# PLASMA PHYSICS AND FUSION ENERGY

There has been an increase in worldwide interest in fusion research over the last decade due to the recognition that a large number of new, environmentally attractive, sustainable energy sources will be needed to meet the ever increasing demand for electrical energy. This has led to an international agreement to build a large, $4 billion, reactor scale device known as the "International Thermonuclear Experimental Reactor" (ITER).

Based on a series of course notes from graduate courses in plasma physics and fusion energy at MIT, the text begins with an overview of world energy needs, current methods of energy generation, and the potential role that fusion may play in the future. It covers energy issues such as production of fusion power, power balance, the design of a simple fusion reactor and the basic plasma physics issues faced by the developers of fusion power – macroscopic equilibrium and stability, transport, and heating.

This book is suitable for graduate students and researchers working in applied physics and nuclear engineering. A large number of problems accumulated over two decades of teaching are included to aid understanding.

JEFFREY P. FREIDBERG is a Professor and previous Head of the Nuclear Science and Engineering Department at MIT. He is also an Associate Director of the Plasma Science and Fusion Center, which is the main fusion research laboratory at MIT.

# PLASMA PHYSICS AND FUSION ENERGY

Jeffrey P. Freidberg

*Massachusetts Institute of Technology*

CAMBRIDGE
UNIVERSITY PRESS

# CAMBRIDGE
### UNIVERSITY PRESS

University Printing House, Cambridge CB2 8BS, United Kingdom

Published in the United States of America by Cambridge University Press, New York

Cambridge University Press is part of the University of Cambridge.

It furthers the University's mission by disseminating knowledge in the pursuit of education, learning and research at the highest international levels of excellence.

www.cambridge.org
Information on this title: www.cambridge.org/9780521733175

© J. Freidberg 2007

First published 2007
First paperback edition 2008
Reprinted 2010

A catalogue record for this publication is available from the British Library

ISBN 978-0-521-85107-7 Hardback
ISBN 978-0-521-73317-5 Paperback

For Karen

# Contents

# Preface

*Plasma Physics and Fusion Energy* is a textbook about plasma physics, although it is plasma physics with a mission – magnetic fusion energy. The goal is to provide a broad, yet rigorous, overview of the plasma physics necessary to achieve the half century dream of fusion energy.

The pedagogical approach taken here fits comfortably within an Applied Physics or Nuclear Science and Engineering Department. The choice of material, the order in which it is presented, and the fact that there is a coherent storyline that always keeps the energy end goal in sight is characteristic of such applied departments. Specifically, the book starts with the design of a simple fusion reactor based on nuclear physics principles, power balance, and some basic engineering constraints. A major point, not appreciated even by many in the field, is that virtually no plasma physics is required for the basic design. However, one of the crucial outputs of the design is a set of demands that must be satisfied by the plasma in order for magnetic fusion energy to be viable. Specifically, the design mandates certain values of the pressure, temperature, magnetic field, and the geometry of the plasma. This defines the plasma parameter regime at the outset. It is then the job of plasma physicists to discover ways to meet these objectives, which separate naturally into the problems of macroscopic equilibrium and stability, transport, and heating. The focus on fusion energy thereby motivates the structure of the entire book – how can we, the plasma physics community, discover ways to make the plasma perform to achieve the energy mission.

Why write such a book now? Fusion research has increased worldwide over the last several years because of the internationally recognized pressure to develop new reliable energy sources. With the recently signed agreement to build the next generation International Thermonuclear Experimental Reactor (ITER), I anticipate a substantial increase in interest on the part of new students and young scientists to join the fusion program. While fusion still has a long way to go before becoming a commercially viable source of energy, the advent of ITER enhances the already existing worldwide interest and excitement in plasma physics and fusion research. The incredibly challenging science and engineering problems coupled with the dream of an energy system characterized by unlimited fuel, near environmental perfection, and economical competitiveness are still big draws to new students and researchers.

Who is the intended audience? This textbook is aimed at seniors, first year graduate students, and new scientists joining the field. In general, the style of presentation includes in depth physical explanations aimed at developing physical intuition. It also includes many detailed derivations to clarify some of the mathematical mysteries of plasma physics. The book should thus be reasonably straightforward for newcomers to fusion to read in a stand alone fashion. There is also an extensive set of homework problems developed over two decades of teaching the subject at MIT.

With more explanations and detailed derivations something must give or else the book would become excessively long. The answer is to carefully select the material covered. In deciding how to choose which material to include and not to include, there are clearly tough decisions to be made. I have made these choices based on the idea of providing newcomers with a good first pass at understanding all the essential issues of magnetic fusion energy. Consequently, the material included is largely focused on the plasma physics mandated by fusion energy, which for a first pass is most easily described by macroscopic fluid models.

As to what is not included, there is very little discussion of fusion engineering. There is also very little discussion of plasma kinetic theory (e.g. the Vlasov equation and the Fokker–Planck equation). Somewhat surprisingly to me, it was not until the next-to-last chapter in the book that I first actually needed any of the detailed results of kinetic theory (i.e., the collisionless damping rates of RF heating and current drive), which I then derived using a simple, intuitive single-particle analysis. The point is that the first time through, the best way to develop an overall understanding of all the issues involved, with particular emphasis on self-consistent integration of the plasma physics, is to focus on macroscopic fluid models which are more easily tied to physical intuition and experimental reality. Ideally, a follow-on study based on kinetic theory would be the next logical step to master fusion plasma physics. In such a study, many of the topics described here would be analyzed at the more advanced level marking the present state of the art in fusion research.

As is clear from the length of the book, it would take a two semester course to cover the entire material in detail. However, a cohesive one semester course can also be easily constructed by picking and choosing from among the many topics covered. In terms of prerequisites, my assumption is that readers will have a solid foundation in undergraduate physics and mathematics. The specific requirements include: (1) mathematics up to partial differential equations, (2) mechanics, (3) basic fluid dynamics, and (4) electromagnetic theory (i.e., electrostatics, magnetostatics, and wave propagation). Experience has shown that an undergraduate degree in physics or most engineering disciplines provides satisfactory preparation.

In the end it is my hope that the book will help educate the next generation of fusion researchers, an important goal in view of the international decision to build ITER, the world's first reactor-scale, burning plasma experiment.

# Acknowledgements

The material for this book has evolved over many years of research and teaching. Many friends, colleagues, and students, too numerous to mention, have contributed in a significant way to my knowledge of the field, making this book possible. I acknowledge my deep appreciation for their collaboration, cooperation, and comraderie.

A number people at MIT also deserve special thanks. Bob Granetz, Ian Hutchinson, Ron Parker, and Abhay Ram have also all taught the subject upon which the book is based. I am grateful to them for sharing their notes and experiences with me.

Many colleagues at MIT have also been kind enough to read chapters of the book and provide me with me valuable feedback. I would like to thank Paul Bonoli, Leslie Bromberg, Peter Catto, Jan Egedal, Martin Greenwald, Jay Kesner, Jesus Ramos, and John Wright for their efforts. Other MIT colleagues gave generously of their time by means of intensive discussions. My appreciation to Darin Ernst, Joe Minervini, Kim Molvig, Miklos Porkolab, and Steve Scott.

A number of friends and colleagues from the general fusion community also read sections of the manuscript and provided me with valuable comments, particularly with respect to Chapter 13, which describes many present day fusion concepts. I would like to acknowledge help from Dan Barnes and Dick Siemon (the FRC), Riccardo Betti and Dale Meade (the tokamak and fusion reactors), Alan Boozer and Hutch Neilson (the stellarator), Bick Hooper (the spheromak), Martin Peng (the spherical tokamak), and John Sarff (the RFP).

Special thanks to my colleague Don Spong for producing the striking illustration appearing on the cover of the book.

As one might expect, preparing a manuscript is an ambitious task. I am extremely grateful to a cadre of MIT graduate students (many of them now full-time researchers) for their help in preparing the figures. My thanks to Joan Decker, Eric Edlund, Nathan Howard, Alex Ince-Cushman, Scott Mahar, and Vincent Tang. Special thanks to Vincent Tang who proofread the entire manuscript for content and style. My assistant Liz Parmelee also provided invaluable administrative and organizational support during the entire preparation of the manuscript.

The team at Cambridge University Press has been a great help in publishing the manuscript, from the initial agreement to write the book to the final production. Thanks to

Simon Capelin (publishing director), Lindsay Barnes (assistant editor), Dan Dunlavey (production editor), Emma Pearce (production editor) and Maureen Storey (copy editor).

Last, but most certainly not least, I would like to thank my wife Karen for her unending support and encouragement while I prepared the manuscript. She was also kind enough to proofread a large fraction of the text for which I am most grateful.

# Units

Throughout the textbook standard MKS units are used. The one exception is the temperature. It is now common practice in the field of fusion plasma physics to absorb Boltzmann's constant $k$ into the temperature so that the combination $kT$ always appears as $T$; that is, $kT \rightarrow T$, where $T$ has the units of energy (joules).

There are also a number of relationships expressed in "practical" units, which unless otherwise specified, are given by

| | | |
|---|---|---|
| Number density | $n$ | $10^{20}$ m$^{-3}$ |
| Temperature | $T$ | keV |
| Pressure | $p$ | atmospheres |
| Magnetic field | $B$ | tesla |
| Current | $I$ | megamperes |
| Minor radius | $a$ | m |
| Major radius | $R$ | m |
| Confinement time | $\tau_E$ | s |

# Part I

Fusion power

# 1

# Fusion and world energy

## 1.1 Introduction

It has been well known for many years that standard of living is directly proportional to
energy consumption. Energy is essential for producing food, heating and lighting homes,
operating industrial facilities, providing public and private transportation, enabling com-
munication, etc. In general a good quality of life requires substantial energy consumption
at a reasonable price.

Despite this recognition, much of the world is in a difficult energy situation at present
and the problems are likely to get worse before they get better. Put simply there is a steadily
increasing demand for new energy production, more than can be met in an economically
feasible and environmentally friendly manner within the existing portfolio of options. Some
of this demand arises from increased usage in the industrialized areas of the world such as
in North America, Western Europe, and Japan. There are also major increases in demand
from rapidly industrializing countries such as China and India. Virtually all projections of
future energy consumption conclude that by the year 2100, world energy demand will at
the very least be double present world usage.

A crucial issue driving the supply problem concerns the environment. In particular, there
is continually increasing evidence that greenhouse gases are starting to have an observable
negative impact on the environment. In the absence of the greenhouse problem the energy
supply situation could be significantly alleviated by increasing the use of coal, of which there
are substantial reserves. However, if the production of greenhouse gases is to be reduced
in the future there are limits to how much energy can be generated from the primary fossil
fuels: coal, natural gas, and oil. A further complication is that, as has been well documented,
the known reserves of natural gas and oil will be exhausted in decades. The position taken
here is that the greenhouse effect is indeed a real issue for the environment. Consequently,
in the discussion below, it is assumed that new energy production will be subject to the
constraint of reducing greenhouse gas emissions.

To help better understand the issues of increasing supply while decreasing emissions, a
short description is presented of each of the major existing energy options. As might be
expected each option has both advantages and disadvantages so there is no obvious single

path to the future. Still, once the problems are identified it then becomes easier to evaluate new proposed energy sources.

This is where fusion enters the picture. Its potential role in energy production is put in context by comparisons with the other existing energy options. The comparisons show that fusion has many attractive features in terms of safety, fuel reserves, and minimal damage to the environment. Equally important, fusion should provide large quantities of electricity in an uninterrupted and reliable manner, thereby becoming a major contributor to the world's energy supply. These major benefits have fueled the dreams of fusion researchers for over half a century. However, fusion also has disadvantages, the primary ones being associated with overcoming the very difficult scientific and engineering challenges that are inherent in the fusion process. The world's fusion research program is finding solutions to these problems one by one. The final challenge will be to integrate these solutions into an economically competitive power plant that will allow fusion to fulfill its role in world energy production.

The remainder of this chapter contains comparative descriptions of the various existing energy options and a more detailed discussion of how fusion might fit into the future energy mix.

## 1.2 The existing energy options

### 1.2.1 Background

The primary natural resources used to produce energy fall into three main categories: fossil fuels, nuclear fuels, and sunlight, which is the driver for most renewables. In general these resources can be used either directly towards some desired end purpose or indirectly to produce electricity which can then be utilized in a multitude of ways. The direct uses include heating for homes, commercial buildings, and industrial facilities and as fuel for transportation. Electricity is used in manufacturing and construction, as well as home, commercial, and industrial lighting and cooling.

One issue applicable to all sources of energy is efficiency of utilization, which directly impacts fuel reserves and/or cost. Clearly high efficiency is desirable and in practical terms this translates into conservation methods. Logically, conservation should be used to the maximal extent possible to help solve the energy problem.

As a simple overview of the current world energy situation consider the end uses of energy. In the year 2001 industrialized countries such as the USA apportioned about 60% of their energy to direct applications and 40% to the production of electricity. See Fig. 1.1.

Electricity is singled out because of its high versatility and the fact that this is the main area where fusion can make a contribution. A detailed breakdown of the relative fuel consumption used to generate electricity in the USA for the year 2001 is illustrated in Fig. 1.2. Observe that fossil fuels are the dominant contributor, providing about 70% of the electricity with 51% generated by coal. Nuclear, gas, and hydroelectric generation also made substantial contributions while wind, solar, and other renewable sources had very little impact (i.e. 0.4%).

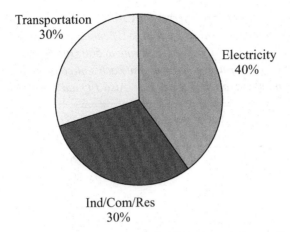

Figure 1.1 Apportionment of energy in the USA in 2001 (Annual Energy Review, 2001 Energy Information Administration, US Department of Energy).

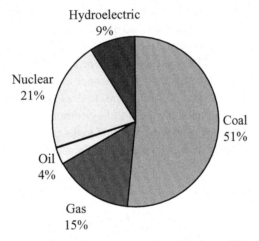

Figure 1.2 Breakdown of fuel consumption to generate electricity in the USA in 2001 (Annual Energy Review, 2001, Energy Information Administration, US Department of Energy).

What are the conclusions from these facts? First, most of the world's energy, including electricity, is derived from fossil fuels. Second, all fossil fuels produce greenhouse gases. Third, if greenhouse emissions are to be reduced in the future, even though energy demand is increasing, new energy capacity will have to be met by a combination of nuclear, hydro-electric, renewable (e.g. wind, solar, geothermal) sources, and conservation. Fourth, some major direct energy usages, such as heating by fossil fuels, could be replaced by electricity, although at an increased cost because of lower efficiency. Fifth, transportation is a special problem because of the need for a mobile fuel. As discussed shortly electricity may be

Table 1.1. *Estimate of energy reserves for various primary fuels. These are very approximate and should be viewed as guidelines. The total usage assumes that the source is used to supply the entire world's energy at a rate of 500 Quads per year (slightly higher than the 2001 rate). The self-usage assumes that each source is used to supply energy at its own individual 2001 usage rate. Also 1 Quad ≈ $10^{18}$ joules.*

| Resource | Energy reserves (Quads) | Total usage (y) | Self-usage (y) |
|---|---|---|---|
| Coal | $10^5$ | 200 | 900 |
| Oil | $10^4$ | 20 | 60 |
| Natural gas | $10^4$ | 20 | 100 |
| U235 (standard) | $10^4$ | 20 | 300 |
| U238, Th232 (breeder) | $10^7$ | 20 000 | |
| Fusion (D–T) | $10^7$ | 20 000 | |
| Fusion (D–D) | $10^{12}$ | $2 \times 10^9$ | |

able to help here through the production of synthetic fuels, ethanol, or hydrogen, which ultimately may be used to replace gasoline and diesel fuel.

To summarize, increasing electricity production in an economic and environmentally friendly way is a vital step in addressing the world's energy problems now and in the future. Fusion is one new energy source that has the potential to accomplish this mission. It is, however, a long term solution (i.e., 30–100 years). In the interim, fossil fuels will remain the primary natural resources producing the world's electricity.

With this as background, one is now in a position to describe in more detail the various existing energy options, particularly with respect to electricity, in order to put fusion in a proper context.

### 1.2.2 Coal

Coal is the main fossil fuel used to generate electricity (51% in the USA). One major advantage of coal is that there are substantial reserves in many countries capable of supplying the world with electricity at the current usage rate for hundreds of years. See Table 1.1 for a list of approximate reserves of various types of fuel. If fuel availability was the only energy issue, coal would be the solution for the foreseeable future. However, when environmental concerns are considered, coal becomes less desirable.

Coal provides continuous, non-stop electricity by means of large, remotely located power plants. This vital non-stop property is known as "base load" electricity. For reference, note that a large power plant typically produces 1 GW of power, capable of supporting a city with a population of about 250 000 people. Two other important advantages of coal are that it is a well-developed technology and that it is among the lowest-cost producers of electricity.

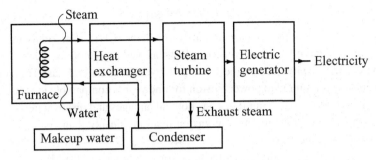

Figure 1.3 Schematic diagram of a fossil fuel power plant.

To help visualize how much coal is required to produce electricity, consider the city of Boston which has a population of about 600 000 people, and whose total rate of electrical energy consumption corresponds to 2.4 GW. The volume of coal required to provide continuous power at this level for one year would completely fill one 70 000 fan football stadium.

Consider next the efficiency of converting coal to electricity. Burning any fossil fuel (i.e., coal, natural gas, or oil) is a chemical process whose main output is heat. As shown in Fig. 1.3, a heat exchanger converts water to steam which then drives a steam turbine connected to an electric generator, thereby producing electricity. The laws of thermodynamics imply that for reasonable operating temperatures, the maximum overall efficiency for converting heat to electricity is about 35–40%. More heat is lost out of the smokestack than is converted to electricity. This unpleasant consequence is unavoidable and occurs whenever a steam cycle is used to produce electricity, as it is for coal and nuclear systems.

The main disadvantage of fossil fuel combustion is environmental in nature. Burning any fossil fuels leads to the unavoidable generation of carbon dioxide ($CO_2$) which is largely responsible for the greenhouse effect. This is a serious disadvantage when considering increased usage of fossil fuels for new electricity generation.

There are also several coal-specific environmental disadvantages. Because of impurities, when coal is burned it also releases fly ash (largely calcium carbonate), sulfur dioxide, nitrous oxide, and oxides of mercury, all of which are harmful to health. These emissions can be reduced, although not completely eliminated, by electrostatic precipitators and scrubbers. However, this increases the cost of electricity.

Interestingly, there are also small amounts of radioactive isotopes contained in natural coal that are released into the atmosphere upon burning. Although the fractional amounts are small, the quantities of coal are large and more radiation is actually released by a coal power plant than by a nuclear power plant. Even so, the level of radioactivity is believed to be sufficiently small not be a concern.

In summary, one can see that coal has both advantages (fuel reserves and cost) and disadvantages (greenhouse gases and emissions). Because of its advantages, and because there are no obviously superior alternatives, coal will remain a major contributor to electricity production for many years to come.

### *1.2.3 Natural gas*

Natural gas is a fossil fuel that consists mainly of methane ($CH_4$). It is widely used to heat homes, commercial buildings, and industrial plants, as well as to produce electricity. About 15% of the electricity produced in the USA is derived from natural gas. The amount of liquefied natural gas required to power Boston for one year is comparable in volume to that of coal. With respect to coal, natural gas has both advantages and disadvantages.

Consider the advantages. First natural gas burns more cleanly than coal. There are far fewer emissions and the amount of $CO_2$ released during combustion is smaller. Second, natural gas plants can be built in smaller units, on the order of 100 MW. This leads to a more rapid construction time and a smaller initial investment, both desirable financial incentives. Third, natural gas powered plants can be operated in a "combined cycle" mode. Here, thermodynamic steam and gas cycles are combined, leading to an increased overall conversion efficiency of gas to electricity of 50–60%. Lastly, many would agree that natural gas, when available, is the most desirable way to heat homes and industrial facilities in terms of convenience and cost.

There are also several disadvantages. First, the amount of $CO_2$ produced per megawatt hour of electricity, while less than for coal, is still very large, as it must be for any fossil fuel. Thus, contributions to the greenhouse effect are considerable. Second, the reserves of natural gas are much less than those of coal. Current estimates are for less than 100 years at the present rate of usage. See Table 1.1. Also, most of the known reserves do not lie within the boundaries of the industrialized nations where the majority of the gas is consumed. Third, high demand coupled with production limits and relatively scarce reserves have led to high and unstable fuel costs. Fourth, it is more difficult and more expensive to transport and store natural gas than coal or oil because of the need for pipelines and high-pressure liquid storage tanks. Fifth, since natural gas is such an ideal fuel for heating, many feel that its use to produce electricity is a poor allocation of a valuable natural resource. The incentive for this poor allocation is largely motivated by short-term economics and energy deregulation with too little thought given to long-term consequences.

To summarize, the use of natural gas to produce electricity has advantages (cleanest burning of any fossil fuel and low short-term cost) and disadvantages (greenhouse gases, limited reserves, and poor allocation of resources). Overall, short-term financial incentives dominate the tradeoffs and will likely lead to the continued use of natural gas for electricity production.

### *1.2.4 Oil*

Oil is the last of the fossil fuels to be discussed. It is an excellent fuel for transportation because of its portability and its large energy content. It is also the fuel of choice for heating when natural gas is not available. A large amount (i.e., 35%) of the energy used in the world is derived from oil, with much of it devoted to transportation usage. It is rarely used to directly produce electricity.

As a measure of energy content note that a 1 gallon milk container filled with gasoline is capable of moving a typical automobile 25 miles, indeed an impressive feat. Furthermore the total weight of a fully loaded 15 gallon fuel tank is only about 120 pounds, a negligible fraction of the total weight of the automobile. A full tank can therefore efficiently move an automobile about 375 miles, again, a truly impressive feat.

The second issue of interest is the cost of gasoline. It is surprisingly inexpensive compared to many other common liquids. In the USA the untaxed price per gallon of gasoline is still less than that of bottled water. Gasoline would appear to be a bargain, even at present higher prices.

Nevertheless, there are disadvantages to the use of gasoline for transportation. First, since gasoline is a fossil fuel it produces a large amount of greenhouse gases, comparable in total magnitude to that of coal. Second, crude oil is only readily available in a few areas of the world. One major source is the Middle East, which is fraught with political instability. Third, the reserves of oil are much less than those of coal, on the order of several decades at present usage rates. The competition for oil from the developing countries will likely increase in the future raising costs and perhaps limiting supplies.

Are there ways to decrease the world's dependency on oil? There are possibilities, but they are not easy. Consuming less oil by using hybrid vehicles could make an important contribution and may be accepted by the public even though it raises the initial cost of an automobile. Consuming less oil by driving smaller automobiles with improved fuel efficiency could also make a large contribution, although many may be reluctant to follow this path, viewing it as a lowering of one's standard of living.

A different approach is based on the fact that gasoline can be produced from coal tars and oil shale, of which there are large reserves. The end product is known as "synfuel," but at present the process is not economical. Also since synfuel is a form of fossil fuel, the production of greenhouse gases still remains an important environmental problem.

Another approach is to use non-petroleum fuels produced by bio-conversion. One method currently in limited use is the conversion of corn to ethanol, a type of alcohol. Although ethanol is a plausibly efficient replacement for gasoline, the economics of production are not. Large amounts of land are required and considerable energy must be expended to produce the ethanol, comparable to and sometimes exceeding the energy content of the final fuel itself.

There has also been considerable interest and publicity in developing the technology of using hydrogen in conjunction with fuel cells to produce a fully electric car, thus completely replacing the need for gasoline. Hydrogen has the advantages of: (1) a large reserve of primary fuel (e.g. water), (2) a high conversion efficiency from fuel to electric power, and (3) most importantly the end product of the process is harmless water vapor rather than $CO_2$. This may be the ultimate transportation solution but there are two quite difficult challenges to overcome.

First hydrogen itself is not a primary fuel. It must be produced separately, for instance by electrolysis, and this requires substantial energy. If the energy for the electrolysis of water is derived from fossil fuels much of the gain in reduced $CO_2$ emissions is canceled. Second, the energy content of hydrogen at atmospheric pressure, including its higher conversion efficiency, is still much lower than that of gasoline, by a factor of about 1200. Therefore, to increase the energy content of hydrogen fuel to a value comparable to gasoline, the hydrogen must be compressed to the very high pressure of 1200 atm. This poses a very difficult fuel tank design problem for on-board storage of hydrogen. Another option is to store the hydrogen in liquid form, but this requires a costly on-board cryogenic system. A third option is to develop room-temperature compounds that are capable of storing and rapidly cycling large quantities of hydrogen. The development of such compounds is a topic of current research, but success is still a long way into the future. One sees that the on-board storage of high-density hydrogen presents a difficult technological challenge.

The conclusions from this discussion are as follows. There is no simple, short-term, attractive alternative to gasoline for transportation. Synthetic fuel, ethanol, and hydrogen are possible long-term solutions, but each has a mixture of unfavorable economic, energy balance, and environmental problems. Providing the energy to produce hydrogen or ethanol by $CO_2$-free electricity (e.g. by nuclear power) would be a big help but would not solve the other problems. In the short term the best strategy may be to increase the use of hybrid vehicles and to evolve towards smaller, more fuel efficient automobiles.

### 1.2.5 Nuclear power

The primary use of nuclear power is the large-scale generation of base load electricity by the fissioning (i.e., splitting) of the uranium isotope $U^{235}$. At present there is still public concern about the use of nuclear power. However, a more careful analysis shows that this form of energy is considerably more desirable than is currently perceived and will likely be one of the main practical solutions for the future production of $CO_2$ free electricity.

There are several comparisons with fossil fuel plants that show why nuclear power has received so much attention as a source of electricity. The first involves the energy content of the fuel. A nuclear reaction produces on the order of one million times more energy per elementary particle than a fossil fuel chemical reaction. The implication is that much less nuclear fuel is required to produce a given amount of energy. Specifically, the total volume of nuclear fuel rods needed to power Boston for one year would just about fit in the back of a pickup truck. This should be compared to the football stadium required for fossil fuels.

A second point of comparison is environmental impact. Nuclear power plants produce neither $CO_2$ nor other harmful emissions. This is a major environmental advantage.

Another issue is safety. Despite public concern, the actual safety record of nuclear power is nothing less than phenomenal. No single nuclear worker or civilian has ever lost his or her life because of a radiation accident in a nuclear power plant built in the Western world. The worst accident in a USA plant occurred at Three Mile Island. This was a financial disaster for the power company but only a negligible amount of radiation was released to the environment. The reason is that Western nuclear power plants are designed with many overlapping layers of safety to provide "defense in depth" culminating with a huge, steel reinforced containment vessel around the reactor to protect the public in case of a "worst" accident. The large loss of life and wide environmental damage resulting from the Chernobyl accident occurred because there was no containment vessel around the reactor. Such a design would never be licensed to operate as a nuclear power plant in the West. Overall, safety is always a major concern in the design and operation of nuclear power plants, but the record shows that for Western power plants the problems are well under control.

Consider next the issue of fuel reserves. This is a complex issue. In the simplest view one can assume that $U^{235}$ is the basic fuel and once most of it has been consumed in the reactor, the resulting "spent fuel" rods are buried in a permanent, non-retrievable repository. In this scenario there is enough $U^{235}$ to provide electricity at the present rate for several hundred years. On the other hand the spent fuel rods contain substantial amounts of plutonium which can be chemically extracted and then used as a new nuclear fuel. In fact, it is possible to use the resulting plutonium in such a way that it actually breeds more plutonium than is being consumed. The use of such "breeder" reactors extends the reserves of nuclear fuels to many thousands of years. Breeders are more expensive than conventional nuclear plants and are not currently used because of the ready availability of low cost $U^{235}$. However, in the long term breeders may be one of the energy sources of choice.

Nuclear waste and how to dispose of it is another important issue. Here too there are subtleties. One point is that many of the radioactive fission byproducts have reasonably short half-lives, on the order of 30 years or less. They need to be stored for about a century during which time they self-destruct by radioactive decay into a harmless form, an ideal end result. It is the long-lived, multi-thousand year wastes that receive much public attention and scrutiny. Several possible solutions have received serious consideration. The waste can be dissolved in glass (i.e., vitrification) and permanently stored. The fuel can be chemically reprocessed for re-use in regular or breeder reactors, thereby transforming much of the long-lived waste into useful electricity. Third, there are techniques that, while currently expensive, transmute long-lived, non-fissioning radioactive waste byproducts into harmless elements. Also, a critical point is that the total volume of nuclear waste is very small. The total nuclear "rubbish" resulting from powering Boston for one year would fill up only a small fraction of a pickup truck. The conclusion is that there are a variety of technological solutions to the waste disposal problem. The main problems are more political than technological.

The last issue of importance is nuclear proliferation, which concerns the possibility that unstable governments or terrorist groups would gain access to nuclear weapons. At first glance one might conclude that reducing the use of nuclear power would obviously reduce the risks of proliferation. This is an incorrect conclusion. The key technical point to recognize is that the spent fuel from a reactor cannot be directly utilized to make a weapon because of the low concentration of fissionable material. Nevertheless, spent fuel is often reprocessed to make new fuel for use in nuclear reactors thereby increasing the fuel reserves as previously discussed. However, one intermediate step in reprocessing is the production of nearly pure plutonium, which at this point could be diverted for use as weapons. A major component of an effective non-proliferation plan should thus involve the detection and prevention of the diversion of plutonium for weapons use by unstable governments. In implementing such a plan two facts should be noted: (1) reprocessing may have valuable energy and economic benefits, and (2) reprocessing technology, while very expensive, is reasonably well established. Consequently any nation can justify the construction of a reprocessing facility based on energy needs, thereby opening up the possibility of a surreptitious diversion of a small amount of plutonium for use in weapons. One approach might be for the major, stable nuclear powers in the world to carry out all the reprocessing in their own countries, and then sell the resulting fuel to smaller countries with legitimate energy needs. This would take away the justification for the proliferation of reprocessing facilities. Ironically, since the Carter administration the USA has had a well-intentioned, but ill-conceived, policy in which it does *no* reprocessing of spent fuel. The hope was that other countries would follow suit. The reality is that reprocessing has expanded in other countries to fill the gap suggesting that USA policy may have made the non-proliferation situation worse rather than better. What are the conclusions from this discussion? First, nuclear non-proliferation is a very serious and important problem that must be addressed. Second, whether or not stable countries like the USA build more nuclear power plants will have little if any direct effect on non-proliferation and may actually divert attention away from the real issues.

To summarize, nuclear power has many underappreciated advantages as well as some disadvantages. Even some well-known environmentalists have started to support nuclear power as the only viable option for producing large quantities of $CO_2$ free electricity. A short term stumbling block to the construction of new nuclear power plants is the fact that while fuel costs are low, the capital costs are high because of the complexity of the reactor. In a deregulated market this is a disincentive to new investment.

### 1.2.6 Hydroelectric power

Hydroelectric power is a widely used renewable source of energy. It provides 2% of the world's energy and about 9% of the electricity in the USA. The idea behind hydroelectric power is conceptually simple. At a geographically and technologically appropriate location along the path of a river, a dam is built creating a huge reservoir lake on the high side of the dam. As reservoir water pours over the dam because of gravity, it turns a turbine which then drives an electric generator, producing electricity.

Hydroelectric power has many attractive advantages. First, no $CO_2$ or other serious pollutants are generated during the production of electricity. Second, large amounts of power are generated in a hydroelectric plant, comparable to that in a coal or nuclear plant. Third, the conversion efficiency of fluid kinetic energy to electricity is high since no thermal steam cycle is involved. Fourth, except in rare cases of extended drought, the power is available continuously for base load electricity. Fifth, the cost of electricity is low, typically comparable to that of coal plants. Sixth, and most importantly, the fuel reserves are effectively infinite. Hydroelectric power is clearly a renewable energy source.

There are two downsides to hydroelectric power. First, most of the suitable rivers already have dams. Therefore, expansion of hydroelectric power is difficult since there are few, if any, unutilized technologically attractive sites available. Second, although not a major problem for early dams, environmental issues will have a much larger impact on any future hydroelectric plants. The main issue is the large amount of land that is flooded to form the reservoir lake. Often this land could be used for agricultural or recreational purposes, so there is a tradeoff that must be evaluated before changing its use to electricity production.

Overall, hydroelectric power will continue to make an important contribution to the supply of electricity although the possibilities for expansion are limited.

### 1.2.7 Wind power

Wind is another renewable energy source that has received much attention in recent years. Even so, it currently provides a negligible fraction of electricity in the USA. Wind should almost certainly be used more than it is at present but for fundamental technological reasons it will not be the ultimate solution to the electricity generation problem.

The idea behind wind power is conceptually easy to understand. Wind striking the blades of a large windmill causes them to rotate. This rotational kinetic energy, by a series of gears, drives an electric generator producing electricity.

Wind has some important advantages. First, wind power is clearly a renewable energy source. Second, it produces electricity in a very clean manner. There is no $CO_2$ , nor are there any harmful pollutants. Third, no steam cycle is involved. Therefore the conversion from wind kinetic energy to electricity is reasonably efficient. Fourth, although the cost of wind power, for reasons described below, is higher than for existing coal plants, it is still within a tolerable range. This is particularly true if one were to add in the additional, often hidden, environmental costs of fossil fuel plants.

There are, however, some disadvantages to wind power. First, the wind does not blow at a constant rate. If it is too weak, not much power is produced. If it is too strong, the blades must turn parallel to the wind to prevent them from spinning too fast and causing mechanical damage. Here too, not much power is produced. On average, a large, modern windmill produces about 35% of its maximum rated power. Much of the gain of not requiring a steam cycle is canceled by the variability of the wind speed.

Second, the 35% availability factor implies that to produce an average of 1 GW of power requires a wind farm whose total power rating is about 3 GW. The problem is that the excess power produced during optimal wind conditions is very difficult and very expensive to store for use during poor wind conditions.

A third disadvantage is that the power intensity of the wind is very low as compared for instance to that in the center of a coal furnace. Therefore producing a significant amount of power requires a large number of windmills spread over a large area. For instance, a modern wind farm, with an optimistic 40% availability factor would need to consist of about 4000 windmills occupying about 400 square miles to produce the 2.4 GW power required to power Boston. Note that Boston has an area of about 50 square miles. Therefore an area 8 times larger than Boston would have to be covered by windmills to produce the required

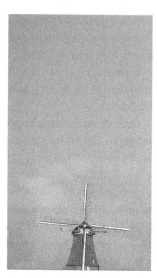

Figure 1.4 Comparison of Big Ben, a modern windmill, and an old fashion Dutch windmill. All the photographs are to the same scale.

power. If the "Stadium" measures coal power, and the "Pickup Truck" measures nuclear power, then the equivalent measure for wind power is the "City plus Suburbs".

Lastly, there are several environmental issues to consider. Windmills tend to be noisy and harmful to birds. There is also the issue of aesthetics. Engineers may find beauty in modern windmills, but the general public tends to view them as unattractive eyesores. Also they are quite large, with mounting towers on the order of 100 m and blades about 50 m in length. The photographs in Fig. 1.4 demonstrate the comparative heights of Big Ben, a modern windmill, and an old fashioned scenic Dutch windmill.

This discussion suggests that wind power faces some extremely difficult challenges if it is ultimately to replace coal as a major source of electricity. A perhaps better role for wind is as a topping source of power, helping to meet peak demand during critical parts of the day and during the more extreme seasons of summer and winter. In this role wind might ultimately provide up to 20% of electricity. It could not provide more because the large fluctuations in wind speed and resulting wind power would likely cause instabilities on the national transmission grid.

### 1.2.8 Solar power

The last renewable source discussed is solar energy. As with wind a negligible amount of USA electricity production is presently derived from solar power. Nevertheless, solar power is often projected to be a potentially attractive alternative to fossil fuels. There are a number of special applications where solar power can be attractive, but for fundamental technological reasons it is unlikely that it will be the ultimate solution for the world's electricity problems.

Understanding how the sun is used to produce electricity involves a detailed knowledge of quantum mechanics and semiconductor theory and is beyond the scope of this book. For present purposes assume that a carefully designed solar cell converts the sun's energy directly into electricity with a daylight averaged efficiency of about 10%.

There are two main advantages of producing electricity from the sun. First, the source of energy is clearly renewable and free. Second, neither $CO_2$ nor other harmful emissions are produced during the energy conversion process. In this sense solar power is very attractive environmentally.

The disadvantages of solar energy are similar to those of wind. First, the sun obviously shines brightly only during periods of cloud-free daytime. Consequently, producing base load power is not possible since there is no simple way to store excess energy during the day for use at night. Second, the sun's intensity is very low compared to that in a coal furnace. Therefore a large area of solar cells is required to produce a significant amount of power. For example, an area of about 50 square miles would have to be covered by solar panels to provide the 2.4 GW required by Boston. Recall that the area of Boston is also about 50 square miles so that the useful measure of energy for solar power is the "City." It takes a City's worth of solar cells to power that same city.

Lastly, solar electricity is inherently expensive. The reason is that a truly large quantity of manufactured material is required to cover a whole city area. The cost of mining, transporting, and manufacturing this material is large and unavoidable.

The conclusion is that solar power faces some very difficult challenges if it is to be used to produce large quantities of electricity. There are other more attractive uses, such as for residential and some commercial heating. Here its contributions can be substantial and should be encouraged.

### 1.2.9 Conservation

Conservation can be defined as the more efficient use of our existing natural resources. Clearly maximizing conservation is an important and worthwhile contribution to help alleviate existing and future energy problems. Although substantial efforts have already been made towards improving conservation, there are still many more opportunities that have yet to be exploited.

There are two ways that conservation can be implemented, one of which has a good chance of acceptance by the public and the other which is on much shakier ground. Although both approaches conserve energy they are separated by a relatively clear line in the sand.

The attractive approach takes advantage of advances in technology to conserve fuel while maintaining performance in appliances, automobiles, and other equipment used in daily living. Examples of this approach include hybrid automobiles, more efficient appliances, additional insulation for older homes, etc.

The second and more difficult approach to conservation requires that citizens directly reduce their use of energy in certain aspects of their daily living. Often this is viewed as a reduction in standard of living. The public is in general much more reluctant to give up something to which they are already accustomed. Examples of this approach to conservation include smaller more gasoline efficient automobiles, smaller houses, increased used of public transportation, less use of air conditioning in summer, lower thermostat settings in the winter, etc.

With the continually increasing demand for new electricity, particularly by some of the developing nations, it is difficult to imagine that conservation can completely solve the world's future electricity generation problems. Nevertheless, it can reduce the magnitude of the problems. This would afford the nations of the world more time to develop and transition to new alternatives.

### *1.2.10  Summary*

The discussion in this section has shown that there are difficult energy problems facing the world that will probably become worse in the future. There is no obvious, single solution. Each of the existing energy options faces a mixture of difficult issues including limited reserves, $CO_2$ production, toxic emissions, waste disposal, excessive land usage, and high costs. In the end energy will be provided by a portfolio of options, hopefully chosen by logic rather than by crisis. One possible new addition to the portfolio that can potentially have a large impact is fusion, which is the next topic for discussion.

## 1.3  The role of fusion energy

### *1.3.1  Fusion energy*

Fusion is a form of nuclear energy. Its main application is the production of electricity in large base load power plants. The basic nuclear processes involved occur at the opposite end of the spectrum of atomic masses than fission. Specifically, fission involves the splitting of heavy nuclei such as $U^{235}$. Fusion involves the merging (i.e., the fusing) of light elements, mainly hydrogen (H) and its isotopes deuterium (D) and tritium (T). The fusion of hydrogen is the main reaction that powers the sun.

There are three main advantages of fusion power: fuel reserves, environmental impact, and safety. Consider first fuel reserves. There are two main reactions of interest that occur at a fast enough rate to produce electricity. These involve pure deuterium and an equal mix of deuterium and tritium. Deuterium occurs naturally in ocean water. There is 1 atom of deuterium for every 6700 atoms of hydrogen. Also deuterium can be easily extracted at a very low cost. If all the deuterium in the ocean were used to power fusion reactors utilizing a standard steam cycle there would be enough energy generated to power the earth for about 2 billion years at the present rate of total world energy consumption! Also, since fusion is a nuclear process, it would take only about a pickup truck full of deuterium laced ocean water (HDO rather than $H_2O$) to power Boston for a year.

The deuterium–tritium (D–T) reaction produces more energy than a pure deuterium (D–D) reaction. However, the main advantage is that D–T reactions occur at a faster rate, thereby making it easier to build such a reactor. Consequently, all first generation fusion reactors will use D–T. In terms of reserves, the multi-billion years of deuterium applies to D–T as well as D–D reactors. However, since tritium is a radioactive isotope with a half-life of only about 12 years, there is no natural tritium to be found on earth. Instead, tritium is obtained by breeding with the lithium isotope $Li^6$ which is one of the components in the fusion blanket. The overall reserves for D–T fusion are thus limited by the reserves of $Li^6$. Geological estimates indicate that there is on the order of 20 000 years of inexpensive $Li^6$ available on earth (assuming total world energy consumption at the present rate). Presumably, well before $Li^6$ is exhausted, the science and technology will have been developed to switch to D–D reactors.

The next advantage is the environmental impact of fusion. Fusion reactions produce no $CO_2$ or other greenhouse emissions. Fusion reactions also do not emit any other harmful chemicals into the atmosphere. The main end product of the fusion reaction is the harmless, inert gas helium. The biggest environmental issue in fusion is that one byproduct of both the D–D and the D–T reaction is a high-energy neutron. These neutrons are captured in the fusion blanket so they pose no threat to the public. However, as they pass through structural material on their way to the blanket, the neutrons cause the structure to become activated. Even so, this radioactive structural material has a short half-life so that the storage time required once it is removed is also short, on the order of 100 years. Overall, when one considers the entire environmental situation, fusion is a very attractive option with respect to fossil, nuclear, and renewable sources.

The last major advantage involves safety. Here, since fusion is a nuclear process, one is concerned about the possibility of a radioactive meltdown such as occurred in the Three Mile Island accident. The basic laws of physics governing fusion reactions make this impossible. Specifically, in a fission reactor the entire energy content corresponding to several years of power production is stored within the reactor core at any instant of time. It is this huge energy content that makes a meltdown possible. A fusion reactor does not depend on maintaining a chain reaction in a large sitting mass of fuel. Instead, fuel must be constantly fed into the reactor at a rate allowing it to be consumed as needed. The end result is that at any instant of time the mass of fuel in a fusion reactor is very small, perhaps corresponding to the weight of several postage stamps. It is this small instantaneous mass of fuel that makes a meltdown impossible in a fusion reactor.

The conclusion from this discussion is that the potential advantages of fusion from the point of view of fuel reserves, environmental impact, and safety are indeed impressive.

As one might expect there are also several disadvantages to fusion that must be considered. These involve scientific challenges, technological challenges, and economics. The key issues are as follows.

The science of fusion is quite complex. Specifically, to burn D–T one is required to heat the fuel to the astounding temperature of $150 \times 10^6$ K, hotter than the center of the sun. At these temperatures the fuel is fully ionized becoming a plasma, a high-temperature collection of independently moving electrons and ions dominated by electromagnetic forces. Once heated some method must be devised to hold the plasma together. The primary method requires a clever configuration of magnetic fields, an admittedly nebulous idea to those unfamiliar with the science of plasma physics. Cleverness is mandatory, not an option. Too simple a configuration allows the plasma to be lost at a rapid rate, thus quenching fusion reactions before sufficient energy can be produced. Even with a clever configuration there are limits to the plasma pressure that can be confined without rapid losses through the magnetic field.

The combined requirement of confining a sufficient quantity of plasma for a sufficiently long time at a sufficiently high temperature to make net fusion power has been the focus of the world's fusion research program for the past 50 years. The unexpected difficulty of these scientific challenges is the primary reason it has taken so long to achieve a net power producing fusion reactor.

There are also engineering challenges, which many believe are of comparable difficulty to the scientific challenges. First, improved low-activation materials need to be developed that can withstand the neutron and heat loads generated by the fusion plasma. Second, large high-field, high-current superconducting magnets need to be developed to confine the plasma. Superconducting magnets on the scale required for fusion have not as yet been built. Third, new technologies to provide heating power have to be developed in order to raise the plasma temperature to the enormously high values required for fusion. This involves a wide variety of techniques ranging from very high-power neutral beams to millimeter wavelength megawatt microwave sources. Clearly a major research and development program is required to make fusion a reality.

The last disadvantage is economics. A fusion reactor is inherently a complex facility. It includes a fuel chamber, a blanket, and a complicated set of superconducting magnets. Also, since the structural material becomes activated, a large remote handling system is required for assembly and disassembly during regular maintenance. The use of tritium plus the structural activation mean that radiation protection is also required. These basic technological requirements imply that the capital cost of a fusion reactor will be larger than that of a fossil fuel power plant, and very likely also that of a fission power plant. This will tend to raise the cost of electricity to consumers. Balancing this are low fuel costs and low costs to protect the environment, both of which tend to reduce the cost of electricity to consumers.

It is clearly difficult to predict the cost of fusion energy as compared to other options 30–50 years in the future. One main complication is that a combination of fuel reserve problems and environmental remediation costs will likely increase the costs of these other options so that comparisons involve a number of simultaneously moving targets. Estimates of future fusion energy costs are in the vicinity of the other options, but because the uncertainties are large, they should be viewed with caution. The main value of these estimates is to show that it makes sense to continue fusion research. Fusion should not be eliminated because of an inherently absurd cost of electricity, nor will it be "too cheap to meter" as one might have hoped in the past.

### *1.3.2 Summary of fusion*

The reality of fusion power is still many years in the future. It is, nonetheless, worth pursuing because of the basic advantages of large fuel reserves, low environmental impact, and inherent safety. Most importantly, fusion should produce large amounts of base load electricity and thus has the potential to have a major impact on the way the nations of the world consume energy.

Two of the main disadvantages of fusion involve mastering the unexpectedly difficult scientific and technological problems. Great progress has been made in solving the scientific problems and large efforts are currently underway to address the technological challenges. Still the outcome is not certain. Many of the critical issues will be addressed in a new experiment known as the International Thermonuclear Experimental Reactor (ITER). This is an internationally funded facility whose construction is anticipated to begin in 2006.

If successful, fusion power should be competitive cost-wise with other energy options although there is a large margin of error in making such predictions. Still the predicted costs are sufficiently reasonable that this should not be a deterrent to completing the research necessary to assess the technological viability of fusion as a source of electricity.

## 1.4 Overall summary and conclusions

The overall summary focuses on the issue of electricity production as it is in this context that fusion could play an important role. The accompanying conclusions are based on the following two realities concerning electricity consumption. First, the demand for electricity is large and is expected to increase in the future. Second, there is increasing evidence that the greenhouse effect is a real problem that must be addressed.

The short-term demand for $CO_2$ free electricity will likely require the increased use of nuclear power to provide large amounts of base load power. Power can also be produced from natural gas, although this seems like a misuse of a fuel that is so ideally suited for heating applications. Hydroelectricity will continue to be an important contributor although, for the reasons discussed, further increases in capacity will be limited. A further important contribution to electricity production can be provided by the wind. However, this form of energy is more appropriate to meet peak demands because of the variable nature of the wind and the fact reserve wind energy cannot be easily stored at low cost. Solar power is currently still too expensive except for special uses such as the heating of water. Conservation can also play an important role in helping to reduce the magnitude of the problem, but by itself will not solve the problem of increasing electricity demand.

In the long term fusion is an excellent new option that ultimately has the potential to become the world's primary source of electricity. This is the main mission of fusion. However, difficult science and technology problems remain and cost may be an issue. Time will tell whether or not fusion research can fulfill its mission.

## Bibliography

There are a large number of books written about the general topic of energy. The ones listed below have been used as primary sources for Chapter 1. One issue has to do with quoted figures for energy usage and energy reserves. Many books give such figures but there are substantial variations among them. The values given in Chapter 1 represent an approximate averaging of these figures, which have been rounded off for simplicity so that they do not imply a false degree of accuracy. The most complete set of data is included in the book by Tester *et al.* and references contained therein.

### *Energy in general*

Hughes, W. L. (2004). *Energy 101*. Rapid City, South Dakota: Dakota Alpha Press.
Rose, D. J. (1986). *Learning About Energy*. New York: Plenum Press.
Tester, J. W., Drake, E. M., Driscoll, M. J., Golay, M. W., and Peters, W. A. (2005). *Sustainable Energy*. Cambridge, Massachusetts: MIT Press.

### *Nuclear power*

Reynolds, A. B. (1996). *Bluebells and Nuclear Energy*. Madison, Wisconsin: Cogito Books.
Waltar, A. E. (1995). *America the Powerless*, Madison, Wisconsin: Cogito Books.

### *Fusion energy*

Fowler, T. K. (1997). *The Fusion Quest*. Baltimore: John Hopkins University Press.
McCracken, G. and Stott, P. (2005). *Fusion, the Energy of the Universe*. London: Elsevier Academic Press.
Wesson, J. (2004). *Tokamaks*, third edn. Oxford: Oxford University Press.

# 2

# The fusion reaction

## 2.1 Introduction

The study of fusion energy begins with a discussion of fusion nuclear reactions. In this chapter this topic is put in context by first comparing the chemical reactions occurring in the burning of fossil fuels with the nuclear reactions that produce the energy in fission and future fusion power plants. The comparison is then taken one level deeper by describing in more detail the basic mechanism of the fission reaction and the reason why this mechanism is not effective for fusion energy. The discussion does, nevertheless, provide the insight necessary to understand the alternative mechanism that must instead be employed to produce large numbers of nuclear fusion reactions. Several fusion reactions, including the deuterium–tritium (D–T) reaction, are described in detail.

Once the analysis of the issues described above has been carried out one is led to the following conclusion. Both the splitting of heavy atoms (fission) and the combining of light elements (fusion) lead to the efficient production of nuclear energy. The opposing energy mechanisms are a direct consequence of the nature of the forces that hold the nuclei of different elements together. The behavior of these nuclear forces is conveniently displayed in a curve of "binding energy" versus atomic number. A simple physical picture is presented that explains the binding energy curve and why it has the shape that it does. This explanation shows why light or heavy elements are good sources of nuclear energy and why intermediate elements are not.

## 2.2 Nuclear vs. chemical reactions

By comparing the energy equivalence of various types of fuel it is easy to see why strong interest exists in the use of nuclear reactions for energy production:

| fossil | fission | fusion |
|--------|---------|--------|
| $10^6$ tonne oil | $=$ 0.8 tonne uranium | $=$ 0.14 tonne deuterium |

There is an advantage of a factor of about one million for nuclear over fossil fuels, which has enormous fuel-reserve and environmental impact. This large separation is associated with the fundamental difference between chemical and nuclear reactions.

To understand the issues, consider first chemical reactions. The burning of a fossil fuel for electricity or transportation is a chemical reaction. A fossil fuel comprising complex hydrocarbons is burned through a reaction with oxygen leading to a release of energy and the formation of new molecules corresponding to different chemical compounds. It is a chemical reaction because the number of atoms of each of the primary element atoms (e.g. H, C, or O) is unchanged by the reaction. The end products, nevertheless, consist of different compounds of H, C, and O than in the original fuel. The new molecules are characterized by a different structure of the electronic bonds that hold the electrons in place around each atom. The dominant force controlling the behavior of the electrons is the electromagnetic force, which is capable of rearranging the electronic structure but leaves the nuclei unchanged.

In a "burning reaction" some of the chemical potential energy of the fuel is converted into kinetic energy of the end products and some is converted into radiation energy as visible light. The chemical potential energy of the end products is therefore lower than that of the fuel. The kinetic energy contribution produced in burning is transformed into heat via randomizing collisions. This heat can ultimately be used to produce steam either to drive a steam turbine or whose increased pressure can drive the pistons in an automobile engine.

It is of great practical interest to determine the amount of energy released by each chemical reaction that takes place. An order of magnitude estimate for this energy can be obtained by noting that an extreme form of chemical reaction is ionization, in which sufficient energy is given to an atom or molecule to completely free one of the electrons bound to its nucleus. Typically the energy released in the burning of a fossil fuel is considerably lower than its ionization energy since very little ionization of the fuel takes place during combustion. As an example, it is well known from quantum theory that the ionization potential of hydrogen, which is similar to that of many substances, is approximately 13.6 eV. Since 1 eV $\approx$ 11 600 K it follows that at the typical temperature of a burning fossil fuel (i.e., a few thousand K) chemical reactions tend to release energies on the order of a fraction of an electron volt per reaction per atom. (Note that this estimate does not apply to the core of a fission reactor, which also operates at about $10^3$ K, but involves nuclear rather than chemical reactions.)

As a practical example consider the burning of gasoline. Although gasoline consists of a broad spectrum of hydrocarbons, a useful approximation for present purposes is to assume that the predominant component is $C_8H_{18}$. The chemical reaction describing the complete combustion of gasoline is then

$$2C_8H_{18} + 25O_2 \rightarrow 16CO_2 + 18H_2O + 94 \, \text{eV}. \tag{2.1}$$

Since there are 102 atoms involved, this averages to 0.9 eV per atom. Macroscopically the 94 eV per reaction is equivalent to an energy release of approximately 40 MJ/kg $\approx$ 100 MJ/gallon of gasoline. This is a good reference number for comparisons between chemical and nuclear reactions. Keep in mind that 40 MJ/kg is an impressive number since the gasoline that would fill only a 1 liter container is capable of moving an automobile 10 km.

Consider next nuclear reactions, which include both fission and fusion. Nuclear reactions produce changes in the basic structure of the nuclei of the atoms involved. A nuclear reaction changes atoms of one element into atoms of another (e.g. in fission uranium is changed into xenon and strontium).

A second major characteristic of nuclear reactions concerns the conservation of elementary particles in the nucleus (described generically as nucleons), which for present purposes can be assumed to involve only protons and neutrons. In some nuclear reactions the numbers of protons and neutrons are each separately conserved. However, in many other nuclear reactions a few of the neutrons may decay into protons by emitting beta particles (i.e. electrons). Thus, in general, a nuclear reaction conserves only the total, but not the individual, number of nucleons in the nucleus.

With regard to energy production, the energy released per nuclear reaction is always enormous compared to that from a chemical reaction. The reason is that it is the nuclear rather than the electromagnetic force that causes the reactions. The nuclear force acts over a short range, comparable to a nuclear diameter, but is much, much stronger than the electromagnetic force over this range.

The energy release in a nuclear reaction represents the decrease in nuclear "potential energy" or equivalently the increase in binding energy of the nuclei between the final and initial states. The final state is more stable with the increase in binding energy appearing as a slight decrease in the final total nuclear mass. It is this difference in mass that is transformed into energy by Einstein's famous relation $E = mc^2$. The released energy is usually in the form of either the kinetic energy of the end particles or gamma rays. Thus, a nuclear reaction, such as fission or fusion, can typically be written as

$$A_1 + A_2 \rightarrow A_3 + A_4 + \cdots + A_k + \text{energy}, \tag{2.2}$$

where the energy $\equiv E$ is given by

$$E = \left[ \left( m_{A_1} + m_{A_2} \right) - \left( m_{A_3} + m_{A_4} + \cdots + m_{A_k} \right) \right] c^2. \tag{2.3}$$

Usually the value of $E$ for nuclear reactions is in the range 10–100 MeV. This difference of a factor of about one million in energy release between nuclear and chemical reactions is responsible for the corresponding difference in macroscopic energy equivalence.

The next point in the discussion is to compare at a deeper level the similarities and differences between the two main nuclear processes of interest, fission and fusion, including some specific examples.

## 2.3 Nuclear energy by fission

The usual method of producing fission energy is to bombard an atom of $_{92}U^{235}$, a relatively rare isotope of uranium, with a slow neutron (i.e., a neutron whose energy is approximately equivalent to room temperature: 0.025 eV). This process has two major advantages with respect to the production of energy, which can be understood by examining a typical fission reaction. Note that there are many different ways in which $_{92}U^{235}$ can be split and the reaction given below is typical in terms of the end products and the energy release:

$$_0n^1 + _{92}U^{235} \rightarrow {}_{54}Xe^{140} + {}_{38}Sr^{94} + 2(_0n^1) + E'. \tag{2.4}$$

In this reaction the xenon and strontium are themselves unstable isotopes and in about two weeks decay to stable elements via several beta emissions. The final reaction becomes

$$_0n^1 + _{92}U^{235} \rightarrow {}_{58}Ce^{140} + {}_{40}Zr^{94} + 2(_0n^1) + 6e^- + E. \tag{2.5}$$

The energy $E$ released can be easily determined from a standard table of nuclear data. The total mass of the initial and final elements is found to be 236.053 u and 235.832 u respectively. Here, u $= 1.660566 \times 10^{-27}$ kg is the atomic mass unit. Using Einstein's relation, this translates into $E = 206$ MeV per nuclear reaction or 0.88 MeV per nucleon of $_{92}U^{235}$. Macroscopically the energy released is equivalent to $84 \times 10^6$ MJ/kg representing an enormous gain of over one million compared to the burning of gasoline.

The two reasons why this powerful fission reaction can be converted into a practical method for producing electricity can now be identified. First, while it takes only one neutron to initiate a fission reaction the end products contain two neutrons. Actually, when averaged over all possible fission reactions, the average number of neutrons produced is slightly higher, approximately 2.4 per reaction. This neutron multiplication allows the buildup of a chain reaction, which has the advantage of making a fission reactor self-sustaining for several years before the fuel has to be replaced. All that is required is a sufficient mass of fuel to minimize the loss of neutrons.

The second reason is associated with the fact that fission reactions are initiated by an electrically neutral particle, the neutron. This is important because a neutron can easily penetrate the electron cloud surrounding the atom and gain close proximity to the nucleus itself. The electromagnetic Coulomb force of the nucleus has no impact on the neutron and cannot repel it from the nuclear interaction region. The result is that it is relatively easy for low-energy neutrons to produce fission reactions with $_{92}U^{235}$, a definite advantage in terms of energy balance, economics, and the ability to operate the reactor in a regime where the temperature is low enough that the fuel remains a solid.

These two reasons are why fission reactions are an effective means of producing nuclear energy for the practical production of electricity.

## 2.4 Nuclear energy by fusion

### 2.4.1 Neutron initiated nuclear reactions in light elements

Following the path of heavy atom fission it is instructive to ask whether large numbers of fission or fusion reactions can be made to occur in light elements by slow neutron bombardment leading to a chain reaction. Unfortunately the answer turns out to be "no"! When the reason for this is understood, it will then become clear what path must instead be chosen to produce fusion energy.

Table 2.1. *Properties of the nuclei of the primary light elements involved in fusion reactions. The mass and charge of each nuclei are given by Au and Ze where u $= 1.6605655 \times 10^{-27}$ kg and e $= 1.6021892 \times 10^{-19}$ coulombs.*

| Element | Symbol | Mass number $A$ | Charge number $Z$ |
|---|---|---|---|
| Electron | $e^-$ or e | 0.000549 | $-1$ |
| Neutron | $_0n^1$ or n | 1.008665 | 0 |
| Hydrogen (proton) | $_1H^1$ or p | 1.007276 | 1 |
| Deuterium | $_1H^2$ or D | 2.013553 | 1 |
| Tritium | $_1H^3$ or T | 3.015501 | 1 |
| Helium-3 | $_2He^3$ | 3.014933 | 2 |
| Helium-4 (alpha) | $_2He^4$ or $\alpha$ | 4.001503 | 2 |
| Lithium-6 | $_3Li^6$ | 6.013470 | 3 |
| Lithium-7 | $_3Li^7$ | 7.014354 | 3 |

The explanation begins by examining two hypothetical nuclear reactions that correspond to the neutron-driven fission or fusion of light elements. The simple calculations below show how neither leads to a successful path. In the case of light element fission consider the bombardment of a deuterium nucleus ($_1H^2$) with a neutron. The relevant reaction can be written as

$$_0n^1 + {_1H^2} \rightarrow {_1H^1} + 2(_0n^1) + E. \tag{2.6}$$

This reaction leads to the desired neutron multiplication. However, the energy released as calculated from the nuclear data in Table 2.1 show that $E = -2.23$ MeV. The negative sign indicates that energy is not actually released but must be supplied as an input to make the reaction take place. Clearly this would be unacceptable as a power source. Other light element fission reactions all have the same undesirable property.

Consider now the alternative, light element fusion by neutron bombardment. As a hypothetical example again assume that a neutron collides with a deuterium nucleus. The resulting nuclear fusion reaction can be written as

$$_0n^1 + {_1H^2} \rightarrow {_2He^3} + e^- + E. \tag{2.7}$$

In this case the nuclear data show that $E = +6.27$ MeV. The fusion reaction is energetically favorable for power production. However, the reaction consumes neutrons. One neutron is needed to initiate the reaction but none remains after fusion takes place. Since there are no readily available sources of neutrons, this reaction is not self-sustainable. Therefore, it too is unacceptable as a practical power source. This same undesirable property characterizes all other neutron-driven light element fusion reactions.

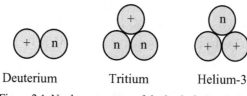

Deuterium              Tritium              Helium-3

Figure 2.1 Nuclear structure of the basic fusion fuels.

### 2.4.2 Light element fusion reactions

The simple examples just described show that the advantages of heavy element fission do not directly carry over into light element fission or fusion. Even so, enough insight has been provided to learn how to exploit light element nuclear reactions for the production of energy. There are two main points to be made. First, from the point of view of energy production one should focus on fusion rather than fission for light element nuclear reactions. These are the reactions that produce rather than consume energy. Second, the reliance on neutrons to initiate fusion reactions must be eliminated since these reactions consume neutrons and no external sources are available.

How then can light element fusion reactions be initiated? The basic idea is to replace the neutron with another light element; that is, generate a nuclear reaction by having two light elements bombard each other, for instance two colliding deuterium nuclei. The advantage of this idea is that the lack of a chain reaction is easily overcome by simply providing a continuous supply of deuterium, which, unlike a neutron supply, is readily and inexpensively available.

The disadvantage is that for two deuterium atoms to undergo a nuclear reaction, their nuclei must be in very close proximity to each other, typically within a nuclear diameter. At these close distances, the inter-particle Coulomb potential produces a strong repulsive force between the two positively charged nuclei, which diverts the particle orbits and greatly reduces the likelihood of a nuclear reaction. If the deuterium nuclei have sufficiently high energies the repulsive Coulomb force can be overcome, and this is indeed the strategy behind all current fusion research. The energies required, and the related physics issues are topics of major importance to fusion. They are discussed in detail in the next chapter. For the moment, however, these issues are not considered. Instead, attention is focused solely on the nuclear energy production of various fusion reactions without regard for how easy or difficult it may be to produce these reactions.

Studies of the nuclear properties of light element fusion indicate that three such reactions may be advantageous for the production of nuclear energy. These involve deuterium, tritium, and helium-3, an isotope of helium. A schematic diagram of the nuclear structure of each of these nuclei is shown in Fig. 2.1. The relevant nuclear fusion reactions are given below. For simplicity of notation, hereafter in the text the following symbol replacements are made: neutron, $_0n^1 \rightarrow n$; hydrogen nucleus (proton), $_1H^1 \rightarrow p$; deuterium nucleus (deuteron), $_1H^2 \rightarrow D$; tritium nucleus (triton), $_1H^3 \rightarrow T$; helium nucleus (alpha), $_2He^4 \rightarrow \alpha$; helium-3 nucleus, $_2He^3 \rightarrow He^3$.

## The D–D reaction

As its name implies, the D–D reaction produces fusion energy by the nuclear interaction of two deuterium nuclei. This is the most desirable reaction in the sense of a virtually unlimited supply of inexpensive fuel, easily extracted from the ocean. However, as will be shown in the next chapter, it is the most difficult of the three reactions to initiate. The D–D reaction actually has two branches, each occurring with an approximately equal likelihood. The relevant reactions are as follows:

$$D + D \rightarrow He^3 + n + 3.27\,MeV, \tag{2.8}$$

$$D + D \rightarrow T + p + 4.03\,MeV. \tag{2.9}$$

In terms of energy content the two reactions produce 0.82 and 1.01 MeV per nucleon respectively. Macroscopically this is equivalent to $78 \times 10^6$ and $96 \times 10^6$ MJ/kg of deuterium, typical of nuclear energy yields. The difficulty of initiating D–D fusion is the reason that this reaction is not the primary focus of current fusion research.

## The D–He$^3$ reaction

This reaction fuses a deuterium nucleus with a helium-3 nucleus. The reaction is also difficult to achieve, but less so than for D–D. However, it requires helium-3 as a component of the fuel and there are no natural supplies of this isotope on earth. Even so, the reaction is worth discussing since the end products are all charged particles. From an engineering point of view charged particles are more desirable than neutrons for extracting energy as they greatly reduce the problems associated with materials activation and radiation damage. They also offer the possibility of converting the nuclear energy directly into electricity without passing through an inefficient steam cycle. The reaction is

$$D + He^3 \rightarrow \alpha + p + 18.3\,Mev. \tag{2.10}$$

The energy released per reaction is impressive, even by nuclear standards. The 18.3 MeV corresponds to 3.66 MeV per nucleon, which is macroscopically equivalent to $351 \times 10^6$ MJ/kg of the combined D–He$^3$ fuel. Note that the reaction is not completely free of neutrons since some D–D and next generation D–T reactions will also occur, producing neutrons. The low He$^3$ availability combined with the difficulty of initiating D–He$^3$ fusion are the reasons that current fusion research is not focused around this reaction.

## The D–T reaction

The D–T reaction involves the fusion of a deuterium nucleus with a tritium nucleus. It is the easiest of all the fusion reactions to initiate (although its initiation still much more difficult than that of $_{92}U^{235}$ fission reactions). In terms of energy desirability issues, D–T reactions produce large numbers of neutrons and require a supply of tritium in order to be capable of continuous operation, but there is no natural tritium on earth. Furthermore, the tritium is radioactive with a half-life of 12.26 years. The D–T reaction, nevertheless, produces a significant amount of nuclear energy. It can be written as

$$D + T \rightarrow \alpha + n + 17.6 \, \text{MeV}. \tag{2.11}$$

This corresponds to 3.52 MeV per nucleon and is macroscopically equivalent to $338 \times 10^6$ MJ/kg. In spite of the problems associated with tritium and neutrons, the D–T reaction is the central focus of worldwide fusion research, a choice dominated by the fact that it is the easiest fusion reaction to initiate.

Having made this choice how does one deal with the tritium and neutron problems? Many years of fission research have taught nuclear engineers how to handle material activation and radiation damage resulting from high-energy neutrons. The same holds true for radioactivity associated with tritium. The solutions are far from simple but they are by now well established.

The one outstanding problem is the tritium supply. The solution is to breed tritium in the blanket surrounding the region of D–T fusion reactions. The chemical element that is most favorable for breeding tritium is lithium. The nuclear reactions of primary interest are

$$_3\text{Li}^6 + n(\text{slow}) \rightarrow \alpha + T + 4.8 \, \text{MeV},$$
$$_3\text{Li}^7 + n(\text{fast}) \rightarrow T + \alpha + n - 2.5 \, \text{MeV}. \tag{2.12}$$

Both reactions produce tritium although the first reaction generates energy while the second one consumes energy. Also natural lithium comprises 7.4% $_3\text{Li}^6$ and 92.6% $_3\text{Li}^7$. Even though there is a much larger fraction of $_3\text{Li}^7$, nuclear data show that the $_3\text{Li}^6$ reaction is much easier to initiate and as a result it is this reaction that dominates in the breeding of tritium.

With respect to the $_3\text{Li}^6$ reaction, if there were no loss of neutrons, then each n consumed in fusion would produce one new T by breeding with the fusion produced neutron: the breeding ratio would be 1.00. In a practical reactor, however, there are always some unavoidable neutron losses. Thus, some form of neutron multiplication is required. Also needed is a method of slowing down the high-energy fusion neutrons since the reaction in Eq. (2.12) is most easily initiated with slow, low-energy neutrons.

A discussion of these issues is deferred until Chapter 5. For present purposes one should assume that the issues have been satisfactorily resolved. Consequently, breeding T from $_3\text{Li}^6$ solves the problem of sustaining the tritium supply, assuming adequate supplies of lithium are available. The known reserves of lithium are sufficiently large to last thousands of years so fuel availability is not a problem. On the longer time scale, the goal would be to develop D–D fusion reactors.

### 2.4.3 Energy partition in fusion reactions

The large amounts of energy released in fusion reactions appear in the form of kinetic energy of the end products. It is important, particularly for the D–T reaction where one end product is electrically charged and the other is not, to determine how the energy is apportioned between the two end products. The apportionment can be easily determined by making use of the well-satisfied assumption that the energy and momentum of each end product far exceeds that of the initial fusing nuclei.

The calculation proceeds as follows. The two end products are designated by the subscripts 1 and 2. The initial fusing particles are assumed to be essentially at rest (with respect to the end product velocities). The conservation of energy and momentum relations before and after the fusion reaction involve only the end products and thus have the forms

$$\frac{1}{2}m_1v_1^2 + \frac{1}{2}m_2v_2^2 = E, \tag{2.13}$$

$$m_1v_1 + m_2v_2 = 0. \tag{2.14}$$

These equations can easily be solved simultaneously for $v_1$ and $v_2$ and the corresponding kinetic energies. The results are

$$\frac{1}{2}m_1v_1^2 = \frac{m_2}{m_1 + m_2}E, \tag{2.15}$$

$$\frac{1}{2}m_2v_2^2 = \frac{m_1}{m_1 + m_2}E. \tag{2.16}$$

Observe that the kinetic energies are apportioned inversely with the mass; in other words the lighter particle carries most of the energy. As an example consider the D–T reaction where $E = 17.6$ MeV. The end products consist of an alpha particle and a neutron with $m_\alpha / m_n = 4$. Thus the kinetic energy of the alpha particle is equal to $(1/5)E = 3.5$ MeV while that of the neutron is equal to $(4/5)E = 14.1$ MeV. The neutron energy is four times larger than that of the alpha particle. One can then rewrite the D–T fusion reaction in the slightly more convenient form

$$D + T \rightarrow \alpha(3.5 \text{ MeV}) + n(14.1 \text{ MeV}). \tag{2.17}$$

This completes the discussion of the basic properties of fusion reactions. The remainder of the book is primarily concerned with the D–T reaction, as it is this reaction that dominates the world's fusion research program.

## 2.5 The binding energy curve and why it has the shape it does

In this section we provide a physical explanation of the observation that nuclear reactions are most readily initiated for either heavy elements (i.e., fission) or light elements (i.e., fusion) but not with intermediate elements. The explanation is presented in two parts. First, an examination of the curve of binding energy vs. atomic mass, as obtained from experimental measurements, shows that the binding forces holding the nuclei of either light or heavy elements together are weaker than those of intermediate elements. This is the basic explanation of why it is easier to initiate nuclear reactions with elements at the extreme ends of atomic mass. Having established this conclusion, one can then address the second part of the explanation, which is concerned with why the binding energy curve actually has

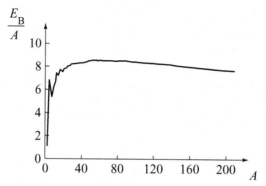

Figure 2.2 Binding energy per nucleon vs. mass number for the dominant form of each chemical element.

the shape it does. It is shown that the shape arises from a geometric competition between the strong short-range nuclear force and the weak long-range Coulomb force.

### 2.5.1 The binding energy curve

Consider a primary chemical element whose nucleus contains $N$ neutrons and $Z$ protons. Note that the integer sum $N + Z$ is very nearly, but not exactly, equal to the actual experimentally measured mass number $A$: $N + Z \approx A$, where $m_A = A$ u is the nuclear mass. A comparison of the actual nuclear mass with the total mass of the isolated individual particles making up the nucleus shows that

$$N m_n + Z m_p > m_A. \tag{2.18}$$

The difference in mass can be thought of as being converted into binding energy to hold the nucleus together. Specifically, the binding energy is defined as

$$E_B \equiv (N m_n + Z m_p - m_A)c^2. \tag{2.19}$$

An amount of energy equal to $E_B$ would have to be added to the nucleus to break it apart into its separate components. A somewhat more convenient quantity is the binding energy per nucleon, defined as $E_B/A$. This quantity is a measure of the average energy binding each nucleon to the nucleus.

As an example one can calculate the binding energy per nucleon for fluorine which has $N = 10$, $Z = 9$, and $A = 18.998\,40 \approx 19$. The relevant masses are $m_n = 1.008\,66$ u, $m_p = 1.007\,28$ u, and $m_A = 18.998\,40$ u. This corresponds to a mass differential $N m_n + Z m_p - m_A = 0.154$ u. Substituting into Eq. (2.19) yields $E_B = 143\,\mathrm{MeV}$ and $E_B/A = 7.5$ MeV/nucleon. Observe that the energies binding the particles together in the nucleus are quite large.

This calculation can be repeated for all the elements. The results are plotted as a curve of binding energy per nucleon vs. mass number as shown in Fig. 2.2 for the dominant form of each element. Note that $E_B/A$ is small for both light and heavy elements and is maximized at an intermediate value corresponding to iron (i.e., $A \approx 56$). The conclusion is that relative

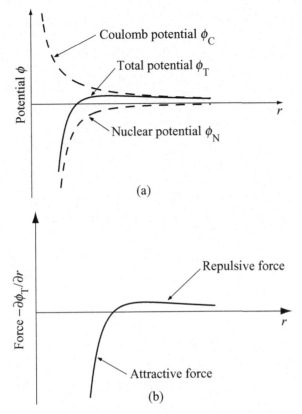

Figure 2.3 (a) The Coulomb, nuclear, and total potentials as functions of radius, and (b) the corresponding total force $F = -\partial\phi_T/\partial r$ as a function of radius.

to intermediate elements, the nuclear forces binding the nucleons together in either light or heavy elements are weaker. Thus, it is easier to initiate nuclear reactions in heavy or light elements. Furthermore, fusing light elements together pushes one up the curve to more strongly bound heavier elements implying that energy will be released by such reactions. Similarly, splitting heavy elements apart by fission also pushes one up the curve to lighter elements, implying a release of energy.

The shape of the binding energy curve explains why light and heavy elements dominate the nuclear reactions for energy production. The final question to be asked is why does the $E_B/A$ vs. $A$ curve have the shape that it does.

### 2.5.2 *The shape of the binding energy curve*

The shape of the binding energy curve is a consequence of a competition between the strong short-range nuclear force and the weak long-range Coulomb force. Schematic diagrams of the combined nuclear and Coulomb potential energies and the corresponding forces felt by a charged particle are shown in Fig. 2.3. Based on this figure a simple model is presented

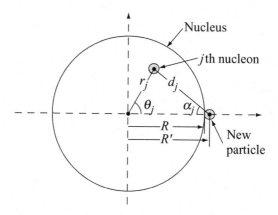

Figure 2.4 Geometry of the simple binding energy model. Note that $R' = R + r_0$.

below that predicts the qualitative shape of the binding energy curve. At the outset it should be emphasized that the model neglects all nuclear and quantum effects. The nucleus is assumed to consist of small hard-sphere protons and neutrons of radius $r_0$, behaving in accordance with classical Newtonian mechanics. This is all that is required to obtain the desired results.

The basic idea of the model is to assume the existence of a spherical nucleus comprising $N + Z \approx A$ nucleons and to then calculate the net attractive force (i.e., the nuclear force minus the Coulomb force) on a new charged particle placed on the surface of the sphere. The geometry is illustrated in Fig. 2.4. The goal of the calculation is to determine the dependence of this force on $A$. Since the magnitude of the force is a good measure of how strongly the new particle would be bound to the nucleus, its dependence on $A$ should be qualitatively similar to the behavior of the binding energy curve.

The calculation proceeds as follows. Consider first the repulsive Coulomb force (proportional to $1/r^2$) on the new particle due to the $j$th particle in the nucleus. The component of force along $R$ is given by

$$\mathbf{F}_j^{(C)} = \frac{e^2}{4\pi\varepsilon_0} \frac{\cos\alpha_j}{d_j^2} \mathbf{e}_R. \tag{2.20}$$

The total Coulomb force along $R$ is obtained by summing over $j$ (note that the tangential components of the force cancel by symmetry after summing):

$$\mathbf{F}_R^{(C)} = \frac{1}{2} \frac{e^2}{4\pi\varepsilon_0} \sum_j \frac{\cos\alpha_j}{d_j^2} \mathbf{e}_R. \tag{2.21}$$

The additional factor $1/2$ is a reflection of the fact that only about one half of the nucleons are charged particles.

The sum over particles in Eq. (2.21) can be converted into an integral by recognizing that the volume occupied by each particle must be set approximately equal to the size of a differential volume element.

$$d\mathbf{r} \equiv r^2 \sin\theta \, dr \, d\theta \, d\phi \approx (4/3)\pi r_0^3. \tag{2.22}$$

The expression for the force becomes

$$\mathbf{F}_R^{(C)} \approx \left( \frac{e^2}{8\pi \varepsilon_0} \frac{3}{4\pi r_0^3} \int \frac{\cos\alpha}{d^2} d\mathbf{r} \right) \mathbf{e}_R, \tag{2.23}$$

where the subscript $j$ has been dropped from all particle quantities since they are now considered to be continuous variables. After some slightly tedious algebra and geometry, one can evaluate this integral analytically. In view of the simplicity of the model it makes sense to leave these details to the problem section rather than present them in the text. Also, the final form for the force can be expressed in terms of $A$ rather than $R$ by noting that the total volume of the nucleus is approximately equal to $A$ times the volume per nucleon:

$$\frac{4\pi}{3} A r_0^3 \approx \frac{4\pi}{3} R^3. \tag{2.24}$$

The desired expression for the $R$ component of the Coulomb force is now

$$\mathbf{F}_R^{(C)} \equiv F_R^{(C)} \mathbf{e}_R = \frac{F_0}{2} \frac{A}{(A^{1/3} + 1)^2} \mathbf{e}_R, \tag{2.25}$$

where $F_0 = e^2/4\pi \varepsilon_0 r_0^2$.

Since the "attractive" direction for the force is along $-R$, the "attractive" Coulomb force is $-F_R^{(C)}$, and is plotted as a function of $A$ in Fig. 2.5(a). As intuitively expected, the magnitude of the Coulomb force increases monotonically with $A$. More charges produce a larger force. Also for large $A$, the force scales as $A^{1/3}$. This is a consequence of the fact that the total charge enclosed in the nucleus increases with $R^3$ while the geometric behavior of the force decreases with $R^{-2}$. The net effect is proportional to the product of these contributions given by $R \sim A^{1/3}$.

Consider next the nuclear force. Perhaps the simplest model (in terms of evaluating integrals) that one can construct to demonstrate the effects of a strong short-range force is that of a central force that scales as $-K/r^4$. The value of $K$ can be determined by assuming that the nuclear and Coulomb forces are equal to each other in magnitude at a critical distance $r = r_c$ leading to $K = e^2 r_c^2/4\pi \varepsilon_0$. Here, $r_c$ is typically several nucleon radii; that is, $r_c = kr_0$ with $k$ assumed to be a known dimensionless number of order unity. Observe that for $r < r_c$ the nuclear force dominates the Coulomb force, while for $r > r_c$ the reverse is true. These are the desired qualitative properties of a strong short-range force.

Under these assumptions, the $R$ component of the nuclear force can be written as

$$\mathbf{F}_R^{(N)} \equiv F_R^{(N)} \mathbf{e}_R \approx - \left( \frac{e^2}{4\pi \varepsilon_0} \frac{3}{4\pi r_0^3} \int \frac{k^2 r_0^2 \cos\alpha}{d^4} d\mathbf{r} \right) \mathbf{e}_R. \tag{2.26}$$

The minus sign indicates that the nuclear force is in the negative $R$ (i.e., attractive) direction. After some additional algebra this integral can also be evaluated analytically. The desired

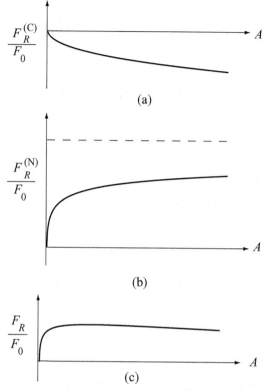

Figure 2.5 Normalized attractive forces vs. mass number $A$: (a) the Coulomb force $-F_R^{(C)}$; (b) the nuclear force $-F_R^{(N)}$; and (c) the total attractive force: $F_R = -(F_R^{(C)} + F_R^{(N)})$.

expression for the nuclear force as a function of $A$ has the form

$$F_R^{(N)} = -F_0 \frac{k^2 A}{(A^{1/3} + 1)^2 (2A^{1/3} + 1)}. \tag{2.27}$$

The "attractive" nuclear force $-F_R^{(N)}$ is plotted in Fig. 2.5(b) for $k = 3.4$. Note that the force increases with $A$ for small $A$, eventually leveling off at large $A$. This behavior is consistent with the following physical picture. For small nuclei, corresponding to small $A$, the attractive nuclear force increases as each new nucleon is added. However, after sufficient nucleons have been added, the nucleus becomes so large that new particles no longer feel the effect of distant nucleons because of the short-range nature of the nuclear force. In other words, the short-range nature of the force restricts new particles to feeling the effects of only a limited fixed number of nucleons (i.e., those lying approximately within a distance $r_c$), regardless of how large the total number of nucleons in the nucleus may be.

The total attractive nuclear force

$$F_R = -\left(F_R^{(C)} + F_R^{(N)}\right) \tag{2.28}$$

is plotted in Fig. 2.5(c). Observe that there is a qualitatively similar behavior between this curve and the binding energy curve ($E_B/A$ vs. $A$). The conclusion is that the shape of the binding energy curve is determined by a strong nuclear force whose attraction increases for small nuclei but which eventually saturates for large nuclei, ultimately becoming dominated by a weaker, but continually increasing repulsive Coulomb force. The result is that the binding energies for very light and very heavy elements are weaker than for intermediate mass elements. This explains why energy production from nuclear reactions is most effective at the ends of the spectrum of atomic masses.

## 2.6 Summary

The chemical reactions occurring in the burning of fossil fuels involve the rearrangement of the electron structure of the molecules involved, but conserve the individual chemical elements. The reactions are dominated by the electromagnetic force and tend to release energies on the order of a fraction of an electron volt per atom per reaction. Nuclear reactions on the other hand involve either the splitting or combining of the nuclei of the atoms involved, causing the transmutation of the initial fuel into new chemical elements. In these reactions the nuclear force dominates and energy releases on the order of mega-electron volts per nucleon per reaction are typical. Thus, nuclear reactions normally release about one million times more energy per reaction than chemical reactions.

There are two types of nuclear reactions, fission and fusion. Fission is easiest to achieve with very heavy elements, $_{92}U^{235}$ in particular. In these reactions, fission is initiated by a slow neutron and the resulting reactions produce on average 2.4 new neutrons. This generates a chain reaction, which allows a self-sustaining mode of operation.

Neutrons are not effective initiators for light element fusion reactors as they are consumed rather than produced in fusion reactors. Instead fusion results from the direct interaction of two positively charged nuclei. The need to overcome the repulsive Coulomb barrier in such reactions makes them more difficult to initiate than fission reactions. Since no chain reaction occurs, new fuel must be continually added to keep the operation continuous. The easiest way to initiate fusion is by the D–T reaction, which releases 17.6 MeV, 14.1 MeV in the neutron and 3.5 MeV in the alpha particle. The world's fusion energy research program is focused on the D–T reaction, the ease of initiation being of more importance than the problem of breeding tritium.

## Bibliography

Much of the material in Chapter 2 has been well known for many years. Various slightly different ways of understanding the material can be found in the references below, all of which include a strong focus on fusion energy.

### *The fusion reaction*

Dolan, T. J. (1982). *Fusion Research*. New York: Pergamon Press.

Glasstone, S. and Loveberg, R. H. (1960). *Controlled Thermonuclear Reactions.* Princeton, New Jersey: Van Nostrand.

Gross, R. (1984). *Fusion Energy.* New York: John Wiley & Sons.

Miyamoto, K. (2001). *Fundamentals of Plasma Physics and Controlled Fusion,* revised edn. Toki City: National Institute for Fusion Science.

Rose, D. J. and Clark, M. (1961). *Plasmas and Controlled Fusion.* Cambridge, Massachusetts: MIT Press.

Stacey, W. M. (1981). *Fusion Plasma Analysis.* New York: John Wiley & Sons.

Stacey, W. M. (2005). *Fusion Plasma Physics.* Weinheim: Wiley-VCH.

Wesson, J. (2004). *Tokamaks,* third edn. Oxford: Oxford University Press.

## Problems

2.1 Consider the energy released in fully "catalyzed D–D" fusion. In this process the $H^3$ and T produced by the pure D–D reaction react with other D nuclei leaving only alpha particles, protons, and neutrons. Calculate the average energy released per deuteron corresponding to the complete burn and compare it to that of the pure D–D reaction.

2.2 Calculate the energy content of the following fuels. Express each answer in joules.
   (a) A barrel of coal (200 lb) 1 lb coal $= 1.3 \times 10^4$ BTU.
   (b) 50 gallons of gasoline 1 gal $= 3 \times 10^7$ cal.
   (c) 50 gallons of sea water – fully catalyzed D–D.
   (d) 25 gallons of sea water plus an equal amount of T for D–T burn.
   (e) A barrel of uranium ore (4000 lb) (0.2% of the ore is uranium and of the uranium 0.7% is in the form of $_{92}U^{235}$). The fission of one $_{92}U^{235}$ nucleus releases about 200 MeV of energy.
   (f) One chocolate sundae, every day for one year (1 sundae $= 500$ food calories).

2.3 Some fusion engineering students decide to travel by car from Boston to Fort Lauderdale for their holiday break, a distance of 1500 miles. The students are driving a new fusion-powered automobile which burns fully catalyzed D–D fuel. If 0.0153% of the atoms in sea water are deuterium atoms how many gallons of sea water are required for the drive? For comparison, how many gallons of gasoline would be required for a car which averages 25 miles/gal.

2.4 One advantage of the D–D reaction as compared to the D–T reaction is that a larger fraction of the energy is delivered to charged particles. Assuming both branches of the D–D reaction have equal probability, calculate the fraction of energy in charged particles if the tritium is completely burned but the $He^3$ escapes before interacting with the D. Compare this to the D–T value.

2.5 Carry out the algebraic steps required to verify Eqs. (2.25) and (2.27).

2.6 To test the sensitivity of the nuclear force model to the shape of the binding energy curve, repeat the derivation of Eq. (2.27) using the following central force:

$$\mathbf{F}_j^{(N)} = -\frac{1}{4\pi\varepsilon_0 r_c^2}\mathrm{cosech}^2\left(\frac{r^2}{r_c^2}\right)\mathbf{e}_R.$$

As in the text write $r_c = kr_0$ and adjust $k$ so that the maximum of the binding energy curve occurs for iron. This will probably involve a numerical calculation. Compare the result with the one obtained in the text.

# 3

# Fusion power generation

## 3.1 Introduction

The laws of physics have shown that a large amount of kinetic energy is released every time a nuclear fusion reaction occurs. Determining the conditions under which this energy can be converted into useful societal applications, such as the production of electricity or hydrogen, requires a substantial amount of analysis and is discussed in Chapters 3–5. The logic of the presentation, starting from the end goal and working backwards, is as follows. The desirability of fusion ultimately depends upon the design of practical, economical reactors that have a favorable power balance: $P_{out} \gg P_{in}$. Two qualitatively different concepts have been proposed to achieve this goal, magnetic fusion and inertial fusion. This book focuses on magnetic fusion.

The end goal of Chapters 3–5 is to present a simple design for a magnetic fusion reactor. In order to develop the design, knowledge of the macroscopic power balance in a magnetic fusion system is required as input. Power balance is discussed in Chapter 4. As might be expected, this analysis involves a variety of physical phenomena representing various sources and sinks of power. Many of these phenomena will be familiar to readers, including thermal conduction, convection, and compression. However, the macroscopic power generated by nuclear fusion reactions, which is clearly the most crucial source term in the system, will probably be less familiar. Similarly, the radiation losses due to the Coulomb interactions between charged particles may also be less familiar. The fusion power generation and radiation losses are the subjects of Chapter 3 and serve as inputs to the discussion on power balance.

Chapter 3 begins with the calculation of the fusion power, and requires the introduction of several microscopic quantities: cross section, mean free path, and collision frequency. These quantities effectively determine the probability of initiating fusion reactions in a given D–T fuel mixture. A transition is then made from the micro to the macro world through the concept of the reaction rate. This is essential if one wants to be able to calculate the total power produced in a macroscopic volume of D–T fuel as would exist in a reactor.

Next, the power lost due to radiation is calculated. This loss is known as Bremsstrahlung radiation and results from the Coulomb interaction between charged particles. A simple

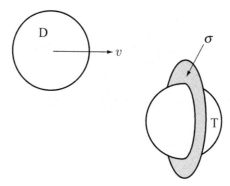

Figure 3.1 Cross sectional area $\sigma$ for a "hard-sphere" collision.

heuristic derivation is presented that yields the correct scaling dependence on density and temperature along with a semi-quantitatively accurate scaling coefficient.

As stated, knowledge of the fusion power generation and radiation losses is essential in determining the overall power balance in a magnetic fusion reactor.

## 3.2 The concepts of cross section, mean free path, and collision frequency

The first step along the path of calculating the power generated by D–T fusion reactions is to investigate the basic microscopic physics of nuclear collisions. Once this knowledge is in hand it is relatively straightforward to move to the macroscopic scale in order to determine the power generated in a magnetic fusion reactor. The microscopic concepts involve include the cross section, mean free path, and collision frequency.

### 3.2.1 Cross section

The cross section characterizes in some quantitative form the probability that a pair of D and T nuclei will undergo a nuclear fusion reaction. To be specific, assume the T nucleus is stationary and the D nucleus is moving towards it with a velocity $v$ as shown in Fig. 3.1. The T is considered to be the target particle while the D is the incident particle. Imagine now that a spherically symmetric force field surrounds the target particle. The projection of this sphere perpendicular to the motion of the incident particle is a circle of cross sectional area $\sigma$. If the incident particle passes through the area $\sigma$, then the force exerted on it by the target particle is sufficiently strong that a nuclear reaction takes place. For obvious reasons such an interaction is called a "collision". If on the other hand the incident particle does not pass through the area $\sigma$, then the target force is sufficiently weak that there is no collision.

While the concept of a cross section is appealingly simple, it is not at all obvious what magnitude and functional dependence on $v$ and geometry to choose for $\sigma$. This depends crucially upon the nature of the forces acting between the two particles. For fusion reactions

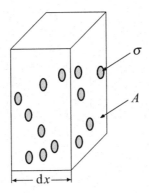

Figure 3.2 Volume of target particles. Each target particle blocks out a cross sectional area $\sigma$.

dominated by the short-range nuclear force one expects, and indeed finds, that the cross section is on the order of a nuclear diameter. The short-range nature of the force also implies that incident particles passing outside this area would experience at most a very weak interaction, not sufficient to produce fusion. Thus the sharp boundary between having and not having a collision as illustrated in Fig. 3.1 provides a reasonable approximation for nuclear collisions and the corresponding cross section is often referred to as a "hard-sphere" cross section. As might be expected the actual behavior of the cross section is more complicated and must in general be determined experimentally. Actual fusion cross sections are presented shortly. For the moment it suffices to assume that $\sigma$ is a known constant quantity, independent of $v$ and geometry, characterizing hard-sphere collisions. These assumptions will be relaxed shortly.

### 3.2.2 Mean free path

Intuitively one expects the value of $\sigma$ to be related to the probability of having a fusion collision. Large $\sigma$ implies a large target. Hence it should be relatively easy to initiate collisions – the probability is high. Small $\sigma$ represents a small target making collisions less likely – the probability is low. The cross section $\sigma$ is related to the probability of a collision through the concept of the "mean free path."

To understand the concept, consider a volume of target particles of width $dx$ and macroscopic cross sectional area $A$ as shown in Fig. 3.2. The volume of interest is $V = A dx$. Assume that the number density of target particles is denoted by $n_1$. The total number of particles in the target volume is then given by $N_1 = n_1 V = n_1 A dx$.

Each of these target particles blocks out an area $\sigma$ within which it can collide with incident particles. Assume now that the target density is sufficiently low and the cross section sufficiently small that from the point of view of an incident particle, the $\sigma$s do not overlap. This is typically a good approximation for most materials including fusion plasmas since the nuclei occupy only a small portion of the available volume. Under this

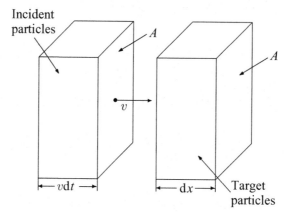

Incident
particles

*A*

*A*

*v*

— *v*d*t* —

— d*x* —

Target
particles

Figure 3.3 Flux of incident particles with velocity $v$ impinging on a volume of stationary target particles.

approximation the fraction of the total area blocked by target particles, d$F$, can be written as

$$dF = \frac{\sigma N_1}{A} = \frac{\sigma n_1 A dx}{A} = \sigma n_1 dx. \qquad (3.1)$$

The quantity d$F$ can be interpreted as the probability that one incoming particle will have a nuclear collision as it traverses a distance d$x$. Equivalently, the quantity

$$\frac{dF}{dx} = \sigma n_1 \qquad (3.2)$$

is the probability of nuclear collision per unit length.

This probability can be used to calculate a very useful quantity known as the mean free path $\lambda_m$. The calculation proceeds as follows. Assume that there is not one, but many incident particles, with a density $n_2$, all approaching the target with a velocity $v$ as shown in Fig. 3.3. The flux of incident particles, $\Gamma$, is by definition, the total number of particles crossing an area $A$ per unit time, per unit area:

$$\Gamma = \frac{N_2}{A dt} = \frac{n_2 A v dt}{A dt} = n_2 v. \qquad (3.3)$$

The portion of this flux, d$\gamma$, having a nuclear collision as the particles move along a distance d$x$ = $v$d$t$ is just

$$d\gamma = dF\,\Gamma = \sigma n_1 \Gamma dx. \qquad (3.4)$$

Equivalently, the quantity $-d\gamma$ is equal to the decrease in the total flux of particles not having a collision after traversing a distance d$x$: $-d\gamma = \Gamma(x + dx) - \Gamma(x) = d\Gamma$. The change in non-colliding flux can thus be written as

$$d\Gamma = -\sigma n_1 \Gamma dx. \qquad (3.5)$$

This is a simple ordinary differential equation describing the spatial evolution of the incident flux. Under the reasonable assumptions that $\sigma$ and $n_1$ are approximately constant over the length scales of interest, Eq. (3.5) can be easily integrated, yielding

$$\Gamma = \Gamma_0 e^{-x/\lambda_\mathrm{m}}, \tag{3.6}$$

where $\lambda_\mathrm{m} = 1/n_1\sigma$ is known as the mean free path. The mean free path can be interpreted in two ways. First, it clearly represents the characteristic e-folding decay length of the incident flux that has not yet undergone a collision. Second, it can also be interpreted as the average distance that an incident particle travels before it has a collision. Note that $\lambda_\mathrm{m}$ is inversely proportional to $n_1$ and $\sigma$. This makes intuitive sense in that a high density of target particles and/or a large cross section imply that an incident particle will not have to travel very far before undergoing a collision.

### 3.2.3 Collision frequency

The final quantities of interest are the collision frequency $\nu_\mathrm{m}$ and its inverse $\tau_\mathrm{m}$, the mean time between collisions, often simply called the collision time. These are defined as follows. The previous analysis has shown that a typical incident particle will travel a distance $\lambda_\mathrm{m}$ before having a collision. If the particle is moving with a velocity $v$ then the corresponding time before having a collision is

$$\tau_\mathrm{m} = \frac{\lambda_\mathrm{m}}{v} = \frac{1}{n_1 \sigma v}. \tag{3.7}$$

The collision frequency is defined as the inverse of the collision time:

$$\nu_\mathrm{m} = \frac{1}{\tau_\mathrm{m}} = n_1 \sigma v. \tag{3.8}$$

Physically, on average $\nu_\mathrm{m}$ particles will undergo a nuclear collision each second. Clearly, for a nuclear interaction an incident particle undergoes only one collision, after which the particle no longer exists in its original form. However, there are many other types of collisions, Coulomb collisions between charged particles in particular, wherein a single particle can make repeated collisions without losing its identity. For these types of multiple interactions, the quantity $\nu_\mathrm{m}$ can also be interpreted as the average number of collisions per second made by a single particle.

The concepts of cross section, mean free path, collision time and collision frequency are used extensively in many different aspects of plasma physics and fusion energy and will appear frequently throughout the text. In the context of evaluating fusion power generation, the collision frequency, which is proportional to $\sigma v$ and which represents the number of collisions per second, is a good measure of the ease or difficulty of producing the desired D–T reactions.

## 3.3 The reaction rate

### 3.3.1 The hard-sphere reaction rate

The concepts just discussed involve the microscopic physics of nuclear collisions between individual particles. The next step on the path of calculating the power produced in a fusion reactor is to convert from the micro to the macro world. This task is accomplished by the introduction of the reaction rate $R_{12}$, which determines the number of fusion collisions per unit volume per unit time. Once the value of $R_{12}$ is known, it is then straightforward to calculate the fusion power produced per unit volume.

The calculation of the reaction rate is an easy task if one again focuses on hard-sphere collisions. To begin, note that in a time $dt = dx/v$, $n_2 A dx$ incident particles will pass through the target volume. Of these, the number having a collision is given by $dF(n_2 A dx)$. The reaction rate is the number of particles having a collision per unit volume per unit time – the number of collisions per cubic meter per second. Thus,

$$R_{12} = \frac{dF n_2 A \, dx}{A \, dx \, dt} = \sigma n_1 n_2 \frac{dx}{dt} = n_1 n_2 \sigma v. \tag{3.9}$$

The fusion power density can now easily be calculated. If each fusion collision generates an energy $E_f$, then the total energy produced per unit volume per second is $E_f R_{12}$. This quantity is equal to the fusion power density as measured in watts per cubic meter and can be written as

$$S_f = E_f n_1 n_2 \sigma v \quad \text{W/m}^3. \tag{3.10}$$

This is the desired expression. It is important to emphasize that the energy $E_f$ in Eq. (3.10) may correspond to the total fusion energy, the alpha particle energy, or the neutron energy depending upon the context in which the power density is being calculated. The distinctions are straightforward but nonetheless crucial, and are discussed in detail in the next chapter. Lastly, one should recognize that while Eq. (3.10) is conceptually correct, it is in an overly simplified form because of the focus on hard-sphere collisions.

### 3.3.2 Reaction rate generalized to include multiple velocities

#### The distribution function

The first important generalization of the reaction rate concept is to take into account that the target and incident particles both have a distribution of random velocities; that is, not all the target particles are stationary and not all the incident particles are moving with the same velocity $v$. The generalization is implemented by introducing a higher-level density function that contains far more information about the particles than the simple number density $n$. This higher-level function is called the distribution function and is denoted by the symbol $f$.

To understand the difference between $n$ and $f$ consider a small cubic volume of physical space $d\mathbf{r} = dx \, dy \, dz$. The total number of particles in this volume at a given instant of time is just $n(\mathbf{r}, t) d\mathbf{r}$, and includes all particles, regardless of their velocities. Suppose now that one instead wants to know the total number of a certain subclass of particles in this volume – those particles moving only within a restricted velocity range. Specifically, assume the subclass is defined as those particles moving with a velocity in the range between $v_x$ and $v_x + dv_x$, $v_y$ and $v_y + dv_y$, and $v_z$ and $v_z + dv_z$. The distribution function is defined by first generalizing the concept of a 3-D physical volume $d\mathbf{r}$ into a 6-D phase space volume $d\mathbf{r}\,d\mathbf{v}$, where $d\mathbf{v} = dv_x dv_y dv_z$. The distribution function is the generalization of the number density. The total number of particles in the 6-D differential phase space volume (i.e., those located within a physical volume $d\mathbf{r}$ centered about the point $\mathbf{r}$ moving within the range of velocities $d\mathbf{v}$ centered about the velocity $\mathbf{v}$) at a given instant of time is just the product of the density times the volume: $f(\mathbf{r}, \mathbf{v}, t) d\mathbf{r}\,d\mathbf{v}$.

Clearly $f$ contains much more information than $n$. In fact $n$ can be easily obtained from $f$. If one sums (i.e., integrates) over all velocities at a given $\mathbf{r}$, then one obtains the total number of particles in the physical volume $d\mathbf{r}$ including all particle velocities. Dividing this number by the volume $d\mathbf{r}$ then yields $n$, the number of particles per unit volume.

$$n(\mathbf{r}, t) = \frac{dx\,dy\,dz \int f\,dv_x dv_y dv_z}{dx\,dy\,dz} = \int f\,d\mathbf{v}. \tag{3.11}$$

Note that even though $d\mathbf{r}$ and $d\mathbf{v}$ are treated as differential quantities, they should be viewed as small quantities, varying negligibly on the length and velocity scales of interest but yet still containing a sufficiently large number of particles so that a statistical interpretation makes sense.

Similarly, if one wants to calculate the average velocity $\mathbf{u}$ of all the particles in $d\mathbf{r}$ centered at the location $\mathbf{r}$, the procedure is to choose a velocity $\mathbf{v}$, multiply by the number of particles moving with this $\mathbf{v}$, sum (i.e., integrate) over all possible values of $\mathbf{v}$, and then divide by the total number of particles in $d\mathbf{r}$:

$$\mathbf{u}(\mathbf{r}, t) = \frac{d\mathbf{r} \int \mathbf{v} f\,d\mathbf{v}}{n\,d\mathbf{r}} = \frac{1}{n} \int \mathbf{v} f\,d\mathbf{v}. \tag{3.12}$$

In fact the mean value of any quantity $W$ can easily be found by calculating the distribution function weighted average as follows:

$$\langle W \rangle = \frac{1}{n} \int W f\,d\mathbf{v}. \tag{3.13}$$

### First generalization of the reaction rate

The reaction rate $R_{12} = n_1 n_2 \sigma v$ is generalized to include a distribution of velocities through several simple steps. First, the densities $n_1$ and $n_2$ must be replaced by $n_1 \rightarrow f_1(\mathbf{r}, \mathbf{v}_1, t) d\mathbf{v}_1$ and $n_2 \rightarrow f_2(\mathbf{r}, \mathbf{v}_2, t) d\mathbf{v}_2$. Second, since the choice of labels "target" and "incident" is arbitrary, the velocity $v$ appearing in the expression for $R_{12}$ actually represents the relative velocity between the two particles: $v = |\mathbf{v}_2 - \mathbf{v}_1|$. Finally, one must in general allow the cross section to be a function of relative velocity. For example, in a D–T collision if the relative velocity between the two particles is very small, the Coulomb repulsion will strongly

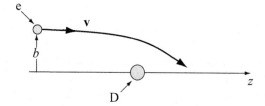

Figure 3.4 Electron with velocity **v** and impact parameter $b$ "colliding" with a stationary deuterium nucleus.

deflect the orbits preventing the particles from getting very close to one another. In other words, for small relative velocities the cross section for a nuclear collision will be very, very small. As the relative velocity increases, the Coulomb repulsion has a smaller effect and the cross section increases. Thus, the cross section is in general a function of relative velocity: $\sigma = \sigma(|\mathbf{v}_2 - \mathbf{v}_1|)$. Combining these modifications leads to the following generalization of the reaction rate:

$$R_{12} = \int f_1(\mathbf{v}_1) f_2(\mathbf{v}_2) \sigma(|\mathbf{v}_2 - \mathbf{v}_1|) |\mathbf{v}_2 - \mathbf{v}_1| \, d\mathbf{v}_1 d\mathbf{v}_2$$
$$= n_1 n_2 \langle \sigma v \rangle. \tag{3.14}$$

Here, the definition of $\langle W \rangle$ from Eq. (3.13) has been used to obtain the second form of $R_{12}$. Lastly, note that for like particle collisions

$$R_{11} = \tfrac{1}{2} \int f_1(\mathbf{v}_1) f_1(\mathbf{v}_2) \sigma(|\mathbf{v}_2 - \mathbf{v}_1|) |\mathbf{v}_2 - \mathbf{v}_1| \, d\mathbf{v}_1 d\mathbf{v}_2$$
$$= \tfrac{1}{2} n_1^2 \langle \sigma v \rangle. \tag{3.15}$$

The factor $\frac{1}{2}$ appears because when integrating over the two velocities $\mathbf{v}_1$, $\mathbf{v}_2$ each collision is counted twice.

### Second generalization of the reaction rate

The expression for the reaction rate just presented works well for calculating the fusion power produced by short-range nuclear collisions in which each collision produces an identical amount of energy $E_f$. In other words, the generalization of Eq. (3.10) can be written as

$$S_f = E_f n_1 n_2 \langle \sigma v \rangle. \tag{3.16}$$

The second generalization corresponds to the situation where the interaction force is long range in nature and the particle quantity of interest itself depends upon the relative velocity and the geometry of the collision. This is typical of Coulomb collisions, which are very important in fusion physics.

To understand the issue, consider the Coulomb interaction of an electron (the incident particle) with a deuterium nucleus (the target) as shown in Fig. 3.4. Note that the geometry of

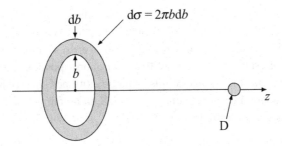

Figure 3.5 Illustration of the differential cross section $d\sigma = 2\pi b\,db$.

the orbit is characterized by the "impact parameter" $b$. To be specific assume it is of interest to calculate the change in $z$ directed momentum of the electron, $\Delta m v_z$, as it interacts with the D nucleus. For a Coulomb collision the role of the quantity $\Delta m v_z$ is analogous to that of $E_f$ for a nuclear collision. Intuitively, one expects that for a fixed value of $b$ the D will minimally perturb the orbit of a fast electron since it passes by so quickly. There is very little time to interact. A slower electron with the same $b$ will have a much greater deflection of its orbit. This implies that $\Delta m v_z$ is a function of relative velocity $v$. Likewise, two electrons with the same $v$ but different values of $b$ will have different orbit deflections. The electron with larger $b$ will be deflected less since the Coulomb force is weaker at larger distances. Thus, $\Delta m v_z$ is also a function of the impact parameter $b$. The implication is that for Coulomb collisions it is incorrect to first calculate the reaction rate and then, after integrating over velocity, to multiply by the change $\Delta m v_z$ since $\Delta m v_z$ is itself a function of velocity. It must be brought in under the integral sign and become part of the integral.

The second issue for Coulomb collisions involves the impact parameter $b$ and its relation to the cross section. Because of the long-range nature of the Coulomb force the cross section in some sense seems infinite. All particles, regardless of their $b$ and $v$, will feel the Coulomb force. This difficulty is addressed by sub dividing the "infinite" cross section into a series of differential cross section rings of area $d\sigma = 2\pi b\,db$ as shown in Fig. 3.5. As one integrates $d\sigma$ over all values of the impact parameter in the range $0 < b < \infty$ the total area is indeed infinite. However, since $\Delta m v_z$ is a function of $b$ that decreases for large $b$, the differential cross section is weighted with a decreasing function of $b$ in the integrand and the overall integral remains finite[1] when integrating over $b$. Note that $d\sigma$ is independent of $v$ since any particle at the given value of $b$ will by definition pass through a differential ring, regardless of its velocity.

This behavior in a long-range force field is generalized into the reaction rate concept as follows. The cross section $\sigma$ is replaced by the differential cross section $d\sigma = 2\pi b\,db$ and an additional integral over $b$ is now required. Next, assume that there is a certain quantity

---

[1] The actual situation is somewhat more complex. Even with the weighted cross section there still remains a weak logarithmic divergence for Coulomb collisions which is remedied by integrating over a finite range $0 < b < b_{max}$. This issue is discussed in detail in Chapter 9. For the moment readers should just assume that the integral remains finite.

of interest $W = W(v, b)$ characterizing the collision, where $v = |\mathbf{v}_2 - \mathbf{v}_1|$. This could be the change in $z$ directed momentum, the change in kinetic energy, the loss of energy due to radiation, etc. due to a single collision. In other words, $W$ represents the change in some quantity of interest per particle per collision. The corresponding macroscopic change, denoted by $w$, represents the change in this same quantity per unit volume per unit time. The generalized reaction rate allows one to calculate $w$ assuming one knows $W$ and the distribution functions. The desired relation is

$$w = 2\pi \int W(v, b) f_1(\mathbf{v}_1) f_2(\mathbf{v}_2) v b \, db \, d\mathbf{v}_1 d\mathbf{v}_2. \tag{3.17}$$

Note that this expression reduces to the short-range collision form for nuclear reactions by choosing $W = E_f H(b_0 - b)$, where $H$ is the Heaviside step function and $b_0^2 = \sigma(v)/\pi$.

## 3.4 The distribution functions, the fusion cross sections, and the fusion power density

After a lengthy discussion, an appropriate formulation, given by Eq. (3.16), has been presented that allows one, at least conceptually, to evaluate the fusion power density in a reactor. To actually carry out this evaluation, however, requires an explicit knowledge of the distribution functions and the cross section which are as determined as follows.

### 3.4.1 The distribution functions

As is shown in Chapter 9, the Coulomb cross section is much larger than the D–T fusion cross section. The implication is that random Coulomb collisions will cause the particle distribution functions to relax to thermodynamic equilibrium on a time scale much shorter than the nuclear fusion collision time. Furthermore, it is well known from statistical mechanics that thermodynamic equilibrium is described by a Maxwellian distribution function. Specifically, it is assumed that the distribution functions for D and T are given by

$$f_j(\mathbf{r}, \mathbf{v}, t) = n_j \left( \frac{m_j}{2\pi T_j} \right)^{3/2} e^{-m_j v^2 / 2T_j}. \tag{3.18}$$

Here, $j$ denotes D or T, $n(\mathbf{r}, t)$, $T(\mathbf{r}, t)$ are the number density and temperature respectively, and $v^2 = v_x^2 + v_y^2 + v_z^2$. These are the distribution functions that will be used to evaluate the fusion power.

### 3.4.2 The fusion cross section and power density

The fusion cross section is developed in several steps. First, a simple estimate is given using the hard-sphere model. Second, an attempt is made to include the Coulomb velocity dependence of $\sigma$ using purely classical physics. This estimate turns out to be highly pessimistic and a third evaluation is made that includes nuclear quantum mechanical effects.

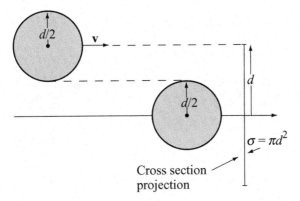

Figure 3.6 Hard-sphere collision cross section $\sigma = \pi d^2$.

The quantum model is accurate although the final cross sections are ultimately determined from experimental measurements.

### The hard-sphere cross section

As its name implies the hard-sphere cross section assumes that each of the colliding particles acts like a billiard ball as shown in Fig. 3.6. The diameter of each particle is assumed to be a nuclear diameter: $d \approx 5 \times 10^{-15}$ m. This picture implies that the cross section $\sigma = \pi d^2$. The usual units for nuclear cross sections are "barns," where 1 barn $= 10^{-28}$ m$^2$. Thus, the hard sphere model predicts a fusion cross section given by

$$\sigma = \pi d^2 \approx 0.8 \times 10^{-28} \text{ m}^2 \approx 1 \text{ barn.} \tag{3.19}$$

This is a reasonable estimate of $\sigma$ except that it fails to take into account the fact that slow particles will not undergo nuclear collisions because of the repulsive Coulomb force. A first attempt to correct this shortcoming is the next topic.

### The classical cross section

The classical model attempts to determine the velocity dependence of $\sigma$ by taking into account the competition between the repulsive Coulomb force and the attractive nuclear force using a combination of classical physics and the hard-sphere model. The idea is straightforward. Assume a D nucleus is moving with a velocity $v$ directly at a stationary T nucleus. This is shown in both the laboratory and center of mass frames in Fig. 3.7. In order for a hard-sphere collision to occur, conservation of energy requires that in the center of mass frame, the sum of the initial kinetic energies of the D and T when they are far from each other must exceed the Coulomb potential energy evaluated at the surface of the particles when they just touch each other. If this is not true, the D and T are repelled away from each other before they collide. In mathematical terms, a collision occurs when

$$\frac{m_D}{2}\left(\frac{m_T}{m_D + m_T}v\right)^2 + \frac{m_T}{2}\left(\frac{m_D}{m_D + m_T}v\right)^2 \geq \frac{e^2}{4\pi\varepsilon_0 d}, \tag{3.20}$$

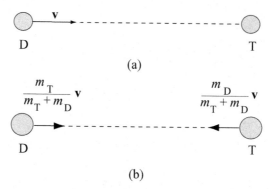

(a)

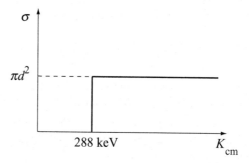

(b)

Figure 3.7  A D–T collision in (a) the laboratory frame and (b) the center of mass frame.

Figure 3.8  The D–T cross section as a function of energy assuming a classical interaction.

which simplifies to

$$\frac{1}{2}m_r v^2 \geq \frac{e^2}{4\pi \varepsilon_0 d} \tag{3.21}$$

where $m_r = m_D m_T/(m_D + m_T)$ is the reduced mass. If one substitutes the value $d = 5 \times 10^{-15}$ m and defines the center of mass kinetic energy $K_{cm} = m_r v^2/2$, then the condition for a collision to occur reduces to

$$K_{cm} \geq 288 \text{ keV}. \tag{3.22}$$

The classical picture of the dependence of $\sigma$ on $v$, or equivalently $K_{cm}$, is illustrated in Fig. 3.8. The crucial feature is the hard barrier at 288 keV, a quite high value. Any particle with a lower energy will not undergo a nuclear fusion collision. As compared to the actual situation, this is a pessimistic result.

### Nuclear quantum mechanical effects

The correct cross section for fusion collisions involves nuclear quantum mechanical effects. The reason is that the region of strongest nuclear interactions occurs on the scale length of

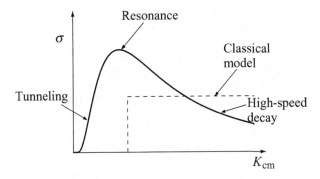

Figure 3.9 Nuclear quantum mechanical modifications to the cross section.

the nucleus. On such small-scale lengths classical physics does not apply and it is necessary to include nuclear quantum mechanical effects.

The most important effect is that on nuclear scale lengths nuclei exhibit both particle like and wavelike properties. The wavelike properties introduce three qualitatively new modifications to the simple classical hard-sphere model. First, in the quantum mechanical regime, the "tunneling effect" is present. A simple example of tunneling occurs when a sound wave is reflected off an insulating acoustic tile. Even if the insulator is perfectly lossless, sound waves still penetrate into the material, although with exponentially decreasing amplitude. If the tile is thin enough some sound energy will emerge from the back of the insulator; energy will have "tunneled" its way through. In terms of the cross section, "tunneling" is equivalent to barrier penetration. In other words, even for kinetic energies below the Coulomb cutoff, there is still a finite probability of interaction. The 288 keV is not a hard cutoff. Intuitively, the further below the cutoff, the lower is the probability of interaction.

The second wavelike effect is that at nuclear distances two nuclei can actually pass through one another. A nuclear reaction can thus be thought of as the interaction of two closely coupled waves. If the relative velocity of the particles (waves) is very large (i.e., $K_{cm} \gg 288\,\text{keV}$), the time available for a closely coupled interaction is very short, thereby reducing the probability of a fusion collision. Therefore the fusion cross section should decrease for large increasing relative velocities.

The final wavelike effect concerns the possibility of resonance. Under certain conditions of geometry and relative velocity the combined potential energies of the two colliding nuclei can exhibit a resonance. The net result is that under such conditions there is an enhanced probability of a nuclear reaction occurring, corresponding to an increased cross section. This is in fact the situation for the D–T interaction.

These three wavelike modifications are sketched schematically in Fig. 3.9 and compared with the classical picture. The actual cross sections for fusion reactions are determined experimentally, usually by directing a beam of mono-energetic particles (deuterons in the present case) at a stationary target. The experimental cross sections are plotted vs. deuteron energy (not relative mass energy) in Fig. 3.10 for the main fusion reactions of interest. All

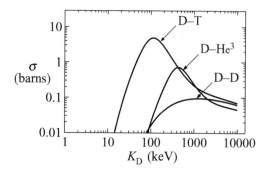

Figure 3.10 Experimentally measured cross sections for the D–T, D–He³, and D–D fusion reactions as a function of deuteron energy $K_D = m_D v_D^2/2$ (Glasstone, S., Lovberg, R. (1975). *Controlled Thermonuclear Reactions.* New York: Robert E. Krieger Publishing).

exhibit the qualitative behavior sketched in Fig. 3.9. Observe that the D–T cross section has a peak of about 5 barns at $K_D \approx 120\,\text{keV}$. In terms of cross section D–T is clearly favorable in comparison to D–D or D–He³. As a specific comparison observe that $\sigma_{DT} \approx 100\,\sigma_{DD}$ at 20 keV, which as is shown later, is a typical value for a fusion reactor. For a final point of comparison, note that the fission cross section for a thermal neutron (i.e., $K_n \approx 0.025\,\text{eV}$) colliding with $_{92}U^{235}$ is about 600 barns. With respect to cross section, fission has a big advantage compared to fusion.

Assuming that $\sigma(v)$ is known, it is then straightforward in principle to calculate the reaction rate as given by Eq. (3.14). This has been done numerically for the main fusion reactions using equal temperature Maxwellian distribution functions. The results are illustrated in Fig. 3.11 as curves of $\langle \sigma v \rangle$ vs. temperature $T$. Observe that for D–T, $\langle \sigma v \rangle$ has a peak value of $9 \times 10^{-22}\,\text{m}^3/\text{s}$ at a temperature of 70 keV. Knowing $\langle \sigma v \rangle$ one can then easily calculate the fusion power density as determined by Eq. (3.16) and repeated here for convenience

$$S_f = E_f n_1 n_2 \langle \sigma v \rangle. \tag{3.23}$$

Consider now the optimum ratio of D to T in the D–T reaction. For this case write $n_1 = n_D$ and $n_2 = n_T$. The optimum ratio is found by noting that overall charge neutrality requires that the total number of electrons in the fuel $n_e$ must equal the total sum of all positive charges: $n_D + n_T = n_e$ (assuming that the number of alpha particles is small). Thus, defining $k$ as the fraction of deuterium one can write $n_D = k n_e$ and $n_T = (1 - k)n_e$ implying that $n_D \cdot n_T = k(1 - k)n_e^2$. The product is maximized when $k = \frac{1}{2}$ and has the value $n_D \cdot n_T = \frac{1}{4}n_e^2$. The optimum fuel mixture is a 50%–50% combination of D and T leading to the final desired expression for the fusion power density:

$$S_f = \tfrac{1}{4}E_f n_e^2 \langle \sigma v \rangle. \tag{3.24}$$

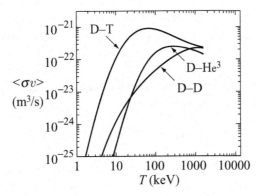

Figure 3.11 Velocity averaged cross section (i.e., $\langle \sigma v \rangle = R_{ij}/n_i n_j$) for the D–T, D–He$^3$, and D–D fusion reactions as a function of temperature.

Equation (3.24) represents the dominant source of power in the overall power balance of a fusion system.

### Some closing comments on fusion power generation

A further point of interest concerns the fact that one can obtain an analytic approximation to the evaluation of $\langle \sigma v \rangle$ by using an analytic model for $\sigma$, which is valid for low energies. This derivation is not essential to the present discussion since accurate curves of $\langle \sigma v \rangle$ have been presented in Fig. 3.11. However, readers may find it of interest to see how a $\langle \sigma v \rangle$ calculation can be carried out from beginning to end. Also, the derivation is helpful for some of the problems at the end of the chapter. Since the derivation is lengthy and not essential for the continuity of the discussion it is presented in Appendix A.

There are two additional points to be made. First, the particle energies required to initiate fusion reactions are on the order of 70 keV. This exceeds the ionization potential by a factor of more than 1000. The conclusion is that a burning D–T fuel is a fully ionized gas, hereafter referred to as a plasma. The second point is that, as is shown shortly, Coulomb collisions combined with power balance requirements result in an optimum operating temperature on the order of 15 keV, well below the 70 keV maximum of the $\langle \sigma v \rangle$ curve. The implication is that for a Coulomb-induced Maxwellian distribution function, most of the fusion reactions occur for particles on the tail of the distribution function.

## 3.5 Radiation losses

### 3.5.1 Overview of radiation losses

An important, although usually not dominant energy loss mechanism affecting power balance in a fusion reactor is that due to radiation. There are in fact several types of radiation losses that can occur: line radiation due to impurities, cyclotron radiation due to particle motion in a magnetic field, and Bremsstrahlung radiation due to Coulomb collisions. Of these, Bremsstrahlung radiation usually produces the largest losses. Therefore the present subsection focuses on calculating these losses, which are irreducible and unavoidable in any fusion reactor.

The calculation proceeds by using the general form of the reaction rate formalism given by Eq. (3.17), repeated here for convenience,

$$w = 2\pi \int W(v, b) f_1(\mathbf{v_1}) f_2(\mathbf{v_2}) vb \, db \, d\mathbf{v_1} d\mathbf{v_2}. \tag{3.25}$$

Recall, that in Eq. (3.25) $W$ is the microscopic change per particle per collision for the process under consideration. For the present situation, $W$ corresponds to the radiation energy loss per particle during a single Coulomb collision. This loss occurs because, as is well known from electromagnetic theory, a charged particle undergoing an accelerated motion radiates energy. A non-accelerating charged particle traveling in a straight line at constant velocity will have its orbit altered when it passes in close proximity to another charged particle because of the Coulomb interaction. This Coulomb collision deflects the particle from its straight-line orbit. The changing direction of the orbit corresponds to an accelerated motion. During the period of orbit deflection, the charged particle loses some of its energy by radiation.

The first task of the present calculation is to derive an approximate expression for $W$. Once $W$ is known it is then straightforward to substitute into Eq. (3.25) to obtain the value of $w$, corresponding to the loss of radiation energy due to Coulomb collisions per cubic meter per second. In other words $w \equiv S_B$, where $S_B$ is the Bremsstrahlung radiation power density loss measured in watts per cubic meter.

In practical fusion situations photons emitted by Bremsstrahlung radiation occur over a continuous range of frequencies, usually in the ultraviolet or soft x-ray region. Their mean free path for reabsorption is enormous, implying that once a photon is emitted, it is lost from the system. It is in this sense that the losses are unavoidable and irreducible. The calculation proceeds as follows.

### 3.5.2 Calculation of W, the energy lost per particle per Coulomb collision

Coulomb collisions play a very important role in fusion physics with respect to particle and heat transport as well as radiation losses. Transport phenomena are dominated by distant, multiple, small-deflection collisions, which require a somewhat sophisticated analysis that is presented in Chapter 9. In contrast, Bremsstrahlung radiation losses are dominated by close, single, large-deflection collisions. This fact enables one to make a set of reasonably accurate approximations, leading to a simple expression for the radiation losses. The expression has the correct scaling dependence on density and temperature and a semi-quantitatively accurate scaling coefficient. A further point worth noting is that the dominant collisions contributing to Bremsstrahlung radiation are due to electrons colliding with ions. Detailed calculations, not presented here, show that the radiation fields generated by each particle in a like-particle Coulomb collision cancel each other out, implying no loss due to radiation.

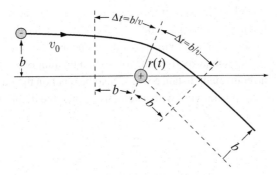

Figure 3.12 Geometry of a Coulomb collision.

The evaluation of $W$ requires two steps. First, the power $P$ radiated during a Coulomb collision is calculated. Second, an approximation is made to estimate the collision duration $\Delta t$ during which this radiation is emitted. The product $P\Delta t$ is the desired quantity $W$, representing the amount of energy lost by radiation per Coulomb collision.

The derivation begins with the formula for the power radiated by a charged particle undergoing an accelerated motion. This is a well-known formula from the theory of electromagnetic radiation that will be familiar to many readers. However, for those unfamiliar with the formula, a derivation is presented in Appendix B including a brief discussion of the principles of radiating electromagnetic fields. The formula for the power radiated by a non-relativistic accelerating electron is

$$P = \frac{\mu_0 e^2 \dot{v}^2}{6\pi c} \quad \text{W},$$
(3.26)

where $\dot{v}$ is the acceleration. The task now is to apply this formula to a Coulomb collision as illustrated in Fig. 3.12.

During the collision the electron experiences an accelerated motion due to the Coulomb interaction force, which can be written as follows:

$$\dot{v} = \frac{F_C}{m_e} = \frac{Ze^2}{4\pi \varepsilon_0 m_e r^2(t)}.$$
(3.27)

Here $Z$ is the charge number of the ion. An examination of Fig. 3.12 suggests that it is reasonable to approximate $r(t)$ during the collision interaction time by the original impact parameter $b$: $r(t) \approx b$. Under this approximation,

$$\dot{v} = \frac{F_C}{m_e} \approx \frac{Ze^2}{4\pi \varepsilon_0 m_e b^2}.$$
(3.28)

Substituting Eq. (3.28) into the expression for $P$ yields

$$P = \left( \frac{Z^2 e^6}{96\pi^3 \varepsilon_0^3 c^3 m_e^2} \right) \frac{1}{b^4} \quad \text{W}.$$
(3.29)

Here, use has been made of the relation $c^2 = 1/\mu_0 \varepsilon_0$.

The next step is to calculate the time interval $\Delta t$ during which the electron is actually accelerating. The Coulomb collision shown in Fig. 3.12 indicates that the electron begins to accelerate when it is a distance of approximately $b$ from the ion. The electron continues its acceleration until it is far enough away from the ion so that the mutual force again becomes small. By symmetry, the equivalent distance would also be $b$ but on the far side of the orbit. Consequently, if the electron is moving with a velocity $v$, then the duration of the collision is

$$\Delta t \approx \frac{2b}{v} \ \mathrm{s}. \tag{3.30}$$

The expression for $W$ can now be easily evaluated:

$$W \approx p\Delta t \approx \left( \frac{Z^2 e^6}{48\pi^3 \varepsilon_0^3 c^3 m_e^2} \right) \frac{1}{vb^3} \ \mathrm{J}. \tag{3.31}$$

This is the desired expression.

The last point of interest is to compare the energy radiated with the initial kinetic energy of the electron. The comparison is obtained by noting that a strong, large-angle collision is characterized by an electron kinetic energy that is of the same order of magnitude as the Coulomb potential energy evaluated at a separation $r \sim b$. A simple calculation yields

$$\frac{W}{m_e v^2/2} \sim \frac{v^3}{c^3} \ll 1. \tag{3.32}$$

The energy radiated by an electron during a single Coulomb collision is a very small fraction of its initial kinetic energy.

### 3.5.3 Calculation of $S_B$

The next step in the calculation is to substitute the expression for $W$ into the equation determining $S_B$ (Eq. (3.25)) and then carry out the integrations. The calculation is simplified by taking advantage of the fact that the ions are far more massive than the electrons. The consequences of this are that at equal temperatures: (1) the ions move much more slowly than the electrons, and (2) during a collision the ion orbit is negligibly perturbed. Thus, if one associates species "1" with electrons and species "2" with ions it is an excellent approximation to assume that the relative velocity of the particles is given by $v = |\mathbf{v}_1 - \mathbf{v}_2| = |\mathbf{v}_e - \mathbf{v}_i| \approx |\mathbf{v}_e|$. This approximation decouples the ions from the remainder of the integrand. The integration over ion velocities then yields the simple contribution

$$\int f_i(\mathbf{v}_i) \, d\mathbf{v}_i = n_i. \tag{3.33}$$

A second simplification in the remaining electron integration results from the fact that the relative velocity depends only on the magnitude of $\mathbf{v}_e$. This can be exploited by introducing standard spherical velocity variables. The 3-D velocity integration reduces to a 1-D integration since none of the quantities in the integrand depends upon the spherical angle variables. In particular it follows that

$$d\mathbf{v_e} = dv_x dv_y dv_z = v^2 \sin\theta \, dv \, d\theta \, d\phi = 4\pi v^2 dv, \tag{3.34}$$

where the integrations over $\theta$ and $\phi$ yield the factor $4\pi$. Under these simplifications the expression for $S_B$ (Eq. (3.25)) using a Maxwellian distribution function for electrons reduces to

$$S_B = 8\pi^2 n_i n_e \left(\frac{Z^2 e^6}{48\pi^3 \varepsilon_0^3 c^3 m_e^2}\right) \left(\frac{m_e}{2\pi T_e}\right)^{3/2} \int \frac{v^2 e^{-m_e v^2/2T_e}}{b^2} \, dv \, db. \tag{3.35}$$

Consider now the integration over the impact parameter $b$. One's first intuition might be to integrate over all values of $b$: $0 < b < \infty$. This, however, leads to a divergent integral. The divergence arises from small values of $b$ corresponding to strong, large-angle collisions. The difficulty is that as $b$ becomes smaller and smaller, the implicit assumption of a continuous range of emitted frequencies breaks down and the formula for the radiated power $P$ is no longer valid. Quantum mechanical effects become important and radiation is allowed only at a discrete set of quantized frequencies; that is, at small $b$ the power radiated approaches a finite constant and is not infinite. A simple approximation that accounts for the quantum transition is to change the lower limit on $b$ from $b = 0$ to $b = b_{\min}$, where $b_{\min}$ is the transition impact parameter corresponding to the onset of quantum mechanical effects. The transition occurs when the impact parameter becomes so small that the quantum mechanical uncertainty principle starts to become violated. Avoiding this problem requires that $\Delta x \Delta p \geq h/2\pi$. If one identifies $\Delta x \to b$ and $\Delta p \to m_e v$ then

$$b_{\min} \approx \frac{h}{2\pi m_e v}. \tag{3.36}$$

Note that for an electron velocity corresponding to 1 keV, $b_{\min}$ is somewhat smaller than an atomic diameter: $b_{\min}$ (1 keV) $= 6 \times 10^{-12}$ m $< 10^{-10}$ m. The $b$ integration is now straightforward.

$$\int_{b_{\min}}^{\infty} \frac{db}{b^2} = \frac{1}{b_{\min}} = \frac{2\pi m_e v}{h}. \tag{3.37}$$

Using this result in Eq. (3.35) leads to a simple velocity integral, which can be easily evaluated analytically, yielding the following expression for $S_B$:

$$S_B = \left(\frac{2^{1/2}}{6\pi^{3/2}}\right) \left(\frac{e^6}{\varepsilon_0^3 c^3 h \, m_e^{3/2}}\right) Z^2 n_i n_e T_e^{1/2} \text{ W/m}^3. \tag{3.38}$$

A more exact calculation that treats both the Coulomb interaction and the quantum mechanical effects self-consistently leads to an expression for $S_B$ that is identical to Eq. (3.38) except for the numerical coefficient. The more exact expression is obtained by replacing

$$\frac{2^{1/2}}{6\pi^{3/2}} \to \frac{2^{1/2}}{3\pi^{5/2}}. \tag{3.39}$$

Numerically, the change is given by $0.0423 \rightarrow 0.0269$. Hereafter, all calculations involving Bremsstrahlung radiation will use the more accurate coefficient.

### 3.5.4 *The effect of multiple ion species*

Equation (3.38) represents the Bremsstrahlung power radiated by electrons colliding with a single species of ions. In many situations there are multiple ion species present: D, T, alphas, and impurities. The total radiated power is obtained by simply summing over the individual ion species. The final sum for $S_B$ is usually written in terms of a quantity known as $Z_{eff}$, the "effective charge" of all the ions. The definition of $Z_{eff}$ makes use of the overall charge neutrality of the system as given by

$$\sum_j Z_j n_j = n_e, \tag{3.40}$$

where the sum is over all ion species. Using this relation one defines $Z_{eff}$ as follows:

$$Z_{eff} = \frac{\sum_j Z_j^2 n_j}{\sum_j Z_j n_j} = \frac{\sum_j Z_j^2 n_j}{n_e}. \tag{3.41}$$

For a pure D–T plasma $Z_{eff} = 1$. The final, desired expression for $S_B$ is now written as

$$S_B = \left( \frac{2^{1/2}}{3\pi^{5/2}} \right) \left( \frac{e^6}{\varepsilon_0^3 c^3 h \, m_e^{3/2}} \right) Z_{eff} \, n_e^2 \, T_e^{1/2} \quad \text{W/m}^3. \tag{3.42}$$

In practical units

$$S_B = C_B \, Z_{eff} \, n_{20}^2 \, T_k^{1/2} \quad \text{W/m}^3,$$
$$C_B = 5.35 \times 10^3. \tag{3.43}$$

Observe that Bremsstrahlung radiation, like fusion power, increases with the square of the density. However, its increase with temperature is much weaker.

## 3.6 Summary

Chapter 3 has described the first steps along the path to the design of a fusion reactor. These steps involve the calculations of: (1) the dominant heating source, fusion power, and (2) an important sink of energy, Bremsstrahlung radiation. Both contributions are essential in assessing the overall power balance in a fusion reactor.

The calculation of the fusion power density requires the introduction of microscopic nuclear physics including the concepts of cross section, mean free path, and collision time. The D–T reaction has the largest fusion cross section, about 5 barns at a center of mass kinetic energy of 120 keV. High kinetic energies are required in order to overcome the repulsive

Coulomb force. Coulomb collisions, which are much more frequent than fusion collisions, also cause the plasma to relax to a Maxwellian distribution function. The combination of a Maxwellian distribution function and the experimentally measured D–T cross section enables one to evaluate the velocity-averaged cross-section $\langle \sigma v \rangle$ as a function of $T$. The maximum value of $\langle \sigma v \rangle$ is approximately $9 \times 10^{-22}$ m$^3$/s at $T = 70$ keV. Combining these results leads to the desired expression for the fusion power density in a 50%–50% D–T plasma:

$$S_f = \tfrac{1}{4} E_f n^2 \langle \sigma v \rangle \quad \text{W/m}^3. \tag{3.44}$$

The dominant radiation loss in a fusion plasma is due to Coulomb interactions between electrons and ions. It is known as Bremsstrahlung radiation and occurs during the orbit deflection period while the electron is experiencing an accelerated motion. During this period the electron loses energy by radiation emission. Using the generalized reaction rate formalism leads to the following expression for radiation power density loss:

$$S_B = \left( \frac{2^{1/2}}{3\pi^{5/2}} \right) \left( \frac{e^6}{\varepsilon_0^3 c^3 h\, m_e^{3/2}} \right) Z_{\text{eff}} n^2 T_e^{1/2} \quad \text{W/m}^3. \tag{3.45}$$

Both the fusion power and Bremsstrahlung radiation scale as $n^2$. However, the fusion power increases much more rapidly with $T$ in the regime of fusion interest ($T > 10$ keV).

## Bibliography

The material in Chapter 3 covers the basic nuclear physics of fusion reactions plus an explanation and simple derivation of Bremsstrahlung radiation. This material has been well known for many years. The references below present earlier descriptions of the material, sometimes with more rigor and detail than presented here. In particular, the derivation of Bremsstrahlung radiation by Hutchinson is both rigorous and complete.

### *The nuclear physics of fusion reactions*

Dolan, T. J. (1982). *Fusion Research*. New York: Pergamon Press.
Glasstone, S., and Loveberg, R. H. (1960). *Controlled Thermonuclear Reactions*. Princeton, New Jersey: Van Nostrand.
Gross, R. (1984). *Fusion Energy*. New York: John Wiley & Sons.
Kammash, T. (1975). *Fusion Reactor Physics*. Ann Arbor, Michigan: Ann Arbor Science
Miyamoto, K. (2001). *Fundamentals of Plasma Physics and Controlled Fusion*, revised edn. Toki City: National Institute for Fusion Science.
Rose, D. J., and Clark, M. (1961). *Plasmas and Controlled Fusion*. Cambridge, Massachusetts: MIT Press.
Stacey, W. M. (1981). *Fusion Plasma Analysis*. New York: John Wiley & Sons.
Stacey, W. M. (2005). *Fusion Plasma Physics*. Weinheim: Wiley-VCH.
Wesson, J. (2004). *Tokamaks*, third edn. Oxford: Oxford University Press.

### Bremsstrahlung radiation

Dolan, T. J. (1982). *Fusion Research*. New York: Pergamon Press.

Glasstone, S. and Loveberg, R. H. (1960). *Controlled Thermonuclear Reactions*.
    Princeton, New Jersey: Van Nostrand.

Hutchinson, I. H. (1987). *Principles of Plasma Diagnostics*. Cambridge, England:
    Cambridge University Press.

Rose, D. J. and Clark, M. (1961). *Plasmas and Controlled Fusion*. Cambridge,
    Massachusetts: MIT Press.

Wesson, J. (2004). *Tokamaks*, third edn. Oxford: Oxford University Press.

### Problems

3.1 This problem compares the reaction rates for D–T fusion for several different distribution functions. For reference note that a good analytic approximation for $\langle \sigma v \rangle$ when both D and T are equal temperature Maxwellians is given by (Hively, H. M. (1977). Convenient computational forms for Maxwellian reactivities. *Nuclear Fusion*, **17**, 873).

$$\langle \sigma v \rangle = 10^{-6} \exp\left( \frac{a_{-1}}{T_i^\alpha} + a_0 + a_1 T_i + a_2 T_i^2 + a_3 T_i^3 + a_4 T_i^4 \right) \text{ m}^3/\text{s},$$

where $T_i = T_i$ (keV) and

| $\alpha$ | $a_{-1}$ | $a_0$ | $a_1$ | $a_2$ | $a_3$ | $a_4$ |
|---|---|---|---|---|---|---|
| 0.2935 | −21.38 | −25.20 | −7.101 × 10⁻² | 1.938 × 10⁻⁴ | 4.925 × 10⁻⁶ | −3.984 × 10⁻⁸ |

   (a) Calculate $\langle \sigma v \rangle$ assuming $f_T(\mathbf{v}) = n_T \delta(\mathbf{v})$ and $f_D(\mathbf{v})$ is a Maxwellian for the cases $T = 5$ keV, 10 keV, 15 keV, 20 keV. Compare the results for the case in which both species are Maxwellians.

   (b) Calculate $\langle \sigma v \rangle$ assuming $f_T(\mathbf{v}) = n_T \delta(\mathbf{v})$ and $f_D(\mathbf{v}) = n_D \delta(\mathbf{v} - v_0 \mathbf{e}_x)$. Choose $v_0$ such that $m_D v_0^2 / 2 = 3T/2$ for the following cases: $T = 5$ keV, 10 keV, 15 keV, 20 keV. Compare the results for the case in which both species are Maxwellians. Are beam distribution functions better or worse for fusion? Explain.

3.2 Determine the optimum operating temperature $T_{pp}$ (keV) by maximizing the alpha power density at fixed pressure. This leads to an expression of the form $S_\alpha = C_{pp} p^2$, where $S_\alpha$ is in watts per cubic meter and $p$ is in atmospheres. Consider now the situation where the density rather than the pressure is held constant. Using the expression for $\langle \sigma v \rangle$ in Problem 3.1 determine the optimum temperature $T_{nn}$ that maximizes $S_\alpha$ at fixed $n$ and calculate the corresponding value of $S_\alpha = K_{nn} n^2 = K_{nn}(p/2T_n)^2 = C_{nn} p^2$. Lastly repeat the calculation for the case where the product $np$ is constant and calculate the corresponding value of $S_\alpha = K_{np} np = K_{np} p^2/2n = C_{np} p^2$. Compare the values of $C_{nn}$, $C_{np}$ and $C_{pp}$. Note that usually holding the pressure fixed makes the most physical sense. However, in a few situations the other options can also make sense. Even so, it is always the pressure that is limited by macroscopic instabilities, thereby providing the motivation for always writing $S_\alpha = C p^2$.

3.3 The purpose of this problem is to investigate the effect of plasma profiles on the alpha power density. The idea behind the calculation is to replace the 0-D model where all quantities are equal to their average value by a volume-averaged 1-D model where the density and temperature have known profiles that vary in space. Specifically, in a plasma with a circular cross section the volume-averaged alpha power density is defined as

$$\bar{S}_\alpha = \frac{2}{a^2} \int_0^a \left( \frac{E_\alpha}{4} n^2 \langle \sigma v \rangle \right) r \, dr.$$

Assume now that the density and temperature profiles are given by

$$n = (1 + \nu_n)\bar{n}(1 - r^2/a^2)^{\nu_n},$$
$$T = (1 + \nu_T)\bar{T}(1 - r^2/a^2)^{\nu_T},$$

where $\bar{n}$ and $\bar{T}$ are the volume-averaged density and temperature respectively. To determine the effect of temperature profile on alpha power density numerically evaluate $\bar{S}_\alpha$ for $\nu_n = 0$ and $0 \le \nu_T \le 4$ using $\langle \sigma v \rangle$ from Problem 3.1. For each $\nu_T$ find the optimum value of $\bar{T}$ that maximizes $\bar{S}_\alpha$ at fixed average pressure: $\bar{p} = 2\left[(1 + \nu_n)(1 + \nu_T)/(1 + \nu_n + \nu_T)\right]\bar{n}\bar{T} = \text{const}$. Plot the optimum $\bar{T}(\text{keV})$ and corresponding $\bar{S}_\alpha/\bar{p}^2 \equiv C_p$ as a function of $\nu_T$ showing the 0-D limit $\nu_n = \nu_T = 0$ for reference. To determine the variation of $\bar{S}_\alpha$ with density repeat the above calculation for $\nu_T = 2$ and $0 \le \nu_n \le 4$. Are peaked profiles good, bad, or unimportant in maximizing $\bar{S}_\alpha$?

# 4

# Power balance in a fusion reactor

## 4.1 Introduction

Based on the results derived in Chapter 3 it is now possible to assemble all the sources and sinks that contribute to the overall power balance in a fusion reactor. Chapter 4 describes the construction and analysis of such a power balance model. The goal is to determine quantitatively the requirements on pressure, density, temperature, and energy confinement of the D–T fuel so as to produce a favorable overall power balance in a reactor: $P_{out} \gg P_{in}$.

Clearly, power balance plays a crucial role in determining the desirability of magnetic fusion as a source of electricity. An analysis of power balance determines the ease or difficulty of initiating fusion reactions. Specifically, how much external power must be supplied, either initially or continuously, to produce a given amount of steady state fusion power? The input power required must be sufficiently low in comparison to the output power in order that a large net power is produced – this is the basic requirement of a power reactor.

Basic power balance for a magnetic fusion reactor involves the analysis of the 0-D form of the law of conservation of energy from fluid dynamics. The general procedure for deriving the 0-D energy equation is the first topic discussed in Chapter 4. This is followed by a detailed analysis of the 0-D model, leading to quantitative conditions on the pressure, temperature, density, and energy confinement that must be satisfied in order to achieve a favorable power balance. Knowledge of the power balance constraints provides the necessary input for the design of a simple magnetic fusion reactor as developed in Chapters 5.

## 4.2 The 0-D conservation of energy relation

The approach used throughout the book to explain the physical behavior of a fusion plasma is based on the analysis of a 3-D fluid model, encompassing conservation of mass, momentum, and energy, plus Maxwell's equations. These are a set of non-linear, coupled, partial differential equations, made even more complicated by the fact that separate sets of fluid equations are required for each of the different species (i.e., electrons, ions, alphas, etc.). Furthermore, an accurate description of certain fusion phenomena requires knowledge of yet even more sophisticated microscopic physics aimed at determining the actual particle distribution functions. Microscopic plasma physics is, however, beyond the scope of the present book. Even so, the full 3-D fluid model is more than complicated enough to challenge one's mathematical skills.

A simpler model is needed as an introduction to fusion energy, one that leads to an overview of the overall power balance requirements in a fusion reactor. This model is 0-D conservation of energy. The overview provided by such a model helps to bracket the parameter regimes in terms of pressures, densities, temperatures, and energy confinement for magnetic fusion.

The 0-D model is obtained from the full set of 3-D fluid equations by making a number of simplifying assumptions and approximations as follows. First, the fuel is assumed to consist of a 50%–50% mixture of D–T with a negligible concentration of alpha particles. The implication is that $2n_D = 2n_T = n_e \equiv n$ and $n_\alpha \ll n$, where $n_j$ is the number density of the $j$th species of particles. Second, each component of the fuel is assumed to be at the same temperature: $T_D = T_T = T_e \equiv T$. Note that while the alpha number density is small its energy density $(n_\alpha E_\alpha)$ is usually substantial because $E_\alpha \gg T$. Third, the fuel is assumed to be in the form of a fully ionized gaseous plasma near thermodynamic equilibrium, corresponding to Maxwellian distribution functions. Under this assumption the internal energy density and corresponding particle pressure for each species $j$ are given by $U_j = (3/2)n_j T_j$ and $p_j = n_j T_j$. The total internal energy density and pressure of the fuel are now easily calculated: $U = U_D + U_T + U_e = 3nT$ and $p = p_D + p_T + p_e = 2nT$. Note that $U = \frac{3}{2}p$. These assumptions are reasonably well satisfied in fusion reactors and reduce the complexity from a multi-species model to a single one species model.

Using these simplifications one can now write the well-known conservation of energy relation from fluid dynamics. There are several standard forms of this relation and the one given below is convenient for present purposes:

$$\frac{3}{2}\frac{\partial p}{\partial t} + \frac{3}{2}\nabla \cdot p\mathbf{v} + p\nabla \cdot \mathbf{v} + \nabla \cdot \mathbf{q} = S. \tag{4.1}$$

The interpretation is as follows. For a small fixed volume in the laboratory reference frame, the first term represents the time rate of change of internal energy density within this volume. The second term describes the net flux of energy density leaving the volume by convection. The third term corresponds to the loss of energy density due to expansion of the fluid (and corresponds to the familiar "$p\ dV$" term in thermodynamics). The fourth term represents the loss of energy density due to diffusive processes. The most common diffusive process is heat conduction in which $\mathbf{q} = -\kappa\nabla T$. The last term $S$ describes the various sources and sinks of power density contributing to the energy balance. This is a crucial term consisting of three contributions

$$S = S_f - S_B + S_h. \tag{4.2}$$

Here, $S_f$ is the fusion heating power density produced by nuclear reactions, $S_B$ is the radiation loss per unit volume due to Bremsstrahlung, and $S_h$ is the external heating power density supplied to the system (e.g. ohmic heating power or external RF power). Also, care must be taken, depending upon application, as to whether to include the alpha power, the neutron power, or both in evaluating $S_f$. There is additional discussion on this point later in the chapter.

For present purposes assume the source terms are known. The 0-D power balance relation is now derived as follows. Equation (4.1) is integrated over the volume of the plasma and then divided by the plasma volume $V$. The 0-D power balance relation is thus given by

$$\frac{1}{V} \int \left[ \frac{3}{2} \left( \frac{\partial p}{\partial t} + \nabla \cdot p\mathbf{v} \right) + p\nabla \cdot \mathbf{v} + \nabla \cdot \mathbf{q} - S \right] d\mathbf{r} = 0. \tag{4.3}$$

The remainder of the chapter is focused on the simplification and analysis of this critical equation.

## 4.3 General power balance in magnetic fusion

The plan to examine power balance is as follows. First, the general 0-D power balance relation is simplified, leading to an explicit form directly applicable to a magnetic fusion reactor. Second, this relation is examined with respect to power balance within the plasma itself. In particular, the conditions for the plasma to be self-sustained in steady state equilibrium are determined. Third, global power balance is investigated. Once the plasma is in steady state equilibrium, one needs to guarantee that the total electric power output of the reactor greatly exceeds the electric power input in order to have a desirable source of electricity.

The fourth and final topic concerns the dynamical behavior of the plasma. Here, there are two important issues, thermal stability and the minimum external power required to reach steady state. First, for thermal stability one assumes that the desired steady state reactor operating parameters are achieved. It is essential to show that this operating point is stable in order to avoid a thermal runaway. Second, the issue of a minimum external heating power arises even in a fully ignited reactor where no such power is required in steady state. However, external heating is required during the startup transient phase in order to heat the plasma from its low initial temperature to the desired final ignition temperature. This power does not affect the steady state operating costs of running the reactor but does contribute to the capital cost. The corresponding constraints resulting from each of these dynamic phenomena are important inputs to the design of a reactor.

## 4.4 Steady state 0-D power balance

The analysis begins by recalling the general 0-D plasma power balance relation, repeated here for convenience

$$\frac{1}{V} \int \left[ \frac{3}{2} \left( \frac{\partial p}{\partial t} + \nabla \cdot p\mathbf{v} \right) + p\nabla \cdot \mathbf{v} + \nabla \cdot \mathbf{q} - S \right] d\mathbf{r} = 0. \tag{4.4}$$

One can now introduce several approximations that reduce Eq. (4.4) to a much simpler form. First, a magnetic fusion reactor will almost certainly be a steady state system with small or negligible flows. Consequently the time derivative term is zero and the convection and compression terms can be neglected.

The next simplification involves the source terms. Consider the fusion power contribution. Since Eq. (4.4) applies to the plasma, only the alpha particle energy should be included. The alphas are charged particles that are confined by the magnetic field and thus remain within the plasma, providing a source of heat. The neutrons, which have no charge, escape from the plasma. They provide the main source of heat in the blanket and are ultimately responsible for producing electricity. Even so, since they are not confined in the plasma they make no contribution to the power balance within the plasma. The conclusion is that the fusion power contribution $S_f \rightarrow S_\alpha$, where

$$S_\alpha = \tfrac{1}{4} E_\alpha n^2 \langle \sigma v \rangle .$$  (4.5)

Here, recall that $E_\alpha = 3.5$ MeV. Also it has been assumed that the temperatures of all species are equal and that the fuel is a 50%–50% mixture of D and T, each with density $n/2$. Observe that as written $S_\alpha = S_\alpha(n, T)$. From a physics point of view it makes more sense to rewrite $S_\alpha$ in the form $S_\alpha = S_\alpha(p, T)$. The reason is the misconception that arises because of the experimental ease of raising $n$ by simply injecting more gas into the system. However, this is not very useful since the temperature would certainly decrease as the density increases, thereby reducing the number of fusion reactions. The implication is that it is not the number density that is fundamental but the energy density, which is proportional to the pressure. In the expression for $S_\alpha$, and all other contributions to the power balance that follow, the quantities $p$ and $T$ will be treated as the fundamental variables. Recall that the total pressure $p = 2nT$, allowing one to rewrite $S_\alpha$ as follows:

$$S_\alpha = \frac{1}{16} E_\alpha p^2 \frac{\langle \sigma v \rangle}{T^2}.$$  (4.6)

The next source term to consider is the Bremsstrahlung radiation. This contribution is easily expressed in terms of $p$ and $T$:

$$S_B = C_B Z_{\text{eff}} n^2 T^{1/2} = \frac{1}{4} C_B Z_{\text{eff}} \frac{p^2}{T^{3/2}}.$$  (4.7)

For simplicity, a high degree of plasma purity is assumed implying that $Z_{\text{eff}} \approx 1$.

The last contribution to $S$ is the external heating term. This, in general, comprises an ohmic heating term plus sources of auxiliary power such as microwave heating. The ohmic heating term is a function of temperature, current, and geometry. In the regime of steady state fusion power production this contribution is zero. The reason is that the ohmic plasma current is the secondary current of a transformer and a DC transformer is not physically possible. The ohmic power can only have an impact during the initial transient. On the other hand, the auxiliary heating power is operational during both transients and steady state. This is the dominant source of external heating power, and is assumed to be deposited in the plasma with a known profile, independent of $p$ and $T$. Consequently, the heating power can be written as

$$S_h = S_h(\mathbf{r}, t).$$  (4.8)

All the terms, with the exception of the heat flux have now been defined. Leaving this term aside temporarily, one is next faced with the task of integrating the remaining contributions over the plasma volume. Since the profiles of $p$ and $T$ are not known, a plausible approximation must be made to carry out this step. The approximation is as follows. In general, the pressure and temperature are monotonically decreasing profiles that are always positive (i.e., there are no negative regions that can lead to a cancellation when integrating). Therefore, a qualitatively and semi-quantitatively accurate approximation is to assume that the $p$ and $T$ profiles are constant across the whole profile with magnitudes equal to their average value. It is this approximation that converts the multi-dimensional model into a useful 0-D model. The uniform profile assumption simplifies general integral contributions as follows:

$$\frac{1}{V}\int G(p,T)\,d\mathbf{r} \approx \frac{1}{V}\int G(\bar{p},\bar{T})\,d\mathbf{r} = G[\bar{p}(t),\bar{T}(t)],$$
$$\frac{1}{V}\int S(\mathbf{r},t)\,d\mathbf{r} = \bar{S}(t).$$
(4.9)

The quantities $\bar{p},\bar{T},\bar{S}$ are average values over space. For convenience the "over bars" will be suppressed hereafter.

Under the assumptions introduced, the 0-D power balance relation simplifies to

$$\left(\frac{E_\alpha}{16}\right)p^2\frac{\langle\sigma v\rangle}{T^2} - \left(\frac{C_B}{4}\right)\frac{p^2}{T^{3/2}} + S_h - \frac{1}{V}\int_A \mathbf{q}\cdot d\mathbf{A} = 0.$$
(4.10)

The term involving the heat flux, obtained using the divergence theorem, is an integral over the plasma surface area. The last step in the derivation is to simplify this term. As an illustrative example consider a circular cylindrical plasma of radius $a$ in which the heat flux is given by Fourier's law: $\mathbf{q} = -\kappa\nabla T$ with $\kappa$ the thermal conductivity. Under this assumption the heat flux contribution to Eq. (4.10) reduces to

$$\frac{1}{V}\int_A \mathbf{q}\cdot d\mathbf{A} = -2\frac{\kappa}{r}\frac{\partial T}{\partial r}\bigg|_{r=a}.$$
(4.11)

To evaluate this term one needs to know the unknown temperature gradient at the plasma edge. The uniform profile assumption obviously does not work. Equally complicating is the fact that $\kappa$ for a fusion plasma is not accurately known. The thermal conductivity in a fusion plasma is often noticeably higher than the value calculated by the classical theory of collisions. The reason is that most plasmas experience a variety of small-scale micro-turbulence, which leads to anomalously large values of transport coefficients. The usual procedure used in fusion research to circumvent this problem is to define a 0-D energy confinement time $\tau_E$, valid for general geometry, as follows:

$$\frac{1}{V}\int_A \mathbf{q}\cdot d\mathbf{A} \equiv \frac{3}{2}\frac{p}{\tau_E}.$$
(4.12)

The quantity $\tau_E$ represents the e-folding relaxation time of the plasma energy due to heat conduction. In practice it is determined experimentally by regression analysis of a large database of plasma discharges from different devices. For present purposes one should assume that $\tau_E$ is a known quantity. Note that, in general, $\tau_E = \tau_E(p, T)$. However, for the moment $\tau_E$ is treated as being independent of $p$ and $T$ to simplify the analysis. This assumption is relaxed in future chapters after plasma transport phenomena have been discussed.

Substituting the simplified expression for the heat flux contribution leads to the desired form of the 0-D steady state power balance relation in a magnetic fusion reactor.

$$S_\alpha + S_h = S_B + S_\kappa,$$

$$\left(\frac{E_\alpha}{16}\right) p^2 \frac{\langle \sigma v \rangle}{T^2} + S_h = \left(\frac{C_B}{4}\right) \frac{p^2}{T^{3/2}} + \frac{3}{2} \frac{p}{\tau_E}. \tag{4.13}$$

The goal now is to analyze this equation in order to determine the conditions on $p$, $T$, and $\tau_E$ that lead to a favorable power balance.

## 4.5 Power balance in the plasma

The first topic of interest is to investigate the conditions that determine how to maintain the plasma in steady state power balance without, for the moment, being concerned about the overall power balance in the reactor. There are two relevant subtopics. The first is called "ideal ignition" and as will be shown sets a lower limit for the operating temperature of the plasma. The second subtopic is simply called "ignition" and leads to constraints on $p$, $\tau_E$, and $T$ for the plasma to be in steady state equilibrium under more realistic conditions.

### 4.5.1 Ideal ignition

Ideal ignition corresponds to the condition of steady state power balance in the plasma assuming negligible heat conduction losses and no external heating. In other words, the fusion produced alpha power must be large enough to overcome the irreducible Bremsstrahlung radiation losses:

$$S_\alpha = S_B. \tag{4.14}$$

Substituting for $S_\alpha$ and $S_B$ leads to a condition involving only the plasma temperature. The dependence on $p$ cancels out:

$$\frac{\langle \sigma v \rangle}{T_k^{1/2}} = \frac{4C_B}{E_\alpha} = 3.8 \times 10^{-24} \frac{\mathrm{m^3/s}}{\mathrm{keV^{1/2}}}. \tag{4.15}$$

This relationship is illustrated in Fig. 4.1. Observe that for the plasma to satisfy the ideal ignition condition the temperature must satisfy

$$T \geq 4.4 \text{ keV}. \tag{4.16}$$

Equation (4.16) sets a lower limit on the plasma temperature in a magnetic fusion reactor. The corresponding value for the D–D reaction is much higher, approximately 30 keV. Note

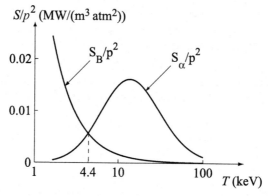

Figure 4.1 Alpha power and Bremsstrahlung radiation as a function of temperature. The intersection point corresponds to ideal ignition.

also that once the ideal ignition temperature is exceeded, the alpha power increases much more rapidly than the Bremsstrahlung losses. For example at $T = 10$ keV, one finds that $S_\alpha/S_B \approx 8$ for D–T fuel.

### 4.5.2 Ignition

The ignition condition represents a more realistic power balance situation. The assumption here is that in steady state the alpha power heating is sufficiently large to balance the combined Bremsstrahlung and thermal conduction losses, without the need for any external power:

$$S_\alpha = S_B + S_\kappa. \tag{4.17}$$

This clearly is a very desirable power balance goal since once the plasma is ignited, steady state fusion power is produced with no external power required to sustain the plasma. Equation (4.17) sets limits on $p$, $T$, and $\tau_E$. To determine these limits quantitatively it is convenient to express the various power densities in practical units as follows:

$$S_\alpha = \frac{E_\alpha}{16} \frac{\langle \sigma v \rangle}{T^2} p^2 = K_\alpha \frac{\langle \sigma v \rangle}{T_k^2} p^2 \quad \text{MW/m}^3,$$

$$S_B = \frac{C_B}{4} \frac{p^2}{T^{3/2}} = K_B \frac{p^2}{T_k^{3/2}} \quad \text{MW/m}^3, \tag{4.18}$$

$$S_\kappa = \frac{3}{2} \frac{p}{\tau_E} = K_\kappa \frac{p}{\tau_E} \quad \text{MW/m}^3.$$

Here $K_\alpha = 1.37$, $K_B = 0.052$, and $K_\kappa = 0.15$. The units are $T_k$ (keV), $\tau_E$ (s), $\langle \sigma v \rangle$ $(10^{-22} \text{ m}^3/\text{s})$, and $p$ $(10^5 \text{ Pa})$. The reason for the choice of pressure units is that $10^5$ Pa $= 1$ bar $\approx 1$ atm. Hence, numerical values of $p$ should provide some sense of physical intuition for the reader.

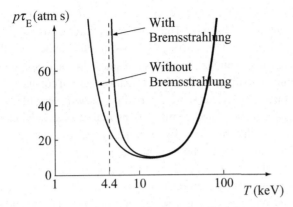

Figure 4.2 Critical $p\tau_E$ for ignition as a function of temperature.

These definitions of power density are substituted into the ignition power balance condition. After some simple algebra one obtains a condition on the product $p\,\tau_E$ as a function of $T$:

$$p\,\tau_E \geq \frac{K_\kappa T_k^2}{K_\alpha \langle \sigma v \rangle - K_B T_k^{1/2}} \approx K_I \frac{T_k^2}{\langle \sigma v \rangle} \quad \text{atm s.} \tag{4.19}$$

The quantity $p\tau_E = 2T(n\tau_E)$, where $n\tau_E$ is known as the Lawson parameter. The approximate form follows from the reasonably good assumption of neglecting the Bremsstrahlung losses in the regime of interest. In this case $K_I = K_\kappa/K_\alpha = 0.11$.

The exact and approximate forms of Eq. (4.19) are plotted in Fig. 4.2. Observe that $T$ must exceed the ideal ignition temperature in order for $p\,\tau_E$ to be positive. More importantly, $p\,\tau_E$ must exceed a certain minimum value for ignition to occur. The minimizing value of $T$ and the corresponding value of $p\,\tau_E$ are given by

$$T_{\min} = 15 \text{ keV},$$
$$(p\,\tau_E)_{\min} = 8.3 \text{ atm s.} \tag{4.20}$$

For $p\tau_E < (p\tau_E)_{\min}$ ignition is not possible. For $p\tau_E > (p\tau_E)_{\min}$ two solutions corresponding to two different temperatures are possible. The significance of these two solutions is discussed shortly and is related to the thermal stability of the system. Note that for a 15 keV plasma with an energy confinement time of 1 s, a pressure of about 8 atm is required for the plasma to be ignited; that is, it is sustained purely by the self-heating of the fusion alpha particles.

The existence of a minimum $p\,\tau_E$ has very important practical implications. In general, increasing either $p$ or $\tau_E$ requires an increase in either the size of the device or the magnetic field, both of which lead to an increase in the capital cost of the reactor. Therefore, the ease or difficulty of satisfying Eq. (4.20) is a crucial factor in distinguishing the relative desirability of various proposed magnetic fusion configurations (e.g. tokamak, stellarator, etc.).

There is one final issue, related to the use of a non-zero external heating power to consider with respect to plasma power balance. There are two situations where the use of external heating power is essential. First, in existing D–T burning fusion devices, the $p \tau_E$ criterion cannot be satisfied because their size or field is too small. In such cases additional external heating power is required to maintain the plasma against thermal conduction losses. Second, even in larger reactor-scale devices some external power will very likely be needed. The reason is that many magnetic fusion configurations require a toroidal current to hold the plasma in equilibrium. Such currents are driven in steady state by external sources of directed beam power or microwave power. The unavoidable power necessary to drive this current also contributes to the heating power and sets a lower bound on $S_h$.

The amount of external heating power required is very important in the overall power balance of the reactor. When $S_h > 0$, the plasma is subignited. The reactor becomes a power amplifier with the gain (defined as $P_{out}/P_{in}$) being the critical measure of performance. The use of external power has both an advantage and a disadvantage. The advantage is that external power reduces the requirement on $p \tau_E$ since less alpha power is now required to sustain the plasma. The disadvantage is that too much external power degrades the overall gain of the reactor. As an extreme example, a device heated solely by external power requires $p \tau_E = 0$. It can sustain a low-temperature plasma but produces zero fusion power. The quantitative impact of external heating power is discussed in the context of plasma power balance immediately below, and in the context of overall reactor power balance in the next subsection.

The approach taken with respect to plasma power balance is to assume that a certain amount of external heating power is required without actually focusing on any particular application. The goal is to determine the impact that this power has on the minimum value of $p \tau_E$ required to maintain the plasma in steady state equilibrium and the corresponding reduction in gain. A convenient way to introduce the external heating power is to assume that the alpha power provides only a fraction $f_\alpha$ of the total heating power:

$$f_\alpha \equiv \frac{S_\alpha}{S_\alpha + S_h}. \tag{4.21}$$

Observe that $f_\alpha = 1$ corresponds to ignition, $f_\alpha = \frac{1}{2}$ to equal alpha and external power, and $f_\alpha = 0$ to no alpha power.

Next, one can easily express $S_h$ in terms of $S_\alpha$ and $f_\alpha$:

$$S_h = \frac{1 - f_\alpha}{f_\alpha} S_\alpha. \tag{4.22}$$

This relation is substituted into the full steady state power balance relation

$$S_\alpha + S_h = S_\kappa + S_B. \tag{4.23}$$

A short calculation gives the following modified form of the $p \tau_E$ condition:

$$p \tau_E \geq \frac{K_\kappa T_k^2}{(1/f_\alpha) K_\alpha \langle \sigma v \rangle - K_B T_k^{1/2}} \approx f_\alpha K_I \frac{T_k^2}{\langle \sigma v \rangle} \quad \text{atm s.} \tag{4.24}$$

As expected, the addition of external heating reduces the demands on the amount of alpha heating required to maintain the plasma. Specifically, the minimum $p\tau_E$ for steady state power balance is reduced by a factor of $f_\alpha$. Even so, this is a mixed blessing since too much external heating power has a detrimental effect on the overall power balance in the reactor. This is the next topic of interest.

## 4.6 Power balance in a reactor

Assume that some combination of alpha power and external heating power is maintaining the plasma in steady state equilibrium at a temperature of about 15 keV. The critical issue then is to calculate the ratio of the output power to the input power (i.e., the gain) to determine the conditions under which the system indeed makes sense as a power reactor. To address this issue two dimensionless "gain parameters" are introduced. The first parameter $Q$ is widely used in the fusion community and is based primarily on physics considerations. The second parameter $Q_E$ is somewhat more realistic and attempts, in a simple manner, to include some basic engineering constraints. The goal of the analysis is to determine the dependence of both $Q$ and $Q_E$ on the value of $p\tau_E$ when $p\tau_E$ is smaller than the value required for ignition.

### 4.6.1 The physics gain factor Q

The analysis proceeds by carefully defining $Q$ in terms of the various power density sources and sinks. Some simple algebra then allows one to write this definition in the form $Q = Q(p\tau_E, T)$.

The definition of the physics gain factor $Q$ is as follows:

$$Q = \frac{\text{net thermal power out}}{\text{heating power in}}$$

$$= \frac{\text{total thermal power out} - \text{heating power in}}{\text{heating power in}}$$

$$= \frac{P_{out} - P_{in}}{P_{in}}. \tag{4.25}$$

The motivation for the definition is as follows. Ultimately, in a reactor electricity is generated by the total amount of thermal power produced in the plasma. For a net gain in output thermal power, the total output thermal power must exceed the input heating power required to maintain the plasma. Thus, in the limit where no fusion reactions take place, all the input heating power is converted into the total output thermal power in the form of thermal conduction and radiation power losses; that is, $P_{out} = P_{in}$, and $Q = 0$. At the other extreme corresponding to full ignition, no heating power is required to maintain the plasma ($P_{in} = 0$), since this task is carried out by the alphas. In this limit $Q = \infty$. The conclusion is that with the definition of $Q$ given by Eq. (4.25) the interesting regime for a steady state power reactor is defined by $0 < Q < \infty$.

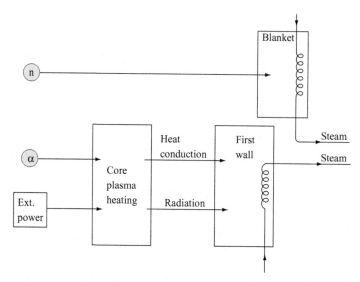

Figure 4.3 Schematic diagram of power flow in a fusion reactor showing how the input sources of fusion neutrons, fusion alphas, and external heating power are ultimately converted into steam for electricity.

A simpler, explicit form for the physics gain factor of the form $Q = Q(p\tau_E, T)$ is now derived by substituting the appropriate contributions for total output thermal power and input heating power into the definition given by Eq. (4.25). To begin, note that the input power is easily identifiable as $P_{in} = S_h V$, where $V$ is the plasma volume. This is external power that must be continuously absorbed by the plasma to keep it heated to the right temperature and to drive any steady state toroidal current that may be required.

Consider next the total output thermal power. What precise contributions must be included here? To answer this question, think again of a reactor where the output electric power is produced by means of a thermal conversion system (e.g. a heat exchanger–steam turbine system) that converts the heat produced by the plasma into steam and then electricity. See Fig. 4.3. From the point of view of physics the total thermal output power should therefore include all the sources of heat that leave the plasma and are available for thermal conversion. There are three such sources. First, the 14.1 MeV fusion neutrons that escape from the plasma are the primary source of heat. This heat is converted to steam in the surrounding blanket. Second, Bremsstrahlung radiation escapes from the plasma and is deposited on the first wall. To prevent radiation-induced melting, the first wall must be cooled. The heat carried away by the coolant is also available for conversion to steam. Third, plasma energy is continuously carried to the first wall by thermal conduction. Here too, to prevent melting this thermal heat must be carried away by the first wall coolant, after which it becomes available for steam production. As for the alpha particles, it is assumed that they give their heat to the plasma before they diffuse out. Thus, their energy is not directly available to produce external heat. The alpha power contribution is only

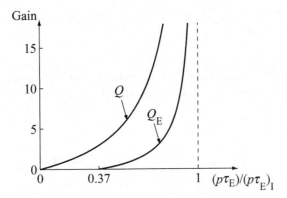

Figure 4.4 The physics gain factor $Q$ and engineering gain factor $Q_E$ as functions of $(p\tau_E)/(p\tau_E)_I$.

indirect through heating of the plasma. It ultimately appears as output thermal power in the form of radiation and heat conduction. A similar argument applies to the external heating power.

On the basis of this reasoning, the output power is given by $P_{out} = (S_n + S_B + S_\kappa)V$, where $S_n = (E_n/E_\alpha)S_\alpha = 4S_\alpha$. Combining these results leads to the following definition of $Q$:

$$Q = \frac{4S_\alpha + S_B + S_\kappa - S_h}{S_h}. \tag{4.26}$$

The simplified form of $Q$ can be obtained by substituting $S_h$ from Eq. (4.23) into the numerator of Eq. (4.26). This yields

$$Q = \frac{S_f}{S_h}, \tag{4.27}$$

where $S_f = S_n + S_\alpha = 5S_\alpha$. Equation (4.27) shows that $Q$ is simply the ratio of the total fusion power produced to the power input.

The desired relation giving $Q = Q(p\tau_E, T)$ is obtained by eliminating $S_h$ from the denominator of Eq. (4.27), again by means of Eq. (4.23), and substituting for the separate contributions. A short calculation, in which for simplicity Bremsstrahlung is neglected, then leads to the following expression for $Q$:

$$Q = 5\frac{p\tau_E}{(p\tau_E)_I - p\tau_E}, \tag{4.28}$$

$$(p\tau_E)_I = K_1\frac{T_k^2}{\langle\sigma v\rangle} \quad \text{atm s}$$

Note that $(p\tau_E)_I$ denotes the value of $p\tau_E$ required for ignition. The curve of $Q$ vs. $p\tau$ is illustrated in Fig. 4.4. As expected when $p\tau = (p\tau)_I$, then $Q = \infty$. When some external heating power is used, the required $p\tau < (p\tau)_I$ although the gain $Q$ is lowered. For example, if one desires ten times the fusion power as compared to the input power (i.e., $Q = 10$), then $p\tau_E$ is smaller by a factor of $\frac{2}{3}$ from the case of pure ignition.

Finally, a comparison of Eqs. (4.24) and (4.28) shows that $f_\alpha$ and $Q$ are related to

$$f_\alpha = \frac{Q}{5+Q}. \tag{4.29}$$

Thus, if entry into the "burning plasma regime" is defined as plasma power balance in which alpha power just equals external power, then this corresponds to $f_\alpha = \frac{1}{2}$ or equivalently $Q = 5$.

### 4.6.2 The engineering gain factor $Q_E$

The gain factor $Q$ provides a reasonable accounting of the various power sources and sinks that contribute to power balance as viewed from the physics of the fusion process. Even so it is not entirely satisfactory in the sense that the various contributions to $Q$ represent different types of power. For example $S_n$ and $S_\kappa$ are thermal power densities while $S_h$ is, for instance, a microwave power density. None of the contributions is an actual electric power density. The engineering gain factor $Q_E$ remedies this situation by converting all contributions to electric power densities by the introduction of appropriate power conversion efficiencies. The more realistic figure of merit $Q_E$ is defined as

$$
\begin{aligned}
Q_E &= \frac{\text{net electric power out}}{\text{electric power in}} \\[2mm]
&= \frac{\text{total electric power out} - \text{electric power in}}{\text{electric power in}} \\[2mm]
&= \frac{P_{\text{out}}^{(E)} - P_{\text{in}}^{(E)}}{P_{\text{in}}^{(E)}}. \tag{4.30}
\end{aligned}
$$

The separate terms are evaluated as follows. First, $P_{\text{in}}^{(E)}$ is the actual electric power required to drive the external heating sources. A fraction $\eta_e$ of this power is converted into a form suitable for plasma heating, say, microwave power. Furthermore, the plasma itself only absorbs a fraction $\eta_a$ of the microwave power, the remainder being reflected away. The overall conversion efficiency of electric power input to power absorbed in the plasma is thus given by the product $\eta_e \eta_a$, implying that

$$P_{\text{in}}^{(E)} = S_h V / \eta_e \eta_a. \tag{4.31}$$

Typically, $\eta_e \approx 0.7$ and $\eta_a \approx 0.7$.

Consider next the electric power output. The argument here concerning which terms actually produce heat is similar to that presented with the definition of $Q$ with two exceptions. First, one should also include the heat produced by each fusion neutron when breeding tritium in the lithium blanket. Thus, in addition to the $E_n = 14.1$ MeV contribution one should add $E_{Li} = 4.8$ MeV. Second, the microwave power reflected off the plasma is absorbed by the wall producing heat and it too should be included. This contribution is given by $(1 - \eta_a)\eta_e P_{\text{in}}^{(E)}$. Adding these contributions leads to the following expression for the total thermal power out: $P_{out} = (S_n + S_{Li} + S_B + S_\kappa)V + (1 - \eta_a)\eta_e P_{in}^{(E)}$, where

$$S_n + S_{Li} = [(E_n + E_{Li})/E_\alpha] S_\alpha = 5.4 S_\alpha. \tag{4.32}$$

Assume that the thermal heat is converted to electricity through a steam cycle and turbine with a conversion efficiency $\eta_t$, where typically $\eta_t \approx 0.4$. The electric power out can then be written as

$$P_{out}^{(E)} = \eta_t \left[ 5.4 S_\alpha + S_B + S_\kappa + \frac{(1 - \eta_a)}{\eta_a} S_h \right] V. \tag{4.33}$$

Combining terms leads to the following expression for the engineering power gain $Q_E$:

$$Q_E = \frac{\eta_t \eta_e \eta_a (5.4 S_\alpha + S_B + S_\kappa) - [1 - (1 - \eta_a)\eta_t \eta_e] S_h}{S_h}. \tag{4.34}$$

After substituting for the various sources and sinks, and again neglecting Bremsstrahlung for simplicity, one can easily rewrite this expression in the desired form:

$$Q_E = \frac{(6.4\eta_t \eta_e \eta_a + 1 - \eta_t \eta_e) p\tau - (1 - \eta_t \eta_e)(p\tau)_I}{(p\tau)_I - p\tau}$$

$$\approx 2.0 \frac{p\tau - 0.37(p\tau)_I}{(p\tau)_I - p\tau}. \tag{4.35}$$

A plot of $Q_E$ vs. $p\tau$ is also shown in Fig 4.4. Observe that as with $Q$, the value of $Q_E = \infty$ when $p\tau = (p\tau)_I$. Similarly, $p\tau_E$ is reduced from its fully ignited value when external heating is used. However, the reduction is not as great. Specifically, a simple calculation shows that the relationship between $Q$ and $Q_E$ is given by

$$Q = \frac{E_n + E_\alpha}{E_n + E_\alpha + E_{Li}} \frac{Q_E + 1 - \eta_t \eta_e}{\eta_t \eta_e \eta_a} = 4.0(Q_E + 0.72). \tag{4.36}$$

Some interesting numerical values are as follows. For electric power breakeven $(P_{out}^{(E)} = P_{in}^{(E)})$ corresponding to $Q_E = 0$, a value of $p\tau_E/(p\tau_E)_I = 0.37$ is required. This corresponds to $Q = 2.9$. If the goal is an electric power gain factor of $Q_E = 10$, then $p\tau_E = 0.90(p\tau_E)_I$ is required and corresponds to $Q \approx 43$. Finally, note that a physics gain factor of $Q = 10$ is equivalent to an engineering gain factor of only $Q_E = 1.8$.

What are the conclusions to be drawn from this analysis? They can be summarized as follows. First, full self-ignition ($Q = Q_E = \infty$) requires $p\tau_E = 8.3$ atm s at $T = 15$ keV. Second, if some external heating is required the value of $p\tau_E$ is reduced since this power relieves some of the heating burden from the alpha particles ($Q_E$ is now finite). Third, for reasonable choices of the efficiencies $\eta_e$, $\eta_a$, and $\eta_t$ the reduction is not very large. For $Q_E = 10$, a reasonable value in a reactor, the value of $p\tau_E$ is reduced by a factor of only about 0.9. Fourth, even electric power breakeven ($Q_E = 0$) requires a value of $p\tau_E$ equal to about 0.4 of its ignition value. Finally, a typical $Q_E = 10$ reactor producing 1000 MW of electric power combined with an overall heating efficiency of $\eta_e \eta_a \approx 0.5$ implies that no more than 50 MW of absorbed microwave power is available for external heating and current drive.

## 4.7 Time dependent power balance in a fusion reactor

The analysis above describes important constraints with respect to the steady state operation of a fusion reactor. This section focuses on additional constraints that arise during the transient phase as the plasma is heated from its initial low temperature to its final operating temperature. There are two topics of interest: thermal stability and the minimum external heating power. To address these issues the 0-D power balance relation is generalized to include time dependence. The generalized relation is discussed first and then applied to the two topics of interest.

### 4.7.1 Time dependent 0-D power balance relation

The goal here is to add time dependence plus the convection and compression terms into the 0-D steady state power balance relation. Once accomplished, this allows an analysis of transient effects. The difficulty in carrying out this task is the appearance of the velocity **v** and, particularly, its spatial derivatives, which are unknown quantities. A convenient way to circumvent this problem, which exploits the uniform (in space) profile assumption, is as follows.

Recall that the terms of interest in the energy equation Eq. (4.1) are

$$\frac{3}{2}\left(\frac{\partial p}{\partial t} + \nabla \cdot p\mathbf{v}\right) + p\nabla \cdot \mathbf{v}. \tag{4.37}$$

This expression can be rewritten more conveniently as

$$\frac{3}{2}\left(\frac{\partial p}{\partial t} + \nabla \cdot p\mathbf{v}\right) + p\nabla \cdot \mathbf{v} = \frac{3}{2}\frac{\partial p}{\partial t} + \frac{5}{2}\nabla \cdot p\mathbf{v} - \mathbf{v} \cdot \nabla p. \tag{4.38}$$

Equation (4.38) is now averaged over the plasma volume to obtain the corresponding 0-D form of the terms. The $\nabla \cdot p\mathbf{v}$ terms integrate to zero by means of the divergence theorem, assuming the plasma boundary is stationary; that is, assuming $\mathbf{n} \cdot \mathbf{v} = \mathbf{0}$ on the plasma boundary. Next, under the assumption of spatially uniform profiles, one can neglect the last term in Eq. (4.38). This is not a very good approximation (because the actual profiles are not flat), but the term is usually not a dominant contribution and the approximation greatly simplifies the analysis by eliminating the appearance of **v**. Combining terms leads to the desired time dependent form of the 0-D power balance equation:

$$\frac{3}{2}\frac{dp}{dt} = S_\alpha + S_h - S_B - S_\kappa$$

$$= \left(\frac{E_\alpha}{16}\right) p^2 \frac{\langle \sigma v \rangle}{T^2} + S_h - \left(\frac{C_B}{4}\right)\frac{p^2}{T^{3/2}} - \frac{3}{2}\frac{p}{\tau_E}. \tag{4.39}$$

This equation can now be used to analyze thermal stability and the minimum external heating power.

### 4.7.2 *Thermal stability*

Consider a plasma in steady state thermal equilibrium at a certain desired operating temperature $T_0$. The alpha and external heating powers balance the heat conduction and Bremsstrahlung losses. Now assume the plasma experiences a small, random temperature fluctuation. Thermal stability asks whether the heating dynamics of the plasma are such that the temperature returns to its original value $T_0$ (stable) or if instead the temperature experiences a thermal runaway (unstable). The answer, as shown below, is closely associated with the relative rates at which the heating power and the power loss change with temperature.

In order to address the temperature evolution by means of Eq. (4.39), one must introduce an assumption that relates the pressure change to the temperature change. This is necessary to reduce the problem from that of a single equation with two unknowns $(T, p)$ to one of a single equation with one unknown $(T)$. A simple plausible approximation is to assume that the density remains constant as the temperature is perturbed from equilibrium. This assumption will serve the present purpose of illustrating the basic problem of thermal stability. Thus, recalling that $p = 2nT$ and assuming that $n = n_0 = $ const, one can rewrite Eq. (4.39) as follows:

$$3n_0 \frac{dT}{dt} = S_H(T) - S_L(T)$$

$$= \left( \frac{E_\alpha}{4} \right) n_0^2 \langle \sigma v \rangle + S_h - C_B n_0^2 T^{1/2} - 3 \frac{n_0 T}{\tau_E}. \qquad (4.40)$$

Here, $S_H = S_\alpha + S_h$ is the total heating power density while $S_L = S_B + S_\kappa$ is the total power loss density. At thermal equilibrium $S_H(T_0) = S_L(T_0)$ by definition.

An intuitive picture of thermal stability can be obtained by examining a plot of $\dot{T}$ vs. $T$ in the vicinity of $T \approx T_0$ as illustrated in Fig. 4.5. Note that there are two possibilities depending on the relative rate of temperature change of the heating and loss terms. The first case, corresponding to a more rapid change of losses as compared to heating, has a negative slope and is thermally stable. To see this, observe that a small positive temperature perturbation moves one to a higher value of $T$ with respect to $T_0$. The negative slope of the curve implies that $\dot{T} < 0$ at this increased value of $T$. Since $\dot{T} < 0$ the dynamics of the system causes the temperature to decrease. In other words, since the losses increase faster than the heating, the net loss of power tends to lower the temperature back towards its equilibrium value. The system is thermally stable. A similar argument holds for a negative temperature perturbation.

In contrast, a positive slope of the curve, corresponding to a more rapid change of heating as compared to losses, implies a thermal runaway. A small positive temperature perturbation places one on a portion of the curve where $\dot{T} > 0$. The dynamics, dominated by the heating, tends to further increase the temperature. This is a thermal runaway, often referred to as a thermal instability.

This intuitive picture can be quantified mathematically by the well-known procedure of linear stability analysis. This method is also widely used in many other aspects of fusion

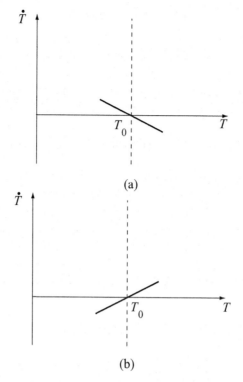

Figure 4.5 $\dot{T}$ as a function of $T$ for: (a) a thermally stable system and (b) a thermally unstable system.

science. It works as follows. Assume the temperature is perturbed only a very small amount away from equilibrium:

$$T(t) = T_0 + T_1(t). \tag{4.41}$$

The "small" assumption implies that $T_1 \ll T_0$. Equation (4.41) is substituted into the equation for $T$, which is then Taylor expanded about the temperature $T_0$. Since $T_0 = \text{const}$, the expanded temperature equation becomes

$$
\begin{aligned}
3n_0 \frac{\mathrm{d}T_1}{\mathrm{d}t} = {} & S_H(T_0) - S_L(T_0) \\
& + \left[ S_H'(T_0) - S_L'(T_0) \right] T_1 \\
& + \left[ S_H''(T_0) - S_L''(T_0) \right] \frac{T_1^2}{2} + \cdots
\end{aligned}
\tag{4.42}
$$

Here, the prime denotes differentiation with respect to $T_0$.

The equation simplifies as follows. The first two terms on the right hand side exactly cancel by virtue of the equilibrium condition. All the terms on the third line can be neglected because of the smallness assumption of $T_1/T_0$. In particular, if $T_1/T_0 \ll 1$, then $T_1/T_0 \gg$

$(T_1/T_0)^2 \gg (T_1/T_0)^3 \ldots$ These simplifications reduce Eq. (4.42) to a first order linear ordinary differential equation in the unknown variable $T_1$:

$$3n_0 \frac{\mathrm{d}T_1}{\mathrm{d}t} = \left[ S_{\mathrm{H}}'(T_0) - S_{\mathrm{L}}'(T_0) \right] T_1. \tag{4.43}$$

The goal now is to examine the time dependence of $T_1$. If $T_1$ decays to zero, the temperature returns to its original value $T_0$ and the plasma is thermally stable. If on the other hand $T_1$ grows without bound, the temperature is moving further and further from its equilibrium value and the plasma is thermally unstable. The enormous advantage of the small perturbation assumption is that the stability equation is linear, allowing a simple analytic solution. Linearity implies that the solutions have exponential behavior. Thus, the solution to Eq. (4.43) can be written as

$$T_1(t) = T_1(0)e^{\gamma t}. \tag{4.44}$$

The growth rate $\gamma$ is found by direct substitution and is given by

$$\gamma = \frac{1}{3n_0} \left[ \frac{\mathrm{d}S_{\mathrm{H}}(T_0)}{\mathrm{d}T_0} - \frac{\mathrm{d}S_{\mathrm{L}}(T_0)}{\mathrm{d}T_0} \right]. \tag{4.45}$$

As predicted intuitively, when the losses increase faster than the heating, the growth rate is negative, implying thermal stability. Conversely, when the heating increases faster than the losses, the growth rate is positive, indicating exponential growth and thermal instability. The condition for thermal stability can therefore be written as

$$\frac{\mathrm{d}S_{\mathrm{L}}(T_0)}{\mathrm{d}T_0} > \frac{\mathrm{d}S_{\mathrm{H}}(T_0)}{\mathrm{d}T_0}. \tag{4.46}$$

Equation (4.46) can easily be applied to magnetic fusion by substituting the appropriate expressions appearing in $S_{\mathrm{L}}$ and $S_{\mathrm{H}}$. The algebra is simplified by neglecting the Bremsstrahlung radiation, which, as it turns out, is a good approximation in the temperature regime near the stability–instability transition. A short calculation yields the following condition for thermal stability:

$$3\frac{n_0}{\tau_{\mathrm{E}}} \geq \frac{1}{4} E_\alpha n_0^2 \frac{\mathrm{d}\langle \sigma v \rangle}{\mathrm{d}T_0}. \tag{4.47}$$

This equation is valid even in the presence of external heating since $S_{\mathrm{h}}$ is assumed to be a fixed quantity independent of $T$. The stability condition can be further simplified by substituting $S_\kappa = S_\alpha + S_{\mathrm{h}}$ and recalling that at the equilibrium temperature $T_0$, $S_{\mathrm{h}}(T_0) = [(1 - f_\alpha)/f_\alpha]S_\alpha(T_0)$. The final form of the stability condition is then

$$\frac{\mathrm{d}}{\mathrm{d}T_0} \left( \frac{T_0^{1/f_\alpha}}{\langle \sigma v \rangle} \right) \geq 0. \tag{4.48}$$

The implications of the stability criterion on power balance can be seen by re-examining the curve of $p\tau_{\mathrm{E}}$ vs. $T$. This is illustrated in Fig. 4.6 for several values of $f_\alpha$. The bold portions of the curves correspond to stability. Observe that for a given $p\tau_{\mathrm{E}}$ thermal stability occurs for the high-temperature solution to power balance. However, while natural thermal

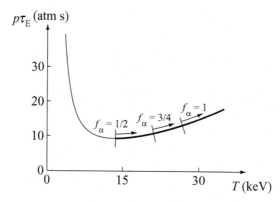

Figure 4.6 Plot of the thermally stable portion of the $p\tau_E$ vs. $T$ curve for various $f_\alpha$. The darker portions of the curve are stable to the right of the $f_\alpha$ points.

stability is desirable, there are other plasma stability and engineering issues that often require operation at the lower-temperature solution. Since the low-temperature solution is thermally unstable, the conclusion is that some form of feedback, usually referred to as "burn control" must be employed in this regime.

The conclusion regarding the need for burn control has been derived assuming that $\tau_E$ is a constant. A more accurate model based on experimental observations would have to include a crucial new effect, that of a temperature dependence on $\tau_E$. Typically $\tau_E$ degrades as the temperature increases. This makes power balance more difficult, but, as will be shown in Chapter 14, substantially improves the situation with respect to thermal stability. Specifically, the qualitative physics of burn control remains unchanged but, when the effect of the experimental $\tau_E = \tau_E(T)$ is included, the undesirable problem of burn control is eliminated.

### 4.7.3 The minimum external power

Assume an optimized reactor situation in which steady state power balance is achieved without the need for any external heating, either to maintain the temperature or to drive plasma current. Even in this situation, however, external power is still needed during the start up transient phase in order to heat the plasma from its initial low temperature to a high enough value for the alpha heating to become dominant. The goal here is to calculate the minimum value of the external heating power and to compare it to the reactor output power. As stated earlier the external power does not affect the steady state operating costs of a fully ignited reactor. It does, nonetheless, contribute directly to the capital cost.

The analysis proceeds by investigating the time evolution of the plasma from its cold initial state to its final ignited state. This, in general, requires the numerical solution of the non-linear, time dependent, energy balance equation. However, the desired information can

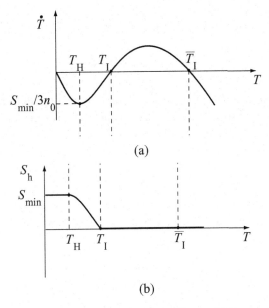

Figure 4.7 (a) $\dot{T}$ vs. $T$ showing the low ($T_I$) and high ($\bar{T}_I$) equilibrium points plus the negative minimum $\dot{T} = [S_\alpha(T_H) - S_\kappa(T_H)]/3n_0 \equiv -S_{min}/3n_0$. (b) Typical $S_h(T)$ vs. $T$ profile that vanishes at $S_h(T_I) = 0$ and whose value at $T = T_H$ satisfies $S_h(T_H) \geq S_{min}$.

be extracted analytically by examining the complete $\dot{T}$ vs. $T$ curve (i.e. Eq. (4.40)) for the case $S_h = 0$ (and neglecting Bremsstrahlung) as shown in Fig. 4.7(a). For simplicity, the entire time evolution is again assumed to take place at fixed $n = n_0 = $ const. Also the value of $p\tau_E$ along the curve is assumed to be sufficiently large that two steady state equilibria exist (i.e. at the two temperatures at which $\dot{T} = 0$).

There are several points to be made. First, the slope at the lower ignition point $T_I$ is positive, indicating thermally instability and the need for some form of burn control. Second, the slope at the upper ignition point $\bar{T}_I$ is negative, corresponding to thermal stability. Since $\dot{T} > 0$ between $T_I$ and $\bar{T}_I$ the thermal instability at $T = T_I$ automatically drives the plasma to the higher stable temperature $\bar{T}_I$. In either case it is necessary to heat the plasma from its initial low value up to $T = T_I$.

The difficulty is now apparent. In the region $0 < T < T_I$, the value of $\dot{T}$ satisfies $\dot{T} < 0$. With $\dot{T}$ negative the plasma temperature cannot increase unless enough external heating is added to raise the curve a sufficient amount above the axis to make $\dot{T}$ positive everywhere. A typical external heating profile $S_h(T)$ is shown in Fig. 4.7(b). The profile must satisfy two strict requirements. First, at $T = T_I$, $S_h(T_I) = 0$ corresponding to the original requirement of a fully ignited plasma. Second, at $T = T_H$ the external power $S_h(T_H)$ must be sufficiently large that the minimum of the $\dot{T}$ vs. $T$ curve is positive. This corresponds to the minimum external power: $S_{min} \equiv S_h(T_H)$.

In mathematical terms $S_{\min}$ is defined as follows. First, at $T = T_{\mathrm{I}}$, the assumptions made imply that $\dot{T} = 0$, $S_h = 0$ and $S_\alpha = S_\kappa$. Specifically, the last condition is given by

$$\frac{1}{4} E_\alpha n_0^2 \langle \sigma v \rangle_{\mathrm{I}} = 3 \frac{n_0 T_{\mathrm{I}}}{\tau_{\mathrm{E}}}. \tag{4.49}$$

In Eq. (4.49), assume that $T_{\mathrm{I}}$ is the known, desired, operating temperature and that the reactor has been designed so that $n_0 \tau_{\mathrm{E}}$, or equivalently $p \tau_{\mathrm{E}}$, is sufficiently large to achieve plasma power balance without external heating.

Second, the temperature $T_{\mathrm{H}}$ is found by calculating the minimum of the $\dot{T}$ vs. $T$ curve:

$$3 n_0 \dot{T} = \frac{1}{4} E_\alpha n_0^2 \langle \sigma v \rangle - 3 \frac{n_0 T}{\tau_{\mathrm{E}}} + S_h. \tag{4.50}$$

In the vicinity of $T = T_{\mathrm{H}}$ assume that $S_h = \mathrm{const}$ as shown in Fig. 4.7(b). Setting $(\partial \dot{T} / \partial T)_{T_{\mathrm{H}}} = 0$ then yields

$$\frac{1}{4} E_\alpha n_0^2 \frac{\mathrm{d} \langle \sigma v \rangle}{\mathrm{d} T_{\mathrm{H}}} = 3 \frac{n_0}{\tau_{\mathrm{E}}}. \tag{4.51}$$

This expression can be simplified by eliminating $n_0/\tau_{\mathrm{E}}$ by means of Eq. (4.49):

$$\frac{\mathrm{d} \langle \sigma v \rangle}{\mathrm{d} T_{\mathrm{H}}} = \frac{\langle \sigma v \rangle_{\mathrm{I}}}{T_{\mathrm{I}}}. \tag{4.52}$$

Equation (4.52) defines $T_{\mathrm{H}}$. The final step is to calculate $S_{\min}$ by setting $S_h$ equal to the imbalance in heating powers at $T = T_{\mathrm{H}}$: $S_{\min} = (S_\kappa - S_\alpha)_{T_{\mathrm{H}}}$.

$$S_{\min} = 3 \frac{n_0 T_{\mathrm{H}}}{\tau_{\mathrm{E}}} - \frac{1}{4} E_\alpha n_0^2 \langle \sigma v \rangle_{\mathrm{H}}. \tag{4.53}$$

It is perhaps of more importance to calculate the ratio of $S_{\min}$ to the alpha power at the final ignition temperature: $S_{\min}/(S_\alpha)_{\mathrm{I}}$. The ratio can be easily converted into electric power units, which then determines the fraction of the total electric power out that is required to drive the external heating power for startup. This information is clearly of great practical importance. A short calculation shows that the ratio $S_{\min}/(S_\alpha)_{\mathrm{I}}$ is given by

$$\frac{S_{\min}}{(S_\alpha)_{\mathrm{I}}} = \frac{T_{\mathrm{H}}}{T_{\mathrm{I}}} - \frac{\langle \sigma v \rangle_{\mathrm{H}}}{\langle \sigma v \rangle_{\mathrm{I}}}. \tag{4.54}$$

Upon introducing the heating and thermal efficiencies as described in Subsection 4.6.2, one obtains the desired power ratio

$$\frac{P_{\mathrm{in}}^{(\mathrm{E})}}{P_{\mathrm{out}}^{(\mathrm{E})}} = \frac{1}{6.4 \, \eta_t \eta_e \eta_a} \left( \frac{T_{\mathrm{H}}}{T_{\mathrm{I}}} - \frac{\langle \sigma v \rangle_{\mathrm{H}}}{\langle \sigma v \rangle_{\mathrm{I}}} \right) \approx 0.80 \left( \frac{T_{\mathrm{H}}}{T_{\mathrm{I}}} - \frac{\langle \sigma v \rangle_{\mathrm{H}}}{\langle \sigma v \rangle_{\mathrm{I}}} \right). \tag{4.55}$$

Before plotting the results, some intuition can be obtained by noting that in the normal temperature regime of a fusion reactor (6 keV $< T_{\mathrm{I}} <$ 30 keV), the velocity-averaged cross

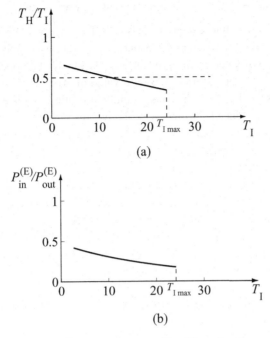

Figure 4.8 (a) The ratio $T_H/T_I$ vs. $T_I$ and (b) the ratio $P_{in}^{(E)}/P_{out}^{(E)}$ vs. $T_I$ corresponding to the minimum external heating for start up.

section is reasonably well approximated by $\langle \sigma v \rangle \approx K_\sigma T^2$. Using this approximation in Eqs. (4.52),(4.54), and (4.55) yields

$$
\begin{aligned}
\frac{T_H}{T_I} &= \frac{1}{2}, \\
\frac{S_{min}}{(S_\alpha)_I} &= \frac{1}{4}, \\
\frac{P_{in}^{(E)}}{P_{out}^{(E)}} &\approx 0.2.
\end{aligned}
\tag{4.56}
$$

The actual curves of $T_H/T_I$ and $P_{in}^{(E)}/P_{out}^{(E)}$ are plotted in Fig. 4.8 as a function of $T_I$.

Observe that during start up considerable power is required to heat the plasma to the ignition temperature. The simple 0-D model predicts that about 20% of the steady state electric power is required as independent startup power. This result is highly model dependent and is a direct consequence of the assumption that $\tau_E$ is a constant. When the experimentally determined $\tau_E = \tau_E(T)$ is used in the analysis the situation is greatly improved. It is shown in Chapter 14 that in this case the qualitative picture of the minimum power is unchanged. However, the required auxiliary power is reduced by about one order of magnitude, again helping to avoid an undesirable problem.

## 4.8 Summary of magnetic fusion power balance

Analysis of the 0-D power balance relation leads to the following conclusions with respect to a magnetic fusion reactor. In a reactor the plasma must operate at a temperature $T \approx 15$ keV and achieve a minimum value of the product $p\tau_E$. In steady state operation in the fully ignited regime (i.e., with no external power) a value of $p\tau_E \approx 8.3$ atm s is required. If external power is utilized to produce some additional heating or current drive, the value of the required $p\tau_E$ is reduced. In this mode the reactor operates as an amplifier with a physics gain factor $Q$ or equivalently, a more realistic engineering gain factor $Q_E$. Typically economic considerations require $Q_E \approx 10$. Under this condition the required $p\tau_E$ is reduced by only a small amount, to 90% of its ignited value.

The transient dynamics leading to steady state operation introduces two phenomena of importance. First is the question of thermal stability. For values of $p\tau_E$ above the minimum required for steady state operation there are, in general, two equilibrium solutions corresponding to two different temperatures. The lower-temperature solution is thermally unstable and requires some form of burn control. The higher-temperature solution is thermally stable but when $\tau_E$ is constant, $T$ has too high a value and is therefore undesirable for other physics reasons not yet discussed. In Chapter 14 it is shown that for the realistic case $\tau_E = \tau_E(T)$ the high-temperature solution occurs at a lower value of $T$ and thermally stable operation becomes accessible.

Second, even if the plasma is fully ignited in steady state ($Q_E = \infty$) a substantial amount of external heating power is still required during the transient phase to heat the plasma to its ignited state. The simple 0-D model predicts a value $P_{in}^{(E)} \approx 0.2 P_{out}^{(E)}$ when $\tau_E$ is a constant. Here too, the situation is improved for the realistic case $\tau_E = \tau_E(T)$. In Chapter 14 it is shown that the minimum heating power is reduced by about a factor of about 10 as compared to the constant $\tau_E$ case.

## Bibliography

Chapter 4 describes the power balance issues in a fusion reactor. Much of this material is quite basic and has been described in a number of textbooks over the years. Several textbooks describing fusion reactor power balance are listed below. Also given is the original reference for the famous Lawson criterion.

### *Power balance in a fusion reactor*

Dolan, T. J. (1982). *Fusion Research*. New York: Pergamon Press.
Gross, R. (1984). *Fusion Energy*. New York: John Wiley & Sons.
Rose, D. J. and Clark, M. (1961). *Plasmas and Controlled Fusion*. Cambridge, Massachusetts: MIT Press.
Stacey, W. M. (1981). *Fusion Plasma Analysis*. New York: John Wiley & Sons.
Stacey, W. M. (2005). *Fusion Plasma Physics*. Weinheim: Wiley-VCH.
Wesson, J. (2004). *Tokamaks*, third edn. Oxford: Oxford University Press.

### The Lawson criterion

Lawson, J. D. (1957). Some criteria for a power producing thermonuclear reactor, *Proceedings of the Physical Society* **B70**, 6.

## Problems

4.1 This problem involves the derivation of a generalized version of the Lawson criterion. Consider a subignited reactor in which $S_\alpha < S_k$. In the plasma power balance include alpha heating, external heating, and thermal conduction losses. Also, include the power produced by breeding tritium from $Li^6$. However, assume that of the total alpha power only a fraction $f$ deposits its energy in the plasma while $1 - f$ is immediately lost to the first wall and converted to heat. Assume a thermal conversion efficiency $\eta_t$ and an input electricity to plasma heating conversion efficiency $\eta_h$ (i.e., $\eta_h = \eta_e$, $\eta_a = 1$).
   (a) Derive an expression for $p\tau_E = G(Q_E, f)$ for steady state operation.
   (b) Assume $T = 15$ keV, $\eta_t = 0.35$, $\eta_h = 0.5$. Plot curves of $p\tau_E$ vs. $Q_E$ for $f = 0$, 0.5, 1. Compare the required $p\tau_E$ values at $Q_E = 20$ with the fully ignited value ($Q_E = \infty$, $f = 1$) and the Lawson breakeven criterion ($Q_E = 1$, $f = 0$).

4.2 An axisymmetric toroidal D–T fusion reactor requires a toroidal current as well as a toroidal magnetic field to provide good confinement. In most present day experiments this current is driven by a transformer. However, in a steady state reactor the toroidal current must be driven non-inductively. One method of doing this is by means of "RF current drive." The efficiency of current drive is defined as $\eta = n_{20} I R_0 / P_{CD}$, where $P_{CD}$ is the total RF power in watts and $I$ is the plasma current in amperes. About the best efficiency thus far achieved is $\eta \approx 0.3$ A/(m$^2$ W). Consider now a reactor with a circular cross section plasma, major radius $R_0 = 8$ m, minor radius $a = 2.5$ m, and toroidal magnetic field $B_0 = 6$ T. Assume the density and temperature profiles are uniform. In steady state the toroidal $\beta$ is limited to $\beta = 2\mu_0 p / B_0^2 = 0.05$. The toroidal current is $I = 20$ MA. For the purpose of this problem assume $I$ is driven entirely non-inductively.
   (a) Derive an expression for the circulating power quality factor, defined by $Q_{CD} = P_{fusion} / P_{CD}$.
   (b) Using the analytic formula for the cross section given in Problem 3.1 calculate the value of $T_k$ that maximizes $Q_{CD}$.
   (c) Calculate the corresponding value of $Q_{CD}$. Will current drive be a major or minor problem for a reactor?

4.3 For uncatalyzed D–D fusion estimate the minimum $p\tau_E$ required for ignition and the corresponding value of $T$. Repeat the calculation for fully catalyzed D–D fusion. Use the $\langle \sigma v \rangle$ curves in the text.

4.4 The purpose of this problem is to investigate the effect of plasma profiles on the critical value of $p\tau_E$ for ignition assuming that alpha power and thermal conduction losses are the dominant contributions to power balance. Consider a 1-D plasma with a circular cross section. The volume-averaged alpha power density and thermal conduction losses are defined as

$$\overline{S}_\alpha = \frac{2}{a^2} \int_0^a \left( \frac{E_\alpha}{4} n^2 \langle \sigma v \rangle \right) r \, dr,$$

$$\overline{S}_1 = \frac{2}{a^2} \int_0^a \left( 3 \frac{nT}{\tau_E} \right) r \, dr.$$

Assume now that the density and temperature profiles are given by

$$n = (1 + v_n)\bar{n}(1 - r^2/a^2)^{v_n},$$
$$T = (1 + v_T)\bar{T}(1 - r^2/a^2)^{v_T},$$

where $\bar{n}$ and $\bar{T}$ are the volume-averaged density and temperature respectively. To determine the effect of temperature profile on $\bar{p}\tau_E$ numerically evaluate $\bar{S}_\alpha$ and $\bar{S}_l$ for $v_n = 0$ and $0 \leq v_T \leq 4$ using $\langle \sigma v \rangle$ from Problem 3.1. For each $v_T$ find the optimum value of $\bar{T}$ that minimizes $\bar{p}\tau_E$ at fixed average pressure: $\bar{p} = 2\left[(1 + v_n)(1 + v_T)/(1 + v_n + v_T)\right]\bar{n}\bar{T} = \text{const}$. Plot the optimum $\bar{T}$ (keV) and corresponding $\bar{p}\tau_E$ as a function of $v_T$ showing the 0-D limit $v_n = v_T = 0$ for reference. To determine the variation of $\bar{p}\tau_E$ with density repeat the above calculation for $v_T = 2$ and $0 \leq v_n \leq 4$. Are peaked profiles good, bad, or unimportant in minimizing $\bar{p}\tau_E$?

# 5

# Design of a simple magnetic fusion reactor

## 5.1 Introduction

Power balance considerations have shown that a magnetic fusion reactor should operate at a temperature of about 15 keV and be designed to achieve a value of $p\tau_E > 8.3$ atm s. Even so, these considerations do not shed any light on the optimum tradeoff between $p$ and $\tau_E$. Nor do they provide any insight into the geometric scale and magnetic field of a fusion reactor. This is the goal of Chapter 5, which presents the design of a simple magnetic fusion reactor. All geometric and magnetic quantities are calculated as well as the critical plasma physics parameters.

Remarkably, the design requires virtually no knowledge of plasma physics even though for nearly half a century the field has been dominated by the study of this new branch of science. The design is actually driven largely by basic engineering and nuclear physics constraints. These constraints determine the geometric scale of the reactor as well as the size of the magnetic field. Equally important, they make "demands" on the plasma parameters. Plasma physicists must learn how to create plasmas that satisfy these demands (e.g. pressure and confinement time) in order for fusion to become a commercially viable source of energy. Knowledge of the desired plasma parameters is crucial as it defines the end goals of fusion-related plasma physics research, and serves as the guiding motivation for essentially all of the discussion of plasma physics in the remainder of the book.

The plan in this chapter to design a reactor is as follows. First, a description of a generic magnetic fusion reactor is presented. Second, a definition is given of what actually constitutes a design; that is, which parameters specifically need to be calculated. Third, the basic engineering and nuclear physics constraints as well as the design goals are introduced. Finally, the constraints and goals, combined with power balance considerations, are utilized to produce a design. In general, the design parameters calculated are reliable to within factors of 2–3 with respect to several far more detailed and comprehensive reactor design studies.

## 5.2 A generic magnetic fusion reactor

A generic magnetic fusion reactor is illustrated in Fig. 5.1. It has the following features. A basic requirement is that the reactor must operate as a steady state device. Pulsed reactors

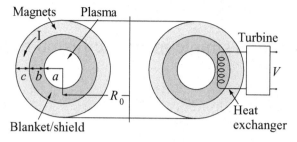

Figure 5.1 Generic toroidal fusion reactor showing the plasma, blanket-and-shield, and magnets.

are less desirable because cyclical thermal and mechanical stresses on the structure lead to increased maintenance and more frequent replacement of components due to fatigue failure. The reactor itself is in the shape of a torus. This is not an ideal shape from an engineering point of view because of its mechanical complexity. A cylinder or sphere would be more desirable. However, as will be shown, the basic properties of magnetic fields require that the geometry be toroidal. Also, the plasma cross section is assumed to be circular. This is not the case in actual designs. However, for present purposes this is a useful assumption to simplify the mathematics and leads to a design that falls within the desired accuracy.

The purpose of the magnetic field is to confine the hot plasma away from the "first wall." Outside the first wall is the blanket where the energy conversion takes place. Surrounding the blanket is a shield to protect the magnets and the workers in the plant from neutron and gamma ray radiation. Finally, the coils producing the magnetic field are located outside the shield.

The magnets must, in general, be superconducting. The reason is that copper magnets dissipate substantial amounts of ohmic power during steady state operation even though the electrical conductivity of copper is quite high. This dissipated power would be sufficient to seriously impair the overall power balance of the reactor. In other words, the ohmic power would be comparable to the fusion produced electrical power. Superconducting magnets dissipate literally zero power in steady state and require only a small (as compared to the power output) amount of cooling power to keep the magnets superconducting. Another feature of the magnets is that they must be located outside the blanket-and-shield because an internal superconducting magnet could not withstand the bombardment of the fusion neutrons. Not only would it sustain material damage but it would also quickly return to its non-superconducting state.

Thus, a generic fusion reactor consists of a toroidal plasma surrounded by a first wall, a blanket-and-shield, and superconducting magnets. For perspective the generic reactor illustrated in Fig. 5.1 can be compared to a far more realistic, detailed design produced as part of the ARIES reactor program and illustrated in Fig. 5.2.

## 5.3 The critical reactor design parameters to be calculated

This section describes the critical parameters that define the design of the reactor. These parameters include the geometry of the reactor, the magnetic field, and various plasma

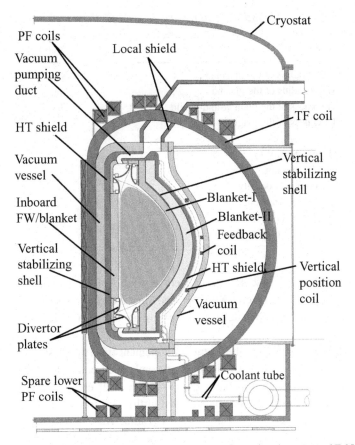

Figure 5.2 Cross section of the ARIES-AT power core configuration (courtesy of F. Najmabadi).

physics parameters. Clearly, because of the simplicity of the model, one can only calculate a limited number of quantities but these will be sufficient to provide the desired overview of a magnetic fusion reactor. The quantities to be determined are listed in Table 5.1.

All of these quantities, with the exception of $\beta$, have been previously defined or are self-explanatory. The quantity, $\beta$, is the normalized plasma pressure defined as follows:

$$\beta \equiv \frac{p}{B^2/2\mu_0}. \tag{5.1}$$

Note that $\beta$ is the ratio of the plasma pressure to magnetic pressure. It is a measure of the efficiency with which the magnetic field confines the plasma. High $\beta$ is desirable for economic power balance in a reactor but is difficult to achieve experimentally because of various plasma instabilities. Low $\beta$ is easier to achieve but represents a lower confinement efficiency.

A substantial part of fusion research has been devoted to the discovery of particular magnetic geometries that give stable confinement at high $\beta$. In the context of the present simple reactor design, the actual details of the magnetic geometry are unimportant and the

Table 5.1. *Quantities defining the design of a fusion reactor*

| Quantity | Symbol |
| --- | --- |
| Minor radius of the plasma | $a$ |
| Major radius of the plasma | $R_0$ |
| Thickness of the blanket-and-shield | $b$ |
| Thickness of the magnets | $c$ |
| Plasma temperature | $T$ |
| Plasma density | $n$ |
| Plasma pressure | $p$ |
| Fusion power density | $S_f$ |
| Energy confinement time | $\tau_E$ |
| Magnetic field | $B$ |
| Normalized plasma pressure | $\beta$ |

magnetic field is simply represented by an average value $B$. Even so, the engineering and economic constraints on the reactor will "demand" that a certain value of $\beta$ be achievable and it is then up to the plasma physicists to discover magnetic geometries that can meet this demand.

A similar argument applies to the energy confinement time $\tau_E$. Plasma turbulence and the resulting anomalous transport must be sufficiently well controlled that the value of $\tau_E$ required by technological constraints can be achieved. If either the required $\beta$ or $\tau_E$ cannot be achieved by clever plasma physics, then further constraining the reactor design to accommodate these deficiencies leads to a much less attractive reactor in terms of economic viability. A major part of the discussion of plasma physics in the remainder of the book is aimed at discovering ways to achieve the necessary values of these parameters.

## 5.4 Design goals, and basic engineering and nuclear physics constraints

### 5.4.1 Design goals

The overriding goal of the design is to minimize the capital cost of the nuclear reactor. As will be shown, a magnetic fusion reactor is inherently a large, rather complex facility (as compared for instance to a natural gas power plant). Consequently, the initial funds required to construct the facility will be relatively high. It is, therefore, critical that the reactor be designed, to the extent possible, to minimize this capital cost in order that fusion can become economically competitive when the capital, fuel, and operating costs are all combined to yield the cost-of-electricity (CoE).

Plasma physics considerations enter the design only in the sense of the following guidelines. In Chapters 13–14 it will be shown that the basic physics governing the behavior of fusion grade plasmas is such that it is difficult in practice to achieve large $\tau_E$ and high $\beta$. The achievable values from a plasma physics point of view do not leave much, if any, margin with respect to the typical values demanded from the engineering and nuclear physics constraints. Consequently, it is desirable, whenever possible, to design the reactor so as to minimize the demands on $\tau_E$ and $\beta$.

The design goals can thus be summarized as follows: minimize the cost and wherever possible minimize the demands on $\tau_E$ and $\beta$.

### 5.4.2 Engineering constraints

There are four basic engineering constraints that directly enter into the design of the reactor. They are as follows.

First, a typical large-scale power plant generates approximately 1000 MW of electricity. This is the value used in the present design. It is denoted by $P_E = 1000$ MW.

Second, there is a wall loading limit on the first wall; that is, there is an upper limit on the amount of power per unit area than can safely pass through the first wall without causing unacceptable damage to the wall material. There are two effects contributing to this limit. One is the heat load due to thermal conduction losses plus the somewhat smaller Bremsstrahlung radiation losses. There are a number of techniques (e.g. limiters, divertors) which address the thermal conduction problem. All techniques obviously involve some form of cooling. The heat loss problem is still not completely resolved at present, although progress is being made. In a reactor heat loss is assumed to pose the less severe wall loading limit.

The second effect results from the fact that all the 14.1 MeV neutrons pass through the first wall. They do so with a more or less uniform angular power deposition profile around the circumference. This neutron flux can lead to various types of radiation-induced damage to the first wall material including sputtering, embrittlement, erosion, cracking, etc. The 14.1 MeV neutrons are assumed to be responsible for the most stringent wall loading limit. Studies of existing materials suggest that limits in the range 1–6 MW/m$^2$ are possible. Here, a reasonable, perhaps slightly optimistic value for the maximum neutron wall loading limit is chosen to be $P_W = 4$ MW/m$^2$.

The third engineering constraint results from the electrical properties of superconducting magnets. In order for a magnet to remain superconducting its temperature $T$, current density $J$, and magnetic field $B$ must lie beneath a dome-shaped curve in a 3-D $T, J, B$ space as shown in Fig. 5.3. Once operation crosses the boundary of the curve, the magnet reverts to its normal, relatively poorly conducting state. Magnetic fusion, as its name implies, requires high magnetic fields and high currents. At the present time the highest-field, highest-current-density magnets use niobium–tin as the superconducting material. Large-scale magnets with magnetic fields on the order of 10–15 T, such as the "ITER model coil" located in Japan, have been successfully built. See Fig. 5.4. As high-temperature superconducting magnets are developed, this value will likely increase. For present purposes, however, the largest magnetic field allowed is chosen to be $B_{max} = 13$ T.

The fourth engineering constraint is also associated with the magnetic field. It is related to the fact that high-field magnets produce enormous forces because of the self-generated

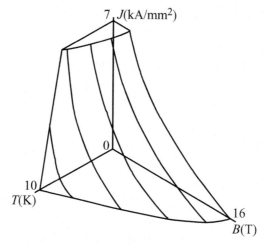

Figure 5.3 The 3-D *T, J, B* dome for niobium–tin under which a superconducting magnet must operate to be in its superconducting state (Wilson, M. N. (1983). *Superconducting Magnets*. Oxford: Clarendon Press).

Figure 5.4 The ITER superconducting model coil built by an international team and assembled and tested in Japan (courtesy of K. Okuno).

magnetic pressure. If a sufficiently strong structural support system is not provided, the magnetic forces could quite literally blow the magnet apart. Indeed, the size and cost of a high-field superconducting magnet is frequently dominated by the requirements of the structural support system. The critical parameter driving the design is the maximum allowable stress on the support structure, often stainless steel. The actual value of the maximum stress depends on the choice of structural material, the geometry of the coil, the stress distribution around the coil (which, in general, will not be uniform), and various mandated safety margins. For present structural materials a reasonable value for the maximum allowable average stress is $\sigma_{max} = 300$ MPa $\approx 3000$ atm.

The basic engineering constraints on a magnetic fusion reactor thus involve the desired output power, the maximum neutron wall loading, the maximum magnetic field of a superconducting magnet, and the maximum allowable stress on the structural support system.

### 5.4.3 Nuclear physics constraints

There are three basic nuclear physics constraints that directly impact the design of a fusion reactor as described below.

The first constraint is the size of the D–T fusion cross section. Since the cross section is related to the number of fusion reactions taking place per second, this ultimately determines the required plasma pressure for a given volume in order to produce the desired amount of power. From the analysis in Chapter 3, the relevant cross section constraint can be written as $\langle \sigma v \rangle = 3 \times 10^{-22}$ m$^3$/s at a temperature of $T = 15$ keV.

The second nuclear physics constraint is related to the blanket. Obtaining good tritium breeding in lithium requires that the 14.1 MeV neutrons slow down to thermal speed. This is accomplished by combining a moderating material with the lithium in the blanket. The mean free path for slowing down the neutrons in the moderator is a critical parameter, which largely determines the size of the blanket. The value of the mean free path depends somewhat on the moderating material and the energy of the neutrons. Typically, over the range of neutron energies of interest the slowing down cross section is on the order of a barn. For simplicity in the present calculation, the cross section is assumed to be a constant given by $\sigma_{sd} = 1$ barn.

The third nuclear physics constraint is also related to the blanket. Once the neutrons are slowed down by the moderator they are readily captured by the Li$^6$ to breed tritium. The cross section for this process is quite large thereby providing the motivation for slowing down the neutrons. This neutron capture cross section is therefore the third constraint. A 0.025 eV thermal neutron has a tritium breeding cross section of $\sigma_{br} = 950$ barns.

This completes the list of constraints. A summary of all the limiting values is given in Table 5.2. In the section that follows, it is shown how this set of seemingly disconnected constraints essentially determines all the parameters of the basic reactor design.

## 5.5 Design of the reactor

### 5.5.1 Outline of the design calculation

The constraints described above are used in the following order to calculate the design parameters for the simple magnetic fusion reactor under consideration.

- The cross section constraints related to the blanket determine the thickness of the blanket.
- The minimum cost constraint and the magnet constraints combine to determine the coil thickness and plasma minor radius.

Table 5.2. *Basic engineering and nuclear physics constraints*

| Quantity | Symbol | Limiting value |
|---|---|---|
| Electric power output | $P_E$ | 1000 MW |
| Maximum wall loading | $P_W$ | 4 MW/m$^2$ |
| Maximum magnetic field | $B_{max}$ | 13 T |
| Maximum mechanical stress | $\sigma_{max}$ | 300 MPa $\approx$ 3000 atm |
| Velocity-averaged cross section | $\langle \sigma v \rangle$ | $3 \times 10^{-22}$ m$^3$/s |
| Fast neutron slowing down cross section | $\sigma_{sd}$ | 1 barn |
| Slow neutron breeding cross section in $Li^6$ | $\sigma_{br}$ | 950 barns at 0.025 eV |

- The output power and wall loading constraints determine the major radius of the plasma.
- The D–T fusion cross section and the power output constraints determine the plasma pressure.
- The need for plasma power balance determines the required value for the energy confinement time.
- The need to minimize the normalized pressure $\beta$ determines the size of the magnetic field and the corresponding required value of $\beta$.

The details of the design are given below.

### 5.5.2 *The blanket-and-shield thickness*

The blanket-and-shield is a complex, highly engineered component of a fusion reactor with a choice of materials for each of its components. It has several critical functions as shown in the highly simplified diagram in Fig. 5.5. The simplification arises because each function is treated as separate and independent from all others whereas in reality there is considerable overlap. Even so, the simplified model is convenient for providing a qualitative understanding of the operation of the blanket-and-shield. Starting from the first wall and progressing outwards, the function of each component is as follows.

First, a neutron multiplier is required to create excess neutrons to replace various unavoidable losses in the blanket (e.g. lost space due to the coolant and the presence of structural material, etc.). The 14.1 MeV fusion neutrons all pass through the relatively thin multiplier region. A fraction of these undergoes a multiplying reaction with each corresponding original neutron leading to the production of two new neutrons.

Second, a moderator is required to slow down the fast neutrons to thermal energy levels so they can be readily captured by $Li^6$. A variety of lithium compounds meet this need, depending upon whether the blanket is in the form of a solid, liquid metal, or molten salt. There is no universally agreed upon optimum choice and each option involves a series of engineering trade offs whose discussion is beyond the scope of the present book. Even so, the slowing down process is relatively insensitive to these choices, which is fortunate since it is in fact a dominant driver determining the thickness of the blanket.

Third the blanket must include a tritium breeding region containing a substantial amount of $Li^6$. Natural lithium consists of 7.5% $Li^6$ and 92.5% $Li^7$. Recall that the $Li^7$ can also

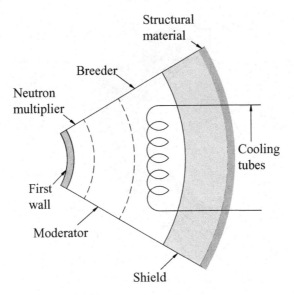

Figure 5.5 Simple model for the components of the blanket-and-shield. Note that the cooling tubes actually extend into the first wall, neutron multiplier, and moderator.

react with a fusion neutron, resulting in tritium breeding without the loss of a neutron, but losing rather than gaining energy:

$$\mathrm{Li}^7 + \mathrm{n} \rightarrow \mathrm{T} + \alpha + \mathrm{n} - 2.5 \,\mathrm{MeV}. \tag{5.2}$$

However, the $\mathrm{Li}^7$ cross section is much smaller than that of $\mathrm{Li}^6$, resulting in only a small gain in tritium and a small loss in energy. Furthermore, when breeding does actually take place in the $\mathrm{Li}^6$, it does so over a relatively narrow region because of the large cross section and corresponding short mean free path. Consequently, if one attempts to improve blanket performance (in the form of reduced blanket thickness) by means of an enriched $\mathrm{Li}^6$ breeding material, this may not yield a large return on investment. The reason is that the breeding region is already only a relatively small part of the overall blanket. In the end the gains in performance must be balanced against the cost of enrichment.

Fourth, the blanket itself is surrounded by a shield. Its purpose is to absorb any neutrons that escape the blanket as well as any gamma rays that are produced by subsidiary nuclear reactions. The absorption must be almost perfect. The reason is that just outside the shield are the superconducting coils that operate at about 5–10 K. Thus, for a reasonably sized cryogenic cooling system, the magnets can withstand only a very small heat load.

Fifth, embedded in the blanket is a set of cooling tubes with either a liquid or gas acting as the coolant. The role of the coolant is to carry away the fusion-produced heat resulting from the slowing down of energetic particles. It is this heat that is ultimately converted into electricity through an appropriate thermal conversion system. This, after all, is the main goal of a fusion power reactor.

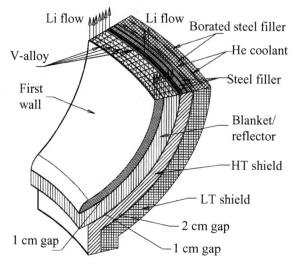

Figure 5.6 Blanket module from the AIRES design project (courtesy of F. Najmabadi).

Finally, there is structural material surrounding the blanket-and-shield. This allows a modular construction, which is important for engineering purposes. A blanket-and-shield constructed of modules can be more easily maintained since individual modules can be readily extracted, repaired, and replaced as required.

This completes the qualitative description of the blanket-and-shield. The simple design presented here can be compared to the more realistic one used in the ARIES project and illustrated in Fig. 5.6.

Attention is now focused on calculating the combined thickness of the moderator and breeding region, which is the largest part of the blanket-and-shield. A reasonable estimate is obtained from the admittedly simple model described below. The only subtlety involved is to recognize that two processes are actually taking place – slowing down and breeding. For the inner part of the moderator, slowing down dominates. Further out, extending into the breeding region, breeding becomes more important.

Consider first the slowing down of fast neutrons. Neutrons lose energy through a sequence of elastic collisions. The energy $E$ of a typical neutron decays with distance into the moderator–breeder region in accordance with the simple relation

$$\frac{dE}{dx} + \frac{E}{\lambda_{sd}} = 0. \tag{5.3}$$

The choice for the mean free path $\lambda_{sd}$ is related to the slowing down cross section by the relation $\lambda_{sd} = 1/n_L \sigma_{sd} \approx 0.055$ m. Here, $n_L = 4.5 \times 10^{28}$ m$^{-3}$ is the number density of natural lithium and $\sigma_{sd} = 1$ barn, the constraint value previously discussed. For simplicity the mean free path is assumed to be at most only weakly dependent on energy. In other words, the dependence of $\lambda_{sd}$ on $E$ is neglected. Under these assumptions Eq.(5.3) can be easily integrated yielding

$$E = E_n e^{-x/\lambda_{sd}}, \tag{5.4}$$

where $E_n = 14.1$ MeV. As expected, the energy decays exponentially with distance.

Next, consider the breeding of tritium. This involves a nuclear reaction with lithium and each such reaction results in the loss of a neutron. Therefore, the flux of neutrons penetrating into the moderator–breeding region decreases with distance. The flux of neutrons is defined as $\Gamma_n \equiv n_n v_n$ and satisfies a decay-type of equation given by

$$\frac{d\Gamma_n}{dx} + \frac{\Gamma_n}{\lambda} = 0. \tag{5.5}$$

The mean free path for breeding, $\lambda$, is related to the corresponding cross section by the usual form of relation $\lambda = 1/n\sigma$. The quantities $n$ and $\sigma$ are determined as follows. For breeding $n$ is equivalent to the density of $Li^6$. Given that 7.5% of natural lithium is $Li^6$, one must set $n = 0.075 \, n_{Li}$. This assumes no enrichment. The breeding cross section $\sigma$ can be related to the thermal breeding cross section $\sigma_{br}$ by noting that at low energies experimental measurements show $\sigma$ to be inversely proportional to the neutron velocity where $v_n \sim E^{1/2}$. Thus, a plausible model for the breeding cross section, valid in the region where it is dominant, is given by

$$\sigma \approx \sigma_{br} \left(\frac{E_t}{E}\right)^{1/2}, \tag{5.6}$$

where $\sigma_{br} = 950$ barns is the constraint value and $E_t = 0.025$ eV. Combining these results yields an expression for $\lambda$ which can be written as

$$\lambda = \lambda_{br} \left(\frac{E}{E_t}\right)^{1/2}. \tag{5.7}$$

Here, $\lambda_{br} = 1/(0.075 \, n_L \sigma_{br}) = 0.0031$ m.

Substituting this expression, plus the expression for $E$, into the equation for $\Gamma_n$ leads to the following simple differential equation:

$$\frac{d\Gamma_n}{dx} + \left(\frac{E_t}{E_f}\right)^{1/2} \frac{e^{x/2\lambda_{sd}}}{\lambda_{br}} \Gamma_n = 0. \tag{5.8}$$

Equation (5.8) can easily be solved yielding the neutron flux as a function of distance into the moderator–breeding region:

$$\frac{\Gamma_n}{\Gamma_{n0}} = \exp\left[-2\left(\frac{E_t}{E_f}\right)^{1/2} \frac{\lambda_{sd}}{\lambda_{br}} \left(\exp\frac{x}{2\lambda_{sd}} - 1\right)\right]. \tag{5.9}$$

Here, $\Gamma_{n0} = \Gamma_n(0)$. Observe that the flux decreases slowly near $x \sim 0$, the entrance to the moderator, but then very rapidly decreases once breeding begins.

The desired expression for the moderator–breeding region thickness $\Delta x$ can now be obtained by inverting this relationship and then asking how thick the region must be in order for the flux to be attenuated by a given amount. For instance, when the flux is reduced by a factor of 100, then 99% of the fusion neutrons have slowed down and undergone a breeding reaction. The thickness of the moderator–breeding region required to accomplish this reduction is given by

$$\Delta x = 2\lambda_{sd} \ln \left[ 1 - \frac{1}{2} \left( \frac{E_f}{E_t} \right)^{1/2} \frac{\lambda_{br}}{\lambda_{sd}} \ln \left( \frac{\Gamma_n}{\Gamma_{n0}} \right) \right] \approx 0.88 \, \text{m}. \qquad (5.10)$$

Note that the thickness is about 0.88 m. The value is relatively insensitive to the choice of reduction factor (100) because of the double logarithm. Also, note that the approximate boundary separating the moderator and breeding regions, defined as the point where $\lambda_{sd} = \lambda_{br}(E/E_t)^{1/2}$, is given by $x \approx 0.79$ m. While breeding is taking place over the entire region it dominates only over a relatively narrow region. However, within this region it is very effective because the breeding mean free path is very short.

The value $\Delta x \approx 0.88$ m for the moderator–breeder is a reasonable estimate. Highly sophisticated numerical neutronics modeling of various designs, including neutron multiplication, moderation, breeding, the shield, the cooling tubes, and the structural support indicates that the complete thickness of the blanket-and-shield system should typically be in the range $1 \, \text{m} < b < 1.5 \, \text{m}$. For the design of the simple fusion reactor under consideration, the choice made for the thickness $b$ is

$$b = 1.2 \, \text{m} \qquad (5.11)$$

After a lengthy discussion, this is the desired value of $b$, which will be shown shortly to play a dominant role in determining the remainder of the reactor geometry.

### 5.5.3 The plasma radius and coil thickness

The plasma radius $a$ and the coil thickness $c$ are determined by the simultaneous consideration of two major design issues: minimizing the cost and satisfying magnet stress limitations. A simple model is presented that addresses each issue, leading to a set of coupled algebraic relations for $a$ and $c$. These are then solved simultaneously yielding the desired values.

#### The minimum cost

As one might imagine, calculating the cost of a fusion reactor is quite a complex task. In general one must include the capital cost, the operating cost, and the fuel cost in addition to the sophisticated rules of financing. As will be shown, a fusion reactor is a rather large facility compared to either fossil or fission reactors suggesting that the capital cost is a major consideration. This point of view is the one adopted here. A reasonably accurate but simple approximation to the goal of minimizing the overall cost is to focus solely on the capital cost.

The capital cost itself comprises two main components: the fixed cost and the nuclear island cost:

$$\text{total capital cost} = \text{fixed cost} + \text{nuclear island cost}. \qquad (5.12)$$

The fixed costs are essentially the same regardless of the source of heat (e.g. fusion, fission, fossil). These costs include those of the turbines, generators, buildings, etc. For a fusion reactor detailed reactor design studies show that the fixed costs are typically somewhat less than half of the total capital cost. The nuclear island costs are those costs directly associated with the fusion device. They are dominated by the large, highly engineered components including the blanket-and-shield and the magnets. The first wall also makes a finite, but not dominant, contribution to the cost. For simplicity this contribution is neglected here.

The capital cost model used here is based on the following two simple but plausible assumptions. First, the fixed costs are assumed to be proportional to the electric power output; higher power output requires larger buildings, larger turbines, larger generators, etc. If one denotes the fixed costs by $K_F$, then this assumption implies that

$$K_F = C_F P_E, \tag{5.13}$$

where $C_F$ is a constant. The second assumption is that the cost of the nuclear island is proportional to the volume of highly engineered material: the blanket-and-shield and magnets in the present case. The bigger these components, the more material, the more construction, and the more engineering required. If the nuclear island costs are denoted by $K_I$ then the cost assumption implies that

$$K_I = C_I V_I. \tag{5.14}$$

Here, $C_I$ is a constant and the engineered volume $V_I$ is given by (see Fig. 5.1)

$$V_I = 2\pi^2 R_0[(a + b + c)^2 - a^2]. \tag{5.15}$$

In minimizing the capital cost of the reactor it is not the actual cost itself that is most relevant but instead the cost per watt. This avoids the trivial result of zero cost for zero power. Consequently, if one defines the cost per watt as $C = (K_F + K_I)/P_E$, then the quantity to be minimized can be expressed as

$$C = C_F + C_I \frac{V_I}{P_E}. \tag{5.16}$$

Since $C_F$ and $C_I$ are constants, minimizing the cost is equivalent to minimizing the volume of material required to produce a given power output:

$$\frac{V_I}{P_E} = \frac{2\pi^2 R_0[(a + b + c)^2 - a^2]}{P_E}. \tag{5.17}$$

This form is convenient because the quantity to be minimized is expressed directly in terms of the reactor geometry without the need to explicitly specify the difficult to calculate cost coefficients.

One further step is required to transform Eq. (5.17) into a more useful form for determining $a$ and $c$. This step requires the elimination of $R_0$ by means of the relation between $P_E$ and $P_W$ which can be carried out as follows. First, recall from Chapter 4 that in a fully ignited plasma the electric power out is related to the thermal fusion power produced by

$$P_E = \tfrac{1}{4}\eta_t(E_\alpha + E_n + E_{Li})n^2\langle\sigma v\rangle(2\pi^2 R_0 a^2). \tag{5.18}$$

Second, the product of the neutron wall loading and the surface area of the plasma is, by definition, equal to the total neutron power produced in the plasma; that is, all the fusion produced neutron power passes through the first wall. This can be expressed as

$$P_W(4\pi^2 R_0 a) = \tfrac{1}{4}E_n n^2\langle\sigma v\rangle(2\pi^2 R_0 a^2). \tag{5.19}$$

A simple elimination yields the required expression for $R_0$:

$$R_0 = \left(\frac{1}{4\pi^2\eta_t}\frac{E_n}{E_\alpha + E_n + E_{Li}}\right)\frac{P_E}{a P_W} = 0.04\frac{P_E}{a P_W} \text{ m}. \tag{5.20}$$

In Eq. (5.20) the value chosen for the thermal conversion efficiency is $\eta_t = 0.4$.

The quantity $R_0$ is now substituted into Eq. (5.17) leading to

$$\frac{V_I}{P_E} = \left(\frac{1}{2\eta_t}\frac{E_n}{E_\alpha + E_n + E_{Li}}\right)\frac{(a+b+c)^2 - a^2}{a P_W} = 0.79\frac{(a+b+c)^2 - a^2}{a P_W}. \tag{5.21}$$

Equation (5.21) is the first of the two desired relations involving the unknowns $a$ and $c$. Observe the appearance of $P_W$ in the denominator, confirming the intuition that the minimum cost reactor should operate at the maximum possible wall loading: $P_W = 4 \text{ MW/m}^2$ under the assumed constraint.

### The minimum coil thickness

The minimum coil thickness is estimated by means of a simple stress analysis that assumes the coil comprises a homogeneous structural superconducting material whose maximum stress limit is given by the constraint value $\sigma_{max}$. The calculation is simplified by unbending the magnet from its toroidal configuration into an equivalent straight solenoid of length $2\pi R_0$. The geometry is illustrated in Fig. 5.7.

A reasonable estimate of the coil thickness is obtained by focusing on the tensile forces. The basic idea is to split the coil into two halves and then calculate the upward magnetic force trying to explode the magnet. The two tensile forces that develop at the midplane must balance this force. See Fig. 5.7.

The net upward magnetic force is calculated by first noting that there is a differential outward radial force $dF^{(M)}\mathbf{e}_r$ acting on the current flowing along the differential line element $dl$ of the coil. This force is given by $dF^{(M)}\mathbf{e}_r = BI\,dl\,\mathbf{e}_r$. Here, $B$ is the average magnetic field acting on the current. Since the magnetic field varies from $B_c$ just inside the solenoid to 0 outside the solenoid a reasonable approximation for the average value is just $B_c/2$. Next, the current $I$ is expressed in terms of the magnetic field in the solenoid through Ampère's law, leading to the familiar result $I = 2\pi R_0 B_c/\mu_0$. Lastly, the geometry illustrated in Fig. 5.7 implies that the differential line element along the average coil radius can be written as

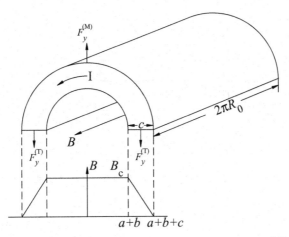

Figure 5.7 Straight model of the magnet coil showing the magnetic force $F_y^{(M)}$ and the two tensile forces $F_y^{(T)}$.

$dl = (a + b + c/2)\, d\theta$. Combining results yields

$$dF^{(M)}\mathbf{e}_r = \frac{\pi R_0 B_c^2}{\mu_0}\left(a + b + \frac{c}{2}\right) d\theta\, \mathbf{e}_r. \tag{5.22}$$

The upward component of the force is calculated by forming the dot product of Eq. (5.22) with $\mathbf{e}_y$ and noting that $\mathbf{e}_r \cdot \mathbf{e}_y = \sin\theta$.

$$dF_y^{(M)} = \frac{\pi R_0 B_c^2}{\mu_0}\left(a + b + \frac{c}{2}\right)\sin\theta\, d\theta. \tag{5.23}$$

The net upward magnetic force $F_y^{(M)}$ is obtained by integrating over $\theta$ in the range $0 \leq \theta \leq \pi$. Thus, $F_y^{(M)}$ is given by

$$F_y^{(M)} = \frac{2\pi R_0 B_c^2}{\mu_0}\left(a + b + \frac{c}{2}\right). \tag{5.24}$$

This represents the first half of the tensile force balance. Note, that the value of $B_c$ has not as yet been set to the constraint value $B_{\max}$. Some subtle reasoning is required to reach this conclusion and is discussed shortly. For the moment, however, focus on the second half of the force balance, the calculation of the internal tensile forces in the structural material. Returning to Fig. 5.7, one sees by symmetry that two equal tensile forces $F_y^{(T)}$ are developed at the midplane. Clearly, the coil thickness is minimized when the structure is stressed to the maximum allowable constraint value $\sigma_{\max}$. Under the simplifying assumption that the stress is uniform across the coil thickness, the tensile force developed at each half of the midplane is just the product of the maximum stress with the cross sectional area:

$$F_y^{(T)} = \sigma_{\max}\left(2\pi R_0 c\right). \tag{5.25}$$

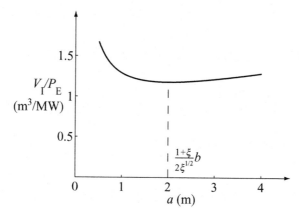

Figure 5.8 The cost function $V_I/P_E$ as a function of plasma radius $a$ for $P_W = 4 \text{ MW/m}^2$ and $\xi = 0.11$.

This is the second half of the tensile force balance. For the magnet to be in equilibrium the two tensile forces must balance the upward magnetic force

$$2F_y^{(T)} = F_y^{(M)}. \tag{5.26}$$

Upon substituting the appropriate expressions for the forces and then solving for the coil thickness $c$ one obtains

$$c = \frac{2\xi}{1-\xi}(a+b), \tag{5.27}$$

where $\xi = B_c^2/4\mu_0\sigma_{\max}$. As expected, the higher the magnetic field, the thicker the coil. Equation (5.27) is the second desired relation involving the unknowns $a$ and $c$.

The next step in the derivation is to substitute the expression for $c$ into the expression for $V_I/P_E$. A short calculation leads to

$$\frac{V_I}{P_E} = \frac{0.79}{P_W(1-\xi)^2}\left[(4\xi)a + (1+\xi)^2\frac{b^2}{a} + 2(1+\xi)^2b\right]. \tag{5.28}$$

The cost parameter $V_I/P_E$ is a function of $a$, which has a minimum as shown in Fig. 5.8. Values of $a$ much smaller than the optimum value lead to an inefficient use of the blanket as shown in Fig. 5.9. Larger values of $a$ result in a large size and low power density and thus a higher cost, also shown in Fig. 5.9. The optimum value of $a$ lies in the intermediate region and is obtained by setting $\partial(V_I/P_E)/\partial a = 0$. The result is

$$a = \frac{1+\xi}{2\xi^{1/2}}b. \tag{5.29}$$

The values of $c$ and $V_I/P_E$ are obtained by back substitution:

$$c = \frac{\xi^{1/2}(1+\xi^{1/2})}{1-\xi^{1/2}}b,$$

$$\frac{V_I}{P_E} = 1.58\frac{1+\xi}{(1-\xi^{1/2})^2}\frac{b}{P_W}. \tag{5.30}$$

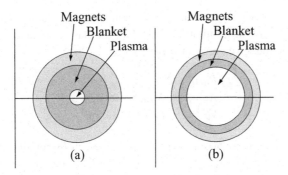

Figure 5.9 Low efficiency of the blanket because of: (a) too small a plasma, and (b) too large a plasma leading to low power density.

Before substituting numerical values a short discussion is needed with respect to the choice of $\xi \sim B_c^2$. Obviously the coil thickness and cost are minimized as $B_c \to 0$. However, a small magnetic field implies high $\beta$ which is difficult to achieve from a plasma physics point of view. One can ask if there is an optimum tradeoff between $\beta$ and $B_c$. In fact, for most fusion configurations high $\beta$ is sufficiently difficult to achieve experimentally that the optimum tradeoff occurs at the boundary corresponding to the maximum allowable magnetic field, even though the higher field increases the cost. The conclusion is that plasma physics considerations require that $B_c$ be set equal to its maximum allowable constraint value: $B_c = B_{max} = 13$ T. With this choice the value of $\xi$ is given by $\xi = B_{max}^2/4\mu_0\sigma_{max} = 0.11$.

Using this value for $\xi$, along with the other constraint values, leads to the following values for $a$, $c$, and $V_I/P_E$:

$$a = 2.0 \text{ m},$$
$$c = 0.79 \text{ m}, \qquad\qquad\qquad (5.31)$$
$$V_I/P_E = 1.2 \text{ m}^3/\text{MW}.$$

These values are reasonably compatible with those obtained from more detailed and rigorous studies. As a practical example again examine the model coil built as part of the ITER design project shown in Fig. 5.4. Finally, it is worth pointing out that the value of $V_I/P_E$ is approximately a factor of 2.5 larger with the maximum magnetic field ($\xi = 0.11$) as compared to the very small magnetic field limit ($\xi \to 0$). There is indeed a substantial cost penalty resulting from the plasma physics difficulty of achieving high $\beta$.

### 5.5.4 The major radius and plasma surface area and volume

The last geometric quantity to be calculated is the major radius. This is found by substituting the value of $a$ back into Eq. (5.20) yielding

$$R_0 = 0.04\frac{P_E}{a P_W} = 5.0 \text{ m}. \qquad\qquad\qquad (5.32)$$

The geometry has now been fully determined. Several related geometric quantities of interest can be evaluated by direct substitution. These include the plasma aspect ratio, plasma surface area and plasma volume, given respectively by

$$
\begin{aligned}
R_0/a &= 2.5, \\
A_P &= 4\pi^2 R_0 a \approx 400 \text{ m}^2, \\
V_P &= 2\pi^2 R_0 a^2 \approx 400 \text{ m}^3.
\end{aligned} \tag{5.33}
$$

At this point a useful comparison can be made between fission and fusion. A typical pressurized water reactor (PWR) may contain on the order of 65 000 fuel rods, each with a diameter of 1.3 cm and length 3.7 m. The total surface area of these fuel rods is approximately 10 000 m$^2$, a factor of 25 greater than in a fusion reactor. This greater surface area makes it much easier for the coolant to carry away heat and thereby enables a fission reactor to operate at a much higher power density than a fusion reactor. Similarly, the total volume of the fuel rod assemblies (rods, structure, coolant), equivalent to the plasma volume, fits into a cylinder of about 3.7 m diameter and 4 m length. This corresponds to a volume of approximately 40 m$^3$, which is 10 times smaller than for fusion.

The conclusions are as follows. Heat can be removed from a fusion reactor only through the bounding toroidal surface area of the plasma. Cooling tubes cannot be inserted into the plasma as they will cool the plasma and probably be melted themselves. Consequently, to produce a given amount of power, a fusion reactor must have a large volume since heat can be removed only through the relatively small bounding surface area. The large volume suggests that the capital cost of a fusion reactor will be inherently higher than for fission or fossil fuels. The high cost has to be balanced against the following: the lower level of radioactivity, the ease of waste disposal, the reduced risk of proliferation in comparison to a fission reactor. Also, the low cost and high reliability of the fuel supply, and the absence of greenhouse gases and other pollutants, are advantages of fusion whose tradeoffs have to be measured with respect to fossil fuels.

### 5.5.5 *Power density and plasma pressure*

Recall that high power density leads to a small reactor core and a lower cost to produce a given amount of power. The converse is true for low power density. A convenient measure of the power density is the ratio of the total power produced in the plasma (from alphas and neutrons) divided by the plasma volume. This quantity is calculated by noting that overall power balance implies that

$$
\begin{aligned}
P_\alpha + P_n &= \frac{E_\alpha + E_n}{E_\alpha + E_n + E_{\text{Li}}} (P_\alpha + P_n + P_{\text{Li}}) \\
&= \frac{E_\alpha + E_n}{E_\alpha + E_n + E_{\text{Li}}} \frac{P_E}{\eta_t}.
\end{aligned} \tag{5.34}
$$

The plasma power density is thus given by

$$\frac{P_\alpha + P_n}{V_P} = \frac{E_\alpha + E_n}{E_\alpha + E_n + E_{Li}} \frac{P_E}{\eta_t V_P} = 4.9 \text{ MW/m}^3. \tag{5.35}$$

The equivalent value for a fission reactor is approximately $100 \text{ MW/m}^3$. As previously suggested, the relatively small surface-to-volume ratio of fusion, compared to fission, leads to low power density, a larger reactor, and an inherently higher capital cost.

Consider next the plasma pressure. Recall that the alpha-plus-neutron power is related to the fusion cross section by

$$P_\alpha + P_n = \tfrac{1}{4}(E_\alpha + E_n) n^2 \langle \sigma v \rangle V_P. \tag{5.36}$$

Since $p = 2nT$ it then follows that the plasma pressure can be written as

$$p = \left( \frac{16}{E_\alpha + E_n} \frac{P_\alpha + P_n}{V_P} \right)^{1/2} \left( \frac{T^2}{\langle \sigma v \rangle} \right)^{1/2} = 8.4 \times 10^{-12} \left( \frac{T_k^2}{\langle \sigma v \rangle} \right)^{1/2} \text{ atm.} \tag{5.37}$$

In the second formula the units are $T$ (keV) and $\langle \sigma v \rangle$ (m$^3$/s). Clearly, in order to minimize the required $\beta$ one must operate at the temperature that minimizes the function $T^2/\langle \sigma v \rangle$; that is, at 15 keV. Using this result and the corresponding value of $\langle \sigma v \rangle = 3 \times 10^{-22}$ m$^3$/s yields the following set of plasma parameters for a typical fusion reactor:

$$
\begin{aligned}
p &= 7.2 \text{ atm,} \\
T &= 15 \text{ keV,} \\
n &= 1.5 \times 10^{20} \text{ m}^{-3}.
\end{aligned}
\tag{5.38}
$$

Observe that as compared to room temperature air (i.e. approximately $N_2$), a fusion plasma has a temperature ratio $(T_e/T_{N_2})$ about 600 000 times higher and a number density ratio $(n_e/n_{N_2})$ about 170 000 times smaller. In terms of number density, a plasma makes a respectable vacuum.

### 5.5.6 The plasma physics quantities $\beta$ and $\tau_E$

The last two quantities of interest in the design are the plasma $\beta$ and the energy confinement time $\tau_E$. These are particularly important quantities since their values set the stage for the entire second part of the book. Specifically, it is the task of plasma physicists to create plasmas that simultaneously achieve the values of $\beta$ and $\tau_E$ required in a fusion reactor. The second part of the book focuses on the plasma physics and corresponding methods to accomplish these goals.

Consider first the energy confinement time. The required value is easily calculated from the minimum ignition requirement $p\tau_E = 8.3$ atm s. For $p = 7.2$ atm this yields

$$\tau_E = 1.2 \text{ s.} \tag{5.39}$$

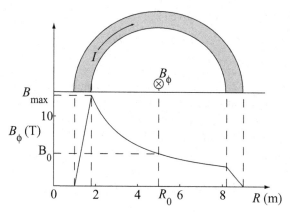

Figure 5.10 Magnetic field $B_\phi$ as a function of major radius $R$, showing the $1/R$ reduction moving away from the inside of the coil where $B_\phi = B_{max}$.

The magnetic field configuration must be able to provide confinement of the plasma thermal energy so that its natural thermal conduction decay time is slightly over a second.

Next, the minimum required plasma $\beta$ can be determined from knowledge of the plasma pressure and the maximum allowable $B$ field. The only subtlety is that in a torus, the magnetic field in the center of the plasma (where $\beta$ is defined) is lower than $B_{max}$ because of geometric effects. To see this, examine Fig. 5.10. Here, application of the integral form of Ampère's law shows that the magnetic field decreases inversely with the major radius $R$:

$$\mu_0 I_c = \oint \mathbf{B} \cdot d\mathbf{l} = \oint B_\phi R \, d\phi = 2\pi R B_\phi \tag{5.40}$$

or

$$B_\phi = \frac{\mu_0 I_c}{2\pi R} = B_0 \left( \frac{R_0}{R} \right). \tag{5.41}$$

In these expressions $I_c$ is the total current flowing in the toroidal field coils, $B_\phi$ is the toroidal magnetic field, and $B_0$ is the value of $B_\phi$ at the location of the major radius $R = R_0$, the radius where $\beta$ is defined.

The $1/R$ scaling implies that the maximum magnetic field occurs on the inside of the coil: $B = B_{max}$ at $R = R_0 - a - b$. Therefore, it follows that

$$B_0 = \frac{R_0 - a - b}{R_0} B_{max} = 4.7 \text{ T}. \tag{5.42}$$

Geometric effects substantially reduce the field at the center of the plasma. Using this value one can easily calculate the required value of $\beta$:

$$\beta = \frac{p}{B_0^2/2\mu_0} = 0.082. \tag{5.43}$$

A value of $\beta = 8.2\%$ is about the maximum that has been reliably achieved in the most successful current fusion device, the tokamak.

Table 5.3. *Summary of parameters for a generic fusion reactor*

| Quantity | Symbol | Value |
|---|---|---|
| Blanket-and-shield thickness | $b$ | 1.2 m |
| Coil thickness | $c$ | 0.79 m |
| Minor radius | $a$ | 2.0 m |
| Major radius | $R_0$ | 5.0 m |
| Aspect ratio | $R_0/a$ | 2.5 |
| Plasma surface area | $A_P$ | 400 m$^2$ |
| Plasma volume | $V_P$ | 400 m$^3$ |
| Power density | $(P_\alpha + P_n)/V_P$ | 4.9 MW/m$^3$ |
| Magnetic field at $R = R_0$ | $B_0$ | 4.7 T |
| Plasma pressure | $p$ | 7.2 atm |
| Plasma temperature | $T$ | 15 keV |
| Plasma number density | $n$ | $1.5 \times 10^{20}$ m$^{-3}$ |
| Energy confinement time | $\tau_E$ | 1.2 s |
| Normalized plasma pressure | $\beta$ | 8.2% |

## 5.6 Summary

Chapter 5 has presented the design of a simple, generic fusion reactor. Typical values have been obtained for all the geometric quantities of interest as well as the critical plasma parameters. The design parameters are listed in Table 5.3. Interestingly, the design is driven by a combination of engineering and nuclear physics constraints, with plasma physics playing only a very small role. The key contribution of plasma physics is the recognition that the achievement of high $\beta$ is a difficult task, implying that high magnetic fields are desirable, even though this leads to an increase in cost.

An examination of the design indicates that a relatively large, low-power density core is a characteristic of a magnetic fusion reactor. The fundamental reason for this is the small surface-to-volume ratio, a necessity driven by the fact that heat can only be extracted through the toroidal surface area of the plasma. Cooling tubes cannot be inserted into the plasma for the purpose of increasing the surface-to-volume ratio. The combination of large size and low power density means that the capital cost of a fusion reactor will be relatively high. This has to be balanced against the low cost and reliable supply of fuel as well as some potentially large environmental advantages. The evolution of the world's energy usage, supply, and environmental impact over the next few decades will ultimately determine the economic competitiveness of fusion power.

Finally, it is worth stating that the critical plasma physics requirements in a reactor correspond to the simultaneous achievement of $\beta \approx 8\%$ and $\tau_E \approx 1$ s at a temperature of $T \approx 15$ keV. These guidelines serve as the main focus for the discussion of magnetic fusion plasma physics discussed in the second part of the book.

## Bibliography

Over the years the international fusion community has developed many designs for large-scale ignition experiments and fusion reactors. Below are some useful references which give an overview of reactor design and a sampling of more detailed conceptual and engineering designs for specific devices.

### *Reactor design overviews*

Dolan, T. J. (1982). *Fusion Research*. New York: Pergamon Press.
Gross, R. (1984). *Fusion Energy*. New York: John Wiley & Sons.
Rose, D. J. and Clark, M. (1961). *Plasmas and Controlled Fusion*. Cambridge, Massachusetts: MIT Press.
Stacey, W. M. (1981). *Fusion Plasma Analysis*. New York: John Wiley & Sons.
Stacey, W. M. (2005). *Fusion Plasma Physics*. Weinheim: Wiley-VCH.

### *Specific designs for ignition experiments and fusion reactors*

Badger, B., Abdou, M. A., *et al.* (1974). *UWMAK-I – A Wisconsin Toroidal Fusion Reactor Design*, UWFDM-68, University of Wisconsin Report. Madison Wisconsin: University of Wisconsin.
Coppi, B., Airoldi, A., Bombarda, F., Cenacchi, G., Defragiache, P., and Sugiyama, L. E. (2001). Optimal regimes for ignition and the Ignitor Experiment, *Nuclear Fusion* **41**, 1253.
ITER Team (2002). *ITER Technical Basis*. ITER EDA Documentation Series Number 24. Vienna: IAEA.
Maisonnier, D., Cook, L., *et al.* (2005). *A Conceptual Study of Commercial Fusion Power Plants*, EFDA-RP-RE-5.0, EFDA Report.
Meade, D. M. (2000). Mission and design of the Fusion Ignition Research Experiment (FIRE), *Proceedings of the Eighteenth IAEA International Conference on Fusion Energy, Sorrento, Italy*. Vienna: IAEA.
Najmabadi, F., Conn, R. W., *et al.* (1991). *The Aries-I Tokamak Fusion Reactor Study*. *Fusion Technology*, p. 253 Amsterdam: North Holland.
Najmabadi, F. and the ARIES Team (1997). *Overview of the ARIES-RS Reverse-Shear Tokamak Power Plant Study*, Fusion Engineering and Design Vol 38, P. 3. Amsterdam, North Holland.
Najmabadi, F. Jardin, S. C. *et al.* (2000). ARIES-AT: An advanced Tokamak, advanced technology fusion power plant, *Proceedings of the Eighteenth IAEA International Conference on Fusion Energy, Sorrento, Italy*. Vienna: IAEA.

## Problems

5.1 A resistive magnet tokamak neutron source is illustrated in Fig. 5.11. The central leg of the magnet fills the entire center of the torus as shown. For simplicity assume that the total ohmic power dissipated in the magnet is equal to twice the power dissipated in the central leg of length $L$. The plasma density corresponds to the maximum operational value (known as the Greenwald limit) which for present purposes can be simplified to $n_{20} = 0.8 B_0 / R_0$.

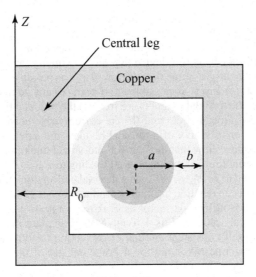

Figure 5.11

(a) Show that for known $R_0$ and $b$ the ratio of neutron power to magnet dissipation power has a maximum as a function of minor radius $a$.
(b) For the case $T = 15$ keV, $R_0 = 3$ m, $b = 1$ m calculate the optimum $a$, the optimum aspect ratio $R_0/a$, and the corresponding maximized value of the power ratio. Also calculate the total neutron power and neutron wall loading if $B_{max} = 15$ T, where $B_{max}$ is the maximum value of the field at the coil.

5.2 A DC magnetic field is generated by a uniform current flowing in a metallic cylinder of inner radius $a/2$ and outer radius $a$. The current flows parallel to the longitudinal axis. The coil is cooled by water flowing in the center of the coil (i.e., $0 < r < a/2$) which maintains a uniform temperature $T(a/2)$ on the inner surface. The coil is well insulated externally so that no heat escapes from the outer surface at $r = a$. Assume the thermal conductivity $\kappa$ and electrical resistivity $\eta$ are constants.
(a) Derive an expression for the temperature difference between the inner and outer surfaces when the magnetic field on the surface is $B(a)$.
(b) Evaluate the temperature difference for $\kappa = 400$ W/(m K), $\eta = 2 \times 10^{-8}$ $\Omega$ m, and $B(a) = 2$ T.

5.3 Consider the problem of maximizing the alpha particle power density in the simple superconducting fusion reactor discussed in the text. Assume the maximum allowable magnetic field on the toroidal field (TF) magnet is $B_{max} = 13$ T. The reactor is to operate at the maximum permissible neutron wall loading of $P_W = 4$ MW/m$^2$. The plasma pressure in the reactor corresponds to the maximum MHD stability limit known as the "Troyon limit" which for present purposes can be written in the following simplified form: $\beta = 0.12 (a/R_0)$. Here, $\beta = 2\mu_0 p/B_0^2 = 2\mu_0 (p_e + p_D + p_T)/B_0^2$. Also assume that $b = 1.2$ m.
(a) Show that there is an optimum temperature that maximizes the alpha power density and calculate this temperature.
(b) Show that there is an optimum $R_0$ that maximizes the alpha power density. Note that the optimum $R_0 = R_0(a) = R_0[a(S_\alpha)]$.

(c) Calculate the optimized $R_0$, the maximum alpha power density, the maximum neutron power density, and the corresponding value of $a$.

5.4 The purpose of this problem is to test the economic desirability of the Perfect Plasma Power Reactor (PPPR). This is a device that has maximally favorable plasma physics performance. Specifically, assume that: (a) any $\beta \leq 1$ is possible at no cost; (b) any $I$ is possible at no cost; (c) any $\tau_E$ is possible at no cost; (d) any auxiliary heating is available at no cost; (e) the toroidal magnets are free; (f) the poloidal field magnets are free; (g) the ohmic transformer is free; and (h) the current drive power is free. In essence, the fusion reactor consists only of a toroidal blanket-and-shield surrounding the first wall. Neutronic studies show that the blanket-and-shield thickness is about $b = 1.5$ m. The first wall material limits the neutron wall loading to $P_W = 4$ MW/m$^2$. The electric power out is required to be $P_E = 1$ GW at a thermal conversion efficiency of $\eta_t = 0.35$. Assume that each fusion reaction ultimately produces about 22 MeV of thermal energy. Now, define the mass utilization factor as $F \equiv$ reactor mass/electric power. For a fission reactor a typical value of $F$ is 1 tonne/MW. Calculate the values of $R_0$ and $a$ for a topologically self-consistent circular cross section fusion reactor that minimizes $F$ (i.e., is most economic). Assume the average density of the blanket-and-shield is $3 \times 10^3$ kg/m$^3$. Discuss the results.

5.5 This problem investigates the effect of a non-circular plasma cross section on the design of a simple fusion reactor. The idea is to repeat the design calculation presented in the text assuming the plasma has an elliptic shape with horizontal diameter $2a$ and vertical diameter $2\kappa a$. The quantity $\kappa$ is known as the plasma elongation. The blanket-and-shield and magnets are also elliptic in shape but their thicknesses $b$ and $c$ are uniform around the cross section. Calculate the minimum value of $V_I/P_E$ and the corresponding reactor parameters in Table 5.3 for $0.5 \leq \kappa \leq 2$, assuming that all constraints are unchanged. Based on this simple model what conclusions can be drawn about the desirability of a non-circular plasma? Two points to note: first, in the regime of interest, the circumference of an ellipse can be approximated by $C = 2\pi a(1 + \kappa^2)/2$. Second, when calculating the maximum magnet stress note that the largest magnetic force occurs on the flattened side of the ellipse.

# Part II

The plasma physics of fusion energy

# 6

# Overview of magnetic fusion

## 6.1 Introduction

The analysis presented in the previous chapters has established the plasma properties necessary for a magnetic fusion reactor. In particular, a combination of engineering and nuclear physics constraints has shown that a fusion plasma must achieve a temperature $T \sim 15$ keV, a pressure $p \sim 7$ atm, a plasma beta $\beta \sim 8\%$, and an energy confinement time $\tau_E \sim 1$ s. Furthermore, the plasma must be confined in the shape of a torus with minor radius $a \sim 2$ m and major radius $R_0 \sim 5$ m. The challenge to the fusion plasma physics community is to discover ways to simultaneously achieve these parameters.

Because the behavior of a fusion plasma can be quite complicated and subtle, as well as being far from our everyday intuitive experiences, it is perhaps not surprising that this has led to the development of a new subfield of physics known as "plasma physics." Only after knowledge of this new state of matter has been mastered will it be possible to produce robust fusion plasmas suitable for a fusion reactor.

The need to master plasma physics is the motivation for the second part of the book. Presented in these chapters is a description of the plasma physics necessary to produce a fusion plasma. The goal is to provide a reasonably rigorous introduction to the field of plasma physics as viewed from the perspective of a nuclear engineer. That is, the topics and order of presentation are always chosen by keeping in mind the final goal of producing a practical fusion reactor. One consequence of this approach is that a wide variety of topics is covered in order to give the reader a relatively complete overview of the field. A second consequence is that most of the understanding results from an analysis of a set of fluid models of varying complexity. More accurate models involving kinetic descriptions that determine the particle distribution functions are available, but these are advanced topics, and are not necessary for an introduction to fusion.

Over many years of research a natural logic has developed to introduce new scientists to the field of plasma physics and fusion energy. The basis for this logic is the need to create plasmas that satisfy the basic condition for ignition and the accompanying production of large quantities of fusion power. The ignition condition is given by

$$p\tau_E = \frac{24}{E_\alpha} \left( \frac{T^2}{\langle \sigma v \rangle} \right) = 0.11 \times 10^{-22} \frac{T_k^2}{\langle \sigma v \rangle} \quad \text{atm s} \tag{6.1}$$

The material presented in the book follows the natural logic required to achieve this goal and is organized as follows. Chapter 7 presents a semi-quantitative overview of the definition of a plasma and certain global properties of fusion plasma behavior. The aim is to provide some insight, admittedly at an elementary level, into how to think about a plasma in the context of its behavior in electric and magnetic fields. Chapters 8 and 9 describe how individual charged particles behave in prescribed electric and magnetic fields. No attempt is made to calculate the fields self-consistently. Even this simple description sheds considerable light on the structure of a fusion reactor, in particular why the plasma must be in the shape of a torus.

To proceed further, self-consistent models are required and this is the subject of Chapter 10. As previously stated, these are a set of fluid models of varying complexity. Self-consistency is crucial because the electrical currents and charges resulting from the motion of the plasma particles can often make large modifications to the applied fields, which then in turn alter the motion of the charged particles, and so on.

Next, one of the simpler fluid models, known as magnetohydrodynamics (MHD) is employed (Chapters 11 and 12) to investigate the macroscopic equilibrium and stability of various simple fusion configurations. MHD stability limits set the maximum achievable value of $p$ appearing in the ignition condition. The MHD results are applied in Chapter 13 to describe the main magnetic concepts currently under investigation for the production of fusion power. While there are a number of such concepts, the current leader in terms of performance is the "tokamak". The details of the magnetic geometry are shown to be critical in determining the maximum value of $\beta$ (and hence $p$) that can be confined in a stable manner.

Following this is a description of the transport properties of a fusion plasma (Chapter 14). Knowing how rapidly particles and energy can be lost from a plasma by means of transport processes such as diffusion and heat conduction is a critical issue. These losses determine the energy confinement time $\tau_E$ of the plasma which appears as a critical parameter in the ignition condition.

Plasma heating and current drive are discussed in Chapter 15. Several heating methods are described that produce plasmas with fusion grade temperatures: $T \sim 15$ keV $\approx 170 \times 10^6$ K. Achieving $T = 15$ keV is important in order to minimize $T^2/\langle \sigma v \rangle$, which in turn minimizes the requirements on $p\tau_E$ in the ignition condition. In addition, a method is presented for driving steady state toroidal currents without a transformer. Chapter 16 briefly discusses the future direction of the international fusion program with a strong emphasis on ITER.

In summary, the order of the presentation is as follows: (1) basic description of a plasma, (2) single-particle behavior, (3) self-consistent models, (4) MHD equilibrium and stability, (5) fusion configurations, (6) transport, (7) heating and current drive, and (8) the future of fusion research. Before proceeding with the detailed discussions it is useful to collect in one place the main issues and conclusions that will be drawn from each of these chapters.

## 6.2 Basic description of a plasma

A fusion plasma is a fully ionized gas whose behavior is dominated by long-range electric and magnetic fields (as opposed to short-range nearest-neighbor Coulomb collisions). A major consequence of this behavior is that a plasma is an exceptionally good conductor of electricity. The high conductivity may be somewhat surprising since the number density of electrons (i.e., the number of charge carriers) is typically eight orders of magnitude smaller than that found in a good electrical conductor such as copper. Even so, the electrical conductivity of a fusion plasma is about 40 times larger than that of copper. The reason is associated with the fact that at high temperatures and low densities there are rarely any Coulomb collisions between electrons and ions and thus there is very little resistance to the flow of current.

The high conductivity implies that the inside of the plasma is shielded from DC electric fields to a very large degree. On the other hand, DC magnetic fields can penetrate, although somewhat slowly, and it is these fields that provide plasma confinement. This is the reason it is known as "magnetic fusion" rather than "electric fusion" or "electromagnetic fusion."

Explicit criteria are derived that determine the conditions under which a plasma is a good conductor of electricity, dominated by long-range collective effects. These criteria are well satisfied in fusion plasmas. In terms of a fusion reactor, it is the high conductivity that opens the possibility of confinement by magnetic fields.

## 6.3 Single-particle behavior

The discussion here assumes that a prescribed DC magnetic field has penetrated the plasma. Its task is to provide plasma confinement. The simplest way to learn how magnetic confinement works is to use Newton's law of motion to determine the trajectories of charged particles in the prescribed field. No attention is devoted to the problem of how the current and charge densities associated with the motion of the charged particles alter the prescribed field. The goal here is to see, on a microscopic level, whether or not the applied magnetic field confines individual charged particles within the plasma or instead causes them to be lost to the first wall.

The analysis shows that there are three qualitatively different classes of particle motion. The fastest and most dominant motion concerns the behavior of charged particles in the long-range magnetic field, which varies slowly, in time and space. This is the so-called "collisionless behavior." It is shown that charged particles spiral around the magnetic field lines along helical trajectories. The perpendicular radius of the helix, called the gyro radius, is quite small, less than 1 cm. It is in this sense that particles are confined perpendicular to the field. Parallel to the field there is free motion and particles would be very quickly lost if the magnetic field lines directly intersected the first wall. The need to prevent parallel losses is the main reason why fusion configurations are toroidal with magnetic lines wrapping around the surface as on a barber pole, but not intersecting the wall. In addition to the perpendicular "gyro" motion, electrons and ions develop slow drift velocities across the field lines due to various field gradients, but the direction of these drifts usually does not lead to a loss of particles. The conclusion from the collisionless behavior is that in a wide variety of toroidal magnetic geometries all particles are confined – there is no loss of particles or energy.

The second class of single-particle motions involves the close, nearest-neighbor electric interaction between two charged particles. These are Coulomb collisions. They occur relatively infrequently since it is rare in a low-density, high-temperature plasma that two particles are in close enough proximity for a sufficiently long time that their paths are significantly altered from their collisionless trajectories. Even so, Coulomb collisions are very important since they lead to a loss of particles and energy through diffusion like processes. The goal of the analysis is to determine various like and unlike particle collisional loss rates for momentum and energy. These loss rates are used in later chapters to determine the macroscopic particle and energy transport coefficients to be used in the self-consistent fluid models.

The last, and unfortunately most infrequent, class of particle motions is nuclear fusion collisions. These collisions have already been discussed in earlier chapters. Ironically, even though fusion is a source of nuclear energy, in most of the plasma physics required to understand present-day experiments it is a good approximation to neglect all nuclear effects, or at most include them perturbatively. Nuclear effects will become important in the overall plasma power balance once experiments are built with substantial amounts of alpha heating power, although even in this regime the number of fusion collisions remains relatively small.

In summary, the hierarchy of single-particle motions is as follows: (1) dominant motion – collisionless behavior in the long-range electromagnetic fields; (2) Coulomb collisions – infrequent interactions in the nearest-neighbor inter-particle Coulomb potential; and (3) nuclear fusion collisions – even less frequent interactions in the very short-range nuclear potential. The collisionless behavior provides essentially perfect single-particle confinement. The Coulomb collisions are responsible for transport. The fusion collisions produce the nuclear power. With respect to a fusion reactor, the excellent confinement of charged particles in the long-range fields is essential in demonstrating the principle of "magnetic fusion." The transport losses have to be kept sufficiently small to allow an achievable path to ignition and an overall favorable power balance.

## 6.4 Self-consistent models

In order to be able to predict quantitatively the performance of fusion plasmas, one needs self-consistent models. The study of plasma physics as presented in this book is based largely on fluid models. Such models can be derived from first principles from various kinetic models. However, these derivations require a substantial amount of mathematics and are somewhat formal in nature. Instead, the fluid models presented are derived on the basis of conservation of mass, momentum, and energy combined with some simple intuition. The resulting equations look similar to the equations of standard fluid dynamics although there are some crucial differences.

The differences can be summarized as follows. First, a separate set of fluid equations is required for both electrons and ions. These equations are coupled through Maxwell's equations. Second, in addition to the pressure gradient force of fluid dynamics, one must also include the bulk fluid electric and magnetic forces. Finally, although the equations look similar to those of fluid dynamics, much of the basic physics contained in the model is qualitatively different. This difference reflects the fact that plasma behavior is dominated by the long-range collective effects associated with the macroscopic electric and magnetic fields rather than nearest-neighbor collisions.

To illustrate this point, assume that a group of electrons bunch together over a localized region of plasma causing a slight charge imbalance. This imbalance generates a macroscopic electric field, which because of its long-range nature, is felt by all other charged particles far from the bunch. In particular, distant particles have their orbits altered in response to this field (and in general will move to shield out the field). The point is that the electric field far from the bunched particles is felt essentially instantaneously (i.e., the field effectively moves at the speed of light) even if all the particles are behaving in a collisionless manner. Contrast this with the fluid behavior of a neutral gas. A local pressure perturbation is felt far away only because of the propagation of a succession of nearest-neighbor collisions – the effect travels at the sound speed.

The resulting set of two-fluid equations is quite useful for the study of various plasma wave-heating techniques in which substantial geometric simplifications can be made without much loss in accuracy. When real geometric effects are important the two-fluid equations are often too complicated to solve. In these cases, by restricting attention to long-wavelength, low-frequency phenomena, the two-fluid model can be reduced to a single-fluid model known as MHD. Various forms of the MHD model are useful for the study of macroscopic equilibrium and stability as well as transport.

To summarize, a set of two-fluid and single-fluid models is presented based on the conservation of mass, momentum, and energy plus Maxwell's equations. These models provide a self-consistent description of plasma behavior. In contrast to the standard fluid behavior of a neutral gas, long-range collective effects associated with the macroscopic electric and magnetic fields dominate the behavior of a plasma. In the context of a fusion reactor, the development of self-consistent fluid models is crucial since such models enable a reasonably accurate, tractable analysis of all the major plasma physics issues.

## 6.5 MHD equilibrium and stability

The first application of the self-consistent models involves the macroscopic equilibrium and stability of the plasma. The analysis makes use of the single-fluid MHD model. The goal is to understand and discover various magnetic geometries that are capable of confining high-$\beta$ (corresponding to high $p$) plasmas in stable equilibrium. It is largely MHD considerations that determine the magnet systems required for any given concept and in this sense they are responsible for its outward appearance, the first image that meets the eye.

First it is shown that even though a fusion plasma is toroidal in shape, it cannot be held in equilibrium by a purely toroidal magnetic field. A transverse magnetic field, known as the poloidal magnetic field, is required. Equilibrium configurations, therefore, require a combination of toroidal and poloidal magnetic fields. The optimum combination is based on an analysis of macroscopic stability – the optimum being defined in terms of the maximum achievable stable $\beta$.

The next point to note is that this optimization is somewhat subtle. For example, the presence of a perfectly conducting wall near the surface of the plasma has a strong stabilizing effect on certain classes of MHD instabilities. If, however, the wall is resistive (as it must be for practical reasons), instability returns although with much slower growth rates. If one believes that these slow growing modes can be feedback stabilized, then one class of optimized configuration results. If feedback is not considered to be a viable option, say for various technological reasons, then wall stabilization is not effective and a different class of optimized configurations is obtained.

The net result, depending upon the assumed constraints made with respect to stability, is that there is a wide variety of "optimized" magnetic configurations currently under consideration for ultimate use as a magnetic fusion reactor. At present, in terms of actual experimental performance, the "tokamak" is the most successful concept. Its maximum theoretical $\beta$ limit is lower than those of some of the other configurations but is, nonetheless, still in the right range for a fusion reactor. The higher performance of the tokamak turns out to be closely associated with a substantially lower level of transport, MHD performance being adequate for experimental purposes.

## 6.6 Magnetic fusion concepts

The major magnetic fusion concepts under consideration are the tokamak, the spherical tokamak, the stellarator, the reversed field pinch, the spheromak, the field reversed configuration and the levitated dipole. All but the stellarator are 2-D axisymmetric toroidal configurations. The stellarator is an inherently 3-D configuration.

Each of these concepts is described primarily from the point of view of macroscopic MHD equilibrium and stability. Also, various experimental operational limitations are discussed that affect performance but are not as yet well understood from a first-principles theory.

It is shown that there is a very large variation in the evolution and corresponding progress of the various concepts. Tokamaks have been studied the most and have achieved the best overall performance: next is the stellarator followed by the spherical tokamak, which is actually a very tight aspect ratio tokamak. These configurations all have relatively strong toroidal magnetic fields and reasonable transport losses. Each is capable of MHD stable operation at acceptable values of $\beta$ without the need for a conducting wall close to the plasma. The stellarator has the advantage of being the only concept not requiring toroidal current drive in a reactor but has a noticeably more complicated magnet configuration which increases complexity and cost. When any of these configurations is scaled to a reactor, the presence of the toroidal magnetic field results in a relatively large device, leading to increased capital cost.

The reversed field pinch, the spheromak, and the field reversed configuration, all have much smaller or zero toroidal magnetic fields. Scaling any of these to a reactor would lead to a compact, higher-power-density reactor which would reduce the capital cost. However, the smaller toroidal field leads to poorer plasma performance. Transport losses are higher and each configuration would be macroscopically unstable without the presence of a perfectly conducting wall near the plasma surface. Thus, even if transport losses could be reduced, steady state operation with a realistic finite conductivity wall, would require the use of feedback, which adds to the technological complexity.

To summarize, over many years of research a large number of configurations have been considered for use as a magnetic fusion reactor. So far, the tokamak has achieved the best performance. Other configurations aim to overtake the tokamak by either improving the physics but sacrificing technological attractiveness or vice versa. At present, no non-tokamak configuration seems obviously destined to overtake the tokamak. It is for this reason that the world's first burning plasma experiment, ITER, will be a tokamak.

## 6.7 Transport

A plasma confined in stable MHD equilibrium will, by definition, not lose its energy by means of a rapid, coherent, macroscopic motion to the wall. Even so, it will lose energy by means of various transport processes. The dominant transport mechanism is thermal diffusion, although particle and magnetic field diffusion can also play a finite role. Thermal losses have to be compensated (mainly by alpha power) in order to maintain the steady production of fusion energy. Clearly, the magnitude of these losses cannot be too large or else the overall power balance becomes unfavorable.

The confining magnetic field dramatically alters the energy transport properties of the plasma. Specifically, the plasma becomes enormously anisotropic. For example, the thermal conductivity perpendicular to the magnetic field can be ten orders of magnitude smaller than that parallel to the field. This is the main reason why fusion plasmas must be toroidal with the magnetic field lines wrapping around the torus. It is only in such a toroidal magnetic geometry that the large parallel losses can be eliminated. That is, the continuous wrapping around of magnetic field lines geometrically eliminates any direct parallel loss path along **B** to the first wall. The only loss path to the first wall is across the magnetic field where the transport losses are much smaller.

An accurate evaluation of the transport coefficients due to Coulomb collisions requires a great deal of mathematics using sophisticated kinetic models. These difficulties are overcome in this book by the use of much simpler "random walk" arguments. Using such arguments transport coefficients are easily calculated in a straight cylindrical geometry. The calculations are generalized to toroidal geometry where it is shown that the losses increase substantially because of a special class of particles whose orbits are trapped on the outside of the torus.

While these calculations provide a good basic understanding of the transport processes in a fusion plasma they, nevertheless, almost always result in optimistic predictions of energy confinement in actual experiments. The reason is that the excitation of a variety of microscopic plasma instabilities, driven by the plasma temperature and density gradients, leads to anomalously large losses of plasma energy. The theory of plasma micro-instabilities is quite complicated and a fully definitive predictive capability is not yet available. To circumvent this problem fusion researchers have developed empirical scaling relations based on a large experimental database that predict energy transport, and these relations are discussed in the book.

In summary, a fully ionized magnetized fusion plasma is characterized by a high electrical conductivity and highly anisotropic transport coefficients. The cross-field energy transport is in general many orders of magnitude smaller than the parallel transport. In practice, experimentally observed cross-field transport is more rapid than that predicted by classical Coulomb collisions. Microscopic plasma instabilities drive anomalous transport, which at present is most often modeled by empirical scaling relations. For a fusion reactor the conclusions are as follows. The tokamak currently has the best performance with respect to anomalous transport and the corresponding value of $\tau_E$ is probably adequate when scaled to a reactor. The performance of other configurations, with the possible exception of the stellarator, is at present marginal at best. If, however, techniques can be developed to suppress turbulence causing micro-instabilities other configurations may become more attractive.

## 6.8 Heating and current drive

The first topic discussed here is the heating of a plasma to the enormously high temperatures required for a fusion reactor: $T = 15 \, \text{keV}$. Several methods have been successfully used for this purpose, including ohmic heating, neutral beam heating, and RF heating. The end result is that temperatures approaching 30 keV have been achieved in large-scale tokamak experiments. The external power levels required to drive the heating sources are substantial but not excessive in the sense that scaling to a reactor does not prevent a favorable overall power balance. The heating methods are as follows.

The simplest way to heat a plasma is by ohmic heating. The toroidal current that flows in most magnetic configurations ohmically heats the plasma, in much the way that a current heats a wire. However, in contrast to a wire, the resistivity of a plasma decreases with temperature so that heating becomes less efficient as $T$ increases. In practice, it is shown that ohmic heating is effective up to several keV but not much higher.

To increase $T$ from several keV to 15 keV requires additional sources of external power. One method is to inject high-energy neutral particle beams into the plasma. The beam particles must be of the same species as the plasma particles (e.g. deuterium or tritium). The neutral particles undergo charge-exchange collisions with low-energy plasma ions; that is, the high-energy neutral particle transfers an electron to a low-energy ion producing a high-energy ion and a low-energy neutral particle. This method has been very effective in experiments but will be more difficult in a reactor because of technological problems associated with scaling the beams to higher energies.

A second method of achieving high plasma temperatures is by using RF heating and there are several natural resonant frequencies in the plasma that make this an efficient process – somewhat like heating food in a microwave oven. Both the electrons and ions in the plasma can be driven resonantly, although the external sources for ions are more readily available at high power levels and lower costs. Even so, resonant electron heating has some attractive physics features because, as is shown in Section 15.7, heat can be deposited in narrow localized regions of the plasma, thereby providing an external method for controlling the temperature profile.

In summary, heating represents a success story for fusion. The temperatures required for a reactor should be achievable without the need for an excessive amount of external power.

The second topic of this section is current drive. Since a magnetic fusion reactor is expected to be a steady state device, those concepts that require a toroidal current for equilibrium must have some non-inductive means to drive this current. A transformer, which can drive current during startup and for finite duration pulsed operation, cannot operate in steady state. As with heating, several current drive methods have been proposed and are being developed in laboratory experiments.

One method is to inject neutral beams tangentially into the plasma, generating a toroidal momentum parallel to the direction of injection. Part of this momentum is transferred to electrons through collisions. Electrons flowing with a preferred momentum in the toroidal direction constitute an electric current in the opposite direction (because of the negative charge).

A second method is to launch RF waves at an appropriate microwave frequency. The launching system (i.e., antennas or waveguides) must be designed so that the waves propagate preferentially in the toroidal direction. When this occurs, the waves drag electrons in the wave-troughs. The speeding up of slower electrons in the wave as it travels around the torus produces an electric current opposite to the direction of propagation. The situation is somewhat similar to a surfer catching a wave and then moving with the front.

Each of these methods is successful in driving currents non-inductively in present-day experiments, opening up the possibility of steady state operation with toroidal current. The main difficulty is that current drive is not as efficient as heating. It takes substantial power to drive a modest amount of current. The net result is that a reactor in which all the current must be driven non-inductively would likely lead to an unfavorable overall power balance. Fortunately, in configurations such as the tokamak, there is a natural self-induced toroidal current arising from toroidal transport phenomena. This rather complex process, which has been observed in all high-performance tokamaks, can generate up to 90% of the total current. Consequently, if only a small fraction of the total current needs to be driven non-inductively, then the overall power balance may become acceptable. The transport-induced current is known as the "bootstrap current" and is discussed in Chapter 14.

In summary, a variety of methods involving neutral beams and RF power sources can be used to heat and non-inductively drive current in fusion experiments. Both types of methods have been successful for both tasks. From the reactor point of view, the current drive problem is more difficult because of the relatively low efficiency of converting power into toroidal current.

## 6.9  The future of fusion research

The final topic discussed in the book is the future plans for fusion research. This research will largely be aimed at learning about the physics of burning plasmas where alpha heating substantially exceeds external heating. It will be centered on the construction of a large, next generation fusion device involving an international collaboration. This device is ITER. The experimental reactor will be a tokamak, the size of which is comparable to that of a full scale reactor. Its primary goal is to create for the first time a strongly burning fusion plasma with $Q = 10$. ITER will open a new frontier in the study of fusion plasma physics – the study of high-temperature plasmas dominated by alpha particle heating.

## Bibliography

Chapter 6 contains a brief overview of the plasma physics required to understand and produce a fusion reactor. There are two other somewhat longer books that also provide such an overview including an historical perspective.

### *Overview of Fusion*

Fowler, T. K. (1997). *The Fusion Quest*. Baltimore: John Hopkins University Press.
McCracken, G. and Stott, P. (2005). *Fusion, the Energy of the Universe*. London: Elsevier Academic Press.

# 7

# Definition of a fusion plasma

## 7.1 Introduction

In previous chapters a plasma has been defined qualitatively as an ionized gas whose behavior is dominated by collective effects and by possessing a very high electrical conductivity. The purpose of this chapter is to derive explicit criteria that allow one to quantify when this qualitative definition is valid. In particular, for a general plasma, not restricted to fusion applications, three basic parameters are derived: a characteristic length scale (the Debye length $\lambda_D$), a characteristic inverse time scale (the plasma frequency $\omega_p$), and a characteristic collisionality parameter (the "plasma parameter" $\Lambda_D$). The sizes of these parameters allow one to distinguish whether certain partially or fully ionized gases are indeed plasmas, for example a thin beam of electrons, the ionosphere, the gas in a fluorescent light bulb, lightening, a welding arc, or especially a fusion "plasma."

It is shown in this chapter that for an ionized gas to behave like a plasma there are two different types of criterion that must be satisfied. One type involves the macroscopic lengths and frequencies as compared to $\lambda_D$ and $\omega_p$. The macroscopic length is defined as the typical geometric dimension of the ionized gas ($L \sim a \sim R_0$), while the macroscopic frequency is defined as the inverse thermal transit time for a particle to move across the plasma ($\omega_T \equiv v_T/L$). Plasma behavior requires a small Debye length $\lambda_D \ll L$ and a high plasma frequency $\omega_p \gg \omega_T$. In this regime the plasma is remarkably effective in shielding out electric fields, implying a high electrical conductivity.

The second type of criterion involves the effects of microscopic Coulomb collisions. Such collisions determine whether or not the behavior of an ionized gas is dominated by long-range collective effects as opposed to short-range Coulomb interactions. Collisions are also closely connected to the property of high electrical conductivity. For plasma behavior low collisionality is required, which is shown to correspond to large values of the plasma parameter $\Lambda_D$. This parameter represents the number of charged particles located within a sphere whose radius is equal to the Debye length (and often referred to as a Debye sphere). Low collisionality implies a large number of particles in a Debye sphere.

The criteria described above must be satisfied by any type of ionized gas for it to behave like a plasma. When one further specializes to fusion plasmas, there are additional constraints that must be satisfied, related to the presence of the magnetic field. The new

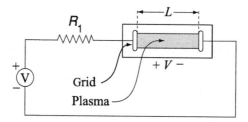

Figure 7.1 Idealized circuit diagram to examine Debye shielding.

constraints are as follows. When a charged particle is confined by a magnetic field, it spirals around a given magnetic line with a characteristic frequency, proportional to $B$, known as the gyro frequency ($\omega_c$). Both electron and ion gyro frequencies must be large compared to the corresponding inverse thermal transit times in a fusion plasma. There is also a characteristic radius of the orbit associated with the gyro motion, proportional to $B$, and known as the gyro radius ($r_L$). In a fusion plasma both electron and ion gyro radii must be small compared to the typical macroscopic dimensions of the plasma. The frequency and length conditions must be fulfilled in order for the plasma to be magnetically confined.

The discussion below focuses on determining quantitative criteria that define a fusion plasma. A main conclusion is that for the typical parameters describing a fusion reactor derived in Section 5.5, the criteria are very well satisfied.

## 7.2 Shielding DC electric fields in a plasma – the Debye length

This section discusses the ability of a plasma to shield DC electric fields. A plasma does this extremely effectively, implying that it has a high electrical conductivity. The calculation presented below makes use of a simple, intuitive fluid model for the plasma in which collective effects dominate. The result of the analysis is the derivation of a critical length, known as the Debye length. For plasmas much bigger in dimension than a Debye length, the shielding is extremely effective.

### 7.2.1 A physical picture of Debye shielding

To understand the shielding effect, recall that in general a plasma has an equal number of positively and negatively charged particles. However, the particles are not bound together as in atoms, but are free to move about through each other, consistent with Maxwell's equations. With this in mind, consider the idealized plasma circuit illustrated in Fig. 7.1. The plasma, characterized by a density $n$ and temperatures $T_e$, $T_i$, is located between two grids separated by a distance $L$. The source corresponds to a DC voltage. In addition the value of $R_1$ is assumed to be sufficiently high that the current flowing in the plasma is low and for simplicity can be neglected in the simple calculation that follows.

The physical picture of Debye shielding follows from an analysis of the motion of electrons and ions when a voltage is applied across the grids. Electrons are attracted to the positive grid and ions to the negative grid. Consider first the electrons. As the electrons flow towards the positive grid, a thin surface charge builds up near the grid surface. The sign of the charge is opposite that of the grid potential. Consequently, the direction of the electric field arising from the surface charge is opposite that of the applied electric field. A similar argument holds for the ions at the negative grid. This opposing induced electric field produces a shielding effect known as "Debye shielding."

In practice, the surface charge layer is not infinitesimally thin but has a finite thickness, known as the Debye length. The reason for the finite thickness is associated with the non-zero temperature of each species. Qualitatively what happens is as follows. Focus first on the electrons. Clearly, electrons are repelled away from the negative grid. However, because of the finite thermal spread in velocities, some high-energy electrons can overcome the negative potential and approach the negative grid. The presence of these electrons prevents perfect shielding in an infinitesimally thin layer – there will always be a small amount of negative charge near the negative grid contributing to the local value of potential. Intuitively, one can see that as the temperature increases, the shielding distance also increases as a larger portion of electrons are able to reach the grid.

At the positive grid an analogous phenomenon occurs. Electrons are attracted to the positive grid and the positive potential traps a thin layer of negative charge. However, because of the finite thermal spread in energies some electrons can escape the trapping potential preventing complete shielding from occurring in an infinitesimally thin layer. Thus, at both grids finite temperature effects lead to incomplete shielding and a finite sheath thickness. By symmetry a similar analysis applies to ions at both grids.

With this qualitative picture in hand, the next task is to derive an explicit expression for the sheath thickness $\lambda_D$.

### 7.2.2 Derivation of the Debye length

The derivation of the Debye length follows from an analysis of a simple fluid model for the plasma. The model is a straightforward generalization of the well-known equations of fluid dynamics. The plasma model is derived below and then coupled to Maxwell's equation. An analysis of these coupled equations demonstrates the shielding effect and leads to an explicit expression for the Debye length.

The starting point is the familiar momentum equation of fluid dynamics

$$\rho \left( \frac{\partial \mathbf{v}}{\partial t} + \mathbf{v} \cdot \nabla \mathbf{v} \right) = -\nabla p + \rho \mathbf{g}. \tag{7.1}$$

To generalize this equation to a plasma one needs to take into account two effects. First, there are two species, electrons and ions, so that a separate momentum equation is needed for each species. Second, there is an additional force acting on the plasma due to the electric field. To determine this force, assume that the plasma behaves like a fluid, dominated by the smooth long-range electric field, and not the short-range collisions. A small volume $V$ of electrons thus experiences a Lorentz force $\mathbf{F_E} = Q\mathbf{E}$, where $Q$ is the total electron charge located in $V$: $Q = -en_e V$. Since the momentum equation involves the force density, the required addition to Eq. (7.1) is simply $\mathbf{f_E} = \mathbf{F_E}/V = -en_e\mathbf{E}$. A similar expression holds for ions.

Several further approximations can be made. First, since the electric current is assumed to be small, the effects of the magnetic field can be neglected. Second, in all terrestrial laboratory experiments the effect of gravity can be ignored since it is infinitesimal compared to the electric field force. Under these assumptions the two-fluid model for plasma momentum conservation becomes

$$m_e n_e \left( \frac{\partial \mathbf{v_e}}{\partial t} + \mathbf{v_e} \cdot \nabla \mathbf{v_e} \right) = -en_e\mathbf{E} - \nabla p_e,$$
$$m_i n_i \left( \frac{\partial \mathbf{v_i}}{\partial t} + \mathbf{v_i} \cdot \nabla \mathbf{v_i} \right) = en_i\mathbf{E} - \nabla p_i. \tag{7.2}$$

The model is now applied to the problem of Debye shielding as follows. The assumption of negligible current implies that in DC steady state $\mathbf{v_e} \approx \mathbf{v_i} \approx 0$. There is no flow of charge. Next, for convenience the electron and ion temperatures are assumed constant: $T_e = $ const, $T_i = $ const. Also, the assumption of steady state implies that $\nabla \times \mathbf{E} = 0$ so that $\mathbf{E} = -\nabla\phi$. Lastly, the geometry under consideration is idealized to be 1-D: all quantities are functions only of $x$. Combining these simplifications leads to the following equations relating $n_e$ and $n_i$ to $\phi$:

$$\frac{d\phi}{dx} - \frac{T_e}{en_e}\frac{dn_e}{dx} = 0,$$
$$\frac{d\phi}{dx} + \frac{T_i}{en_i}\frac{dn_i}{dx} = 0, \tag{7.3}$$

which can be easily integrated, yielding

$$n_e = n_0 \exp(e\phi/T_e),$$
$$n_i = n_0 \exp(-e\phi/T_i). \tag{7.4}$$

Here, $n_0$ is the unperturbed number density of electrons and ions far from each grid; that is, at the midpoint between them. The unperturbed densities must be equal in order to maintain overall charge neutrality. For simplicity, the arbitrary constant associated with the potential has been chosen so that $\phi = 0$ at the midpoint, defined as $x = 0$.

The system of equations is now closed by combining Eq. (7.4) with Poisson's equation:

$$\frac{d^2\phi}{dx^2} = \frac{e}{\varepsilon_0}(n_e - n_i). \tag{7.5}$$

After substituting for the densities one obtains

$$\frac{d^2\phi}{dx^2} = \frac{en_0}{\varepsilon_0}[\exp(e\phi/T_e) - \exp(-e\phi/T_i)]. \tag{7.6}$$

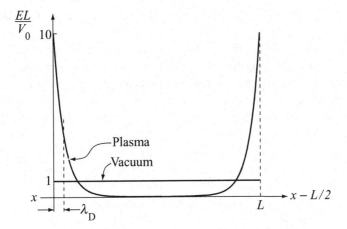

Figure 7.2 Electric field versus distance for the case $L/\lambda_D = 20$.

Note that even this simplest of problems leads to a non-linear equation. The easiest and traditional way to proceed is to assume a low voltage such that $|e\phi|/T_{e,i} \ll 1$. This allows Eq. (7.6) to be transformed into one that is linear but still captures the essential physics. Taylor expanding $\exp(\pm e\phi/T_{e,i}) \approx 1 \pm e\phi/T_{e,i}$ leads to

$$\frac{d^2\phi}{dx^2} - \frac{e^2 n_0}{\varepsilon_0}\left(\frac{1}{T_e} + \frac{1}{T_i}\right)\phi = 0. \tag{7.7}$$

The boundary conditions are: (1) the normalizing condition $\phi(0) = 0$, and (2) the circuit condition $\phi(-L/2) - \phi(L/2) = V_0$. The solution is straightforward and is given by

$$\phi = -\frac{V_0}{2}\frac{\sinh(x/\lambda_D)}{\sinh(L/2\lambda_D)},$$

$$E = -\frac{d\phi}{dx} = \frac{V_0}{2\lambda_D}\frac{\cosh(x/\lambda_D)}{\sinh(L/2\lambda_D)}. \tag{7.8}$$

Here, the Debye length $\lambda_D$ is defined by

$$\frac{1}{\lambda_D^2} = \frac{1}{\lambda_{De}^2} + \frac{1}{\lambda_{Di}^2} = \frac{e^2 n_0}{\varepsilon_0 T_e} + \frac{e^2 n_0}{\varepsilon_0 T_i}. \tag{7.9}$$

A plot of $E = -d\phi/dx$ vs. $x$ is illustrated in Fig. 7.2 for the case $L/\lambda_D = 20$; also shown for comparison is the vacuum solution. Observe the enormous shielding of the electric field by the plasma. As might be expected the shielding layer thickness $\lambda_D$ increases with temperature since at high $T$ there are more energetic particles contributing to incomplete shielding. Also $\lambda_D$ decreases with density as more particles are available to provide shielding thereby reducing the sheath thickness.

An important conclusion from this analysis is the following. Since one of the conditions for an ionized gas to be a plasma is that it shield out DC electric fields, then this implies that the geometric dimension of the gas must be much larger than a Debye length: $L \gg \lambda_D$.

To apply the Debye length criterion to a fusion plasma one simply needs to substitute numerical values. Before doing this, note that in the literature often only the electron Debye length is used ($\lambda_D \rightarrow \lambda_{De}$) as a mathematical convenience. This is sometimes erroneously justified by the large mass of the ions. The assumption is that the ions are so heavy that they do not move, and enter the problem only as a constant neutralizing background of charge. However, it is the ion temperature and not its mass that enters the derivation so that technically the ion Debye length is negligible only when $T_i \gg T_e$ which rarely occurs in practice. In any event for a fusion plasma with $T_i \approx T_e$, neglecting the ion Debye length increases the value of the Debye length by a numerical factor of order unity that is of little consequence for present purposes. Still, to keep with tradition, a numerical value given in terms of the electron Debye length is:

$$\lambda_{De} = \left( \frac{\varepsilon_0 T_e}{e^2 n_0} \right)^{1/2} = 2.35 \times 10^{-5} \left( \frac{T_k}{n_{20}} \right)^{1/2} \text{ m} \tag{7.10}$$

where the units are $T_k$ (keV) and $n_{20}$ ($10^{20} \text{ m}^3$). For a fusion plasma with $T_k = 15$ and $n_{20} = 2$, this yields $\lambda_{De} = 6.4 \times 10^{-5}$ m. Clearly, a fusion plasma with $a \sim 2$ m satisfies the Debye length criterion by over four orders of magnitude, indeed a large margin.

The final point to make in this subsection involves the concept of "quasi-neutrality". In the main bulk of the plasma the shielding of the electric field is extremely effective implying that the local number densities of electrons and ions are barely perturbed from their background value $n_0$. Consequently the *local* electron density is almost exactly equal to the *local* ion density: $n_e(x) \approx n_i(x)$ over almost the entire plasma volume. For example, at a typical point one quarter of the distance away from either grid ($x = \pm L/4$) the difference between the electron and ion number densities for equal temperatures is

$$\left| \frac{n_e - n_i}{n_0} \right| = \frac{e V_0}{2T} \frac{1}{\cosh(L/4\lambda_D)} \ll 1. \tag{7.11}$$

The local property $n_e(x) \approx n_i(x)$ is known as quasi-neutrality and is a basic characteristic of a plasma. Beware, however, of the following trap. The quasi-neutrality condition does not imply that $\mathbf{E} = 0$ or $\nabla \cdot \mathbf{E} = 0$, only that $\varepsilon_0 \nabla \cdot \mathbf{E} \ll e n_e$.

In summary, it has been shown that the critical shielding distance against DC electric fields is the Debye length. Furthermore, for an ionized gas to behave like a plasma its geometric dimensions must be much larger than a Debye length. The next task is to determine the characteristic frequency associated with plasma behavior.

## 7.3 Shielding AC electric fields in a plasma – the plasma frequency

There is a characteristic frequency, known as the electron plasma frequency ($\omega_{pe}$), that defines a second criterion for determining when an ionized gas behaves like a plasma.

The criterion requires that the macroscopic frequencies and inverse time scales be much smaller than the electron plasma frequency. This criterion is shown to be equivalent to the requirement that a plasma screen out low-frequency AC, as well as DC, electric fields. The calculation presented below leads to an explicit expression for $\omega_{pe}$ by means of an analysis of the plasma response to an AC voltage source of frequency $\omega$ and amplitude $V_0$ in the circuit shown in Fig. 7.1. The analysis is based on a straightforward generalization of the simple fluid model described above. Specifically, the conservation of mass must be added to the model as well as allowing for time dependence.

### 7.3.1 A physical picture of the screening of AC fields

The existence of a critical frequency for plasma behavior follows from the dynamics of the screening process. If a voltage is suddenly applied across the grids shown in Fig. 7.1, both the electrons and the ions move to shield this field. Because of their small mass the electrons respond much more rapidly than the ions. Even so, there is still some delay because of their inertia.

If the applied frequency is sufficiently low, one expects that electrons will have more than enough time to complete their motion and shield out the field. In other words for low frequencies the electron motion essentially instantaneously tracks the time variation of the applied field. However, for sufficiently high frequencies, the applied field is oscillating so rapidly that the electrons, even with their small mass, can barely move before the field changes sign. In this case, the plasma is unable to shield out the field and screening no longer occurs; the presence of the plasma is barely noticeable.

The transition frequency is known as the electron plasma frequency and its derivation is presented below.

### 7.3.2 Derivation of the electron plasma frequency

The derivation of the electron plasma frequency follows from an analysis similar to that of DC Debye shielding with three modifications. First, the calculation is greatly simplified by exploiting the fact that the heavy mass of the ions prevents them from moving very much in comparison to the electrons under the action of an AC field for all but the lowest frequencies. It is therefore a good approximation to treat the ions as having an infinite, rather than large finite mass. The infinite mass assumption implies that the ions enter the calculation only as a fixed neutralizing background of positively charged particles. Even after the AC field is applied, the ions stay fixed in place with a constant density. In terms of the two-fluid model the density and velocity of the ions are given by $n = n_0$ and $\mathbf{v}_i = 0$ respectively.

The second modification again makes use of the assumption that the current generated by the motion of electrons is small. This allows one to neglect the effect of the induced magnetic field. The implication is that the inductive part of the electric field is small. Hence, the electric field is approximately electrostatic:

$$\mathbf{E}(x, t) = -\nabla\phi(x, t) - \partial\mathbf{A}(x, t)/\partial t \approx -\nabla\phi(x, t).$$

The third modification takes into account the fact that the electrons are constantly moving under the action of the AC field: $\mathbf{v}_e \neq 0$. While the electron current density is negligible, the effects on the charge density are important and must be included. In terms of the model, there is now an additional unknown ($\mathbf{v}_e = v_e\mathbf{e}_x$). To close the system of equations it is necessary to include the well-known mass conservation relation in the model in addition to the momentum equation. For the 1-D geometry with constant $T_e$ electrons, the self-consistent AC shielding model becomes

$$\frac{\partial n_e}{\partial t} + \frac{\partial}{\partial x}(n_e v_e) = 0,$$

$$\frac{\partial v_e}{\partial t} + v_e \frac{\partial v_e}{\partial x} = \frac{e}{m_e} \frac{\partial\phi}{\partial x} - \frac{T_e}{m_e n_e} \frac{\partial n_e}{\partial x}, \qquad (7.12)$$

$$\frac{\partial^2\phi}{\partial x^2} = \frac{e}{\varepsilon_0}(n_e - n_0).$$

The unknowns are $n_e$, $v_e$ and $\phi$. Observe that the model consists of three non-linear coupled partial differential equations. To proceed further two additional simplifications are required. First, the same assumption that was made for DC Debye shielding is made here. The amplitude of the potential is assumed to be small compared to the temperature: $e\phi \ll T_e$. This allows one to linearize the equations about the background state $n_e = n_0$, $\phi = v_e = 0$. It then follows that $n_e(x, t) = n_0 + n_1(x, t)$ with $n_1 \ll n_0$. Similarly, $\phi$ and $v_e$ are also considered to be small, first order quantities.

The second simplification is to focus the calculation solely on the sinusoidal steady state response of the plasma. One has to imagine that a very small amount of dissipation is present in the system that damps out all initial transients. Under this assumption, the principles of AC circuit analysis imply that the time dependence of all perturbed quantities has the following mathematical form: $Q_1(x, t) = Q_1(x) \exp(-i\omega t)$.

From these simplifications one can easily show that the AC shielding model reduces to

$$n_1 = -i\frac{n_0}{\omega}\frac{dv_e}{dx},$$

$$v_e = i\frac{T_e}{\omega m_e}\frac{d}{dx}\left(\frac{e\phi}{T_e} - \frac{n_1}{n_0}\right), \qquad (7.13)$$

$$\frac{d^2\phi}{dx^2} = \frac{e}{\varepsilon_0}n_1.$$

The quantities $n_1$ and $v_e$ can be easily eliminated in terms of $\phi$. The elimination introduces two free integration functions which for sinusoidal steady state solutions must be set to zero. The end result is a single ordinary differential equation for $\phi$:

$$\frac{d^2\phi}{dx^2} - \frac{\left(1 - \omega^2/\omega_{pe}^2\right)}{\lambda_{De}^2}\phi = 0, \qquad (7.14)$$

where the plasma frequency $\omega_{pe}$ is defined as

$$\omega_{pe}^2 = \frac{n_0 e^2}{m_e \varepsilon_0}. \tag{7.15}$$

As before, the boundary conditions on $\phi$ are $\phi(-L/2) - \phi(L/2) = V_0$ and $\phi(0) = 0$. The desired solution is

$$\phi(x,t) = -\frac{V_0}{2} \frac{\sinh(x/\hat{\lambda}_D)}{\sinh(L/2\hat{\lambda}_D)} \exp(-i\omega t),$$

$$\frac{n_1(x,t)}{n_0} = -\frac{e V_0}{2 T_e} \left(1 - \frac{\omega^2}{\omega_{pe}^2}\right) \frac{\sinh(x/\hat{\lambda}_D)}{\sinh(L/2\hat{\lambda}_D)} \exp(-i\omega t). \tag{7.16}$$

Here, $\hat{\lambda}_D$ is the effective Debye length of the electrons defined by

$$\hat{\lambda}_D^2 = \frac{\lambda_{De}^2}{1 - \omega^2/\omega_{pe}^2}. \tag{7.17}$$

An examination of Eq. (7.16) leads to the following conclusions with respect to the shielding of AC fields. For frequencies below the plasma frequency ($\omega^2 < \omega_{pe}^2$) shielding is very similar to the DC case except that the effective Debye length is somewhat longer. The continual lagging behind of the electrons caused by their inertia reduces the amount of shielding that occurs, leading to a larger Debye length.

As the frequency approaches the plasma frequency ($\omega^2 \to \omega_{pe}^2$), the effective Debye length becomes infinite: $\hat{\lambda}_D \to \infty$. In this limit the potential and perturbed density reduce to

$$\phi(x,t) \approx -V_0 \frac{x}{L} \exp(-i\omega t),$$

$$\frac{n_1(x,t)}{n_0} \approx -\frac{e V_0}{2 T_e} \left(1 - \frac{\omega^2}{\omega_{pe}^2}\right) \frac{x}{L} \exp(-i\omega t) \to 0. \tag{7.18}$$

Observe that the potential approaches the vacuum solution and the perturbed density vanishes. At this critical value of frequency, and above, the shielding effect has completely vanished.

The main conclusion from this analysis is that $\omega_{pe}$ is the critical transition frequency for the shielding of AC fields. Below the plasma frequency there is good shielding. Above the plasma frequency there is no shielding. Therefore, for an ionized gas to exhibit plasma behavior all characteristic frequencies must be much lower than the plasma frequency.

In practice, in terms of numbers the plasma frequency can be written as

$$\omega_{pe} = \left(\frac{n_0 e^2}{m_e \varepsilon_0}\right)^{1/2} = 5.64 \times 10^{11} \, n_{20}^{1/2} \, \text{s}^{-1}. \tag{7.19}$$

For a fusion plasma with $n_{20} = 2$, it follows that $\omega_{pe} = 8.0 \times 10^{11} \, \text{s}^{-1}$. This is a very high frequency. It greatly exceeds all collision frequencies as well as the macroscopic stability frequencies, and inverse transport and confinement times. If the shortest of these macroscopic time scales is defined as the very fast thermal transit time of an electron across a macroscopic length, then $\tau_{Te} \equiv 1/\omega_{Te} \sim L/v_{Te}$, and the requirement $\omega_{pe} \gg \omega_{Te}$ is equivalent to $\lambda_{De} \ll L$. This last equivalence is due to the basic relationship between the Debye length and the plasma frequency:

$$\omega_{pe} = \sqrt{2} \, \frac{\lambda_{De}}{v_{Te}}. \tag{7.20}$$

The conclusion is that the AC shielding properties of a fusion plasma are therefore excellent.

The next topic when defining a fusion plasma concerns low collisionality and the dominance of collective effects.

## 7.4 Low collisionality and collective effects

In the discussion so far the plasma has been treated as a two-fluid model dominated by long-range collective effects. The collective effects have been implicitly built into the model by the way the electric field is calculated. Specifically, the charge density generating the shielding fields is assumed to be accurately approximated by a smooth, continuous distribution of electrons and ions. The field is *not* calculated by summing the contributions of a large number of individual point charges, one charge at a time.

Smoothing out the charge distribution would intuitively seem to be a reasonable approximation for charges located far from a given point. However, for charges located very near the point, one might think that the inter-particle Coulomb potential would be the dominant factor because of the divergence of the $1/r$ dependence of $\phi_{Coul}$ for small $r$. In a plasma this is *not* the case. The reason is that in a plasma the particle density is sufficiently low that two charges rarely get close enough to each other for the two-particle Coulomb potential to dominate the long-range electric field generated by the smoothed out distribution of the entire charge population.

There are several different ways to derive the critical criterion that separates the regimes where long-range collective effects dominate short-range Coulomb interactions. Each way has a different physical interpretation, but all lead to the same criterion. Below, three different calculations are presented: one based on a statistical analysis, another on comparing the average spacing between particles to the Coulomb interaction distance, and the last comparing the Coulomb collision frequency to the plasma frequency.

Each derivation concludes that the critical parameter is the number of particles contained within a sphere whose radius is equal to the Debye length. This parameter is denoted by $\Lambda_D$ and for an ionized gas to behave like a plasma it is necessary that $\Lambda_D \gg 1$.

### 7.4.1 A statistical picture of long-range collective effects

A statistical description of plasma behavior implies that a continuum fluid model is an accurate representation of the physics. For such a model to be valid, two criteria must be satisfied. On the one hand, imagine that the plasma is subdivided into a large number of fluid volume elements. These elements must be small enough to provide the desired spatial resolution. On the other hand, the volume elements must be large enough that each contains a large number of particles. This is necessary in order for "averaging" the charge density and current density into a smooth function to make statistical sense. A volume element containing one or two particles would clearly display quite poor statistics.

For shielding problems, the characteristic length describing plasma behavior is the Debye length. Ideally, one would like the size (i.e., linear dimension) of a volume element to be small compared to $\lambda_{De}$ in order to achieve good spatial resolution. The largest size element that would be just barely tolerable occurs when the linear dimension of the volume element is comparable to the Debye length.

From the discussion above it follows that for a statistical fluid description to accurately model the physics, a volume element with characteristic linear dimension equal to $\lambda_{De}$ must contain a large number of particles. To within a numerical factor of order unity, which is of little consequence, it does not matter whether a volume element is viewed as a cube, sphere, or other comparable shape in terms of precisely defining its linear dimension. Traditionally, the volume element is chosen to be a sphere. Consequently, the condition that the number of particles in a "Debye sphere" (denoted by $\Lambda_D$) be large can be written as

$$\Lambda_D \equiv \frac{4\pi}{3} n_e \lambda_{De}^3 = \frac{4\pi}{3} \frac{\varepsilon_0^{3/2}}{e^3} \frac{T_e^{3/2}}{n_e^{1/2}} \gg 1. \tag{7.21}$$

Observe that $\Lambda_D$ increases with temperature since higher $T_e$ implies a larger Debye length. In terms of density dependence, the competition between large $n_e$ for a large number of particles and small $n_e$ for a large Debye length is dominated by cubic dependence of $\lambda_{De}$. Hence the density appears in the denominator – large $\Lambda_D$ is easiest to achieve with low $n_e$. Finally, note that the expression for $\Lambda_D$ is independent of mass implying that for comparable temperatures, Eq. (7.21) is valid for ions as well as electrons.

### 7.4.2 The inter-particle spacing versus the Coulomb interactive distance

A second way to determine when long-range collective effects dominate short-range interactions is to calculate and compare the average distance between particles, $b_{part}$, with the characteristic Coulomb interaction distance, $b_{Coul}$. Collective effects dominate when particles rarely are close enough to each other to experience a short-range Coulomb interaction: $b_{part} \gg b_{Coul}$.

The distances $b_{part}$ and $b_{Coul}$ are easily calculated as follows. Consider first $b_{part}$. Assume there are $n_e$ particles in a cubic meter of plasma and divide this volume into $n_e$ equally sized micro-cubes. By definition the linear dimension of each micro-cube is $n_e^{-1/3}$. Since each micro-cube contains on average one particle, the average distance between particles is simply $b_{part} = 1/n_e^{1/3}$.

Consider next the characteristic Coulomb interaction distance. This is defined as the largest distance between particles that results in a strong Coulomb interaction. Its calculation can be deduced from an examination of Fig. 7.3, which shows one electron moving with a velocity $v$ towards a second stationary electron. Intuitively, the transition condition for a strong interaction to occur corresponds to the situation where the separation distance $b$ is such that the kinetic energy of the moving particle is comparable to its Coulomb potential energy:

$$\frac{1}{2}m_e v^2 \approx \frac{e^2}{4\pi\varepsilon_0 b}. \tag{7.22}$$

Small values of $b$ imply a strong interaction while large values of $b$ imply a weak interaction. Next, note that the kinetic energy of a typical electron is comparable to the thermal energy of the plasma: $m_e v^2 \approx T_e$. Substituting this result into Eq. (7.22) yields an expression for the Coulomb interaction distance $b = b_{\text{Coul}}$:

$$b_{\text{Coul}} = \frac{e^2}{2\pi\varepsilon_0 T_e}. \tag{7.23}$$

The condition for long-range collective effects to dominate short-range Coulomb interactions, $b_{\text{part}} \gg b_{\text{Coul}}$, can now be written as

$$\frac{1}{n_e^{1/3}} \gg \frac{e^2}{2\pi\varepsilon_0 T_e}. \tag{7.24}$$

A simple rearrangement of terms shows that Eq. (7.24) can be rewritten as

$$\Lambda_D \gg \frac{4\pi}{3(2\pi)^{3/2}} \approx 0.27. \tag{7.25}$$

Except for an inconsequential numerical factor, this criterion is identical to the one given by Eq. (7.21).

### 7.4.3 The plasma frequency vs. the Coulomb collision frequency

The third way to investigate whether or not long-range collective effects dominate plasma behavior is to calculate the collision frequency $\nu_{ee}$ and mean free path $\lambda_{ee}$ and directly compare them to the plasma frequency and Debye length respectively. Based on the shielding arguments one requires $\omega_{pe} \gg \nu_{ee}$ and $\lambda_{De} \ll \lambda_{ee}$ for long-range effects to dominate.

In order to quantify these conditions, expressions are needed for $\nu_{ee}$ and $\lambda_{ee}$. These are easily estimated from the value of $b_{\text{Coul}}$. A re-examination of Fig. 7.3 indicates that the cross section corresponding to a strong two-particle Coulomb interaction is given by

$$\sigma_{ee} \approx \pi b_{\text{Coul}}^2 = \frac{e^4}{4\pi\varepsilon_0^2 m_e^2 v^4}. \tag{7.26}$$

Recall from the general discussion on collisions that the collision frequency and mean free path are related to the cross section as follows: $\nu_{ee} = n_e \sigma_{ee} v$ and $\lambda_{ee} = 1/n_e\sigma_{ee}$. Again assuming that for a typical particle $m_e v^2 \approx T_e$, it follows that

$$\nu_{ee} \approx \frac{n_e e^4}{4\pi\varepsilon_0^2 m_e^{1/2} T_e^{3/2}} = \frac{1}{3}\frac{\omega_{pe}}{\Lambda_D},$$

$$\lambda_{ee} \approx \frac{4\pi\varepsilon_0^2 T_e^2}{n_e e^4} = 3\lambda_{De}\Lambda_D. \tag{7.27}$$

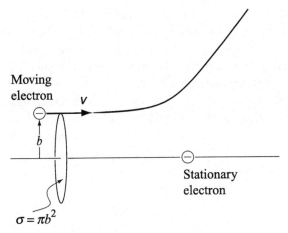

Figure 7.3 Strong Coulomb interactions occurring when $m_e v^2/2 \approx e^2/4\pi\varepsilon_0 b$.

Observe that the criterion for long-range effects to dominate short-range collisions reduces to

$$\Lambda_D \gg 1/3, \tag{7.28}$$

which is essentially the same as previously obtained.

The last step is to substitute numerical values and then determine whether or not a fusion plasma satisfies the condition $\Lambda_D \gg 1$. A simple calculation yields[1]

$$\Lambda_D \equiv \frac{4\pi}{3} n_e \lambda_{De}^3 = \frac{4\pi}{3} \frac{\varepsilon_0^{3/2}}{e^3} \frac{T_e^{3/2}}{n_e^{1/2}} = 5.4 \times 10^6 \frac{T_k^{3/2}}{n_{20}^{1/2}}. \tag{7.29}$$

For a fusion plasma with $T_k = 15$ and $n_{20} = 2$ it follows that $\Lambda_D = 2.2 \times 10^8$. The condition for long-range collective effects to dominate behavior is satisfied by a large margin in a fusion plasma.

## 7.5 Additional constraints for a magnetic fusion plasma

In addition to the $\lambda_{De}$, $\omega_{pe}$, and $\Lambda_D$ constraints required for a general plasma, there are additional constraints for a fusion plasma because of the presence of the magnetic field and the need for magnetic confinement. Specifically, as will be shown in the next chapter, when a charged particle is placed in a magnetic field its motion perpendicular to the field corresponds to a circular orbit characterized by a frequency $\omega_c$ (the gyro frequency) and a

---

[1] In the chapter on Coulomb collisions it is shown that the quantity that appears is not actually $\Lambda_D$ but $\Lambda = 9\Lambda_D$, but this has little effect on the numerical validity of the criterion. It is also shown that Coulomb interactions are dominated by multiple small-angle collisions rather than the large-angle interactions considered here. The net result is a small change in the scaling relations with the more accurate ones being given by $\nu_{ee} \sim (\omega_{pe} \ln \Lambda)/\Lambda$ and $\lambda_{ee} \sim \lambda_{De} \Lambda / \ln \Lambda$. However, here too, the weak behavior of the new logarithmic term does not have any substantial effect on the numerical values justifying the dominance of long-range effects.

radius $r_L$ (the gyro radius). For good confinement the gyro radius must be much smaller than the minor radius of the plasma so the particles do not collide with the first wall.

The small gyro radius requirement is also equivalent to the condition that the gyro frequency be much larger than the inverse thermal transit time. Particles spiral around a field line before having time to collide with the wall.

Without going into great detail one can easily estimate the characteristic gyro frequency and gyro radius of electrons and ions by dimensional analysis and then make the necessary comparisons. The gyro frequency follows directly from Newton's law for a charged particle moving under the influence of the Lorentz force. For the case of finite magnetic field and zero electric field, Newton's law has the form

$$m\frac{d\mathbf{v}}{dt} = q\mathbf{v} \times \mathbf{B}. \tag{7.30}$$

From straightforward dimensional analysis it follows that the gyro frequencies for electrons and ions scale as

$$\omega_{ce} = \frac{eB}{m_e} = 1.76 \times 10^{11} B = 8.8 \times 10^{11} \text{ s}^{-1},$$
$$\omega_{ci} = \frac{eB}{m_i} = 4.79 \times 10^{7} B = 2.4 \times 10^{8} \text{ s}^{-1}. \tag{7.31}$$

Here, $B$ is measured in teslas and the numerical values correspond to $B = 5$ and a deuterium mass.

The gyro radius is determined from the relation between position and velocity:

$$\mathbf{v} = \frac{d\mathbf{r}}{dt}. \tag{7.32}$$

Dimensional analysis implies that the gyro radius scales as $r_L = v/\omega_c$. For a typical thermal particle with $v \approx v_T$ one finds

$$r_{Le} = \frac{v_{Te}}{\omega_{ce}} = \frac{(2m_e T_e)^{1/2}}{eB} = 1.07 \times 10^{-4} \frac{T_k^{1/2}}{B} = 8.3 \times 10^{-5} \text{ m},$$
$$r_{Li} = \frac{v_{Ti}}{\omega_{ci}} = \frac{(2m_i T_i)^{1/2}}{eB} = 6.46 \times 10^{-3} \frac{T_k^{1/2}}{B} = 5.0 \times 10^{-3} \text{ m}, \tag{7.33}$$

with $B = 5$, $T_k = 15$, and a deuterium mass.

Consider now the necessary comparisons. In terms of the macroscopic lengths the condition on the ion gyro radius is more stringent than that on the electron gyro radius because of the mass dependence. Thus, a fusion plasma requires that

$$r_{Li} \ll L \tag{7.34}$$

in order to have well-confined particles, a condition easily satisfied for typical reactor parameters.

The equivalent frequency comparison requires that the ion gyro frequency be large compared to the inverse ion thermal transit time $\omega_{Ti} = L/v_{Ti}$:

$$\omega_{ci} \gg \omega_{Ti}. \tag{7.35}$$

This condition follows from Eq. (7.34) after dividing by $v_{Ti}$ and rearranging terms. It is thus automatically satisfied in a fusion plasma when Eq. (7.34) is satisfied. Physically, particles must make many gyrations before there is any macroscopic motion of the plasma. If this were not the case the effect of the magnetic field would be unimportant and magnetic confinement would be lost

## 7.6 Macroscopic behavior vs. collisions

Observe that the definitions of a plasma always compare characteristic frequencies and lengths of the plasma with macroscopic frequencies and lengths. Comparisons are not directly made with respect to collision frequencies and mean free paths, although one might expect a similar set of inequality requirements to hold in order for an ionized gas to behave like a plasma. The reason is that for typical fusion plasma parameters it follows that

$$a \ll \lambda_{ee} \approx \lambda_{ii},$$
$$\omega_{Ti} \gg \nu_{ii}, \tag{7.36}$$
$$\omega_{Te} \gg \nu_{ee}.$$

The conclusion is that the macroscopic requirements are more severe than the collisional requirements. Satisfying the macroscopic requirements guarantees that the collisional requirements will be satisfied by an even larger margin.

## 7.7 Summary

In this chapter several characteristic parameters describing the behavior of a plasma have been derived. They are the Debye lengths, the gyro radii, the plasma frequencies, the gyro frequencies, and the plasma parameter. These are summarized below:

$$\lambda_{De} = \left( \frac{\varepsilon_0 T_e}{e^2 n_0} \right)^{1/2} = 2.35 \times 10^{-5} \left( \frac{T_k}{n_{20}} \right)^{1/2} \text{ m},$$

$$\lambda_{Di} = \left( \frac{\varepsilon_0 T_i}{e^2 n_0} \right)^{1/2} = 2.35 \times 10^{-5} \left( \frac{T_k}{n_{20}} \right)^{1/2} \text{ m},$$

$$r_{Le} = \frac{(2m_e T_e)^{1/2}}{eB} = 1.07 \times 10^{-4} \frac{T_k^{1/2}}{B} \text{ m},$$

$$r_{Li} = \frac{(2m_i T_i)^{1/2}}{eB} = 6.46 \times 10^{-3} \frac{T_k^{1/2}}{B} \text{ m},$$

$$\omega_{pe} = \left( \frac{n_0 e^2}{m_e \varepsilon_0} \right)^{1/2} = 5.64 \times 10^{11} n_{20}^{1/2} \text{ s}^{-1}, \tag{7.37}$$

$$\omega_{pi} = \left( \frac{n_0 e^2}{m_i \varepsilon_0} \right)^{1/2} = 9.33 \times 10^{9} n_{20}^{1/2} \text{ s}^{-1},$$

$$\omega_{ce} = \frac{eB}{m_e} = 1.76 \times 10^{11} B \text{ s}^{-1},$$

$$\omega_{ci} = \frac{eB}{m_i} = 4.79 \times 10^7 B \quad s^{-1},$$

$$\Lambda_D \equiv \frac{4\pi}{3} n_e \lambda_{De}^3 = \frac{4\pi}{3} \frac{\varepsilon_0^{3/2}}{e^3} \frac{T_e^{3/2}}{n_e^{1/2}} = 5.4 \times 10^6 \frac{T_k^{3/2}}{n_{20}^{1/2}}.$$

Here, all the ion terms correspond to a deuterium mass.

In general, a plasma is a partially or fully ionized gas whose behavior is dominated by long-range collective effects. It is an excellent conductor of electricity, shielding both DC and AC electric fields, and exhibiting local quasi-neutral behavior: $n_e(\mathbf{x}) \approx n_i(\mathbf{x})$. The requirements on the characteristic plasma parameters for these conditions to be met are as follows:

- shielding of DC electric fields:            $\lambda_{De} \ll L$;
- shielding of AC electric fields:            $\omega_{pe} \gg v_{Te}/L$;
- collective effects dominate collisions:      $\Lambda_D \gg 1$.

If attention is focused on the special subclass of plasmas corresponding to fusion applications, there are additional constraints that must be satisfied that are more severe for the ions than the electrons because of the mass dependence:

- magnetic confinement of orbits:            $r_{Li} \ll L$;
- many gyro orbits before macro-motion:      $\omega_{ci} \gg v_{Ti}/L$.

A fusion plasma, which is fully ionized because of its high temperature, satisfies all of these criteria by a large margin.


## Bibliography

This chapter provides a working definition of a plasma in terms of certain fundamental parameters including the Debye length and the plasma frequency. In the literature there are derivations of this fundamental length and time scale based on: (1) a qualitative heuristic discussion, (2) a fluid derivation such as presented in the present textbook, and (3) a more rigorous treatment based on plasma kinetic theory. Several useful treatments are listed below.

Boyd, T. J. M. and Sanderson, J. J. (2003). *The Physics of Plasmas*. Cambridge, England: Cambridge University Press.

Chen, F. F. (1984). *Introduction to Plasma Physics and Controlled Fusion*, second edn. New York: Plenum Press.

Dendy, R. O. (1990). *Plasma Dynamics*. Oxford: Clarendon Press.

Goldston, R. J. and Rutherford, P. H. (1995). *Introduction to Plasma Physics*. Bristol, England: Institute of Physics Publishing.

Hutchinson, I. H. (1987). *Principles of Plasma Diagnostics*. Cambridge, England: Cambridge University Press.

Krall, N. A. and Trivelpiece, A. W. (1973). *Principles of Plasma Physics*. New York: McGraw Hill Book Company.

Spitzer, L. (1962). *The Physics of Fully Ionized Gases*, second edn. New York: Interscience.

Wesson, J. (2004). *Tokamaks*, third edn. Oxford: Oxford University Press.

## Problems

7.1 This problem concerns Debye shielding. Consider electron and ion distribution functions of the form

$$f_j(x, \mathbf{v}) = K_j/(1 + E_j/W_j)^2,$$

where $j$ stands for electrons or ions, $K_j$ and $W_j$ are constants, and $E_j = m_j v^2/2 + q_j\phi(x)$ is the total particle energy.
  (a) Calculate $n_e(\phi)$ and $n_i(\phi)$ choosing $K_j$ such that $n_e = n_i \equiv n_0$ when $\phi = 0$.
  (b) Assume that $\phi \ll W_e$, $W_i$ and evaluate the 1-D Debye shielding distance in terms of $n_0$, $W_e$, $W_i$.

7.2 A plasma has electrons and ions of equal temperature: $T_e = T_i \equiv T$. Their background densities are $n_e = n_i \equiv n_0$. The corresponding non-linear equation describing Debye shielding in a 1-D slab geometry is

$$\frac{d^2\phi}{dx^2} = \frac{en_0}{\varepsilon_0}[\exp(e\phi/T) - \exp(-e\phi/T)].$$

The boundary conditions are $\phi(0) = 0$ and $\phi(-L/2) - \phi(L/2) = V_0$. Solve this equation numerically for $eV_0/T = 0.1, 1, 10$, and compare the profiles. Make a plot of $x_c/\lambda_D$ vs. $eV_0/T$, where $x_c$ is defined as the point where $\phi(x_c) = V_0/2$.

7.3 A spherical electrode is charged up to a potential $V_0$ with respect to the potential at infinity which is assumed to be zero. The electrode is situated in a plasma with equal electron and ion temperatures $T$ and equal electron and ion densities $n_0$ far from the electrode. Calculate the potential profile as a function of $r$ in the limit $eV_0/T \ll 1$.

7.4 An electron-emitting planar grid is placed midway between two collecting planar anodes. The spacing between the two anodes is equal to $2a$. The 1-D electron distribution function is $f_e = (n_0/\pi^{1/2}v_{Te})\exp(-E/T_e)$, where $E = m_e v^2/2 - e\phi(x)$. The ion density is zero. The boundary conditions on the potential are $\phi(0) = 0$ and $\phi(\pm a) = V_0$. There exists a special value of $V_0$ that removes all the surface charge at $x = 0$; that is, a certain $V_0$ makes $\phi'(0) = 0$. Find this value of $V_0$.

7.5 This problem shows how to derive Child's law for electron flow between two electrodes. Consider a planar cathode and planar anode separated by a distance $a$, each with cross sectional area $A$. The cathode is constructed of a thermionic emitting material, which releases an effectively infinite supply of cold electrons at the location of the cathode $x = 0$. The cathode potential is assumed to be zero: $\phi(0) = 0$. The anode is maintained at a potential $\phi(a) = V$. The current flowing between the electrodes is denoted by $I$ and is at this point unknown.
  (a) Using the conservation of current and the conservation of energy for each electron derive an expression for the local charge density $\sigma$ as a function of the local potential $\phi$ and $I$. Neglect collisions.
  (b) Substitute this expression into the 1-D Poisson equation and solve for $\phi$. The boundary conditions are $\phi(0) = 0$ and $\phi(a) = V$. There is an additional constraint that arises because even though the supply of electrons at the cathode is infinite, they cannot all flow to the anode because the negative charge between the electrodes acts as a repelling force to newly emitted electrons which partially compensates the attractive force due to the anode voltage. The limiting current that can flow occurs when $E_x(0) = -\phi'(0) = 0$. This is the limiting case since the electric field (which is negative) cannot become positive since if it did electrons would be drawn from the anode and driven to the cathode. The constraint $\phi'(0) = 0$ determines a relation between $I$ and $V$ known as Child's law. Derive this relationship and show that $I = K V^{3/2}$. Find $K$.

7.6 A cylindrical conducting vacuum chamber of radius $r = a$ is filled with a uniform plasma of density $n_0$ and temperature $T_0$. A finite diameter electron beam with a diffuse radial profile propagates along the axis of the cylinder. The goal of the problem is to investigate the shielding effects of the plasma on the DC electrostatic potential inside the chamber. To simplify the analysis assume the problem is 1-D with all quantities depending on the radial coordinate: $Q = Q(r)$. Also, assume the density profile of the electron beam is given by $n_b(r) = n_{b0} J_0(\alpha r/a)$, where $J_0(\alpha) = 0$ and $J_0$ is the Bessel function.

(a) For the first part of the problem ignore the effects of the background plasma. Calculate the potential $\phi(r)$ assuming the vacuum chamber is grounded: $\phi(a) = 0$. What is the magnitude of the peak potential?

(b) Now consider the effects of the plasma which fills the entire waveguide; that is, the plasma overlaps the beam. Assume the ions are infinitely massive so that $n_i = n_0 = \text{const.}$ and that the electron density is given by the familiar relation $n_e(r) = n_0 \exp(e\phi/T_0)$. For simplicity treat the case $e\phi/T_0 \ll 1$. Recalculate the potential $\phi(r)$ including both the beam and the plasma. What is the magnitude of the peak potential in this case and how does it compare to the case without plasma?

# 8

# Single-particle motion – guiding center theory

## 8.1 Introduction

A major goal of this book is to provide an understanding of how magnetic fields confine charged particles in a fusion plasma. As such, one would like to develop an intuition about the detailed behavior of particle orbits in self-consistent magnetic fields. In particular, it must be demonstrated that charged particles stay confined within the plasma and do not become lost drifting across the field and hitting the first wall.

As a first step towards this goal this chapter focuses on the motion of charged particles in prescribed magnetic and electric fields. No attempt is made at self-consistency – for example, to include the currents and corresponding induced magnetic fields resulting from the flow of charged particles. The fields are simply specified as known quantities. They are assumed to be smooth, slowly varying functions in order to be compatible with the requirement that plasmas be dominated by long-range collective effects. The question of self-consistent fields is deferred to future chapters after appropriate models have been developed.

In the process of studying single-particle motion it will become apparent that there is a well-separated hierarchy of frequencies that characterize the different types of motion that can occur. The fastest and dominant behavior corresponds to gyro motion in which particles move freely along magnetic field lines and rotate in small circular orbits perpendicular to the magnetic field. This motion provides perpendicular confinement of charged particles and makes a toroidal geometry necessary in order to avoid parallel losses.

The next contribution to the hierarchy of frequencies involves slow spatial and time variations in the fields, which lead to important modifications of the basic gyro motion. This regime is known as "guiding center motion." Of particular interest is the development of guiding center drifts ($\mathbf{v}_g$) across the magnetic field. These drifts are, in general, slow compared to the thermal speed ($|\mathbf{v}_g| \ll v_T$) but are nevertheless very important for several reasons. First, one must check the direction of $\mathbf{v}_g$ to make sure that particles do not drift directly into the wall – they do not, although it is by no means obvious at the outset. Second, these drifts are largely responsible for the currents that flow in the plasma and are therefore essential for the ultimate development of self-consistent models.

The study of guiding center motion in slowly varying fields is the main topic of this chapter. Here, the key word is "slowly." Guiding center theory exploits the assumptions that the fields vary slowly in space with respect to the gyro radius and slowly in time with respect to the inverse gyro frequency. The primary motivation for the development of guiding center theory is that the theory provides the basic intuition necessary to understand particle confinement in fusion plasmas.

Continuing, the third regime in the hierarchy of frequencies is the Coulomb collision frequency $\nu_{Coul}$. While such collisions are rare, they are nevertheless crucial for the understanding of magnetic confinement. The reason is that Coulomb collisions are the primary mechanism by which particles and energy diffuse across a magnetic field (ignoring for the moment plasma turbulence) thereby reducing confinement. Even though collisions are infrequent, $\nu_{Coul} \ll |\mathbf{v}_g|/r_L \equiv \omega_g$, they represent the first appearance of a physical mechanism that leads to confinement losses.

The last term in the hierarchy corresponds to nuclear fusion collisions, which unfortunately are very rare. These are basically hard-sphere collisions, which were discussed in Section 3.2. Fusion collisions have little direct effect on particle motion. Indirectly they affect plasma confinement through alpha particle heating and D–T fuel depletion.

In summary, the hierarchy of frequency scales is

$$\omega_c \gg \omega_g \gg \nu_{Coul} \gg \nu_{fus}. \tag{8.1}$$

This chapter describes gyro motion and then focuses on guiding center theory, which correspond to the first two terms in the hierarchy. Coulomb collisions are discussed in the next chapter.

This chapter is organized as follows. The discussion begins with the basic building block of magnetic fusion – gyro motion in a uniform, time independent magnetic field. The gyro orbits are derived exactly starting from Newton's law and the Lorentz electromagnetic force.

Next, a sequence of modifications is made to the magnetic field to model more realistic magnetic geometries. For each modification, attention is focused on calculating the resulting guiding center drift. The analysis makes use of straightforward perturbation theory, which exploits the assumptions of slow space and time variation of the applied fields. This allows each guiding center drift to be calculated by superposition.

There are a number of drifts to include. First the $\mathbf{E} \times \mathbf{B}$ drift arising from perpendicular electric and magnetic fields is calculated. Although it may seem counter-intuitive at present, the non-zero electric field does not violate the plasma's shielding ability and in fact this drift is mandatory in order for the shielding effect to be maintained. Next, perpendicular gradients in a straight magnetic field are introduced leading to the $\nabla B$ drift. Following this, the straight field assumption is relaxed. It is shown that the curvature of a magnetic field leads to a drift appropriately known as the curvature drift.

The next modification is time dependence in both the magnetic and electric field. The dominant effect is the development of an inertia-driven drift, known as the polarization drift. The final modification involves gradients parallel to the magnetic field. This generates a parallel mirroring force that tends to keep particles with a high perpendicular velocity confined between regions of high magnetic field and gives rise to the mirror concept. However, while the mirroring force improves parallel confinement, in the end collisions destroy the effect and the need for toroidicity persists.

The descriptions above indicate that there are a substantial number of modifications to include, and one may wonder whether or not the list is complete. In terms of the guiding center drifts the list is indeed complete – there are no additional guiding center drifts within the order to which the theory is carried out.

The main conclusion from this chapter is that a magnetic field can quite effectively confine charged particles in the perpendicular direction. There is no long-time confinement parallel to the field and this leads to the requirement for a toroidal geometry. While a number of slower cross-field particle drifts do develop because of modifications and additions to the constant, uniform magnetic field, the direction of these drifts does not lead to a flow of particles directly to the first wall. In terms of fusion, guiding center theory predicts good confinement of charged particles for a wide range of toroidal magnetic geometries.

## 8.2 General properties of single-particle motion

The development of guiding center theory begins with a discussion of several general properties of single-particle motion in magnetic and electric fields. Included in the discussion are the statement of the exact equations of motion to be solved and the derivation of general conservation laws leading to the identification of exact constants of the motion.

### 8.2.1 Exact equations of motion

The starting point for the development of guiding center theory is the exact equations of motion as determined from Newton's law. For plasma physics applications only the magnetic and electric forces, given by the Lorentz force are required. Gravity is a very small effect and can be neglected. The equations to be solved are thus

$$m\frac{d\mathbf{v}}{dt} = q(\mathbf{E} + \mathbf{v} \times \mathbf{B}),$$

$$\frac{d\mathbf{r}}{dt} = \mathbf{v}.$$

(8.2)

In general, $\mathbf{B} = \mathbf{B}(\mathbf{r}, t)$ and $\mathbf{E} = \mathbf{E}(\mathbf{r}, t)$ are functions of three dimensions plus time. Equation (8.2) is thus a set of coupled, non-linear, ordinary differential equations for the unknowns $\mathbf{v}$ and $\mathbf{r}$ as functions of $t$. They will be solved for a wide variety of cases by exploiting the underlying assumptions of guiding center theory, namely that the spatial variations of $\mathbf{B}$ and $\mathbf{E}$ occur on a length scale long compared to a gyro radius and that time variations occur on a time scale slow compared to the inverse gyro frequency.

## 8.2.2 General conservation relations

Several general conservation relations can be derived from Eq. (8.2). These involve the conservation of energy and momentum. When applicable the conservation relations lead to "exact constants of the motion," which strongly constrain the particle's orbit.

Consider first the situation in which $E = 0$ and $B$ is independent of time: $B = B(r)$. Forming the dot product of Eq. (8.2) with $v$ leads to

$$m v \cdot \frac{dv}{dt} = \frac{d}{dt} \left( \tfrac{1}{2} m v^2 \right) = 0 \tag{8.3}$$

or

$$\tfrac{1}{2} m v^2 = \text{const.} \tag{8.4}$$

The conclusion is that the kinetic energy of a particle in a static magnetic field is a constant. In other words, a static magnetic field can do no work on a charged particle. Another basic related result is that a static magnetic field produces no force parallel to $B$, a result that follows trivially from the relation $B \cdot (v \times B) = 0$.

This relation can be generalized to include a static electric field. Since the fields are assumed static, Faraday's law implies that $E(r) = -\nabla \phi(r)$. The dot product of Eq. (8.2) with $v$ is again formed. One now makes use of the identity (for a static field)

$$\frac{d\phi}{dt} = \frac{\partial \phi}{\partial t} + v \cdot \nabla \phi = v \cdot \nabla \phi, \tag{8.5}$$

from which it immediately follows that

$$W \equiv \tfrac{1}{2} m v^2 + q \phi = \text{const.} \tag{8.6}$$

The sum of kinetic and potential energy is a constant.

A simple prescription exists for the determination of exact constants of the motion. In general the fields are functions of $x, y, z, t$. Consider the special cases where one or more of these variables is ignorable (i.e., the fields do not depend on these variables). For each ignorable variable, there is one exact constant of the motion. The time independent case above led to the conservation of energy. As another example assume the fields are independent of the coordinate $y$ but not $x, z, t$. Introduce the scalar and vector potential in the usual way: $E = -\nabla \phi - \partial A / \partial t$ and $B = \nabla \times A$. Forming the dot product of the momentum equation with $e_y$ leads to

$$\frac{d}{dt} m v_y = q(E_y - v_x B_z + v_z B_x)$$

$$= -q \left( \frac{\partial A_y}{\partial t} + \frac{dx}{dt} \frac{\partial A_y}{\partial x} + \frac{dz}{dt} \frac{\partial A_y}{\partial z} \right)$$

$$= -q \frac{dA_y}{dt}, \tag{8.7}$$

where in the last step use has been made of the fact that $\partial A_y/\partial y = 0$. It thus follows that

$$p_y \equiv mv_y + qA_y = \text{const.} \tag{8.8}$$

The quantity $p_y$ is the $y$ component of canonical momentum. In a similar way it can be shown (see Problem 8.1) that in a cylindrical geometry with azimuthal symmetry (i.e., $\partial/\partial\theta = 0$) the $\theta$ component of canonical angular momentum is also a constant of the motion:

$$p_\theta \equiv mrv_\theta + qrA_\theta = \text{const.} \tag{8.9}$$

The existence of exact constants of the motion often proves useful in understanding the behavior of particle motion in complex electric and magnetic fields. In the discussion that follows, relatively simple forms for **B** and **E** are chosen that allow for a complete analytic solution of the particle orbits, and that explicitly demonstrate the existence of exact constants of motion.

## 8.3 Motion in a constant B field

The basic building block of magnetic confinement is the behavior of a charged particle in a uniform, time independent, magnetic field. The orbit of such a particle exhibits good confinement perpendicular to the direction of the magnetic field and no confinement parallel to the magnetic field. This behavior can be explicitly demonstrated by solving Newton's laws of motion assuming $\mathbf{E} = 0$ and $\mathbf{B} = B\mathbf{e}_z$, where $B = \text{const}$.

In component form, the full set of Newton's laws reduces to

$$
\begin{aligned}
dv_x/dt &= \omega_c v_y & v_x(0) &= v_{x0} \equiv v_\perp \cos\phi, \\
dv_y/dt &= -\omega_c v_x & v_y(0) &= v_{y0} \equiv v_\perp \sin\phi, \\
dv_z/dt &= 0 & v_z(0) &= v_{z0} \equiv v_\|, \\
dx/dt &= v_x & x(0) &= x_0, \\
dy/dt &= v_y & y(0) &= y_0, \\
dz/dt &= v_z & z(0) &= z_0.
\end{aligned}
\tag{8.10}
$$

Here $\omega_c = qB/m$ is the gyro frequency (sometimes also called the cyclotron or Larmor frequency) and $v_\perp, \phi, v_\|, x_0, y_0, z_0$ are constants representing the initial velocity and position of the particle.

### 8.3.1 Parallel motion

Focus first on the motion parallel to the field. The relevant subset of equations is

$$
\begin{aligned}
dv_z/dt &= 0 & v_z(0) &= v_{z0} \equiv v_\|, \\
dz/dt &= v_z & z(0) &= z_0.
\end{aligned}
\tag{8.11}
$$

The solution is easily found and is

$$
\begin{aligned}
v_z(t) &= v_\|, \\
z(t) &= z_0 + v_\| t.
\end{aligned}
\tag{8.12}
$$

The behavior corresponds to a constant uniform motion. There are no parallel forces providing confinement and particles simply proceed unimpeded. The motion is therefore unconfined along a given magnetic line.

### 8.3.2 Perpendicular motion

In the $x,y$ plane the force is always perpendicular to $\mathbf{v}$. Intuition from classical mechanics suggests that this will lead to a circular-type motion and this is indeed the case. Consider first the relevant equations for the velocity:

$$
\begin{aligned}
dv_x/dt &= \omega_c v_y & v_x(0) &= v_{x0} \equiv v_\perp \cos\phi, \\
dv_y/dt &= -\omega_c v_x & v_y(0) &= v_{y0} \equiv v_\perp \sin\phi.
\end{aligned}
\tag{8.13}
$$

Eliminating $v_x$ yields

$$
\begin{aligned}
& d^2 v_y/dt^2 + \omega_c^2 v_y = 0, \\
& v_y(0) = v_\perp \sin\phi, \\
& dv_y(0)/dt = -\omega_c v_x(0) = -\omega_c v_\perp \cos\phi.
\end{aligned}
\tag{8.14}
$$

Equation (8.14) is a linear, ordinary differential equation with constant coefficients. Its general solution is easily found, and applying the initial conditions leads to

$$
\begin{aligned}
v_y(t) &= -v_\perp \sin(\omega_c t - \phi), \\
v_x(t) &= v_\perp \cos(\omega_c t - \phi).
\end{aligned}
\tag{8.15}
$$

Observe that the particles rotate with an angular frequency equal to the gyro frequency. Also, for a uniform magnetic field, not only is the total kinetic energy conserved, but the separate parallel and perpendicular energies are individually conserved: $v_z^2(t) = v_\parallel^2 = \text{const.}$ and $v_x^2(t) + v_y^2(t) = v_\perp^2 = \text{const.}$

The solution for the perpendicular motion is completed by integrating the velocity, yielding expressions for the particle trajectory $x(t)$, $y(t)$. One obtains

$$
\begin{aligned}
x(t) &= x_g + r_L \sin(\omega_c t - \phi), \\
y(t) &= y_g + r_L \cos(\omega_c t - \phi).
\end{aligned}
\tag{8.16}
$$

Here, the gyro radius (sometimes called the Larmor radius) is given by $r_L = v_\perp/\omega_c = mv_\perp/qB$. The quantities $x_g$, $y_g$ are defined as the guiding center position of the particle:

$$
\begin{aligned}
x_g &\equiv x_0 + r_L \sin\phi, \\
y_g &\equiv y_0 - r_L \cos\phi.
\end{aligned}
\tag{8.17}
$$

This nomenclature is motivated by the trajectory relationship

$$
(x - x_g)^2 + (y - y_g)^2 = r_L^2,
\tag{8.18}
$$

which is illustrated in Fig. 8.1. Observe that the orbit of the particle is circular with a radius equal to the gyro radius. The center of the orbit is located at $x_g$, $y_g$ and hence the

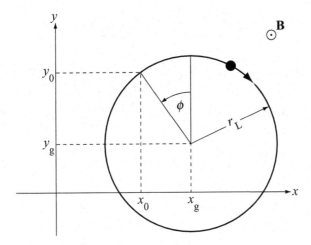

Figure 8.1 Gyro orbit of a positively charged particle in a magnetic field. Shown are the guiding center $x_g$, $y_g$ and the initial position $x_0$, $y_0$.

name "guiding center." Since the gyro radius is, in general, quite small in comparison to the plasma radius, one can conclude that there is good confinement perpendicular to the magnetic field.

The concept of the guiding center is, as its name implies, the basis for "guiding center theory." By following the velocity and position of the guiding center for more general fields one obtains an accurate picture of the average particle location, differing from the exact orbit by only a small deviation of order of the gyro radius. Guiding center motion indeed provides a powerful intuition into the motion of charged particles in slowly varying magnetic and electric fields, a very common practical situation.

A further property of gyro motion is the direction of rotation. Because the electrons and ions have opposite sign charges, they rotate in opposite directions. The actual rotation direction is determined in Fig. 8.2 by calculating the direction of the force $\pm |q|\, \mathbf{v} \times \mathbf{B}$. An easy way to remember the rotation direction is to note that the magnetic field generated by the electric current of a gyrating particle always opposes the applied magnetic field; that is, the gyro motion is diamagnetic. The sign of the charge can be easily taken into account in the description of gyro motion by defining the gyro frequency and gyro radius to always be positive, $\omega_c = |q|\, B/m$, $r_L = mv_\perp/|q|\, B$, and rewriting the solutions as follows:

$$
\begin{aligned}
v_x(t) &= v_\perp \cos(\omega_c t \pm \phi), \\
v_y(t) &= \pm v_\perp \sin(\omega_c t \pm \phi), \\
x(t) &= x_g + r_L \sin(\omega_c t \pm \phi), \\
y(t) &= y_g \mp r_L \cos(\omega_c t \pm \phi),
\end{aligned}
\tag{8.19}
$$

where the upper sign corresponds to a negative charge. Hereafter, the oscillating parts of these solutions is abbreviated to $\mathbf{v}_{\text{gyro}}(t)$ and $\mathbf{r}_{\text{gyro}}(t)$.

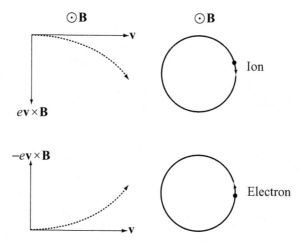

Figure 8.2 Force on a charged particle showing that the rotation is in the diamagnetic direction. For ions $q = +e$, while for electrons $q = -e$.

Lastly, consider the scaling consequences of gyro motion. Note that the gyro frequency increases with the magnetic field $B$: high $B \rightarrow$ high $\omega_c$. Also the electron gyro frequency is much larger than the ion gyro frequency by the ratio $m_i/m_e$. The gyro radius increases with the perpendicular velocity $v_\perp$ and decreases as the magnetic field $B$ increases: high $v_\perp$, low $B \rightarrow$ large $r_L$. For a typical thermal particle with $v_\perp = v_T \equiv (2T/m)^{1/2}$ the ion gyro radius is larger than the electron gyro radius by the ratio $(m_i/m_e)^{1/2}$. Typical numerical values have been given in Chapter 7 and are repeated here for convenience:

$$\omega_{ce} = \frac{eB}{m_e} = 1.76 \times 10^{11} B = 8.8 \times 10^{11} \, s^{-1},$$

$$\omega_{ci} = \frac{eB}{m_i} = 4.79 \times 10^{7} B = 2.4 \times 10^{8} \, s^{-1},$$

$$r_{Le} = \frac{(2m_e T_e)^{1/2}}{eB} = 1.07 \times 10^{-4} \frac{T_k^{1/2}}{B} = 8.3 \times 10^{-5} \, m, \qquad (8.20)$$

$$r_{Li} = \frac{(2m_i T_i)^{1/2}}{eB} = 6.46 \times 10^{-3} \frac{T_k^{1/2}}{B} = 5.0 \times 10^{-3} \, m.$$

These values correspond to $T_k = 15 \, keV$, $B = 5 \, T$, and a deuterium mass.

### 8.3.3 Consequences of gyro motion

The combined perpendicular and parallel motion of a charged particle corresponds to a helical trajectory as shown in Fig. 8.3. Particles spiral unimpeded along field lines with a small perpendicular excursion equal to the gyro radius. This has important implications for the geometry of a magnetic fusion reactor. Specifically, the magnetic geometry must be toroidal. A technologically simpler, linear geometry does not work, as shown in Fig. 8.4(a).

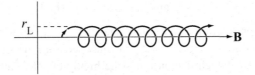

Figure 8.3 Helical trajectory of a charged particle in a uniform magnetic field.

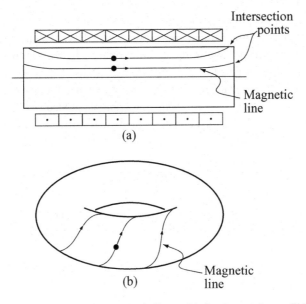

Figure 8.4 (a) Particles streaming along a magnetic line and being lost as they collide with the wall. (b) Magnetic lines wrapping around a torus preventing free streaming end loss.

Observe that in a finite length linear geometry all magnetic field lines must eventually make contact with the first wall as they leave the system. The particles therefore free stream along the field lines directly colliding with the wall in a very short time. In other words, the particles are not confined and there is no time for fusion reactions to occur. This crucial problem is avoided in a toroidal geometry as illustrated in Fig. 8.4(b). Here, particles again spiral continuously along field lines. However, they never make contact with the first wall since the field lines do not leave the chamber in a toroidal geometry and the particle's perpendicular excursions are very small: $r_{Li} \ll a$.

It should be noted that various ingenious configurations have been invented to "plug" the ends of open ended systems. These configurations are based on the "mirror" effect, which is discussed shortly. Even so, in practice, the end losses are just too great to overcome and it is for this reason that all the leading magnetic geometries for fusion applications are toroidal.

To reiterate, the gyro motion of charged particles in a static, homogeneous magnetic field serves as the basic building block for magnetic confinement of fusion plasma.

## 8.4 Motion in constant B and E fields: the E × B drift

The first additional contribution to the static magnetic field to consider corresponds to a constant (in space and time) electric field. This may seem a little strange in view of the discussion in Chapter 7, which demonstrates the highly effective ability of a plasma to shield electric fields. The compatibility of Debye shielding with the existence of electric fields is discussed as the analysis proceeds, and in fact it is shown that no contradictions arise.

For the moment, in keeping with the spirit of "single-particle motion", it is simply assumed that constant electric and magnetic fields are prescribed. The challenge then is to determine the motion of a charged particle in the combined set of fields. The modifications to the original gyro motion separate into two contributions, one due to the parallel electric field and the other due to the perpendicular component. It is shown that the parallel component leads to a constant acceleration and the perpendicular component leads to an initially surprising drift perpendicular to both **E** and **B** known as the **E** × **B** drift.

### 8.4.1 Effect of a parallel electric field

In addition to the constant magnetic field $\mathbf{B} = B\mathbf{e}_z$, assume a constant electric field $\mathbf{E} = \mathbf{E}_\perp + E_\parallel \mathbf{e}_z$ exists in the plasma. Consider first the effect of the parallel electric field. The parallel component of Newton's law reduces to

$$m\frac{\mathrm{d}v_z}{\mathrm{d}t} = qE_\parallel \qquad v_z(0) = v_\parallel. \tag{8.21}$$

The solution is easily found and is given by

$$v_z(t) = v_\parallel + \frac{q}{m}E_\parallel t. \tag{8.22}$$

In addition to the free streaming motion associated with $v_\parallel$ there is a constant acceleration due to the parallel electric field. Hypothetically the particle velocity would continue to increase monotonically and indefinitely until it became relativistic.

In practice, there is a reason why this does not often occur. The ability of electrons and ions to free stream along the magnetic field implies that the parallel electric field that can be generated in a plasma is in general very small, in accordance with the principles of Debye shielding. The actual parallel electric field is not, however, quite as small as predicted by Debye shielding because of the presence of Coulomb collisions. These collisions produce a small frictional drag on the parallel motion leading to a small, but finite plasma resistivity. This resistivity generates a small (but still higher than the Debye value), parallel electric field, similar to the small, but finite, voltage drop across a length of copper wire. The combination of a small electric field and the frictional drag force limits the maximum velocity achievable by a charged particle to non-relativistic values. The frictional drag force due to collisions is discussed in detail in the next chapter.

A final interesting point concerning parallel motion is that under certain conditions, the frictional drag due to collisions is too weak to prevent the slowing down of a certain class of electrons in the plasma. In this situation the electrons do indeed accelerate to relativistic velocities. These electrons are appropriately called "runaway electrons" and this phenomenon is also discussed in the next chapter.

### 8.4.2 Effect of a perpendicular electric field

The next topic concerns the effect of a perpendicular electric field on gyro motion. Consider first the mathematical solution to the problem. To simplify the analysis assume that $\mathbf{E}_\perp = E_x \mathbf{e}_x$, where $E_x = $ const. The perpendicular equations of motion become

$$\frac{dv_x}{dt} = \omega_c v_y + \frac{q}{m} E_x,$$
$$\frac{dv_y}{dt} = -\omega_c v_x. \tag{8.23}$$

Eliminating $v_x$ by means of the second equation yields

$$\frac{d^2 v_y}{dt^2} + \omega_c^2 \left( v_y + \frac{E_x}{B} \right) = 0. \tag{8.24}$$

The solution is easily found by introducing a new velocity variable $v_y' = v_y + E_x/B$. The equation for $v_y'$ simplifies to

$$\frac{d^2 v_y'}{dt^2} + \omega_c^2 v_y' = 0 \tag{8.25}$$

and corresponds to the gyro motion previously discussed.

The solution for the original velocity thus becomes

$$\mathbf{v}_\perp(t) = \mathbf{v}_{\text{gyro}}(t) - (E_x/B)\mathbf{e}_y. \tag{8.26}$$

Note the addition of a new drift perpendicular to both **E** and **B**. This result is easily generalized to an arbitrary perpendicular electric field $\mathbf{E}_\perp = E_x \mathbf{e}_x + E_y \mathbf{e}_y$, where $E_x$, $E_y$ are constants. For the general case one introduces a new perpendicular velocity variable

$$\mathbf{v}_\perp' = \mathbf{v}_\perp - \mathbf{E}_\perp \times \mathbf{B}/B^2. \tag{8.27}$$

The basic equation of motion for the perpendicular $(x, y)$ components, given by

$$m\frac{d\mathbf{v}_\perp}{dt} = q(\mathbf{E}_\perp + \mathbf{v}_\perp \times \mathbf{B}), \tag{8.28}$$

reduces to

$$\frac{d\mathbf{v}_\perp'}{dt} = \omega_c \mathbf{v}_\perp' \times \mathbf{e}_z. \tag{8.29}$$

Equation (8.29) corresponds to gyro motion in a uniform magnetic field. The general form for the original velocity can, therefore, be written as

$$\mathbf{v}_\perp(t) = \mathbf{v}_{\text{gyro}}(t) + \mathbf{V}_E,$$
$$\mathbf{V}_E = \frac{\mathbf{E} \times \mathbf{B}}{B^2}. \tag{8.30}$$

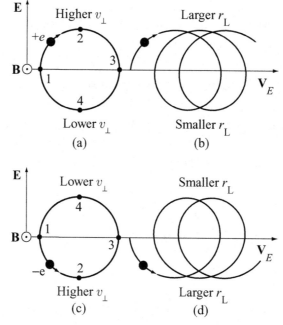

Figure 8.5 (a) Effect of $\mathbf{E}_\perp$ on a positive charge and (b) the resulting perpendicular $\mathbf{E} \times \mathbf{B}$ drift. (c) Effect of $\mathbf{E}_\perp$ on a negative charge and (d) the resulting perpendicular $\mathbf{E} \times \mathbf{B}$ drift.

The conclusion is that the addition of a uniform perpendicular electric field superimposes a constant drift velocity ($\mathbf{V}_E$) on the gyro motion. This drift is known as the $\mathbf{E} \times \mathbf{B}$ drift and is perpendicular to both $\mathbf{E}$ and $\mathbf{B}$. It is also independent of the mass and charge. In other words, electrons and ions drift with the same velocity.

The next step is to develop a physical picture of the origin of the $\mathbf{E} \times \mathbf{B}$ drift and then lastly to address the issue of how the existence of a perpendicular electric field can be compatible with Debye shielding. A physical picture of the $\mathbf{E} \times \mathbf{B}$ drift can be obtained by examining Fig. 8.5 and recalling that the gyro radius increases with the perpendicular velocity: $r_L \sim v_\perp / B$. Consider the motion of a positive charge located in an electric and magnetic field as shown in Fig. 8.5(a). This illustration shows the gyro motion of the charge *without* the electric field. As the particle moves from point 1 to point 2, the effect of the electric field, because of its direction, is to accelerate the charge – increase its velocity. As it moves from point 2 to point 3 it slows down returning to its original velocity. Note that at every point along the top part of the trajectory the velocity is larger than the original velocity without the electric field, implying that on average its gyro radius has increased in size.

The opposite is true on the lower portion of the curve. From point 3 to point 4 the charge is decelerated and slows down. From point 4 back to point 1 the charge accelerates back to its original velocity. Over the bottom portion of the trajectory the average velocity and therefore the average gyro radius is smaller than without the electric field.

The combination of these effects is shown in Fig. 8.5(b). A higher $v_\perp$ on the top portion of the trajectory and a lower $v_\perp$ on the bottom portion lead to a drift perpendicular to both **E** and **B** resulting from the different sizes of the average gyro radius. A similar picture holds for negatively charged electrons as shown in Figs. 8.5(c) and 8.5(d). Observe that the direction of the drift is independent of the size of the charge.

Lastly the simultaneous existence of a perpendicular electric field and Debye shielding has to be reconciled. There are two points to consider. First note that since both electrons and ions have the same $\mathbf{E} \times \mathbf{B}$ drift velocity this corresponds to a macroscopic fluid flow $\mathbf{u}_\perp = \mathbf{V}_E$ without the generation of any electric current $\mathbf{J}_\perp = en\,(\mathbf{u}_{\perp i} - \mathbf{u}_{\perp e}) = 0$. The expression for the $\mathbf{E} \times \mathbf{B}$ drift velocity can thus be rewritten as $\mathbf{E}_\perp + \mathbf{u}_\perp \times \mathbf{B} = 0$. Now recall from the theory of low-frequency electromagnetism that the electric and magnetic fields in a fluid moving with a velocity $\mathbf{u}_\perp$ can be transformed to the reference frame moving with the fluid by the relations $\mathbf{E}'_\perp = \mathbf{E}_\perp + \mathbf{u}_\perp \times \mathbf{B}$ and $\mathbf{B}' = \mathbf{B}$. Consequently, in the reference frame where the fluid is stationary it follows that $\mathbf{E}'_\perp = 0$, which is consistent with the principles of Debye shielding.

The second point is slightly more subtle. In future chapters it will be shown that small perpendicular electric fields (but still larger than the predicted Debye value) can exist in a plasma. This involves the development and solution of self-consistent plasma models. Qualitatively, such electric fields arise because perpendicular to **B** the electrons and ions are magnetically confined and therefore are not free to flow and shield out any local charge imbalances that may develop. It will be shown that these imbalances are a consequence of the different size electron and ion gyro radii and ultimately lead to potentials on the order of $e\phi \sim T$.

In summary, the $\mathbf{E} \times \mathbf{B}$ drift is one of the fundamental cross-field drift velocities appearing in the guiding center theory of charged particle motion.

## 8.5  Motion in fields with perpendicular gradients: the $\nabla B$ drift

The second modification to gyro motion to be investigated involves inhomogeneities in the fields. Specifically, this section includes the effects of gradients in **B** and **E** perpendicular to the magnetic field. Although the **B** field is inhomogeneous, its direction nevertheless remains straight; that is, **B** is assumed to be of the form $\mathbf{B} = B(x, y)\mathbf{e}_z$. For the electric field, the gradients allowed are given by $\mathbf{E} = E_x(x)\mathbf{e}_x + E_y(y)\mathbf{e}_y$. Note that $\nabla \times \mathbf{E} = 0$. Were this not the case, then from Faraday's law, a time dependence would have to be included in **B**. The time dependence issues are discussed in Section 8.7. The form of field gradients considered here appear in plasmas created in long, straight, solenoidal coils.

The analysis presented below demonstrates that the magnetic field gradient produces a particle drift perpendicular to both **B** and $\nabla B$ known as the $\nabla B$ drift. The gradient in the electric field is shown to produce a small shift in the gyro frequency, which is of no great consequence for present purposes. The analysis is carried out using a straightforward perturbation expansion. The small parameter in the expansion is the ratio of the gyro radius to the

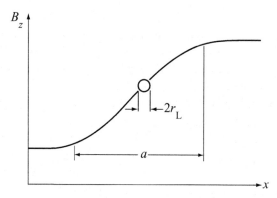

Figure 8.6 Width of the $x$ dimension of a gyro orbit in a field with a weak gradient: $r_L \ll a$.

scale length characterizing the field inhomogeneities: $r_L \nabla B / B \sim r_L \nabla E / E \sim r_L / a \ll 1$. The magnetic and electric fields vary slowly compared to the gyro radius. The details of the perturbation expansion proceed as follows.

### 8.5.1 Perpendicular gradient in B with E = 0

To simplify the calculation assume initially that $\mathbf{E} = 0$ and $B(x, y) \to B(x)$. These assumptions are relaxed shortly. The perpendicular equations of motion can then be written as

$$
\begin{aligned}
m dv_x / dt &= q B(x) v_y, \\
m dv_y / dt &= -q B(x) v_x, \\
dx / dt &= v_x, \\
dy / dt &= v_y.
\end{aligned}
\tag{8.31}
$$

These equations are complicated non-linear differential equations because of the $x$ dependence of $B$. The equations are simplified by exploiting the small gyro radius approximation. The key step is to Taylor expand $B$ about its guiding center. The implication is that a weak field gradient only allows a particle's $x$ position to deviate slightly from its guiding center trajectory. See Fig. 8.6. Under this assumption, the perpendicular equations of motion can be written as

$$
\begin{aligned}
dv_x / dt &\approx \omega_c(x_g) \left[ 1 + \frac{\partial B(x_g)}{\partial x_g} \frac{x - x_g}{B(x_g)} \right] v_y, \\
dv_y / dt &\approx -\omega_c(x_g) \left[ 1 + \frac{\partial B(x_g)}{\partial x_g} \frac{x - x_g}{B(x_g)} \right] v_x, \\
dx / dt &= v_x, \\
dy / dt &= v_y.
\end{aligned}
\tag{8.32}
$$

Note that the magnitutude of second term in the square bracket is smaller by the ratio $r_L / a$. The solution to Eq. (8.32) is found by a straightforward perturbation expansion:

$$
\begin{aligned}
\mathbf{v} &= \mathbf{v}_0 + \mathbf{v}_1 + \cdots, \\
\mathbf{r} &= \mathbf{r}_0 + \mathbf{r}_1 + \cdots.
\end{aligned}
\tag{8.33}
$$

The expansion is substituted into the equations of motion. Setting the leading order contribution to zero yields

$$d\mathbf{v}_0/dt = \omega_c \mathbf{v}_0 \times \mathbf{e}_z,$$
$$d\mathbf{r}_0/dt = \mathbf{v}_0. \tag{8.34}$$

Since $\omega_c = \omega_c(x_g) = \text{const.}$, the solution to Eq. (8.34) is simply the basic gyro motion given by Eqs. (8.15) and (8.16).

The zero order solution is now substituted into the first order contribution to the perturbation equations, which can be written as

$$\frac{dv_{x1}}{dt} - \omega_c v_{y1} = -\frac{v_\perp^2}{2B}\frac{\partial B}{\partial x_g}[1 - \cos 2(\omega_c t - \phi)],$$
$$\frac{dv_{y1}}{dt} - \omega_c v_{x1} = -\frac{v_\perp^2}{2B}\frac{\partial B}{\partial x_g}\sin 2(\omega_c t - \phi). \tag{8.35}$$

These are linear inhomogeneous differential equations. Observe that there are two types of driving terms – a constant term and a term oscillating at twice the gyro frequency. Since the equations are linear the response to each type of driving term can be determined by superposition. It is shown in Problem 8.2 that the second harmonic terms give rise to a small shift in the location of the guiding center plus a small correction to the size of the gyro radius. Neither of these effects is of any consequence since they do not result in a guiding center drift. They can thus be ignored for present purposes. Under this assumption the equations for the velocity components reduce to

$$\frac{dv_{x1}}{dt} - \omega_c v_{y1} = -\frac{v_\perp^2}{2B}\frac{\partial B}{\partial x_g},$$
$$\frac{dv_{y1}}{dt} + \omega_c v_{x1} = 0. \tag{8.36}$$

These equations are identical in form to Eq. (8.23), which produced the $\mathbf{E} \times \mathbf{B}$ drift. By direct comparison it follows that any driving term representing a constant acceleration $\mathbf{F}/m \rightarrow q\mathbf{E}/m$ gives rise to an equivalent $\mathbf{E} \times \mathbf{B}$ drift of the form

$$\mathbf{V}_F = \frac{1}{q}\frac{\mathbf{F} \times \mathbf{B}}{B^2}. \tag{8.37}$$

Applying this result to Eq. (8.36) leads to the $\nabla B$ drift

$$\mathbf{V}_{\nabla B} = \frac{v_\perp^2}{2\,\omega_c}\frac{1}{B}\frac{\partial B}{\partial x_g}\mathbf{e}_y. \tag{8.38}$$

In Problem 8.2 it is shown that this result can be easily generalized to the 2-D case $B = B(x, y)$. The result is a generalized form of the $\nabla B$ drift given by

$$\mathbf{V}_{\nabla B} = \mp\frac{v_\perp^2}{2\,\omega_c}\frac{\mathbf{B} \times \nabla B}{B^2}, \tag{8.39}$$

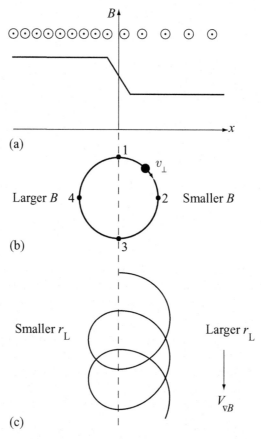

Figure 8.7 (a) Magnetic field gradient due to $\mathbf{B} = B(x)\mathbf{e}_z$. (b) Gyro motion ignoring the $B$ field gradient. (c) Gyro motion plus the $\nabla B$ drift.

where the upper sign corresponds to a negative charge and it is understood that the field is evaluated at the guiding center.

Observe the following properties of the $\nabla B$ drift. (1) The drift is perpendicular to both $\mathbf{B}$ and $\nabla B$. (2) For a typical thermal particle ($v_{\perp} \sim v_T$) the $\nabla B$ drift is small compared to the thermal velocity: $|\mathbf{V}_{\nabla B}/v_T| \sim r_L/a$. (3) Since the drift is proportional to $mv_{\perp}^2$ it has the same (velocity-averaged) magnitude for electrons and ions when $T_e = T_i = T$. (4) Since $\mathbf{V}_{\nabla B}$ is proportional to $1/q$ the direction of the drift is opposite for electrons and ions, causing a net flow of current.

A physical picture of the $\nabla B$ drift can be obtained by examining Fig. 8.7. A magnetic field profile with an admittedly exaggerated magnetic field gradient is shown in Fig. 8.7(a) and Fig. 8.7(b) illustrates the zeroth order gyro motion for a positive particle in which the effects of the gradient are ignored. Recalling that the gyro radius scales as $r_L \sim v_{\perp}/B$, it follows that along the trajectory from point 1 to point 2 to point 3 the gyro radius will be slightly larger because the magnetic field is slightly smaller. Similarly, from point 3 to point 4 and back to point 1 the gyro radius is slightly smaller because of the increased magnetic field. These modifications to gyro motion are combined in Fig. 8.7(c) demonstrating the existence of the $\nabla B$ drift. A similar picture holds for negative charges.

The $\nabla B$ drift makes an important contribution to the flow of current and the corresponding self-consistent magnetic field in a fusion plasma.

### 8.5.2 Perpendicular gradient in E with uniform B

The next topic involves the effects of a weak perpendicular gradient in the electric field. The magnetic field can be considered to be uniform ($B = $ const.) since the effects of a weak gradient in $B$ have already been calculated and, as has been shown, can be easily included by means of superposition. The derivation below demonstrates that the main effect of the electric field gradient is to produce a small correction to the gyro frequency which is of no great significance.

The analysis is carried out assuming the following form for the electric field: $\mathbf{E} = E_x(x)\mathbf{e}_x$. This form satisfies $\nabla \times \mathbf{E} = 0$ so that no time dependence need be included in the magnetic field. As with the $\nabla B$ drift, the mathematical solution is obtained by a straightforward perturbation technique in which the electric field is expanded about the guiding center position of the particle. The relevant equations for the velocity components become

$$\frac{dv_x}{dt} - \omega_c \left[ v_y + \frac{E_x(x_g)}{B} \right] \approx \frac{q}{m} \frac{\partial E_x}{\partial x_g}(x - x_g),$$
$$\frac{dv_y}{dt} + \omega_c v_x = 0. \tag{8.40}$$

Note, that if there were no gradient in the electric field the solution would be given by the sum of the gyro motion plus $\mathbf{E} \times \mathbf{B}$ drift as expected. When the gradient is included one must be careful before simply substituting the zeroth order solutions into the correction term on the right hand side of Eq. (8.40). The reason is that this term might oscillate at the fundamental gyro frequency, thereby appearing as a potentially resonant driving term in the equation. As is well known, a resonant driving term often leads to solutions that grow linearly with time. In other words, the solutions become linearly divergent with $t$ and the perturbation procedure breaks down.

A more careful examination of Eq. (8.40) shows, however, that resonant growth does not occur and the solutions remain bounded. To see this, differentiate the first equation and then eliminate $dv_y/dt$ by means of the second equation. A short calculation yields

$$\frac{d^2 v_x}{dt^2} + \omega_c^2 \left( 1 - \frac{1}{\omega_c B} \frac{\partial E_x}{\partial x_g} \right) v_x = 0. \tag{8.41}$$

Equation (8.41) shows that the main effect of a perpendicular gradient in the electric field is to generate a small correction to the gyro frequency. There is no new particle drift or resonance. In other words, the effect is of no great consequence and is ignored hereafter.

Finally, it is worth noting that if one carries out the expansion one order higher in the ratio $r_L/a$ a drift does develop known as the "finite gyro radius" drift. This drift is in the same direction for electrons and ions but is larger in magnitude for the ions. However, because of its small magnitude ($r_L/a$ smaller than the other guiding center drifts) it does not play an important role for much of the fusion plasma physics discussed in this book. For this reason it is ignored hereafter, although it is discussed in Problem 8.3.

## 8.6 Motion in a curved magnetic field: the curvature drift

The spatial dependence of the magnetic fields thus far considered has been either uniform or possessing a perpendicular gradient. In all cases, however, the direction of the field has been straight, along the $\mathbf{e}_z$ direction. The present section relaxes this constraint and allows for a curved magnetic field. It is shown that the field line curvature leads to a new guiding center drift perpendicular to both the magnetic field and the curvature vector. The drift is driven by the centrifugal force felt by a particle due to its free streaming, parallel motion along a curved field line. Hence, it is known as the "curvature drift."

The analysis is first carried out for a simple curvilinear geometry in which the fields are assumed to be of the form $\mathbf{B} = B(r)\mathbf{e}_\theta$ and $\mathbf{E} = E_r(r)\mathbf{e}_r$. A perturbation expansion is again used. Once the drift has been calculated, the derivation is extended to a generalized curvilinear geometry. The first derivation begins by noting that in a cylindrical coordinate system the position, velocity, and acceleration are related by

$$\mathbf{r}(t) = r(t)\mathbf{e}_r + z(t)\mathbf{e}_z,$$

$$\mathbf{v}(t) = \frac{d\mathbf{r}}{dt} = \frac{dr}{dt}\mathbf{e}_r + r\frac{d\theta}{dt}\mathbf{e}_\theta + \frac{dz}{dt}\mathbf{e}_z = v_r\mathbf{e}_r + v_\theta\mathbf{e}_\theta + v_z\mathbf{e}_z, \qquad (8.42)$$

$$\mathbf{a}(t) = \frac{d\mathbf{v}}{dt} = \left(\frac{dv_r}{dt} - \frac{v_\theta^2}{r}\right)\mathbf{e}_r + \left(\frac{dv_\theta}{dt} + \frac{v_r v_\theta}{r}\right)\mathbf{e}_\theta + \frac{dv_z}{dt}\mathbf{e}_z.$$

Here, use has been made of the fact that the directions of two of the unit vectors change with $\theta$:

$$\frac{d\mathbf{e}_r}{dt} = \frac{\partial \mathbf{e}_r}{\partial\theta}\frac{d\theta}{dt} = \frac{v_\theta}{r}\mathbf{e}_\theta,$$

$$\frac{d\mathbf{e}_\theta}{dt} = \frac{\partial \mathbf{e}_\theta}{\partial\theta}\frac{d\theta}{dt} = -\frac{v_\theta}{r}\mathbf{e}_r. \qquad (8.43)$$

The equations of motion for the velocity components can now be written as

$$\frac{dv_r}{dt} - \frac{v_\theta^2}{r} = \frac{q}{m}(E_r - v_z B),$$

$$\frac{dv_z}{dt} = \frac{q}{m}v_r B, \qquad (8.44)$$

$$\frac{dv_\theta}{dt} + \frac{v_r v_\theta}{r} = 0.$$

The dominant behavior again corresponds to gyro motion plus an $\mathbf{E} \times \mathbf{B}$ drift. This can be seen by introducing a perturbation expansion similar to the $\nabla B$ drift analysis: $\mathbf{v}(t) \approx \mathbf{v}_0(t) + \mathbf{v}_1(t)$. Here, $\mathbf{v}_0(t)$ consists of $\mathbf{v}_{\perp 0}(t) = \mathbf{v}_{\text{gyro}} + \mathbf{V}_E$ and $v_{\theta 0}(t) = v_\| = \text{const}$. Note that parallel now refers to the $\theta$ direction. The next step is to substitute into Eq. (8.44) and to expand all quantities about the guiding center position $r_g$. A short calculation yields an equation for $\mathbf{v}_1(t)$:

$$\frac{d\mathbf{v}_{\perp 1}}{dt} - \omega_c \mathbf{v}_{\perp 1} \times \mathbf{e}_\theta = \frac{\omega_c (r - r_g)}{B} \left[ \frac{\partial B}{\partial r_g} \mathbf{v}_{\perp 0} \times \mathbf{e}_\theta + \frac{\partial E_r}{\partial r_g} \mathbf{e}_r \right] + \frac{v_\|^2}{r_g} \mathbf{e}_r,$$

$$\frac{d v_{\| 1}}{dt} = -\frac{v_{r0} v_{z0}}{r_g},$$

(8.45)

where $\omega_c = q B(r_g)/m$.

The solution has the following properties. The parallel velocity $v_{\| 1}(t)$ develops a small, unimportant, second harmonic modulation, a consequence of the fact that both $v_{r0}(t)$ and $v_{z0}(t)$ are oscillatory at the fundamental frequency and are $\pi/2$ out of phase. The first two terms on the right hand side of the $\mathbf{v}_{\perp 1}$ equation represent the $\nabla B$ drift and the $\mathbf{E}_\perp(\mathbf{r}_\perp)$ gyro frequency correction already discussed. Only the last term represents a new contribution. Because the perturbation expansion essentially linearizes the first order equations, the effect of the new term can again be calculated using superposition.

Physically, this term represents the centrifugal force acting on the particle because of its free streaming parallel motion along a curved magnetic field line. Mathematically, the term has the form of a constant external force. Therefore, in accordance with Eq. (8.37) a guiding center drift develops that is perpendicular to both the magnetic field and the centrifugal force. It is known as the curvature drift and is given by

$$\mathbf{V}_\kappa = \frac{v_\|^2}{\omega_c r} \mathbf{e}_z$$

(8.46)

with all quantities evaluated at the guiding center. The drift has a similar scaling as $\mathbf{V}_{\nabla B}$ except that $v_\perp^2$ is replaced with $2v_\|^2$. It is small compared to the thermal velocity ($|\mathbf{V}_\kappa|/v_T \sim r_L/a$) and comparable in magnitude for electrons and ions of similar temperatures. The direction of the curvature drift for electrons is opposite to that for ions and therefore generates a current.

The expression for $\mathbf{V}_\kappa$ can be generalized to an arbitrary curvilinear magnetic geometry by introducing the radius of curvature vector $\mathbf{R}_c$. Several steps are required. First, the unit vector parallel to the magnetic field is introduced: $\mathbf{b}(\mathbf{r}) \equiv \mathbf{B}/B$. Second, the velocity vector is decomposed into a perpendicular and a parallel component: $\mathbf{v}(t) = \mathbf{v}_\perp + v_\| \mathbf{b}$. Next, the perpendicular components of the equations of motion (with $\mathbf{E} = 0$ for simplicity) are extracted by forming the operation

$$\mathbf{b} \times \left\{ \left[ \frac{d}{dt} (\mathbf{v}_\perp + v_\| \mathbf{b}) - \omega_c (\mathbf{v}_\perp + v_\| \mathbf{b}) \times \mathbf{b} \right] \times \mathbf{b} \right\} = 0,$$

(8.47)

where $\omega_c = qB(\mathbf{r})/m$. The various terms are simplified as follows:

$$\mathbf{b} \times \{[\omega_c(\mathbf{v}_\perp + v_\parallel \mathbf{b}) \times \mathbf{b}] \times \mathbf{b}\} = -\omega_c \mathbf{v}_\perp \times \mathbf{b},$$

$$\mathbf{b} \times \left\{ \left[\frac{d\mathbf{v}_\perp}{dt}\right] \times \mathbf{b} \right\} = \left(\frac{d\mathbf{v}_\perp}{dt}\right)_\perp,$$

$$\mathbf{b} \times \left\{ \left[\frac{d}{dt}(v_\parallel \mathbf{b})\right] \times \mathbf{b} \right\} = v_\parallel \mathbf{b} \times \left[\left(\frac{d\mathbf{b}}{dt}\right) \times \mathbf{b}\right] \qquad (8.48)$$

$$= v_\parallel \left[(\mathbf{b} \cdot \mathbf{b})\frac{d\mathbf{b}}{dt} - \left(\mathbf{b} \cdot \frac{d\mathbf{b}}{dt}\right)\mathbf{b}\right].$$

The last term can be further simplified by noting that for a unit vector $\mathbf{b} \cdot \mathbf{b} = 1$ and therefore $\mathbf{b} \cdot d\mathbf{b}/dt = (1/2)d(\mathbf{b} \cdot \mathbf{b})/dt = 0$. Also, the term $d\mathbf{b}/dt$ can be rewritten as

$$\frac{d\mathbf{b}(\mathbf{r})}{dt} = \left(\frac{\partial}{\partial t} + \frac{d\mathbf{r}}{dt} \cdot \nabla\right)\mathbf{b} = \mathbf{v} \cdot \nabla\mathbf{b}. \qquad (8.49)$$

Combining results leads to a simpler form of the perpendicular equations of motion:

$$\left(\frac{d\mathbf{v}_\perp}{dt}\right)_\perp - \omega_c \mathbf{v}_\perp \times \mathbf{b} = -v_\parallel \mathbf{v}_\perp \cdot \nabla\mathbf{b} - v_\parallel^2 \mathbf{b} \cdot \nabla\mathbf{b}. \qquad (8.50)$$

The left hand side of this equation represents the familiar gyro motion. In the context of a perturbation expansion the right hand side of the equation represents two inhomogeneous driving terms, both smaller by $r_L/a$. The term with $v_\parallel \mathbf{v}_\perp$ oscillates at the gyro frequency with zero average value. It thus makes small modifications to the gyro motion as previously discussed, but does not lead to a drift of the guiding center. Only the last term has the form of a constant external force. It represents the generalization of the centrifugal force and leads to the curvature drift.

The last step in the analysis is to determine a relation between the magnetic curvature vector $\mathbf{b} \cdot \nabla\mathbf{b}$ and the radius of curvature vector $\mathbf{R}_c$. This relationship is easily established by examining Fig. 8.8. Observe that the change in $\mathbf{b}$ along a curved magnetic line is given by

$$d\mathbf{b} = \mathbf{b}(\mathbf{r}_\perp, l + dl) - \mathbf{b}(\mathbf{r}_\perp, l) = \frac{\partial \mathbf{b}}{\partial l}dl = (\mathbf{b} \cdot \nabla\mathbf{b})dl,$$

$$|d\mathbf{b}| = d\theta = \frac{dl}{R_c}. \qquad (8.51)$$

Here use has been made of the fact that the change along the magnetic field is equivalent to taking the parallel gradient: $\partial/\partial l = \mathbf{b} \cdot \nabla$. From the geometry and the definition of the radius of curvature vector it is clear that $\mathbf{R}_c$ is anti-parallel to $\mathbf{b} \cdot \nabla\mathbf{b}$. Therefore $\mathbf{R}_c = -K\mathbf{b} \cdot \nabla\mathbf{b}$. The scale factor $K$ is found by noting that $|\mathbf{b} \cdot \nabla\mathbf{b}| = |d\mathbf{b}|/dl = 1/R_c$. Combining results

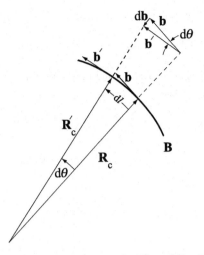

Figure 8.8 Geometry showing the relation between $\mathbf{b} \cdot \nabla \mathbf{b}$ and $\mathbf{R}_c$. Here, $\mathbf{b} = \mathbf{b}(\mathbf{r}_\perp, l)$ and $\mathbf{b}' = \mathbf{b}(\mathbf{r}_\perp, l + dl)$. Similarly for $\mathbf{R}_c$ and $\mathbf{R}'_c$.

leads to

$$\mathbf{b} \cdot \nabla \mathbf{b} = -\frac{\mathbf{R}_c}{R_c^2}. \tag{8.52}$$

The generalized form of the curvature drift can now be calculated. Equation (8.52) is substituted into the centrifugal force term in Eq. (8.50). Then, using the relation between a constant external force and the resulting guiding center drift given by Eq. (8.37), one obtains the desired generalization:

$$\mathbf{V}_\kappa = \mp \frac{v_\parallel^2}{\omega_c} \frac{\mathbf{R}_c \times \mathbf{B}}{R_c^2 B}. \tag{8.53}$$

Again the top sign refers to electrons.

Like the $\nabla B$ drift, the curvature drift makes an important contribution to the flow of current in a plasma and the determination of the self-consistent magnetic fields.

## 8.7 Combined $\mathbf{V}_{\nabla B}$ and $\mathbf{V}_\kappa$ drifts in a vacuum magnetic field

In a steady state fusion plasma with $\mathbf{E} = 0$, an inhomogeneous, curved magnetic field produces two guiding center drifts – the $\nabla B$ drift and the curvature drift. For the special situation where the plasma currents are small, the magnetic field becomes approximately a vacuum magnetic field and a simplifying relationship exists between $\mathbf{V}_{\nabla B}$ and $\mathbf{V}_\kappa$. The goal of this section is to derive this relationship. It is shown that for vacuum fields $\mathbf{V}_{\nabla B}$ and $\mathbf{V}_\kappa$ are both in the same direction, implying that there is no way for their resulting currents to cancel.

The derivation follows from the well-known vector identity

$$\nabla(\mathbf{B} \cdot \mathbf{B}) = 2\mathbf{B} \times (\nabla \times \mathbf{B}) + 2\mathbf{B} \cdot \nabla\mathbf{B}. \tag{8.54}$$

For a vacuum magnetic field $\nabla \times \mathbf{B} = 0$. One now forms the cross product of Eq. (8.54) with $\mathbf{b}$. A short calculation yields

$$\mathbf{B} \times \nabla B = B\mathbf{b} \times [\mathbf{b}(\mathbf{b} \cdot \nabla B) + B\mathbf{b} \cdot \nabla\mathbf{b}] = -B\frac{\mathbf{B} \times \mathbf{R}_c}{R_c^2}. \tag{8.55}$$

Using this relation in the expression for the $\nabla B$ drift (Eq. (8.39)) leads to a simple expression for $\mathbf{V}_{\nabla B} + \mathbf{V}_\kappa$:

$$\mathbf{V}_\kappa + \mathbf{V}_{\nabla B} = \mp\frac{1}{\omega_c}(v_\parallel^2 + \frac{v_\perp^2}{2})\frac{\mathbf{R}_c \times \mathbf{B}}{R_c^2 B}. \tag{8.56}$$

As stated, each drift is obviously in the same direction and hence the resulting currents cannot cancel. This leads to the following interesting question. If the guiding center currents always add, and if $\mathbf{V}_{\nabla B}$ and $\mathbf{V}_\kappa$ are the only current-producing guiding center drifts associated with an inhomogeneous, curved magnetic field, how then can $\mathbf{B}$ correspond to a vacuum field? The answer lies in the development of an additional macroscopic fluid like current, known as the "magnetization current", which exactly cancels the $\mathbf{V}_{\nabla B} + \mathbf{V}_\kappa$ contribution to $\mathbf{J}$. A discussion of the magnetization current is deferred until Chapter 10, where self-consistent fluid models are developed.

## 8.8 Motion in time varying E and B fields: the polarization drift

The next contribution to the theory of single-particle guiding center motion involves the effects of slow time varying electric and magnetic fields. Specifically, attention is focused on fields of the form $\mathbf{E}(\mathbf{r}, t) = E_x(\mathbf{r}_\perp, t)\mathbf{e}_x + E_y(\mathbf{r}_\perp, t)\mathbf{e}_y$ and $\mathbf{B} = B(\mathbf{r}_\perp, t)\mathbf{e}_z$. It is shown that the main consequences of the time variation are the development of a new guiding center drift known as the "polarization drift" and the identification of a new approximate constant of the motion known as the "adiabatic invariant."

The polarization drift arises from the effects of particle inertia in a time varying electric field. As $\mathbf{E}_\perp$ changes slowly in time the particle motion tracks the time evolution of the field, although lagging slightly behind because of particle inertia. The analysis demonstrates that the resulting polarization drift is in the direction of $\mathbf{E}_\perp$ (and not $\mathbf{E}_\perp \times \mathbf{B}$) and is larger for ions than electrons because of the heavier ion mass.

The adiabatic invariant predicts how the perpendicular energy of a charged particle evolves in time in the presence of a slowly varying magnetic field. It is shown that an increasing $\mathbf{B}$ field causes a corresponding increase in $v_\perp^2$. The invariant is not an exact constant of the motion in the sense that its value remains unchanged only after time averaging over the gyro motion.

The analysis is separated into two parts. In the first part the magnetic field is assumed to be uniform in space and time ($B$ = const.) and the electric field is assumed to vary only with time ($\mathbf{E} = E_x(t)\mathbf{e}_x + E_y(t)\mathbf{e}_y$). This simplified model captures the essential features of the polarization drift. The mathematical solution is obtained by a straightforward iteration procedure.

The second part of the analysis allows the magnetic field to also be a function of time. This slightly complicates the calculation because a time varying $\mathbf{B}$ field generates a spatially varying electric field in accordance with Faraday's law. These effects are treated by introducing a special mathematical time transformation into the analysis. Two results follow. First, there is a slight modification to the polarization drift. Second, the new approximate constant of the motion is derived. This constant is known as the adiabatic invariant $\mu$.

### 8.8.1  *The polarization drift for* $\mathbf{E}_\perp = E_x(t)\mathbf{e}_x + E_y(t)\mathbf{e}_y$ *and* $B$ = const.

This subsection focuses on the simple form of the fields given above. The mathematical analysis of the polarization drift is presented first, followed by a simple physical picture.

#### *Mathematical derivation*

For the fields under consideration the equations of motion for the perpendicular particle velocity are given by

$$dv_x/dt - \omega_c v_y = \omega_c E_x(t)/B,$$
$$dv_y/dt + \omega_c v_x = \omega_c E_y(t)/B. \tag{8.57}$$

A formal exact mathematical solution to these equations is readily obtainable for arbitrary $E_x$, $E_y$. However, the solutions are not very insightful since they involve a variety of complicated integrals. Insight can ultimately be obtained by making use of the slow time variation assumption, which then allows an approximate evaluation of the integrals.

For present purposes, it is more convenient mathematically to assume slow variation from the outset. With this assumption, one can obtain an accurate approximation to the solution by means of a straightforward iteration procedure. The basis for the procedure is the introduction of a small parameter that measures the slowness of the time variation. Specifically, the characteristic frequency $\omega$ associated with the time variation of the electric fields is assumed to be low compared to the gyro frequency: $|\dot{\mathbf{E}}_\perp|/|\mathbf{E}_\perp| \sim \omega \ll \omega_c$. The low-frequency assumption guarantees that each new term in the iteration is smaller by $\omega/\omega_c$ than the previous term.

The first step in the iteration procedure is to introduce a new velocity variable $\mathbf{v}'_\perp$ that subtracts out the $\mathbf{E} \times \mathbf{B}$ drift.

$$v_x = v'_x + E_y(t)/B,$$
$$v_y = v'_y - E_x(t)/B. \tag{8.58}$$

The equations of motion for $\mathbf{v}'_{\perp}$ become

$$\frac{dv'_x}{dt} - \omega_c v'_y = -\frac{1}{B}\frac{dE_y}{dt},$$

$$\frac{dv'_y}{dt} + \omega_c v'_x = \frac{1}{B}\frac{dE_x}{dt}. \tag{8.59}$$

Note that the right hand side of Eq. (8.59) is smaller by $\omega/\omega_c$ than the corresponding terms in Eq. (8.57).

The next step in the iteration is to treat the terms on the right hand side of Eq. (8.59) as a new "constant" (actually slowly varying) external force. In analogy with the $\mathbf{E} \times \mathbf{B}$ drift, these terms can be explicitly separated out from the solution by introducing a new velocity variable $\mathbf{v}''_{\perp}$ as follows:

$$v'_x = v''_x + \frac{1}{\omega_c B}\frac{dE_x}{dt},$$

$$v'_y = v''_y + \frac{1}{\omega_c B}\frac{dE_y}{dt}. \tag{8.60}$$

The equations for $\mathbf{v}''_{\perp}$ are now given by

$$\frac{dv''_x}{dt} - \omega_c v''_y = -\frac{1}{\omega_c B}\frac{d^2 E_x}{dt^2} \approx 0,$$

$$\frac{dv''_y}{dt} + \omega_c v''_x = -\frac{1}{\omega_c B}\frac{d^2 E_x}{dt^2} \approx 0. \tag{8.61}$$

The terms on the right hand side of Eq. (8.61) can be neglected since they involve the same components of electric field as the starting equations and are smaller by $(\omega/\omega_c)^2$. In principle, one could continue with the iteration procedure to higher and higher order, although it is obvious by construction that each new right hand side driving term is smaller by $\omega/\omega_c$ than the previous iteration. Once the higher order terms in Eq. (8.61) are neglected it is clear that the solution for $\mathbf{v}''_{\perp}$ is just the familiar gyro motion.

The conclusion from the analysis is that in a constant $B$ field with a slowly varying perpendicular electric field the combined orbit of the particle is accurately approximated by

$$\mathbf{v}_{\perp}(t) = \mathbf{v}_{gyro} + \frac{\mathbf{E}_{\perp} \times \mathbf{B}}{B^2} + \mathbf{V}_p, \tag{8.62}$$

where (with the upper sign corresponding to electrons)

$$\mathbf{V}_p = \mp\frac{1}{\omega_c B}\frac{d\mathbf{E}_{\perp}}{dt}. \tag{8.63}$$

Observe the following properties of the solution. The velocity consists mainly of gyro motion plus the instantaneous value of the $\mathbf{E} \times \mathbf{B}$ drift. This is what one might expect from a slowly varying electric field. There is, however, a small additional drift velocity $\mathbf{V}_p$ in the direction of the electric field and this is the polarization drift. It flows in opposite direction for electrons and ions (tending to cause a charge "polarization" in the direction

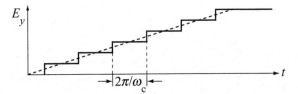

Figure 8.9 Dashed curve: linearly rising electric field that levels off after a period of time. Solid curve: step function model of the electric field evolution.

of the electric field) and is much larger for ions because of their heavier mass. In terms of its magnitude, the polarization drift is small compared to the $\mathbf{E} \times \mathbf{B}$ drift. In particular, $\mathbf{V}_p/\mathbf{V}_E \sim \omega/\omega_c \ll 1$. One might ask if $\mathbf{V}_p$ is small why keep it at all? The reason is that while it is small, it is still the first non-zero perpendicular drift in the direction of $\mathbf{E}_\perp$. There is clearly no contribution in this direction from the $\mathbf{E} \times \mathbf{B}$ drift.

The difference in direction is important. In terms of currents flowing in the direction of $\mathbf{E}_\perp$ it makes more sense to compare the polarization drift with the displacement current which also points in the same direction. This comparison is easily made by calculating

$$\mathbf{J}_p \approx qn\mathbf{V}_{pi} = \frac{nm_i}{B^2}\frac{d\mathbf{E}_\perp}{dt},$$
$$\mathbf{J}_d = \varepsilon_0\frac{\partial\mathbf{E}_\perp}{\partial t}. \tag{8.64}$$

The ratio of polarization to displacement currents is thus given by

$$\frac{\mathbf{J}_p}{\mathbf{J}_d} = \frac{c^2}{v_A^2}, \tag{8.65}$$

where $v_A = (B^2/\mu_0 nm_i)^{1/2}$ is known as the Alfvén speed. For typical reactor parameters, this ratio is about $3 \times 10^3 \gg 1$. In the comparison, the polarization current is dominant.

### A physical picture

The physical origin of the polarization drift is associated with the inertia of the particles. To understand how the drift arises consider the motion of a positively charged particle in a constant $B$ field and a linearly time varying $E_y$ as shown in Fig. 8.9. Now, for simplicity, approximate the time behavior of the electric field as a series of increasing steps with the duration of each step corresponding to one gyro period.

A qualitative picture of the orbit under the action of these fields is illustrated in Fig. 8.10. The dashed curve is a reference circular gyro orbit with no electric field. The solid curve is the orbit during the first step of the electric field. Note that in addition to the $\mathbf{E} \times \mathbf{B}$ shift of the guiding center to the right, both the top and bottom points of the trajectory (i.e., points 1 and 2), are shifted slightly upward because of the different average gyro radius size in the upper and lower portions of the orbit. This difference in gyro radius is associated with inertia which causes the particle motion to lag behind the changing electric field.

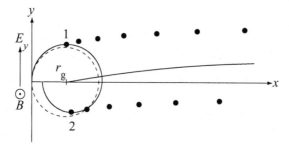

Figure 8.10 Locus of the maximum, minimum, and guiding center location of the particle orbit for the step model of the electric field.

During the second step the process repeats itself with the following modifications due to the larger value of the electric field. The $\mathbf{E} \times \mathbf{B}$ shift of the guiding center to the right is slightly larger. Similarly, the upward shifts of point 1 and point 2 are also both larger. Figure 8.10 plots the envelopes of point 1, point 2 and $\mathbf{r}_g$ for consecutive steps in the electric field. There is clearly a drift of the guiding center in the $y$ direction as long as the electric field is varying in time. This is the polarization drift. Once the electric field levels off the polarization drift vanishes and all that remains is a constant $\mathbf{E} \times \mathbf{B}$ drift.

### 8.8.2 The polarization drift for $\mathbf{E}_\perp = E_x(\mathbf{r}_\perp, t)\mathbf{e}_x + E_y(\mathbf{r}_\perp, t)\mathbf{e}_y$ and $\mathbf{B} = B(t)\mathbf{e}_z$

In this subsection the analysis of the polarization drift is generalized to include the effect of a time varying magnetic field. A further result is the identification of the adiabatic invariant $\mu$ as an approximate constant of the motion.

Note that for simplicity the perpendicular spatial dependence of $B$ is ignored as these effects have already been investigated. Even so, a time varying magnetic field complicates the analysis by requiring a time and spatially varying electric field because of Faraday's law. These effects are treated by means of a mathematical transformation of the time variable which greatly simplifies the analysis.

#### Mathematical analysis

The calculation begins by assuming that the perpendicular electric field is of the form $\mathbf{E}_\perp = E_y(x, t)\mathbf{e}_y$, a simplification that helps to keep the algebra tractable but still captures the essential physics under consideration. With some straightforward additional work the calculation can be easily generalized to the case $\mathbf{E}_\perp = E_x(\mathbf{r}_\perp, t)\mathbf{e}_x + E_y(\mathbf{r}_\perp, t)\mathbf{e}_y$. The starting model corresponds to the equations of motion with the electric field expanded about the guiding center of the particle:

$$\frac{dv_x}{dt} - \omega_c v_y = 0,$$

$$\frac{dv_y}{dt} + \omega_c v_x = \frac{\omega_c}{B}\left[E_y + \frac{\partial E_y}{\partial x_g}(x - x_g)\right]. \tag{8.66}$$

Here, $\omega_c(t) = q B(t)/m$ and all the electric field terms are functions of $(x_g, t)$. Note that even the pure gyro motion is difficult to calculate in the present form of the equations because of the time dependent gyro frequency. The equations are greatly simplified by introducing a new time variable $\tau$ defined by

$$\tau = \int_0^t \omega_c(t')dt', \tag{8.67}$$

implying that $d\tau = \omega_c dt$. Under this transformation the model reduces to

$$\frac{dv_x}{d\tau} - v_y = 0,$$

$$\frac{dv_y}{d\tau} + v_x = \frac{1}{B}\left[E_y + \frac{\partial E_y}{\partial x_g}(x - x_g)\right],$$

$$\frac{dx}{d\tau} = \frac{v_x}{\omega_c}, \tag{8.68}$$

$$\frac{dy}{d\tau} = \frac{v_y}{\omega_c}.$$

An accurate approximate solution to these equations can be obtained by introducing the iteration procedure of the previous subsection and rewriting $v_x$, $v_y$ in terms of cylindrical velocity coordinates:

$$v_x = v_\perp(\tau)\cos[\tau + \varepsilon(\tau)] + \frac{E_y}{B},$$

$$v_y = -v_\perp(\tau)\sin[\tau + \varepsilon(\tau)] + \frac{d}{d\tau}\left(\frac{E_y}{B}\right). \tag{8.69}$$

The variables $v_x$, $v_y$ have been replaced by new unknowns $v_\perp(\tau)$, $\varepsilon(\tau)$. Both the amplitude and phase of the gyro motion are assumed to be functions of time and, in fact, they turn out to be slowing varying functions of time. The form given by Eq. (8.69) already demonstrates the slight modification to the polarization drift in which the $B$ field must be included in the time derivative. The remainder of the analysis focuses on solving for $v_\perp(\tau)$ leading to the identification of the new approximate constant of the motion. The solution for $\varepsilon(\tau)$ can also be easily found but no new important information is contained therein and hence the corresponding analysis is suppressed.

To find the solution for $v_\perp(\tau)$ one additional step is required before substituting Eq. (8.69) into Eq. (8.68). An expression is required for $x - x_g$ in the velocity equations. Since this expression appears only in the small, expanded term, the leading order gyro motion contribution is all that is required. From the second two trajectory equations in Eq. (8.68) one finds that

$$x - x_g \approx \frac{v_\perp(\tau)}{\omega_c(\tau)}\sin[\tau + \varepsilon(\tau)]. \tag{8.70}$$

Equations (8.69) and (8.70) are now substituted into the velocity components of Eq. (8.68). The resulting two equations can easily be solved simultaneously for $dv_\perp/d\tau$ and $d\varepsilon/d\tau$. A short calculation yields the desired equation for $v_\perp(\tau)$.

$$\frac{dv_\perp}{d\tau} + \frac{v_\perp}{2\omega_c B}\left[\frac{\partial E_y}{\partial x_g} + \frac{\partial E_y}{\partial x_g}\cos 2(\tau + \varepsilon)\right] = \frac{d^2}{d\tau^2}\left(\frac{E_y}{B}\right)\sin(\tau + \varepsilon) \approx 0. \quad (8.71)$$

As in the previous subsection the term on the right hand side is a higher order iteration correction and can be neglected.

The next step is to simplify Eq. (8.71) by using Faraday's law to replace $\partial E_y/\partial x_g = -dB/dt = -\omega_c dB/d\tau$. Equation (8.71) reduces to

$$\frac{1}{\mu}\frac{d\mu}{d\tau} = \frac{1}{B}\frac{dB}{d\tau}\cos 2(\tau + \varepsilon), \quad (8.72)$$

where

$$\mu \equiv mv_\perp^2/2B \quad (8.73)$$

is known as the magnetic moment (for reasons to be discussed shortly). This expression can be further simplified. Upon integrating Eq. (8.72) over one gyro period ($\tau_0 \le \tau + \varepsilon \le \tau_0 + 2\pi$), one finds that the right hand side almost exactly averages to zero, except for a very small, negligible correction of order $(\omega/\omega_c)^2$. Thus, to a very high degree of accuracy it follows that $\langle d\ln\mu/d\tau\rangle = 0$. The implication is that $\mu$ is a constant of the motion when averaged over one gyro period.

$$\mu = \frac{mv_\perp^2(t)}{2B(t)} = \text{const.} \quad (8.74)$$

### Significance of $\mu$

The quantity $\mu$ is known as the first adiabatic invariant and is equal to the gyro-averaged magnetic moment of the charged particle. This can be easily seen by recalling that the usual definition of the magnetic moment is $\mu = IA$, where $I$ is the current flowing in a circular loop and $A$ is the area of the loop. For a particle gyrating in a magnetic field the current averaged over one gyro period is given by $I = q/\tau_c = q\omega_c/2\pi$, while the area is given by $A = \pi r_L^2 = \pi(mv_\perp/qB)^2$. Since the product $IA = mv_\perp^2/2B$, the quantity $\mu$ is indeed the magnetic moment.

The fact that $\mu$ is constant when averaged over a gyro period can be interpreted as follows. The magnetic flux enclosed by a particle over one gyro orbit is just $\psi = \pi r_L^2 B = (2\pi m/q^2)\mu \sim \mu$. Therefore, as the $B$ field changes slowly in time the perpendicular velocity and corresponding gyro radius also change slowly in time in such a way that the flux contained within the orbit is a constant.

## Summary of generalized results

A charged particle moving in time varying electric and magnetic fields experiences an additional guiding center drift known as the polarization drift. This drift, for $\mathbf{B} = B(t)\mathbf{e}_z$ and the general case $\mathbf{E}_\perp = E_x(\mathbf{r}_\perp, t)\mathbf{e}_x + E_y(\mathbf{r}_\perp, t)\mathbf{e}_y$, follows from Eq. (8.69) and is given by (top sign for electrons)

$$\mathbf{V}_p = \mp \frac{1}{\omega_c} \frac{d}{dt} \left( \frac{\mathbf{E}_\perp}{B} \right). \tag{8.75}$$

The second new result is the identification of an approximate constant of the motion known as the adiabatic invariant. It is only "approximately" a constant since the derivation requires averaging over a gyro period assuming that the magnetic field is varying slowly (i.e., adiabatically). The adiabatic invariant is given by

$$\mu = \frac{mv_\perp^2(t)}{2B(t)} = \text{const.} \tag{8.76}$$

In terms of fusion applications the polarization drift plays an important role in setting the time scale for macroscopic plasma instabilities. As is shown in Chapter 12 which describes macroscopic macroscopic equilibrium and stability, the time scale associated with the polarization drift is very fast compared to experimental times. If a given magnetic configuration is unstable the plasma is rapidly lost to the wall because of the fast time scale. The conclusion is that for fusion the magnetic configurations must be designed to avoid such instabilities.

The adiabatic invariant plays an important role in two different ways. First it is the basis for a magnetic confinement configuration known as the "mirror machine," which is discussed shortly. Second, the magnetic moment plays an important role in many toroidal magnetic geometries leading to a surprisingly enhanced collisional transport of energy and particles across the magnetic field. This behavior is known as "neoclassical transport theory" and is discussed in Chapter 14. While both of these applications depend on the adiabatic invariant, they are more connected to the result that $\mu$ is a constant in slow spatially varying magnetic fields as opposed to slow time varying fields. This spatial result has not as yet been demonstrated but is a major topic in the next section.

## 8.9 Motion in fields with parallel gradients: the magnetic moment and mirroring

The last topic concerning guiding center motion involves the effect of a parallel gradient in the magnetic field, which can arise in configurations such as those illustrated in Fig. 8.11. Two important results are obtained in the limit where the gyro radius is small compared to the spatial gradient length of the field. First, the quantity $\mu = mv_\perp^2/2B$ is again shown to be an adiabatic invariant. Second, a gyro-averaged force develops parallel to the magnetic field gradient which can have a large impact on the parallel guiding center motion. This force gives rise to the "mirror" effect and provides the basis for one of the earliest

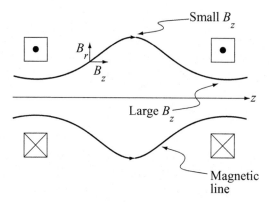

Figure 8.11 Coil configuration giving rise to a parallel gradient in $B$.

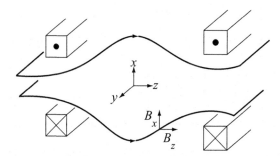

Figure 8.12 Slab model of a magnetic geometry with a parallel gradient.

fusion experiments. The mirror effect and the simple mirror machine are also discussed in this section.

In terms of the mathematics, parallel magnetic field gradients complicate the analysis because the geometry is inherently 2-D. For example, even in the simplest case where $B_z = B_z(z)$, there must be an additional transverse component of $\mathbf{B}$ in order to satisfy $\nabla \cdot \mathbf{B} = 0$. For mathematical tractability in the analysis these transverse components are chosen to satisfy the "long-thin" approximation which assumes that the parallel gradient length is large compared to the transverse gradient length. Although not essential, this approximation greatly simplifies the calculation while still capturing the essential physics.

### 8.9.1 The mathematical formulation

Consider first the prescribed fields. The electric field is assumed to be zero: $\mathbf{E} = 0$. The magnetic field geometry, for simplicity, is taken to be a slab version of the cylindrical configuration in Fig. 8.11. The slab model is illustrated in Fig. 8.12. The magnetic field is static in time and has the following non-zero components: $\mathbf{B} = B_x(x, z)\mathbf{e}_x + B_z(x, z)\mathbf{e}_z$. For the moment no long–thin approximation is made, implying that $B_x \sim B_z$. Under these assumptions the equations of motion for the particle velocities become

$$\frac{dv_x}{dt} = \frac{q}{m} v_y B_z,$$

$$\frac{dv_y}{dt} = -\frac{q}{m}(v_x B_z - v_z B_x), \qquad (8.77)$$

$$\frac{dv_z}{dt} = -\frac{q}{m} v_y B_x.$$

As in earlier calculations these equations will be solved by expansion techniques. A potential difficulty that arises is that the simple coordinate $z$ no longer corresponds to the parallel direction. Thus, while $v_z$ may be nearly the parallel velocity and $v_x, v_y$ the perpendicular velocities they are not exactly so and these deviations introduce a number of small additional terms that compete with the small gyro radius corrections. These geometric complications can be eliminated at the outset by the introduction of a set of three orthogonal unit vectors and corresponding velocity components that exactly distinguish between the perpendicular and parallel directions. The new unit vectors $\mathbf{e}_1, \mathbf{e}_2, \mathbf{b}$, and their inverse relations are given by

$$
\begin{aligned}
\mathbf{b} &= b_x \mathbf{e}_x + b_z \mathbf{e}_z, & \mathbf{e}_z &= b_z \mathbf{b} - b_x \mathbf{e}_1, \\
\mathbf{e}_2 &= \mathbf{e}_y, & \mathbf{e}_y &= \mathbf{e}_2, \qquad (8.78) \\
\mathbf{e}_1 &= \mathbf{e}_2 \times \mathbf{b} = b_z \mathbf{e}_x - b_x \mathbf{e}_z, & \mathbf{e}_x &= b_z \mathbf{e}_1 + b_x \mathbf{b},
\end{aligned}
$$

where $b_x = B_x/B$ and $b_z = B_z/B$. Observe that $\mathbf{b}$ points along the magnetic field while $\mathbf{e}_1, \mathbf{e}_2$ are exactly perpendicular to $\mathbf{B}$. The corresponding velocity components $v_1, v_2, v_\parallel$ and their inverses can now be written as

$$
\begin{aligned}
v_\parallel &= b_z v_z + b_x v_x, & v_z &= b_z v_\parallel - b_x v_1, \\
v_2 &= v_y, & v_y &= v_2, \qquad (8.79) \\
v_1 &= b_z v_x - b_x v_z, & v_x &= b_x v_\parallel + b_z v_1.
\end{aligned}
$$

Using these transformations, one can show after a short calculation that the equations of motion are substantially simplified and can be rewritten as follows:

$$\frac{dv_1}{dt} - \omega_c v_2 = K v_\parallel,$$

$$\frac{dv_2}{dt} + \omega_c v_1 = 0, \qquad (8.80)$$

$$\frac{dv_\parallel}{dt} = -K v_1,$$

where $\omega_c = q B/m$, $B = (B_x^2 + B_z^2)^{1/2}$ and

$$K = K[x(t), z(t)] = b_x \frac{db_z}{dt} - b_z \frac{db_x}{dt}. \qquad (8.81)$$

The equations are now in the desired form.

### 8.9.2 Solution to the equations

The mathematical solution to the problem requires two steps. First a new time variable is introduced, similar to the transformation used in the generalized polarization drift analysis. Second, an explicit model is introduced for the magnetic field enabling the introduction of the long–thin approximation.

The analysis begins with the time transformation which is given by

$$\tau = \int_0^t \omega_c dt \tag{8.82}$$

with $\omega_c(t) = \omega_c[x(t), z(t)]$. Note that this transformation is formally identical to the one used for the polarization drift (i.e., Eq. (8.67)). However, it is inherently implicit in nature since $x(t), z(t)$ are unknown functions. Even so, as is shown, this does not lead to any difficulties in the analysis. Substituting the transformation into the equations of motion yields

$$dv_1/d\tau - v_2 = \hat{K} v_{\|},$$
$$dv_2/d\tau + v_1 = 0, \tag{8.83}$$
$$dv_{\|}/d\tau = -\hat{K} v_1,$$

with

$$\hat{K} = K/\omega_c = b_x db_z/d\tau - b_z db_x/d\tau. \tag{8.84}$$

The next step is to introduce an explicit model for the magnetic field. The simplest model containing a parallel field gradient has the form $B_z = B_z(z)$. Perpendicular gradients in $B_z$ have already been discussed, are not necessary for the present calculation, and are thus not included. The condition that $\nabla \cdot \mathbf{B} = 0$ requires the existence of a non-zero transverse magnetic field. For the slab geometry under consideration this implies a non-zero $B_x(x, z)$. A simple calculation then shows that the explicit magnetic field under consideration is given by

$$B_z = B_z(z),$$
$$B_x = -x dB_z/dz. \tag{8.85}$$

It is now straightforward to introduce the long–thin approximation into the model. The primary motivation for introducing the approximation is to obtain a simplified expression for $\hat{K}$. One assumes that the transverse scale of the configuration is characterized by $x \sim a$ and that the parallel gradient length is defined by $B'_z/B_z \sim 1/L$. The long–thin approximation requires that $a/L \ll 1$ and implies that $B_x/B_z \sim a/L \ll 1$.

After a short calculation one can show that substitution of the model magnetic field and the long–thin approximation results in the following leading order contribution to $\hat{K}$:

$$\hat{K} = b_x \frac{db_z}{d\tau} - b_z \frac{db_x}{d\tau} = \frac{B_z^2}{B^2} \frac{d}{d\tau} \left( \frac{x}{B_z} \frac{dB_z}{dz} \right) \approx \frac{dx}{d\tau} \left( \frac{1}{B_z} \frac{dB_z}{dz} \right)$$
$$\approx \frac{v_1}{\omega_c B_z} \frac{dB_z}{dz} \approx \frac{v_1}{v_{\|} B_z} \frac{dB_z}{d\tau}. \tag{8.86}$$

In the last two expressions $z$ and $\tau$ are used interchangeably as independent variables by the one-to-one implicit relationship $dz = (v_z/\omega_c)d\tau \approx (v_\parallel/\omega_c)d\tau$. Note that there are many more terms contributing to Eq. (8.86) but they are all smaller by at least $a/L$ or $r_L/a$.

It is now straightforward to solve the equations. Consider first, the adiabatic invariant. As in the analysis of the generalized polarization drift it is useful to introduce cylindrical velocity variables with slowly varying coefficients:

$$
\begin{aligned}
v_1 &= v_\perp(\tau)\cos[\tau + \varepsilon(\tau)], \\
v_2 &= -v_\perp(\tau)\sin[\tau + \varepsilon(\tau)].
\end{aligned}
\tag{8.87}
$$

One substitutes into the perpendicular components of the equations of motion obtaining a set of simultaneous equations for $\dot{v}_\perp$ and $\dot{\varepsilon}$. The unknown $\dot{\varepsilon}$ can easily be eliminated yielding the following equation for $\dot{v}_\perp$:

$$
\frac{dv_\perp}{d\tau} = \frac{v_\perp}{2B_z}\frac{dB_z}{d\tau}[1 + \cos 2(\tau + \varepsilon)],
\tag{8.88}
$$

which can straightforwardly be rewritten as

$$
\frac{1}{\mu}\frac{d\mu}{d\tau} = \left(\frac{1}{B_z}\frac{dB_z}{d\tau}\right)\cos 2(\tau + \varepsilon).
\tag{8.89}
$$

After averaging over a gyro period one again finds that

$$
\mu = \frac{mv_\perp^2(z)}{2B(z)} = \text{const.}
\tag{8.90}
$$

The quantity $\mu$ is an adiabatic invariant, although in this case for a slow spatially rather than time varying magnetic field.

The second part of the mathematical solution involves the parallel component of the equations of motion which in the long–thin approximation reduces to

$$
\frac{dv_\parallel}{d\tau} = -\frac{v_1^2}{\omega_c B_z}\frac{dB_z}{dz} = -\frac{v_\perp^2}{2\omega_c B_z}\frac{dB_z}{dz}[1 + \cos(\tau + \varepsilon)].
\tag{8.91}
$$

After averaging over the gyro motion and converting back to the real time independent variable, one can rewrite this expression as

$$
m\frac{dv_\parallel}{dt} = -\mu\frac{dB_z}{dz} = -\mu\nabla_\parallel B.
\tag{8.92}
$$

Observe that there is a gyro-averaged force acting on the parallel guiding center motion of the particle. The force is driven by the parallel gradient in the magnetic field. Two forms are given for the force. The first is the direct result of the calculation, while the second is a generalization that does not make use of the long–thin approximation.

At this point one might think that a paradox has arisen. It has been shown in Section 8.2 that the parallel magnetic force acting on a charged particle is exactly and instantaneously zero. How then can there be an average force parallel to the magnetic field as derived above in Eq. (8.92)? The answer is subtle and can be understood by examining Fig. 8.13, which shows a particle with perpendicular velocity $v_\perp$ gyrating around a magnetic line in a field with a parallel gradient. The key point is that when $v_\perp \neq 0$, the particle has a finite gyro

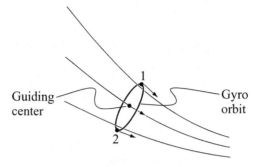

Figure 8.13 Comparison of field directions between the guiding center and the actual particle trajectory (point 1 steeper and point 2 shallower).

radius which produces a small excursion of the orbit (i.e., gyro motion) perpendicular to the guiding center trajectory. Observe that as the particle gyrates, the top of the orbit (point 1) lies on a magnetic line that is slightly steeper than the magnetic line of the guiding center. Similarly, at the bottom of the orbit (point 2) the particle lies on a shallower magnetic line. To leading order the steepness and shallowness average out and the average parallel motion of the particle is parallel to the guiding center. However, to first order the cancellation is not perfect and there is a small correction leading to the "parallel" force given by Eq. (8.92).

The resolution of the paradox can thus be summarized as follows. In a magnetic field with a parallel gradient there is indeed an average parallel force acting on the guiding center motion of the particle. It should be emphasized that the force acts at the guiding center and not the instantaneous position of the particle. Furthermore, the direction of the field at the guiding center is slightly different from the average direction of the actual field experienced by the particle as it gyrates along its orbit. In other words, the field at the guiding center is not exactly parallel to the actual average field experienced by the particle. Therefore, while the guiding center motion feels a parallel force along the gradient, this force is actually in the perpendicular direction when viewed in terms of the instantaneous position of the particle.

In conclusion a parallel magnetic field gradient produces a force that acts on the parallel guiding center motion of the particle. This force produces an important mirroring effect on the particles which is the topic of the next subsection.

### 8.9.3 The mirror effect and the mirror machine

The combination of $\mu = $ const. and $F_\parallel = -\mu \nabla_\parallel B$ can have a dramatic impact on the parallel motion of the guiding center. In particular, the direction of the parallel motion can be completely reversed so that a particle moving to the right along a given field line at a certain instant of time can be moving to the left a short time later. In fact there is a critical

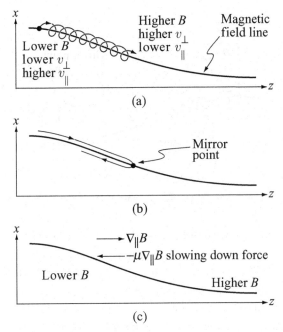

Figure 8.14 The mirror effect (a) as a particle moves into a region of higher $B$, $v_\perp$ increases and $v_\parallel$ decreases; (b) parallel guiding center velocity reflected at the mirror point where $v_\parallel = 0$; (c) the parallel guiding center force.

*A qualitative picture of the mirror effect*

point along the trajectory where the particle is reflected. Not surprisingly, this point is called the "mirror point" and the whole reversal process, the "mirror effect."

The phenomenon of mirroring can be understood qualitatively by examining Fig. 8.14. The trajectory of a particle moving to the right into a region of higher magnetic field is shown in Fig. 8.14(a). The particle starts off in a region of lower field with a certain value of $v_\perp$ and $v_\parallel$. As the particle gyrates and moves parallel to **B** into the high-field region, the value of $B$ along the guiding center increases. Since $\mu = mv_\perp^2/2B = $ const. this implies that $v_\perp$ must also increase. Next, recall that in a static magnetic field the kinetic energy of a particle is an exact constant of the motion: $E = m(v_\perp^2 + v_\parallel^2)/2 = $ const. Consequently, an increase in $v_\perp$ must be accompanied by a decrease in $v_\parallel$. If the increase in $B$ is sufficiently large, the particle eventually reaches a point along its trajectory where $v_\parallel = 0$. This is the reflection point as shown in Fig. 8.14(b). Once reflected, the parallel velocity of the particle reverses direction and the guiding center motion starts moving to the left. The force causing this behavior of the parallel motion is just $F_\parallel = -\mu \nabla_\parallel B$. As can be seen in Fig. 8.14(c) it acts to slow down parallel guiding center motion as a particle enters a high-field region.

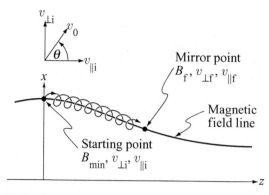

Figure 8.15 Conditions for reflecting a particle at the mirror point $B = B_f$.

### *The quantitative conditions for mirroring*

The conditions for mirroring to occur can easily be quantified using the constants of the motion $\mu$ and $E$. The goal of the calculation is to determine the relation between $v_\perp$ and $v_\parallel$ necessary to reflect a particle at a given point along the parallel field gradient. To begin, consider a particle moving in a mirror field as illustrated in Fig. 8.15. Assume the particle starts initially at the midplane, where the magnetic field is weakest. At this point $B = B_{min}$, $v_\perp = v_{\perp i}$, $v_\parallel = v_{\parallel i}$. The corresponding magnetic moment and energy are given by $\mu = m v_{\perp i}^2/2B_{min}$ and $E = m(v_{\perp i}^2 + v_{\parallel i}^2)/2$.

Assume now that the particle moves to the right and is reflected at the point where $B = B_f > B_{min}$. At this point $v_\perp = v_{\perp f}$ and by definition of the reflection point $v_\parallel = v_{\parallel f} = 0$. The corresponding energy and magnetic moment then have the values $E = m v_{\perp f}^2/2$ and $\mu = m v_{\perp f}^2/2B_f$.

The reflection condition can now be easily calculated by equating the initial and final values of $E$ and $\mu$. To proceed it is convenient to define a normalized energy $E = m v_0^2/2$. The initial velocity can then be expressed in terms of a pitch angle $\theta$ as follows (see Fig. 8.15):

$$\begin{aligned} v_{\perp i} &= v_0 \sin\theta, \\ v_{\parallel i} &= v_0 \cos\theta. \end{aligned} \tag{8.93}$$

Conservation of energy clearly implies that

$$v_{\perp f}^2 = v_{\perp i}^2 + v_{\parallel i}^2 = v_0^2. \tag{8.94}$$

Next conservation of $\mu$ is applied leading to

$$\frac{v_{\perp i}^2}{B_{min}} = \frac{v_{\perp f}^2}{B_f}, \tag{8.95}$$

which simplifies to

$$\sin^2\theta_c = \frac{B_{min}}{B_f}. \tag{8.96}$$

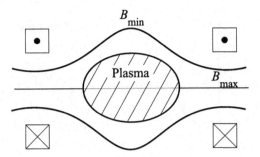

Figure 8.16 Geometry of the simple mirror machine.

Here, $\theta = \theta_c$ is the critical pitch angle for mirror reflection at the point where $B = B_f$. A particle with a higher initial perpendicular velocity, corresponding to a pitch angle $\theta > \theta_c$, will be reflected sooner. Conversely, a particle with a smaller initial perpendicular velocity, $\theta < \theta_c$, will pass the point where $B = B_f$ and may or may not be reflected later, depending upon how large the magnetic field becomes.

In summary, the analysis has shown that for a given parallel gradient in the magnetic field, it is easier to reflect particles with a large pitch angle (i.e., high perpendicular and low parallel initial velocities).

### The simple mirror machine

The mirror effect just described forms the basis for one of the earliest magnetic fusion configurations, appropriately known as the "mirror machine." Its simplest form is illustrated in Fig. 8.16. Two coils with current flowing in the same direction create a magnetic field with a maximum just under each coil and a local minimum midway between. Assume now that plasma initially fills the volume between the coils. Using the guiding center theory of the mirror effect one wishes to determine which, if any, particles remain confined in the prescribed magnetic geometry. Within the context of the theory, it is shown that a large fraction of the particles remain confined, and this fact provided the motivation for the early consideration of the mirror machine as a fusion device.

The analysis is straightforward. Particles with a sufficiently large pitch angle (i.e., large $v_\perp/v_\parallel$) at the center of the configuration where $B = B_{min}$ reflect off the mirror point somewhere along the gradient where $B = B_f$. The particle with the smallest initial pitch angle that is still reflected is the one that is reflected at the mirror throat where $B = B_{max}$. The corresponding critical pitch angle is given by

$$\sin^2 \theta_c = \frac{B_{min}}{B_{max}} \equiv \frac{1}{R}. \tag{8.97}$$

Here, $R = B_{max}/B_{min} > 1$ is defined as the mirror ratio. Particles with a pitch angle $\theta > \theta_c$ (i.e., a high $v_\perp$) will be reflected sooner, before reaching the mirror throat. These particles then reverse direction and reflect off the opposite mirror. In this way, the particles remain

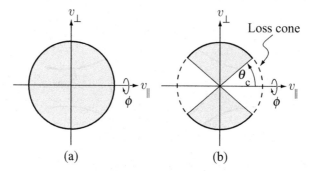

Figure 8.17 Velocity phase space showing: (a) a full, isotropic Maxwellian and (b) a Maxwellian with a loss cone.

confined indefinitely, continually bouncing between mirror reflection points. In contrast, particles with $\theta < \theta_c$ (i.e., a high $v_\parallel$) pass the mirror throat without being reflected. They are quickly lost to the first wall.

This analysis shows that the subset of particles confined in a mirror machine is defined by the range of pitch angles

$$\theta_c \leq \theta \leq \pi - \theta_c. \tag{8.98}$$

Pitch angles outside this range form a "loss cone" in velocity space in which all the particles have been lost. The concept of the loss cone is illustrated in Fig. 8.17, which depicts the density of particles in $v_\perp$, $v_\parallel$, $\phi$ space. Figure 8.17(a) corresponds to an isotropic distribution function such as a Maxwellian, with no loss cone. The shaded region represents a sphere with a uniform distribution of particles. Figure 8.17(b) shows the effect of losing particles with a small pitch angle. A cone of particles is removed from opposite poles of a sphere and only the remaining shaded region contains mirror confined particles.

The fraction of confined particles $f$ of an initially Maxwellian distribution function $F_M(v)$ is equal to the ratio of the number of particles outside the loss cone divided by the total number of particles. This fraction is easily calculated in terms of the mirror ratio as follows:

$$f = \frac{\displaystyle\int_{\theta_c}^{\pi-\theta_c} \sin\theta \, d\theta \int_0^{2\pi} d\phi \int_0^{2\pi} v^2 F_M(v) dv}{\displaystyle\int_0^{\pi} \sin\theta \, d\theta \int_0^{2\pi} d\phi \int_0^{2\pi} v^2 F_M(v) dv}$$

$$= \frac{\displaystyle\int_{\theta_c}^{\pi-\theta_c} \sin\theta \, d\theta}{\displaystyle\int_0^{\pi} \sin\theta \, d\theta} = \left(1 - \frac{1}{R}\right)^{1/2}. \tag{8.99}$$

Observe that for a mirror ratio $R = 2$, about 70% of the particles are confined, quite a substantial fraction.

In practical experiments, the simple mirror machine did not work as well as predicted. Both macroscopic and microscopic instabilities were observed, leading to anomalously fast losses of particles. Careful analysis and several very clever ideas ultimately were able to mitigate these effects. However, there still remained one irreducible problem. Coulomb collisions scattered confined particles into the loss cone, after which they were immediately lost out of the ends of the device. The rate at which particles were lost was just too fast to achieve a favorable power balance in a mirror machine fusion reactor. This topic will be revisited in more detail after the discussion of Coulomb collisions in the next chapter.

## 8.10 Summary – putting all the pieces together

This chapter has described the motion of a charged particle in a prescribed set of smooth magnetic and electric fields. A wide choice of fields has been investigated allowing for perpendicular and parallel spatial gradients as well as time variation. A useful intuition has been developed by assuming that the spatial gradient length is long compared to the gyro radius and the characteristic frequency associated with the time variation is low compared to the gyro frequency. This is very often the situation of practical importance.

The analysis has shown that the perpendicular particle motion can be decomposed into two components: the fast gyro motion and the slower guiding center motion comprising primarily the guiding center drifts. In the parallel direction, only guiding center motion is important. Thus, the trajectory of a particle can be accurately approximated by

$$
\begin{aligned}
\mathbf{v}(t) &= \mathbf{v}_{\text{gyro}} + \mathbf{v}_g + v_{\parallel}\mathbf{b}, \\
\mathbf{r}(t) &= \mathbf{r}_{\text{gyro}} + \mathbf{r}_g + l\,\mathbf{b}.
\end{aligned}
\tag{8.100}
$$

The velocity and position are given in terms of the magnetic and electric fields, which are assumed to be of the form

$$
\begin{aligned}
\mathbf{B} &= B(\mathbf{r}, t)\mathbf{b}, \\
\mathbf{E} &= \mathbf{E}_{\perp}(\mathbf{r}, t) + E_{\parallel}(\mathbf{r}, t)\mathbf{b}.
\end{aligned}
\tag{8.101}
$$

The gyro motion, expressed in terms of a local perpendicular, rectangular coordinate system whose axis corresponds to the guiding center of the particle, is given by

$$
\begin{aligned}
\mathbf{v}_{\text{gyro}} &= v_{\perp} \cos \tau\, \mathbf{e}_x \pm v_{\perp} \sin \tau\, \mathbf{e}_y, \\
\mathbf{r}_{\text{gyro}} &= r_L \sin \tau\, \mathbf{e}_x \mp r_L \cos \tau\, \mathbf{e}_y.
\end{aligned}
\tag{8.102}
$$

where the upper sign here and below corresponds to electrons, $r_L = v_{\perp}/\omega_c$, $\omega_c = |q|B(\mathbf{r}_g, l, t)/m$, and

$$
\tau = \int_0^t \omega_c(\mathbf{r}_g, l, t)\mathrm{d}t.
\tag{8.103}
$$

In these expressions $v_\perp, \mathbf{r}_g, l$ are slowly varying functions of time determined from the solution of the guiding center trajectories.

The guiding center motion is described by a closed set of equations for the unknowns $v_\perp, \mathbf{v}_g, v_\parallel, \mathbf{r}_g, l$. Also, each guiding center particle is characterized by a magnetic moment $\mu$ as well as a charge $q$ and mass $m$, all of which are assumed to be known quantities. Consider first the perpendicular guiding center drift velocity, which comprises the following contributions:

$$\mathbf{v}_g = \mathbf{V}_E + \mathbf{V}_{\nabla B} + \mathbf{V}_\kappa + \mathbf{V}_p. \tag{8.104}$$

The individual drift velocities, expressed in terms of the local rectangular coordinate system $(\mathbf{r}_g = x_g \mathbf{e}_x + y_g \mathbf{e}_y)$ can be written as

$$\mathbf{V}_E = \frac{\mathbf{E}_\perp \times \mathbf{B}}{B^2} \quad \mathbf{E} \times \mathbf{B} \text{ drift,}$$

$$\mathbf{V}_{\nabla B} = \mp \frac{v_\perp^2}{2\omega_c} \frac{\mathbf{B} \times \nabla B}{B^2} \quad \nabla B \text{ drift,}$$

$$\mathbf{V}_\kappa = \mp \frac{v_\parallel^2}{\omega_c} \frac{\mathbf{R}_c \times \mathbf{B}}{R_c^2 B} \quad \text{curvature drift,} \tag{8.105}$$

$$\mathbf{V}_p = \mp \frac{1}{\omega_c} \mathbf{b} \times \frac{d\mathbf{V}_E}{dt} \quad \text{polarization drift,}$$

Here and below, all fields are evaluated at the guiding center.

The perpendicular velocity is expressed in terms of the adiabatic invariant

$$v_\perp^2 = 2\mu B / m, \tag{8.106}$$

while the parallel velocity is obtained by solving the differential equation

$$m \frac{dv_\parallel}{dt} = q E_\parallel - \mu \frac{\partial B}{\partial l}. \tag{8.107}$$

Finally, the guiding center position is obtained by solving

$$d\mathbf{r}_g / dt = \mathbf{v}_g,$$
$$dl / dt = v_\parallel. \tag{8.108}$$

Equations (8.104)–(8.108) form a closed set of coupled ordinary differential equations for determining the guiding center motion. Often when the fields are static or possess geometric symmetry one can solve the equations analytically. Qualitatively, the guiding center motion represents the gyro-averaged trajectory of the particle. In the perpendicular direction the motion consists of the combination of drifts given above. In the parallel direction the velocity is determined by: (1) the parallel gradient in the magnetic field coupled with the fact that $\mu$ is an adiabatic invariant, as well as (2) the parallel electric field if one exists. Focusing on the guiding center motion often provides a much better intuition of plasma behavior than examining the details of the exact particle trajectory. This intuition, as will be shown, is of great help in understanding the confinement of fusion plasmas.

## Bibliography

Single-particle motion in magnetic and electric fields has many applications including fusion, space plasma physics, and astrophysics. Various treatments appear in the literature, some simple and intuitive, others more formal and rigorous. Most treatments focus on deriving the guiding center motion of charged particles, ignoring collisions, and assuming slow variation in the length and time scale of the applied fields. Several references for additional reading are listed below.

Boyd, T. J. M. and Sanderson, J. J. (2003). *The Physics of Plasmas*. Cambridge, England: Cambridge University Press.

Chen, F. F. (1984). *Introduction to Plasma Physics and Controlled Fusion*, second edn. New York: Plenum Press.

Dolan, T. J. (1982). *Fusion Research*. New York: Pergamon Press.

Goldston, R. J. and Rutherford, P. H. (1995). *Introduction to Plasma Physics*. Bristol, England: Insititute of Physics Publishing.

Helander, P. and Sigmar, D. J. (2002). *Collisional Transport in Magnetized Plasmas*. Cambridge, England: Cambridge University Press.

Miyamoto, K. (2001). *Fundamentals of Plasma Physics and Controlled Fusion*, revised edn. Toki City: National Institute for Fusion Science.

Northrup, T. G. (1966). *Adiabatic Charged particle Motion* (Kunkel, W. B. editor). New York: McGraw Hill Book Company.

Stacey, W. M. (2005). *Fusion Plasma Physics*. Weinheim: Wiley-VCH.

Wesson, J. (2004). *Tokamaks*. third edn. Oxford: Oxford University Press.

## Problems

8.1 Consider a plasma with azimuthal symmetry: $\partial/\partial\theta = 0$. Express the fields in terms of a scalar potential $\phi(r, z, t)$ and vector potential $\mathbf{A}(r, z, t)$. Form the dot product of the single-particle momentum equation with the $\mathbf{e}_\theta$ vector. Show that the canonical angular momentum $p_\theta = mrv_\theta + q\psi$ is an exact constant of the motion. Here, $\psi = rA_\theta$.

8.2 This problem investigates several points arising in connection with the derivation of the $\nabla B$ drift. Specifically, the calculation in the text is generalized to a 2-D magnetic field and the consequences of the second harmonic terms appearing in the derivation are investigated.

(a) Consider a magnetic field of the form $\mathbf{B} = B(x, y)\mathbf{e}_z$. Taylor expand about the guiding center position in both the $x$ and the $y$ direction. Following the derivation in section 8.5.1 show that the general form of the $\nabla B$ drift is given by

$$\mathbf{V}_{\nabla B} = \mp \frac{v_\perp^2}{2\omega_c} \frac{\mathbf{B} \times \nabla B}{B^2}.$$

(b) Next, consider the contributions due to the second harmonic terms. Find the first order corrections to both the particle velocity and position by calculating a particular solution to the equations and then satisfying the initial conditions by an appropriate choice of homogeneous solution. Show that the modified trajectory remains circular but with a slightly different location for the guiding center and a slightly modified size for the gyro radius as given by the primed quantities below:

$$r_L'^2 = r_L^2 \left( 1 + \frac{\mathbf{v}_{\perp 0} \cdot \mathbf{B} \times \nabla B}{\omega_c B^2} \right),$$

$$\mathbf{r}_g' = \mathbf{r}_g - \frac{\mathbf{v}_{\perp 0} \times (\mathbf{v}_{\perp 0} \times \nabla B)}{2\omega_c^2 B},$$

$$\mathbf{v}_{\perp 0} = v_\perp (\mathbf{e}_x \cos \phi + \mathbf{e}_y \sin \phi).$$

Note: The algebra involved in part (b) is straightforward but somewhat tedious.

8.3 This problem involves calculating the *second* order corrections to the guiding center motion assuming a uniform magnetic field and an electric field with a perpendicular gradient. Of particular interest is the derivation of the second order "finite gyro radius" drift. Assume the fields are given by $\mathbf{B} = B\mathbf{e}_z$ with $B = \text{const.}$ and $\mathbf{E} = -\nabla \Phi(x, y)$. Expand the equations including all second order terms.

(a) Calculate the generalized corrections to the gyro frequency by assuming that

$$v_x = -\frac{1}{B}\frac{\partial \Phi}{\partial y_g} + v_\perp \cos \Omega t + a_1 \sin \Omega t + v_{2x}(t),$$

$$v_y = +\frac{1}{B}\frac{\partial \Phi}{\partial x_g} + c_1 \sin \Omega t + v_{2y}(t).$$

Note the implied special choice of initial conditions to make the problem slightly simpler. Find $a_1$ and $c_1$ and show that the generalized shift in gyro frequency, correct to second order in $r_L/a$, is given by

$$\Omega^2 \approx \omega_c^2 \mp \omega_c \frac{\nabla^2 \Phi}{B} + \frac{1}{B^2}\left[ \left(\frac{\partial^2 \Phi}{\partial x \partial y}\right)^2 - \left(\frac{\partial^2 \Phi}{\partial x^2}\right)\left(\frac{\partial^2 \Phi}{\partial y^2}\right) \right] + \cdots.$$

(b) Show that the dominant contribution to the second order velocity $\mathbf{v}_2(t)$ is the finite gyro radius drift. The total drift thus can be written as

$$\mathbf{V}_D = -\left( 1 + \frac{r_L^2}{4}\nabla^2 \right)\frac{\nabla \Phi \times \mathbf{B}}{B^2}.$$

8.4 A 1-D magnetic field with a reversal at the origin can be modeled in a slab geometry by $\mathbf{B} = B_0 \tanh(x/a)\mathbf{e}_z$ with $-\infty < x < \infty$.
  (a) Why are the guiding center formulas for the particle drifts derived in the text invalid?
  (b) Sketch the orbit of a proton with initial conditions $x(0) = y(0) = \dot{y}(0) = 0$ and $\dot{x}(0) = v_\perp$.
  (c) Expand about $x = 0$ and derive an expression for the turning point of the orbit $x_{\max}$. Show that $x_{\max} = Cr_L^\alpha a^\beta$, where $r_L = mv_\perp/eB_0$. Find $C, \alpha, \beta$.

8.5 A cylindrical plasma is immersed in a longitudinal magnetic field given by $\mathbf{B} = B_0[1 - \beta_0 \exp(-r^2/a^2)]\mathbf{e}_z$. For $B_0 = 6$ T, $\beta_0 = 0.75$, and $T_e = T_i = 1$ keV:
  (a) Calculate the electron and ion gyro frequency and average thermal gyro radius at $r = a$;
  (b) Calculate the magnitude and sign of the electron and ion $\nabla B$ drifts at $r = a$;

(c) Are the guiding center assumptions $r_{Le}/a \ll 1$, $r_{Li}/a \ll 1$, $V_{\nabla B}/v_\perp \ll 1$ satisfied?

(d) Calculate the direction and sign of the macroscopic current required to produce the dip in the $B_z$ field. Is this compatible with step (b)? Explain.

8.6 A plasma has a constant uniform magnetic field $\mathbf{B} = B_0 \mathbf{e}_z$. Superimposed is an electrostatic electric field of the form $\mathbf{E} = E_0 \cos(\omega t - kz)\mathbf{e}_z$, where $\omega$ and $k$ are known constants. Assume a positively charged particle is initially located at $z(0) = 0$ with a parallel velocity $v_z(0) = v_\parallel$. Show that for a sufficiently large value of $E_0$ the particle is trapped in the wave. Calculate the critical $E_0$.

8.7 This problem involves a generalization of the previous electrostatic trapping problem. Consider a positively charged particle acted upon by a magnetic field $\mathbf{B} = B_0 \cos(ky - \omega t)\mathbf{e}_x$.

(a) Prove that the electric field is given by

$$\mathbf{E} = -(\omega B_0/k)\cos(ky - \omega t)\mathbf{e}_z.$$

(b) The trajectory of the particle is defined as $\mathbf{r}(t) = x(t)\mathbf{e}_x + y(t)\mathbf{e}_y$. The initial position and velocity of the particle are as follows: $\dot{y}(0) = v_0$ and $y(0) = x(0) = \dot{x}(0) = 0$. Derive a pair of coupled differential equations for $x(t)$ and $y(t)$. One equation should be integrable with the result then substituted into the other equation. The final result should be a single, second order, differential equation involving only one dependent variable. The goal of this part of the problem is to derive this equation.

(c) Derive a relationship between $v_0, \omega, k, B_0$ that defines the boundary between trapped and un trapped particles.

8.8 The magnetic field due to an infinitely long wire carrying a current $I$ is given by $\mathbf{B} = (\mu_0 I/2\pi R)\mathbf{e}_\phi$, where $\phi$ is the toroidal angle.

(a) Explain why this configuration is not able to successfully confine individual electrons and ions in the $R, Z$ plane.

(b) As an extreme example calculate how long it would take for a 10 keV ion to escape from a toroidal chamber whose minor radius is $b = 0.1$ m if the particle is initially located at $R = R_0 = 100$ m, $Z = 0$.

8.9 This problem has a somewhat unintuitive answer. Consider the motion of a charged particle in combined magnetic and electric fields $\mathbf{B} = B_0 \mathbf{e}_z$ and $\mathbf{E} = -\nabla\phi$ with $\phi(x, y) = Kxy$. The goal is to find the exact orbit of the particle.

(a) Write down the exact equations of motion for the trajectory $x(t)$, $y(t)$ of a positively charged particle. These equations should have the form of two coupled second order ODEs. For convenience define $K = \varepsilon e B_0^2/m$, where $\varepsilon$ is an equivalent parameter representing the normalized electric field.

(b) Find the general solution to the equations. For simplicity assume $\varepsilon$, is small but finite. Describe the qualitative behavior of the orbit for large time.

8.10 A positive ion is situated in a uniform magnetic field $\mathbf{B} = B_0 \mathbf{e}_z$. A time varying, spatially uniform electric field is applied of the form $\mathbf{E} = E_0(1 - e^{-t/\tau})\mathbf{e}_x$.

(a) Calculate the exact perpendicular velocity of the particle for an ion with the following initial conditions: $v_x(0) = 0$ and $v_y(0) = v_\perp$.

(b) Calculate the guiding center velocity $\mathbf{v}_g(t')$ in the limit $\varepsilon \equiv 1/\omega_c\tau \ll 1$ by averaging over one gyro period as follows:

$$\mathbf{v}_g(t') = \frac{\omega_c}{2\pi} \int_{t'}^{t'+2\pi/\omega_c} \mathbf{v}(t)dt.$$

Are there any transient or steady state guiding center drifts in the $x$ or $y$ direction?

8.11 Draw a picture of the earth and its dipole magnetic field. Describe and calculate the orbit of an electron starting off at the equatorial plane with $v_\parallel \gg v_\perp$. Repeat for an electron with $v_\parallel \ll v_\perp$.

8.12 Consider a hollow cylindrical copper tube. Along the axis is a copper wire. A current $I$ flows in the wire and a low-frequency AC voltage is applied across the tube and the wire.

(a) Sketch the electric and magnetic fields as a function of $r$. For simplicity ignore the AC magnetic field.

(b) Describe and calculate the orbit of a typical electron and ion placed in this combined magnetic and electric field.

8.13 A positive ion is placed in a sheared magnetic field given by

$$\mathbf{B} = B_0[\mathbf{e}_z + (x/L)\mathbf{e}_y].$$

(a) Write down the exact equations of motion describing the orbit of the particle.

(b) Find a relation between $v_z(t)$ and $x(t)$ assuming the following initial conditions: $v_y(0) = v_z(0) = x(0) = y(0) = z(0) = 0$ and $v_x(0) = v_0$.

(c) Using this relation derive a single, second order ODE for $x(t)$.

(d) Calculate the $x$ location of the turning point of the orbit.

8.14 An ion in a cylindrical plasma column moves under the action of a combined magnetic field and electric potential given by $\mathbf{B} = B_0\mathbf{e}_z$ and $\phi = \phi_0(r/a)^2$. Assume that at $t = 0$ the particle passes through the origin $r(0) = 0$ with a velocity $\dot{r}(0) = (2T_i/m_i)^{1/2}$. Calculate and sketch the exact trajectory of the ion as a function of time for various positive values of the parameter $\alpha = \phi_0/a^2 B_0\omega_{ci}$. Can the radial extent of the orbit ever be much smaller than an ion gyro radius? Explain.

8.15 In a simple, azimuthally symmetric ($\partial/\partial\theta = 0$) mirror machine, the magnitude of the longitudinal magnetic field near the axis is approximately given by $B_z(r, z) \approx B_0(1 + z^2/L^2)$. Here $L$ is a constant and $z = 0$ is the reflection plane of symmetry.

(a) Evaluate the magnitude and direction of the curvature vector $\kappa = \mathbf{b} \cdot \nabla\mathbf{b}$ as a function of $r$, $z$ for small but finite values of $r$.

(b) A mirror trapped ion with total kinetic energy $mv^2/2$ is reflected at the point where $|B| = 2B_0$. Find the particle's $v_\parallel$ at the point $z = 0$, $r = r_0$.

(c) Calculate the magnitude and direction of the guiding center drift velocity at $z = 0$, $r = r_0$.

# 9

# Single-particle motion – Coulomb collisions

## 9.1 Introduction

Coulomb collisions are the next main topic in the study of single particle motion. The theory of Coulomb collisions is a basic building block in the understanding of transport processes in a fusion plasma. This understanding is important since the transport of energy and particles directly impacts the power balance in a fusion reactor. An overview of the topics covered in Chapter 9 and their relation to fusion energy is presented below.

An analysis of Coulomb collisions shows that there are two qualitatively different types of transport: velocity space transport and physical space transport. In velocity space transport, Coulomb collisions lead to a transfer of momentum and energy between particles in $v_\perp$, $v_\parallel$ space that tends to drive any initial distribution function towards a local Maxwellian. Generally there are no accompanying direct losses of energy or particles from the plasma (the mirror machine being an exception). In physical space transport, Coulomb collisions lead to a diffusion of energy and particles out of the plasma. These are direct losses affecting the plasma power balance, with energy transport almost always the more serious of the two.

A comparison of the two types of transport shows that velocity space transport almost always occurs on a much faster time scale. Typically, the transport time in velocity space is approximately $(r_L/a)^2$ shorter than for physical space. This chapter focuses on velocity space transport. The important issue of physical space transport is addressed in Chapter 14 after self-consistent plasma models have been developed.

Consider now some of the general properties of velocity space transport. An interesting feature resulting from the analysis is that one cannot define a simple, unique "Coulomb cross section" as was done for the nearly hard-sphere nuclear fusion collisions. The reason is associated with the long-range nature of the Coulomb potential (i.e., long compared to the size of a nucleus but still short with respect to inter-particle spacing). For example, if one wants to calculate the change in electron momentum due to collisions with ions, it is *not* correct to first calculate the total number of collisions per second by summing over different particle velocities and impact parameters and *then* multiplying by the change in momentum per collision. Because of the long-range nature of the Coulomb potential, the change in momentum per collision is itself a function of velocity and impact parameter and this dependence must be included in the summation. Assuming the summation has been carried out correctly, one could then, if desired, define an equivalent cross section, but this cross section would be different for different processes: momentum exchange, energy exchange, etc. The cross section is not unique. The usual procedure with Coulomb collisions is to define a set of collision frequencies (or equivalently collision times), one for each process.

A second property of interest is that two different types of collision frequencies are needed to describe velocity space transport. The first corresponds to a "test particle" collision frequency. In this case one focuses on a typical single test particle, for instance an electron, and asks how much directed momentum or energy does this particle lose from undergoing Coulomb collisions with all other electrons or ions in the plasma. This collision frequency is a function of the velocity $v$ of the test particle and the density $n$ and temperature $T$ of the target particles. Test particle collision frequencies are important in understanding the particle losses in a mirror machine, the energy transfer of high-energy ions, such as alphas, to the background plasma, and the interesting phenomenon of runaway electrons.

The second type of velocity space collision frequency corresponds to a "net exchange" collision frequency. Here, one is interested, for instance, in the net exchange of energy that occurs when electrons are hotter than ions, or the net exchange of momentum when electrons are flowing through ions. Note that since Coulomb collisions are elastic, the net exchange of energy or momentum for like particle collisions is always zero when summing over all particles. Net exchange collisions take place only between different species. The resulting collision frequencies are functions of $n$ and $T$ but not $v$ since summations are carried out over both test and target species.

Net exchange collisions are important in the development of self-consistent fluid models describing the behavior of fusion plasmas. Specific applications include the temperature equilibration between electrons and ions when one species is preferentially heated, and the friction felt by electrons as they are driven through the ions by an electric field. This last phenomenon gives rise to electrical resistivity and cross-field particle diffusion, which are discussed in Chapters 10 and 14.

Another interesting and general feature of the analysis is that when summing over different particle velocities and impact parameters, it is the distant, small-angle collisions that dominate in determining all velocity space collision frequencies. The long-range nature of the Coulomb potential shows that even though each distant collisional deflection angle is small, there are many, many more such collisions than near-encounter, large-angle deflections.

The specific goals of this chapter are to derive the collision frequencies for various plasma processes, and to apply these results to important fusion plasma phenomena. The analysis begins with a mathematical derivation of the scattering angle of a particle's trajectory due to a Coulomb collision as a function of the particle's velocity and impact parameter. This critical piece of knowledge is used to derive the "test particle" and "net exchange" collision frequencies for momentum and energy transfer. The results are applied to problems of fusion interest including a re-examination of the mirror concept, the heating of a plasma by high-energy alphas or a high-energy neutral beam, and the phenomenon of runaway electrons.

## 9.2 Coulomb collisions – mathematical derivation

### 9.2.1 Formulation of the problem

The goal of this section is to calculate the scattering angle of a particle's trajectory resulting from a Coulomb collision as a function of its velocity and impact parameter. Obtaining this critical information requires the solution of Newton's equations of motion for two particles interacting under the influence of the Coulomb potential.

Several approximations are implicit in the analysis that greatly simplify the calculation. First, the Coulomb collision takes place over such a short distance that the curvature of the orbit due to gyro motion can be neglected. In other words, it is not necessary to include the magnetic field in the derivation. Second, because of the long-range nature of the Coulomb potential one might think that it would be necessary to solve a many body problem which simultaneously takes into account the multiple Coulomb interactions from a large number of nearby particles. This is not the case although the reason is slightly complex. Although not obvious at the outset, it is shown that the dominant collisions affecting transport correspond to small-angle deflections. Since each such deflection represents a small perturbation to the trajectory, the combined result can be calculated as the superposition of two-body interactions. The effect of one perturbation on another perturbation is small and can be neglected. The validity of superposition is the reason why only two-body Coulomb collisions need to be studied.

The formulation of the problem begins by considering the interaction of two particles with charges and masses $q_1$, $q_2$, $m_1$, $m_2$ moving with velocities $\mathbf{v}_1$, $\mathbf{v}_2$. The only force acting on the particles is due to the Coulomb electric field:

$$\mathbf{E} = -\nabla\phi$$
$$\phi = \frac{q_1 q_2}{4\pi\varepsilon_0 r}. \tag{9.1}$$

The geometry of the interaction is illustrated in Fig. 9.1. Using Newton's law the equations describing the trajectories of the particles are

$$m_1 \frac{d\mathbf{v}_1}{dt} = -\frac{q_1 q_2}{4\pi\varepsilon_0} \frac{\mathbf{r}_2 - \mathbf{r}_1}{|\mathbf{r}_2 - \mathbf{r}_1|^3},$$
$$m_2 \frac{d\mathbf{v}_2}{dt} = -\frac{q_1 q_2}{4\pi\varepsilon_0} \frac{\mathbf{r}_1 - \mathbf{r}_2}{|\mathbf{r}_1 - \mathbf{r}_2|^3},$$
$$\frac{d\mathbf{r}_1}{dt} = \mathbf{v}_1, \tag{9.2}$$
$$\frac{d\mathbf{r}_2}{dt} = \mathbf{v}_2.$$

The starting model consists of 12 coupled non-linear ordinary differential equations. This ambitious model is reduced to a single differential equation by transforming to the center of mass frame and making use of several conservation relations.

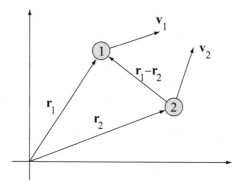

Figure 9.1 Geometry of a two-body Coulomb collision in the laboratory frame.

The first step in the simplification procedure is to note the conservation of momentum, obtained by adding the two separate momentum equations together:

$$\frac{d}{dt}(m_1\mathbf{v}_1 + m_2\mathbf{v}_2) = 0. \tag{9.3}$$

This motivates the introduction of two new velocity vectors as dependent variables: the center of mass velocity $\mathbf{V}$ and the relative velocity $\mathbf{v}$. The definitions and inverse relations are as follows:

$$\mathbf{V} = \frac{m_1\mathbf{v}_1 + m_2\mathbf{v}_2}{m_1 + m_2}, \qquad \mathbf{v}_1 = \mathbf{V} + \frac{m_2}{m_1 + m_2}\mathbf{v},$$

$$\mathbf{v} = \mathbf{v}_1 - \mathbf{v}_2, \qquad \mathbf{v}_2 = \mathbf{V} - \frac{m_1}{m_1 + m_2}\mathbf{v}. \tag{9.4}$$

Clearly, $d\mathbf{V}/dt = 0$ implying that $\mathbf{V} = \text{const.}$ and that the position of the center of mass is given by

$$\mathbf{R}(t) = \frac{m_1\mathbf{r}_1 + m_2\mathbf{r}_2}{m_1 + m_2} = \mathbf{R}(0) + \mathbf{V}t. \tag{9.5}$$

The equations for $\mathbf{v}$ and the relative position $\mathbf{r} = \mathbf{r}_1 - \mathbf{r}_2$ simplify to

$$\frac{d\mathbf{v}}{dt} = \frac{q_1 q_2}{4\pi\varepsilon_0 m_r}\frac{\mathbf{r}}{r^3},$$

$$\frac{d\mathbf{r}}{dt} = \mathbf{v}, \tag{9.6}$$

where

$$m_r = \frac{m_1 m_2}{m_1 + m_2} \tag{9.7}$$

is the reduced mass and $\mathbf{r} = r\,\mathbf{e}_r$. Observe that in the center of mass frame there are only six unknowns $(\mathbf{v}, \mathbf{r})$. The problem now corresponds to the motion of an equivalent single

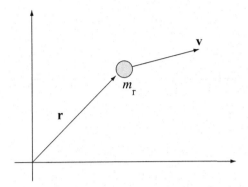

Figure 9.2 Center of mass geometry showing the central force acting on a particle with reduced mass $m_r$.

particle with mass $m_r$ acted upon by a central force whose origin is at the center of mass. See Fig. 9.2 for the center of mass geometry.

The next step in the analysis is to derive the conservation of energy relation by forming the dot product of the momentum equation with $\mathbf{v}$. A short calculation yields

$$\frac{d}{dt}\left(\frac{1}{2}m_r v^2 + \frac{q_1 q_2}{4\pi\varepsilon_0}\frac{1}{r}\right) = 0 \tag{9.8}$$

or

$$\frac{1}{2}m_r v^2 + \frac{q_1 q_2}{4\pi\varepsilon_0}\frac{1}{r} = E_0 = \text{const.} \tag{9.9}$$

The sum of kinetic plus potential energies is a constant. A similar conservation relation applies for the conservation of angular momentum, which is obtained by forming the cross product of the momentum equation with $\mathbf{r}$. One finds

$$\frac{d}{dt}m_r(\mathbf{r} \times \mathbf{v}) = 0 \tag{9.10}$$

or

$$m_r \mathbf{r} \times \mathbf{v} = \mathbf{L}_0 = \text{const.} \tag{9.11}$$

One consequence of angular momentum conservation is that if the initial position and velocity in the center of mass frame are given by $\mathbf{r}(0)$, $\mathbf{v}(0)$, then the particle trajectory remains in the initial plane (i.e., the plane perpendicular to $\mathbf{r}(0) \times \mathbf{v}(0)$) during the entire collision process.

The problem is thus reduced to a 2-D problem where, without loss in generality, the center of mass coordinate system can be rotated such that there are only four unknowns corresponding to the positions and velocities lying in the initial $\mathbf{r}(0)$, $\mathbf{v}(0)$ plane.

The problem can be further simplified by obtaining expressions for the conservation constants $E_0$, $L_0$ in terms of the initial relative velocity $\mathbf{v}(0) \equiv v_0 \mathbf{e}_x$ and impact parameter $b$ as shown in Fig. 9.3. Again, without loss in generality the coordinate system can be

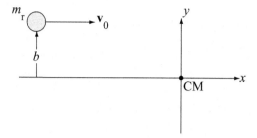

Figure 9.3 Center of mass (CM) geometry showing how to calculate $E_0$ and $L_0$ in terms of $b$ and $v_0$.

oriented so that $\mathbf{v}(0)$ lies in the $x$ direction and is displaced by an impact parameter $b$ in the $y$ direction. The initial conditions correspond to the situation where the particle is far from the center of mass so that $\phi \to 0$. It thus follows that

$$E_0 = \tfrac{1}{2} m_r v_0^2. \tag{9.12}$$

Similarly $\mathbf{L}_0 = L_0 \mathbf{e}_z = m_r y(0) \mathbf{e}_y \times v_0 \mathbf{e}_x$ or

$$L_0 = -m_r b v_0. \tag{9.13}$$

The final simplification is obtained by introducing cylindrical coordinates $x(t) = r(t)\cos\theta(t)$, $y(t) = r(t)\sin\theta(t)$ and recalling that $\mathbf{r} = r\,\mathbf{e}_r$ and $\mathbf{v} = \dot{r}\,\mathbf{e}_r + r\dot{\theta}\mathbf{e}_\theta$. The conservation of energy and angular momentum relations reduce to

$$\dot{r}^2 + r^2\dot{\theta}^2 + \frac{q_1 q_2}{2\pi\varepsilon_0}\frac{1}{r} = v_0^2,$$
$$r^2\dot{\theta} = -bv_0. \tag{9.14}$$

After eliminating $\dot{\theta}$ one obtains a single first order ODE for the unknown $r(t)$

$$\frac{dr}{dt} = \mp v_0 \left(1 - \frac{b^2}{r^2} - 2\frac{b_{90}}{r}\right)^{1/2}, \tag{9.15}$$

The choice of sign depends upon whether the particle is approaching the center of mass just before the collision or leaving it just after. The quantity $b_{90}$ is defined by

$$b_{90} = \frac{q_1 q_2}{4\pi\varepsilon_0 m_r v_0^2}. \tag{9.16}$$

The reason for the subscript will become apparent shortly.

Equation (9.15) is the desired relation.

### 9.2.2 Solution to the problem

It is now a relatively straightforward matter to derive an expression for the scattering angle as a function of initial velocity and impact parameter. The geometric relations necessary for the calculation are illustrated in Fig. 9.4. Observe that the scattering angle is denoted

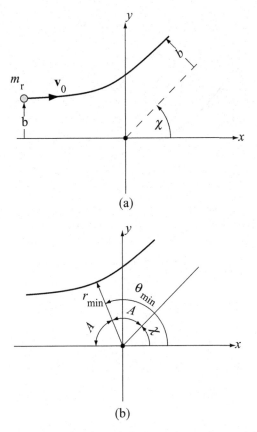

Figure 9.4 Center of mass geometry showing: (a) the scattering angle $\chi$ and (b) the relation between $\chi$, $\theta_{min}$, and $A$.

by $\chi$. Also, the point of closest approach of the orbit, denoted by $r = r_{min}$, occurs where $\dot{r} = 0$ and $\theta = \theta_{min}$. The goal is to calculate $\chi = \chi(v_0, b)$.

The relations among the various angles needed to carry out the calculation follow from the symmetry of the orbit: $\chi + A = \theta_{min}$ and $\chi + 2A = \pi$. Eliminating $A$ yields a relation between $\chi$ and $\theta_{min}$:

$$\chi = 2\,\theta_{min} - \pi. \tag{9.17}$$

The angle $\theta_{min}$ is determined from Eq. (9.15) using the relation

$$\theta_{min} \equiv \pi - \int_{\theta_{min}}^{\pi} d\theta = \pi - \int_{r_{min}}^{\infty} \frac{d\theta}{dr} dr. \tag{9.18}$$

Now, note that

$$\frac{d\theta}{dr} = \frac{\dot{\theta}}{\dot{r}} = -\frac{bv_0}{r^2 \dot{r}}. \tag{9.19}$$

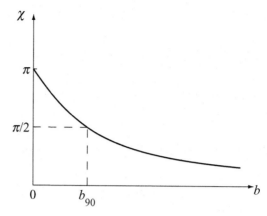

Figure 9.5 The scattering angle $\chi$ versus the impact parameter $b$. Note that $\chi = \pi/2$ for $b = b_{90}$.

Substituting for $\dot{r}$ (using the negative sign since $r$ is decreasing as it moves towards the center of mass) yields

$$\theta_{min} = \pi - b \int_{r_{min}}^{\infty} \frac{dr}{r(r^2 - b^2 - 2b_{90}r)^{1/2}}. \tag{9.20}$$

This integral can be easily evaluated by letting

$$\frac{b}{r} = -\frac{b_{90}}{b} + \left(1 + \frac{b_{90}^2}{b^2}\right)^{1/2} \sin\alpha. \tag{9.21}$$

Equation (9.20) becomes

$$\theta_{min} = \pi - \int_{\alpha_1}^{\alpha_2} d\alpha = \pi - (\alpha_2 - \alpha_1), \tag{9.22}$$

where $\alpha_2 = \pi/2$ and

$$\sin\alpha_1 = \frac{b_{90}}{\left(b^2 + b_{90}^2\right)^{1/2}}. \tag{9.23}$$

The overall goal of the calculation is accomplished by substituting the value of $\theta_{min}$ into Eq. (9.17). A short calculation yields the desired relation $\chi = \chi(v_0, b)$:

$$\cot\frac{\chi}{2} = \frac{b}{b_{90}} = \frac{4\pi\varepsilon_0 m_r}{q_1 q_2} v_0^2 b. \tag{9.24}$$

A plot of the deflection angle $\chi$ versus impact parameter $b$ is illustrated in Fig. 9.5. As expected, a small impact parameter, which corresponds to a very close encounter, leads to a large scattering angle. Conversely, when the impact parameter is large, implying a weak distant collision, the scattering angle is small. Also, when $b = b_{90}$ the deflection angle is 90°, thus motivating the subscript. Observe also that as the particle velocity increases, the deflection angle decreases since the particle spends less time in the vicinity of the center of mass, corresponding to a weaker collision.

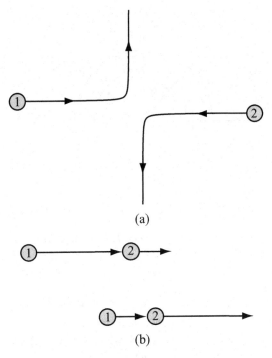

(a)

(b)

Figure 9.6 (a) Particle 1 loses directed momentum, but not energy, in the transverse plane, (b) particle 1 loses directed momentum and energy along the original direction of motion.

Equation (9.24) represents the critical property of Coulomb collisions necessary for the various applications that follow.

## 9.3 The test particle collision frequencies

### 9.3.1 The general formulation

Consider the momentum relaxation of a test electron with momentum $m_e \mathbf{v}_e$, injected into a stationary plasma of electrons and ions. The problem being posed is to determine how long it takes, on average, for this electron to lose its directed momentum (i.e., momentum along its initial direction) by colliding, for instance, with the ions in the plasma. The directed momentum can be lost either by being transferred to the transverse direction or to the energy of the target particle, as shown in Fig. 9.6. It does not matter which – in both cases directed momentum is lost.

The relaxation process is characterized by a relaxation time $\tau_{ei}(v_e)$ or equivalently a collision frequency $\nu_{ei}(v_e) = 1/\tau_{ei}(v_e)$. Similar collision frequencies exist for the test electron slowing down on plasma electrons ($\nu_{ee}$), or a test ion slowing down on plasma ions ($\nu_{ii}$) or electrons ($\nu_{ie}$). Knowledge of these collision frequencies provides a good estimate of the time required for an initial distribution of particles to relax to a Maxwellian. That is,

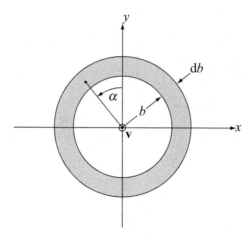

Figure 9.7 Transverse plane showing the angle $\alpha$.

after a few collision times the particle has essentially lost all memory of its initial state and becomes part of the background plasma.

The analysis below focuses on the calculation of $\nu_{ei}$. However, the derivation is general and a simple switching of particle subscripts yields the other relevant collision frequencies. The calculation is, in principle, relatively straightforward, making use of the generalized definition of collision frequency given in Chapter 3 and the critical information $\chi = \chi(v, b)$ just derived. Even so, substantial algebra is required and there is one subtlety involving the actual range of the "long-range" Coulomb potential.

The derivation begins by recalling from Chapter 3 that the "hard-sphere" collision frequency for a typical test electron with initial directed velocity $\mathbf{v}_e = v_e \mathbf{e}_x$ colliding with a fixed background of zero velocity ions of density $n_i$ is given by $\nu = n_i \sigma v_e$. Consider now the generalization of this relation to the present situation. Assume the target ions do not all have zero velocity but have a distribution function. Then, $n_i \rightarrow f_i(\mathbf{v}_i) d\mathbf{v}_i$. Next, the velocity appearing in the definition of $\nu$ is actually the relative velocity implying that $v_e \rightarrow |\mathbf{v}_e - \mathbf{v}_i|$. For the long-range Coulomb potential the hard-sphere cross section is replaced by $\sigma \rightarrow 2\pi b \, db$. Actually, a more accurate generalization that will be of use in future sections replaces the "$2\pi$" with the explicit angular dependence of the particle's initial position in the transverse plane: $\sigma \rightarrow 2\pi b \, db \rightarrow b \, db \, d\alpha$. See Fig. 9.7. Finally, we focus on momentum relaxation by introducing the quantity $\Delta m_e v_e$, which represents the loss in the initial $x$ directed momentum of the electron due to a single collision. This quantity will be related to the scattering angle $\chi$ shortly.

The total loss in $x$ directed momentum of the test electron is obtained by summing the loss per collision over all collisions. Thus, in simple and generalized form

$$\frac{d}{dt}(m_e v_e) = -(\Delta m_e v_e) n_i \sigma v_e$$

$$= -\int (\Delta m_e v_e) f_i(\mathbf{v}_i) |\mathbf{v}_e - \mathbf{v}_i| b \, db \, d\alpha \, d\mathbf{v}_i. \tag{9.25}$$

The next step is to define the test particle momentum relaxation collision frequency:

$$\frac{d}{dt}(m_e v_e) \equiv -v_{ei}(m_e v_e).$$

(9.26)

The desired expression for $v_{ei}$ is thus

$$v_{ei}(v_e) = \frac{1}{m_e v_e} \int (\Delta m_e v_e) f_i(\mathbf{v_i})|\mathbf{v_e} - \mathbf{v_i}| b \, db \, d\alpha \, d\mathbf{v_i}.$$

(9.27)

The task now is to evaluate $\Delta m_e v_e$ and then the five-fold integral.

### 9.3.2 Evaluation of $\Delta m_e v_e$ and the integration over the impact parameter

The evaluation of $\Delta m_e v_e$ proceeds as follows. The velocity of the electron at any instant of time is given in terms of the center of mass coordinates as follows (see Eq. (9.4)):

$$\mathbf{v_e} = \mathbf{V} + \frac{m_i}{m_i + m_e}\mathbf{v}.$$

(9.28)

Since $\mathbf{V} = $ const., the loss in the $x$ directed component of electron momentum as a result of a collision with an ion can be written as

$$\Delta m_e v_e = m_r \Delta \mathbf{v} \cdot \mathbf{e_x} = m_r(\mathbf{v}_{initial} - \mathbf{v}_{final}) \cdot \mathbf{e_x}.$$

(9.29)

If one writes the initial ion velocity as $\mathbf{v_i} = v'_x \mathbf{e_x} + v'_y \mathbf{e_y} + v'_z \mathbf{e_z}$, then from Fig. 9.3 it follows that $\mathbf{v}_{initial} = v \mathbf{e_X}$, where $v = [(v_e - v'_x)^2 + v'^2_y + v'^2_z]^{1/2}$ is the initial relative speed and $\mathbf{e_X} = [(v_e - v'_x)\mathbf{e_x} - v'_y \mathbf{e_y} - v'_z \mathbf{e_z}]/v$ is a unit vector along the direction of the initial relative velocity. Similarly, $\mathbf{v}_{final} = v \cos \chi \, \mathbf{e_X} + v \sin \chi \, \mathbf{e_\perp}$, where $\mathbf{e_\perp} = \cos \alpha \, \mathbf{e_Y} - \sin \alpha \, \mathbf{e_Z}$ is a unit vector perpendicular to the initial relative velocity (see Fig. 9.7) and use has been made of the fact that $v(t_{final}) = v(t_{initial}) \equiv v$ from conservation of energy. The loss in momentum $\Delta m_e v_e$ thus becomes

$$\Delta m_e v_e = m_r v[(1 - \cos \chi)\mathbf{e_X} \cdot \mathbf{e_x} - \sin \chi \, \mathbf{e_\perp} \cdot \mathbf{e_x}].$$

(9.30)

The last term can be ignored since it averages to zero when integrated over $\alpha$: $\int (\mathbf{e_\perp} \cdot \mathbf{e_x}) d\alpha = \int (\mathbf{e_Y} \cdot \mathbf{e_x} \cos \alpha + \mathbf{e_Z} \cdot \mathbf{e_x} \sin \alpha) d\alpha = 0$. Physically, the reason is that by symmetry, the numbers of upward and downward deflections are equal; that is, the product $\mathbf{e_\perp} \cdot \mathbf{e_x}$ has an equal probability of being positive or negative. The remaining non-vanishing contribution now has the form

$$\Delta m_e v_e = m_r(v_e - v'_x)(1 - \cos \chi).$$

(9.31)

In the final simplification this expression is rewritten in terms of the impact parameter by using the scattering angle relation derived in Section 9.2 and given by Eq. (9.24). The result is

$$\Delta m_e v_e = 2m_r(v_e - v'_x)\frac{b^2_{90}}{b^2 + b^2_{90}}.$$

(9.32)

One can now carry out the integrations over $b$ and $\alpha$ in Eq. (9.27) although a subtle difficulty arises with the upper limit of integration on $b$. For the moment define this limit as $b_{max}$. The integral over $b$ and $\alpha$ becomes

$$\int_0^{2\pi} d\alpha \int_0^{b_{max}} \frac{b_{90}^2}{b^2 + b_{90}^2} b \, db = \pi b_{90}^2 \ln\left(1 + \frac{b_{max}^2}{b_{90}^2}\right). \tag{9.33}$$

The problem is now apparent. Nominally, one would think that to include all possible collisions with ions, the impact parameter must cover the full range $0 < b < \infty$. However, setting $b_{max} = \infty$ leads to a logarithmic divergence.

The full mathematical resolution of this difficulty requires extensive analysis, well beyond the scope of the present discussion. Nevertheless, the difficulty can be resolved by a simple physical argument associated with Debye shielding. The logic is as follows. The weak $1/r$ dependence of the Coulomb potential implies that its effect extends over a very large distance: hence, the divergence for $b_{max} \to \infty$. However, the test particle actually feels a much smaller potential from any target particle whose inter-particle distance exceeds a critical value. The reason is that *in a plasma* the test particle is shielded from the electric field of all target particles further away than a Debye length. The consequence is that one can set $b_{max} \approx \lambda_{De}$ thereby eliminating the divergence. The numerical coefficient in front of $\lambda_{De}$ is not very important since the divergence appears only in the insensitive logarithmic term.

Using this resolution of the difficulty, one finds that the ratio of $b_{max}/b_{90}$ appearing in the integral is given by

$$\frac{b_{max}}{b_{90}} \approx \frac{\lambda_{De}}{b_{90}} = \left(\frac{\varepsilon_0 T_e}{n_e e^2}\right)^{1/2} \frac{4\pi \varepsilon_0 m_r v^2}{e^2}. \tag{9.34}$$

Equation (9.34) is still somewhat unsatisfactory in that it mixes the single-particle velocity $v$ with the velocity averaged quantities $T_e$ and $n_e$. This second difficulty is easily resolved by assuming that for a typical thermal electron $m_r v^2/2 \approx 3T_e/2$. Again the weak dependence of the logarithmic term makes the final answer insensitive to the exact details of the approximation. The expression for $b_{max}/b_{90}$ reduces to

$$\frac{b_{max}}{b_{90}} \equiv \Lambda \approx \frac{12\pi \varepsilon_0^{3/2} T_e^{3/2}}{n_e^{1/2} e^3} = 9\left(\frac{4\pi}{3} n_e \lambda_{De}^3\right). \tag{9.35}$$

The quantity $\ln \Lambda$ is known as the Coulomb logarithm. In practical units

$$\Lambda = 4.9 \times 10^7 \frac{T_k^{3/2}}{n_{20}^{1/2}} = 2.0 \times 10^9, \tag{9.36}$$

where the numerical values correspond to $T_k = 15$ and $n_{20} = 2$. This implies that

$$\ln \Lambda \approx 20 \tag{9.37}$$

in a fusion reactor. For just about all fusion plasmas, both those in present day experiments and those in future fusion reactors, $\ln \Lambda$ lies between 15 and 20 demonstrating its relative insensitivity to substantial variations in plasma density and temperature.

The net result of this discussion is that the integration over impact parameters reduces to

$$\int_0^{2\pi} d\alpha \int_0^{b_{max}} \frac{b_{90}^2}{b^2 + b_{90}^2} \, b \, db \approx \pi b_{90}^2 \ln(1 + \Lambda^2) \approx 2\pi b_{90}^2 \ln \Lambda, \tag{9.38}$$

where the last simplification follows from the fact that $\Lambda \gg 1$. Also, note that replacing $m_r v^2 \approx 3T_e$ in $b_{90}^2$ is reasonable for the insensitive logarithmic term, but not for the $b_{90}^2$ coefficient in front in which the stronger velocity dependence must be maintained.

This result is now substituted into the expression for $v_{ei}$ which reduces to

$$v_{ei}(v_e) = \frac{4\pi m_r}{m_e v_e} \ln \Lambda \int (v_e - v_x') b_{90}^2 f_i(\mathbf{v}_i) |\mathbf{v}_e - \mathbf{v}_i| \, d\mathbf{v}_i. \tag{9.39}$$

### 9.3.3 Integration over target velocities

The integration over target velocities is accomplished by introducing spherical velocity variables $(v, \theta, \phi)$ as follows:

$$\begin{aligned} v_x' &= v_e + v\cos\theta, \\ v_y' &= v\sin\theta\sin\phi, \\ v_z' &= v\sin\theta\cos\phi. \end{aligned} \tag{9.40}$$

This transformation gives $d\mathbf{v}_i = dv_x' dv_y' dv_z' = v^2 \sin\theta \, dv \, d\theta \, d\phi$, $|\mathbf{v}_e - \mathbf{v}_i| = v$, and $b_{90}^2 = (e^2/4\pi\varepsilon_0 m_r)^2/v^4$ with $q_i = -q_e = e$ for singly charged ions.

Assume now that $f_i$ corresponds to a Maxwellian with temperature $T_i$ and density $n_i$. Substituting into Eq. (9.39) and carrying out the trivial integration over $\phi$ leads to a simplified expression for $v_{ei}$ given by

$$v_{ei}(v_e) = \left( \frac{2^{1/2}}{8\pi^{3/2}} \frac{e^4 n_i}{\varepsilon_0^2 T_i^{3/2}} \frac{m_i^{3/2}}{m_e m_r} \ln \Lambda \right) I(w_e),$$

$$I(w_e) = -\frac{e^{-w_e^2}}{w_e} \int_0^\infty dw \int_0^\pi d\theta \sin\theta \cos\theta \, e^{-w^2 - 2w_e w \cos\theta}, \tag{9.41}$$

where $w = v/v_{Ti}$ and $w_e = v_e/v_{Ti}$ are normalized forms of the velocity and $v_{Ti} = (2T_i/m_i)^{1/2}$. The quantity $I$ is evaluated by first integrating over $\theta$:

$$\int_0^\pi e^{-2w_e w \cos\theta} \sin\theta \cos\theta \, d\theta = -\frac{1}{w_e} \frac{\partial}{\partial w} \left( \frac{\sinh 2w_e w}{2w_e w} \right). \tag{9.42}$$

The integral over $w$ is evaluated by integrating by parts and carrying out some simple algebraic manipulations. One obtains

$$I(w_e) = \frac{1}{w_e^2} \left[ \frac{\pi^{1/2}}{2w_e} \Phi(w_e) - e^{-w_e^2} \right], \tag{9.43}$$

where $\Phi(w_e)$ is the familiar probability (or error function) integral

$$\Phi(w_e) = \frac{2}{\pi^{1/2}} \int_0^{w_e} e^{-w^2} dw. \tag{9.44}$$

The desired expression for $\nu_{ei}$ can finally be written as

$$\nu_{ei}(v_e) = \left( \frac{2^{1/2}}{8\pi^{3/2}} \frac{e^4 n_i}{\varepsilon_0^2 T_i^{3/2}} \frac{m_i^{3/2}}{m_e m_r} \ln \Lambda \right) \left[ \frac{\pi^{1/2}}{2w_e^3} \Phi(w_e) - \frac{e^{-w_e^2}}{w_e^2} \right]. \tag{9.45}$$

A more transparent form of $\nu_{ei}$ is obtained using a simple approximation for $I$ that makes use of the fact that

$$\Phi \approx \begin{cases} 1, & w_e \gg 1, \\ \dfrac{2}{\pi^{1/2}} e^{-w_e^2} \left( w_e + \dfrac{2}{3} w_e^3 \right), & w_e \ll 1. \end{cases} \tag{9.46}$$

A reasonable approximation for $I$, valid for both large and small $w_e$, is thus

$$I(w_e) \approx \frac{\pi^{1/2}}{2} \frac{1}{w_e^3 + 3\pi^{1/2}/4}. \tag{9.47}$$

The expression for $\nu_{ei}$ reduces to

$$\nu_{ei}(v_e) = \left( \frac{1}{4\pi} \frac{e^4 n_i}{\varepsilon_0^2 m_e m_r} \ln \Lambda \right) \frac{1}{v_e^3 + 1.3 v_{Ti}^3}. \tag{9.48}$$

### 9.3.4 Properties of $\nu_{ei}$ and other collision frequencies

After a lengthy calculation one is now in a position to investigate the properties of $\nu_{ei}$. The first point to note is that for electron–ion collisions the velocity of a typical electron satisfies $v_e \sim v_{Te} \gg v_{Ti}$. Consequently the second term in the denominator of $\nu_{ei}$ can be neglected for all but the very slowest electrons. Also, for electron–ion collisions $m_r \approx m_e$. The collision frequency simplifies to

$$\nu_{ei}(v_e) \approx \left( \frac{1}{4\pi} \frac{e^4 n_i}{\varepsilon_0^2 m_e^2} \ln \Lambda \right) \frac{1}{v_e^3}. \tag{9.49}$$

Observe that $\nu_{ei} \sim 1/v_e^3$. High-velocity electrons have a smaller collisional loss of directed momentum because they move so fast that they spend very little time in the vicinity of a target ion where the interaction takes place. As will be shown, the $1/v_e^3$ dependence plays an important role in the phenomenon of runaway electrons.

The next point to note is that the dominant contribution to $\nu_{ei}$ arises from particles with large impact parameters, corresponding to small-angle collisions. In fact, the $\nu_{ei}$ integral would diverge because of small-angle collisions except for the physically motivated argument that sets an upper limit of $b_{max} \approx \lambda_{De}$, resulting in a finite bound. Physically, each large-$b$ collision produces only a small deflection, but there are so many of these collisions because of the long-range nature of the Coulomb potential that they dominate the behavior.

As a quantitative comparison consider the cross section for large-angle deflections greater than or equal to $90°$: $\sigma_{90} = \pi b_{90}^2$. The corresponding collision frequency is $\nu_{90} = n_i \sigma_{90} v_e$. Comparing this value with Eq. (9.49) shows that

$$\nu_{ei}(v_e) = (4 \ln \Lambda) \nu_{90}(v_e). \tag{9.50}$$

The actual collision frequency is about a factor of 80 larger than the $90°$ collision frequency. Consider now a typical value of $\nu_{ei}$ obtained by setting $v_e = \hat{v}_e \equiv (3T_e/m_e)^{1/2}$. One finds

$$\hat{\nu}_{ei} = \nu_{ei}|_{\hat{v}_e} \approx \frac{1}{\sqrt{3}} \frac{\omega_{pe}}{\Lambda} \ln \Lambda = 1.33 \times 10^5 \frac{n_{20}}{T_k^{3/2}} \text{ s}^{-1}. \tag{9.51}$$

In the practical formula, $n_e = n_i$ and $\ln \Lambda = 20$. Observe, as might be expected, that the collision frequency increases with density and decreases with temperature. For a fusion reactor with $n_{20} = 2$, $T_k = 15$, the collision frequency $\nu_{ei} \approx 4.6 \times 10^3 \text{ s}^{-1}$. Also, since $\Lambda \gg 1$, it follows that $\nu_{ei} \ll \omega_{pe} \sim \omega_{ce}$. Coulomb collisions are indeed rare compared to the basic characteristic frequencies defining a fusion plasma.

Similar expressions for the other directed momentum collision frequencies are easily obtained from Eq. (9.45) by switching the appropriate species subscripts and taking the correct limit for the reduced mass. The practical expressions below are calculated assuming $n_e = n_i$, and $\ln \Lambda = 20$. Also the typical particle velocities have been chosen to satisfy $m_e v_e^2 = 3T_e$ and $m_i v_i^2 = 3T_i$. Some simple algebra yields

$$\nu_{ee} \approx \left( \frac{1}{2\pi} \frac{n_e e^4}{\varepsilon_0^2 m_e^2} \ln \Lambda \right) \frac{1}{v_e^3 + 1.3v_{Te}^3} \approx 1.2\hat{\nu}_{ei},$$

$$\nu_{ii} \approx \left( \frac{1}{2\pi} \frac{n_i e^4}{\varepsilon_0^2 m_i^2} \ln \Lambda \right) \frac{1}{v_i^3 + 1.3v_{Ti}^3} \approx 1.2 \left( \frac{m_e}{m_i} \right)^{1/2} \left( \frac{T_e}{T_i} \right)^{3/2} \hat{\nu}_{ei}, \tag{9.52}$$

$$\nu_{ie} \approx \left( \frac{1}{4\pi} \frac{n_e e^4}{\varepsilon_0^2 m_e m_i} \ln \Lambda \right) \frac{1}{v_i^3 + 1.3v_{Te}^3} \approx 1.4 \left( \frac{m_e}{m_i} \right) \hat{\nu}_{ei}.$$

The value of $\hat{\nu}_{ei}$ in these relations is the practical one given by Eq. (9.51).

In terms of scaling, note that if one defines the mass ratio $\mu = m_e/m_i$, then

$$\nu_{ee} \sim \nu_{ei},$$

$$\nu_{ii} \sim \mu^{1/2} \nu_{ei}, \tag{9.53}$$

$$\nu_{ie} \sim \mu \, \nu_{ei}.$$

The interpretation is as follows. A test electron will lose a comparable amount of momentum whether it scatters off an electron or an ion. Hence $\nu_{ee} \sim \nu_{ei}$. A test ion scattering off other ions is similar to $\nu_{ee}$ except that the frequency of collisions is slowed down by the heavier mass and corresponding slower motion. The net result is that the collision frequency is reduced by $\nu_{ii} \sim \mu^{1/2} \nu_{ei}$. Finally, when ions scatter off electrons they barely lose any momentum at all because of their heavy mass. The collisional slowing down time is very long or, conversely, the collision frequency is very low: $\nu_{ie} \sim \mu \, \nu_{ei}$.

The last point concerns the relative size of the Coulomb collision frequency with respect to the D–T fusion collision frequency. Recall that the cross section for D–T fusion reactions is about 5 barns, which corresponds to a fusion collision frequency of approximately $\nu_{DT} \approx n_i \sigma_{DT} \nu_{Ti} \approx 0.1$ s$^{-1}$. On the other hand Eq. (9.53) implies that for D–T Coulomb collisions $\nu_{ii} \approx 91$ s$^{-1}$, corresponding to an equivalent cross section $\sigma_{ii} \approx 3800$ barns. The equivalent Coulomb cross section is about a factor of 1000 larger. The ions (and electrons) in a fusion plasma relax to a Maxwellian distribution function well before many fusion reactions can take place.

The collision frequencies just calculated are used throughout the remainder of the book to help understand the behavior of plasmas in a variety of situations including macroscopic equilibrium and stability, transport, and heating. Also, it is worth pointing out that one could, if desired, calculate other test particle collision frequencies corresponding to total or transverse energy relaxation. However, these are not essential for present purposes. Instead, attention is focused on using the momentum test particle collision frequencies to investigate three applications wherein they dominate the behavior: the mirror machine, the slowing down of high energy ions, and runaway electrons.

## 9.4 The mirror machine revisited

The presence of Coulomb collisions leads to an irreducible limit on the maximum value of $Q$ (i.e., the physics energy gain) that can be achieved in a simple mirror machine fusion reactor. The limit results from the fact that Coulomb collisions cause confined mirror particles to scatter into the loss cone, after which they are lost virtually instantaneously from the plasma; that is, the test particle Coulomb collision frequency directly determines the energy confinement time $\tau_E$, thereby setting a limit on the plasma power balance. The limit is irreducible in that there is no way to eliminate Coulomb collisions. The analysis below shows that the maximum value of $Q$ in a simple mirror machine is too low to be of interest for fusion energy applications.

The goal of the analysis is to derive an expression for $Q$ applicable to a simple mirror machine. Two interesting features of the calculation are that: (1) $Q$ turns out to be independent of density, and (2) the optimum temperature is not 15 keV, but somewhat over 100 keV. The analysis is simplified by making several slightly optimistic assumptions. First, it is assumed that neither macroscopic nor microscopic plasma instabilities are present – there is no anomalously large particle or energy loss from the plasma. Second, the mirror ratio is assumed to be large enough so that for a particle to diffuse into the loss cone, it must lose essentially all of its perpendicular (to **B**) momentum.

### 9.4.1 Calculation of $\tau_E$

The critical issue in the analysis is the estimate of the energy loss time $\tau_E$. The estimate is determined as follows. The test particle collision analysis has shown that $\nu_{ei} \sim \nu_{ee} \gg \nu_{ii}$.

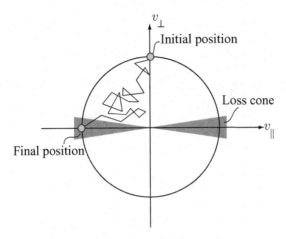

Figure 9.8 A particle starting from its initial position in velocity space is lost instantaneously once it diffuses into the loss cone as a result of many small-angle collisions.

Thus, at first glance, one would expect electrons to diffuse into the loss cone much faster than ions. This, however, is not the case. If the fast diffusing electrons were lost first, this would leave a large positive charge on the plasma. The resulting electric field is in the direction to draw the electrons back into the plasma and drive ions out of the plasma. The ions, because of their heavy mass, hardly move at all. The net result is that it is the electrons that are drawn back into the plasma and, therefore, do not leave at the fast rate associated with electron collisions. The loss rate is therefore actually determined by the ions, which also scatter into the loss cone, but at the slower rate related to $\nu_{ii}$. Once they do so, both the ions and the electrons (which are electrostatically held to the ions) are lost together.

The actual estimate for $\tau_E$ is obtained by solving for $v_i(t)$ from the differential equation that defines $\nu_{ii}(v_i)$ (i.e., the equivalent of Eq. (9.26) for ions). Intuitively, one expects that $\tau_E \sim 1/\nu_{ii}(v_{Ti})$. Solving the differential equation yields a reasonable approximation for the multiplicative constant. Specifically, the goal is to calculate how long it takes a typical test ion to lose essentially all of its perpendicular momentum (and corresponding perpendicular energy) assuming it starts off with $v_\perp = \hat{v}_i \equiv (3T_i/m_i)^{1/2}$, $v_\parallel = 0$. See Fig. 9.8. This is the time required for a typical ion to enter the loss cone.

The governing equation is obtained by substituting the simplified expression for $\nu_{ii}(v_i)$ given by Eq. (9.52) into the differential equation for $v_i(t)$ and introducing a normalized velocity variable $w_i(t) = v_i(t)/(3T_i/m_i)^{1/2}$. A short calculation yields

$$\frac{dw_i}{dt} = -\hat{\nu}_{ii}\left(\frac{1+k}{w_i^3 + k}\right)w_i,$$

$$w_i(0) = 1.$$

(9.54)

Here $\hat{\nu}_{ii} = \nu_{ii}|_{\hat{v}_i} = 2.3 \times 10^3 n_{20}/T_k^{3/2}\ \mathrm{s}^{-1}$, the ion mass $m_i = 2.5 m_{\mathrm{proton}}$ corresponding to a 50%–50% D–T fuel, and $k = 1.3\,(2/3)^{3/2} = 0.71$. Equation (9.54) can be simply solved

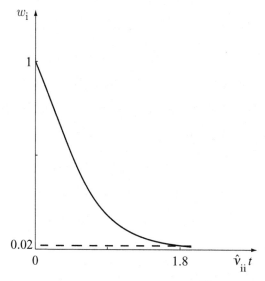

Figure 9.9 The normalized velocity $w_i = v_i/(3T_i/m_i)^{1/2}$ as a function of time.

leading to the following implicit solution for $w_i(t)$:

$$w_i^3 + 3k \ln w_i = 1 - 3(1+k)\hat{v}_{ii}t. \tag{9.55}$$

The solution is illustrated in Fig. 9.9. Observe that $w_i$ is a monotonically decreasing function of time. For practical purposes one can assume that all the perpendicular momentum is lost when $w_i$ is reduced to approximately 98% of its initial value: $w_i = 0.02$. This is equivalent to about 4 e-folding times. The actual fraction chosen is not very critical since it is the insensitive logarithmic term that dominates the behavior for small $w_i$. The perpendicular momentum loss time, which as discussed, is equal to $\tau_E$, is obtained by solving Eq. (9.55) for the time $t$ at which $w_i = 0.02$. The solution is given by $\hat{v}_{ii}\tau_E \approx 1.8$, or in practical units

$$\tau_E = 7.8 \times 10^{-4} \frac{T_k^{3/2}}{n_{20}} \text{ s.} \tag{9.56}$$

### 9.4.2 Power balance in a simple mirror machine

One is now in a position to calculate power balance in a mirror machine. Following the discussion in Chapter 4 it follows that steady state plasma power balance (neglecting Bremsstrahlung) is governed by the relation

$$S_\alpha + S_h - S_\kappa = 0. \tag{9.57}$$

Here $S_h$ is the external heating power density, $S_\alpha$ is the alpha power density, and $S_\kappa$ is the heating loss power density. Explicit expressions for $S_\alpha$ and $S_\kappa$ are

$$S_\alpha = \tfrac{1}{4} E_\alpha n^2 \langle \sigma v \rangle = K_\alpha n_{20}^2 \langle \sigma v \rangle \quad \text{W/m}^3,$$

$$S_\kappa = \frac{3nT}{\tau_E} = K_\kappa \frac{n_{20}^2}{T_k^{1/2}} \quad \text{W/m}^3,$$

$$(9.58)$$

where $K_\alpha = 1.4 \times 10^{27}$ and $K_\kappa = 6.2 \times 10^7$.

Next, recall that the physics gain factor $Q$ is defined as

$$Q = \frac{S_f}{S_h} = \frac{E_n + E_\alpha}{E_\alpha} \frac{S_\alpha}{S_h}. \tag{9.59}$$

Here, $S_f$ is the total thermal fusion output power density resulting from the alphas and the neutrons. Substituting from Eq. (9.58) yields

$$Q = 5 \frac{S_\alpha}{S_\kappa - S_\alpha} = \frac{5}{\left( K_M / T_k^{1/2} \langle \sigma v \rangle \right) - 1}, \tag{9.60}$$

where $K_M = K_\kappa / K_\alpha = 4.4 \times 10^{-20}$. Observe that the density dependence has canceled out because of the density dependence of the collision frequency. Consequently, for a simple mirror machine $Q = Q(T)$.

For positive $Q$ the function $Q(T)$ is maximized when $T_k^{1/2} \langle \sigma v \rangle$ has a maximum. After consulting the $\langle \sigma v \rangle$ curves in Fig. 3.11 one finds that $T_k^{1/2} \langle \sigma v \rangle$ has a relatively broad maximum at a temperature somewhat above 100 keV. At $T_k = 100$ the product $T_k^{1/2} \langle \sigma v \rangle = 8.1 \times 10^{-21}$. It then follows that for a simple mirror machine

$$Q_{max} = 1.1. \tag{9.61}$$

More sophisticated calculations than the one just described have been carried out but the conclusion remains the same. A simple mirror machine can achieve at most a value of $Q$ slightly above unity. Since a value $Q \sim 50$ is required for a power reactor, this is too difficult a hurdle to overcome for the simple mirror machine. Finally, it is worth noting that several ingenious ideas have been suggested to improve the $Q$ value in devices based on the mirror effect. However, in these devices some of the optimistic idealized assumptions made above breakdown, and the conclusion remains the same. One way or another, the problems associated with end loss are too severe to allow a high power gain in a steady state mirror reactor.

## 9.5 The slowing down of high-energy ions

### 9.5.1 The high-energy-ion slowing down model

An important application of the test particle collision analysis involves the slowing down of high-energy ion beams with the background plasma, a situation that typically arises in one of two ways. First, the alpha particles produced by fusion reactions have a very high energy (3.5 MeV) compared to the background plasma (15 keV). The alpha energy must be transferred to the plasma in order to sustain the plasma against thermal conduction transport losses. It is therefore of interest to learn how long it takes for this transfer to occur as compared to the energy confinement time, and whether the transfer is preferentially to electrons or ions.

A second application involves the initial heating of the plasma to ignition conditions, before the alphas are created. One method to provide this heating is by injecting a high-energy neutral beam of either D or T into the plasma. The beam is ionized as it passes through the plasma and then transfers its energy via Coulomb collisions. Typical present day beams have energies on the order of 100 keV. In a reactor, 1 MeV beams will be required. Again, it is of interest to know how long it takes to transfer the energy and which species is preferentially heated.

In both cases one is faced with the problem of determining how a high, nearly mono-energetic population of ions transfers its momentum and energy to the background plasma. Strictly speaking, momentum and energy transfers between alphas and the plasma involve net exchange collisions. However, because of the mono-energetic property, the test particle collision analysis provides a good method for answering the questions of interest.

The specific goals are to calculate $v_{be}$ and $v_{bi}$ and then analyze the slowing down trajectory of a beam ion. Here the subscript "b" refers to the beam. By comparing the magnitudes of these collision frequencies, one can determine whether beam energy is preferentially transferred to plasma electrons or ions and how long the transfer takes. The answer is not immediately obvious. Beam ions colliding with electrons lose very little energy per collision. On the other hand, the higher thermal velocity of the electrons results in many more beam–electron collisions than beam–ion collisions. The analysis shows that for sufficiently high beam energies the electrons are preferentially heated. Below a critical value of beam energy it is the ions that are preferentially heated. Lastly, a simple calculation is presented that determines the actual fraction of beam energy that is transferred to both the electrons and ions. The analysis proceeds as follows.

The basic equation governing the slowing down of beam ions is

$$m_b \frac{dv_b}{dt} = -m_b(v_{be} + v_{bi})v_b. \tag{9.62}$$

The collision frequencies are easily obtained from the general results derived in Section 9.3.4 with one simple modification, affecting the charge number $Z_b$. All the collision frequencies are proportional to $e^4$, which arises from the more basic form $q_1^2 q_2^2$. For the plasma $q_1 = -q_2 = e$. However, for beam–plasma interactions the plasma species have a charge $e$ while the beam particles, in general, have a charge $Z_b e$ (e.g. for alpha particles $Z_b = 2$). Therefore, for beam–plasma collisions one must replace $e^4 \rightarrow Z_b^2 e^4$.

With this modification, focus first on the calculation of the beam–electron collision frequency. From Eq. (9.48) it follows that

$$v_{be}(v_b) = \left( \frac{1}{4\pi} \frac{Z_b^2 e^4 n_e}{\varepsilon_0^2 m_e m_b} \ln \Lambda \right) \frac{1}{v_b^3 + 1.3 v_{Te}^3}. \tag{9.63}$$

Note that $v_b = 1.3 \times 10^7$ m/s for 3.5 MeV alphas, while $v_{Te} = 7.3 \times 10^7$ m/s for 15 keV electrons: $v_{Te}^3 \gg v_b^3$. The inequality is even stronger for neutral beam heating. Therefore, in the regime of interest $v_{be}$ can be accurately approximated by (replacing $1.3 \rightarrow 3\pi^{1/2}/4$)

$$v_{be}(v_b) \approx \frac{1}{3(2\pi)^{3/2}} \frac{Z_b^2 e^4 m_e^{1/2} n_e}{\varepsilon_0^2 m_b T_e^{3/2}} \ln \Lambda = 100 \frac{n_{20}}{T_k^{3/2}} \ \text{s}^{-1}. \tag{9.64}$$

The numerical value corresponds to alpha heating.

Consider next the beam–ion collision frequency. Equation (9.48) implies that $v_{bi}$ is given by

$$v_{bi}(v_b) = \left( \frac{1}{4\pi} \frac{Z_b^2 e^4 n_i}{\varepsilon_0^2 m_r m_b} \ln \Lambda \right) \frac{1}{v_b^3 + 1.3 v_{Ti}^3} \tag{9.65}$$

For 15 keV fusion ions, $v_{Ti} = 1.1 \times 10^6$ m/s implying that $v_{Ti}^3 \ll v_b^3$. Thus, it is reasonable to approximate

$$v_{bi}(v_b) \approx \frac{1}{4\pi} \frac{Z_b^2 e^4 n_i}{\varepsilon_0^2 m_r m_b v_b^3} \ln \Lambda = 0.94 \frac{n_{20}}{\left(m_b v_b^2/2\right)^{3/2}} \ \text{s}^{-1}. \tag{9.66}$$

The numerical values correspond to alpha heating and the units of $m_b v_b^2/2$ are megaelectronvolts.

### 9.5.2 *Which species is preferentially heated?*

One can now determine whether beam ions preferentially transfer their momentum and energy to electrons or ions by plotting $v_{be}$ and $v_{bi}$ as a function of $v_b$ as illustrated in Fig. 9.10. Observe that for $v_b > v_c$, the beam–electron collision frequency is higher than the beam–ion frequency. The beam is transferring more energy to electrons than ions. Once the beam has slowed down such that $v_b < v_c$, then the beam–ion collision frequency dominates and the remaining beam energy is transferred to the ions. The transition velocity occurs when $v_b = v_c$ corresponding to $v_{be} = v_{bi}$. For $n_e = n_i$ the critical velocity is

$$\frac{m_b v_c^2}{2 T_e} = \left( \frac{3\pi^{1/2}}{4} \right)^{2/3} \left( 1 + \frac{m_b}{m_i} \right)^{2/3} \left( \frac{m_b}{m_e} \right)^{1/3} = 44. \tag{9.67}$$

The numerical value again corresponds to alpha heating.

The conclusion is that in a 15 keV plasma the alphas will predominantly transfer their energy to the electrons starting from the initial alpha energy of 3.5 MeV down to 660 keV.

### 9.5.3 *The alpha particle slowing down time*

The slowing down time of the alphas can be estimated by solving the differential equation determining $v_b(t)$ (Eq. (9.62)). If one introduces a normalized beam velocity $w_b = v_b/v_c$ and substitutes the simplified forms of the collision frequency described above, then Eq. (9.62)

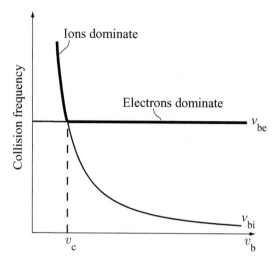

Figure 9.10 Collision frequencies $\nu_{\mathrm{be}}$ and $\nu_{\mathrm{bi}}$ as a function of beam velocity $v_{\mathrm{b}}$.

for $w_{\mathrm{b}}$ becomes

$$\frac{dw_{\mathrm{b}}}{dt} = -\nu_{\mathrm{be}}\left(1 + \frac{1}{w_{\mathrm{b}}^3}\right)w_{\mathrm{b}}, \tag{9.68}$$

$$w_{\mathrm{b}}(0) \equiv w_0 = \left(\frac{E_\alpha}{44T}\right)^{1/2} = 2.30.$$

The numerical value in the initial condition corresponds to $E_\alpha = 3.5$ MeV and $T = 15$ keV. Also, at $T = 15$ keV the beam–electron collision frequency has the value $\nu_{\mathrm{be}} = 3.45$ s$^{-1}$.

The solution is

$$w_{\mathrm{b}} = \left[\left(1 + w_0^3\right)e^{-3\nu_{\mathrm{be}}t} - 1\right]^{1/3}. \tag{9.69}$$

The function $w_{\mathrm{b}}(t)$ is plotted in Fig. 9.11. As expected the beam velocity monotonically decreases with time. During early times the beam is preferentially heating electrons. The critical transition time at which electron and ion heating rates are equal is found by setting $w_{\mathrm{b}} = 1$ in Eq. (9.69) and solving for $t_{\mathrm{c}}$. One finds

$$t_{\mathrm{c}} = \frac{1}{3\nu_{\mathrm{be}}}\ln\frac{1 + w_0^3}{2} = 0.18 \text{ s}. \tag{9.70}$$

This time is noticeably shorter than the required energy confinement time $\tau_{\mathrm{E}}$. The implication is that the alphas can give most of their energy to the electrons before the plasma loses its energy by heat conduction; the alphas do indeed have time to heat the electrons.

For $t > t_{\mathrm{c}}$ most of the remaining alpha energy is given to the ions. The time for this transfer can be estimated as follows. First, set the beam energy equal to the background

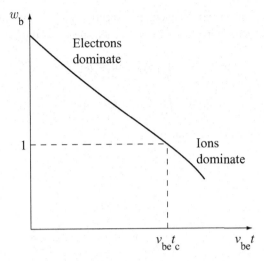

Figure 9.11 Normalized beam velocity $w_b = v_b/v_c$ as a function of time.

plasma thermal energy: $m_b v_b^2/2 \approx T$, which is equivalent to $w_b^2 = 1/44 \ll 1$. When this occurs, the beam, for all practical purposes, has transferred all of its energy to the plasma. The corresponding time $t_f$ is

$$t_f \approx \frac{1}{3v_{be}} \ln(1 + w_0^3) = 0.25 \text{ s.} \tag{9.71}$$

The heating time for ions is just $t_f - t_c = 0.07$ s, again a small fraction of the energy confinement time.

### 9.5.4 The fraction of alpha energy transferred to electrons and to ions

One now wants to know how the alpha energy is actually apportioned between electrons and ions. The answer is found by utilizing the conservation of energy as follows. The loss of alpha energy is determined by multiplying the alpha momentum equation (Eq. (9.62)) by $v_b$ and introducing $U_b = m_b v_b^2/2$:

$$\frac{dU_b}{dt} = -2(v_{be} + v_{bi})U_b. \tag{9.72}$$

Conservation of energy implies that the alpha energy is transferred to electrons and ions in accordance with the following relations:

$$\frac{dU_e}{dt} = 2\, v_{be} U_b,$$
$$\frac{dU_i}{dt} = 2\, v_{bi} U_b, \tag{9.73}$$

where $U_e$, $U_i$ are the electron and ion energies respectively. Coulomb collisions convert this energy into heat. The total energy transferred to each species is thus

$$U_e = 2 \int_0^{t_f} \nu_{be} U_b \, dt = m_b v_c^2 I_e,$$

$$U_i = 2 \int_0^{t_f} \nu_{bi} U_b \, dt = m_b v_c^2 I_i.$$

(9.74)

Here, the integrals $I_e$, $I_i$ can be written as

$$I_e = \nu_{be} \int_0^{t_f} w_b^2 \, dt,$$

$$I_i = \nu_{be} \int_0^{t_f} \frac{\nu_{bi}}{\nu_{be}} w_b^2 \, dt = \nu_{be} \int_0^{t_f} \frac{1}{w_b} \, dt.$$

(9.75)

The electron integral is easily evaluated in the important limit of large beam energy: $w_0^3 \gg 1$. If one introduces $y = \exp(-3\,\nu_{be} t)$, then

$$I_e \approx \frac{w_0^2}{3} \int_{1/w_0^3}^1 \frac{(y - 1/w_0^3)^{2/3}}{y} \, dy \approx \frac{w_0^2}{3} \int_0^1 \frac{1}{y^{1/3}} dy = \frac{w_0^2}{2}.$$

(9.76)

Similarly, the ion integral is evaluated by introducing $z = w_0^3 \exp(-3\,\nu_{be} t)$:

$$I_i \approx \frac{1}{3} \int_1^{w_0^3} \frac{dz}{z(z-1)^{1/3}} \approx \frac{1}{3} \int_1^\infty \frac{dz}{z(z-1)^{1/3}} = \frac{1}{3} \Gamma(1/3)\Gamma(2/3) = 1.21.$$

(9.77)

The desired energy fractions $F_e$, $F_i$ are now calculated as follows:

$$F_e \equiv \frac{I_e}{I_e + I_i} = \frac{T_c}{T_c + T},$$

$$F_i \equiv \frac{I_i}{I_e + I_i} = \frac{T}{T_c + T},$$

(9.78)

where $T_c$ is given by

$$T_c = 0.342 \frac{(m_e/m_b)^{1/3}}{(1 + m_b/m_i)^{2/3}} E_\alpha = 33 \text{ keV}.$$

(9.79)

The energy fractions are plotted as a function of temperature in Fig. 9.12. Observe that for alpha heating of a D–T plasma the energy is divided equally among electrons and ions if the plasma temperature is 33 keV. At the typical reactor temperature of 15 keV the electrons receive about 70% of the alpha energy, while the ions receive 30%. In a fusion reactor the alpha particles preferentially heat the electrons by more than a factor of 2:1.

### 9.5.5 Discussion of beam heating

The conclusion from this analysis is that Coulomb collisions allow the alpha particles to transfer their energy to the plasma electrons and ions before the energy is lost by thermal

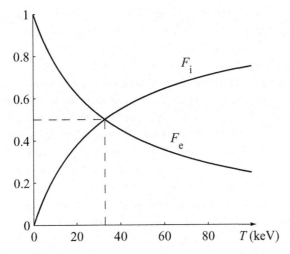

Figure 9.12 Fraction of electron energy $F_e$ and fraction of ion energy $F_i$ transferred by alpha particles as a function of temperature.

conduction. This assumes that there are no plasma instabilities that lead to an anomalous loss of alpha particles. Since most of the alpha energy is transferred to electrons, the one remaining question is to determine how long it takes the electrons to equilibrate with the ions. This is an important issue since it is the ions that ultimately must be heated to produce fusion reactions. Electron–ion equilibration is discussed in Section 9.7.3, where it is shown that the corresponding equilibration time is on the order of 0.4 s.

The last point of interest concerns plasma heating by neutral beams. A similar analysis to that for alpha particles applies here as well. For example, if one wants to heat a $T = 3$ keV D–T plasma with a beam of 100 keV deuterium ions, then the critical beam energy, given by Eq. (9.67), satisfies $m_b v_c^2 / 2T = 27.6$, which is equivalent to $m_b v_c^2 / 2 = 83$ keV. In this case the critical beam energy is not too different from the starting beam energy. Also, from Eq. (9.79), the corresponding critical temperature for equal heating of electrons and ions is $T_c = 1.5$ keV. Thus the division of beam energy between electrons and ions is almost opposite that for alpha heating. In the neutral beam example the ions receive 2/3 of the beam energy and the electrons 1/3. Lastly, the beam slowing down time for $n_{20} = 1$ and a 100 keV beam is approximately $t_f \approx 0.018$ s, which is considerably shorter than a typical energy confinement time in a tokamak experiment; the conclusion is that from a physics point of view, neutral beams are a good way to heat a plasma, and this is indeed borne out experimentally.

## 9.6 Runaway electrons

Another application of the test particle collision frequency theory concerns the phenomenon of runaway electrons. This is a situation in which a small, special population of electrons is accelerated from an initially modest velocity (compared to $c$) to relativistic speeds by a

DC electric field. The acceleration occurs even though the electric field may be small and Coulomb collisions may be slowing down the motion. In general runaway electrons have a deleterious effect on plasma performance and one must therefore learn how to minimize the population of such particles.

In experiments, runaway electrons are sometimes generated as follows. During the start up and subsequent approximately steady state phase of many fusion plasma discharges a toroidal current is induced in the plasma by means of transformer action with the plasma being the secondary of the transformer. There is an accompanying electric field with a component parallel to $\mathbf{B}$. The magnitude of this slowly varying, nearly DC $E_\parallel$ is small, but not zero, since Coulomb collisions provide a friction against the flow of electrons equivalent to an electrical resistivity; that is, collisions slightly reduce the effect of Debye shielding.

The reason why a small $E_\parallel$ can accelerate electrons to relativistic speeds is connected to the fact that both $\nu_{ei} \sim \nu_{ee} \sim 1/v_e^3$ for high $v_e$. The collision frequency decreases rapidly with increasing velocity. Therefore, an electron with a sufficiently high initial velocity can gain more momentum from the electric field between collisions than it will lose after each collision. The faster the electron moves, the more time it has to gain momentum from $E_\parallel$ before its next collision. This is an unstable situation leading to electron runaway. Ultimately, runaway electrons (1) reach relativistic speeds, (2) are lost to the first wall, or (3) are sometimes scattered by several possible plasma instabilities.

Runaway electrons can be important in fusion plasmas for several reasons. First, runaways escaping from the plasma as a result of plasma instabilities mostly strike the first wall. Because of their high energy having too many runaways may cause material damage to the wall. Second, if one is trying to ohmically heat a plasma, this technique will not be very effective if a large number of runaways are present. A substantial fraction of the input power is absorbed by the runaways and cannot be transferred back to the bulk electrons or ions because the collision frequency is too small.

In the calculation presented below a simple criterion is derived that determines the condition for the onset of runaway electrons. The condition is a function of $n$, $T$, and $E_\parallel$. With this relation it is possible to estimate the number density, energy density, and current density associated with the runaway population. The theory shows that the runaway population is a very strong function of number density and a practical way to minimize the effects of runaways is to operate at sufficiently high $n$. Runaways should not be too serious a problem in a fusion reactor although one must be careful during startup. They can, however, be more of a problem in certain experiments.

### 9.6.1 The threshold condition for runaway electrons

The threshold condition for the onset of runaway electrons can be easily derived by adding the effect of a macroscopic DC electric field to the collisional slowing down equation. Attention is focused on electrons since they are much more likely to runaway than the heavier ions. Both electron–ion and electron–electron collisions must be considered as both slow down a test particle at comparable rates: $\nu_{ei} \sim \nu_{ee}$. Also, only motion parallel to **B** is important since the magnetic field strongly inhibits motion in the perpendicular direction. Lastly, it is assumed for mathematical convenience, without loss of generality, that the parallel electric field is negative: $E_\parallel = -|E_\parallel|$. Under these conditions the equation describing the parallel motion of a test electron is

$$\frac{dv_e}{dt} = -\frac{e}{m_e}E_\parallel - (\nu_{ei} + \nu_{ee})v_e. \tag{9.80}$$

The goal now is to solve this differential equation and determine the conditions under which the electron velocity exhibits runaway behavior.

To begin, note that the electrons most likely to runaway are those which just happen to have a relatively large initial velocity in the direction of the electric field acceleration. In this regime the collision frequencies can be approximated by their large velocity $1/v_e^3$ limit. Also, for convenience the velocity is normalized as follows: $v_e = (2T_e/m_e)^{1/2} w_e = v_{Te}w_e$. The differential equation now has the form

$$\frac{dw_e}{dt} = \hat{E}_\parallel - \frac{\nu_R}{w_e^2}, \tag{9.81}$$

where

$$\hat{E}_\parallel = \frac{e|E_\parallel|}{(2m_e T_e)^{1/2}} = 9.4 \times 10^3 \frac{|E_\parallel|}{T_k^{1/2}} \text{ s}^{-1},$$

$$\nu_R = \frac{3}{8\sqrt{2\pi}} \frac{ne^4 \ln\Lambda}{\varepsilon_0^2 m_e^{1/2} T_e^{3/2}} = 7.3 \times 10^5 \frac{n_{20}}{T_k^{3/2}} \text{ s}^{-1}. \tag{9.82}$$

Here and below, the units for $|E_\parallel|$ are volts per meter.

The behavior of the differential equation can be easily understood by examining a plot of $\dot{w}_e$ vs. $w_e$ as shown in Fig. 9.13. Observe that the value of $\dot{w}_e$ can be either negative (i.e., decelerating) or positive (i.e., accelerating), the transition occurring at $w_c = (\nu_R/\hat{E}_\parallel)^{1/2}$. Therefore, for any particle with an initial velocity $w_e(0) < w_c$, then $\dot{w}_e < 0$, implying that the particle will slow down continuously, reaching a very small velocity (where the low-velocity corrections to the collision frequencies must be included). Clearly these particles are not runaways. On the other hand, particles with an initial velocity $w_e(0) > w_c$ are characterized by $\dot{w}_e > 0$. These particles accelerate indefinitely until relativistic effects become important or they are lost from the plasma. These are the runaway electrons.

Based on this discussion, one sees that runaway electrons occur for $w_e \geq w_c$ or in unnormalized units

$$\frac{v_e^2}{v_{Te}^2} \geq \frac{v_c^2}{v_{Te}^2} = \frac{3}{8\pi} \frac{ne^3 \ln\Lambda}{|E_\parallel|\varepsilon_0^2 T_e} = 78 \frac{n_{20}}{|E_\parallel|T_k}. \tag{9.83}$$

This is the desired relation. The next task is to investigate the properties of runaway electrons to determine when they may be important in fusion experiments.

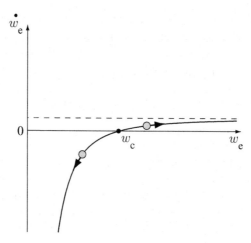

Figure 9.13 Plot of $w_e$ vs. $\dot{w}_e$ showing that a particle with initial velocity $w_e > w_c$ has $dw_e/dt > 0$ and therefore accelerates to relativistic speeds.

### 9.6.2 Properties of runaway electrons

Some insight into the importance of runaway electrons can be obtained by calculating the fraction of (1) the number density, (2) the energy density, and (3) the parallel current density associated with the runaway population. This is straightforward once the onset condition is known. To carry out the calculation one assumes that the plasma electrons are described by a Maxwellian distribution function. The quantities above can then be directly calculated using the onset condition as the transition point between the background plasma and the runaway population. Note that it is also of interest to calculate the fraction of runaways lost from the plasma and the rate at which new runaways are born by diffusion across the onset boundary, but these problems are beyond the scope of the present discussion.

The key simplifying point in the derivation is the recognition that for most present day plasmas and projected future reactors the quantity $n_{20}/|E_\parallel|T_k \sim 1$. The implication is that the onset velocity $v_c/v_{Te} \gg 1$; the runaway electrons are mainly generated from the high-energy tail of the distribution function. Consequently, when the probability integral $\Phi(v_c/v_{Te})$ appears in the evaluation of various quantities, one can obtain an accurate analytic approximation by using the large argument expansion: $\Phi(\zeta) \approx 1 - [\exp(-\zeta^2)/\pi^{1/2}\zeta]$ with $\zeta \gg 1$.

The first quantity of interest is the fraction of runaway electrons $f_n = n_R/(n_R + n)$, where $n_R/n$ is calculated as follows:

$$\frac{n_R}{n} = \frac{1}{\sqrt{\pi}\,v_{Te}} \int_{v_c}^{\infty} e^{-v_\parallel^2/v_{Te}^2} dv_\parallel = \frac{1}{2}(1 - \Phi) \approx \frac{1}{2\sqrt{\pi}} \frac{e^{-\zeta^2}}{\zeta}. \qquad (9.84)$$

Here $\Phi = \Phi(\zeta)$ and $\zeta = v_c/v_{Te} = 8.8(n_{20}/|E_\parallel|T_k)^{1/2}$.

Consider next the fraction of the total electron energy density carried by runaways. As a simple approximation assume that all the runaways accelerate to near relativistic

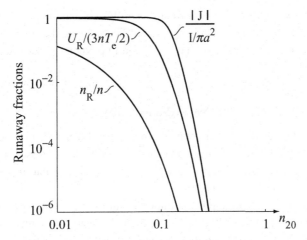

Figure 9.14 Contributions of runaway electrons to the number density, energy density, and current density.

speeds so that their energy density is given by $n_R m_e c^2/2$. The energy density fraction $f_U = U_R/(U_R + 3nT_e/2)$ is determined by noting that

$$\frac{U_R}{3nT_e/2} \approx \frac{n_R m_e c^2/2}{3nT_e/2} = \left(\frac{m_e c^2}{3T_e}\right)\left(\frac{n_R}{n}\right). \qquad (9.85)$$

The last quantity of interest is the fraction of the total current density carried by runaways. Again, if one assumes acceleration to relativistic speeds, then the current density carried by runaways is just $J_R = -en_R c$. The corresponding current density fraction $f_J = |J_R|/(|J_R| + I/\pi a^2)$ can thus be determined from

$$\frac{|J_R|}{I/\pi a^2} = \left(\frac{\pi a^2 enc}{I}\right)\left(\frac{n_R}{n}\right). \qquad (9.86)$$

Note that all the quantities of interest are proportional to $n_R/n$. These quantities are plotted as a function of $n_{20}$ in Fig. 9.14 for a typical present day experiment characterized by $T_k = 2$, $|E_\parallel| = 0.5$ V/m, $a = 1$ m, and $I = 1$ MA. The most striking feature of Fig. 9.14 is the very strong dependence of all the quantities of interest on the density. For densities $n_{20} \sim 1$, the fraction of runaways is negligible: $n_R/n \sim 8 \times 10^{-35}$. For densities $n_{20} \sim 0.1$, the runaways are starting to make a substantial contribution to the energy density and already dominate the current density. For densities $n_{20} \sim 0.01$, the runaways also dominate the current density.

There are several conclusions to be drawn from the analysis. First, runaway electrons are an interesting physical phenomenon arising because of the $1/v_e^3$ dependence of the Coulomb collision frequency. Second, for typical experimental parameters in current and future experiments, the contribution of runaways to the number density, energy density, and current density is a very strong function of the background plasma density. Third, runaways are unimportant when $n_{20} \sim 1$. They can dominate behavior when $n_{20} \sim 0.01$. Finally, the way to avoid the undesirable effects of runaways in fusion experiments and reactors is to always operate at high densities. This is not too difficult during the steady state portion of the discharge but some caution is needed during start up when the densities are lower and the electric fields are higher.

## 9.7 Net exchange collisions

The discussion so far has focused on momentum slowing down collisions between a single test particle and a population of target particles. This section addresses the issues of net momentum and net energy exchanges between an entire test population (not just a single test particle) with an entire target population.

Knowledge of these exchange rates is important in the development of self-consistent fluid models. For example, a parallel electric field drives electrons through ions. The friction caused by Coulomb collisions causes the electrons to transfer some of their momentum to the ions, leading to the macroscopic phenomenon of resistivity. As another example, consider an external heating source that preferentially heats electrons. Coulomb collisions cause the electrons to transfer some of their energy to the ions, leading to temperature equilibration. The rate and effectiveness of this transfer is important since it is the ions that ultimately must be heated to 15 keV for fusion reactions to occur.

In terms of the analysis, recall that the derivation of a test particle collision frequency requires the specification of the test particle velocity $v_e$ and a sum (i.e., integration) over target particles. The result is a collision frequency of the form $v_{ei}(v_e)$. At first glance it might appear that the net exchange collision rates simply require some appropriate form of averaging over all test particle velocities; that is, a distribution function weighted integration must be carried out over the test particles. This is essentially true although there are several subtleties.

The first point is that Coulomb collisions are elastic. This implies that the combined momentum and combined energy of the two interacting particles are conserved (although not individually for each particle) before and after a Coulomb collision. Therefore if both species have a Maxwellian distribution function with the same temperature, there can be no net exchange of momentum or energy when summing over all test and target particles; by symmetry there is no preferred direction for exchanges to occur. For net exchanges to take place the species must have different distribution functions. In the analysis the electrons and ions are assumed to have shifted Maxwellian distribution functions with slightly different temperatures. The shift refers to the location of the peak of $f(\mathbf{v})$, which generates a net macroscopic momentum for each species.

Second, net exchange collisions can only occur between unlike species (e.g. electrons and ions). Since total momentum and energy are conserved during a Coulomb collision, electrons colliding with electrons or ions with ions produce no net exchanges when summing over all like particles.

The third point involves the parameter regime. For mathematical simplicity attention is focused on plasmas where the temperature difference between electrons and ions is small:

specifically $(T_e - T_i)/(T_e + T_i) \ll 1$. Similarly the average macroscopic flow velocities $\mathbf{u}_e$, $\mathbf{u}_i$ are assumed to be small compared to the thermal velocities: $|\mathbf{u}_e| \ll v_{Te}$ and $|\mathbf{u}_i| \ll v_{Ti}$. This regime is characteristic of many present day experiments.

The goal then of this section is to calculate the net momentum and net energy exchange collision frequencies due to Coulomb collisions in a plasma where the electrons and ions have near Maxwellian distribution functions.

### 9.7.1 Formulation of the problem

The formulation of the problem is similar to one presented in Section 9.3 for test particle collision frequencies. It is also similar to the early discussion of generalized reaction rates described in Section 3.3. The goal of the formulation is to obtain a meaningful definition, and corresponding mathematical relationship, enabling the evaluation of the net momentum and net energy exchange collision frequencies.

To begin consider the electron–ion momentum exchange collision frequency. Without loss in generality, one can assume the ions are the target particles and the electrons the test particles (since the summation is carried out over both species). Assume that the initial velocity of the electron is $\mathbf{v}_e = v_{ex}\mathbf{e}_x$ while the ion is at rest. As before, the directed momentum lost by a single test electron undergoing a Coulomb collision with a single target ion is denoted by $\Delta m_e v_{ex}$.

Next, starting with the simple hard sphere collision model, recall that the reaction rate, which is the total number of collisions per second per unit volume, is given by $R_{ei} = n_e n_i \sigma v_e$. Consequently, for a small physical volume element $d\mathbf{r} = dx\,dy\,dz$, the total momentum lost per second by the electrons to the ions is just $(\Delta m_e v_{ex}) n_e n_i \sigma v_e d\mathbf{r}$. This expression can be easily generalized to Coulomb collisions, including a distribution function for electrons and ions, by the following substitutions: $n_e \rightarrow f_e(\mathbf{v}_e)d\mathbf{v}_e$, $n_i \rightarrow f_i(\mathbf{v}_i)d\mathbf{v}_i$, $v_e \rightarrow |\mathbf{v}_e - \mathbf{v}_i|$, and $\sigma \rightarrow b\,db\,d\alpha$.

One additional step is needed to close the formulation, which is related to the definition of the "total electron momentum." Observe that for a pure Maxwellian distribution function the total momentum density of the electrons is zero because of odd symmetry in the integrand: $\int m_e \mathbf{v}_e f_e d\mathbf{v}_e = 0$. However, for a shifted Maxwellian the net momentum of all the electrons in $d\mathbf{r}$ is not zero and is in fact defined in terms of a macroscopic velocity $\mathbf{u}_e(\mathbf{r}, t)$ as follows: $d\mathbf{r} \int m_e \mathbf{v}_e f_e d\mathbf{v}_e \equiv m_e n_e \mathbf{u}_e d\mathbf{r}$. Therefore, when there is a net loss of total electron momentum to the ions, it is the quantity $m_e n_e \mathbf{u}_e$ that is reduced in value.

Combining these results leads to the following relationship governing momentum exchange between electrons and ions:

$$\frac{d}{dt}(m_e n_e \mathbf{u}_e d\mathbf{r}) = -d\mathbf{r}\int (\Delta m_e v_{ex})\,f_e\,(\mathbf{v}_e)\,f_i\,(\mathbf{v}_i)\,|\mathbf{v}_e - \mathbf{v}_i|\,b\,db\,d\alpha\,d\mathbf{v}_e\,d\mathbf{v}_i. \quad (9.87)$$

Here, for mathematical simplicity it has been assumed that $\mathbf{u}_e = u_e\mathbf{e}_x$. The final result can be easily generalized to an arbitrary direction of velocity. A further simplification to Eq. (9.87) results from the recognition that the total number of electrons in the volume $d\mathbf{r}$ remains fixed during the collision process because of the fast time scale and short length scale of each collision. Particles move around a great deal in velocity space but not in physical space. Stated mathematically (and proved in detail in the next chapter), this argument implies that

$$\frac{d}{dt}(n_e d\mathbf{r}) = 0. \tag{9.88}$$

Equation (9.87) reduces to

$$m_e n_e \frac{du_e}{dt} = -\int (\Delta m_e v_{ex}) f_e(\mathbf{v_e}) \, f_i(\mathbf{v_i}) |\mathbf{v_e} - \mathbf{v_i}| b \, db \, d\alpha \, d\mathbf{v_e} \, d\mathbf{v_i}. \tag{9.89}$$

One is now in a position to define the net momentum exchange collision frequency between electrons and ions, $\bar{v}_{ei}(n, T)$. The frequency is defined in terms of the difference in macroscopic velocities as follows:

$$m_e n \frac{du_e}{dt} \equiv -m_e n \, \bar{v}_{ei}(u_e - u_i), \tag{9.90}$$

where $n_e = n_i = n$. Note that there can be a net exchange of momentum only when the macroscopic velocities of electrons and ions are different. Conversely, when $u_e = u_i$ there is no reason or preferred direction for one species to transfer momentum to the other. The combination of Eqs. (9.89) and (9.90) provides an explicit formulation for the calculation of $\bar{v}_{ei}(n, T)$.

A similar analysis applies to net energy exchange collisions between electrons and ions. In this case, the macroscopic thermal energies of the electrons and ions in a volume $d\mathbf{r}$ are given by $U_e = (3/2)n_e T_e \, d\mathbf{r}$ and $U_i = (3/2)n_i T_i \, d\mathbf{r}$. Energy is exchanged between electrons and ions only when there is a temperature differential between the species: $T_e \neq T_i$. In analogy with momentum exchange, this situation suggests the following definition for the temperature equilibration time $\bar{\tau}_{eq}(n, T)$, or equivalently, the temperature equilibration collision frequency $\bar{v}_{eq}(n, T) = 1/\bar{\tau}_{eq}(n, T)$

$$\frac{dU_e}{dt} \equiv -\bar{v}_{eq}(U_e - U_i) \tag{9.91}$$

or, noting that $n_e d\mathbf{r} = \text{const.}$,

$$\frac{3}{2}n \frac{dT_e}{dt} \equiv -\frac{3}{2}n\bar{v}_{eq}(T_e - T_i). \tag{9.92}$$

The left hand side of Eq. (9.92) represents the time rate of change of the electron energy density due to energy exchange collisions with the ions. If one now denotes the loss in electron energy due to a single electron–ion collision as $\Delta m_e v_e^2 / 2$, then, again in analogy with momentum exchange, the left hand side can be calculated as follows

$$\frac{3}{2}n \frac{dT_e}{dt} = -\int \left(\Delta \frac{m_e v_e^2}{2}\right) f_e(\mathbf{v_e}) \, f_i(\mathbf{v_i}) |\mathbf{v_e} - \mathbf{v_i}| b \, db \, d\alpha \, d\mathbf{v_e} \, d\mathbf{v_i}. \tag{9.93}$$

Equations (9.92) and (9.93) provide an explicit set of relations to determine $\bar{v}_{eq}$.

There are two remaining points to make regarding the formulation. First, since the combined momentum and combined energy of an electron–ion pair are conserved during a

Coulomb collision the macroscopic momentum and energy exchange rates for the ions are related to those of the electrons by

$$m_i n \frac{du_i}{dt} = -m_e n \frac{du_e}{dt} = m_e n \, \bar{\nu}_{ei}(u_e - u_i),$$

$$\frac{3}{2} n \frac{dT_i}{dt} = -\frac{3}{2} n \frac{dT_e}{dt} = \frac{3}{2} n \bar{\nu}_{eq}(T_e - T_i).$$

$$(9.94)$$

What one species loses, the other species gains.

The second point is that the actual evaluation of the net exchange collision frequencies is a somewhat ambitious task, involving an eight-fold integral. Even so, for slightly shifted Maxwellian distribution functions with a small temperature differential, the quantities $\bar{\nu}_{ei}$ and $\bar{\nu}_{eq}$ can be evaluated analytically.

### 9.7.2 The net momentum exchange collision rate

The net momentum exchange collision frequency $\bar{\nu}_{ei}$ is evaluated by a procedure similar to that used for the test particle collision frequencies. The details are given below starting with the specification of the distribution functions, followed by the straightforward but somewhat tedious evaluation of the multi-dimensional collision integral.

#### The distribution functions

For net exchange collisions, the appropriate distribution functions correspond to shifted Maxwellians with slightly different temperatures. Specifically, $f_e$ and $f_i$ are assumed to be of the forms:

$$f_e(\mathbf{v}_e) = n_e \left(\frac{m_e}{2\pi T_e}\right)^{3/2} \exp\left(-\frac{m_e}{2T_e}\left[(v_{ex} - u_e)^2 + v_{ey}^2 + v_{ez}^2\right]\right),$$

$$f_i(\mathbf{v}_i) = n_i \left(\frac{m_i}{2\pi T_i}\right)^{3/2} \exp\left(-\frac{m_i}{2T_i}\left[(v_{ix} - u_i)^2 + v_{iy}^2 + v_{iz}^2\right]\right).$$

$$(9.95)$$

For simplicity the macroscopic velocities $\mathbf{u}_e$, $\mathbf{u}_i$ are assumed to flow in the $x$ direction; that is, one can easily show that

$$\int \mathbf{v}_e f_e \, d\mathbf{v}_e = n_e u_e \mathbf{e}_x,$$

$$\int \mathbf{v}_i f_i \, d\mathbf{v}_i = n_i u_i \mathbf{e}_x.$$

$$(9.96)$$

The calculation is made tractable by introducing two assumptions. First, the flow velocity is assumed small compared to the thermal velocity: $u \ll v_T$ for each species. Second the temperature differential is assumed to be small. Thus, the temperatures can be written as

$$T_e = T + \Delta T/2,$$

$$T_i = T - \Delta T/2,$$

$$(9.97)$$

where $T = (T_e + T_i)/2$, $\Delta T = T_e - T_i$, and $\Delta T \ll T$. These assumptions are reasonably well satisfied in most present day experiments.

The next step is to exploit the small $u$, small $\Delta T$ assumptions by Taylor expanding the distribution functions. A short calculation yields

$$f_e(\mathbf{v}_e) \approx n_e \left(\frac{m_e}{2\pi T}\right)^{3/2} \exp\left(-\frac{m_e v_e^2}{2T}\right)\left[1 + \frac{m_e v_{ex} u_e}{T} - \left(3 - \frac{m_e v_e^2}{T}\right)\frac{\Delta T}{4T}\right],$$

$$f_i(\mathbf{v}_i) \approx n_i \left(\frac{m_i}{2\pi T}\right)^{3/2} \exp\left(-\frac{m_i v_i^2}{2T}\right)\left[1 + \frac{m_i v_{ix} u_i}{T} + \left(3 - \frac{m_i v_i^2}{T}\right)\frac{\Delta T}{4T}\right].$$

(9.98)

This approximation for the distribution functions is sufficient to determine both $\bar{\nu}_{ei}$ and $\bar{\nu}_{eq}$.

### New velocity variables

Having specified the distribution functions one must now start to evaluate the multi-dimensional collision integral. This task is most easily carried out by transforming the velocity variables $\mathbf{v}_e$, $\mathbf{v}_i$ to the center of mass variables $\mathbf{v}$, $\mathbf{V}$ using the previously defined relations

$$\mathbf{V} = \frac{m_e \mathbf{v}_e + m_i \mathbf{v}_i}{m_e + m_i}, \qquad \mathbf{v}_e = \mathbf{V} + \frac{m_i}{m_e + m_i}\mathbf{v},$$

$$\mathbf{v} = \mathbf{v}_e - \mathbf{v}_i, \qquad \mathbf{v}_i = \mathbf{V} - \frac{m_e}{m_e + m_i}\mathbf{v}.$$

(9.99)

The calculation is further simplified by introducing spherical velocity coordinates for both the center of mass velocity $\mathbf{V}$ and the relative velocity $\mathbf{v}$:

$$\mathbf{v} = v(\cos\theta\, \mathbf{e}_x + \sin\theta\sin\phi\, \mathbf{e}_y + \sin\theta\cos\phi\, \mathbf{e}_z),$$

$$\mathbf{V} = V(\cos\theta'\mathbf{e}_x + \sin\theta'\sin\phi'\mathbf{e}_y + \sin\theta'\cos\phi'\mathbf{e}_z).$$

(9.100)

A short calculation shows that the collision integral for $\bar{\nu}_{ei}$ given by Eq. (9.89) can now be written as

$$m_e n_e \frac{du_e}{dt} = -\int (\Delta m_e v_{ex}) f_e\, f_i\, v(b\, db\, d\alpha)(v^2 \sin\theta dv d\theta d\phi)(V^2 \sin\theta' dV d\theta'\phi').$$ (9.101)

### Calculation of $\Delta m_e v_{ex}$ and the integration over $b$ and $\alpha$

One must now calculate $\Delta m_e v_{ex}$ and begin to evaluate the multi-dimensional integral. The quantity $\Delta m_e v_{ex}$ has already been evaluated in connection with the test particle collision frequencies and is given here for convenience in terms of the present notation:

$$\Delta m_e v_{ex} = m_r v[(1 - \cos\chi)\mathbf{e}_X \cdot \mathbf{e}_x - \sin\chi\, \mathbf{e}_\perp \cdot \mathbf{e}_x].$$ (9.102)

As before the second term averages to zero when integrating over $\alpha$ and can be ignored. The remaining non-vanishing term can be expressed as

$$\Delta m_e v_{ex} = 2m_r v \cos\theta \frac{b_{90}^2}{b^2 + b_{90}^2}.$$ (9.103)

Since the only explicit appearance of $b$ and $\alpha$ in the integrand occurs in $\Delta m_e v_{ex}$ it is now straightforward to carry out the integration over these variables:

$$\int_0^{2\pi} d\alpha \int_0^{b_{max}} (\Delta m_e v_e x)\, b\, db = (4\pi m_r \ln \Lambda)\, b_{90}^2\, v \cos \theta. \tag{9.104}$$

### Integration over the velocities

The next step is to substitute the velocity transformation into the distribution functions and carry out the integrations over $d\mathbf{V}$ and $d\mathbf{v}$. The details are as follows. First, note that

$$f_e(\mathbf{v}_e) f_i(\mathbf{v}_i) \approx n_e n_i \left( \frac{m_e}{2\pi T} \right)^{3/2} \left( \frac{m_i}{2\pi T} \right)^{3/2} \exp \left( -\frac{m_e v_e^2 + m_i v_i^2}{2T} \right) F(\mathbf{v}_e, \mathbf{v}_i), \tag{9.105}$$

$$F(\mathbf{v}_e, \mathbf{v}_i) = 1 + \frac{m_e v_{ex} u_e + m_i v_{ix} u_i}{T} + \frac{m_e v_e^2 - m_i v_i^2}{T} \frac{\Delta T}{4T}.$$

The exponential factor transforms into a relatively simple form given by

$$\exp \left( -\frac{m_e v_e^2 + m_i v_i^2}{2T} \right) = \exp \left( -\frac{(m_e + m_i)V^2 + m_r v^2}{2T} \right). \tag{9.106}$$

The function $F$, when expanded, consists of a substantial number of terms. The algebra can be simplified by focusing on the $\theta$ dependence of $F$ and noting that the integral over $b$ and $\alpha$ has produced a term proportional to $\cos \theta$. See Eq. (9.104). Therefore only those terms in $F$ that are of the form $\hat{F} \cos \theta$ lead to a contribution that does not average to zero when integrating over $0 \le \theta \le \pi$. These terms are easily selected and can be written as

$$F = \frac{m_r v (u_e - u_i)}{T} \cos \theta. \tag{9.107}$$

One now combines all these results and introduces normalized velocity variables $W$ and $w$ as follows: $V = [2T/(m_e + m_i)]^{1/2} W$, $v = (2T/m_r)^{1/2} w$. The basic momentum exchange relation given by Eq. (9.101) reduces to

$$m_e n_e \frac{du_e}{dt} = -\frac{\sqrt{2}I}{8\pi^4} \frac{n_e n_i e^4 m_e^{1/2} \ln \Lambda}{\varepsilon_0^2 (1 + m_e/m_i)^{3/2} T^{3/2}} (u_e - u_i), \tag{9.108}$$

where the multi-dimensional integral $I$ is the product of six, easily evaluated, separable integrals

$$I = \int_0^{2\pi} d\phi' \int_0^{2\pi} d\phi \int_0^{\pi} \sin \theta' d\theta' \int_0^{\pi} \sin \theta \cos^2 \theta \, d\theta \int_0^{\infty} W^2 e^{-W^2} dW \int_0^{\infty} w e^{-w^2} dw$$

$$= \frac{2\pi^{5/2}}{3}. \tag{9.109}$$

Finally, by comparing Eq. (9.108) with Eq. (9.90) and assuming that $m_e \ll m_i$, one obtains the desired expression for $\bar{\nu}_{ei}$:

$$\bar{\nu}_{ei} = \frac{\sqrt{2}}{12\pi^{3/2}} \frac{e^4 n_i}{\varepsilon_0^2 m_e^{1/2} T^{3/2}} \ln \Lambda = \sqrt{\frac{2}{\pi}} \frac{\omega_{pe}}{\Lambda} \ln \Lambda = 1.8 \times 10^5 \frac{n_{20}}{T_k^{3/2}} \, \text{s}^{-1}. \tag{9.110}$$

Observe that $\bar{\nu}_{ei} \sim \hat{\nu}_{ei}$. The net momentum exchange collision frequency, to within a numerical factor on the order of unity, is equal to the typical value of the test particle slowing down frequency $\bar{\tau}_{ei}(v_i)$. In terms of numerical values, $\bar{\tau}_{ei} = 1/\bar{\nu}_{ei} \sim 0.16 \times 10^{-3}$ s for a fusion plasma, which is much shorter than the required energy confinement time. Momentum will be exchanged well before energy is lost.

The primary application of the quantity $\bar{\nu}_{ei}$ is in its contribution to the self-consistent fluid equations describing momentum conservation in fusion plasmas. This important topic is discussed in detail in the next chapter. It is shown that $\bar{\nu}_{ei}$ gives rise to the small but finite resistivity of the plasma and thus directly impacts the ohmic heating of a plasma. In Chapter 14 it is also shown that $\bar{\nu}_{ei}$ is closely connected to the particle diffusion coefficient in physical space.

### 9.7.3 The net energy exchange collision rate

The calculation of the net energy exchange collision frequency is very similar to the one just presented. Two modifications must be made. First, one must calculate the loss in electron energy $\Delta m_e v_e^2 / 2$ per collision as opposed to the loss in directed momentum $\Delta m_e v_{ex}$. Second, a different term in the function $F$, defined in Eq. (9.105), now makes the non-vanishing contribution to the collision integral.

Consider first the electron energy loss per collision. The electron energy at any instant of time, expressed in terms of the center of mass variables, is

$$\frac{1}{2} m_e v_e^2 = \frac{m_e}{2} \left[ V^2 + \left( \frac{m_i}{m_e + m_i} \right)^2 v^2 \right] + m_r \mathbf{V} \cdot \mathbf{v}. \tag{9.111}$$

Since $\mathbf{V}$ and $v^2$ are conserved before and after the collision, it follows that the loss in electron energy can be written as

$$\Delta m_e v_e^2 / 2 = m_r \mathbf{V} \cdot \Delta \mathbf{v} = m_r \mathbf{V} \cdot (\mathbf{v}_{\text{initial}} - \mathbf{v}_{\text{final}}). \tag{9.112}$$

Next, recall that $\mathbf{v}_{\text{initial}} = v \mathbf{e}_X$ and $\mathbf{v}_{\text{final}} = v(\cos \chi \, \mathbf{e}_X + \sin \chi \, \mathbf{e}_\perp)$, implying that

$$\Delta m_e v_e^2 / 2 = m_r v \mathbf{V} \cdot [(1 - \cos \chi) \mathbf{e}_X - \sin \chi \, \mathbf{e}_\perp]. \tag{9.113}$$

As before the second term averages to zero when integrating over $\alpha$. The non-vanishing contribution reduces to

$$\Delta \frac{m_e v_e^2}{2} = m_r \mathbf{V} \cdot \mathbf{v} (1 - \cos \chi) = \frac{2b_{90}^2}{b^2 + b_{90}^2} m_r \mathbf{V} \cdot \mathbf{v}. \tag{9.114}$$

The second modification involves the non-vanishing contributions arising in the function $F$. If one focuses on the $\theta$ and $\theta'$ dependence of the various terms, comparing them with the corresponding dependence in Eq. (9.114), it follows that the only non-vanishing contribution is given by

$$F = \left(\frac{m_r \mathbf{V} \cdot \mathbf{v}}{T}\right) \frac{\Delta T}{T}. \tag{9.115}$$

Combining terms leads to the following simplification of Eq. (9.93), the defining equation for the $\bar{\nu}_{eq}$ collision integral:

$$\frac{3}{2} n \frac{dT_e}{dt} = -(T_e - T_i) \int f_{Me} f_{Mi} v \left(\frac{m_r \mathbf{V} \cdot \mathbf{v}}{T}\right)^2 \frac{2b_{90}^2}{b^2 + b_{90}^2} b \, db \, d\alpha \, d\mathbf{V} \, d\mathbf{v}. \tag{9.116}$$

Here, $f_{Me}$, $f_{Mi}$ are unperturbed Maxwellians (i.e., no shifts and equal temperatures) and, in spherical velocity variables,

$$\mathbf{V} \cdot \mathbf{v} = Vv[\cos\theta \cos\theta' + \sin\theta \sin\theta' \cos(\phi - \phi')]. \tag{9.117}$$

The integral in Eq. (9.116) can now be evaluated in a straightforward manner. After a lengthy calculation one finds for $m_e \ll m_i$

$$\frac{3}{2} n \frac{dT_e}{dt} = -\frac{\sqrt{2}}{4\pi^{3/2}} \frac{n_e n_i m_e^{1/2} e^4 \ln\Lambda}{m_i T^{3/2} \varepsilon_0^2} (T_e - T_i). \tag{9.118}$$

A comparison with Eqs. (9.92) and (9.110) leads to the desired expression for $\bar{\nu}_{eq}$:

$$\bar{\nu}_{eq} = 2 \frac{m_e}{m_i} \bar{\nu}_{ei} = 78 \frac{n_{20}}{T_k^{3/2}} \quad s^{-1}. \tag{9.119}$$

The numerical value corresponds to a 50%–50% D–T mixture.

As expected, the collision frequency for energy exchange is much lower than that for momentum exchange. It is smaller by the mass ratio $m_e/m_i$. An electron colliding with a heavy ion can have a large change in the direction of its velocity but its energy remains largely unchanged – a familiar analogy is a ping pong ball bouncing off a bowling ball. For a fusion plasma $\bar{\tau}_{eq} = 1/\bar{\nu}_{eq} \sim 0.4$ s. The time is shorter, but not by much, than the required energy confinement time. Still, heat preferentially supplied for instance to electrons, should have time to transfer and equilibrate with the ions, before it is lost by heat conduction.

The energy equilibration collision frequency plays an important role in the conservation of energy relations for electrons, ions, and alphas. Clearly, energy must be transferred from the electrons and the alphas to the ions in a timely fashion in order to produce fusion power. An energy balance between alphas and electrons with cold ions makes no physical sense since it is the ions and not the electrons that produce fusion reactions.

## 9.8 Summary

Coulomb collisions cause transport in both velocity space and physical space. Chapter 9 focuses on velocity space transport. Almost all of the useful results concerning various collision frequencies are based on a fundamental property of two-particle Coulomb collisions that relates the scattering angle to the relative velocity and impact parameter: $\cot(\chi/2) = b/b_{90}$, where $b_{90} = q_1 q_2 / 4\pi\varepsilon_0 m_r v^2$.

The first class of problems considered involves single test particle collision frequencies. Here, attention is focused on the loss of directed momentum of a single test particle colliding with the entire populations of plasma ions and electrons. It is shown that small-angle collisions dominate the loss process and in fact it requires the introduction of Debye shielding to set a finite bound on the collision frequencies. Typically the electron–ion slowing down time scales as $\nu_{ei} \sim (\omega_{pe}/\Lambda) \ln \Lambda \propto n/T^{3/2}$ and is a fraction of a millisecond for a fusion plasma. For a mass ratio $\mu = m_e/m_i$ the other test particle collision frequencies scale as $\nu_{ee} \sim \nu_{ei}$, $\nu_{ii} \sim \mu^{1/2}\nu_{ei}$, $\nu_{ie} \sim \mu\nu_{ei}$.

Several applications of the test particle collision analysis are investigated. First, the simple mirror machine is revisited. Its $Q$ value is limited to values of the order of unity because of the rapid loss of particles on the ion–ion collision time scale due to Coulomb collisions. The problem of alpha slowing down is also investigated. The results show that alphas slow down predominantly on the electrons for alpha energies in the range 3.5 MeV $> m_b v_b^2/2 >$ 660 keV. Overall, about 70% of the alpha energy is apportioned to electrons and 30% to ions. The electron energy is then transferred to the ions on the electron–ion equilibration time scale. The whole process takes place in about 0.5 s, which is faster, although not by a large factor, than the required energy confinement time, typically 1 s. The last application involves runaway electrons. Here, a small population of electrons can be accelerated to relativistic speeds by a small DC electric field, a phenomenon made possible by the $1/v_e^3$ dependence of $\nu_{ei}$ and $\nu_{ee}$ for large velocities. Runaways have a damaging effect on experiments and can be avoided by operation at sufficiently high densities: $n_{20} > 0.1$.

A second form of velocity space transport is investigated involving the net exchange of momentum and energy between electrons and ions. Here, the total populations of both species are included when summing up the consequences of Coulomb collisions. Net exchange collisions are important only between different species. Furthermore, the species must have different distribution functions, for example shifted Maxwellians with different temperatures. It has been shown that the net exchange collision frequency for momentum transfer scales as $\bar{\nu}_{ei} \sim (\omega_{pe}/\Lambda) \ln \Lambda \propto n/T^{3/2}$. A similar calculation has shown that the net exchange collision frequency for energy equilibration is much lower: $\bar{\nu}_{eq} \sim \mu\bar{\nu}_{ei}$. The main application of the net exchange collision frequencies is in the formulation of self-consistent plasma models. This is the topic of the next chapter.

## Bibliography

Very early in the fusion program researchers appreciated the fact that Coulomb collisions represented a basic irreducible minimum limit for various types of transport losses in a plasma including heat, particles, and flux. Therefore, a great deal of effort has been devoted to understanding Coulomb collisions, a more difficult task than one might have imagined because of the weak decay of the Coulomb potential and the necessity of including the effects of Debye shielding. Several useful references are as follows.

Boyd, T. J. M. and Sanderson, J. J. (2003). *The Physics of Plasmas*. Cambridge, England: Cambridge University Press.

Braginskii, S. I. (1965). *Reviews of Plasma Physics* (Leontovich, M. A. editor) Vol. 1. New York: Consultants Bureau.

Dolan, T. J. (1982). *Fusion Research*. New York: Pergamon Press.

Dreicer, H. (1960). Electron and ion runaway in a fully ionized gas, *Physical Review* **117**, 329.

Goldston, R. J. and Rutherford, P. H. (1995). *Introduction to Plasma Physics*. Bristol, England: Insititute of Physics Publishing.

Helander, P. and Sigmar, D. J. (2002). *Collisional Transport in Magnetized Plasmas*. Cambridge, England: Cambridge University Press.

Krall, N. A. and Trivelpiece, A. W. (1973). *Principles of Plasma Physics*. New York: McGraw Hill Book Company.

Miyamoto, K. (2001). *Fundamentals of Plasma Physics and Controlled Fusion*, revised edn. Toki City: National Institute for Fusion Science.

Spitzer, L. (1962). *The Physics of Fully Ionized Gases*, second edn. New York: Interscience.

Stacey, W. M. (2005). *Fusion Plasma Physics*. Weinheim: Wiley-VCH.

Wesson, J. (2004). *Tokamaks*, third edn. Oxford: Oxford University Press.

## Problems

9.1 Consider two particles acting upon each other with a repulsive central force potential given by $\phi(r) = \phi_0(r_0^2/r^2)$, where $\phi_0$ and $r_0$ are positive constants.

   (a) Using a heuristic argument estimate the cross section for a 90° momentum collision.

   (b) Using the exact particle trajectories, calculate the relationship between the scattering angle $\chi$ and the impact parameter $b$.

   (c) Calculate the directed momentum collision frequency and corresponding scattering cross section by integrating over all scattering angles $\chi$. (One integral will have to be evaluated numerically.) Compare this result with that obtained in part (a).

9.2 Consider two equal but oppositely charged particles with different masses undergoing a binary collision under the action of a central force potential $\phi = -\phi_0(r_0/r)^4$, $\phi_0 > 0$.

   (a) Write down the equations of motion in the center of mass reference frame.

   (b) Calculate the smallest impact parameter that the center of mass particle can have and still ultimately escape the central force field. In other words the collision should deflect the orbit but not inextricably draw the particle into the origin $r(t) \rightarrow 0$. Express your answer as $b = b(v_0, \phi_0, r_0, q, m_1, m_2)$, where $v_0$ is the relative speed.

9.3 This goal of this problem is to determine the time it takes for an anisotropic distribution function to relax to an isotropic Maxwellian. Consider a plasma in which the electron distribution function is

$$f_e = \frac{n_0}{\pi^{3/2}} \left(\frac{m_e}{2T_\perp}\right) \left(\frac{m_e}{2T_\parallel}\right)^{1/2} \exp\left(-\frac{m_e v_\perp^2}{2T_\perp} - \frac{m_e v_\parallel^2}{2T_\parallel}\right).$$

Assume that $T_\perp = T + \Delta T/2$ and $T_\parallel = T - \Delta T/2$, where $T = (T_\perp + T_\parallel)/2$ and $\Delta T = T_\perp - T_\parallel$. Assume that $\Delta T/T \ll 1$. Following the analysis in the text for energy exchange collisions define

$$\frac{3}{2}n\frac{dT_\perp}{dt} = -\frac{3}{2}n\bar{v}_\perp(T_\perp - T_\parallel),$$

$$\frac{3}{2}n\frac{dT_\parallel}{dt} = -\frac{3}{2}n\bar{v}_\parallel(T_\parallel - T_\perp).$$

Show that $\bar{v}_\perp = \bar{v}_\parallel$, thus demonstrating conservation of total energy for like particle collisions. Calculate $\bar{v}_\perp$, the characteristic frequency for relaxing the anisotropy in the distribution function.

9.4 A beam of 100 keV deuterons collides head-on with a beam of 100 keV tritons. The distribution function for each species can be modeled as

$$f_D(\mathbf{v}) = n_D \delta(v_x - v_D)\delta(v_y)\delta(v_z),$$
$$f_T(\mathbf{v}) = n_T \delta(v_x - v_T)\delta(v_y)\delta(v_z),$$

where $n_D = n_T = n_e/2$ and $m_D v_D^2/2 = m_T v_T^2/2 = 100$ keV.

(a) Calculate the momentum exchange collision frequency $\bar{v}_{DT}$ following the procedure outlined in the text.

(b) Evaluate $\bar{v}_{DT}$ assuming that $n_e = 2 \times 10^{20}$ m$^{-3}$.

(c) Assume now that the momentum exchange time is comparable to the time for the distribution function to relax to a Maxwellian. Compare this time to the characteristic time for D–T fusion collisions. Would most of the fusion collisions occur when the D and T are beams or would they occur on the tail of the distribution function after they have relaxed to Maxwellians?

9.5 Consider the situation in which the electrons in a plasma develop a high-energy tail due to cyclotron heating. Furthermore, assume that all the current in the plasma is carried by these hot electrons. This suggests that the electron and ion distribution functions can be modeled as follows:

$$f_e(\mathbf{v}) = \frac{n_B}{\pi^{3/2} v_B^3} \exp\left(-\frac{v^2}{v_B^2}\right) + \frac{n_H}{\pi^{3/2} v_H^3} \exp\left[-\frac{(v_x - u)^2 + v_y^2 + v_z^2}{v_H^2}\right],$$

$$f_i(\mathbf{v}) = \frac{n_i}{\pi^{3/2} v_i^3} \exp\left(-\frac{v^2}{v_i^2}\right),$$

where $v_B = (2T/m_e)^{1/2}$, $v_i = (2T/m_i)^{1/2}$, and $v_H = (2T_H/m_e)^{1/2}$. Note that $T$ is the temperature of the bulk electrons and ions, while $T_H \gg T$ is the temperature of the high-energy tail. Also, charge neutrality requires that $n_e \equiv n_B + n_H = n_i$. An important point is that the current density is given by $\mathbf{J} \equiv -en_e u_e = -en_H u$. Following the analysis related to momentum exchange collisions define

$$m_e n_e \frac{du_e}{dt} \equiv -m_e n_e \bar{v}_{ei}(u_e - u_i).$$

Calculate $\bar{v}_{ei}$ for the above distribution function and compare it with the one derived in the text. Does the result agree with your intuition? Explain.

# 10

# A self-consistent two-fluid model

## 10.1 Introduction

The discussion so far has focused on single-particle motion in prescribed, long-range electric and magnetic fields as well as short-range Coulomb collisions. No attempt has been made at self-consistency. That is, no attempt has been made to determine how the current density and charge density generated by single-particle motion feeds back and alters the original applied electric and magnetic field. The development of a self-consistent plasma model is the goal of Chapter 10.

Self-consistency is a critical issue. It is important in: (1) providing the physical understanding of the macroscopic forces that hold a plasma together; (2) determining the transport of energy, particles, and magnetic flux, across the plasma; (3) understanding how electromagnetic waves propagate into a plasma to provide heating and non-inductive current drive; and (4) learning how small perturbations in current density and charge density can sometimes dramatically affect the macroscopic and microscopic stability of a plasma.

In developing self-consistent models one should be aware that various levels of description are possible. The most accurate models involve kinetic theory. These strive to determine the particle distribution functions $f_e(\mathbf{r}, \mathbf{v}, t)$ and $f_i(\mathbf{r}, \mathbf{v}, t)$. Kinetic models are very accurate as well as being inclusive of a wide variety of physical phenomena. They are also more complicated to solve and tend to be somewhat abstract with respect to physical intuition. Consequently, with respect to the introductory nature of the book, kinetic theory is considered to be an advanced topic, awaiting study at a future time.

The next level of description, and the one focused on in this book, corresponds to macroscopic fluid models. Here, the basic unknowns in the model are easily recognizable physical quantities, such as density, temperature, pressure, etc. The great simplification in comparison to kinetic theory is that all of the unknowns are functions of only space and time: $Q = Q(\mathbf{r}, t)$. In general, a fluid model is not as accurate or complete as a kinetic model but is considerably simpler to solve, leading to a more easily developed physical intuition.

With respect to fusion research, fluid models provide a reasonably accurate description of all the important phenomena: macroscopic equilibrium and stability, transport, and heating and current drive. Although occasionally missing some important physics contained only in a kinetic model, fluid modeling is nevertheless ideal for an introduction to fusion research.

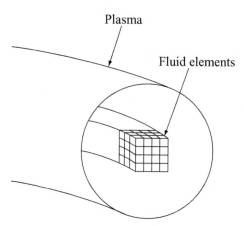

Figure 10.1 Subdividing a plasma into small fluid elements.

It should be noted that kinetic and fluid models are not independent descriptions. Fluid models can be rigorously derived by taking various velocity moments of a kinetic model. The moment procedure always leads to a system with more unknowns than equations. Some form of mathematical expansion, based on the physical regime of interest must be introduced to obtain closure of the system: number of equations = number of unknowns.

Rather than utilizing the more formalistic, mathematical moment approach, one can directly derive a self-consistent two-fluid model of a plasma using physical intuition and some simple mathematics, based on the conservation of mass, momentum and energy. This is the approach taken in Chapter 10. Specifically, a general two-fluid model is derived based on these conservation laws, which serves as the foundation for understanding all of the important fusion-related plasma phenomena described in the remainder of this book.

## 10.2 Properties of a fluid model

The discussion begins with a description of the basic properties of a fluid model. This knowledge is important in developing the framework upon which such models can be developed. The discussion focuses on the assumptions inherent in any fluid description and whether or not they are valid for a fusion plasma

### 10.2.1 Macroscopic averages

The basic idea in a fluid model is to take the medium of interest, in this case a plasma, and subdivide it into a large number of small, moving fluid elements as shown in Fig. 10.1. Each element is assumed to contain a large number of particles. The behavior of each fluid element is described by certain *average* macroscopic properties of the particles contained therein. An example of such a property is the electron number density $n_e$ defined as follows. Assume that at time $t$ the fluid element centered at point $\mathbf{r}$ has a volume $\Delta V$ and contains a total of $N_e$ electrons. The number density of electrons is then defined as

$$n_e(\mathbf{r}, t) = \frac{\text{number of particles}}{\text{volume}} = \frac{N_e}{\Delta V} = \int f_e \, d\mathbf{v}. \tag{10.1}$$

The last form connects the intuitive counting of single particles with the integration of the smoothed out distribution function over velocity. Observe that the number density represents the average number of particles per unit volume. It is not concerned with the microscopic details of the velocity distribution, only the integration over all particles.

Similarly, in a fluid model one defines the macroscopic velocity $\mathbf{u}_e$ as the average velocity of all the electrons contained in the fluid element.

$$\mathbf{u}_e(\mathbf{r}, t) = \frac{\text{sum of the particle velocities}}{\text{number of particles}}$$

$$= \frac{\mathbf{v}_1 + \mathbf{v}_2 + \mathbf{v}_3 + \cdots \mathbf{v}_n}{N_e} = \frac{1}{n_e} \int \mathbf{v} \, f_e \, d\mathbf{v}. \tag{10.2}$$

Here too, only information about the average and not the individual velocities is contained in $\mathbf{u}_e$.

A fluid description of a plasma is thus concerned with developing a model, in this case a set of coupled partial differential equations, describing the evolution of important, macroscopic properties of the plasma. For present purposes, these properties include the number densities $n_e(\mathbf{r}, t)$, $n_i(\mathbf{r}, t)$, macroscopic flow velocities $\mathbf{u}_e(\mathbf{r}, t)$, $\mathbf{u}_i(\mathbf{r}, t)$, pressures $p_e(\mathbf{r}, t)$, $p_i(\mathbf{r}, t)$, and temperatures $T_e(\mathbf{r}, t)$, $T_i(\mathbf{r}, t)$. The as yet undefined macroscopic pressures and temperatures are defined as they appear naturally in the analysis.

### 10.2.2 Size of a fluid element

Another property concerning the validity of fluid models is the size of a fluid element. For a fluid model to be valid it must be possible to define a range of sizes for each element that satisfies two conflicting requirements. On one hand, the element cannot be too small. If it is too small, then only a small number of particles are contained within the element and averaging makes little sense from a statistical point of view.

On the other hand, if the element is too large, then spatial resolution is lost and this may be unacceptable in terms of reduced accuracy. For a fusion plasma, there is a wide range of fluid sizes that are consistent with these two constraints implying that, at least from the averaging point of view, a fluid model makes good sense. As an example consider a fusion plasma with $n_e = 10^{20}$ m$^{-3}$. For most phenomena of interest a fluid element in the shape of a cube whose linear dimension is $\Delta x = 10^{-5}$ m offers a high degree of resolution (keeping in mind that the macroscopic plasma dimension is on the order of 1 m ). This corresponds to a fluid element volume $\Delta V = (\Delta x)^3 = 10^{-15}$ m$^3$. The total number of electrons within this volume element is thus $N_e = n_e \Delta V = 10^5 \gg 1$; there is a wide window where the conflicting requirements are well satisfied.

### 10.2.3 When is a plasma fluid model useful?

Even if subdividing the plasma into small fluid elements and computing various average properties of each element makes sense, this still does not guarantee that a fluid model will be useful. To understand the issue, consider a very familiar fluid, air at atmospheric pressure which also easily satisfies the two conflicting constraints. The air satisfies another crucial property – the molecules within each fluid element are collision dominated. Collisions keep the molecules closely confined together in physical space. A molecule, whose instantaneous velocity can vary in magnitude and direction over a wide range during its orbit, still cannot easily move over long distances with respect to its neighbors. Multiple collisions lead to many random changes in direction, implying that a given particle is more or less confined to a region whose size is comparable to the mean free path. The conclusion is that the molecules in each element form a well-defined cluster of particles, whose identity is maintained as the system evolves in time. This coherence, caused by a high collisionality, is the major reason why fluid models are very useful for air. Each fluid element essentially corresponds to a "super particle" with mass $mn\Delta V$ and velocity $\mathbf{u}$.

Based on this discussion one may wonder why fluid models are useful for fusion plasmas, which have been shown to be nearly collisionless? The answer is slightly subtle. Perpendicular to the magnetic field, the small size of the gyro radius acts to keep particles, even those with widely varying perpendicular velocities, close to one another. The magnetic field replaces collisions in providing perpendicular coherence to the particles in a fluid element. In contrast, particles move freely parallel to the magnetic field. Two particles in the same fluid element with different parallel velocities can easily separate by a large distance along $\mathbf{B}$ thereby preventing parallel coherence from developing.

The conclusion is that plasmas exhibit fluidlike behavior perpendicular, but not parallel, to the magnetic field. In general, the parallel motion must be treated kinetically, which is a more difficult task. The kinetic difficulty is often avoided by the convenient but worrisome procedure of extending the fluidlike treatment to the parallel direction, even though this is an invalid assumption.

The situation is not as bad as it sounds. For many fusion plasma phenomena, it will be shown that the parallel motion is a minor effect. In these situations the fluid model is incorrect only when it is unimportant. In other phenomena, the fluid model captures part, but not all, of the physics. Here, one can obtain a qualitatively correct overall physical picture, although a more sophisticated kinetic model is required for detailed, accurate understanding. Finally, there are some phenomena where the parallel fluid motion is simply treated incorrectly. Fortunately, for present purposes these are few in number.

The end result is that, for an introductory book, the simplifications of fluid vs. kinetic are so large that they dominate the decision as to which approach should be taken. Consequently, this chapter and the remainder of the book focus on pure fluid models to describe the physics of fusion plasmas. However, it will be pointed out whenever the improper treatment of the parallel motion is important.

The task now is to develop a self-consistent two-fluid model, based on the conservation of mass, momentum, and energy, that describes the evolution of the plasma fluid variables – density, velocity, pressure, and temperature. Once obtained, this model is coupled to Maxwell's equations to form a closed system.

## 10.3 Conservation of mass

Consider first the conservation of mass. The goal is to derive an equation for each species that reflects mass conservation. In general, the equation should involve only the basic unknown fluid variables plus the electric and magnetic fields. Although the final conservation relation may be familiar to many readers, the calculation is carried out in detail to demonstrate the procedure, which may be somewhat less intuitive for the momentum and energy relations.

A further point to note is that here, and for the other conservation laws, the derivations are carried out in Lagrangian coordinates which move with the fluid, and are then transformed at the end into the laboratory frame represented by Eulerian coordinates. The reason is that it is intuitively simpler to derive the conservation laws in Lagrangian coordinates, but more convenient to apply them in Eulerian coordinates. The convenience is associated with the ease of specifying $\mathbf{E}$ and $\mathbf{B}$ as well as the boundary conditions in a fixed Eulerian coordinate system.

The derivation begins with a statement of the underlying physical assumption upon which mass conservation is based: a plasma is an ionized gas in which particles are neither created nor destroyed. This assumption neglects ionization and recombination, which are negligible at fusion temperatures since the plasma is already fully ionized by a wide margin. Also neglected are fusion collisions, which are very infrequent in existing experiments. Even in a reactor, fusion collisions only gradually deplete the D–T plasma fuel. This effect can be easily included by introducing a third species of alpha particles and adding particle sources and sinks to the conservation relations. For simplicity, however, alpha particle effects on the electron and ion populations are ignored at present.

Under these assumptions the conservation of mass relation can be easily expressed in Lagrangian coordinates. The relation is valid for both electrons and ions and for convenience the species subscripts have been suppressed. Assume that at time $t = 0$ a small fluid element of volume $\Delta V = \Delta x \Delta y \Delta z$ contains a total number of particles $N = n \Delta V$. See Fig. 10.2. Consistent with the fluid picture, the particles stay close together and move along a smooth trajectory. Since particles are being neither created nor destroyed, conservation of mass requires that

$$\frac{d}{dt}(\text{number of particles in a fluid element}) = \frac{dN}{dt} = 0, \tag{10.3}$$

where the time derivative is taken along the trajectory. With a little work, Eq. (10.3) can be cast in a more familiar form. Observe that

$$\frac{dN}{dt} = \Delta V \frac{dn}{dt} + n \frac{d\Delta V}{dt} \tag{10.4}$$

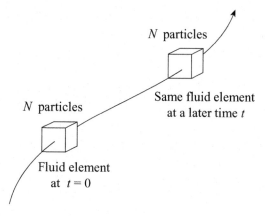

Figure 10.2 Lagrangian trajectory of a fluid element conserving particles.

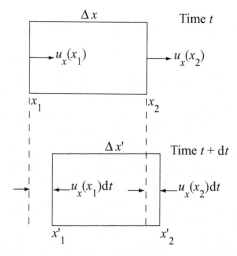

Figure 10.3 Change in $\Delta x$ over a time $dt$. The motion shown corresponds to compression: $\Delta x' < \Delta x$.

with

$$\frac{d\Delta V}{dt} = \Delta y \Delta z \frac{d\Delta x}{dt} + \Delta x \Delta z \frac{d\Delta y}{dt} + \Delta x \Delta y \frac{d\Delta z}{dt}. \tag{10.5}$$

Although the total number of particles in the fluid element is constant, both the number density and the volume of the element can change separately.

The change in the volume element can be rewritten in a more convenient form by examining Fig. 10.3, which for simplicity focuses only on the $x$ direction. At a given time $t$ the left and right hand boundaries of the fluid element are located at $x_1 = x - \Delta x/2$ and $x_2 = x + \Delta x/2$. The corresponding velocities of the boundaries are $u_x(x_1)$ and $u_x(x_2)$. A short time $dt$ later the boundaries have moved to new positions given by $x_1' = x_1 + u_x(x_1)dt$

and $x_2' = x_2 + u_x(x_2)dt$. The new width of the element $\Delta x' = x_2' - x_1'$ is related to the original width $\Delta x = x_2 - x_1$ by

$$\Delta x' = \Delta x + [u_x(x_2) - u_x(x_1)]\,dt \approx \Delta x + \frac{\partial u_x}{\partial x}\Delta x\,dt, \tag{10.6}$$

where the last form is obtained by Taylor expanding about small $\Delta x$. The time rate of change of $\Delta x$ can thus be expressed as

$$\frac{d\Delta x}{dt} = \frac{\Delta x' - \Delta x}{dt} = \frac{\partial u_x}{\partial x}\Delta x. \tag{10.7}$$

Similar expressions apply to the $y$ and $z$ directions.

Combining these results leads to the following form for $dN/dt$:

$$\frac{dN}{dt} = \Delta V\left(\frac{dn}{dt} + n\frac{\partial u_x}{\partial x} + n\frac{\partial u_y}{\partial y} + n\frac{\partial u_z}{\partial z}\right) = \Delta V\left(\frac{dn}{dt} + n\nabla\cdot\mathbf{u}\right) = 0. \tag{10.8}$$

The final form of the conservation of mass is obtained by transforming the Lagrangian time derivative of $n$ in Eq. (10.8) to an Eulerian coordinate system by the usual relation

$$\left(\frac{dn}{dt}\right)_{\text{Lagrangian}} = \left(\frac{\partial n}{\partial t} + \mathbf{u}\cdot\nabla n\right)_{\text{Eulerian}}. \tag{10.9}$$

Equation (10.8), written separately for electrons and ions in Eulerian coordinates, reduces to

$$\frac{\partial n_e}{\partial t} + \nabla\cdot(n_e\mathbf{u}_e) = 0,$$
$$\frac{\partial n_i}{\partial t} + \nabla\cdot(n_i\mathbf{u}_i) = 0. \tag{10.10}$$

This is the desired form for the conservation of mass relations.

## 10.4 Conservation of momentum

### 10.4.1 The basic principle

The next relation involves the conservation of momentum. The basic physical principle involved is Newton's law of motion applied to the particles in the moving fluid element. In schematic form, conservation of momentum requires

$$\frac{d}{dt}(\text{momentum}) = \text{force on the fluid element}. \tag{10.11}$$

For a plasma the forces that need to be included are: (1) the electric field force, (2) the magnetic field force, (3) the pressure gradient force, and (4) the net momentum exchange collisional force. The gravitational force is negligible. The force due to viscosity is also neglected. The reason is that in most cases of interest the viscosity coefficient is not very large and the plasma does not move very fast or in general form narrow boundary layers, except perhaps at the very edge. Therefore, viscosity rarely plays a dominant role in determining plasma behavior.

The terms in Eq. (10.11) are now evaluated one by one for electrons to obtain the conservation of momentum relation. A simple switching of subscripts then yields the corresponding equation for ions.

### 10.4.2 The inertial force

To begin consider a small fluid element containing $N_e = n_e \Delta V$ electrons. The average momentum per electron is $m_e \mathbf{u}_e$. Consequently, the total momentum in the element is $m_e \mathbf{u}_e N$. The left hand side of Eq. (10.11), which represents the inertial forces, can now be written as

$$\frac{d}{dt}(\text{momentum}) = \frac{d}{dt}(m_e n_e \Delta V \, \mathbf{u}_e). \tag{10.12}$$

Since $n_e \Delta V = \text{const.}$ from the conservation of mass, it follows that

$$\frac{d}{dt}(\text{momentum}) = \Delta V m_e n_e \left( \frac{\partial}{\partial t} + \mathbf{u}_e \cdot \nabla \right) \mathbf{u}_e, \tag{10.13}$$

where the transformation to Eulerian coordinates has been made.

### 10.4.3 The electric field force

One must next evaluate the various forces appearing on the right hand side of Eq. (10.11). The first of these is due to the electric field. Since each electron has a charge $-e$, the total charge in the fluid element is just $Q_e = -eN_e$. The Lorentz force acting on this charge due to the electric field is simply $Q_e \mathbf{E}$, which can be rewritten as

$$\text{electric field force} = Q_e \mathbf{E} = -\Delta V e n_e \mathbf{E}. \tag{10.14}$$

Implicit in this expression is the assumption that $\mathbf{E}(\mathbf{r}, t)$ is the long-range, smooth electric field arising from collective effects. The short-range electric field associated with Coulomb collisions is treated explicitly when the collisional force contribution is evaluated.

### 10.4.4 The magnetic field force

The second force is due to the magnetic field. Since the fluid element has a charge $Q_e$ and moves with an average velocity $\mathbf{u}_e$, the Lorentz force due to the magnetic field is given by $Q_e \mathbf{u}_e \times \mathbf{B}$, which can be expressed as

$$\text{magnetic field force} = Q_e \mathbf{u}_e \times \mathbf{B} = -\Delta V e n_e \mathbf{u}_e \times \mathbf{B}. \tag{10.15}$$

### 10.4.5 The pressure gradient force

The next force that contributes to momentum conservation is due to the pressure gradient. To derive this macroscopic force one must actually look microscopically at the flow of electrons

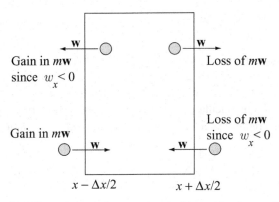

Figure 10.4 Flow of thermal momentum in the $\mathbf{u} = 0$ reference frame.

in and out of each boundary. The force arises because the electrons in the fluid element do not all move with the exact macroscopic velocity $\mathbf{u}_e$. This is only the average value of all their velocities. The actual velocity of any given electron can be written as $\mathbf{v} = \mathbf{u}_e + \mathbf{w}$, where $\mathbf{w}$ is the thermal component whose value is randomly distributed among the different electrons. When averaging over the distribution function it follows by definition that $\langle \mathbf{v} \rangle = \mathbf{u}_e$ and $\langle \mathbf{w} \rangle = 0$.

The pressure gradient force can be understood by examining the flow of thermal momentum across the boundaries of the fluid element as shown in Fig. 10.4. Consider the right hand boundary in the reference frame in which the element is stationary (i.e., the frame where $\mathbf{u}_e = 0$). Electrons just inside the boundary with a thermal momentum $m_e\mathbf{w}$ will escape if the $x$ component of $\mathbf{w}$ is positive: $w_x > 0$. This represents a loss of $x$ directed momentum from the fluid element. Similarly particles just outside the boundary with $w_x < 0$ enter the fluid element. Note that they also decrease the $x$ directed momentum of the fluid element because they add negative momentum ($w_x < 0$). An analogous effect occurs at the left hand boundary although in this case both entering and exiting particles add to the $x$ directed momentum. If the gains and losses at each boundary do not exactly balance there is a net increase (or decrease) in the momentum of the fluid element. This change in momentum gives rise to the pressure gradient force.

The details of the derivation proceed by focusing on the $x$ direction and noting that the number of electrons with a positive $x$ directed velocity $w_x$ leaving the right hand boundary of the fluid element per second at $x + \Delta x/2$ is given by

$$\text{no. of electrons leaving/s} = \{[\text{particle flux}](\text{area})\}$$
$$= \{[(f_e (\mathbf{r}, \mathbf{w}, t) \, d\mathbf{w}) (w_x)] (\Delta y \Delta z)\}_{x+\Delta x/2}. \quad (10.16)$$

If each of these electrons has a momentum $m_e\mathbf{w}$, then the loss of momentum per second from the right hand boundary due to exiting particles is just the product of the number of particles leaving per second times the momentum lost per particle; that is, (particle flux)(area)$(m_e\mathbf{w})$. The total momentum lost by the exiting particles is obtained by summing (i.e., integrating) over all $\mathbf{w}$ subject to the constraint $w_x > 0$:

$$\text{loss of momentum/s} = \left[ \Delta y \Delta z \int_{w_x > 0} m_e \mathbf{w} w_x f_e(\mathbf{r}, \mathbf{w}, t) d\mathbf{w} \right]_{x + \Delta x/2} . \tag{10.17}$$

A completely analogous calculation applies to the particles entering with $w_x < 0$. Their negative velocity also represents a loss of momentum which can be written as

$$\text{loss of momentum/s} = \left[ \Delta y \Delta z \int_{w_x < 0} m_e \mathbf{w} w_x f_e(\mathbf{r}, \mathbf{w}, t) d\mathbf{w} \right]_{x + \Delta x/2} . \tag{10.18}$$

The sum of the integrals is easily evaluated by assuming the distribution function is a Maxwellian ($f_M$) in $\mathbf{w}$ (i.e., a shifted Maxwellian in the laboratory frame). One finds

$$\int_{w_x < 0} + \int_{w_x > 0} m_e \mathbf{w} w_x f_e(\mathbf{r}, \mathbf{w}, t) d\mathbf{w} = \mathbf{e}_x \int m_e w_x^2 f_M d\mathbf{w} = \mathbf{e}_x p_e, \tag{10.19}$$

where $p_e = n_e T_e$ is the electron pressure. The total loss of momentum per second from exiting and entering particles can now be written as

$$\text{total loss of momentum/s} = \mathbf{e}_x \, \Delta y \Delta z \, p_e|_{x + \Delta x/2} , \tag{10.20}$$

implying that, as might be expected, the force is just the pressure times the area. A similar calculation at the left hand boundary shows that the total gain in momentum due to thermal motion is given by

$$\text{total gain in momentum/s} = \mathbf{e}_x \, \Delta y \Delta z \, p_e|_{x - \Delta x/2} . \tag{10.21}$$

The pressure gradient force is obtained by calculating the net gain in momentum/per second and Taylor expanding for small $\Delta x$. The result is

$$\begin{aligned}
\text{pressure gradient force} &= \Delta y \Delta z ( p_e|_{x - \Delta x/2} - p_e|_{x + \Delta x/2} ) \mathbf{e}_x \\
&= -\Delta V \frac{\partial p_e}{\partial x} \mathbf{e}_x \\
&= -\Delta V (\nabla p_e).
\end{aligned} \tag{10.22}$$

The last form is the obvious generalization to three dimensions.

### 10.4.6 The collisional friction force

The final force is the result of momentum exchange collisions. If the electrons have a larger fluid velocity than the ions ($\mathbf{u}_e > \mathbf{u}_i$), then friction with the ions due to Coulomb collisions produces a drag force on the electrons. The drag force per unit volume for electrons and ions has already been calculated in Chapter 9 and by simply multiplying by $\Delta V$ one obtains the required results:

$$\text{electron drag force} = \Delta V m_e n \frac{d\mathbf{u}_e}{dt}\bigg|_{\text{collisions}} = -\Delta V m_e n\, \bar{\nu}_{ei}(\mathbf{u}_e - \mathbf{u}_i),$$

$$\text{ion drag force} = \Delta V m_i n \frac{d\mathbf{u}_i}{dt}\bigg|_{\text{collisions}} = +\Delta V m_e n\, \bar{\nu}_{ei}(\mathbf{u}_e - \mathbf{u}_i),$$

(10.23)

where

$$\bar{\nu}_{ei} = \frac{\sqrt{2}}{12\pi^{3/2}} \frac{e^4 n_i}{\varepsilon_0^2 m_e^{1/2} T^{3/2}} \ln \Lambda = 1.8 \times 10^5 \frac{n_{20}}{T_k^{3/2}} \quad \text{s}^{-1}.$$

(10.24)

### 10.4.7 The conservation of momentum equations

All of the forces have now been calculated. Combining these contributions leads to the desired set of macroscopic fluid equations describing conservation of momentum for electrons and ions:

$$m_e n_e \left( \frac{\partial}{\partial t} + \mathbf{u}_e \cdot \nabla \right) \mathbf{u}_e = -e n_e \left( \mathbf{E} + \mathbf{u}_e \times \mathbf{B} \right) - \nabla p_e - m_e n_e \bar{\nu}_{ei} \left( \mathbf{u}_e - \mathbf{u}_i \right),$$

$$m_i n_i \left( \frac{\partial}{\partial t} + \mathbf{u}_i \cdot \nabla \right) \mathbf{u}_i = e n_i (\mathbf{E} + \mathbf{u}_i \times \mathbf{B}) - \nabla p_i - m_e n_e \bar{\nu}_{ei} (\mathbf{u}_i - \mathbf{u}_e).$$

(10.25)

There is one further point of physics worth discussing. Where exactly in the derivation did the issue of high collisionality vs. low collisionality appear, an issue that directly affects the validity of a fluid description? The answer is in the assumption of using a Maxwellian distribution function to calculate the pressure gradient force. For plasma phenomena occurring on a time scale slow compared to the collisional time scale, the Maxwellian assumption is valid. Collisions would rapidly drive the plasma to a Maxwellian.

However, many, perhaps most, plasma phenomena occur on a more rapid time scale. For example, macroscopic equilibrium and stability, RF heating and current drive, and a variety of plasma micro-instabilities all involve more rapid time scales. Thus, even if the initial distribution of the plasma is Maxwellian, its time response to any of these effects will in general be non-Maxwellian in nature; that is, the plasma will not respond as if it has an isotropic scalar pressure as assumed in the derivation, a consequence of the low collisionality and inherent anisotropy caused by the magnetic field. In general treating these effects correctly requires a kinetic model. Even so, this incorrect assumption concerning high plasma collisionality and the resulting isotropy of the pressure is often not important, or at least is not a dominant effect. Fluid descriptions therefore provide a reasonably good basis for understanding the behavior of fusion plasmas.

## 10.5  Conservation of energy

### 10.5.1  The basic principle

In this subsection attention is focused on the thermal energy of the plasma. Note that the conservation of kinetic energy can easily be obtained by forming the dot product of Eq. (10.25) with either $\mathbf{u}_e$ or $\mathbf{u}_i$ as appropriate. However, no new information is obtained by this procedure.

The basic principle governing conservation of internal energy is given in schematic form as follows:

$$\frac{d}{dt}(\text{internal energy}) = \text{net heating power.} \tag{10.26}$$

There are a number of terms contributing to the net heating power, which is defined as the net power that increases the temperature of the plasma (as opposed to driving orderly macroscopic motion). The net heating power consists of a set of sources and sinks as follows: (1) the rate of work done to compress or expand the plasma fluid element, (2) the rate of energy lost due to thermal conduction, (3) ohmic heating power, (4) external auxiliary heating power, (5) alpha power, (6) Bremsstrahlung radiation losses, and (7) the rate of energy lost (or gained) from one species to another by Coulomb temperature equilibration collisions. This somewhat lengthy list of contributions is now evaluated, one by one.

### 10.5.2  The rate of change of internal energy

The easiest way to focus on the internal energy is to move to the reference frame where the fluid element of interest is at rest: $\mathbf{u}_e = 0$ (as before, the derivation is carried out for electrons after which a simple switching of subscripts gives the relation for ions). To begin, recall that a shifted Maxwellian in the laboratory reference frame becomes a stationary Maxwellian in the random velocity $\mathbf{w}$ in the stationary frame. Note that each electron only has random thermal energy $(m_e w^2/2)$ in the stationary frame. Therefore, the average internal energy per particle in the fluid element is given by

$$\text{internal energy/particle} = \frac{1}{n_e} \int \frac{m_e w^2}{2} f_M \, d\mathbf{w} = \frac{3}{2} T_e. \tag{10.27}$$

The total internal energy in a small fluid element is then $(3T_e/2) n_e \Delta V$. The time rate of change of this quantity corresponds to the left hand side of the power balance equation:

$$\frac{d}{dt}(\text{internal energy}) = \frac{d}{dt}\left(\frac{3}{2} n_e \Delta V T_e\right)$$

$$= \frac{3}{2} n_e \Delta V \left(\frac{\partial}{\partial t} + \mathbf{u}_e \cdot \nabla\right) T_e. \tag{10.28}$$

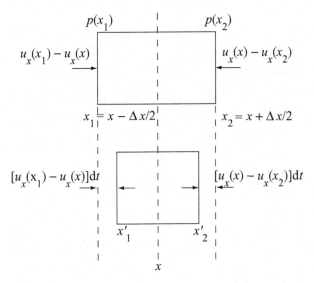

Figure 10.5 Compression of a fluid element in the $u = 0$ reference frame due to a gradient in $u_x$.

### 10.5.3 The compression work

In Subsection 10.4.5 it was shown that microscopic random motion leads to the development of a pressure force on each of the surfaces of the fluid element. An imbalance in these pressure forces gives rise to the macroscopic pressure gradient force. It is shown below that even if there is no pressure gradient, work can be supplied to compress the fluid element if there is an imbalance in the macroscopic fluid velocity on opposing surfaces of the element. Since there is no pressure gradient, none of this work causes a change in the macroscopic velocity ($\mathbf{u_e}$) and corresponding macroscopic energy ($m_e u_e^2/2$) of the fluid element. The compression work is entirely converted into internal energy. This contribution to the power balance represents the familiar "$pdV$" term in thermodynamics.

The compression term is easily calculated by examining Fig. 10.5. Focus first on the left hand boundary. At a time $t$ in the stationary frame, the left hand boundary moves with a macroscopic velocity $u_x(x_1) - u_x(x)$, where $x_1 = x - \Delta x/2$. An infinitesimal time $dt$ later the surface has moved to the right to a new location $x_1' = x_1 + [u_x(x_1) - u_x(x)]\,dt$. If the pressure at $x_1$ is denoted by $p(x_1)$, then the work done moving this surface is just the product of force times distance.

$$\text{work on left boundary} = [p(x_1)\Delta y \Delta z](x_1' - x_1)$$
$$= [p(x_1)\Delta y \Delta z][u_x(x_1) - u_x(x)]\,dt. \qquad (10.29)$$

A similar calculation on the right hand surface shows that the work done compressing the element is given by

$$\text{work on right boundary} = -[p(x_2)\Delta y \Delta z](x_2' - x_2)$$
$$= -[p(x_2)\Delta y \Delta z][u_x(x_2) - u_x(x)]\,dt, \qquad (10.30)$$

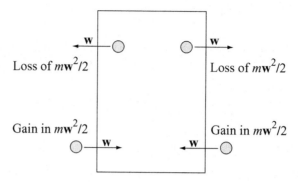

Figure 10.6 Flow of thermal energy in the $\mathbf{u} = 0$ reference frame.

where $x_2 = x + \Delta x/2$. Note the presence of the minus sign since "compression" of the right hand boundary corresponds to motion to the left which is the negative $x$ direction.

The power that heats the plasma by compression is then equal to the rate of change per second of the combined work done on both surfaces:

$$\text{compression power} = \frac{(\text{work on left} + \text{work on right})}{dt}$$
$$= \Delta y \Delta z [p(x_1)][u_x(x_1) - u_x(x)]$$
$$- \Delta y \Delta z [p(x_2)][u_x(x_2) - u_x(x)]. \tag{10.31}$$

Taylor expanding for small $\Delta x$ leads to

$$\text{compression power} = -\Delta V \left( p \frac{\partial u_x}{\partial x} \right) = -\Delta V (p_e \nabla \cdot \mathbf{u}_e). \tag{10.32}$$

Equation (10.32) is the desired expression, the last form representing the obvious generalization to three dimensions. Observe, as expected, that the compression force depends on the inhomogeneity in the flow velocity but not on the gradient in the pressure.

### 10.5.4 Thermal conduction

The thermal conduction power is a microscopic effect in the power balance relation, analogous to the pressure gradient force in the momentum equation, although, as is shown, with one major subtlety. Thermal conductivity involves the flow of thermal energy carried by the random motion of the particles across the boundaries of the fluid element. The geometry is illustrated in Fig. 10.6 in the stationary frame.

Focus first on the left hand boundary. Consider a particle just inside the surface with a thermal energy $mw^2/2$ (where for convenience the species subscript has been temporarily suppressed). The number of these particles leaving per second is equal to the flux times the area assuming the particle has a negative random velocity in the $x$ direction: $dN_{\text{out}} = (w_x f \, dw)\Delta y \Delta z$ for $w_x < 0$. The thermal energy lost per second by the exiting particles is thus given by $(mw^2/2)dN_{\text{out}} = (mw^2/2)(w_x f \, dw)\Delta y \Delta z$ for $w_x < 0$. Similarly particles entering the fluid element from the left hand side add to the thermal energy at a rate $(mw^2/2)dN_{in} = (mw^2/2)(w_x f d\mathbf{w})\Delta y \Delta z$ for $w_x > 0$. Note that in contrast to the calculation of the pressure gradient force, the outgoing and ingoing thermal energy contributions on a given boundary tend to cancel. For the pressure force they add together since momentum has a positive or negative vector direction.

The net gain in thermal energy per second entering the fluid element from the left is just equal to the total (gain + loss) of the two contributions above summed over all particle velocities (i.e., including both positive and negative $w_x$):

$$\text{gain in energy/s} = \Delta y \Delta z \int (mw^2/2) \, w_x f \, d\mathbf{w}\big|_{x-\Delta x/2}. \tag{10.33}$$

It is at this point that a subtlety arises. For a Maxwellian distribution function the integral exactly averages to zero because of odd symmetry in the integrand. In reality, however, the thermal conduction effect represents the dominant loss mechanism in most fusion plasmas. This apparent contradiction is resolved by noting that the actual distribution function is not a pure Maxwellian, but is instead a Maxwellian plus an additional contribution which has odd symmetry in velocity. The odd symmetry component of $f$ produces an integrand with even symmetry and a non-vanishing contribution to the thermal conduction losses. The odd component has no effect on the pressure gradient where it averages to zero.

Having explained the mathematical difficulty qualitatively, one is still faced with the problem of calculating the odd symmetry component of $f$. This is a complicated calculation involving the solution of a kinetic model of the plasma including Coulomb collisions. It is beyond the scope of the present book. A more convenient approach is adopted by introducing the heat flux vector $\mathbf{q}(\mathbf{r}, t)$. Its $x$ component is defined by

$$q_x(\mathbf{r}, t) = \int (mw^2/2)w_x f \, d\mathbf{w}. \tag{10.34}$$

The specific form of $q_x(\mathbf{r}, t)$ for a fusion plasma is derived in Chapter 14 by means of a simple heuristic argument that eliminates the need for the complicated calculation. It is shown that $q_x \approx -\kappa_{xx}(\partial T/\partial x)$, and an explicit approximation for $\kappa_{xx}$, the thermal conductivity, is also derived. For the moment, the thermal conduction contribution to the two-fluid energy equation is simply expressed in terms of the as yet to be calculated heat flux vector $q_x(\mathbf{r}, t)$.

Based on this discussion the gain in thermal energy per second at the left hand surface can be rewritten as

$$\text{gain in energy/s} = \Delta y \Delta z (q_x)\big|_{x-\Delta x/2}. \tag{10.35}$$

Similarly, the gain in energy per second at the right hand surface is given by

$$\text{gain in energy/s} = -\Delta y \Delta z (q_x)\big|_{x+\Delta x/2}. \tag{10.36}$$

The minus sign occurs because gain and loss switch roles from left to right surfaces, a result of the change in sign for $w_x$.

Combining Eqs. (10.35) and (10.36) leads to the desired expression for the heat conduction power gain

$$\text{heat conduction power} = \Delta y \Delta z [(q_x)|_{x-\Delta x/2} - (q_x)|_{x+\Delta x/2}]$$
$$= -\Delta V \frac{\partial q_x}{\partial x} = -\Delta V (\nabla \cdot \mathbf{q}_e). \tag{10.37}$$

The last form is the generalization to three dimensions.

### 10.5.5 The alpha power

The alpha power density has been calculated in Chapter 3 and used extensively in the discussion on power balance. In addition the apportionment of the power between electrons and ions has been derived in Chapter 9 in the discussion of the slowing down of high-energy ions. Combining these results leads to the following expressions for the alpha power:

$$\text{alpha power to electrons} = \Delta V \left( \frac{F_e^{(\alpha)}}{4} E_\alpha n_e^2 \langle \sigma v \rangle \right),$$
$$\text{alpha power to ions} = \Delta V \left( \frac{1 - F_e^{(\alpha)}}{4} E_\alpha n_e^2 \langle \sigma v \rangle \right), \tag{10.38}$$

where $F_e^{(\alpha)} = T_c / (T_c + T)$ is the fraction of power absorbed by electrons, $T_c = 33$ keV, and in the expression for $F_e^{(\alpha)}$ it has been assumed that $T_e \approx T_i \equiv T$.

### 10.5.6 The external auxiliary heating power

The external heating power has also been discussed in detail in connection with power balance. The total external auxiliary heating power, primarily arising from neutral beams or microwaves, is defined as $\Delta V S_a$. Here, $S_a(\mathbf{r}, t)$ is assumed to be a known function of space and time. This power is also divided between electrons and ions with $F_e^{(a)}$ representing the fraction going to electrons. The actual fraction cannot be determined until the type of external heating source is specified. This is discussed in more detail in Chapter 15, which covers heating and current drive. For the moment, one should just assume that $S_a(\mathbf{r}, t)$ and $F_e^{(a)}$ are known quantities. Under this assumption, the external heating power contributions can be written as

$$\text{heating power to electrons} = \Delta V F_e^{(a)} S_a,$$
$$\text{heating power to ions} = \Delta V (1 - F_e^{(a)}) S_a. \tag{10.39}$$

### 10.5.7 The ohmic heating power

The ohmic power is generated by the frictional force acting on the electrons as they are driven through ions by means of an electric field. Since the power is generated by a collisional friction force it produces heat which intuitively one might expect to go preferentially to the electrons. This is indeed the case. The simple calculation below, based on a reduced set of fluid momentum equations, shows that the ohmic power is given by $\eta J^2$. The calculation also leads to a value for the resistivity $\eta$ in terms of the collision frequency as well as determining the power distribution between electrons and ions.

Consider a simple 0-D, two-fluid model where the only forces acting are due to inertia, momentum exchange collisions, and a constant electric field turned on at time $t = 0$. The model reduces to

$$m_e \frac{du_e}{dt} = -eE - m_e \bar{\nu}_{ei}(u_e - u_e),$$

$$m_i \frac{du_i}{dt} = eE + m_e \bar{\nu}_{ei}(u_e - u_i). \tag{10.40}$$

Assume now that the electrons and ions start off at rest. Also, for simplicity in the mathematics, assume that $\bar{\nu}_{ei} = \text{const}$. In the limit of $m_e/m_i \ll 1$ the solution to Eq. (10.40) is given by

$$u_e = -\frac{eE}{m_e \bar{\nu}_{ei}}(1 - e^{-\bar{\nu}_{ei}t}) \rightarrow -\frac{eE}{m_e \bar{\nu}_{ei}},$$

$$u_i = \frac{eE}{m_i \bar{\nu}_{ei}}(1 - e^{-\bar{\nu}_{ei}t}) \rightarrow \frac{eE}{m_i \bar{\nu}_{ei}}. \tag{10.41}$$

The second forms correspond to the steady state values. As expected, conservation of momentum leads to a much higher velocity for the electrons than for the ions because of the small electron mass.

Next, note that the steady state power delivered by the electric field to the electron fluid element is equal to the product of the force per particle times the average particle velocity times the number of particles in the element: $(-eE)(u_e)(n_e\Delta V)$. It is this power that is converted to heat. A similar expression holds for ions but is smaller by $m_e/m_i$ and can be neglected; that is, the electric field primarily heats electrons. Therefore, in steady state the ohmic heating power can be written as

$$\text{ohmic heating to electrons} = \Delta V m_e n_e \bar{\nu}_{ei} u_e^2,$$

$$\text{ohmic heating to ions} \approx 0. \tag{10.42}$$

The last step is to recognize that the current density $\mathbf{J}$ is defined as the flux of electric charge: $\mathbf{J} = e(n_i\mathbf{u}_i - n_e\mathbf{u}_e)$. For the present model $\mathbf{J} \approx -en_e u_e \mathbf{e}_x$. Eliminating $u_e$ leads to the desired expression for the ohmic heating power to the electrons:

$$\text{ohmic heating to electrons} = \Delta V \eta J^2, \tag{10.43}$$

where the resistivity $\eta$ is given by

$$\eta = \frac{m_e \bar{\nu}_{ei}}{e^2 n_e}. \tag{10.44}$$

The resistivity is discussed in more detail in Chapter 11, including several subtleties that affect only the multiplicative coefficient in front of Eq. (10.44) by a factor of about 2. For

*A self-consistent two-fluid model*

the moment note that substituting the value for $\bar{\nu}_{ei}$ yields the following form for $\eta$:

$$\eta = \frac{\sqrt{2}}{12\pi^{3/2}} \frac{e^2 m_e^{1/2}}{\varepsilon_0^2 T_e^{3/2}} \ln \Lambda = 6.5 \times 10^{-8} \left( \frac{1}{T_k^{3/2}} \right) \quad \Omega \text{ m}. \tag{10.45}$$

The resistivity is independent of density and scales inversely as $T^{-3/2}$. For a fusion plasma $\eta \approx 1.1 \times 10^{-9}$ $\Omega$ m which is approximately a factor of 20 higher than for copper.

### 10.5.8 Bremsstrahlung radiation

The primary radiation loss in the bulk of the plasma is due to Bremsstrahlung radiation which has been calculated in Chapter 3. Since the radiation is due to the acceleration of electrons, Bremsstrahlung radiation appears as a loss term in the electron energy equation. It has a negligible direct effect on the ion equation. From Chapter 3 it then follows that for a small fluid element of plasma the radiation losses are given by

$$\text{electron radiation loss} = \Delta V \left( \frac{2^{1/2}}{3\pi^{5/2}} \right) \left( \frac{e^6}{\varepsilon_0^3 c^3 h m_e^{3/2}} \right) Z_{eff} n_e^2 T_e^{1/2}$$

$$= \Delta V C_B n_{20}^2 T_k^{1/2} \quad \text{W}, \tag{10.46}$$

$$\text{ion radiation loss} \approx 0,$$

where $C_B = 5.35 \times 10^3$ for $Z_{eff} = 1$.

### 10.5.9 Energy equilibration

The last term contributing to the power balance is due to energy exchange collisions. If the electrons are hotter than the ions, then Coulomb collisions cause the electrons to lose energy and the ions to gain energy until the temperatures equilibrate. The equilibration rate has already been calculated in Chapter 9 and by simply multiplying by $\Delta V$ one obtains the desired results

$$\text{electron equilibration power} = -\tfrac{3}{2} \Delta V n_e \bar{\nu}_{eq} (T_e - T_i),$$

$$\text{ion equilibration power} = +\tfrac{3}{2} \Delta V n_e \bar{\nu}_{eq} (T_e - T_i), \tag{10.47}$$

where $\bar{\nu}_{eq} = 2(m_e/m_i)\bar{\nu}_{ei}$.

### 10.5.10 The conservation of energy equations

The many terms involved in the conservation of energy have now been calculated. Combining these contributions leads to the desired set of macroscopic fluid equations describing the conservation of energy:

electrons

$$\frac{3}{2}n_e\left(\frac{\partial}{\partial t}+\mathbf{u}_e\cdot\nabla\right)T_e+p_e\nabla\cdot\mathbf{u}_e+\nabla\cdot\mathbf{q}_e=S_e,$$

$$S_e=\frac{F_e^{(\alpha)}}{4}E_\alpha n_e^2\langle\sigma v\rangle+F_e^{(a)}S_a+\eta J^2-C_B n_e^2 T_e^{1/2}-\frac{3}{2}n_e\bar{\nu}_{eq}(T_e-T_i); \tag{10.48}$$

ions

$$\frac{3}{2}n_i\left(\frac{\partial}{\partial t}+\mathbf{u}_i\cdot\nabla\right)T_i+p_i\nabla\cdot\mathbf{u}_i+\nabla\cdot\mathbf{q}_i=S_i,$$

$$S_i=\frac{1-F_e^{(\alpha)}}{4}E_\alpha n_e^2\langle\sigma v\rangle+\left(1-F_e^{(a)}\right)S_a+\frac{3}{2}n_e\bar{\nu}_{eq}(T_e-T_i). \tag{10.49}$$

## 10.6 Summary of the two-fluid model

The model consists of the conservation of mass, momentum, and energy for electrons and ions plus Maxwell's equations. The coupling of the fluid variables to Maxwell's equations occurs through the definitions that relate the number densities ($n_e$, $n_i$) and flow velocities ($\mathbf{u}_e$, $\mathbf{u}_i$) to the current density $\mathbf{J}$ and charge density $\sigma$. Specifically, $\mathbf{J}=e(n_i\mathbf{u}_i-n_e\mathbf{u}_e)$ and $\sigma=e(n_i-n_e)$. The basic unknowns in the model are $n_e$, $n_i$, $\mathbf{u}_e$, $\mathbf{u}_i$, $T_e$, $T_i$, $\mathbf{E}$, $\mathbf{B}$. Also, the pressures are related to the densities and temperatures by the simple relations $p_e=n_e T_e$ and $p_i=n_i T_i$. The complete two-fluid model is given by the following.

*Conservation of mass*

$$\frac{\partial n_e}{\partial t}+\nabla\cdot(n_e\mathbf{u}_e)=0,$$

$$\frac{\partial n_i}{\partial t}+\nabla\cdot(n_i\mathbf{u}_i)=0. \tag{10.50}$$

*Conservation of momentum*

$$m_e n_e\left(\frac{\partial}{\partial t}+\mathbf{u}_e\cdot\nabla\right)\mathbf{u}_e=-en_e(\mathbf{E}+\mathbf{u}_e\times\mathbf{B})-\nabla p_e-m_e n_e\bar{\nu}_{ei}(\mathbf{u}_e-\mathbf{u}_i),$$

$$m_i n_i\left(\frac{\partial}{\partial t}+\mathbf{u}_i\cdot\nabla\right)\mathbf{u}_i=en_i(\mathbf{E}+\mathbf{u}_i\times\mathbf{B})-\nabla p_i-m_e n_e\bar{\nu}_{ei}(\mathbf{u}_i-\mathbf{u}_e). \tag{10.51}$$

*Conservation of energy*

$$\frac{3}{2}n_e\left(\frac{\partial}{\partial t}+\mathbf{u}_e\cdot\nabla\right)T_e+p_e\nabla\cdot\mathbf{u}_e+\nabla\cdot\mathbf{q}_e=S_e,$$

$$\frac{3}{2}n_i\left(\frac{\partial}{\partial t}+\mathbf{u}_i\cdot\nabla\right)T_i+p_i\nabla\cdot\mathbf{u}_i+\nabla\cdot\mathbf{q}_i=S_i, \tag{10.52}$$

with

$$S_e=\frac{F_e^{(\alpha)}}{4}E_\alpha n_e^2\langle\sigma v\rangle+F_e^{(a)}S_a+\eta J^2-C_B n_e^2 T_e^{1/2}-\frac{3}{2}n_e\bar{\nu}_{eq}(T_e-T_i),$$

$$S_i=\frac{1-F_e^{(\alpha)}}{4}E_\alpha n_e^2\langle\sigma v\rangle+\left(1-F_e^{(a)}\right)S_a-\frac{3}{2}n_e\bar{\nu}_{eq}(T_i-T_e). \tag{10.53}$$

*Maxwell's equations*

$$\nabla \times \mathbf{E} = -\frac{\partial \mathbf{B}}{\partial t},$$

$$\nabla \times \mathbf{B} = \mu_0 e \left( n_i \mathbf{u}_i - n_e \mathbf{u}_e \right) + \frac{1}{c^2} \frac{\partial \mathbf{E}}{\partial t},$$

$$\nabla \cdot \mathbf{E} = \frac{e}{\varepsilon_0} \left( n_i - n_e \right),$$

$$\nabla \cdot \mathbf{B} = 0. \tag{10.54}$$

The two-fluid model describes the self-consistent behavior of a fusion plasma including all of the main phenomena: macroscopic equilibrium and stability, transport, and heating and current drive. The model is a closed system of equations (i.e., number of equations = number of unknowns) except for the as yet undefined heat flux vectors $\mathbf{q}_e$ and $\mathbf{q}_i$. These are derived in Chapter 14. The model presents an accurate description of most plasma phenomena except for those in which behavior parallel to the magnetic field is dominant. Fortunately, this is not a serious limitation for most applications.

The two-fluid model is simpler to solve than any of the more general kinetic models. Even so, it still consists of a complicated set of coupled non-linear partial differential equations. In practice, depending upon the application under consideration, one can often make a number of further simplifications that substantially reduce the complexity of the model. This will be done for all of the applications discussed in the book.

## Bibliography

A good understanding of the basics of plasma physics and fusion energy can be obtained by means of fluid models. Although not as complete as a kinetic description, fluid models do allow a rigorous analysis of a great many phenomena in plasma physics. Most of the references below present a derivation of fluid equations from kinetic theory. One exception is the book by Bird *et al.*, who present a derivation based on the intuitive application of conservation of mass, momentum, and energy.

Boyd, T. J. M., and Sanderson, J. J. (2003). *The Physics of Plasmas*. Cambridge, England: Cambridge University Press.

Bird, R. B., Stewart, W. E., and Lightfoot, E. W. (1960). *Transport Phenomena*. New York: John Wiley & Sons

Chen, F. F. (1984). *Introduction to Plasma Physics and Controlled Fusion*, second edn. New York: Plenum Press.

Freidberg, J. P. (1987). *Ideal Magnetohydrodynamics*. New York: Plenum Press.

Goedbloed, H., and Poedts, S. (2004). *Principles of Magnetohydrodynamics*. Cambridge, England: Cambridge University Press.

Goldston, R. J., and Rutherford, P. H. (1995). *Introduction to Plasma Physics*. Bristol, England: Insititute of Physics Publishing.

Hazeltine, R. D., and Meiss, J. D. (1992). *Plasma Confinement*. Redwood City: Addison-Wesley.

Krall, N. A., and Trivelpiece, A. W. (1973). *Principles of Plasma Physics*. New York: McGraw Hill Book Company.

## Problems

The problems below involve derivations of global conservation relations for the two-fluid model. The general procedure for obtaining such relations requires forming various combinations of the equations and then algebraically manipulating them into a local conservation form. Specifically for scalar and vector relations the local conservation forms are

$$\frac{\partial S}{\partial t} + \nabla \cdot \mathbf{V} = 0 \qquad \text{scalar equation,}$$

$$\frac{\partial \mathbf{V}}{\partial t} + \nabla \cdot \overset{\leftrightarrow}{\mathbf{T}} = 0 \qquad \text{vector equation,}$$

in which tensor $\overset{\leftrightarrow}{\mathbf{T}} = \mathbf{AB}$, where $\mathbf{A}$ and $\mathbf{B}$ are arbitrary vectors. One then integrates the local conservation equation over the plasma volume, making use of appropriate boundary conditions, to obtain the global conservation laws. The following tensor relations are easily derivable and should be of help in solving the problems:

$$\nabla \cdot (\mathbf{AB}) = \mathbf{B}(\nabla \cdot \mathbf{A}) + (\mathbf{A} \cdot \nabla)\mathbf{B},$$

$$\mathbf{r} \times \nabla \cdot (\mathbf{AB}) = \nabla \cdot [\mathbf{A}(\mathbf{r} \times \mathbf{B})] - \mathbf{A} \times \mathbf{B},$$

$$\int_V d\mathbf{r} \nabla \cdot (\mathbf{AB}) = \int_S dS\, (\mathbf{n} \cdot \mathbf{A})\mathbf{B}.$$

10.1 A plasma is surrounded by a rigid perfectly conducting wall. Show that the electrons and ions each satisfy a global conservation of particle relation given by $dN_{e,i}/dt = 0$, where

$$N_{e,i} = \int n_{e,i} d\mathbf{r}.$$

State and justify the boundary conditions used in the derivation.

10.2 A plasma, surrounded by a fixed, perfectly conducting wall, satisfies the two-fluid equations. Show that if (1) collisions are neglected and (2) both species satisfy the simple adiabatic energy equation, the plasma satisfies the following global conservation law: $dW/dt = 0$, where the energy

$$W = \int \left( \frac{\varepsilon_0 E^2}{2} + \frac{B^2}{2\mu_0} + \frac{p_e + p_i}{\gamma - 1} + \frac{m_e n_e u_e^2}{2} + \frac{m_i n_i u_i^2}{2} \right) d\mathbf{r}.$$

Give a physical interpretation of each term. State and justify the boundary conditions used in the derivation.

10.3 This problem is a generalization of Problem 10.2 aimed at determining overall power balance when the dissipative terms are maintained. To derive the power balance relation assume that charge neutrality prevails: $n_e = n_i \equiv n$. Keep all the dissipative and non-adiabatic terms in the momentum and energy equations. Show that the global two-fluid power balance relation can be written as

$$\frac{dW}{dt} = P_f - P_B + P_a - Q_\kappa,$$

where $W$ is given in Problem 10.2 and $P_f$, $P_B$, and $P_a$ are the fusion, Bremsstrahlung, and auxiliary heating powers respectively:

$$P_f = \int \frac{E_\alpha}{4} n^2 \langle \sigma v \rangle \, d\mathbf{r},$$

$$P_B = \int C_B n^2 T^{1/2} \, d\mathbf{r},$$

$$P_a = \int S_a \, d\mathbf{r}.$$

The $Q_\kappa$ term represents the energy loss by thermal conduction through the wall:

$$Q_\kappa = \int (\mathbf{q}_e + \mathbf{q}_i) \cdot \mathbf{n} \, dS.$$

What happened to the ohmic heating term?

10.4 A plasma is surrounded by a rigid conducting wall. Show that for the two-fluid model the global momentum conservation law has the form $d\mathbf{P}/dt = 0$, where the momentum

$$\mathbf{P} = \int (m_e n_e \mathbf{u}_e + m_i n_i \mathbf{u}_i + \mathbf{S}/c^2) \, d\mathbf{r}$$

and $\mathbf{S} = \mathbf{E} \times \mathbf{H}$ is the Poynting vector. Give a physical interpretation of each term. State and justify the boundary conditions used in the derivation. Some careful reasoning is required to justify the vanishing of the surface terms.

10.5 A plasma is surrounded by a rigid conducting wall. Show that for the two-fluid model the global angular momentum conservation law has the form $d\mathbf{L}/dt = 0$, where the angular momentum

$$\mathbf{L} = \int (m_e n_e \mathbf{r} \times \mathbf{u}_e + m_i n_i \mathbf{r} \times \mathbf{u}_i + \mathbf{r} \times \mathbf{S}/c^2) \, d\mathbf{r}$$

and $\mathbf{S} = \mathbf{E} \times \mathbf{H}$ is the Poynting vector. Give a physical interpretation of each term. State and justify the boundary conditions used in the derivation. Some careful reasoning is required to justify the vanishing of the surface terms.

# 11

# MHD – macroscopic equilibrium

## 11.1 The basic issues of macroscopic equilibrium and stability

The first major issue in which self-consistency plays a crucial role is the macroscopic equilibrium and stability of a plasma. One needs to learn how a magnetic field can produce forces to hold a plasma in stable, macroscopic equilibrium thereby allowing fusion reactions to take place in a continuous, steady state mode of operation. This chapter focuses on the problem of equilibrium. The issue of stability is discussed in Chapters 12 and 13.

The analysis of macroscopic equilibrium and stability is based on a single-fluid model known as MHD. The MHD model is a reduction of the two-fluid model derived by focusing attention on the length and time scales characteristic of macroscopic behavior. Specifically, the appropriate length scale $L$ is the plasma radius ($L \sim a$) while the appropriate time scale $\tau$ is the ion thermal transit time across the plasma ($\tau \sim a/v_{Ti}$). This leads to a characteristic velocity $u \sim L/\tau \sim v_{Ti}$, which is the fastest macroscopic speed that the plasma can achieve – the ion sound speed.

The derivation of the MHD model from the two-fluid model is the first topic discussed in this chapter. Also presented is a derivation of MHD starting from single-particle guiding center theory. The purpose is to show that the intuition leading to MHD is indeed consistent with single-particle guiding center motion.

One particular point worth noting in the derivation is the transformation of the electron momentum equation into a single-fluid Ohm's law, which relates the electric field to the current. There are actually three forms of Ohm's law depending upon how many terms are maintained. Consequently, there are three corresponding forms of the MHD model. The most inclusive form is usually referred to as the "generalized" Ohm's law. Neglecting certain terms in this relation leads to a reduced form, known as the "resistive" Ohm's law. Lastly, if the resistivity itself is neglected, one obtains the "ideal" Ohm's law, which corresponds to a plasma with perfect conductivity. Interestingly, for all three forms of Ohm's law, the equilibrium force balance relation is identical.

The next main topic discussed in this chapter is the formulation and analysis of the MHD equilibrium equations. An overview of the results is as follows. There are two qualitatively different types of forces involved in producing an MHD equilibrium. First there are radial expansion forces due to the natural tendency of a hot gas to expand. Both toroidal (i.e., the long way around) and poloidal (i.e., the short way around) magnetic fields are shown to be capable of balancing the radial expansion force. Second, there are toroidal forces, arising from the toroidal geometry, that tend to make the plasma "ring" expand to a larger and larger major radius. Here, only poloidal magnetic fields can counteract the toroidal expansion forces. It is shown that the critical property of the magnetic field required for toroidal force balance is that the magnetic lines continuously wrap around the torus like the stripes on a barber pole. This property is known as "rotational transform." Several simple magnetic configurations are discussed showing how to calculate the rotational transform. In general, axisymmetric toroidal fusion configurations produce rotational transform by means of a toroidal plasma current. However, transform can also be generated without a toroidal current as in a stellarator, although in this case the configuration is inherently 3-D.

A main conclusion from the analysis is that there is a wide variety of magnetic geometries capable of providing the necessary forces to hold a plasma in toroidal equilibrium. The next problem is to examine the MHD stability of these equilibria to determine which is most attractive for a fusion reactor in terms of achieving a high stable $\beta$. This is the focus of the following two chapters.

## 11.2 Derivation of MHD from the two-fluid model

The derivation of the MHD model from the two-fluid model is straightforward although several steps are required. First, clear statements of the length and time scales of interest are required, which establish the basis for comparing the sizes of various terms in the two-fluid model. Second, since MHD focuses on macroscopic scales, there are several obvious simplifications that can be made related to the small mass of the electrons, the non-relativistic velocity of the plasma, and the shortness of the Debye length. Finally, the transition from the two-fluid model to the MHD model is made by introducing single-fluid variables and neglecting those terms that are small, because of the narrowing focus on the particular length and time scales associated with MHD. The derivation proceeds as follows.

### 11.2.1 Basic scaling relations for MHD

The characteristic length, time and velocity scales describing MHD are given by $L \sim a$, $\tau \sim a/v_{Ti}$, and $u_i \sim v_{Ti}$. To proceed further, one needs to know how the various unknowns in the two-fluid model, as well as the other naturally appearing length, time, and velocity scales, compare with the MHD scales. This will determine which terms to keep and which ones to neglect.

Consider first the electric field. For the macroscopic velocities the perpendicular inductive electric field is assumed to be relatively large, implying that the dominant fluid velocity is the $\mathbf{E} \times \mathbf{B}$ drift. Consistency with the assumed characteristic MHD velocity $v_{Ti}$ then requires that $E_\perp \sim v_{Ti} B$. Furthermore, since electrons and ions both move with the same $\mathbf{E} \times \mathbf{B}$ drift velocity it follows that $u_e \sim u_i \sim E_\perp/B \sim v_{Ti}$. The scaling of $E_\parallel$ is slightly complex and is deferred until later in the derivation.

Next, note that the current density $\mathbf{J} = e(n_i\mathbf{u}_i - n_e\mathbf{u}_e)$ cancels to leading order since the $\mathbf{E} \times \mathbf{B}$ drift velocity is identical for both electrons and ions. The implication is that $|\mathbf{J}|/env_{Ti} \ll 1$. This inequality is quantified shortly. Also, the plasma pressure is assumed to be finite compared to the magnetic pressure, which requires that $\beta \equiv p/(B^2/2\mu_0) \sim 1$. Lastly, to satisfy the definition of a fusion plasma two further inequalities must be satisfied: small ion gyro radius $r_{Li} \ll a$ and low collision frequency $\bar{\nu}_{ei} \ll v_{Ti}/a$. Both are well satisfied in current experiments and fusion reactors.

All of the inequalities defining MHD can be combined and summarized in the following compact form:

$$\text{length: } \mathbf{a} \gg r_{Li} \gg [r_{Le} \sim \lambda_{De}];$$
$$\text{frequency: } \bar{\nu}_{ei} \ll \mathbf{v}_{Ti}/\mathbf{a} \ll \omega_{ci} \ll [\omega_{ce} \sim \omega_{pe}]; \qquad (11.1)$$
$$\text{velocity: } \mathbf{v}_{Ti} \sim v_a \ll v_{Te} \ll [c].$$

Here, recall that the Alfvén speed $v_a^2 \equiv B^2/\mu_0 n_i m_i \approx 2v_{Ti}^2/\beta$. Also, in Eq. (11.1) the bold-face symbols refer to the characteristic MHD scales. The terms in the square brackets correspond to inequalities that are extremely well satisfied as compared to the MHD scales. The hierarchy of inequalities is discussed in more detail as the derivation proceeds.

These simple scaling assumptions allow one to compare various terms in the two-fluid model to determine which ones to maintain and which ones to neglect.

### 11.2.2 The "obvious" simplifications

The "obvious" simplifications in the two-fluid model result from the terms in the square brackets of Eq. (11.1) representing extremely well-satisfied inequalities. The first simplification arises from the fact that the characteristic time scale of interest is $\tau \sim a/v_{Ti}$, or in terms of an equivalent MHD frequency $\omega \sim 1/\tau \sim v_{Ti}/a$. This is a very low frequency compared to the natural electron frequencies: $\omega \ll \omega_{ce} \sim \omega_{pe}$. In practice the inequalities are satisfied by many orders of magnitude. The implication is that the electron inertia term can be neglected in the electron momentum equation since the electron response time is essentially infinitely fast compared to the characteristic MHD time. Formally, one can obtain the appropriate mathematical limit by letting $m_e \to 0$.

The second simplification is associated with the non-relativistic scale of the macroscopic plasma flows. Since $v_{Ti} \ll c$ by many orders of magnitude, one can neglect the displacement current in Maxwell's equations. Similarly, the characteristic scale length of MHD easily satisfies the inequality $a \gg \lambda_{De}$. This leads to a third simplification – the $\nabla \cdot \mathbf{E}$ term in Poisson's equation can be neglected leading to the quasi-neutrality relation: $n_e \approx n_i$. (Do not fall into the trap of assuming that quasi-neutrality implies that $\nabla \cdot \mathbf{E} = 0$. It does not. It only implies that $\varepsilon_0 \nabla \cdot \mathbf{E} \ll en_e$.) These simplifications reduce Maxwell's equation from the exact relativistic Lorentz-invariant form to a self-consistent low-frequency Galilean-invariant form. Formally, the low-frequency Maxwell equations can be obtained mathematically by taking the limit $\varepsilon_0 \to 0$.

Under the above simplifications, the two-fluid model reduces to:

$$\frac{\partial n_e}{\partial t} + \nabla \cdot (n_e \mathbf{u}_e) = 0,$$

$$\frac{\partial n_i}{\partial t} + \nabla \cdot (n_i \mathbf{u}_i) = 0,$$

$$0 = -en_e(\mathbf{E} + \mathbf{u}_e \times \mathbf{B}) - \nabla p_e - m_e n_e \bar{\nu}_{ei}(\mathbf{u}_e - \mathbf{u}_i),$$

$$m_i n_i \left( \frac{\partial}{\partial t} + \mathbf{u}_i \cdot \nabla \right) \mathbf{u}_i = en_i (\mathbf{E} + \mathbf{u}_i \times \mathbf{B}) - \nabla p_i - m_e n_e \bar{\nu}_{ei} (\mathbf{u}_i - \mathbf{u}_e),$$

$$\frac{3}{2} n_e \left( \frac{\partial}{\partial t} + \mathbf{u}_e \cdot \nabla \right) T_e + p_e \nabla \cdot \mathbf{u}_e + \nabla \cdot \mathbf{q}_e = S_e,$$

$$\frac{3}{2} n_i \left( \frac{\partial}{\partial t} + \mathbf{u}_i \cdot \nabla \right) T_i + p_i \nabla \cdot \mathbf{u}_i + \nabla \cdot \mathbf{q}_i = S_i, \tag{11.2}$$

$$\nabla \times \mathbf{E} = -\frac{\partial \mathbf{B}}{\partial t},$$

$$\nabla \times \mathbf{B} = \mu_0 e(n_i \mathbf{u}_i - n_e \mathbf{u}_e),$$

$$n_i - n_e = 0,$$

$$\nabla \cdot \mathbf{B} = 0.$$

### 11.2.3 The single-fluid variables

The next step in the procedure is to introduce single-fluid variables into the two-fluid equations. The single-fluid variables are the mass density $\rho$, the macroscopic velocity $\mathbf{v}$, and the pressure $p$. They are defined as follows. First, since the mass of the ions is much greater than the mass of the electrons, the mass density of the single-fluid corresponds to that of the ions:

$$\rho \equiv m_i n, \tag{11.3}$$

where $n_e = n_i \equiv n$ because of quasi-neutrality.

Second, recall that the ion and electron fluid velocities are both nearly equal to the $\mathbf{E} \times \mathbf{B}$ drift. Therefore, since $m_i \gg m_e$ the momentum of the fluid is carried by the ions:

$$\mathbf{v} \equiv \mathbf{u}_i. \tag{11.4}$$

Next, the electron fluid velocity is defined in terms of the current density by reintroducing $\mathbf{J}$ as one of the basic unknowns in the problem: $\mathbf{J} = en(\mathbf{u}_i - \mathbf{u}_e)$ or

$$\mathbf{u}_e \equiv \mathbf{v} - \mathbf{J}/en. \tag{11.5}$$

Finally, the pressure in the single-fluid model is simply the total electron and ion pressure:

$$p \equiv p_e + p_i. \tag{11.6}$$

Equations (11.3)–(11.6) provide the necessary relations between the single-fluid variables and the two-fluid variables.

### 11.2.4  The conservation of mass equations

The information contained in the two conservation of mass equations is easily obtained as follows. First, one simply multiplies the ion conservation relation by $m_i$ and introduces the single fluid variables. This yields

$$\frac{d\rho}{dt} + \rho\nabla \cdot \mathbf{v} = 0, \tag{11.7}$$

where here and below

$$\frac{d}{dt} = \frac{\partial}{\partial t} + \mathbf{v}\cdot\nabla \tag{11.8}$$

is the usual convective derivative moving with the ion fluid.

   The second relation is obtained by multiplying both conservation equations by the charge $e$ and then subtracting the equations. After making use of charge neutrality, one finds

$$\nabla \cdot \mathbf{J} = 0. \tag{11.9}$$

Equation (11.9) is a redundant relation since it also follows from taking the divergence of the low-frequency form of Ampère's law given in Eq. (11.2): $\nabla \cdot (\nabla \times \mathbf{B} - \mu_0\mathbf{J}) = -\mu_0\nabla \cdot \mathbf{J}$ $= 0$. Thus, Eq. (11.7) represents the only independent information from the conservation of mass relations.

### 11.2.5  The conservation of momentum equations

The two conservation of momentum equations are simplified as follows. First, one adds the two equations together and introduces the single-fluid variables. The resulting equation is

$$\rho\frac{d\mathbf{v}}{dt} = \mathbf{J} \times \mathbf{B} - \nabla p. \tag{11.10}$$

Note that the electric field terms have cancelled because of charge neutrality and the collisional terms have cancelled because of conservation of total electron-plus-ion momentum in elastic Coulomb collisions.

   This equation is the most important one in MHD as it describes the basic force balance of the plasma. The physical interpretation of Eq. (11.10) is as follows. The left hand side represents the inertial force, which is important in determining the dynamical behavior of the plasma. The $\mathbf{J} \times \mathbf{B}$ term represents the magnetic field force used to confine the plasma. The $\nabla p$ term represents the pressure gradient force that causes a hot core of plasma to expand outwards. In a steady state situation without flow the inertial force is zero and equilibrium is achieved when the magnetic force balances the pressure gradient force. In terms of scaling one can see that the current density $\mathbf{J}$ must satisfy $J \sim p/aB$ for the two forces to balance. This implies that $J/env_{Ti} \sim r_{Li}/a \ll 1$ as previously asserted. Because of its importance, the single-fluid momentum equation is discussed in more detail in Section 11.5.

   The second piece of information contained in the momentum equations is obtained by simply rewriting the electron momentum equation in the form of an Ohm's law and substituting the single-fluid variables. A short calculation yields

$$\mathbf{E} + \mathbf{v} \times \mathbf{B} = \frac{1}{en} (\mathbf{J} \times \mathbf{B} - \nabla p_e) + \eta \mathbf{J}. \tag{11.11}$$

Equation (11.11) is known as the "generalized" Ohm's law. The left hand side represents the electric field in the reference frame moving with the plasma. The $\mathbf{J} \times \mathbf{B}$ and $\nabla p_e$ terms corresponds to the Hall term and electron diamagnetic term respectively. Situations exist where these terms play an important role but not within the regime of MHD. In fact, a simple estimate of the size of these terms shows that $JB/envB \sim p/aenvB \sim r_{Li}/a \ll 1$. Therefore, these terms can be neglected. The last term describes the resistivity of the plasma and is also small. The MHD scaling relations imply that $\eta J/vB \sim (\bar{v}_{ei}a/v_{Ti})(r_{Le}^2/a^2) \ll 1$. On this basis, the resistivity term can also be neglected. The net result of this discussion is that Eq. (11.11) reduces to the "ideal" Ohm's law:

$$\mathbf{E} + \mathbf{v} \times \mathbf{B} = 0. \tag{11.12}$$

Note that in the reference frame moving with the plasma the electric field is zero; in other words, the plasma behaves like an ideal perfectly conducting material.

Finally, it is worth pointing out that in spite of its smallness, the resistivity term is sometimes maintained in the Ohm's law. This is because: (1) the resistivity is the only dissipative process appearing in the momentum equation (i.e., the Hall and electron diamagnetic terms are non-dissipative), and (2) the "ideal" MHD scaling arguments apply only to the perpendicular direction; that is, since the parallel component of $\mathbf{v} \times \mathbf{B}$ is zero the only term to compare with on the left hand side of the Ohm's law is $E_\parallel$, which has not as yet been scaled. If one thus assumes that $E_\parallel \sim \eta J_\parallel$, then these two terms should be maintained and the resistivity contribution should be altered as follows for self-consistency: $\eta \mathbf{J} \to \eta \mathbf{J}_\parallel$. In practice, for mathematical simplicity the entire $\eta \mathbf{J}$ is usually kept. This gives the correct expression parallel to the field and produces only a negligibly small inconsistency in the perpendicular direction. Maintaining the resistivity term leads to the "resistive" Ohm's law:

$$\mathbf{E} + \mathbf{v} \times \mathbf{B} = \eta \mathbf{J}. \tag{11.13}$$

A final subtlety regarding resistivity involves the value of $\eta$. More accurate calculations involving kinetic plasma models show that the actual resistive Ohm's law is of the form $\mathbf{E} + \mathbf{v} \times \mathbf{B} = \eta_\parallel \mathbf{J}_\parallel + \eta_\perp \mathbf{J}_\perp$. The resistivity is anisotropic because the correct distribution function is more complicated than a simple shifted Maxwellian. Kinetic theory shows that the value of $\eta$ in Eq. (11.13) from the simple shifted Maxwellian derivation actually corresponds to the perpendicular resistivity: $\eta_\perp = \eta$. However, it is $\eta_\parallel$ that is required to correctly model the parallel Ohm's law. The more accurate value of the parallel resistivity is reduced by a factor of about $1/2$: $\eta_\parallel = 0.51\eta$. The anisotropy is unimportant for present purposes. Consequently, for simplicity it is assumed that the resistivity is isotropic with $\eta \to \eta_\parallel$ since it is the parallel direction that is most affected by dissipation. Specifically, the resistive Ohm's law becomes

$$\mathbf{E} + \mathbf{v} \times \mathbf{B} = \eta_\parallel \mathbf{J}, \tag{11.14}$$

where from Chapter 10

$$\eta_\parallel = (0.51) \frac{\sqrt{2}}{12\pi^{3/2}} \frac{e^2 m_e^{1/2}}{\varepsilon_0^2 T_e^{3/2}} \ln \Lambda = 3.3 \times 10^{-8} \left( \frac{1}{T_k^{3/2}} \right) \quad \Omega \text{ m.} \tag{11.15}$$

There are two main ways that resistivity affects plasma behavior. First, resistivity allows a wider range of possible instabilities than ideal MHD. However, these instabilities have much slower growth rates and usually do not lead to a macroscopic loss of plasma, but instead to enhanced transport losses. Second, even in the absence of resistive instabilities, resistivity still represents the only dissipation in the momentum equation. This dissipation gives rise to particle diffusion and magnetic field diffusion, representing two main transport losses. Here too, the phenomena occur on much slower time scales than the characteristic MHD time scale. For this reason much of the discussion on resistive MHD is deferred to Chapter 14, which treats plasma transport.

The end result of the discussion above is that the two-fluid momentum equations are transformed into a single-fluid momentum equation plus an Ohm's law that can be either ideal or resistive. None of the applications considered in the book requires the generalized Ohm's law.

### 11.2.6 The conservation of energy equations

In the MHD regime the energy equations simplify considerably. All the source and sink terms including thermal conductivity are negligible. Heating and cooling of the plasma in general take place on a much slower time scale than that of ideal MHD. To justify this statement recall that the 0-D form of the heat conduction term (which is always among the largest of these contributions) is given by $\nabla \cdot \mathbf{q} \to 3nT/\tau_E$, where $\tau_E$ is the energy confinement time. Since $\tau_E$ for existing large experiments or a reactor typically lies in the range 0.1–1 s it is clear this time is much longer than the characteristic MHD time scale: $\tau_E \gg a/v_{Ti} \sim 1$ µs.

Based on this reasoning the ion energy equation, expressed in terms of single-fluid variables, reduces to

$$\frac{3}{2} n \left( \frac{\partial}{\partial t} + \mathbf{v} \cdot \nabla \right) T_i + p_i \nabla \cdot \mathbf{v} = 0 \tag{11.16}$$

A short calculation in which $\nabla \cdot \mathbf{v}$ is eliminated by means of the mass conservation relation reduces Eq. (11.16) to the familiar adiabatic form

$$\frac{d}{dt} \left( \frac{p_i}{\rho^\gamma} \right) = 0, \tag{11.17}$$

where $\gamma = 5/3$.

Next, since $\mathbf{u}_e = \mathbf{v} - \mathbf{J}/en \approx \mathbf{v}$ in the MHD scaling it follows that the electrons satisfy an identical type of energy equation

$$\frac{d}{dt} \left( \frac{p_e}{\rho^\gamma} \right) = 0. \tag{11.18}$$

Adding the two energy equations together using the definition $p = p_i + p_e$ leads to the desired form for the single-fluid energy equation:

$$\frac{d}{dt}\left(\frac{p}{\rho^\gamma}\right) = 0. \tag{11.19}$$

Note that in MHD one does not have to distinguish between $p_e$ and $p_i$ since only the sum appears in Eq. (11.19).

### 11.2.7 Summary of the MHD equations

The single-fluid equations derived above can now be collected together. The result is a closed set of equations defining the MHD model. The equations are summarized below:

$$
\begin{aligned}
\text{mass} : \quad & \frac{d\rho}{dt} + \rho\nabla\cdot\mathbf{v} = 0; \\[4pt]
\text{momentum} : \quad & \rho\frac{d\mathbf{v}}{dt} = \mathbf{J}\times\mathbf{B} - \nabla p; \\[4pt]
\text{Ohm's law} : \quad & \mathbf{E} + \mathbf{v}\times\mathbf{B} = 0 \quad \text{ideal MHD}, \\[2pt]
& \mathbf{E} + \mathbf{v}\times\mathbf{B} = \eta_\parallel\mathbf{J} \quad \text{resistive MHD}; \\[4pt]
\text{energy} : \quad & \frac{d}{dt}\left(\frac{p}{\rho^\gamma}\right) = 0; \\[4pt]
\text{Maxwell} : \quad & \nabla\times\mathbf{E} = -\frac{\partial\mathbf{B}}{\partial t}, \\[2pt]
& \nabla\times\mathbf{B} = \mu_0\mathbf{J}, \\[2pt]
& \nabla\cdot\mathbf{B} = 0.
\end{aligned} \tag{11.20}
$$

This is the basic model that will be used to analyze the macroscopic equilibrium and stability of a fusion plasma.

## 11.3 Derivation of MHD from guiding center theory

This section presents an alternative derivation of the ideal MHD model based on guiding center theory. Specifically, the perpendicular MHD momentum equation and Ohm's law are derived by combining the single-particle guiding center drifts with a new contribution known as the magnetization current. The purpose is to provide additional physical intuition by showing the relationship between microscopic single-particle behavior and macroscopic fluid behavior.

### 11.3.1 The basic idea

The basic idea behind the derivation is to calculate the perpendicular current density flowing through an arbitrary open surface in the plasma due to guiding center motion and to see how this current is related to the plasma pressure gradient and inertial forces.

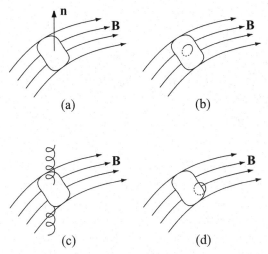

Figure 11.1 (a) Geometry showing a surface with $\mathbf{n} \perp \mathbf{B}$. (b) Cancellation of current due to circular gyromotion. (c) Current due to guiding center drifts. (d) Magnetization current around the boundary.

The geometry is illustrated in Fig. 11.1(a). Since the goal is to calculate $\mathbf{J}_\perp$, the surface of interest is chosen such that its normal vector is perpendicular to $\mathbf{B}$. Some careful "bookkeeping" that takes into account the gyro radius expansion is required to calculate the current. To begin consider the leading order contribution to the particle's motion, which corresponds to the circular gyro motion. As seen in Fig. 11.1(b) this motion makes no contribution to the current since each particle enters and leaves the surface in opposite directions; there is no net flow of charge during a gyro period.

The perpendicular current is a first order effect, generated by two separate mechanisms. The first mechanism corresponds to the flow of charge across the surface resulting from the perpendicular guiding center drifts as shown in Fig. 11.1(c). The second contribution results from the circular gyro motion right near the boundary of the surface. As shown in Fig. 11.1(d), this current, known as the magnetization current, arises because particles cross the boundary in only one direction. Their return path lies outside the boundary and therefore does not cancel the in-surface contribution to the current. Although this may seem like a small effect it is comparable in magnitude to the guiding center drift contribution which is also a small first order effect. Each of these contributions is now calculated.

### 11.3.2 The guiding center drift current and Ohm's law

The guiding center drift velocity of the particles consists of the $\mathbf{E} \times \mathbf{B}$ drift, the $\nabla B$ drift, the curvature drift, and the polarization drift (the last being negligible for electrons). Recall from Chapter 8 that these are given by

$$
\mathbf{v}_{gi} = \mathbf{V}_E + \mathbf{V}_{\nabla B} + \mathbf{V}_\kappa + \mathbf{V}_p
$$

$$
= \frac{\mathbf{E} \times \mathbf{B}}{B^2} + \frac{v_\perp^2}{2\omega_{ci}} \frac{\mathbf{B} \times \nabla B}{B^2} - \frac{v_\parallel^2}{\omega_{ci}} \frac{\boldsymbol{\kappa} \times \mathbf{B}}{B} + \frac{\mathbf{b}}{\omega_{ci}} \times \frac{d}{dt} \frac{\mathbf{E} \times \mathbf{B}}{B^2},
$$

$$
\mathbf{v}_{ge} = \mathbf{V}_E + \mathbf{V}_{\nabla B} + \mathbf{V}_\kappa
$$

$$
= \frac{\mathbf{E} \times \mathbf{B}}{B^2} - \frac{v_\perp^2}{2\omega_{ce}} \frac{\mathbf{B} \times \nabla B}{B^2} + \frac{v_\parallel^2}{\omega_{ce}} \frac{\boldsymbol{\kappa} \times \mathbf{B}}{B},
$$

$$(11.21)$$

where $\boldsymbol{\kappa} = \mathbf{b} \cdot \nabla \mathbf{b} = -\mathbf{R}_c/R_c^2$ is the curvature vector.

Next, note that within the MHD scaling the $\mathbf{E} \times \mathbf{B}$ drift is larger than all the other ion and electron drifts by the ratio $a/r_{Li}$ in the usual situation where the perpendicular and parallel particle velocities are comparable to the thermal velocities: $m_i v_i^2 \sim m_e v_e^2 \sim T$. As in the fluid model the dominant guiding motion for both electrons and ions is the $\mathbf{E} \times \mathbf{B}$ drift velocity. Since each electron and each ion drifts with essentially the same velocity it makes sense to introduce a macroscopic perpendicular fluid velocity as follows: $\mathbf{u}_\perp \approx \mathbf{v}_{gi} \approx \mathbf{v}_{ge}$, where $\mathbf{u}_\perp = \mathbf{E} \times \mathbf{B}/B^2$. This relation can be rewritten as follows

$$
\mathbf{E}_\perp + \mathbf{u}_\perp \times \mathbf{B} = 0, \tag{11.22}
$$

which corresponds to the perpendicular Ohm's law.

Consider now the perpendicular current due to the guiding center drift motion. Although the $\mathbf{E} \times \mathbf{B}$ drift is the largest contribution, the resulting net current cancels because of charge neutrality and the fact that the drift is the same for both species. The easiest way to calculate the net current density is to directly subtract the ion and electron contributions and then sum over all particles. This automatically cancels the larger $\mathbf{E} \times \mathbf{B}$ terms:

$$
\mathbf{J}_{\perp g} = e \langle n_i \mathbf{v}_{gi} \rangle - e \langle n_e \mathbf{v}_{ge} \rangle = e \int (\mathbf{v}_{gi} f_i - \mathbf{v}_{ge} f_e) \, d\mathbf{v}. \tag{11.23}
$$

Substituting for the drift velocities yields

$$
\mathbf{J}_{\perp g} = e \int \left( \frac{v_\perp^2}{2\omega_{ci}} \frac{\mathbf{B} \times \nabla B}{B^2} - \frac{v_\parallel^2}{\omega_{ci}} \frac{\boldsymbol{\kappa} \times \mathbf{B}}{B} + \frac{\mathbf{b}}{\omega_{ci}} \times \frac{d\mathbf{u}_\perp}{dt} \right) f_i \, d\mathbf{v}
$$

$$
+ e \int \left( \frac{v_\perp^2}{2\omega_{ce}} \frac{\mathbf{B} \times \nabla B}{B^2} - \frac{v_\parallel^2}{\omega_{ce}} \frac{\boldsymbol{\kappa} \times \mathbf{B}}{B} \right) f_e \, d\mathbf{v}. \tag{11.24}
$$

One can easily evaluate these integrals assuming stationary Maxwellian distribution functions and noting that in local rectangular velocity coordinates $v_\parallel^2 = v_z^2$ and $v_\perp^2 = v_x^2 + v_y^2$. This yields the desired expression for the current due to the guiding center drift motion:

$$
\mathbf{J}_{\perp g} = p \left( \frac{\mathbf{B} \times \nabla B}{B^3} - \frac{\boldsymbol{\kappa} \times \mathbf{B}}{B^2} \right) + \frac{\rho}{B} \mathbf{b} \times \frac{d\mathbf{u}_\perp}{dt}, \tag{11.25}
$$

where $p = p_e + p_i$ and $\rho = m_i n_i$.

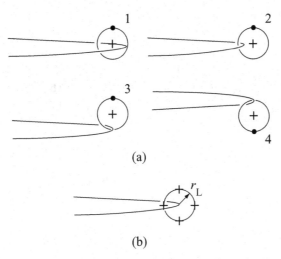

(a)

(b)

Figure 11.2 (a) Several limiting orbits of particles that just intersect the surface one time. The symbol $+$ denotes the center of each orbit. (b) The locus of the centers of limiting orbits is a circle of radius $r_L$ centered on the edge.

### 11.3.3 The magnetization current

The magnetization current arises from particles (near the edge) that intersect the surface under consideration just one time as shown in Fig. 11.1(d). Their contribution to $\mathbf{J}_\perp$ is evaluated by calculating the current carried by a single particle and then multiplying by the number of particles making a single intersection. The calculation proceeds as follows.

Over a gyro period, a particle with a charge $q$ (where $q = \pm e$) produces an average current

$$I_q = -\frac{q}{\tau_c} = -\frac{q\omega_c}{2\pi}. \tag{11.26}$$

The minus sign signifies that the current flows in the diamagnetic direction, tending to cancel the applied field.

The number of particles making just one intersection can be determined by examining Fig. 11.2, which illustrates several limiting cases. The orbit of particle 1 just barely misses the edge of the surface of interest. Particle 2 lies predominantly outside the surface of interest but just manages to intersect it. Similarly, particles 3 and 4 represent limiting orbits producing one intersection. The locus of the centers of these orbits forms a circle whose radius is the gyro radius. Therefore, any particle whose guiding center lies within a circle centered on the surface edge, with radius $r_L$, makes just one intersection. The volume element containing such particles is given by $d\mathbf{r} = \mathbf{A} \cdot d\mathbf{l}$, where $\mathbf{A} = \pi r_L^2 \mathbf{b}$ and $d\mathbf{l}$ is arc length along the surface edge. Observe that the normal vector to the gyro orbit lies along $\mathbf{b}$ which is, in general, not parallel to $d\mathbf{l}$ as shown in Fig. 11.3. This projection reduces the number of single-intersection particles.

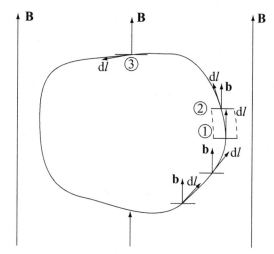

Figure 11.3 Relation between **b** and d**l**. The normal to the circle defining the locus of guiding centers always points upwards along **b**. The direction of d**l** changes around the surface edge. At 1, d**l** and **b** are parallel, maximizing the volume of the single intersection particles (width between dashed lines is maximum). At 2, and then 3, the volume decreases as d**l** becomes perpendicular to **b**.

Next, the number of particles with a velocity **v** contained within this volume element is given by $dN = (f \, d\mathbf{v})(d\mathbf{r})$. They generate a corresponding current $dI = I_q \, (f \, d\mathbf{v})(d\mathbf{r})$. The total magnetization current is obtained by integrating over all velocities and all volume elements along the surface edge:

$$I_M = - \int \int \left(\frac{q\omega_c}{2\pi}\right)\left(\frac{\pi m^2 v_\perp^2}{q^2 B^2}\right) f \, d\mathbf{v} \, \mathbf{b} \cdot d\mathbf{l}. \tag{11.27}$$

For a Maxwellian distribution function the velocity integration is easily evaluated and the expression reduces to

$$I_M = - \oint \frac{p}{B}\mathbf{b} \cdot d\mathbf{l}. \tag{11.28}$$

The next task is to rewrite the magnetization current in a more convenient form in terms of a magnetization current density $\mathbf{J}_M$ as follows.

$$I_M = \int \mathbf{J}_M \cdot d\mathbf{S}, \tag{11.29}$$

where $S$ is the surface. One now introduces a magnetization **M** defined by $\mathbf{J}_M = \nabla \times \mathbf{M}$. Stokes' theorem implies that

$$I_M = \int \nabla \times \mathbf{M} \cdot d\mathbf{S} = \oint \mathbf{M} \cdot d\mathbf{l}. \tag{11.30}$$

Figure 11.4 Physical origin of the magnetization current. There are more downward orbits from the high-density side tangent to the vertical line than upward orbits from the low-density side. The differential causes $\mathbf{J}_M$.

A comparison of Eqs. (11.30) and (11.28) shows that $\mathbf{M} = -(p/B)\,\mathbf{b}$. This leads to the desired expression for the magnetization current density

$$\mathbf{J}_M = -\nabla \times \left(\frac{p}{B}\mathbf{b}\right). \tag{11.31}$$

From a physical point of view one might have expected the magnetization current to cancel out when integrating around the surface edge since for about half the particles the component of $\mathbf{b}$ along $d\mathbf{l}$ is parallel while for the other half it is anti-parallel. However, this cancellation is not perfect. To understand the physical origin of the magnetization current consider the simple situation in which there are no guiding center drifts: $\mathbf{B} = B_0\mathbf{e}_z$ with $B_0 = \text{const}$. For this case $\mathbf{J}_M$ reduces to

$$\mathbf{J}_M = \frac{1}{B_0}\nabla p \times \mathbf{e}_z. \tag{11.32}$$

Equation (11.32) implies that a magnetization current flows if there is a pressure gradient (i.e., $\nabla n$ and/or $\nabla T$) in the plasma. The generation of $\mathbf{J}_M$ from a density gradient is illustrated in Fig. 11.4. Observe that all particles make circular gyro orbits with zero guiding center drifts. However, there are more particles to the left than the right because of the density gradient. Therefore, if one draws a vertical line as shown in Fig. 11.4 it is apparent that there are more downward orbits tangent to the vertical line from particles on the left than there are upward orbits from particles on the right. The magnetization current results from the difference in the number of downward and upward orbits generated by the density gradient. A temperature gradient also causes a magnetization current since hotter particles on average have a higher velocity at the point of tangency than the colder particles. This differential in velocities due to $\nabla T$ also produces a contribution to the magnetization current. The conclusion is that a gradient in the pressure causes a macroscopic magnetization current even though none of the particles needs to have a guiding center drift.

### 11.3.4 The perpendicular MHD momentum equation

The last step in the derivation is to combine the guiding center drift and magnetization current densities to form the total current: $\mathbf{J}_\perp = \mathbf{J}_{\perp_g} + \mathbf{J}_M$. One next forms the cross product $\mathbf{J}_\perp \times \mathbf{B}$:

$$\mathbf{J}_\perp \times \mathbf{B} = \left[ p \left( \frac{\mathbf{B} \times \nabla B}{B^3} - \frac{\kappa \times \mathbf{B}}{B^2} \right) + \frac{\rho}{B} \mathbf{b} \times \frac{d\mathbf{u}_\perp}{dt} \right] \times \mathbf{B} - \left[ \nabla \times \left( \frac{p}{B} \mathbf{b} \right) \right] \times \mathbf{B}. \tag{11.33}$$

The various terms simplify as follows:

$$\left[ \frac{\rho}{B} \mathbf{b} \times \frac{d\mathbf{u}_\perp}{dt} \right] \times \mathbf{B} = \rho \frac{d\mathbf{u}_\perp}{dt} \Big|_\perp ;$$

$$p \frac{(\kappa \times \mathbf{B}) \times \mathbf{B}}{B^2} = -p\,\kappa;$$

$$\left[ \nabla \times \left( \frac{p}{B} \mathbf{b} \right) \right] \times \mathbf{B} = p\,(\nabla \times \mathbf{b}) \times \mathbf{b} + (\nabla p \times \mathbf{b}) \times \mathbf{b} - p \left( \frac{\nabla B}{B^2} \times \mathbf{b} \right) \times \mathbf{B} \tag{11.34}$$

$$= p\,\kappa - \nabla_\perp p - p \left( \frac{\nabla B}{B^3} \times \mathbf{B} \right) \times \mathbf{B}.$$

Here, use has been made of the vector identities $\kappa \cdot \mathbf{b} = 0$ and $(\nabla \times \mathbf{b}) \times \mathbf{b} = \mathbf{b} \cdot \nabla \mathbf{b} - \frac{1}{2}\nabla\,(\mathbf{b} \cdot \mathbf{b}) = \kappa$.

Combining and canceling appropriate terms finally leads to the desired form of the momentum equation as derived from guiding center theory:

$$\rho \frac{d\mathbf{u}_\perp}{dt} \Big|_\perp = \mathbf{J} \times \mathbf{B} - \nabla_\perp p. \tag{11.35}$$

This form is identical to the corresponding equation from ideal MHD. The conclusion is that the fluid and guiding center approaches are equivalent but different ways of obtaining self-consistent descriptions of macroscopic MHD behavior.

## 11.4 MHD equilibrium – a qualitative description

The first important application of the MHD model is concerned with the calculation of equilibrium. How does a combination of externally applied and internally induced magnetic fields act to provide an equilibrium force balance that holds the plasma together at the desired location in the vacuum chamber? This section presents a brief qualitative description of the basic issues arising in the creation of equilibrium force balance.

Although obvious, it should be stated at the outset that the equilibria of interest must correspond to confined equilibria. Specifically, consider a fusion plasma contained within

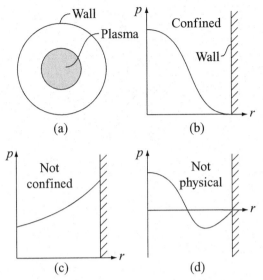

Figure 11.5  (a) Confined plasma isolated from the wall. (b) Corresponding confined pressure profile. (c) An unconfined pressure profile since $p(r)$ is finite at the wall. (d) An unphysical profile since $p(r)$ < 0 over part of the plasma.

a vacuum chamber as shown in Fig. 11.5(a). The purpose of the magnetic field is to isolate the plasma from the first wall vacuum chamber. This keeps the plasma hot and the wall cool. While the physical goal is clear, the MHD equations allow a wide variety of different types of mathematical solutions and one must be certain to focus on those that correspond to confined equilibria. This point is demonstrated Fig. 11.5(b)–(d), which illustrate several possible mathematical solutions arising from the MHD equations. Only one shows good confinement where the pressure is everywhere positive and approaches zero at the wall.

In a qualitative sense, MHD equilibrium in a toroidal geometry separates into two pieces: radial pressure balance and toroidal force balance. The radial pressure balance is illustrated in Fig. 11.6. The plasma is a hot core of gas that tends to expand uniformly along the minor radius $r$. Magnetic fields and currents must exist to balance this radial expansion force. As is shown shortly there are two basic magnetic configurations that can produce radial pressure balance – the "$\theta$-pinch" and "Z-pinch" plus combinations thereof. It is also worth noting that the problem of radial pressure balance is important both in toroidal geometries and in linear geometries, where toroidal effects are neglected.

The second equilibrium issue involves toroidal force balance. As its name implies, this arises solely because of the toroidal geometry. The problem is illustrated in Fig. 11.7. Because of toroidicity, unavoidable forces are generated by both the toroidal (i.e., long way around) and poloidal (i.e., short way around) magnetic fields that tend to push the plasma horizontally outwards along the direction of the major radius $R$. Several methods are described that show how the outward toroidal forces can be balanced to prevent the plasma from striking the first wall. These involve the use of a perfectly conducting wall surrounding

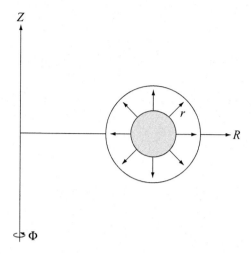

Figure 11.6 Outward pressure expansion force along *r* in both linear and toroidal configurations.

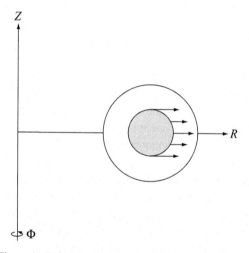

Figure 11.7 Outward force along *R* due to toroidicity.

the plasma, the application of an external "vertical field," and the use of "toroidal-helical" fields.

In the remainder of this chapter the ideal MHD model is simplified to focus on the problem of equilibrium. The discussion begins with a description of several basic properties of MHD equilibria. With this as background, the problem of radial pressure balance is then addressed in the context of a 1-D cylindrical model. Next, the problem of toroidal force balance is discussed and a simple approximate relation is derived showing the conditions required to achieve toroidal equilibrium in a 2-D axisymmetric torus carrying a toroidal current. Lastly, a brief discussion is presented that describes how toroidal force balance is achieved in a 3-D toroidal–helical configuration. These results set the stage for the discussion of stability, which begins in the next chapter.

## 11.5 Basic properties of the MHD equilibrium model

### *11.5.1 The MHD equilibrium model*

The first step in the quantitative analysis is to simplify the MHD model so that it focuses solely on equilibrium. The simplifications result from two basic assumptions that define the MHD equilibria of interest: (1) for a plasma in equilibrium all quantities are independent of time ($\partial/\partial t = 0$); (2) the plasma is assumed to be static ($\mathbf{v} = 0$). Stationary equilibria with time-independent flows are possible ($\mathbf{v} \neq 0$), but such flows normally do not dominate present or future experiments and therefore are not essential for the present discussion.

The simplification now proceeds as follows. Under the above assumptions the MHD conservation of mass and energy relations are trivially satisfied. The ideal Ohm's law requires that $\mathbf{E} = 0$, which in turn implies that Faraday's law is automatically satisfied. For the resistive Ohm's law there remains a small electric field which produces transport on a time scale slow compared to the MHD time scale and can be neglected for present purposes. The remaining non-trivial equations, which are identical for either of the Ohm's laws, define the MHD equilibrium model, given by

$$\mathbf{J} \times \mathbf{B} = \nabla p,$$
$$\nabla \times \mathbf{B} = \mu_0 \mathbf{J}, \tag{11.36}$$
$$\nabla \cdot \mathbf{B} = 0.$$

These equations describe the equilibrium properties of all magnetic configurations of fusion interest.

### *11.5.2 General properties – flux surfaces*

Before proceeding with the quantitative discussion of radial pressure balance and toroidal force balance, one can deduce several general properties characteristic of all MHD equilibria. The first property concerns the concept of flux surfaces. To understand flux surfaces, examine the contours of constant pressure in a well-confined plasma equilibrium, as illustrated in Fig. 11.8. By definition the vector $\nabla p$ is perpendicular to the $p = \text{const.}$ contours. Now, if one forms the dot product of the MHD momentum equation with $\mathbf{B}$ it immediately follows that

$$\mathbf{B} \cdot \nabla p = 0. \tag{11.37}$$

The implication is that the magnetic field lines (i.e., the lines parallel to $\mathbf{B}$) must lie in the surfaces of constant pressure; there is no component of $\mathbf{B}$ perpendicular to the surface.

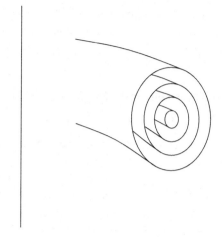

Figure 11.8 Contours of constant pressure in a well-confined toroidal equilibrium.

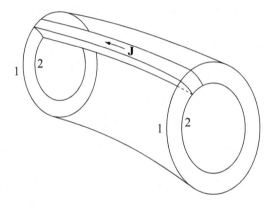

Figure 11.9 Two flux surfaces, 1 and 2, at two different toroidal locations showing that the current flows between and not across them.

Hence, these surfaces are also called flux surfaces. In a confined plasma the pressure and flux contours coincide, forming a set of closed, nested, toroidal surfaces.

### 11.5.3 General properties – current surfaces

Another general property easily derived is obtained by forming the dot product of the momentum equation with $\mathbf{J}$:

$$\mathbf{J} \cdot \nabla p = 0. \tag{11.38}$$

The current lines also lie in the constant pressure surfaces. There is no component of $\mathbf{J}$ perpendicular to the pressure contours. The implication is that as shown in Fig. 11.9 the

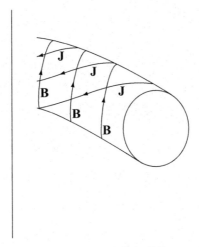

Figure 11.10 Flux surfaces showing that in general the **B** lines and **J** lines are not parallel.

current flows between flux surfaces and not across them. Note that although both the **B** and **J** lines lie in the flux surfaces, this does not imply that **B** and **J** are parallel. This point is illustrated in Fig. 11.10. In general the angle between **B** and **J** is arbitrary with the cases of purely parallel or purely perpendicular lines representing special, but still physically accessible, limits.

### 11.5.4 General properties – magnetic pressure and tension

The last general property concerns the concepts of magnetic pressure and tension. These are the two ways in which the magnetic field can act to hold the plasma in equilibrium force balance. The relevant relation is obtained by using Ampère's law to eliminate **J** from the momentum equation and invoking the vector identity $\nabla(B^2/2) = \mathbf{B} \times (\nabla \times \mathbf{B}) + (\mathbf{B} \cdot \nabla)\mathbf{B}$. A short calculation yields

$$\nabla_\perp \left( p + \frac{B^2}{2\mu_0} \right) - \frac{B^2}{\mu_0}\kappa = 0, \tag{11.39}$$

where $\nabla_\perp = \nabla - \mathbf{b}(\mathbf{b} \cdot \nabla)$ is the perpendicular component of the gradient operator and, as before, $\kappa = \mathbf{b} \cdot \nabla\mathbf{b}$ is the curvature vector.

Equation (11.39) describes pressure balance perpendicular to the magnetic field. The interpretation of the terms is as follows. The quantity $p$ clearly represents the plasma pressure. By analogy, the quantity $B^2/2\mu_0$ represents the magnetic pressure. The last term $(B^2/\mu_0)\kappa$ represents the tension force created by the curvature of the field lines. This force is easy to visualize if one imagines that field lines behave like stretched rubber bands wrapped around a cylindrical piece of wood. Explicit examples of the magnetic pressure and tension forces are described during the discussion of radial pressure balance and toroidal force balance.

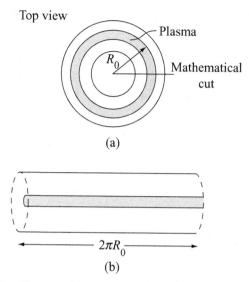

Figure 11.11 (a) Top view of a torus showing the mathematical cut. (b) Equivalent straight cylinder.

In summary, it is useful to think of a well-confined fusion plasma as one in which the pressure contours form a set of closed nested toroidal surfaces. Both the magnetic field lines and current density lines lie on these surfaces. The magnetic forces that hold the plasma together arise from a combination of two different mechanisms: magnetic pressure and magnetic tension.

## 11.6 Radial pressure balance

This section focuses on the problem of radial pressure balance. The analysis is greatly simplified by the following geometric transformation. Start with a toroidal configuration, and make a (mathematical) cut at an arbitrary poloidal plane. Then, straighten out the torus transforming it into an equivalent straight cylinder. See Fig. 11.11. Clearly the problem of toroidal force balance vanishes in a straight cylinder. Also, if the plasma cross section is circular, then the geometry is 1-D, with all quantities depending only upon the minor radius $r$.

Using this model, several basic configurations are investigated: the $\theta$-pinch, the Z-pinch, and the screw pinch. These configurations demonstrate how magnetic pressure and magnetic tension can be used to provide radial pressure balance. Also, the analysis suggests a useful and explicit definition of the plasma $\beta$.

### 11.6.1 The θ-pinch

The geometry and field components of a $\theta$-pinch are illustrated in Fig. 11.12. A current flows in the coil as shown in the figure. This produces an externally applied "toroidal"

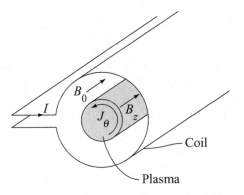

Figure 11.12 Schematic diagram of a $\theta$-pinch.

magnetic field $B_0$ inside the coil. A "poloidal" current is induced in the plasma whose direction is such as to cancel the applied field. The non-trivial unknowns in the problem are as follows: $p = p(r)$, $\mathbf{B} = B_z(r)\mathbf{e}_z$, and $\mathbf{J} = J_\theta(r)\mathbf{e}_\theta$. (The current density in the $\theta$ direction is the origin of the name "$\theta$-pinch.") The goal now is to see whether or not the ideal MHD equations allow this geometric configuration to possess confined equilibria.

The analysis is straightforward. First, by symmetry the condition $\nabla \cdot \mathbf{B} = 0$ is automatically satisfied. Second, Ampère's law reduces to

$$\mu_0 J_\theta = -\frac{dB_z}{dr}. \tag{11.40}$$

Substituting $J_\theta$ into the momentum equation ($p' = J_\theta B_z$) leads to a simple relation between $p$ and $B_z$, which can be written as

$$\frac{d}{dr}\left(p + \frac{B_z^2}{2\mu_0}\right) = 0. \tag{11.41}$$

This equation can be easily integrated:

$$p(r) + \frac{B_z^2(r)}{2\mu_0} = \frac{B_0^2}{2\mu_0}, \tag{11.42}$$

where $B_0^2/2\mu_0 = $ const. is the externally applied magnetic pressure. Equation (11.42) represents the basic radial pressure balance relation for a $\theta$-pinch. It states that at any radial position $r$, the sum of the local plasma pressure plus internal magnetic pressure is equal to the applied magnetic pressure. The $\theta$-pinch pressure balance relation allows a wide variety of confined equilibria in which the pressure is peaked at the center and monotonically decreases to zero at the edge. A typical example is illustrated in Fig. 11.13. Note that since $B_z(0)$ is arbitrary the peak value of $\beta$ defined as $\beta_0 \equiv 2\mu_0 p(0)/[2\mu_0 p(0) + B_z^2(0)]$ can vary over a wide range: $0 < \beta_0 < 1$.

The conclusion is that the $\theta$-pinch configuration is capable of providing radial pressure balance. The confinement mechanism corresponds to magnetic pressure.

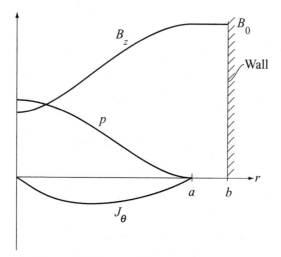

Figure 11.13 Typical $\theta$-pinch profiles.

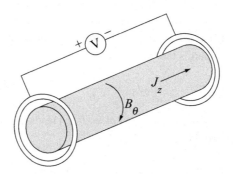

Figure 11.14 Schematic diagram of a Z-pinch.

### 11.6.2 The Z-pinch

The Z-pinch is the complementary configuration to the $\theta$-pinch. Its geometry is illustrated in Fig. 11.14. Here, two ideal electrodes drive a "toroidal" current in the $z$ direction producing a purely "poloidal" magnetic field in the $\theta$ direction. (The current density in the $z$ direction is the motivation for the name Z-pinch.) Therefore, in this configuration the non-trivial unknowns are given by: $p = p(r)$, $\mathbf{B} = B_\theta(r)\mathbf{e}_\theta$, and $\mathbf{J} = J_z(r)\mathbf{e}_z$. The goal now is to determine whether or not the Z-pinch configuration is capable of sustaining confined equilibria.

Here too, the analysis is straightforward. Because of symmetry $\nabla \cdot \mathbf{B} = 0$ is automatically satisfied. Ampère's law yields a relation for $J_z$ given by

$$\mu_0 J_z = \frac{1}{r}\frac{d}{dr}(rB_\theta). \tag{11.43}$$

The pressure balance relation is obtained by substituting this expression into the momentum equation ($p' = -J_z B_\theta$):

$$\frac{dp}{dr} + \frac{B_\theta}{\mu_0 r} \frac{d}{dr}(rB_\theta) = 0. \tag{11.44}$$

A more intuitive form is obtained by expanding the $rB_\theta$ derivative. After a simple calculation the desired form of the Z-pinch pressure balance relation can be written as

$$\frac{d}{dr}\left(p + \frac{B_\theta^2}{2\mu_0}\right) + \frac{B_\theta^2}{\mu_0 r} = 0. \tag{11.45}$$

In order, from left to right, the terms correspond to the plasma pressure, the magnetic pressure, and the magnetic tension. In general, the Z-pinch equation cannot be integrated as simply as the $\theta$-pinch equation. Even so, one can easily demonstrate confined equilibria by means of an example. Assume a physically plausible $J_z(r)$ profile that peaks on axis and monotonically decays to zero at $r = a$:

$$J_z(r) = \begin{cases} \dfrac{2I}{\pi a^2}\left(1 - \dfrac{r^2}{a^2}\right) & 0 < r < a \\ 0 & a < r < b \end{cases}. \tag{11.46}$$

Here, the coefficient in front of $J_z(r)$ has been chosen so that $I = \int J_z dS$ represents the total current flowing in the plasma. An expression for $B_\theta(r)$ is easily obtained by integrating Ampère's law

$$B_\theta(r) = \begin{cases} \dfrac{\mu_0 I}{\pi a}\left(\dfrac{r}{a} - \dfrac{r^3}{2a^3}\right) & 0 < r < a \\ \dfrac{\mu_0 I}{2\pi r} & a < r < b \end{cases}. \tag{11.47}$$

Finally, the pressure profile is determined by integrating the pressure balance relation. A short calculation yields

$$p(r) = \begin{cases} \dfrac{\mu_0 I^2}{12\pi^2 a^2}\left(1 - \dfrac{r^2}{a^2}\right)^2\left(5 - 2\dfrac{r^2}{a^2}\right) & 0 < r < a \\ 0 & a < r < b \end{cases}. \tag{11.48}$$

These profiles are illustrated in Fig. 11.15. The pressure clearly corresponds to a well-confined equilibrium. If one defines the peak value of $\beta$ for a Z-pinch as $\beta_0 \equiv 2\mu_0 p(0)/[2\mu_0 p(0) + B_\theta^2(0)]$, then $\beta_0 = 1$ for all profiles since $B_\theta(0) = 0$. While high $\beta_0$ is potentially good, one sees that a Z-pinch does not have the flexibility in this parameter that a $\theta$-pinch does.

Lastly, it is instructive to plot the three forces in Z-pinch pressure balance as a function of radius as shown in Fig. 11.16. Note that a negative force points radially inwards, and therefore corresponds to a confining force (as opposed to an expansion force). Figure 11.16 indicates that near the outer edge of the plasma the only confining force is due to magnetic tension. Both the particle pressure and magnetic pressure produce outwardly pointing

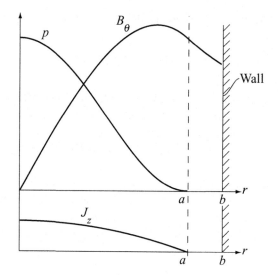

Figure 11.15 Typical Z-pinch profiles.

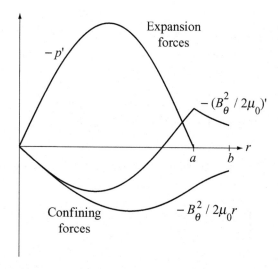

Figure 11.16 The radial forces on a Z-Pinch. Only the tension force is confining in the outer portion of the plasma.

expansion forces. The conclusion is that the Z-pinch configuration is capable of providing radial pressure balance by means of the magnetic tension force.

### 11.6.3 The screw pinch

The screw pinch is a configuration that consists of an arbitrary combination of $\theta$-pinch and Z-pinch fields. In this configuration the magnetic lines twist around the surface giving the appearance of a screw thread. Hence the name "screw pinch." Every magnetic configuration of fusion interest satisfies a form of radial pressure balance relation corresponding to the general screw pinch.

The analysis of the screw pinch is also straightforward. The non-trivial unknowns are as follows: $p$, $\mathbf{B} = B_\theta \mathbf{e}_\theta + B_z \mathbf{e}_z$, and $\mathbf{J} = J_\theta \mathbf{e}_\theta + J_z \mathbf{e}_z$. As before, the condition $\nabla \cdot \mathbf{B} = 0$ is automatically satisfied. Ampère's law yields a relation for the total current density:

$$\mu_0 \mathbf{J} = -\frac{dB_z}{dr}\mathbf{e}_\theta + \frac{1}{r}\frac{d}{dr}(rB_\theta)\mathbf{e}_z. \tag{11.49}$$

Substituting $\mathbf{J}$ into the pressure balance relation ($p' = J_\theta B_z - J_z B_\theta$) yields the desired expression for radial pressure balance in a generalized screw pinch:

$$\frac{d}{dr}\left(p + \frac{B_\theta^2}{2\mu_0} + \frac{B_z^2}{2\mu_0}\right) + \frac{B_\theta^2}{\mu_0 r} = 0. \tag{11.50}$$

Observe that in general one is free to specify two arbitrary functions, for instance $B_\theta(r)$ and $B_z(r)$. MHD then determines the third function, in this case $p(r)$. The option of choosing two functions arbitrarily is the reason why the screw pinch relation is capable of describing such a wide range of configurations. Two free functions also make sense physically, since experimentalists have the possibility of independently programming the coil currents in both the toroidal and poloidal field circuits. Specific examples corresponding to the main fusion configurations are presented in Chapter 13 following the discussion of MHD stability. Once MHD stability is known, the entire MHD picture can be assembled to provide the motivation for each such configuration.

### *11.6.4 General definition of $\beta$ in a screw pinch*

The last topic in this section is the derivation of a general radial pressure balance relation that can be used to define the "$\beta$" of the plasma. The relation is useful because it involves integrated averages of the pressure and the fields and thus is not sensitive to the specific details of the profiles. The relation is obtained by performing the following averaging operation on the pressure balance equation:

$$\frac{1}{a^2}\int_0^a r^2 dr\left[\frac{d}{dr}\left(p + \frac{B_\theta^2}{2\mu_0} + \frac{B_z^2}{2\mu_0}\right) + \frac{B_\theta^2}{\mu_0 r}\right] = 0. \tag{11.51}$$

Here $a$ is the edge of the plasma, the radius at which the pressure and current densities first vanish.

A straightforward integration by parts allows one to simplify the various terms in Eq. (11.51). The pressure term reduces to

$$\frac{1}{a^2}\int_0^a r^2 \frac{dp}{dr}dr = -\frac{2}{a^2}\int_0^a p\,r\,dr \equiv -\langle p\rangle. \tag{11.52}$$

This term clearly represents the average value of the pressure. The "toroidal" field term can be written as

$$\frac{1}{a^2}\int_0^a r^2 \frac{d}{dr}\left(\frac{B_z^2}{2\mu_0}\right)dr = \frac{B_{za}^2}{2\mu_0} - \frac{2}{a^2}\int_0^a \left(\frac{B_z^2}{2\mu_0}\right)r\,dr \equiv -\frac{1}{2\mu_0}\left(B_{za}^2 - \langle B_z^2\rangle\right),$$

$$\tag{11.53}$$

where $B_{za}^2/2\mu_0$ is the applied toroidal magnetic pressure and $\langle B_z^2/2\mu_0\rangle$ is the average value of the internal toroidal magnetic pressure. If $\langle B_z^2/2\mu_0\rangle < B_{za}^2/2\mu_0$, then the toroidal magnetic field is diamagnetic, helping to confine the plasma. This is the situation in a $\theta$-pinch. On the other hand, if $\langle B_z^2/2\mu_0\rangle > B_{za}^2/2\mu_0$, the toroidal magnetic field is paramagnetic and, like the pressure, is a radial expansion force. In this case the poloidal magnetic tension must increase to balance both of these expansion forces. Lastly, the "poloidal" field contributions can be simplified as follows:

$$\frac{1}{a^2}\int_0^a r^2\,dr\left[\frac{d}{dr}\left(\frac{B_\theta^2}{2\mu_0}\right)+\frac{B_\theta^2}{\mu_0 r}\right]=\frac{1}{a^2}\int_0^a dr\,\frac{d}{dr}\left(\frac{r^2 B_\theta^2}{2\mu_0}\right)=\frac{B_{\theta a}^2}{2\mu_0}. \tag{11.54}$$

Here, $B_{\theta a}$ is related to the toroidal plasma current $I$ by the usual relation

$$B_{\theta a}=\frac{\mu_0 I}{2\pi a}. \tag{11.55}$$

Combining these contributions leads to the desired global radial pressure balance relation

$$\langle p\rangle=\frac{1}{2\mu_0}\left(B_{za}^2-\langle B_z^2\rangle+B_{\theta a}^2\right). \tag{11.56}$$

Now, recall that the quantity $\beta$ is supposed to be a measure of how effective the applied fields are at confining plasma pressure. Equation (11.56) therefore suggests the following definition of $\beta$:

$$\beta=\frac{2\mu_0\langle p\rangle}{B_{za}^2+B_{\theta a}^2}. \tag{11.57}$$

Often, there appear in the literature related quantities known as the toroidal $\beta_t$ and the poloidal beta $\beta_p$, whose definitions are

$$\beta_t=\frac{2\mu_0\langle p\rangle}{B_{za}^2},$$

$$\beta_p=\frac{2\mu_0\langle p\rangle}{B_{\theta a}^2}. \tag{11.58}$$

The relationship between these quantities can be written as

$$\frac{1}{\beta}=\frac{1}{\beta_t}+\frac{1}{\beta_p}, \tag{11.59}$$

implying that the total $\beta$ is dominated by the smaller of the two contributions. Observe that by definition $\beta\le 1$. However, either $\beta_t$ or $\beta_p$, but not both, can be greater than unity. These definitions are useful in the comparison of different proposed magnetic fusion configurations. They can also be readily generalized to non-circular cross section, toroidal configurations.

The main conclusion from this section is that both toroidal and poloidal magnetic fields, or combinations thereof, can be used to provide radial pressure balance in magnetic fusion configurations. Furthermore, there is no upper limit on the achievable value of $\beta$ except the obvious one $\beta\le 1$.

## 11.7 Toroidal force balance

### 11.7.1 Introduction

A fusion plasma must be in the shape of a torus in order to avoid end losses. It is shown in this section that bending a straight cylinder into a torus results in the generation of three new toroidal forces all directed outwardly along the direction of the major radius (i.e., along $R$). See Fig. 11.7. These forces and the basic configurations that they affect are as follows:

(a) the hoop force (Z-pinch);
(b) the tire tube force (Z-pinch, $\theta$-pinch);
(c) the $1/R$ force ($\theta$-pinch).

If the plasma is to be held in toroidal force balance, some additional force must be applied to counter the outward forces. The analysis shows that a counterbalancing force can be readily provided for the Z-pinch; that is, a Z-pinch has good toroidal equilibrium and can therefore be easily bent into a torus. However, it is shown in Chapter 12 that a pure Z-pinch has very poor MHD stability properties.

The situation is just the opposite for the pure $\theta$-pinch. There is no simple way that an inward directed restoring force along $R$ can be applied to a $\theta$-pinch; a pure $\theta$-pinch does not have a good toroidal equilibrium and therefore cannot be bent into a torus. On the other hand, a linear $\theta$-pinch is shown in Chapter 12 to have good stability properties. The consequence of this dichotomy between pure $\theta$- and Z-pinches is that the study of fusion MHD has been largely devoted to discovering configurations (e.g. screw pinches) with optimized combinations of toroidal and poloidal magnetic fields that can stably confine plasmas with high $\beta$ in toroidal equilibrium.

In the discussion below a brief qualitative description is given of each of three toroidal forces as well as possible ways to provide a counteracting force. A simple calculation is also presented that semi-quantitatively evaluates the various terms appearing in toroidal force balance. The discussion focuses on 2-D axisymmetric equilibria, although a brief discussion is also presented that describes equilibria in a 3-D toroidal helix.

### 11.7.2 The hoop force

The hoop force is analogous to the outward expansion force generated by the current flowing in a circular loop of wire. In the present case the current corresponds to the toroidal current flowing in the plasma. Since toroidal current is involved, the hoop force is generated by bending a Z-pinch into a torus.

The origin of the force can be understood by examining Fig. 11.17(a). Shown here is a toroidal Z-pinch in which the plasma has been divided into two halves. The inside and outside halves have surface areas $S_1$ and $S_2$ respectively. Clearly $S_1 < S_2$, since the average value of the major radius is smaller on the inside surface than the outside surface.

The toroidal current generates a poloidal magnetic field as shown in the figure. Observe that a given amount of poloidal flux $\psi$ on the outside of the torus must be squeezed into a

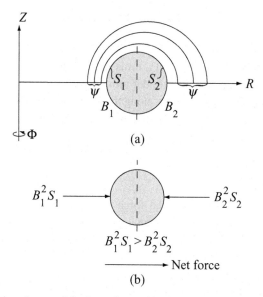

Figure 11.17 Qualitative picture of the hoop force: (a) conservation of flux showing that $B_1 > B_2$; (b) toroidal force balance showing a net outward force along $R$.

smaller cross sectional area on the inside. This implies that the magnitude of the magnetic field is greater on the inside than on the outside: $B_1 > B_2$. The lines are packed more closely together on the inside.

Focus now on the magnetic tension force on each half of the plasma as shown in Fig. 11.17(b). The tension force on the inner surface is just the product of the magnetic tension with the surface area: $F_1 = (B_1^2/2\mu_0)S_1$. It points along $R$ in the outward direction. Similarly, the force on the outer surface is just $F_2 = (B_2^2/2\mu_0)S_2$ and points along $R$ but in the inward direction.

The quadratic dependence of $B$ dominates the expression for the forces. This leads to the conclusion that $F_1 > F_2$. There is a net outward force along the major radius $R$ due to the toroidal current. This is the hoop force.

### 11.7.3 The tire tube force

The tire tube force is so named because it is analogous to the situation in which the internal air pressure stretches the outside surface area of an inflated rubber tire tube more tightly than the inner surface area. Since the plasma pressure is involved, the tire tube force is generated in both a Z-pinch and a $\theta$-pinch.

The origin of the force can be easily understood by examining Fig. 11.18(a), which shows a surface of constant pressure separated into two halves. The pressure produces an expansion force (pressure $\times$ area) equal to $F_1 = pS_1$ on the inner half surface $S_1$. This

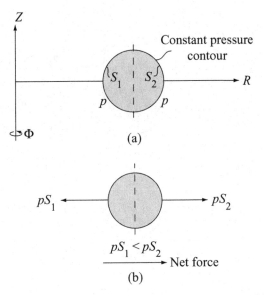

Figure 11.18 Qualitative picture of the tire tube force: (a) constant pressure with $S_2 > S_1$; (b) toroidal force balance showing a net outward force along $R$.

force points inwardly along $R$. There is a similar expansion force $F_2 = pS_2$ on the outer half although in this case the force points outwardly along $R$.

The net force balance picture is shown in Fig. 11.18(b). Clearly, $F_2 > F_1$. On a constant pressure surface, the magnitude of the force is dominated by the size of the surface area. The conclusion is that there is a net tire tube force pointing outwardly along $R$.

### 11.7.4 The $1/R$ force

As its name implies, the "$1/R$" force arises because of the $1/R$ dependence of the toroidal field resulting from the toroidal geometry. Since only the toroidal magnetic field is involved, this force is generated in the $\theta$-pinch configuration but not the Z-pinch.

A simple model demonstrating the $1/R$ force is illustrated in Fig. 11.19(a), which shows a toroidal $\theta$-pinch surrounded by a set of coils carrying current $I_c$. For simplicity the plasma current $I_p$ is assumed to be in the form of a surface current; that is, all the current flows in an infinitesimally thin layer on the plasma surface. Now, recall that the applied field $B_{\phi a}$ has a $1/R$ dependence which follows from integrating Ampère's law around any closed toroidal loop located between the coils and the plasma:

$$\oint \mathbf{B} \cdot d\mathbf{l} = 2\pi R B_{\phi a} = \mu_0 I_c,$$

$$B_{\phi a} = \frac{\mu_0 I_c}{2\pi R}. \qquad (11.60)$$

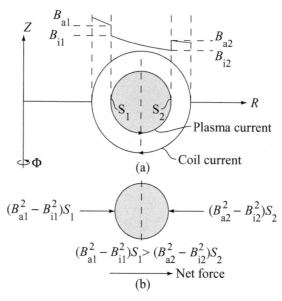

Figure 11.19 Qualitative picture of the $1/R$ force: (a) surface current model showing the $1/R$ dependence of $B_\phi$ along the midplane $Z = 0$; (b) toroidal force balance showing a net outward force along $R$.

The effect of the plasma current is usually diamagnetic, producing a field inside the plasma that partially cancels the applied field. For the surface current model, Ampère's law again yields a $1/R$ dependence for the net toroidal field $B_{\phi i}$ inside the plasma

$$\oint \mathbf{B} \cdot d\mathbf{l} = 2\pi R B_{\phi i} = \mu_0 (I_c - I_p),$$

$$B_{\phi i} = \frac{\mu_0 (I_c - I_p)}{2\pi R}. \tag{11.61}$$

The $1/R$ force can now be understood by dividing the plasma into halves as shown in Fig. 11.19(b). The force on the inner half of the plasma surface is equal to the product of the net pressure times the area: $F_1 = [(B_{\phi a}^2 - B_{\phi i}^2)_{R=R_1}/2\mu_0]S_1$. Similarly, the force on the outer half of the plasma surface can be written as $F_2 = [(B_{\phi a}^2 - B_{\phi i}^2)_{R=R_2}/2\mu_0]S_2$. The combination of the quadratic dependence of the force on $B$ plus the fact that the fields are larger at $R_1$ than at $R_2$ because of the $1/R$ behavior, implies that $F_1 > F_2$. There is a net outward toroidal force along $R$. Note that this conclusion applies to the diamagnetic situation in which $B_{\phi a}^2 > B_{\phi i}^2$. In the opposite paramagnetic situation, $B_{\phi a}^2 < B_{\phi i}^2$, the toroidal force is actually inward. However, as is shown shortly, this force cannot by itself produce toroidal force balance because it prevents the establishment of radial pressure balance.

### 11.7.5 The restoring force due to a perfectly conducting wall

The discussion above has shown that all toroidal plasmas experience an outward toroidal force along $R$. To establish toroidal force balance, an inwardly pointing restoring force is

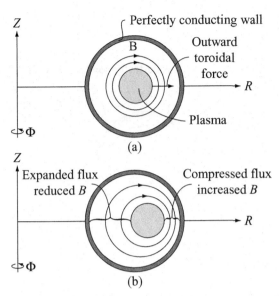

Figure 11.20 (a) A toroidal Z-pinch centered in a perfectly conducting wall. (b) Outward shift of the plasma compresses flux near the outside of the torus providing a toroidal restoring force.

required. In general, for configurations possessing a toroidal current (i.e., the Z-pinch and screw pinch) there are two methods for providing this restoring force: a perfectly conducting wall or the application of an external vertical field. These methods are, however, not effective for a $\theta$-pinch.

This subsection shows how a perfectly conducting wall can provide toroidal force balance. To understand the restoring force, consider a toroidal plasma situated in the center of a perfectly conducting wall as shown in Fig. 11.20(a). Since the plasma experiences a combination of outwardly pointing toroidal forces it naturally starts to move outward along the major radius $R$. As it does so, eddy currents are induced in the wall and the poloidal flux trapped between the plasma and the perfectly conducting wall becomes compressed as shown in Fig. 11.20(b). The compression of the flux implies that the poloidal magnetic field at the outer edge of the plasma has increased in magnitude. As the plasma continues to shift outward, it eventually reaches a point where the magnetic tension on the outer side of the plasma has increased to a large enough value to compensate the outward hoop, tire tube, and $1/R$ toroidal forces. At this point toroidal force balance has been achieved.

While easy to understand in principle, the use of a perfectly conducting wall to achieve toroidal force balance is not very practical. The reason is that in any realistic situation the high temperature of a fusion plasma, coupled with the release of large amounts of neutron power, does not allow the first wall to be superconducting. If the first wall has a finite electrical conductivity, as it must, then the poloidal flux eventually diffuses through this wall as the plasma shifts outward. It is not possible to indefinitely trap the increased flux between the plasma and the wall if the wall has a finite conductivity.

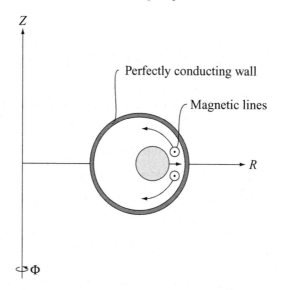

Figure 11.21  Toroidal $\theta$-pinch showing toroidal magnetic lines sliding around the plasma as it moves outward. There is no trapped flux between the plasma and the conducting wall.

A final point to note is that even if the wall were perfectly conducting, this mechanism would not work for the pure $\theta$-pinch. This is because in a pure $\theta$-pinch there is no poloidal field to trap. See Fig. 11.21. Only a toroidal magnetic field is present and the toroidal magnetic field lines simply slide out of the way of the plasma as it moves outward.

### 11.7.6  The restoring force due to a vertical field

A more practical way to achieve toroidal force balance in configurations with a toroidal current is to add an externally applied vertical field as shown in Fig. 11.22. By choosing the magnitude and sign of the vertical field correctly, one can produce an inward restoring force to produce toroidal force balance.

The mechanism can be understood by examining Fig. 11.23. Figure 11.23(a) shows the magnitude and direction of the poloidal magnetic field on the surface of the plasma at the mid-plane $Z = 0$. Observe that the poloidal field on the outside of the torus is smaller in magnitude because of the $1/R$ behavior. Now add a uniform vertical field in the direction shown in Fig. 11.23(b). Note that the vertical field adds to the original poloidal field on the outside of the torus but subtracts from it on the inside. By adjusting the amplitude of the vertical field correctly the magnitude of the combined poloidal fields on the inside and outside of the torus can be balanced as shown in Fig. 11.23(c). When this occurs, toroidal force balance has been achieved.

The use of a vertical field to achieve toroidal force balance in Z-pinches and screw pinches is a practical method used in many experiments. One major reason for its practicality is that the vertical field coils can be located outside the first wall which is convenient for access

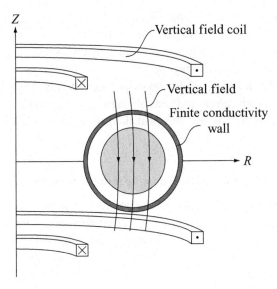

Figure 11.22 Toroidal force balance in a Z-pinch or screw pinch by means of an external vertical field.

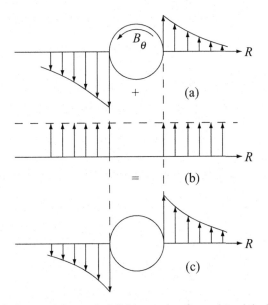

Figure 11.23 Physical picture of the vertical field restoring force: (a) original plasma showing the $B_\theta$ imbalance on the inside and outside; (b) add a pure vertical field in the direction shown; (c) the combination balances $B_\theta$ on the inside and outside providing toroidal force balance.

and preventing radiation damage to the coils. External coils also require that the first wall have a finite conductivity which allows the vertical field to penetrate through to the plasma. If the first wall were a perfect conductor, external coils would be of no use since the fields could not penetrate into the plasma region.

Lastly, observe that a vertical field cannot produce toroidal force balance in a pure $\theta$-pinch. Since the $\theta$-pinch current and vertical magnetic field both lie in the poloidal plane, the magnetic force $\mathbf{J}_\theta \times \mathbf{B}_V$ has no component along the $\mathbf{e}_R$ direction. No force is generated to achieve toroidal force balance.

### 11.7.7 Analytic derivation of toroidal force balance

In this subsection the qualitative ideas discussed above are quantified by a simple calculation which explicitly demonstrates the various forces contributing to toroidal force balance. The main results are the evaluation of the magnitude of the vertical field necessary for toroidal force balance and the demonstration that there is no method for achieving toroidal equilibrium in a pure $\theta$-pinch.

The basic idea behind the calculation is to use the general MHD equilibrium relation to evaluate the net $\mathbf{e}_R$ force acting on the plasma. The calculation is made simpler by assuming the plasma has a circular cross section and a large aspect ratio ($R_0/a \gg 1$). The analysis is further simplified by using a simple model for the magnetic fields which provides a reasonably accurate description of toroidal effects. The derivation proceeds as follows.

#### The model

The basic model consists of a toroidal plasma in which the contours of constant pressure are a set of nested, approximately concentric circles: $p = p(r)$. The geometry is illustrated in Fig. 11.24. Consider now the magnetic field. In a straight cylinder, the magnetic field acting on the plasma is of the form $\mathbf{B} = B_\theta(r)\mathbf{e}_\theta + B_z(r)\mathbf{e}_z$. In a torus the field must be modified to include toroidal effects. It has already been shown that the toroidal field varies as $1/R$ because of toroidicity. Consequently, a good approximation for the toroidal field that combines the radial pressure balance and toroidal force balance characteristics can be written as $B_\phi \approx B_\phi(r)(R_0/R)$.

A similar scaling is assumed for the poloidal magnetic field since it too decreases with major radius: $B_\theta = B_\theta(r)(R_0/R)$. The actual form of the poloidal field in a torus is considerably more complicated (e.g. it includes a $B_r$ component as well as a dipole diamagnetic component). However, the simple form suggested above captures the essential physics and leads to only a small error in the final result. The total poloidal field must also contain an additional contribution representing the applied vertical field, which is required to hold the plasma in toroidal force balance: $\mathbf{B}_V = B_V\mathbf{e}_z$ with $B_V = \text{const}$.

To summarize, the simplified forms of the pressure and magnetic fields to be used in the determination of toroidal force balance are assumed to be

$$p = p(r),$$

$$\mathbf{B} = \frac{R_0}{R}B_\phi(r)\mathbf{e}_\phi + \frac{R_0}{R}B_\theta(r)\mathbf{e}_\theta + B_V\mathbf{e}_z. \tag{11.62}$$

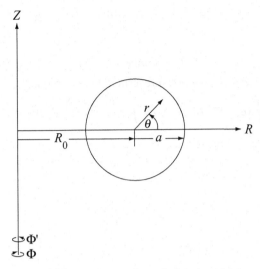

Figure 11.24 Toroidal geometry used to calculate toroidal force balance.

These fields are to be substituted into the general MHD force balance equation. The $\mathbf{e}_R$ component of the force is extracted and then integrated over the entire plasma–vacuum volume to determine the conditions for toroidal force balance. Using the entire volume rather than just the plasma volume is helpful as this avoids the necessity of accurately calculating the poloidal field on the plasma surface, which is considerably more complicated than the expression obtained from the simple model field above. Specifically, for toroidal force balance, one must calculate

$$\int \mathbf{e}_R \cdot [\mathbf{J} \times \mathbf{B} - \nabla p]\, d\mathbf{r} = 0. \tag{11.63}$$

To evaluate the integrand, "cylindrical like" coordinates $(r, \theta)$ are introduced in place of $(R, Z)$ as follows:

$$\begin{aligned} R &= R_0 + r\cos\theta, \\ Z &= r\sin\theta. \end{aligned} \tag{11.64}$$

A short calculation then shows that for the simple model

$$\mu_0 \mathbf{J} = \nabla \times \mathbf{B} = \frac{R_0}{R}\frac{\partial B_\phi}{\partial r}\mathbf{e}_\theta - \frac{1}{r}\frac{\partial}{\partial r}\left(\frac{R_0}{R}rB_\theta\right)\mathbf{e}_\phi \tag{11.65}$$

and

$$\mathbf{e}_R \cdot \mathbf{J} \times \mathbf{B} = -\cos\theta\left[\frac{R_0^2}{R^2}\frac{\partial}{\partial r}\left(\frac{B_\phi^2}{2\mu_0}\right) + \frac{R_0}{R}\frac{B_\theta}{\mu_0 r}\frac{\partial}{\partial r}\left(\frac{R_0}{R}rB_\theta\right)\right] - \frac{B_V}{\mu_0 r}\frac{\partial}{\partial r}\left(\frac{R_0}{R}rB_\theta\right). \tag{11.66}$$

The simple model can now be used to calculate the separate forces contributing to toroidal force balance.

### The tire tube force $F_p$

The tire tube force arises from the pressure term in Eq. (11.63) and is given by

$$F_p = - \int (\mathbf{e}_R \cdot \nabla p) \, d\mathbf{r} = -2\pi \int \left( \frac{\partial p}{\partial r} \cos \theta \right) R r \, dr \, d\theta, \tag{11.67}$$

where use has been made of the relation $d\mathbf{r} = 2\pi R \, r \, dr \, d\theta$ for an axisymmetric torus. Next, recall that $R = R_0 + r \cos \theta$ and note that the $R_0$ term vanishes when integrating over $\theta$. The remaining term can be easily integrated yielding the desired expression for the tire tube force:

$$F_p = -2\pi^2 \int_0^a r^2 \frac{\partial p}{\partial r} \, dr = 2\pi^2 a^2 \langle p \rangle. \tag{11.68}$$

Note that the tire tube force is positive, indicating that it points outward along $R$.

### The $1/R$ force $F_{B_\phi}$

The $1/R$ force arises from the toroidal magnetic field. The $B_\phi$ term from the toroidal force balance relation leads to the following expression for the $1/R$ force:

$$F_{B_\phi} = -\frac{\pi R_0^2}{\mu_0} \int \left( \frac{\partial B_\phi^2}{\partial r} \right) \frac{\cos \theta}{R} r \, dr \, d\theta. \tag{11.69}$$

The integral can be easily evaluated using the large aspect ratio expansion, which assumes that $r/R_0 \ll 1$: specifically

$$\frac{1}{R} \approx \frac{1}{R_0} \left( 1 - \frac{r}{R_0} \cos \theta \right). \tag{11.70}$$

Carrying out the integrations leads to an expression for the $1/R$ force that can be written as

$$F_{B_\phi} = \frac{\pi^2}{\mu_0} \int_0^a r^2 \frac{\partial B_\phi^2}{\partial r} \, dr = 2\pi^2 a^2 \left( \frac{B_{\phi a}^2}{2\mu_0} - \frac{\langle B_\phi^2 \rangle}{2\mu_0} \right). \tag{11.71}$$

Here $B_{\phi a}$ is the applied toroidal magnetic field at $R = R_0$ when no plasma is present. Consistent with the qualitative analysis, one sees that the $1/R$ force points outward along $R$ (i.e., is positive) when the toroidal field is diamagnetic, $B_{\phi a}^2 > \langle B_\phi^2 \rangle$, and points inward when it is paramagnetic.

### The hoop force $F_I$

The hoop force results from the term in the toroidal force balance equation that depends quadratically on the poloidal field $B_\theta$. This term leads to an expression for $F_I$:

$$F_I = -\frac{2\pi R_0}{\mu_0} \int B_\theta \frac{\partial}{\partial r} \left( \frac{R_0}{R} r B_\theta \right) \cos \theta \, dr \, d\theta. \tag{11.72}$$

As for the previous forces, one introduces the aspect ratio expansion and then carries out the integration over $\theta$. A straightforward calculation yields

$$F_1 = \frac{2\pi^2}{\mu_0} \int_0^\infty B_\theta \frac{\partial}{\partial r}(r^2 B_\theta) \, dr = \frac{\pi^2}{\mu_0} \left( r^2 B_\theta^2 \big|_\infty + 2 \int_0^\infty r B_\theta^2 \, dr \right)$$

$$= \frac{\pi^2 a^2}{\mu_0} \left( B_{\theta a}^2 + \frac{2}{a^2} \int_0^\infty r B_\theta^2 \, dr \right), \tag{11.73}$$

where use has been made of the relation $r B_\theta |_\infty = \mu_0 I / 2\pi = a B_{\theta a}$ for a current confined to $r \leq a$. The integral over $r$ is somewhat complex. The reason is that even though the toroidal current density is assumed to vanish at $r = a$ there still exists a vacuum poloidal magnetic field, which is proportional to $I/r$ for $r > a$. This produces a logarithmic divergence in the integral as $r \to \infty$. In practice, the integral is finite because the magnetic energy stored in the vacuum region surrounding a circular loop of wire is finite. Therefore, to obtain the correct expression it is not necessary to let $r \to \infty$ but instead only to let $r \to R_0$ corresponding to the center of the wire. However, in this regime the large aspect ratio expansion breaks down.

The difficulty is circumvented by introducing the macroscopic circuit inductance associated with the toroidal plasma current and then using some well-known results from the theory of magnetostatics. Recall that inductance is defined by

$$\frac{1}{2} L I^2 \equiv \int \frac{B^2}{2\mu_0} \, d\mathbf{r}. \tag{11.74}$$

It is convenient to separate the integration into two regions, an internal one containing the plasma and an external one containing the surrounding vacuum. The internal inductance associated with the plasma is thus defined by

$$\frac{1}{2} L_i I^2 = \int_P \frac{B^2}{2\mu_0} \, d\mathbf{r} = 4\pi^2 R_0 \int_0^a \frac{B_\theta^2}{2\mu_0} r \, dr. \tag{11.75}$$

Similarly, the external inductance is defined by

$$\frac{1}{2} L_e I^2 = \int_V \frac{B^2}{2\mu_0} \, d\mathbf{r} = 4\pi^2 R_0 \int_a^\infty \frac{B_\theta^2}{2\mu_0} r \, dr. \tag{11.76}$$

In fusion research one often utilizes a dimensionless normalized inductance per unit length rather than the actual inductance itself. The definition is $l \equiv (L/2\pi R_0) / (\mu_0/4\pi) = 2L/\mu_0 R_0$.

The desired result is finally obtained by introducing these definitions into Eq. (11.73). A short calculation yields

$$F_I = 2\pi^2 a^2 (l_i + l_e + 1) \frac{B_{\theta a}^2}{2\mu_0}. \tag{11.77}$$

For typical plasmas, $l_i \le 1/2$, depending upon the exact current density profile in the plasma. The quantity $l_e$ requires a surprisingly lengthy calculation from classic magnetostatics. However, the result is well known and is $l_e = 2\ln(8R_0/a) - 4 \sim 2.4$.

For reference one should note that the hoop force can be calculated exactly in the large aspect ratio limit, although this too requires a rather complicated calculation. However, the exact result in the large aspect ratio limit is only slightly modified from the simpler value given above. Hereafter, the more exact relation given below will be used:

$$
\begin{aligned}
F_I &= \frac{I^2}{2}\frac{\partial}{\partial R_0}(L_i + L_e)\\
&= 2\pi^2 a^2 \left(\frac{B_{\theta a}^2}{2\mu_0}\right)\frac{\partial}{\partial R_0}[R_0(l_i + l_e)]\\
&= 2\pi^2 a^2 (l_i + l_e + 2)\frac{B_{\theta a}^2}{2\mu_0}
\end{aligned}
\tag{11.78}
$$

Lastly, note that $F_I > 0$, indicating that the hoop force points radially outwards.

### The vertical field force $F_V$

The last force to calculate arises from the vertical field. The relevant term is

$$
F_V = -\frac{2\pi R_0 B_V}{\mu_0}\int R\frac{\partial}{\partial r}\left(\frac{rB_\theta}{R}\right)dr\,d\theta.
\tag{11.79}
$$

After introducing the aspect ratio expansion and carrying out the integration over $\theta$ one finds that the leading order contribution to Eq. (11.79) reduces to

$$
F_V = -\frac{4\pi^2 R_0 B_V}{\mu_0}(rB_\theta)\Big|_{r\to\infty}.
\tag{11.80}
$$

Since $B_\theta \sim I/r$ as $r \to \infty$, this expression can be rewritten as

$$
F_V = -2\pi^2 a^2 \left(\frac{2R_0 B_V B_{\theta a}}{a\mu_0}\right) = -B_V I\,(2\pi R_0).
\tag{11.81}
$$

The second form represents the familiar result that the force on a "wire" of length $2\pi R_0$, carrying a current $I$ immersed in a uniform field $B_V$ is just the product of the length times the current times the field. Note that if both the current and field are positive, then the force due to the vertical field is negative (i.e. inwards).

### The vertical field for toroidal force balance

By combining the various forces one can easily calculate the vertical field required for toroidal force balance:

$$
\frac{B_V}{B_{\theta a}} = \frac{1}{4}\frac{a}{R_0}\left[\frac{2\mu_0\langle p\rangle}{B_{\theta a}^2} + \frac{B_{\phi a}^2 - \langle B_\phi^2\rangle}{B_{\theta a}^2} + l_i + l_e + 2\right].
\tag{11.82}
$$

In addition to the physical intuition provided by Eq. (11.82), this relatively simple formula, first derived by Shafranov, has provided very useful design guidelines for the vertical field circuits of early tokamak experiments. Since then sophisticated numerical codes have been developed that include the effects of a finite toroidicity and non-circular cross sections. These codes serve as the basis for the design of current experiments. Nevertheless, the forces described by the simple model still represent the basic contributions to toroidal force balance.

In terms of an equilibrium $\beta$ limit for a fusion reactor it would seem at this point, based on Eq. (11.82), that no such limit exists. As one increases the pressure, one simply has to simultaneously increase the applied vertical field. This turns out to be an incorrect conclusion for certain configurations. The issue is subtle and is related to additional equilibrium constraints imposed by stability considerations. These issues, including the equilibrium $\beta$ limit, are examined in Chapter 13.

### Why bending a $\theta$-pinch into a torus doesn't work

It is straightforward to show that a pure toroidal $\theta$-pinch equilibrium cannot exist. The proof comes from examining the simultaneous requirements of radial pressure balance and toroidal force balance for the case $B_{\theta a} = 0$, corresponding to a pure $\theta$-pinch. These relations reduce to

$$\langle p \rangle = \frac{1}{2\mu_0} \left( B_{\phi a}^2 - \langle B_\phi^2 \rangle \right) \qquad \text{radial pressure balance,}$$

$$\langle p \rangle = -\frac{1}{2\mu_0} \left( B_{\phi a}^2 - \langle B_\phi^2 \rangle \right) \qquad \text{toroidal force balance.} \tag{11.83}$$

Clearly, the only possible solution is the trivial one, $\langle p \rangle = 0$.

### 11.7.8 Single particle picture of toroidal force balance

The previous discussion has shown how the single-fluid MHD model can be used to calculate toroidal force balance in a Z-pinch or screw pinch and why such a balance cannot be achieved in a pure toroidal $\theta$-pinch. This sub section provides additional physical intuition by presenting a single-particle guiding center picture of toroidal force balance. In particular, it shows the difficulties that arise in the $\theta$-pinch and how they are resolved by the addition of a toroidal current.

### The pure toroidal $\theta$-pinch

Consider first the pure toroidal $\theta$-pinch. For simplicity assume a low-$\beta$ plasma so that the magnetic field is approximately given by $\mathbf{B} = B_0(R_0/R)\mathbf{e}_\phi$. The geometry is illustrated in Fig. 11.25(a). In terms of guiding center motion note that the $1/R$ dependence of the toroidal field produces both a curvature and $\nabla B$ drift given by

$$\mathbf{v}_g = \mp \frac{1}{\omega_c} \left( v_\parallel^2 + \frac{v_\perp^2}{2} \right) \frac{\mathbf{R}_c \times \mathbf{B}}{R_c^2 B} = \frac{q}{m R_0 B_0} \left( v_\parallel^2 + \frac{v_\perp^2}{2} \right) \mathbf{e}_z. \tag{11.84}$$

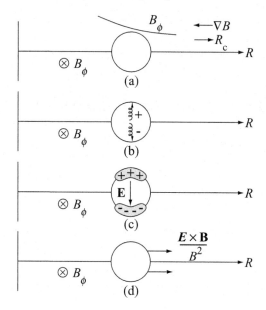

Figure 11.25 Lack of toroidal equilibrium in a $\theta$-pinch: (a) geometry; (b) $\nabla B$ and curvature drifts; (c) induced electric field; (d) plasma moving outward with $\mathbf{E} \times \mathbf{B}/B^2$ drift velocity.

This causes ions to drift vertically upwards and electrons vertically downwards as shown in Fig. 11.25(b).

These guiding center drifts cause accumulation of positive charge on the top of the plasma and negative charge on the bottom of the plasma. The charge separation in turn produces an electric field that points vertically downward through the main bulk of the plasma. See Fig. 11.25(c). Each charged particle in the main bulk of the plasma thus feels the combined influence of the applied toroidal magnetic field and the guiding center induced electric field. Since $\mathbf{E}$ and $\mathbf{B}$ are perpendicular, both electrons and ions develop an $\mathbf{E} \times \mathbf{B}$ drift. This drift is the same for both species and points in the outward direction along $R$ as shown in Fig. 11.25(d). Consequently, the entire plasma moves as a single fluid in the outward $R$ direction towards the first wall; there is no way to achieve toroidal force balance in a pure $\theta$-pinch.

### The Z-pinch and the screw pinch

In an axisymmetric torus the difficulties encountered by the $\theta$-pinch can be overcome by the addition of a poloidal magnetic field, for example in a Z-pinch or screw pinch. For the general case of a screw pinch, the combination of $B_\phi$ and $B_\theta$ produces a magnetic field property known as rotational transform and it is this property that prevents the build up of charge on the top and bottom of the plasma as would occur in a pure $\theta$-pinch.

To understand how this is accomplished one must first understand rotational transform. The property of rotational transform is associated with the fact that in a confined equilibrium,

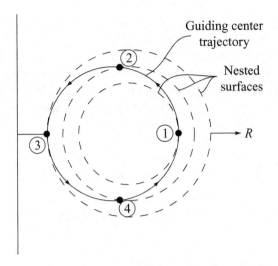

Figure 11.26 Averaging the vertical drift by means of rotational transform.

the combination of toroidal and poloidal magnetic fields causes the magnetic lines to wrap around the plasma, like the stripes on a barber pole bent into a torus. Figure 11.26 shows a poloidal cross section of the plasma in which the location of a given field line marks a point every time it makes one toroidal transit around the torus. After many, many transits the field line puncture points trace out the shape of the flux surface (i.e. a dashed line). Now, if the angle between the $j$ and $j + 1$ transits is denoted by $\Delta\theta_j$, then the rotational transform $\iota$ is defined as the average value of $\Delta\theta_j$ over an infinite number of toroidal transits: $\iota = \langle \Delta\theta \rangle$.

Assuming that rotational transform is present, how does this help to provide toroidal force balance? The answer is that the rotational transform averages out the vertical $\nabla B$ and curvature drifts as the particle freely moves along the field line with $v$. No charge accumulation occurs and therefore no radially outward $\mathbf{E} \times \mathbf{B}$ drift can develop. This averaging occurs even though the particles still possesses the vertical $\nabla B$ and curvature drifts. One is left with the somewhat paradoxical sounding conclusion that a particle that is always drifting upwards does not escape from the plasma.

The physical mechanism of the averaging is illustrated in Fig. 11.26, which depicts three nested pressure contours. Assume a charged particle starts off at point 1 on the inner surface. As it wraps around the torus because of its $v_\parallel$ motion, it would stay on this surface if there were no vertical drifts. However, if the particle has an upward vertical drift its guiding center will drift off this surface and arrive at point 2 on the middle surface after it has moved $\pi/2$ in the poloidal direction. As the particle moves a second $\pi/2$ in the poloidal direction it shifts even further off the original surface arriving at point 3 on the outer surface. The drift process continues during the third $\pi/2$ segment of the orbit although in this case the upward drift causes the particle to move back to the middle surface arriving at point 4. Finally, on the fourth $\pi/2$ segment of the orbit the upward drift of the guiding center brings the particle back to its original starting point. This process is repeated and as one can see there is no buildup of charge at either the top or bottom of the plasma.

The conclusion is that even though a charged particle is always drifting upwards, the presence of rotational transform causes "up" to be *away* from the plasma center half the time and *towards* the plasma center the other half of the time. This is the mechanism by which rotational transform averages out the vertical guiding center drifts and allows toroidal force balance.

### 11.7.9 Calculating the rotational transform

The above discussion shows that rotational transform is an important basic property of the magnetic configuration being used to confine a fusion plasma. This section shows how to calculate the rotational transform in two cases, a straight cylinder of length $2\pi R_0$ and an axisymmetric torus. Knowledge of the transform is also important in the understanding of both equilibrium and stability $\beta$ limits.

#### Rotational transform in a straight cylinder

To evaluate the rotational transform one must be able to calculate the trajectory of a magnetic field line as it wraps around the plasma surface. In a straight cylinder, the equations describing the coordinates of a geometric line parallel to a given magnetic field line are

$$
\begin{aligned}
dr/dz &= B_r/B_z & r(z = 0) &= r_0, \\
d\theta/dz &= B_\theta/r B_z & \theta(z = 0) &= \theta_0.
\end{aligned}
\tag{11.85}
$$

In a straight screw pinch $B_r = 0$ and $B_\theta = B_\theta(r)$, $B_z = B_z(r)$. Because of the cylindrical symmetry the trajectory equations can be easily integrated as follows:

$$
\begin{aligned}
r(z) &= r_0, \\
\theta(z) &= \theta_0 + \left[ \frac{B_\theta(r_0)}{r_0 B_z(r_0)} \right] z.
\end{aligned}
\tag{11.86}
$$

Observe that the magnetic line spirals around the flux surface along a helical trajectory.

The rotational transform is defined as the average value of the change $\Delta\theta$ per toroidal transit. For the straight cylinder the averaging procedure is trivial to carry out because of symmetry. The change in $\Delta\theta$ for any given toroidal transit is exactly the same as for any other toroidal transit. Consequently, the rotational transform in a cylinder, equivalent to a torus of length $2\pi R_0$, is $\iota(r_0) \equiv \Delta\theta = \theta(z + 2\pi R_0) - \theta(z)$, or (suppressing the zero subscript)

$$
\iota(r) = 2\pi \frac{R_0 B_\theta(r)}{r B_z(r)}.
\tag{11.87}
$$

The shape of the profile $\iota(r)$ is one of the critical distinguishing features of different magnetic fusion configurations. Specific examples are presented in Chapter 13.

It is also worth noting that for most axisymmetric toroidal configurations fusion researchers typically introduce a quantity $q(r)$, known as the safety factor, in place of $\iota(r)$. The quantities are inversely related as follows:

$$q(r) \equiv \frac{2\pi}{\iota(r)} = \frac{rB_z(r)}{R_0 B_\theta(r)}. \tag{11.88}$$

Use of the word "safety" is connected to MHD stability and configurations with high $q(r)$ tend to be more stable, (i.e., more safe). The impact of the safety factor is investigated during the discussion of MHD stability in Chapter 13.

### Rotational transform in an axisymmetric torus

The evaluation of the rotational transform in an axisymmetric torus is conceptually similar to that in a straight cylinder although there are two subtleties that arise. First, in a torus, the magnetic fields are functions of $r$ and $\theta$, which is a technical difficulty that makes the derivation of the field line trajectory more difficult. Second, the averaging process to calculate $\iota$ is not as straightforward as for the cylinder. The reason is that the change in $\Delta\theta$ over one toroidal transit differs depending upon the initial poloidal location of the field line. For example, $\Delta\theta$ for one transit is different for a magnetic line starting off on the outside of the torus ($\theta_0 = 0$) as compared to the inside of the torus ($\theta_0 = \pi$). The derivation below overcomes these difficulties although the final answer is in the form of an integral that must, in general, be evaluated numerically. An analytic value is obtained for the simple model field used in the toroidal force balance equation.

The derivation begins with the trajectory equations for a field line in a torus:

$$\begin{array}{ll} dr/d\phi = RB_r(r, \theta)/B_\phi(r, \theta) & r(\phi = 0) = r_0, \\ d\theta/d\phi = RB_\theta(r, \theta)/rB_\phi(r, \theta) & \theta(\phi = 0) = \theta_0, \end{array} \tag{11.89}$$

where $R = R_0 + r\cos\theta$. Assume that the magnetic fields $B_r(r, \theta)$, $B_\theta(r, \theta)$, and $B_\phi(r, \theta)$ are known from an independent solution of MHD equilibrium, usually obtained numerically. Under this assumption, Eq. (11.89) has the form of a set of coupled, non-linear, ordinary differential equations, which can, in principle, easily be solved numerically. Even so, it is not necessary actually to solve these equations. Instead, what is required for the evaluation of the rotational transform is the shape of the flux surface, which is in general non-circular. There are a number of ways to determine the shape of the flux surfaces. One direct way is to just divide the two equations above, leading to

$$dr/d\theta = rB_r(r, \theta)/B_\theta(r, \theta) \qquad r(\theta_0) = r_0. \tag{11.90}$$

This too is an easy equation to solve numerically. Hereafter, it is assumed that Eq. (11.90) has been solved yielding $r = r(\theta; r_0, \theta_0)$.

The next step is to calculate $\Delta\theta$ and carry out the averaging over toroidal transits. The most efficient way to do this is to first calculate how far a magnetic line must travel toroidally ($\delta\phi$) as it makes one full poloidal circuit ($\delta\theta = 2\pi$). Because of axisymmetry, this pattern of field line motion then repeats indefinitely, and a simple use of proportions can be used to determine $\iota$. The angle $\delta\phi$ is easily found from Eq. (11.89):

$$\delta\phi = \int_0^{2\pi} \frac{rB_\phi}{RB_\theta} d\theta. \tag{11.91}$$

Note that in Eq. (11.91) the integration is over the flux surface; that is, the integrand $I(r, \theta) = I[r(\theta), \theta]$. Also it is important to recognize that $\delta\phi = \delta\phi(r_0)$. It is not a function of $\theta_0$ because of axisymmetry and the fact that the integration is over one full poloidal circuit. Thus, without loss in generality one can set $\theta_0 = 0$.

The rotational transform is now found by taking simple proportions

$$\frac{\text{change in } \theta}{\text{change in } \phi} = \frac{2\pi}{\delta\phi} = \frac{\langle \Delta\theta \rangle}{2\pi}, \tag{11.92}$$

where $\langle \Delta\theta \rangle$ is the average poloidal rotation per toroidal transit $\Delta\phi = 2\pi : \langle \Delta\theta \rangle \equiv \iota$. The desired expressions for $\iota$ and $q$ are

$$\frac{\iota(r_0)}{2\pi} = \left[ \frac{1}{2\pi} \int_0^{2\pi} \frac{rB_\phi}{RB_\theta} d\theta \right]^{-1},$$

$$q(r_0) = \frac{1}{2\pi} \int_0^{2\pi} \frac{rB_\phi}{RB_\theta} d\theta. \tag{11.93}$$

In Eq. (11.93) the quantity $r_0$ serves as a label identifying the flux surface on which the transform has been calculated. Specifically, the label corresponds to the radius of the surface on the outside midplane of the cross section $\theta_0 = 0$.

While, in general, the transform has to be calculated numerically, one can demonstrate the procedure using the simple model fields used in the force balance relation. Since $B_r = 0$ for the model fields, this implies that the solution to the flux surface equation is just $r = r_0$. The flux surfaces are circles. The safety factor is then given by

$$q(r_0) = \frac{1}{2\pi} \frac{r_0 B_\phi(r_0)}{B_\theta(r_0)} \int_0^{2\pi} \frac{d\theta}{R_0 + r_0 \cos\theta} = \frac{r_0 B_\phi(r_0)}{R_0 B_\theta(r_0)} \frac{1}{\left(1 - r_0^2/R_0^2\right)^{1/2}}. \tag{11.94}$$

The form is similar to that of a straight cylinder, although there is a small correction due to toroidicity. In the discussion of fusion configurations in Chapter 13 it is shown that there are strong toroidal modifications to $q(r_0)$ when the pressure is increased. In this case the simple model fields are no longer an accurate representation of the equilibrium physics.

### 11.7.10 Toroidal force balance in configurations without toroidal current

The analysis so far has focused on toroidal configurations that are axisymmetric. There is no variation in any of the quantities with the toroidal angle $\phi$. These configurations require a toroidal current and an external vertical field to achieve toroidal force balance. From the point of view of a fusion reactor, the need for a toroidal current is an added complication in the design. Since a transformer cannot produce a steady state DC toroidal current, some

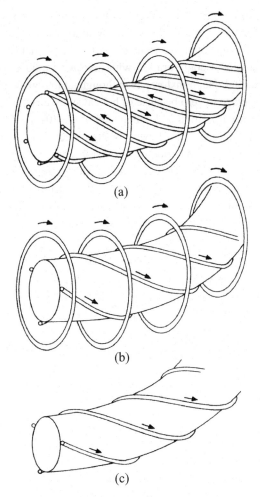

(a)

(b)

(c)

Figure 11.27 Schematic diagrams of three stellarator configurations: (a) simple stellarator; (b) heliotron, and (c) torsatron.

form of non-inductive current drive is required, usually involving complex physics and a substantial amount of expensive auxiliary current drive equipment.

It would clearly be desirable to discover toroidal magnetic configurations that do not require a DC toroidal current in order to maintain toroidal force balance. Such configurations do exist and are known as "stellarators." There are several different types of stellarators under investigation in the world fusion program, but they all share one feature in common. Specifically, all stellarators are 3-D toroidal helices in which the required rotational transform for toroidal force balance is applied externally by means of a helical magnetic field. Simple schematic diagrams of three types of stellarator configurations are illustrated in Fig. 11.27. Although the plasma can, in principle, carry a net DC current this is not a necessity for stellarator equilibrium.

As might be expected, the analysis of 3-D configurations in general requires large amounts of analysis, beyond the scope of the present discussion. Nevertheless, some insight can be gained by calculating the rotational transform of a simple helical configuration. The goal is to show that a helical configuration with zero net toroidal current is capable of externally producing a rotational transform even though, at first glance it might appear that the transform would average to zero because of helical symmetry.

A crucial aspect of the calculation is the inclusion of the effects of a helical modulation to the shape of the flux surface. It is shown, as suggested above, that the leading order contribution of the helical field to the rotational transform does indeed average to zero over each helical period because of symmetry. However, a more careful higher order calculation that takes into account the helical modulation of the flux surface shows that the averaging is not perfect and there remains a small net rotational transform that ultimately provides toroidal force balance. The derivation proceeds as follows.

### The model magnetic field

The existence of the rotational transform does not depend upon toroidicity or finite pressure. The rotational transform is generated even in a straight helical system with vacuum magnetic fields. The model magnetic field thus consists of two contributions: a uniform axial (i.e., toroidal) field $B_z = B_0 = $ const. plus a vacuum helical field $\mathbf{B}_h$. Since $\nabla \times \mathbf{B}_h = 0$ for a vacuum field, $\mathbf{B}_h$ can be represented by a potential: $\mathbf{B}_h = \nabla V$ with $V$ satisfying $\nabla^2 V = 0$. The helical field is assumed to have a helical wavelength $\lambda_h$ and a single helical multi-polarity $l$. This implies that $V(r, \theta, z) = V(r) \sin(l\theta - hz)$, where $\lambda_h = 2\pi/h$. The solution to Laplace's equation then gives

$$V(r) = \hat{C} I_l(hr) \approx C r^l. \tag{11.95}$$

The approximation for the modified Bessel function is valid when $ha \ll 1$ and, while not essential, is used to minimize the complexity of the algebra. The constant $C$ is determined by assuming that the poloidal helical field has a known amplitude $B_h$ at the edge of the plasma. This gives $V(r) = (a B_h/l)(r/a)^l$. The model magnetic field can thus be written as

$$\mathbf{B} = B_0 \mathbf{e}_z + \mathbf{B}_h, \tag{11.96}$$

where

$$\mathbf{B}_h = \nabla V = B_h \left(\frac{r}{a}\right)^{l-1} \left[\mathbf{e}_r \sin \zeta + \mathbf{e}_\theta \cos \zeta - \mathbf{e}_z ha \left(\frac{r}{a}\right) \cos \zeta\right]. \tag{11.97}$$

Here, $\zeta = l\theta - hz$. Note that for the equivalent torus $\zeta = l\theta - N\phi$ with $z = R_0\phi$. Thus, $N = hR_0$ is equal to the number of helical periods around the torus. To simplify the calculation of the rotational transform it is assumed that $N \gg 1$; there are many helical periods around the torus.

The strategy now is to calculate the rotational transform per helical period and then obtain the total transform by multiplying by $N$. The analysis is substantially simplified by assuming that the amplitude of the helical fields is small compared to the toroidal field: $B_h \ll B_0$. This allows one to expand the field line trajectory as follows:

$$r(z) = r_0 + r_1(z) + r_2(z) + \cdots ,$$
$$\theta(z) = \theta_0 + \theta_1(z) + \theta_2(z) + \cdots . \tag{11.98}$$

The trajectory equations, repeated here for convenience, are as follows:

$$dr/dz = B_r/B_z \qquad r(z = 0) = r_0,$$
$$d\theta/dz = B_\theta/r B_z \qquad \theta(z = 0) = \theta_0, \tag{11.99}$$

and must be solved up to and including second order to determine the rotational transform.

### The first order solution

The first non-trivial set of trajectory equations occurs in first order and is

$$\frac{dr_1}{dz} = \frac{B_{hr}(r_0, \theta_0, z)}{B_0} \qquad r_1(z = 0) = 0,$$

$$\frac{d\theta_1}{dz} = \frac{B_{h\theta}(r_0, \theta_0, z)}{r_0 B_0} \qquad \theta_1(z = 0) = 0. \tag{11.100}$$

The equations can easily be integrated yielding

$$r_1(z) = \frac{B_h}{h B_0} \left(\frac{r_0}{a}\right)^{l-1} (\cos \zeta - \cos \zeta_0),$$

$$\theta_1(z) = -\frac{B_h}{h r_0 B_0} \left(\frac{r_0}{a}\right)^{l-1} (\sin \zeta - \sin \zeta_0), \tag{11.101}$$

where, now $\zeta = l\theta_0 - hz$ and $\zeta_0 = l\theta_0$. Physically, this solution corresponds to a small helical modulation of the flux surface. Note that $\theta_1$ is periodic in $z$. Specifically, $\theta_1(2\pi/h) - \theta_1(0) = 0$. As expected, the first non-vanishing contribution to $\theta$ due to the helical fields averages to zero, making no contribution to the rotational transform.

### The second order solution

The second order equations for the trajectory are obtained by a straightforward Taylor expansion:

$$\frac{dr_2}{dz} = \frac{1}{B_0} \left( \frac{\partial B_{hr}}{\partial r_0} r_1 + \frac{\partial B_{hr}}{\partial \theta_0} \theta_1 - \frac{B_{hr} B_{hz}}{B_0} \right) \qquad r_2(0) = 0,$$

$$\frac{d\theta_2}{dz} = \frac{1}{r_0 B_0} \left( \frac{\partial B_{h\theta}}{\partial r_0} r_1 + \frac{\partial B_{h\theta}}{\partial \theta_0} \theta_1 - B_{\theta 1} \frac{r_1}{r_0} - \frac{B_{h\theta} B_{hz}}{B_0} \right) \qquad \theta_2(0) = 0. \tag{11.102}$$

To obtain the rotational transform per helical period one does not actually need the complete solution to these equations. Since the transform per helical period is defined as $\delta\iota = \theta(2\pi/h) - \theta(0) = \theta_2(2\pi/h) - \theta_2(0)$, what is required is the following:

$$\delta\iota = \int_0^{2\pi/h} \frac{d\theta_2}{dz} dz = \frac{1}{h} \int_0^{2\pi} \frac{d\theta_2}{dz} d\zeta . \tag{11.103}$$

A straightforward but slightly tedious calculation using the first order solutions shows that the part of $d\theta_2/dz$ that does not average to zero over $\zeta$ is

$$\frac{d\theta_2}{dz} = \frac{1}{ha^2} \frac{B_h^2}{B_0^2} \left(\frac{r_0}{a}\right)^{2l-4} \left[(l - 2 + h^2 r_0^2) \cos^2 \zeta + l \sin^2 \zeta\right]. \qquad (11.104)$$

It then follows that

$$\delta\iota(r_0) = \frac{\pi}{h^2 a^2} \frac{B_h^2}{B_0^2} \left[2(l - 1) + h^2 r_0^2\right] \left(\frac{r_0}{a}\right)^{2l-4} \approx 2\pi \frac{l - 1}{h^2 a^2} \frac{B_h^2}{B_0^2} \left(\frac{r_0}{a}\right)^{2l-4}. \qquad (11.105)$$

In the second simpler form it has been assumed that $h^2 r_0^2 \ll 1$.

Finally, the total transform is equal to the product of the transform per helical period times the number of helical periods: $\iota = N\delta\iota$. The desired relation can thus be written as

$$\iota(r_0) = 2\pi \frac{l - 1}{N} \left(\frac{R_0 B_h}{a B_0}\right)^2 \left(\frac{r_0}{a}\right)^{2l-4}. \qquad (11.106)$$

Note that for an $l = 2$ stellarator the transform is uniform in space. Actually, the transform increases weakly with radius when the $h^2 r_0^2$ corrections are included. For $l \geq 3$ the transform is zero on axis but increases monotonically with radius. For $l = 1$ the transform would appear to vanish, although a more careful calculation shows that there is a finite transform whose magnitude is reduced by $h^2 a^2$.

The main conclusion is that the combination of a pure toroidal field plus a superimposed helical field produces an external rotational transform capable of providing toroidal force balance without the need for a net DC toroidal current. This is one of the major reasons why the stellarator, despite its technological complexity, is of great interest in fusion research.

## 11.8 Summary of MHD equilibrium

The macroscopic equilibrium of a fusion plasma is described by the MHD model. In general, the equilibria of interest correspond to toroidal configurations in which the confined plasma is described by a set of nested, toroidal pressure contours with circular-like cross sections. Both the magnetic lines and current density lines lie on these contours.

Achieving a toroidal equilibrium requires the solution of two qualitatively different problems. The first involves radial pressure balance where the hot core of plasma tends to expand radially outward along the minor radius $r$. The $\theta$-pinch, $Z$-pinch and the combined screw pinch are capable of confining high values of plasma $\beta$ in radial pressure balance.

The second problem involves toroidal force balance where toroidicity generates a force that tends to expand the plasma outwards along the major radius $R$. In a pure $\theta$-pinch there is no way to counteract this force and thus toroidal equilibrium is not possible. In a $Z$-pinch or a screw pinch the presence of a net DC toroidal current produces a rotational transform that allows toroidal equilibrium by averaging out the vertical $\nabla B$ and curvature drifts. From the fluid point of view, a perfectly conducting wall or an externally applied vertical field is required to balance the forces. Once the forces are balanced, the pressure surfaces and flux surfaces form closed contours, allowing the averaging out of the vertical drifts by the rotational transform.

Lastly, toroidal equilibria can also be generated by 3-D stellarator configurations, which consist of a pure toroidal field plus a superimposed helical field. The helical field generates an externally produced rotational transform which ultimately allows for toroidal force balance. Even though its geometry is more complex than a 2-D axisymmetric system, the stellarator does not require a net DC current, a distinct advantage in terms of reactor desirability.

## Bibliography

The MHD model describes the basic forces that hold the plasma together in macroscopic equilibrium. Almost all textbooks on plasma physics have a section describing MHD equilibrium and stability. The references listed below are almost entirely focused on MHD or at least have very extensive sections devoted to the topic.

Bateman, G. (1978). *MHD Instabilities*. Cambridge, Massachusetts: MIT Press.
Biskamp, D. (1993). *Nonlinear Magnetohydrodynamics*. Cambridge, England: Cambridge University Press.
Freidberg, J. P. (1987). *Ideal Magnetohydrodynamics*. New York: Plenum Press.
Goedbloed, H. and Poedts, S. (2004). *Principles of Magnetohydrodynamics*. Cambridge, England: Cambridge University Press.
Lifschitz, A. E. (1989). *Magnetohydrodynamics and Spectral Theory*. Dordrecht: Kluwer Academic Publishers.
Shafranov, V. D. (1966). *Plasma Equilibrium in a Magnetic Field* (Leontovich, M. A., editor), Vol. 2. New York: Consultants Bureau.
Solov'ev, L. S. and Shafranov, V. D. (1967). *Plasma Confinement in Closed Magnetic Systems* (Leontovich, M. A., editor), Vol. 5. New York: Consultants Bureau.
Wesson, J. (2004). *Tokamaks*, third edn. Oxford: Oxford University Press.
White, R. B. (2001). *Theory of Toroidally Confined Plasmas*. London: Imperial College Press.

## Problems

11.1 Consider a plasma surrounded by a fixed perfectly conducting wall whose behavior is governed by the resistive MHD equations.
   (a) Derive the volume integrated energy balance relation by evaluating the quantity $dW/dt$, where

$$W = \int \left( \frac{\rho v^2}{2} + \frac{B^2}{2\mu_0} + \frac{p}{\gamma - 1} \right) d\mathbf{r}.$$

   (b) The helicity of the plasma is defined as $K = \int \mathbf{A} \cdot \mathbf{B} d\mathbf{r}$, where $\mathbf{A}$ is the vector potential and $\mathbf{B} = \nabla \times \mathbf{A}$. Calculate $dK/dt$ and show that in the limit of ideal MHD (i.e., $\eta = 0$) $K$ is a conserved quantity.

11.2 A cylindrical Z-pinch has profiles satisfying $p(r) = (p_0/J_0^2)J^2(r)$, where $p_0$ and $J_0$ are the pressure and current density on axis respectively. Calculate $B_\theta(r)$ assuming that $J(a) = 0$. Derive a numerical value for $\beta_0 = 2\mu_0 p_0/B_\theta^2(a)$.

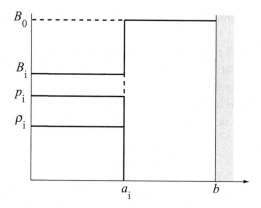

Figure 11.28

11.3  Consider a 1-D cylindrically symmetric screw pinch. Often it is found that the current
density profiles are approximately related by $J_\theta(r) = (\alpha/a)r J_z(r)$, where $\alpha < 1$ is
a constant. Assume that $J_z = K/(a^2 + r^2)^2$. Calculate the pressure profile if the
total $z$ current flowing in the plasma is $I$ and $B_z(\infty) = 0$. The plasma extends over
$0 < r < \infty$. The answer should be written in the form $p = p(r, a, \alpha, I)$. Evaluate
$B_z(0)$.

11.4  This problem treats the adiabatic compression of a $\theta$ pinch in the context of the
surface current model. Consider the simple model illustrated in Fig. 11.28. Here,
$n_i$, $p_i$, $B_i$, $a_i$, $B_0$ represent the initial known state of the plasma. At $t = 0$ the
applied field at $r = b$ is adiabatically increased from its initial value $B_0$ to its final
value $\lambda B_0$ with $\lambda > 1$. Also, the plasma obeys the adiabatic equation of state.
   (a) Using the appropriate conservation laws derive an algebraic expression that
       relates the final beta of the plasma $\beta_f$ to its initial value $\beta_i$. This equation should
       be of the form $F(\beta_f, \beta_i, \lambda, \gamma) = 0$.
   (b) Solve this equation for the special case $\beta_i = 0.5$, $\lambda = 2$, $\gamma = 5/3$.
   (c) If you did the problem correctly you should find that $\beta_f < \beta_i$. Explain why $\beta_f$
       decreases even though the plasma has been compressed.

11.5  Consider a force-free ($\nabla p = 0$) screw pinch of length $2\pi R_0$. Derive an equation for
the $B_z$ profile in terms of the safety factor $q(r)$. Calculate $B_z(r)$ and $B_\theta(r)$ for the
case where $q(r) = q_0 = \text{const.}$ and $B_z(0) = B_0$.

11.6  The goal of this problem is to calculate $B_z(r)$ for a linear $\theta$-pinch confined by an
external magnetic field $B_0$. However, rather than specifying $p(r)$, the pressure is
expressed in terms of the magnetic flux by the relation $p(\psi) = p_0[1 - (\psi/\psi_a)^2]$.
Here, $p_0$ is the pressure on axis, $\psi = r A_\theta$, $A_\theta$ is the vector potential, and $\psi_a$ is
the flux at the plasma edge $r = a$. Calculate the $B_z(r)$ profile and find the value of
$\psi_a = \psi_a(B_0, p_0, a)$.

11.7  Consider radial pressure balance in a 1-D cylindrical screw pinch. The longitudinal
current is given by $\mu_0 J_z(\psi) = -(\alpha^2/R_0 a^2)(\psi - \psi_a)$, where the poloidal flux $\psi$ is
defined such that $B_\theta(r) = (1/R_0)d\psi/dr$. Also, $2\pi R_0$ is the length of the pinch, $a$ is
the minor radius, $\alpha$ is a constant, and $\psi_a = \psi(a)$ is the flux at the edge of the plasma.
The flux is normalized so that $\psi(0) = 0$.
   (a) If the total current flowing in the plasma is $I$ calculate $B_\theta(r)$, $\alpha$, $\psi_a$, assuming
       that $J_z(a) = 0$.

(b) Assume the pressure and longitudinal magnetic field are related by $p(r) = p_0[B_z^2(r) - B_z^2(a)]/[B_z^2(0) - B_z^2(a)]$, where $p_0$ is the pressure on axis. Calculate $B_z(r)$. Can $B_z(a)$ ever reverse sign with the specified boundary conditions?

11.8 A straight 2-D non-circular plasma has an elliptic cross section with horizontal width $2a$ and vertical height $2\kappa a$, where $\kappa$ is the elongation. The plasma is surrounded by a close fitting circular wall of radius $r = 2\kappa b$ with $b \approx a$. For simplicity assume the "toroidal" field $B_z = B_0 = $ const. Now, note that the requirement $\nabla \cdot \mathbf{B} = 0$ implies that the magnetic field in the plasma can be written as $\mathbf{B} = \nabla A \times \mathbf{e}_z + B_0 \mathbf{e}_z$, where $A(r, \theta)$ is the vector potential.

(a) Using the MHD equilibrium equations and Maxwell's equations show that $p = p(A)$ and derive the partial differential equation satisfied by $A$.

(b) Solve the equation for $A$ assuming that $\mu_0 p(A) = (C^2/2)(A_{max}^2 - A^2)$, where $C$ and $A_{max}$ are constants. To obtain an analytic solution assume that $C\kappa a \ll 1$ and solve by expansion.

(c) Magnetic measurements on the wall surface indicate that $B_\theta(\kappa a, \theta) = (\mu_0 I/2\pi \kappa a)(1 + \alpha \cos 2\theta)$, where $I$ and $\alpha$ are measured constants. Solve the equation for $A$ in the vacuum region between the wall and the plasma (where $p = 0$). Match the solutions across the plasma–vacuum interface and derive an expression for $\kappa = \kappa(\alpha)$.

11.9 The purpose of this problem is to derive the Grad–Shafranov equation, a famous partial differential equation describing the MHD equilibrium of configurations possessing toroidal symmetry: $Q(R, Z, \phi) \to Q(R, Z)$.

(a) Using $\nabla \cdot \mathbf{B} = 0$ prove that the magnetic field can be written in terms of a flux function $\psi(R, Z)$ as follows: $\mathbf{B} = \nabla \psi \times \mathbf{e}_\phi/R + B_\phi(R, Z)\mathbf{e}_\phi$.

(b) From Ampères law derive an expression for $\mu_0 J_\phi$ in terms of $\psi$.

(c) From the momentum equation prove that $p(R, Z) = p(\psi)$, where $p(\psi)$ is an arbitrary function.

(d) From the momentum equation prove that $B_\phi(R, Z) \to F(\psi)/R$, where $F(\psi)$ is an arbitrary function.

(e) From the momentum equation derive the Grad–Shafranov equation:

$$R^2 \nabla \cdot (\nabla \psi/R^2) = -\mu_0 R^2 (dp/d\psi) - F(dF/d\psi).$$

11.10 Show that in the large aspect ratio limit of a circular cross section plasma, the Grad–Shafranov equation reduces to the 1-D radial pressure balance relation.

# 12

# MHD – macroscopic stability

## 12.1 Introduction

The second main application of the MHD model concerns the problem of macroscopic stability. The starting point is the assumption that a self-consistent MHD equilibrium has been found that provides good plasma confinement. The stability question then asks whether or not a plasma that has been initially perturbed away from equilibrium would return to its original position as time progresses. If it does, or at worst oscillates about its equilibrium position, the plasma is considered to be stable. On the other hand, when a small initial perturbation continues to grow, causing the plasma to move further and further away from its equilibrium position, then it is considered to be unstable.

For a fusion reactor MHD stability, particularly ideal MHD stability, is crucial. The reason is that ideal MHD instabilities often lead to catastrophic loss of plasma. Specifically, the plasma moves with a rapid, coherent bodily motion directly to the first wall. The resulting loss of plasma combined with the potential damage to the first wall has led to a consensus within the fusion community that ideal MHD instabilities must be avoided in a fusion reactor.

How are such instabilities avoided? In general, plasma stability is improved by limiting the amount of pressure or toroidal current. However, high pressure is desirable in order to achieve high $p\tau_E$ in a reactor, and high current, as will be shown, is desirable for increasing $\tau_E$. MHD stability theory is thus concerned with two basic problems. First, for any given magnetic configuration how does one calculate the actual limits on pressure and current for MHD stability? Second, how does one optimize the magnetic configuration so that the pressure and current limits are as high as possible?

This chapter is primarily concerned with the first problem – developing a procedure to calculate stability limits for a given configuration within the context of the ideal MHD model. Most of the discussion is focused on understanding the basic issues, developing physical intuition, and deriving a general formulation of the stability problem. Several simple applications are presented to demonstrate the procedure for calculating stability. The second and more practical issue of calculating, comparing, and optimizing pressure and current limits in fusion relevant configurations is discussed in Chapter 13.

The specific topics discussed in this chapter are as follows. To begin, several general features of stability are described, including a simple picture of the various types of instabilities that can occur in physical systems and which of these is most applicable to ideal MHD. In addition, an important general property of the ideal model is derived which shows that perpendicular to the magnetic field, plasma and field lines always move together. This property is often referred to as the "frozen-in-field-line" concept and imposes a strong constraint on the types of instabilities that can develop.

The second topic involves a qualitative discussion of the various geometric properties of magnetic fields that are either favorable or unfavorable with respect to stability. Simple physical pictures are presented that demonstrate how, under certain conditions, either parallel or perpendicular plasma currents can drive MHD instabilities. These pictures form the foundation for the development of the physical intuition that has motivated most of the configurations of fusion interest. Also presented is a complementary explanation of MHD stability from a single-particle, guiding center point of view.

The next topic is a derivation of the general formulation of the MHD stability problem, including a description of some of the basic tools of analysis such as linearization and the calculation of eigenvalues that determine stability. The end result is an elegant formulation of the stability problem. Despite its elegance, the model is still quite difficult to solve because, in general, it involves the solution of 3-D partial differential equations.

Lastly, to obtain some actual mathematical experience with MHD stability analysis, several simple special examples are analyzed: the infinite homogeneous plasma, the pure $\theta$-pinch and a subclass of modes for the pure $Z$-pinch. It is shown that a linear $\theta$-pinch has very favorable stability properties while a $Z$-pinch is subject to strong MHD instabilities.

Using the general formulation to examine more interesting, reactor relevant configurations requires huge amounts of analysis and computation, beyond the scope of the present book. To proceed further a very simple, but surprisingly accurate equilibrium model is introduced, known as the surface current model, and this is a primary focus of the next chapter.

## 12.2 General concepts of stability

First the general concepts of ideal MHD stability are discussed, including descriptions of the various types of stability in physical systems and an assessment of which one most closely approximates MHD. Next, a derivation is presented of the frozen-in-field-line property of the ideal model, including a discussion of its important consequences on stability. Lastly, several classification systems are presented that organize the various types of MHD instabilities that can occur in fusion plasmas.

### 12.2.1 Instabilities in physical systems

In order to understand ideal MHD stability it is useful to consider the various types of instabilities than can occur in physical systems. A simple way to do this is to examine a number of mechanical analogs as illustrated in Fig. 12.1. Imagine that the ball represents the plasma and the shape of the curve represents the magnetic geometry.

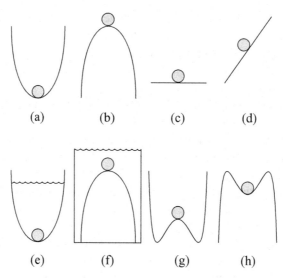

Figure 12.1 Mechanical analogs of stability: (a) stable, (b) unstable, (c) marginally stable, (d) no equilibrium, (e) stable with dissipation, (f) unstable with dissipation, (g) linearly unstable, non-linearly stable, (h) linearly stable, non-linearly unstable.

The ball in Fig. 12.1(a) is considered to be in stable equilibrium. Any perturbation of the ball away from its equilibrium position causes it to oscillate about this position (assuming the well has no friction). In contrast, the ball in Fig. 12.1(b) is in an unstable equilibrium since any small perturbation causes it to continuously mover further and further away from its equilibrium position. The ball in Fig. 12.1(c) is said to be marginally stable. It is on the boundary between stability and instability.

In Fig. 12.1(d) the question of stability makes no sense because the ball is not in an equilibrium position to start with. Figures 12.1(e) and 12.1(f) show qualitatively how some aspects of dissipation may affect stability. Imagine that in both cases the ball is immersed in a viscous fluid such as oil. The stability boundary does not change as compared to the frictionless case; a well remains a well and a hill remains a hill. However, the speed of motion is greatly reduced because of the viscosity. Figure 12.1(g) shows a ball that is linearly unstable but non-linearly stable (i.e., a small perturbation causes the ball to move a finite distance from its initial unstable position to a new stable equilibrium position). Similarly, Fig. 12.1(h) shows a plasma that is linearly stable but non-linearly unstable (i.e., small perturbations are stable but a finite perturbation causes the ball to cross over the hill and become lost).

Which of these analogs corresponds to ideal MHD? For a plasma with or without a perfectly conducting wall the examples in Figs. 12.1(a)–(c) represent accurate analogs for the ideal MHD model. The goal of the stability analysis is to calculate the critical pressure and current corresponding to marginal stability (Fig. 12.1(c)). As $\beta$ and $I$ increase, the well becomes shallower, eventually flattening out and then transforming into a hill. The values

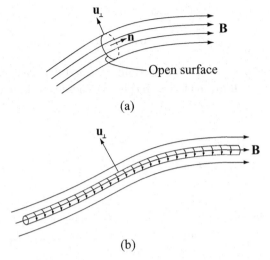

Figure 12.2 The frozen-in-field-line law: (a) geometry showing an open surface; (b) a long thin flux tube.

of $\beta$ and $I$ at marginal stability define the stability limits. The case of a plasma with a resistive wall is qualitatively similar to Figs. 12.1(e) and (f). Note that a plasma that is unstable without a wall (i.e., no oil) is also unstable with a resistive wall (i.e., with oil). However, the wall is still critical because it dramatically slows down the motion of the plasma allowing for the design of a practical feedback system.

### 12.2.2  The frozen-in-field-line concept

The frozen-in-field-line concept refers to the fact that for ideal MHD, the perpendicular motion of the field lines is locked to the perpendicular motion of the plasma. The field lines are "frozen" into the plasma. Below, a derivation of this result is presented, including a discussion of its consequences to ideal MHD stability.

The general proof of the frozen-in-field-line concept for an arbitrary geometry is obtained by making use of an exact, if not very familiar, integral relationship. Consider the flux $\psi(t)$ defined by

$$\psi(t) = \int \mathbf{B} \cdot \mathbf{n} \, dS, \tag{12.1}$$

where the integration is carried out over an open surface area $S$ and $\mathbf{n}$ is the surface normal. See Fig. 12.2(a). The integral relationship shows how to calculate the time rate of change of $\psi(t)$ when $\mathbf{B}$ is a function of time and the boundary of the surface is moving with an arbitrary velocity $\mathbf{u}_\perp$ (which, in general, is not equal to the plasma velocity $\mathbf{v}_\perp$):

$$\frac{d\psi}{dt} = \int \frac{\partial \mathbf{B}}{\partial t} \cdot \mathbf{n} \, dS + \oint \mathbf{B} \times \mathbf{u}_\perp \cdot d\mathbf{l}. \tag{12.2}$$

Now, eliminate $\partial \mathbf{B}/\partial t$ by means of Faraday's law:

$$\frac{d\psi}{dt} = -\int \nabla \times \mathbf{E} \cdot \mathbf{n} \, dS - \oint \mathbf{u}_\perp \times \mathbf{B} \cdot d\mathbf{l}. \tag{12.3}$$

The surface integral is easily converted into a line integral by means of Stokes' theorem. Then setting $\mathbf{E} = -\mathbf{v}_\perp \times \mathbf{B}$ for an ideal MHD plasma leads to the desired flux conservation law:

$$\frac{d\psi}{dt} = \oint [(\mathbf{v}_\perp - \mathbf{u}_\perp) \times \mathbf{B}] \cdot d\mathbf{l}. \tag{12.4}$$

Equation (12.4) shows that in a general magnetic geometry the flux passing through any arbitrary cross section is conserved when the cross sectional area of the flux tube moves with the plasma ($\mathbf{u}_\perp = \mathbf{v}_\perp$).

Consider next applying this result to a long thin tube of magnetic flux as show in Fig. 12.2(b). The tube can be thought of as consisting of a long sequence of connected segments, each with a small cross section. For ideal MHD, as the plasma in the tube moves, the flux in each cross section must move with the plasma. In other words the magnetic lines are "frozen" into the plasma.

The conservation of flux relation has important implications regarding plasma stability. The reason is as follows. Consider a perturbation of the plasma away from equilibrium. For the fluid velocity corresponding to the perturbation to be physically realizable, one requirement is that neighboring plasma fluid elements must remain adjacent to one another; fluid elements are not allowed to tear or break into separate pieces. Now, since the magnetic field lines move with the plasma in ideal MHD, the field-line topology must be preserved during any physical motion. This is a very strong constraint on the types of instabilities that can develop.

There are many magnetic geometries in which it is intuitively clear that it would be energetically favorable for field lines to tear and reconnect, forming a lower-energy state. However, such transitions are not possible in an ideal MHD plasma because of the topological constraint on the field lines. The tearing of a field line requires the tearing of a fluid element and this is not a physically allowable motion. It is for this reason that even a small resistivity can have a large impact on plasma stability. Resistivity allows magnetic field lines to diffuse through the plasma so that the frozen-in topological constraint is removed. A much wider class of plasma motions is now accessible and new instabilities can develop, although admittedly their effects are weaker and they occur on a much slower time scale, associated with the plasma resistivity. Resistive MHD instabilities often, although not always, lead to undesirable enhanced transport and not catastrophic plasma loss. The study of enhanced transport is an advanced topic in plasma physics and is discussed from an empirical point of view in Chapter 14.

In summary, the ideal Ohm's law produces a strong constraint on the allowable motions of the plasma because of the need to preserve field-line topology. This limits the types of instabilities that can develop. However, if such an instability is excited, it is relatively robust since it does not depend on small subtle plasma physics effects. It is this robustness that makes ideal MHD modes so dangerous and has led to the consensus that these instabilities must be avoided in a fusion reactor.

### 12.2.3 Classifications of MHD instabilities

In the literature several different classification schemes have been introduced to distinguish various types of MHD instabilities. Each has its own merits depending upon application. Below is a summary of three main classification schemes.

#### Internal and external modes

Assume the existence of a well-confined plasma equilibrium separated from the first wall by a vacuum region. The first classification scheme distinguishes between internal and external instabilities. This distinction is based on whether or not the surface of the plasma moves as the instability grows. For an internal mode the plasma surface remains fixed in place. These instabilities occur purely within the plasma and place constraints on the shape of the pressure and current profiles. Often they do not lead to catastrophic loss of plasma but can result in important experimental operational limits or enhanced transport. External modes, on the other hand, involve motion of the plasma surface, and hence the entire plasma. Since it is this motion that leads to a plasma striking the first wall, external modes are particularly dangerous in a fusion plasma and must, in general, be avoided.

#### Pressure-driven and current-driven modes

A second way to classify plasma instabilities is by the driving source. In general, a plasma has both perpendicular and parallel currents and each can drive instabilities. The classification system for these instabilities is as follows.

Since $\nabla p = \mathbf{J}_\perp \times \mathbf{B}$ in equilibrium, instabilities driven by perpendicular currents are often called "pressure-driven" modes. Actually, it is a combination of the pressure gradient and the field-line curvature that drives the instabilities. The curvature of the field lines can be favorable, unfavorable, or oscillate with respect to stability. The choice depends upon which way the radius of curvature vector points as compared to the direction of the pressure gradient. Instabilities driven primarily by the pressure gradient are usually further subclassified into one of two forms: the "interchange mode" or the "ballooning mode." The motivation for these names is given in the next section. Pressure-driven instabilities are usually internal modes and set one important limit on the maximum stable $\beta$ that can be achieved in a fusion plasma.

Instabilities driven by the parallel current are often called "current-driven" modes. These instabilities can exist even in the limit of low $\beta$, a regime where all pressure-driven modes are stable. In this regime, current-driven instabilities are often called "kink modes" because the plasma deforms into a kink like shape. Kink modes can be either internal or external. The external kink mode sets an important limit on the maximum toroidal current that can flow in a plasma.

In certain situations, the parallel and perpendicular currents combine to drive an instability, often referred to as the "ballooning-kink" mode. This is usually the most dangerous mode in a fusion plasma. It sets the strictest limits on the achievable pressure and current. Furthermore, it is an external mode, implying that violation of the stability boundary can lead to a rapid loss of plasma energy and plasma current to the first wall. The calculation of the ballooning-kink stability limit is a main focus of the stability analysis described in Chapter 13.

### Conducting wall vs. no wall configurations

The last classification scheme has already been discussed briefly and is based on whether a perfectly conducting wall is required or not. A close fitting perfectly conducting wall can greatly improve the stability of a plasma against external ballooning-kink modes. Since these modes set the strictest stability limits it would be highly desirable to avoid such modes by means of a perfectly conducting wall. The resulting gains in the $\beta$ and current limits due to wall stabilization are substantial, and may be mandatory for reactor viability in certain magnetic configurations.

A real experiment or reactor cannot maintain a superconducting wall close to the plasma. The wall must be resistive, and this subjects the plasma to the resistive wall mode. Based on the simple mechanical analog, the presence of a resistive wall has no effect on the stability boundary of a plasma without a wall. In other words, while a perfectly conducting wall can raise the stability limit above that of the no-wall case, a resistive wall leaves the stability boundary unchanged and only reduces the growth rate.

The possibility of stabilizing the resistive wall mode by feedback or other means is an area of active investigation. This can be a critical issue because certain configurations require a conducting wall even at $\beta = 0$ since they carry a large current. Other configurations with smaller currents are stable at small to moderate $\beta$ even without a wall. It therefore makes sense to classify various magnetic configurations according to whether or not they require a perfectly conducting wall to achieve reactor relevant pressures and currents.

To summarize, MHD instabilities can be classified by several different methods. In terms of the physics one can distinguish internal versus external modes and current-driven versus pressure-driven modes. In terms of experimental practicality, one can distinguish whether a given magnetic configuration does or does not require a perfectly conducting wall to achieve reactor relevant pressures and currents.

### 12.3 A physical picture of MHD instabilities

This section presents several simple physical pictures that show how ideal MHD modes can be excited in a plasma, including interchange modes in a Z-pinch, ballooning modes in a linked set of mirrors, and kink modes in a screw pinch. The need to avoid these basic instability mechanisms has been a major motivation behind the design of essentially all magnetic configurations of fusion interest.

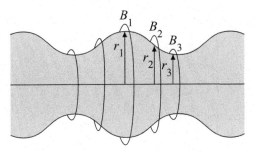

Figure 12.3 Schematic diagram of a Z-pinch interchange instability.

### *12.3.1 Interchange modes*

The first instability considered is the pressure-driven interchange mode. This instability can be most easily understood by considering a pure Z-pinch, which has only perpendicular currents. Assume the initially cylindrically symmetric surface of the plasma is perturbed as shown in Fig. 12.3. Since the total current flowing in the cross section must be constant all along the column, the magnetic field just outside the surface (which scales like $B_\theta \sim 1/r$) and the corresponding radii at the three identified locations satisfy $B_{\theta 1} < B_{\theta 2} < B_{\theta 3}$ and $r_1 > r_2 > r_3$. Consequently, the tension force acting on the surface of the plasma scales as $F_T = (B_\theta^2/\mu_0)(2\pi r L_z) \sim 1/r$, implying that $F_{T1} < F_{T2} < F_{T3}$. The implication is that the force distribution along the column is in the direction to make the perturbation grow since the perturbed force at $r = r_1$ (i.e., $F_{T1} - F_{T2}$) is radially outward while at $r = r_3$ (i.e., $F_{T3} - F_{T2}$) it is radially inward. The perturbed force amplifies the initial perturbation and this corresponds to an unstable driving force. This mode is a form of "interchange instability" since a detailed analysis shows that the perturbation effectively interchanges (i.e., switches the location) of a tube of plasma with a tube of magnetic flux. See Fig 12.3.

A second Z-pinch instability driven by perpendicular currents is illustrated by the perturbation in Fig. 12.4. Here, the magnetic field concentrates in the regions of the plasma where the surface is most tightly curved. The field is reduced in magnitude where the curvature is gentler. Again, the increased tension on the tighter portions of the surface and decreased tension on the gentler portions produce a force distribution that tends to amplify the perturbation, leading to instability.

One sees from these examples that if the field lines curve towards the surface of the plasma, this corresponds to unfavorable curvature. See Fig. 12.5(a). Because of the curvature direction an outward bulge on the surface places the plasma in a lower-field region, which tends to enhance the bulge. Conversely, when the field lines curve away from the surface of the plasma, the curvature is favorable. See the cusp configuration in Fig. 12.5(b). In this case an outward bulge of the surface places the plasma in a region of higher magnetic field. The higher field tends to restore the plasma to its equilibrium position and is thus stabilizing. Note that while the cusp in Fig. 12.5(b) has favorable curvature everywhere, it is not very desirable as a fusion configuration since particles can rapidly flow out through the holes

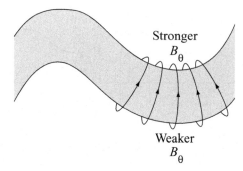

Figure 12.4 Schematic diagram of a twisting Z-pinch instability.

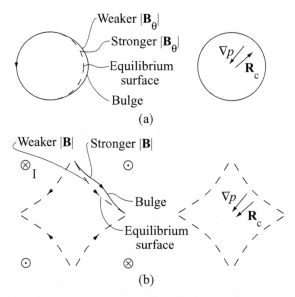

Figure 12.5 (a) Unfavorable curvature in a Z-pinch; (b) favorable curvature in a cusp.

in the cusp. A single-particle picture of the interchange mode that distinguishes favorable from unfavorable curvature is presented in Section 12.4.

### 12.3.2 Ballooning modes

Many magnetic configurations have regions of both favorable and unfavorable curvature. For these cases a somewhat subtle averaging procedure is required to determine whether or not the "average curvature" is favorable or unfavorable. The tokamak has this property. A simpler example to illustrate the point is the series of linked mirrors shown in Fig. 12.6(a). A key property of MHD instabilities in such configurations is the tendency for

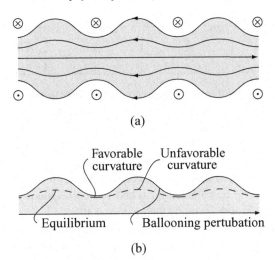

(a)

Favorable    Unfavorable
curvature    curvature

Equilibrium    Ballooning pertubation

(b)

Figure 12.6 Ballooning modes: (a) linked mirror equilibrium; (b) ballooning perturbation in the region of unfavorable curvature.

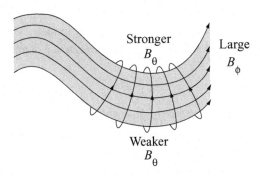

Stronger
$B_\theta$

Large
$B_\phi$

Weaker
$B_\theta$

Figure 12.7 Kink instability in a screw pinch with a large toroidal field. Note that $\mathbf{J} \approx J_\parallel \mathbf{b}$.

the plasma perturbation to concentrate in the regions of unfavorable curvature, causing a "ballooning-like" distortion of the plasma flux surfaces as seen in Fig. 12.6(b). Not surprisingly, these instabilities are often called ballooning modes. Ballooning modes are important because they set one important limit on the maximum $\beta$ that can be stably confined in a fusion plasma. When ballooning modes are excited there is usually either a rapid loss of thermal energy or else greatly enhanced transport losses which prevents the plasma $\beta$ from increasing beyond the ballooning limit. Note also that ballooning instabilities are generally internal plasma instabilities.

### 12.3.3 Current-driven instabilities

Consider now a current-driven instability as illustrated in Fig. 12.7, which shows a Z-pinch with a strong internal axial field (i.e., a screw pinch). Although the current still flows

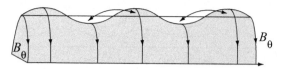

Figure 12.8 Z-pinch column with an interchange perturbation. Note that the plasma and vacuum flux have interchanged positions.

primarily in the axial direction, this now corresponds to the parallel direction. The instability mechanism is similar to that of the pure Z-pinch. However, the toroidal magnetic field lines, which behave somewhat like rubber bands, become bent as the perturbation grows. The tension in the lines tries to restraighten them and this corresponds to a stabilizing force. Instability can still persist if the toroidal field is too small.

If the maximum achievable toroidal magnetic field is determined by magnet engineering considerations, then this sets a limit on the maximum toroidal current that can flow. Specifically, if the current becomes too large with respect to the maximum toroidal field, then the stabilizing effect of toroidal field-line bending is not large enough to stabilize the mode.

In summary, the basic driving mechanisms for the interchange, ballooning, and kink instabilities have been described using simple physical models. Minimizing these driving mechanisms is one major motivation behind the design of essentially all configurations of fusion interest.

### 12.3.4 Single-particle picture of favorable and unfavorable curvature

The distinction between favorable and unfavorable field-line curvature can also be understood from the point of view of single-particle confinement. The analysis below shows how the guiding center curvature drift induces an electric field near the surface of the plasma, which in turn produces an $\mathbf{E} \times \mathbf{B}$ drift. The direction of the $\mathbf{E} \times \mathbf{B}$ drift can be either favorable or unfavorable depending upon the sign of the curvature vector. The analysis proceeds as follows.

Consider a column of plasma with a superimposed interchange perturbation on its surface as shown in Fig. 12.8. Recall that if the field lines are curved a guiding center curvature drift develops given by

$$\mathbf{V}_\kappa = \mp \frac{v_\parallel^2}{\omega_c} \frac{\mathbf{R}_c \times \mathbf{B}}{R_c^2 B}, \tag{12.5}$$

where the plus sign corresponds to a positive charge. Looking at the column end-on as shown in Fig. 12.9(a), one can see that the curvature drift points into or out of the page, depending upon the direction of the curvature vector. For a positive charge, field lines curving towards the plasma produce an inward drift, while field lines curving away produce an outward drift.

Since electrons and ions have opposite sign curvature drifts, the interchange perturbation induces a periodic charge separation along the length of the column as illustrated in

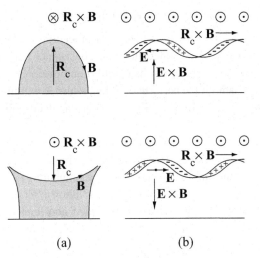

Figure 12.9 Single-particle picture of unfavorable (top) and favorable (bottom) curvature: (a) end view of the plasma showing opposite directions for curvature drifts; (b) side view of plasma showing the induced **E** field and opposing **E** × **B** drifts.

Fig. 12.9(b). This charge separation induces a periodic electric field along the length of the column as shown in the figure. The peak magnitude of the electric field occurs midway in the bulge at the location of the dot on the **E** field direction arrow.

The plasma in the bulge now experiences an **E** × **B** drift. When the field lines curve towards the plasma, the direction of the **E** × **B** drift is such as to enhance the perturbation, leading to instability. Conversely, when the field lines curve away from the plasma, the direction of the **E** × **B** drift is such as to restore the plasma to its equilibrium position, indicating stability.

This simple physical picture provides the intuition to distinguish between favorable and unfavorable curvature.

## 12.4 The general formulation of the ideal MHD stability problem

The previous sections have provided an overview and a description of several general concepts related to MHD stability. The goal now is to quantify these qualitative ideas by means of analysis. This section provides a derivation of the general formulation of the ideal MHD stability problem. The derivation begins with a discussion of the concept of "linear stability." Linear stability is a mathematical procedure that enormously simplifies the analysis of the MHD stability problem. Next, the linear stability procedure is applied to the ideal MHD model. The end result is a set of homogeneous, linear partial differential equations which have the form of an eigenvalue problem, with the eigenvalue being the growth rate of the perturbation. Finally, an important general property of the linear stability equation is derived, which makes the applications in the following sections much simpler to analyze.

### 12.4.1 The concept of linear stability

Linear stability refers to a powerful mathematical procedure that greatly simplifies the stability analysis. The idea is first to calculate a self-consistent MHD equilibrium and then to perturb the plasma *slightly* away from its equilibrium position. By examining the time dependence of the small-amplitude perturbation (i.e., return to equilibrium, oscillate about equilibrium, move further away from equilibrium) one can deduce the (linear) stability of the system.

How does this procedure help? To answer the question, first consider the exact non-linear problem. Here, imagine specifying a set of initial conditions for the plasma. Stability is determined by examining the time evolution of the MHD equations and seeing whether or not the long-time behavior corresponds to a steady state solution exhibiting good confinement. This requires the mathematical solution to a set of coupled, non-linear 3-D partial differential equations in which the unknowns are functions of $R, \phi, Z, t$, a formidable task, even with modern computers.

The mathematics is greatly simplified by separating the problem into two parts: equilibrium and linear stability. Specifically, all dependent variables are written as $Q(\mathbf{r}, t) = Q_0(\mathbf{r}) + Q_1(\mathbf{r}, t)$. The quantity $Q_0(\mathbf{r})$ represents the equilibrium contribution to the total solution. It is an exact solution of the MHD equilibrium equations and therefore, by definition, is not a function of $t$. Also, in most equilibrium configurations of interest there exists some form of geometric symmetry. For example, in an axisymmetric toroidal geometry the dependent variables do not depend upon the toroidal angle $\phi$. In this case $Q_0(\mathbf{r}) = Q_0(R, Z)$, representing a further simplification in the mathematics. The determination of $Q_0(\mathbf{r})$ involves the solution of non-linear differential equations which are much simpler in form than the full model because of the absence of dependence on time and usually at least one spatial coordinate resulting from geometric symmetry. MHD equilibria can be readily obtained numerically and, in certain special cases, analytically. Hereafter, it is assumed that an equilibrium solution has been found representing the configuration of interest.

The quantity $Q_1(\mathbf{r}, t)$ represents the perturbation away from equilibrium. The perturbation is assumed to be small in amplitude implying that $|Q_1(\mathbf{r}, t)| \ll |Q_0(\mathbf{r})|$. It is, in general, a function of all the independent variables $R, \phi, Z, t$ in order to allow for all possible unstable motions. The small-amplitude assumption simplifies the analysis as follows. Consider a typical non-linear term involving the product of two unknowns $J(\mathbf{r}, t) B(\mathbf{r}, t)$. In the context of linear stability these quantities are written as $J = J_0 + J_1$ and $B = B_0 + B_1$. Their product is given by

$$JB = J_0 B_0 + J_0 B_1 + J_1 B_0 + J_1 B_1. \tag{12.6}$$

Now, if one collects all the largest terms in each equation, which correspond to the equilibrium terms (i.e., the $J_0 B_0$ terms), they must exactly cancel since by definition they constitute an exact solution to the MHD model. Of the remaining terms all the non-linear perturbation contributions can be neglected because of the small-amplitude assumption. Specifically,

$|J_1 B_1| \ll |J_1 B_0| \sim |J_0 B_1|$. What survives in the analysis are only the terms that are linear in the unknown perturbations; hence the name "linear stability".

It should also be noted that this conclusion does not depend on the fact that the original non-linearity is quadratic. For a general non-linearity, one simply Taylor expands all unknowns about the equilibrium solution. Thus,

$$F(J) = F(J_0 + J_1) = F(J_0) + \frac{\partial F(J_0)}{\partial J_0} J_1 + \frac{1}{2} \frac{\partial^2 F(J_0)}{\partial J_0^2} J_1^2 + \cdots. \tag{12.7}$$

Again, the leading order terms cancel by virtue of exactly satisfying the equilibrium equations, and all quadratic or higher terms in the Taylor expansion are negligible because of the small-amplitude assumption. Only terms linear in the perturbation survive.

The stability problem has thus been reduced to examining the time evolution of the small-amplitude perturbations. Since the relevant equations are linear, this represents an enormous simplification of the problem. Powerful analytical and numerical techniques are available to solve linear partial differential equations, making the stability analysis of multi-dimensional configurations tractable. This is the primary reason why most MHD stability studies focus on linear stability.

Keep in mind, however, that linear stability does not provide the complete picture. Such an analysis obviously cannot predict the non-linear evolution of a linear instability. Often though knowledge of the non-linear behavior is not too important in the following sense. Experience has shown that plasmas that are linearly MHD unstable usually evolve to a state characterized by a serious or even catastrophic loss of confinement. Therefore, it is not so important to learn the details of how the plasma loses its energy, but instead to learn how to avoid these instabilities in the first place. In other words, one needs to know as accurately as possible the marginal stability limits as determined by linear stability.

### 12.4.2 The MHD linear stability equations

In this subsection the linear stability technique is applied to the ideal MHD equations. The end result is a set of three coupled linear partial differential equations that determine the evolution of the perturbed plasma variables. The equations are cast in the form of an eigenvalue problem for the growth rate by means of a normal mode expansion, a step which further simplifies the determination of stability boundaries.

The analysis begins by expanding all dependent variables as the sum of an equilibrium solution plus a perturbation: $\mathbf{B} = \mathbf{B}_0 + \mathbf{B}_1$, $\mathbf{J} = \mathbf{J}_0 + \mathbf{J}_1$, $p = p_0 + p_1$, and $\rho = \rho_0 + \rho_1$. Note that since the equilibria are static, the appropriate expansion for the velocity is $\mathbf{v} = \mathbf{v}_1$ since $\mathbf{v}_0 = 0$. At this point all equilibrium variables are assumed to be 3-D functions [$Q_0 = Q_0(\mathbf{r})$], while the perturbations depend on three dimensions plus time [$Q_1 = Q_1(\mathbf{r}, t)$]. Substituting into the MHD model leads to the following set of equilibrium and linear stability equations.

- Equilibrium

$$\mathbf{J}_0 \times \mathbf{B}_0 = \nabla p_0;$$
$$\nabla \times \mathbf{B}_0 = \mu_0 \mathbf{J}_0; \qquad (12.8)$$
$$\nabla \cdot \mathbf{B}_0 = 0.$$

- Linear stability

$$\frac{\partial \rho_1}{\partial t} + \nabla \cdot (\rho_0 \mathbf{v}_1) = 0;$$

$$\rho_0 \frac{\partial \mathbf{v}_1}{\partial t} = \mathbf{J}_0 \times \mathbf{B}_1 + \mathbf{J}_1 \times \mathbf{B}_0 - \nabla p_1;$$

$$\frac{\partial p_1}{\partial t} + \mathbf{v}_1 \cdot \nabla p_0 + \frac{\gamma p_0}{\rho_0} \left( \frac{\partial \rho_1}{\partial t} + \mathbf{v}_1 \cdot \nabla \rho_0 \right) = 0; \qquad (12.9)$$

$$\frac{\partial \mathbf{B}_1}{\partial t} = \nabla \times (\mathbf{v}_1 \times \mathbf{B}_0);$$

$$\nabla \times \mathbf{B}_1 = \mu_0 \mathbf{J}_1;$$
$$\nabla \cdot \mathbf{B}_1 = 0.$$

Here, the electric field has been eliminated by means of the ideal Ohm's law and $\rho_0(\mathbf{r})$ is arbitrary. Note that although Eq. (12.9) is linear, it contains a large number of coupled unknowns.

The stability equations are now simplified as follows. First, the time dependence of the perturbations can be explicitly extracted by means of a normal mode expansion. This is possible since the equilibrium quantities are independent of time, and the stability equations are linear. All perturbed quantities can thus be written as

$$Q_1(\mathbf{r}, t) = Q_1(\mathbf{r}) \exp(-i\omega t). \qquad (12.10)$$

The frequency $\omega$, which may be complex, appears as an eigenvalue in the problem. In general, the plasma supports a large, sometimes infinite, number of normal modes, each with its own eigenvalue. Once the normal mode equations are solved and the eigenvalues determined, stability is ascertained by examining the imaginary part of $\omega = \omega_r + i\omega_i$ (i.e., $\omega_i$ is the growth rate) for each mode. If any eigenvalue has $\omega_i > 0$, the system is unstable since the perturbations grow exponentially in time: $Q_1 \sim \exp(\omega_i t)$. If all eigenvalues have $\omega_i \leq 0$, the system is stable as the perturbation either decays to zero or at worst oscillates about the equilibrium position. Note that when introducing the normal mode expansion into the linear stability equations the factor $\exp(-i\omega t)$ cancels in each and every term.

The second step in the simplification process is to introduce the perturbed displacement vector $\boldsymbol{\xi}$ as follows:

$$\mathbf{v}_1 = \frac{\partial \boldsymbol{\xi}}{\partial t} = -i\omega \boldsymbol{\xi}(\mathbf{r}) \exp(-i\omega t). \qquad (12.11)$$

As its name implies, the vector $\boldsymbol{\xi}$ represents the perturbed displacement of the plasma away from its equilibrium position.

The final linearized MHD stability equations are most easily obtained by eliminating all perturbed quantities in terms of $\boldsymbol{\xi}$. A short calculation including all but the linearized momentum equation yields the following relations:

$$\begin{aligned}
\rho_1 &= -\nabla \cdot (\rho \boldsymbol{\xi}); \\
p_1 &= -\boldsymbol{\xi} \cdot \nabla p - \gamma p \nabla \cdot \boldsymbol{\xi}; \\
\mathbf{B}_1 &= \nabla \times (\boldsymbol{\xi} \times \mathbf{B}); \\
\mathbf{J}_1 &= (1/\mu_0) \nabla \times [\nabla \times (\boldsymbol{\xi} \times \mathbf{B})].
\end{aligned}$$
(12.12)

In these expressions the factor $\exp(-i\omega t)$ has been explicitly cancelled from all terms so that the perturbed quantities are functions only of $\mathbf{r}$. Also, for convenience, the "zero" subscript has been dropped from the equilibrium solution.

The quantities in Eq. (12.12) are substituted into the linearized momentum equation. The result is a vector equation for the three components of $\boldsymbol{\xi}$ given by

$$-\omega^2 \rho \boldsymbol{\xi} = \mathbf{F}(\boldsymbol{\xi}),$$
(12.13)

where the force operator $\mathbf{F}$, which is quite complicated, can be written as

$$\mathbf{F}(\boldsymbol{\xi}) = \frac{1}{\mu_0}(\nabla \times \mathbf{B}) \times [\nabla \times (\boldsymbol{\xi} \times \mathbf{B})] + \frac{1}{\mu_0}\{\nabla \times [\nabla \times (\boldsymbol{\xi} \times \mathbf{B})]\} \times \mathbf{B}$$
$$+ \nabla(\boldsymbol{\xi} \cdot \nabla p + \gamma p \nabla \cdot \boldsymbol{\xi}).$$
(12.14)

To complete the formulation of the linear MHD stability problem one must specify boundary conditions. There are two cases of interest corresponding to internal modes and external modes. The internal mode boundary condition is simple to specify and is discussed below. The external mode boundary condition is substantially more complicated and its discussion is deferred until the next chapter.

For internal modes the normal component of the displacement vector must vanish at the plasma surface. If not, a change in the shape of the plasma surface would occur, which violates the definition of an internal mode; that is, an internal mode requires that the plasma surface remain fixed, the instability being confined to the interior of the plasma. Note, however, that the tangential components of the displacement vector need not vanish at the surface. The perturbation is allowed to produce motion wherein the plasma slides tangentially along the unperturbed surface. This physical picture implies that the appropriate boundary condition for internal modes is

$$\mathbf{n} \cdot \boldsymbol{\xi}|_{S_P} = 0,$$
(12.15)

where $S_P$ represents the unperturbed plasma surface and $\mathbf{n}$ is the outward unit normal vector.

Equations (12.13) and (12.15) represent the general formulation of the ideal MHD linear stability problem for internal modes in an arbitrary 3-D equilibrium. The model is usually quite difficult to solve because of the complexity of the force operator $\mathbf{F}$. Several simple 0-D and 1-D geometric configurations are discussed in Sections 12.5–12.8 to demonstrate different procedures for obtaining solutions and eigenvalues. The equations, including an extension of the boundary conditions to external modes, are also solved for the special simple 2-D surface current equilibrium model in the next chapter. This allows a determination of the marginal stability limits for a reasonably wide range of fusion configurations.

### 12.4.3 A general property of linear MHD stability

An important goal of MHD stability analysis is the determination of the pressure and current limits corresponding to marginal stability. Recall that marginal stability is defined as the plasma state in which the growth rate of the most unstable mode approaches zero: $\omega_i \to 0$. For a general physical system, the transition to marginal stability can occur at a non-zero value of real frequency: $\omega_r \neq 0$ as $\omega_i \to 0$. For the ideal MHD system, however, the transition actually does occur at $\omega_r = 0$. The implication is that one can determine marginal stability by setting $\omega^2 \to 0$ at the onset of the analysis. The resulting reduction in mathematical complexity is substantial. Below, a derivation is presented of this important property of marginal stability in ideal MHD.

The proof that marginal stability occurs when $\omega^2 \to 0$ is associated with the fact that the force operator $\mathbf{F}$ has the mathematical property of self-adjointness, which is defined as follows. For any two arbitrary complex vectors $\boldsymbol{\xi}(\mathbf{r})$ and $\boldsymbol{\eta}(\mathbf{r})$ satisfying the boundary conditions, the force operator $\mathbf{F}$ is self-adjoint if it satisfies the integral property

$$\int \boldsymbol{\eta} \cdot \mathbf{F}(\boldsymbol{\xi}) d\mathbf{r} = \int \boldsymbol{\xi} \cdot \mathbf{F}(\boldsymbol{\eta}) d\mathbf{r}. \tag{12.16}$$

The complicated form of $\mathbf{F}$ given by Eq. (12.14) makes this property far from obvious. However, after a surprisingly lengthy calculation the integrals in Eq. (12.16) can be rewritten in a form that is self-adjoint by construction; that is, switching $\boldsymbol{\xi}(\mathbf{r})$ and $\boldsymbol{\eta}(\mathbf{r})$ leaves the integral obviously unchanged. One such self-adjoint form, given without proof, can be written for internal modes as

$$\begin{aligned}
\int \boldsymbol{\eta} \cdot \mathbf{F}(\boldsymbol{\xi}) \, d\mathbf{r} = -\int d\mathbf{r} \Bigg[ &\frac{1}{\mu_0} \nabla \left(\boldsymbol{\xi}_\perp \times \mathbf{B}\right)_\perp \cdot \nabla \times \left(\boldsymbol{\eta}_\perp \times \mathbf{B}\right)_\perp \\
&+ \frac{B^2}{\mu_0} (\nabla \cdot \boldsymbol{\xi}_\perp + 2\boldsymbol{\xi}_\perp \cdot \boldsymbol{\kappa})(\nabla \cdot \boldsymbol{\eta}_\perp + 2\boldsymbol{\eta}_\perp \cdot \boldsymbol{\kappa}) \\
&+ \gamma p(\nabla \cdot \boldsymbol{\xi})(\nabla \cdot \boldsymbol{\eta}) \\
&- \left(\boldsymbol{\xi}_\perp \cdot \nabla_p\right)\left(\boldsymbol{\eta}_\perp \cdot \boldsymbol{\kappa}\right) - \left(\boldsymbol{\eta}_\perp \cdot \nabla_p\right)\left(\boldsymbol{\xi}_\perp \cdot \boldsymbol{\kappa}\right) \\
&- \frac{J_\parallel}{2B} \left(\boldsymbol{\eta}_\perp \times \mathbf{B}\right) \cdot \nabla \times \left(\boldsymbol{\xi}_\perp \times \mathbf{B}\right)_\perp - \frac{J_\parallel}{2B} \left(\boldsymbol{\xi}_\perp \times \mathbf{B}\right) \cdot \nabla \times \left(\boldsymbol{\eta}_\perp \times \mathbf{B}\right)_\perp \Bigg].
\end{aligned} \tag{12.17}$$

Here, $\boldsymbol{\xi}$ and $\boldsymbol{\eta}$ have been written as $\boldsymbol{\xi} = \boldsymbol{\xi}_\perp + \xi_\parallel \mathbf{b}$ and $\boldsymbol{\eta} = \boldsymbol{\eta}_\perp + \eta_\parallel \mathbf{b}$ with $\perp$ and $\parallel$ referring to the equilibrium magnetic field.

Assuming that self-adjointness has been established, one can now easily show that marginal stability in ideal MHD occurs when $\omega^2 \to 0$. The proof is obtained by allowing both $\omega^2$ and $\boldsymbol{\xi}$ to be complex quantities and then forming the dot product of Eq. (12.13) with $\boldsymbol{\xi}^*$ (i.e., set $\boldsymbol{\eta} = \boldsymbol{\xi}^*$). Integrating over the plasma volumes yields

$$\omega^2 \int \rho |\boldsymbol{\xi}|^2 d\mathbf{r} = -\int \boldsymbol{\xi}^* \cdot \mathbf{F}(\boldsymbol{\xi}) \, d\mathbf{r}. \tag{12.18}$$

This step is now repeated using the complex conjugate of Eq. (12.13) and forming the dot product with $\boldsymbol{\xi}$. Using the self-adjoint property of $\mathbf{F}$ it follows that

$$[(\omega^2) - (\omega^2)^*] \int \rho |\boldsymbol{\xi}|^2 d\mathbf{r} = -\int [\boldsymbol{\xi}^* \cdot \mathbf{F}(\boldsymbol{\xi}) - \boldsymbol{\xi} \cdot \mathbf{F}(\boldsymbol{\xi}^*)] d\mathbf{r} = 0. \qquad (12.19)$$

The conclusion is that $(\omega^2) = (\omega^2)^*$. In other words $\omega^2$ is purely real. Therefore, if an eigenfunction is found for which the resulting $\omega^2 > 0$ the system is stable since $\omega_i = 0$ and $\omega_r = \pm\sqrt{\omega^2}$. The modes are oscillatory. On the other hand when $\omega^2 < 0$ the system is unstable since $\omega_r = 0$ and $\omega_i = \pm\sqrt{-\omega^2}$. There is always one mode with a positive growth rate. Marginal stability clearly occurs when $\omega^2 = 0$.

In the simple 0-D and 1-D applications that follow the $\omega^2$ terms are maintained in the analysis to demonstrate how one can actually calculate eigenfunctions and eigenvalues. The more important applications in Chapter 13, which determine pressure and current limits in present day fusion configurations, make use of a 2-D model, and here the assumption $\omega^2 = 0$ is made at the outset to simplify the analysis.

## 12.5 The infinite homogeneous plasma – MHD waves

The first application of ideal MHD stability theory involves the simplest of all geometries, the infinite homogeneous plasma. The equilibrium is 0-D. Since there are no currents flowing to drive instabilities, one expects the "configuration" to be stable. This is indeed the case. The analysis shows that the plasma can support three different types of stable waves: the shear Alfvén wave, the compressional Alfvén wave (sometimes called the fast magnetosonic wave), and the sound wave (sometimes called the slow magnetosonic wave). Below, the general eigenvalue relation encompassing all three modes is derived, followed by a physical discussion of each wave.

### 12.5.1 General derivation of MHD waves

The first step in any stability analysis is to calculate a self-consistent MHD equilibrium. For the infinite homogeneous plasma this is a simple task. The equilibrium is given by

$$\begin{aligned}
\mathbf{B} &= B_0 \mathbf{e}_z, \\
\mathbf{J} &= 0, \\
p &= p_0, \\
\rho &= \rho_0, \\
\mathbf{v} &= 0,
\end{aligned} \qquad (12.20)$$

where $B_0$, $p_0$, and $\rho_0$ are constants. Equations (12.20) obviously satisfy the MHD equilibrium equations.

The next step is to define the most general form of the perturbation. Since the equilibrium is 0-D, the perturbation can be Fourier analyzed in space

$$\boldsymbol{\xi}(\mathbf{r}) = \hat{\boldsymbol{\xi}} \exp(i\mathbf{k} \cdot \mathbf{r}), \qquad (12.21)$$

where $\hat{\xi}$ is a complex amplitude (i.e., $\hat{\xi}$ is a complex constant, and not a function of space). The quantity $\mathbf{k}$ is the wave vector and each of its components must be real in order not to cause a divergence at either $\mathbf{r} = \pm\infty$. Furthermore, without loss in generality the coordinate system can be rotated so that one component of $\mathbf{k}$ lies along $\mathbf{e}_z$ parallel to the magnetic field ($k_\parallel$) while the other component lies along $\mathbf{e}_y$ perpendicular to the field ($k_\perp$). Therefore, the most general form of perturbation for the infinite homogeneous plasma can be written as

$$\xi(\mathbf{r}) = \hat{\xi} \exp(i k_\perp y + i k_\parallel z). \tag{12.22}$$

The advantage of Fourier analyzing in space is that the exponential solutions given by Eq. (12.22) are exact solutions to the linearized partial differential equations describing MHD stability (Eq. (12.13)). The exponential factor cancels from each and every term reducing the set of differential equations in $\xi$ to a set of algebraic equations in $\hat{\xi}$. These algebraic equations are easily derived by noting that each $y$ and $z$ derivative produces a multiplicative factor: $ik_\perp$ and $ik_\parallel$ respectively. Thus, every appearance of the gradient operator reduces to

$$\nabla \rightarrow i\mathbf{k} = ik_\perp \mathbf{e}_y + ik_\parallel \mathbf{e}_z. \tag{12.23}$$

By using this relationship, one can reduce the MHD stability equations to the following simplified algebraic form:

$$\omega^2 \rho_0 \xi = \frac{B_0^2}{\mu_0} \{\mathbf{k} \times [\mathbf{k} \times (\xi \times \mathbf{e}_z)]\} \times \mathbf{e}_z + \gamma p_0 \mathbf{k}(\mathbf{k} \cdot \xi). \tag{12.24}$$

The desired form is then obtained by rewriting Eq. (12.24) as a set of scalar equations for the three vector components. The resulting equations can be cast in matrix form:

$$\begin{vmatrix} \omega^2 - k_\parallel^2 v_a^2 & 0 & 0 \\ 0 & \omega^2 - k^2 v_a^2 - k_\perp^2 v_s^2 & -k_\perp k_\parallel v_s^2 \\ 0 & -k_\perp k_\parallel v_s^2 & \omega^2 - k_\parallel^2 v_s^2 \end{vmatrix} \begin{vmatrix} \xi_x \\ \xi_y \\ \xi_z \end{vmatrix} = 0. \tag{12.25}$$

Here, $v_a = (B_0^2/\mu_0\rho_0)^{1/2}$ is the Alfvén velocity and $v_s = (\gamma p_0/\rho_0)^{1/2}$ is the adiabatic sound speed.

Note that Eq. (12.25) represents three, linear, homogeneous, algebraic equations for the three unknowns $\xi_x$, $\xi_y$, and $\xi_z$. A non-trivial solution to these equations exists only if the determinant of the matrix is zero. A simple inspection shows that the determinant is cubic in $\omega^2$. This implies that there are three separate roots for $\omega^2$ corresponding to the three MHD waves previously mentioned. In the following three subsections the determinant is set to zero and each of the three waves is described in physical terms.

### 12.5.2 The shear Alfvén wave

The first MHD wave of interest is the shear Alfvén wave and is directly determined by setting the $x$ component of the matrix equation to zero. The shear Alfvén eigenvalue is given by

$$\omega^2 = k_\parallel^2 v_a^2 \tag{12.26}$$

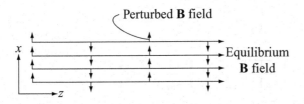

Perturbed **B** field

Equilibrium **B** field

Total **B** field

Figure 12.10 The shear Alfvén wave showing field-line bending.

and clearly corresponds to a stable oscillatory wave. Note that the eigenfrequency is independent of $k_\perp$ even if $k_\perp^2 \gg k_\parallel^2$. The eigenfunction has $\xi_x \neq 0$ and $\xi_y = \xi_z = 0$. The perturbed magnetic field has only an $x$ component given by $B_{1x} = ik_\parallel B_0 \xi_x$. Also, back substitution shows that $\rho_1$, $p_1$, and $\nabla \cdot \boldsymbol{\xi}$ are all zero. The shear Alfvén wave is incompressible, producing neither density nor pressure fluctuations.

Since the perturbed magnetic field and displacement are perpendicular to both **B** and **k**, the wave is purely transverse. Physically, this causes the magnetic field lines to bend as shown in Fig. 12.10. The shear Alfven wave thus describes a basic oscillation between perpendicular plasma kinetic energy (i.e. inertial effects) and line bending magnetic energy (i.e. field line tension). From the stability point of view the shear Alfvén wave is perhaps the most important MHD wave since it is the one that is usually driven unstable by finite currents and geometric effects.

### 12.5.3 The compressional Alfvén wave

The remaining two MHD waves are coupled through the $y$ and $z$ components of the matrix equation. Setting the determinant of this $2 \times 2$ system to zero yields

$$\left(\omega^2 - k^2 v_{\mathrm{a}}^2 - k_\perp^2 v_{\mathrm{s}}^2\right)\left(\omega^2 - k_\parallel^2 v_{\mathrm{s}}^2\right) - k_\perp^2 k_\parallel^2 v_{\mathrm{s}}^4 = 0. \tag{12.27}$$

Observe that there are two roots for $\omega^2$ given by

$$\omega^2 = \frac{1}{2}k^2(v_{\mathrm{a}}^2 + v_{\mathrm{s}}^2)\left[1 \pm (1 - \alpha^2)^{1/2}\right], \tag{12.28}$$

where $k^2 = k_\perp^2 + k_\parallel^2$ and

$$\alpha^2 = 4\frac{k_\parallel^2}{k^2}\frac{v_{\mathrm{s}}^2 v_{\mathrm{a}}^2}{\left(v_{\mathrm{s}}^2 + v_{\mathrm{a}}^2\right)^2} \leq 1. \tag{12.29}$$

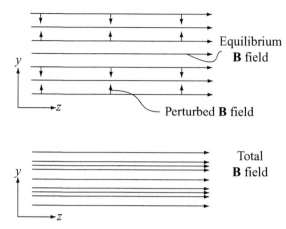

Figure 12.11 The compressional Alfvén wave in the limit $k_\parallel / k_\perp \ll 1$ showing magnetic field compression.

The compressional Alfvén wave (i.e., the fast magnetosonic wave) corresponds to the $+$ sign in Eq. (12.28). The physics can easily be ascertained by noting that $v_s^2/v_a^2 = \gamma \mu_0 p_0/B_0^2 \sim \beta$ and then considering the interesting limit $\beta \ll 1$. This is equivalent to assuming that $\alpha^2 \ll 1$. The eigenfrequency for the compressional Alfvén wave reduces to

$$\omega^2 \approx \left(k_\perp^2 + k_\parallel^2\right) v_a^2. \tag{12.30}$$

Clearly this represents a stable oscillatory wave.

In terms of the eigenfunction one can easily demonstrate that in the low-$\beta$ limit, $\xi_z/\xi_y \approx k_\perp k_\parallel v_s^2/k^2 v_a^2 \sim \beta \ll 1$. The motion is predominantly perpendicular to the magnetic field. Also for $k_\parallel \sim k_\perp$ it follows that $B_{1y}/B_{1z} \approx -k_\parallel/k_\perp \sim 1$. The $B_{1y}$ term is responsible for the $k_\parallel^2 v_a^2$ term in the dispersion relation and corresponds to the previously discussed line bending effect. The $B_{1z}$ term corresponds to the $k_\perp^2 v_a^2$ term and to magnetic compression as illustrated in Fig. 12.11. Usually finite geometric effects require that $k_\parallel \ll k_\perp$, implying that magnetic compressional effects dominate line bending for the compressional Alfvén wave. Finally, a short calculation shows that $\mu_0 p_1/B_0 B_{1z} \approx -\mu_0 \gamma p_0/B_0^2 \sim \beta \ll 1$, indicating that plasma compression plays a negligible role in comparison to magnetic compression in the low-$\beta$ compressional Alfvén wave.

In summary, the compressional Alfvén wave describes a basic oscillation between plasma kinetic energy (i.e., inertia) and magnetic compressional energy (i.e., magnetic pressure). It is very stabilizing for $k_\perp \gg k_\parallel$ and therefore is only weakly excited in most plasma instabilities.

### 12.5.4 The sound wave

The sound wave (i.e., slow magnetosonic wave) is the third MHD wave and is obtained from the negative root in Eq. (12.28). In the low-$\beta$ limit the eigenfrequency reduces to

$$\omega^2 \approx k_\parallel^2 v_s^2. \tag{12.31}$$

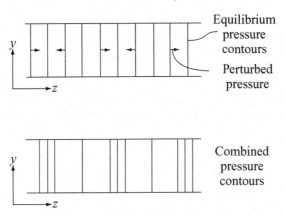

Figure 12.12 The MHD sound wave in the limit $k_\parallel / k_\perp \gg 1$ showing plasma pressure compression.

As for the other MHD modes, the eigenfrequency corresponds to a stable oscillatory wave. For the sound wave the components of the eigenfunction displacement satisfy $\xi_y / \xi_z \approx k_\perp k_\parallel v_s^2 / k^2 v_a^2 \sim \beta \ll 1$. The motion is essentially parallel to the magnetic field. The physics of the sound wave can most easily be seen by considering the regime where $k_\perp \ll k_\parallel$. In this regime a short calculation shows that $B_0 B_{1z} / \mu_0 p_1 \approx k_\perp^2 / k^2 \ll 1$ and $B_0 B_{1y} / \mu_0 p_1 \approx -k_\parallel k_\perp / k^2 \ll 1$. The motion is dominated by the perturbed pressure giving rise to sound waves as shown in Fig. 12.12.

The sound wave represents a basic oscillation between plasma kinetic energy (i.e., inertia) and plasma thermal energy (plasma compression). The sound wave is not usually strongly excited for MHD instabilities. Also, the physics of the sound wave depends on the parallel dynamics of the plasma and the MHD model does not treat this physics very accurately.

### 12.5.5 Summary

The first application of the general MHD stability theory involves the infinite homogeneous plasma. The simplicity of the geometry is used to reduce the problem to a set of linear algebraic equations from which the eigenfrequencies are easily calculated. The results indicate that the infinite homogeneous plasma supports three stable oscillatory waves: the shear Alfvén wave, the compressional Alfvén wave, and the sound wave. Usually it is the shear Alfvén wave that is most strongly excited when MHD instabilities arise.

### 12.6 The linear θ-pinch

The second application of MHD stability theory investigates the stability of the linear θ-pinch. Recall that this configuration has only a perpendicular current and straight magnetic field lines. The absence of field-line curvature eliminates pressure-driven modes, while the absence of parallel current eliminates current-driven modes. The linear θ-pinch should

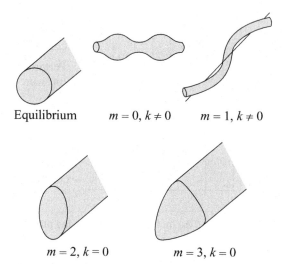

Figure 12.13 Graphical interpretation of perturbations corresponding to various values of $m$ and $k$.

be MHD stable and the goal of this section is to prove this result. The plan is to define the equilibrium, specify the most general form of perturbation, and then analyze the MHD stability equations.

### 12.6.1 The equilibrium and perturbation

The analysis begins with the $\theta$-pinch equilibrium relation, which involves plasma pressure and magnetic pressure, and is repeated here for convenience:

$$p(r) + \frac{B_z^2(r)}{2\mu_0} = \frac{B_0^2}{2\mu_0}. \tag{12.32}$$

Also, $\rho(r)$ is arbitrary.

Consider next the perturbation. Since the equilibrium is symmetric with respect to $\theta$ and $z$, one can Fourier analyze the perturbation as follows:

$$\boldsymbol{\xi}(\mathbf{r}) = \boldsymbol{\xi}(r) \exp\left[i(m\theta + kz)\right]. \tag{12.33}$$

Here, $m$ and $k$ are the "poloidal" and "toroidal" wave numbers respectively. Note that since the equilibrium quantities are functions of radius, it is of no mathematical help to Fourier analyze with respect to $r$ (i.e., a solution of the form $\exp(ikr)$ does not satisfy the differential equations).

A graphical interpretation of the perturbations corresponding to various $m$ and $k$ is presented in Fig. 12.13. Observe that $m = 1$ corresponds to a sideways shift of the plasma cross section. For $m = 2$ the plasma cross section develops an elliptical distortion, while for $m = 3$ the cross section becomes triangular. In addition, when $k \neq 0$ the distortion of the cross section twists around as one moves along the column. For $k = 0$, however, the cross section distortion is constant and does not rotate along the column. The special case $m = 0$, $k \neq 0$ gives rise to a series of bulges along the length of the column. The most unstable external MHD instabilities usually involve modes with low $m$ and small $k$.

Equation (12.33) is the most general form of perturbation for a cylindrically symmetric plasma. Application of Fourier analysis reduces the MHD stability equations from 3-D to 1-D, a major simplification. In general, one must derive and then solve the resulting 1-D MHD differential equations to determine the eigenvalue $\omega^2$ and then see whether $\omega_i > 0$ for any choice of $m$ and $k$.

### 12.6.2 The radial differential equation

The next step is to substitute the equilibrium and perturbation into the generalized MHD stability formulation (Eq. (12.13)) and simplify the resulting 1-D differential equations as much as possible. This is, in principle, straightforward but requires a surprising amount of tedious calculation even for the simplest of equilibria. The result is a single second order differential equation for $\xi_r$ in which the eigenvalue $\omega^2$ appears in a complicated, non-standard form. The equation is displayed below without proof, the purpose being to give the reader an idea of the complexity of MHD stability calculations:

$$\frac{d}{dr}\left[ A \frac{d}{dr}(r\xi_r) \right] - C(r\xi_r) = 0. \tag{12.34}$$

Here

$$A(r) = \frac{\rho\left(v_a^2 + v_s^2\right)\left(\omega^2 - \omega_a^2\right)\left(\omega^2 - \omega_h^2\right)}{r\left(\omega^2 - \omega_f^2\right)\left(\omega^2 - \omega_s^2\right)},$$

$$C(r) = -\frac{\rho}{r}\left(\omega^2 - \omega_a^2\right),$$

$$\omega_a^2(r) = k^2 v_a^2, \tag{12.35}$$

$$\omega_h^2(r) = \frac{v_s^2}{v_a^2 + v_s^2}\omega_a^2,$$

$$\omega_{f,s}^2(r) = \frac{1}{2r^2}(k^2 r^2 + m^2)\left(v_a^2 + v_s^2\right)\left[1 \pm (1 - \alpha^2)^{1/2}\right],$$

$$\alpha^2(r) = \frac{4k^2 r^2}{k^2 r^2 + m^2}\frac{v_a^2 v_s^2}{\left(v_a^2 + v_s^2\right)^2}.$$

Note that even after the tedious algebra has been completed, the eigenvalue $\omega^2$ occurs in such a complicated manner that one cannot easily determine whether or not the θ-pinch is stable. While the answer could be ascertained numerically, for the θ-pinch a simpler, alternative approach yields the desired information.

### 12.6.3 Stability of the θ-pinch

The alternative approach for determining $\theta$-pinch stability involves the derivation of an integral relation which shows that $\omega^2$ must always be positive. The integral relation is obtained by forming the dot product of the general MHD stability equation (Eq. (12.13)) with $\xi^*$ and then integrating over the plasma volume. In carrying out the calculation it is helpful to expand the displacement vector as $\xi = \xi_\perp + \xi_\parallel \mathbf{b}$, where $\xi_\perp = \xi_r \mathbf{e}_r + \xi_\theta \mathbf{e}_\theta$ and $\xi_\parallel = \xi_z$. Also, extensive use is made of the equilibrium relation $B' = -\mu_0 p'/B$ and the expression for the parallel perturbed magnetic field $B_{1z} = \mathbf{b} \cdot \nabla \times (\xi_\perp \times \mathbf{B}) = (\mu_0 p'/B)\xi_r - B\nabla \cdot \xi_\perp$. The various terms appearing in the integrand can now be written as

$$
\begin{aligned}
\xi^* \cdot \nabla(\gamma p \nabla \cdot \xi) &= \nabla \cdot [(\gamma p \nabla \cdot \xi)\xi^*] - \gamma p |\nabla \cdot \xi|^2, \\
\xi^* \cdot \nabla(\xi \cdot \nabla p) &= \nabla \cdot [(\gamma p \nabla \cdot \xi)\xi^*] - p'\xi_r(\nabla \cdot \xi_\perp^* - ik\xi_\parallel^*), \\
\xi^* \cdot \mathbf{J}_1 \times \mathbf{B} &= \frac{1}{\mu_0}\{\nabla \cdot [\mathbf{B}_1 \times (\xi_\perp^* \times \mathbf{B})] - |B_{1r}|^2 - |B_{1\theta}|^2 - |B_{1z}|^2\}, \\
\xi^* \cdot \mathbf{J} \times \mathbf{B}_1 &= -\mu_0 \left(\frac{p'}{B}\right)^2 |\xi_r|^2 + p'\xi_r^*\nabla \cdot \xi_\perp + ikp'\xi_\parallel^*\xi_r.
\end{aligned}
\tag{12.36}
$$

When computing the integral over the plasma volume, note that by Gauss' theorem all the divergence terms integrate to zero for internal mode boundary conditions. The desired stability relation is obtained by collecting terms and solving for $\omega^2$. A short calculation yields

$$
\omega^2 = \frac{\int [|B_{1r}|^2 + |B_{1\theta}|^2 + B^2|\nabla \cdot \xi_\perp|^2 + \gamma p |\nabla \cdot \xi|^2]\,d\mathbf{r}}{\int \rho |\xi|^2\,d\mathbf{r}}.
\tag{12.37}
$$

Since all the terms in the integrand are positive, it immediately follows that $\omega^2 > 0$. The linear $\theta$-pinch is stable to all MHD modes. It is unfortunate in view of this good stability behavior that a linear $\theta$-pinch cannot be bent into a torus because of the problems associated with toroidal force balance. It is worth noting that the procedure of multiplying by $\xi^*$ and integrating over the plasma volume can be applied to arbitrary MHD configurations. However, for all but the $\theta$-pinch equilibrium, the resulting integrand has both positive and negative contributions and therefore one cannot easily determine stability by simple inspection.

## 12.7 The $m = 0$ mode in a linear Z-pinch

The next application of MHD stability theory involves the Z-pinch. Here too, substantial analysis is involved to treat perturbations with arbitrary $m$ and $k$. However, by restricting attention to the $m = 0$ mode the calculation becomes tractable. For this mode a reasonably simple differential equation is derived leading to an analytic expression for the eigenvalue. It is shown that the $m = 0$ mode is unstable for typical Z-pinch profiles unless the pressure decreases sufficiently slowly at large radii. The form of the eigenfunction is similar to that previously discussed in connection with Fig. 12.3, and for obvious reasons is often referred to as the "sausage instability".

### 12.7.1 Derivation of the differential equation

The analysis for the $m = 0$ mode is simplified because for this perturbation $B_{1r} = B_{1z} = 0$ and $\xi_\theta = 0$. Also it is convenient to simplify the right hand side of the general MHD stability equation (Eq. (12.13)) using the vector identity $(\nabla \times \mathbf{B}_1) \times \mathbf{B} + (\nabla \times \mathbf{B}) \times \mathbf{B}_1 = \mathbf{B}_1 \cdot \nabla \mathbf{B} + \mathbf{B} \cdot \nabla \mathbf{B}_1 - \nabla (\mathbf{B} \cdot \mathbf{B}_1)$. Lastly, rather than use $\xi_r$, $\xi_z$ as the basic unknowns, one can further simplify the analysis by introducing two new equivalent unknowns $\xi \equiv \xi_r$, $\nabla \cdot \boldsymbol{\xi} \equiv (r\xi_r)'/r + ik\xi_z$.

Under these assumptions, the perturbed pressure and magnetic field reduce to

$$
\begin{aligned}
p_1 &= -p'\xi - \gamma p \nabla \cdot \boldsymbol{\xi}, \\
B_{1\theta} &= -(B_\theta/r)' r\xi - B_\theta \nabla \cdot \boldsymbol{\xi}.
\end{aligned}
\tag{12.38}
$$

The quantities $p_1$ and $B_{\theta 1}$ are now substituted into the $z$ component of the momentum equation:

$$
-\omega^2 \rho \left[ \nabla \cdot \boldsymbol{\xi} - \frac{(r\xi)'}{r} \right] = k^2 \left( p_1 + \frac{B_\theta B_{1\theta}}{\mu_0} \right).
\tag{12.39}
$$

A short calculation shows that $\nabla \cdot \boldsymbol{\xi}$ appears only algebraically, leading to the relation

$$
\nabla \cdot \boldsymbol{\xi} = \frac{1}{r} \left[ \frac{2k^2 v_a^2 \xi - \omega^2 (r\xi)'}{k^2 (v_s^2 + v_a^2) - \omega^2} \right].
\tag{12.40}
$$

The final equation is obtained by substituting this expression into the $r$ component of the momentum equation:

$$
\omega^2 \rho \xi = \left( p_1 + \frac{B_\theta B_{1\theta}}{\mu_0} \right)' + \frac{2 B_\theta B_{\theta 1}}{\mu_0 r}.
\tag{12.41}
$$

The algebra is carried out by noting that the resulting expression can be written in the following form:

$$
[A_1 (r\xi)']' + A_2 (r\xi)' + A_3 (r\xi) = 0.
\tag{12.42}
$$

A slightly tedious calculation shows that $A_2 = 0$ exactly. Next, assume that the plasma is near marginal stability. Specifically, assume that $\omega^2 \ll k^2(v_a^2 + v_s^2)$. The coefficients $A_1$ and $A_3$ simplify substantially leading to the desired form of the eigenvalue equation:

$$
\omega_i^2 \frac{d}{dr} \left[ \frac{\rho}{r} \frac{d}{dr} (r\xi) \right] - k^2 \left[ \omega_i^2 \frac{\rho}{r} + \frac{2\gamma p K}{r^3} \right] (r\xi) = 0,
\tag{12.43}
$$

where

$$
K(r) = \frac{r}{\gamma p} \frac{dp}{dr} + \frac{2 B_\theta^2}{B_\theta^2 + \gamma \mu_0 p}
\tag{12.44}
$$

and $\omega_i^2 = -\omega^2$ in order to focus on instabilities.

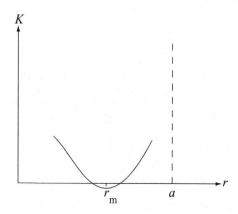

Figure 12.14 $K(r)$ vs. $r$ for a slightly unstable $m = 0$ mode in a Z-pinch.

### 12.7.2 Stability of the $m = 0$ mode

The stability of the $m = 0$ mode is determined by solving Eq. (12.43) and determining the sign of the resulting eigenvalue $\omega_i^2$. The sign of $\omega_i^2$ depends on the profile $K(r)$. For typical profiles, $K(r)$ is usually positive over most of the plasma and may or may or not become negative near the edge where the pressure becomes small.

If $K(r)$ is positive for $0 < r < a$, then one can easily show that the plasma is stable ($\omega_i^2 < 0$) by multiplying by $\xi^*$ and integrating over the plasma volume. If, however, $K(r)$ is negative, even over a very small region, the plasma is unstable. To prove this assume that $K(r)$ has the form shown in Fig. 12.14 and Taylor expand about the minimum point $r = r_m$

$$K(r) \approx -|K_m(r_m)| + \frac{d^2 K_m(r_m)}{dr_m^2} \frac{(r - r_m)^2}{2}. \tag{12.45}$$

Next, introduce normalized variables $\overline{\omega}_i$, $\overline{k}$, and $x$ as follows:

$$\begin{aligned} \omega_i^2 &= (2\,|K_m|\, v_s^2/r_m^2)\overline{\omega}_i^2, \\ k^2 &= \left(K_m''/8\,|K_m|\right)\overline{k}^2, \\ r &= r_m + 2\left(|K_m|\,/K_m''\right)^{1/2}\left(\overline{\omega}_i/\overline{k}\right)^{1/2} x. \end{aligned} \tag{12.46}$$

The eigenvalue equation reduces to

$$\frac{d^2\xi}{dx^2} + \left[\frac{\overline{k}}{2}\left(\frac{1}{\overline{\omega}_i} - \overline{\omega}_i\right) - x^2\right]\xi = 0. \tag{12.47}$$

The boundary conditions in terms of the localized coordinate $x$ are that $\xi(\pm\infty) \to 0$ in order for the eigenfunction to remain well behaved far from the minimum point. Equation (12.47) is easily solved in terms of Hermite functions. For the eigenfunction to vanish at both $x = \pm\infty$ only certain discrete values of the eigenvalue $\bar{\omega}_i$ are allowed, defined by $(\bar{k}/2)(\bar{\omega}_i^{-1} + \bar{\omega}_i) = 2n + 1$, where $n$ is a positive integer. The solution for $\bar{\omega}_i$ for the fastest mode ($n = 0$) is given by

$$\bar{\omega}_i = \frac{\bar{k}}{1 + \left(1 + \bar{k}^2\right)^{\frac{1}{2}}} \to 1. \tag{12.48}$$

The limiting value represents the worst case $\bar{k} \to \infty$. The corresponding eigenfunction can be expressed in terms of Hermite polynomials as follows:

$$\xi = \exp(-x^2/2)H_{2n+1}(x). \tag{12.49}$$

Since $\bar{\omega}_i > 0$ the system is clearly unstable. In terms of the unnormalized variables the maximum growth rate is given by

$$\omega_i = (2|K_m|)^{1/2}\frac{v_s}{r_m}. \tag{12.50}$$

The conclusion is that if the coefficient $K_m$ is negative anywhere in the plasma the Z-pinch is unstable to the $m = 0$ mode. Physically, the mode is driven unstable by the unfavorable curvature of the field lines as previously discussed. There is also, however, a stabilizing force, due to compression of the plasma; one has to do compressional work on the plasma, which is a stabilizing effect, in order to deform the surface into the form of a sausage perturbation. The condition $K_m = 0$ represents the situation in which these two forces balance each other.

### 12.7.3 Profile implications for stabilizing the m = 0 mode

The stability condition $K_m(r) > 0$ is usually most difficult to satisfy near the edge of the plasma, where the pressure is small and the pressure gradient (for a confined plasma) is negative. This intuition follows from writing the stability condition in the way it usually appears in the literature:

$$-\frac{r}{p}\frac{dp}{dr} < \frac{2\gamma B_\theta^2}{B_\theta^2 + \gamma\mu_0 p} \quad \text{(for stability)}. \tag{12.51}$$

Clearly if $p \to 0$ too quickly at the plasma edge, the left hand side of Eq. (12.51) becomes large and the inequality is violated.

The maximum rate at which at which the edge pressure can decrease with radius can be quantified by assuming the plasma is well confined in this region: $\gamma\mu_0 p(r) \ll B_\theta^2(r)$ as $r \to a$. Under this assumption, the stability condition reduces to

$$-\frac{r}{p}\frac{dp}{dr} < 2\gamma, \tag{12.52}$$

implying that the pressure must decay slower than

$$p \sim \frac{1}{r^{2\gamma}} = \frac{1}{r^{10/3}},$$ (12.53)

where the numerical value corresponds to $\gamma = 5/3$. This is a reasonably rapid decay rate, although the pressure at the plasma edge must still be finite for stability.

The $m = 0$ mode is one of two strong MHD instabilities that greatly limit the macroscopic confinement of a simple $Z$-pinch. Specifically, in practical experiments it is usually very difficult to control the pressure profile sufficiently accurately to avoid the $m = 0$ mode, resulting in a catastrophic collapse of the column. This clearly makes the $Z$-pinch a poor choice for a fusion reactor. (The $m = 1$ mode is the other instability and is discussed in the next section.) The $m = 0$ stability criterion also plays a major role in the levitated dipole configuration (LDX). However, as will be shown, the LDX configuration offers a much higher degree of profile control to suppress this potentially catastrophic instability.

## 12.8 The $m = 1$ mode in a linear $Z$-pinch

The $m = 1$ mode is the second macroscopic instability that affects the linear $Z$-pinch. As for the other examples discussed, substantial analysis is required to carry out a complete investigation of this mode for arbitrary wave number $k$. However, since the results have been known for many years, it is possible to make use of these results to present a relatively simple derivation, consistent with the aims of this book. The main results to be exploited are that: (1) only $m = 1$ need be considered since $m \geq 2$ can be shown to be stable; (2) the most unstable wave number is $k \to \infty$; and (3) the analysis should focus immediately on marginal stability by setting $\omega^2 = 0$.

By starting with these assumptions, one can easily derive the marginal stability eigenvalue equation for the $m = 1$, $k \to \infty$ mode. The basic driving mechanism for the unstable perturbation is similar to that shown in Fig. 12.4. The end result of the analysis is the derivation of an infinite set of eigenfunctions and eigenvalue wave numbers $k_n$, where $n$ is an integer. Each solution corresponds to a marginally stable eigenfunction, valid in the limit $k \to \infty$. Note that in this analysis $k^2$ plays the role of the eigenvalue since $\omega^2$ has been set to zero.

When following the derivation, keep in mind that the axial boundary conditions require that $k$ be purely real in order for the eigenfunctions to be bounded at $z = \pm\infty$. Therefore, in principle, complete marginal stability of the $Z$-pinch occurs when the parameters of the plasma are adjusted so that the "worst mode" crosses the stability boundary, which corresponds to the largest positive eigenvalue crossing zero: $k_{max}^2 \to 0$. Once this occurs, there are no further bounded eigenfunctions (i.e., no physical marginal stability states with $k^2 > 0$) and the plasma is stable. Unfortunately, in a simple $Z$-pinch the analysis shows that for the $m = 1$ mode the plasma is always unstable. There always exists a bounded marginal state for sufficiently high values of $k$.

Finally, note that while the calculation presented below is entirely self-consistent, it is by no means obvious that $m = 1$, $k \to \infty$ is the most unstable mode for $m \geq 1$. This issue is addressed below by means of a specific example.

### 12.8.1 The $\theta$ component of the momentum equation

The first step in the derivation is to analyze the $\theta$ component (i.e., the parallel component) of the momentum equation for the $m = 1$ mode in the limit $\omega^2 = 0$:

$$J_z B_{1r} - \frac{1}{r} \frac{\partial p_1}{\partial \theta} = 0. \tag{12.54}$$

Substituting $B_{1r} = (iB_\theta/r)\xi$, $p_1 = -(dp/dr)\xi - \gamma p \nabla \cdot \boldsymbol{\xi}$ (with $\xi_r \equiv \xi$) and making use of the Z-pinch equilibrium relation leads to the simple condition

$$\nabla \cdot \boldsymbol{\xi} = 0. \tag{12.55}$$

The mode is incompressible: that is, $d\rho/dt = dp/dt = 0$.

Under this condition the perturbed pressure reduces to

$$p_1 = -\frac{dp}{dr}\xi, \tag{12.56}$$

which is required as the analysis proceeds.

### 12.8.2 The z component of the momentum equation

The next step in the derivation is the analysis of the $z$ component of the momentum equation, given by

$$J_{1r} B_\theta - \frac{\partial p_1}{\partial z} = 0. \tag{12.57}$$

Substituting $\mu_0 J_{1r} = iB_{z1}/r - ik B_{\theta 1}$ leads to

$$\frac{B_\theta}{\mu_0}\left(\frac{iB_{1z}}{r} - ik B_{1\theta}\right) - ikp_1 = 0. \tag{12.58}$$

The task now is to eliminate $B_{1z}$ and then determine a relation between $B_{1\theta}$ and $\xi$. Only $B_{1\theta}$ is required to obtain the final eigenvalue equation. The quantity $B_{1z}$ is easily eliminated by the condition $\nabla \cdot \mathbf{B}_1 = 0$. For the $m = 1$ mode one finds

$$B_{1z} = -\frac{1}{kr}\left[B_{1\theta} + \frac{d}{dr}(B_\theta \xi)\right]. \tag{12.59}$$

Substituting into Eq. (12.57) leads to the required expression for $B_{1\theta}$:

$$B_\theta B_{1\theta} = \frac{k^2 r^2}{1 + k^2 r^2}\left[\mu_0 \frac{dp}{dr}\xi - \frac{B_\theta}{k^2 r^2}\frac{d}{dr}(B_\theta \xi)\right]. \tag{12.60}$$

Note that at this point the assumption $k \to \infty$ has not as yet been invoked. This assumption is applied in the final form of the eigenvalue equation.

### 12.8.3 The r component of the momentum equation

The last step in the derivation is the simplification of the $r$ component of the momentum equation:

$$J_{z1} B_\theta + J_z B_{\theta 1} + \frac{\partial p_1}{\partial r} = 0, \tag{12.61}$$

which, after a short calculation, can be rewritten as

$$\frac{1}{r^2} \frac{d}{dr} \left[ r^2 \left( B_\theta B_{1\theta} + \mu_0 p_1 \right) \right] - 2\mu_0 \frac{p_1}{r} - i \frac{B_\theta B_{1r}}{r} = 0. \tag{12.62}$$

One next substitutes the expressions for $B_{1\theta}$, $B_{1r}$, and $p_1$ in terms of $\xi$. A straightforward calculation yields

$$\frac{1}{r} \frac{d}{dr} \left[ \frac{r^3 B_\theta^2}{1 + k^2 r^2} \frac{d}{dr} \left( \frac{\xi}{r} \right) \right] + \left[ \frac{d}{dr} \left( r B_\theta^2 \right) \right] \left( \frac{\xi}{r} \right) = 0. \tag{12.63}$$

Equation (12.63) can now be further simplified by considering the limit $k \to \infty$. The desired form of the eigenvalue equation is thus

$$\frac{1}{r} \left( r B_\theta^2 \psi' \right)' + k^2 \left( r B_\theta^2 \right)' \psi = 0, \tag{12.64}$$

where $\psi = \xi/r$.

### 12.8.4 Solution to the m = 1 eigenvalue equation

The solution to the eigenvalue equation clearly depends strongly on the sign of the coefficient $(r B_\theta^2)'$. For typical Z-pinch profiles the current density is flat on axis and decreases to zero at the plasma edge. The corresponding magnetic field first increases linearly with $r$ and then decreases as $1/r$ outside the current edge. A specific example has already been illustrated in Fig. 11.15.

Therefore, in the outer portion of the profile where $B_\theta \sim 1/r$, the coefficient satisfies $(r B_\theta^2)' \sim -1/r^2 < 0$, corresponding to solutions which are monotonic exponentials (i.e., stabilizing). In contrast, in the inner portion of the profile where $B_\theta \sim r$, then $(r B_\theta^2)' \sim +r^2 > 0$. In this region the solutions are oscillatory (i.e., destabilizing). The mathematical intuition is that an unstable eigenfunction should be localized to the inner portion of the plasma. This point is proved below by means of a specific example.

Consider a simple equilibrium where the magnetic field and current density profiles are modeled as shown in Fig. 12.15. The model equilibrium is matched to the realistic diffuse profiles by defining $\bar{a}$ such that the current densities on axis are identical:

$$\bar{a} \equiv \frac{\mu_0 I}{2\pi B_\theta'(0)}. \tag{12.65}$$

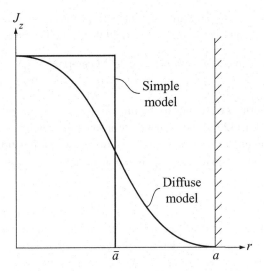

Figure 12.15 Simple model of a diffuse Z-pinch matching the current density on axis and the total current.

Here, $I$ is the total plasma current and $B'_\theta(0)$ is the derivative of the actual diffuse magnetic field.

For this simple model the eigenvalue equation reduces to

$$
\begin{aligned}
\frac{1}{r^3}(r^3\psi')' + 3k^2\psi &= 0 & 0 < r < \bar{a}, \\
r\left(\frac{\psi}{r}\right)' - k^2\psi &= 0 & \bar{a} < r < a.
\end{aligned}
\tag{12.66}
$$

The eigenfunctions are obtained by solving for $\psi$ in each region and choosing the integration constants to satisfy the boundary conditions $\xi(0) = 0$ for regularity at the origin, and $\xi(a) = 0$ for internal modes. The eigenvalue relation is obtained by the matching conditions across the interface $r = \bar{a}$. Specifically, since there are no surface currents in the model the jump conditions require that $[\![\psi]\!]_{\bar{a}} = 0$ and $[\![\psi']\!]_{\bar{a}} = 0$.

The solutions satisfying the boundary conditions are easily expressed in terms of Bessel functions as follows:

$$
\begin{aligned}
\psi_{\text{inner}} &= \hat{C}_1 J_1(\sqrt{3}kr), \\
\psi_{\text{outer}} &= \hat{C}_2 r^2 \left[ K_1(kr) - \frac{K_1(ka)}{I_1(ka)} I_1(kr) \right].
\end{aligned}
\tag{12.67}
$$

At $r = \bar{a}$ in the limit of $k \to \infty$ the asymptotic forms of the solutions are quite accurate and are given by

$$
\begin{aligned}
\psi_{\text{inner}} &\approx \frac{C_1}{\sqrt{r}} \cos(\sqrt{3}kr - 3\pi/4), \\
\psi_{\text{outer}} &\approx \frac{C_2}{\sqrt{r}} (e^{-kr} - e^{-2ka+kr}),
\end{aligned}
\tag{12.68}
$$

where $C_1$, $C_2$ replace $\hat{C}_1$, $\hat{C}_2$ as equivalent integration constants. The two matching conditions on $\xi$ (or equivalently $\psi$) across $r = \bar{a}$ yield the desired eigenvalue condition

$$\cot(\sqrt{3}k\bar{a} - 3\pi/4) = \sqrt{3}e^{-k\bar{a}}\tanh[k(a - \bar{a})]$$
$$\approx 0. \tag{12.69}$$

The second simpler form corresponds to the limit $k \to \infty$.

An examination of Eq. (12.69) shows that there are an infinite number of marginally stable eigenvalues corresponding to each of the zeros of the cotangent function. Specifically, if $n$ is a large positive integer, then the corresponding marginal wave number $k_n$ is given by

$$k_n\bar{a} \approx \left(n + \frac{1}{4}\right)\frac{\pi}{\sqrt{3}}. \tag{12.70}$$

Clearly as $n \to \infty$ the marginal wave numbers become larger and larger. There is no set of plasma parameters for which the "largest" marginal wave number crosses zero; that is, there is always a marginal wave number for which the $m = 1$ mode is on the boundary of instability.

Lastly, it is worth pointing out that a more sophisticated analysis of Eq. (12.64) shows that any diffuse $Z$-pinch is unstable to the $m = 1$, $k \to \infty$ mode if at any point in the profile the coefficient $(rB_\theta^2)'$ becomes positive. In other words, the general condition for $m = 1$ stability requires that everywhere in the plasma

$$\frac{\mathrm{d}}{\mathrm{d}r}\left(rB_\theta^2\right) < 0 \qquad \text{(for stability)}. \tag{12.71}$$

To summarize, the combination of $m = 0$ and $m = 1$ instabilities make the simple $Z$-pinch a poor choice for a fusion reactor. Both modes also play an important role in the stability of the LDX concept but for this configuration a clever modification of the magnetic field has the potential to stabilize these very dangerous macroscopic modes.

## 12.9 Summary of stability

The main thrusts of this chapter have been to: (1) introduce some of the basic concepts related to linear MHD stability; (2) present a simple physical picture of the main driving sources of MHD instabilities; (3) describe a general formulation of the MHD stability problem; and (4) apply the formulation to several simple configurations.

In terms of specifics, the analysis has focused on ideal MHD. External ideal MHD modes usually result in a catastrophic loss of plasma and it is important to learn how to avoid such instabilities. Keeping the pressure and current low help stability but oppose the needs of a fusion reactor. Furthermore, because of the frozen-in-field-line law, ideal MHD modes are robust and likely to be unaffected by more subtle plasma physics phenomena.

Instabilities can be driven by both perpendicular and parallel currents and are closely connected to the relative direction of the pressure gradient and the radius of curvature vector. When the directions are parallel, the curvature is favorable and when they are anti-parallel the curvature is unfavorable.

The general ideal MHD stability problem has been formulated as an elegant set of three coupled partial differential equations for the plasma displacement $\boldsymbol{\xi}$ and eigenvalue $\omega^2$. Even so, the model is difficult to solve in most cases because of the multi-dimensionality of the geometry. Nevertheless, it has been possible to prove that, in general, the marginal stability boundary is crossed when the eigenvalue crosses through zero; that is, marginal stability corresponds to $\omega^2 = 0$.

Several applications have been investigated. It has been shown that the infinite homogeneous plasma is always stable and supports three oscillatory waves: the shear Alfvén wave, the compressional Alfvén wave, and the sound wave. The stability of the linear $\theta$-pinch has also been studied. By means of an integral relation, this configuration has been shown to be always stable, although as previously discussed, it cannot be bent into a toroidal equilibrium. Lastly, the stability of the $m = 0$ and the $m = 1$ mode in a $Z$-pinch has been investigated. In the $m = 0$ case the actual eigenvalue differential equation was derived and solved. The resulting eigenvalues show that the $Z$-pinch can be unstable unless the pressure decreases sufficiently slowly with radius. The $m = 1$ analysis focused on marginal stability in the limit $k \to \infty$. The results show that the center of a linear $Z$-pinch is always unstable to this mode. These instabilities, when excited, lead to a macroscopic loss of plasma on a micro second MHD time scale, thus making the simple $Z$-pinch unattractive as a fusion concept.

The next task is to apply the stability formulation to a simple, but reasonably general 2-D MHD equilibrium model and determine the pressure and current limits in a variety of realistic fusion configurations.

## Bibliography

Many of the books that describe MHD equilibrium also discuss MHD stability and, consequently, the list of references is quite similar to that in Chapter 11. Again, the references below are almost entirely focused on MHD equilibrium and stability.

Bateman, G. (1978). *MHD Instabilities*. Cambridge, Massachusetts: MIT Press.
Biskamp, D. (1993). *Nonlinear Magnetohydrodynamics*. Cambridge, England: Cambridge University Press.
Freidberg, J. P. (1987). *Ideal Magnetohydrodynamics*. New York: Plenum Press.
Goedbloed, H. and Poedts, S. (2004). *Principles of Magnetohydrodynamics*. Cambridge, England: Cambridge University Press.
Kadomstev, B. B. (1966). *Hydromagnetic Stability of a Plasma* (Leontovich, M. A., editor), Vol. 2. New York: Consultants Bureau.
Lifschitz, A. E. (1989). *Magnetohydrodynamics and Spectral Theory*. Dordrecht: Kluwer Academic Publishers.
Wesson, J. (2004). *Tokamaks*, third edn. Oxford: Oxford University Press.
White, R. B. (2001). *Theory of Toroidally Confined Plasmas*. London: Imperial College Press.

## Problems

12.1 This problem investigates the stability of a plasma consisting of a background of cold ions and two interpenetrating beams of cold electrons. For the equilibrium assume

$n_i = 2n_0$, and that the densities of the electron beams are equal: $n_{el} = n_{er} = n_0$. Also, assume $T_i = T_{el} = T_{er} = B_0 = 0$. The first beam moves to the left and the second to the right with equal but opposite velocities $\pm v_0$. Assume the transverse dimensions of the beam are sufficiently large that the plasma can be modeled as a 1-D system with distance parallel to the beam velocities being the dimension of interest.

(a) Derive the dispersion relation for 1-D electrostatic perturbations of the form $Q_1(\mathbf{r}, t) = \hat{Q}_1 \exp(-i\omega t + ikz)$. Use a separate fluid treatment for each electron beam and assume the ions are infinitely massive. Neglect collisions.

(b) Calculate the range of wave numbers (in terms of $n_0$, $v_0$) which are unstable for this "two-stream instability."

12.2 Repeat Problem 12.1 assuming only a single beam of electrons moving with a velocity $v_0$ but treating the ions as having a large but finite mass.

12.3 For those who like algebraic challenges, derive Eq. (12.17).

12.4 For those who continue to like algebraic challenges, derive Eq. (12.35).

12.5 The purpose of this problem is to calculate the eigenvalue relation for shear Alfvén waves propagating along a linear $\theta$-pinch. The geometry of interest has cylindrical symmetry with the following non-zero equilibrium quantities: $p = p(r)$, $\rho = \rho(r)$, and $\mathbf{B} = B(r)\mathbf{e}_z$. All perturbed quantities are assumed to vary as $Q_1 = Q_1(r)\exp(-i\omega t + im\theta + ikz)$. To simplify the algebra assume that the perturbations satisfy the incompressible equation of state, $\nabla \cdot \boldsymbol{\xi} = 0$. Mathematically, this transforms the original problem containing three unknowns $\xi_r, \xi_\theta, \xi_z$ and the adiabatic equation of state to a new problem with four unknowns $\xi_r, \xi_\theta, \xi_z, \hat{p}_1 \equiv p_1 + BB_{1z}/\mu_0$ with the adiabatic equation of state replaced by $\nabla \cdot \boldsymbol{\xi} = 0$.

(a) Derive the eigenvalue equation for $\xi(r) \equiv \xi_r(r)$ and show that it can be written as

$$\frac{d}{dr}\left[ \frac{\omega^2 \mu_0 \rho - k^2 B^2}{m^2 + k^2 r^2} r \frac{d}{dr}(r\xi) \right] - (\omega^2 \mu_0 \rho - k^2 B^2)\xi = 0.$$

(b) Solve this equation for the simple profiles corresponding to the surface current model illustrated in Fig. 12.16. The eigenvalue relation is obtained by solving separately in regions I and II and then matching across the interface at $r = a$ using the jump conditions $[\![\xi]\!]_a = 0$ and $[\![(\omega^2 \mu_0 \rho - k^2 B^2)d(r\xi)/dr]\!]_a = 0$. Obtain a simple answer by taking the limit $ka \ll kb \ll 1$. Hint: Let $\xi = d\psi/dr$.

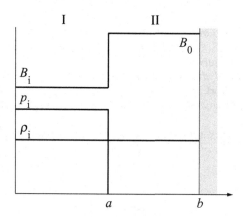

Figure 12.16

12.6 Calculate the eigenfrequency $\omega = \omega(k_\perp, k_\parallel)$ for shear Alfvén waves propagating in an infinite homogeneous resistive MHD plasma characterized by the following equilibrium: $\mathbf{B} = B_0\mathbf{e}_z$, $p = p_0$, $\rho = \rho_0$. Assume Ohm's law includes a small resistivity: $\mathbf{E} + \mathbf{v} \times \mathbf{B} = \eta\mathbf{J}$. To make the algebra a little simpler define $\gamma_D = \eta k^2/\mu_0$ and assume $\gamma_D \ll k_\parallel v_A$.

12.7 This problem investigates the effect of viscosity on the $m = 0$ mode in a cylindrical Z-pinch. A simple way to estimate the effect is to modify the MHD momentum equation as follows:

$$\rho\frac{d\mathbf{v}}{dt} = \mathbf{J} \times \mathbf{B} - \nabla p + \nabla \cdot (\mu\nabla\mathbf{v}),$$

where $\mu$ is the viscosity coefficient. To simplify the stability analysis assume that $\mu = \text{const.}$ and consider short wavelength perturbations (i.e., large $k$) such that $\nabla \cdot (\mu\nabla\mathbf{v}_1) \approx -\mu k^2\mathbf{v}_1$. Calculate the eigenvalue near marginal stability following the analysis carried out in the text. Sketch $\omega_i$ as a function of $k$ and compare it with the no-viscosity case. Does viscosity alter the marginal stability boundary or just change the value of the growth rate?

12.8 The goal here is to determine the effect of thermal conductivity on the $m = 0$ mode in a Z-pinch. The main modification to the analysis presented in the text is that the energy equation is replaced by

$$\frac{1}{\gamma - 1}\left(\frac{\partial p}{\partial t} + \mathbf{v} \cdot \nabla p + \gamma p\nabla \cdot \mathbf{v}\right) = \nabla \cdot (\kappa\nabla T).$$

The addition of the thermal conduction term greatly increases the complexity of the analysis. However, a simple analytic answer can be obtained by carrying out the analysis as described below.

(a) Derive the modified form of the perturbed pressure $p_1$ taking into account the thermal conductivity. To simplify the calculation consider the limit of large $k$ and assume that the growth rate scales linearly with $k$: $\omega_i/k \to \text{const.}$ as $k \to \infty$. Show that in this limit the perturbed pressure reduces to

$$p_1 = -p\nabla \cdot \boldsymbol{\xi} - T\frac{dn}{dr}\xi.$$

(b) The remainder of the stability analysis closely follows that in the text. Calculate the condition for marginal stability. Show that thermal conduction changes the stability condition. By comparing with the $\kappa = 0$ case show that thermal conduction improves stability for a sufficiently gradual density profile and decreases stability for a rapidly decaying density profile.

12.9 The goal here is to determine the MHD stability of a rotating $\theta$-pinch. The calculation involves several steps: (a) calculate the equilibrium fields; (b) derive the eigenvalue differential equation; (c) form an integral relation for the eigenvalue; (d) estimate the eigenvalue using a trial function "guess" for the eigenfunction; and (e) draw your conclusions. Proceed as follows.

(a) The non-trivial equilibrium field quantities for a 1-D, cylindrically symmetric, rotating $\theta$-pinch are $B_z(r)$, $p(r)$, $\rho(r)$, $v_\theta(r)$. Assume that

$$B_z^2(r) = B_0^2 - (B_0^2 - B_i^2)(1 - r^2/a^2)^2,$$
$$p = p_0(1 - r^2/a^2)^2,$$
$$v_\theta = \Omega r,$$

where $B_0$, $B_i$, $p_0$, $\Omega$ are constants and $a$ is the edge of the plasma. Find $p(r)$ and give an expression for $p_0$, the density on axis, in terms of the other equilibrium constants.

(b) Carry out a linear MHD stability analysis for the rotating $\theta$-pinch for arbitrary equilibrium profiles. The following three assumptions make the derivation reasonably simple. First, assume that for a system with equilibrium flow, the relation between the displacement vector and the perturbed velocity is given by $\mathbf{v}_1 = -i\omega\boldsymbol{\xi} + \mathbf{v}_0 \cdot \nabla\boldsymbol{\xi} - \boldsymbol{\xi} \cdot \nabla\mathbf{v}_0$. Second, assume the most unstable perturbations are incompressible: $\nabla \cdot \boldsymbol{\xi} = 0$. Third, assume all perturbed quantities vary as $Q_1(\mathbf{r}, t) = Q_1(r)\exp[-i(\omega t - m\theta - kz)]$ and treat the case $k = 0$. The reasoning is that finite $k$ produces line bending, which is stabilizing, and therefore setting $k = 0$ corresponds to the most unstable mode. (This turns out not to be true for the $m = 1$ mode but make the assumption anyway to keep the algebra tractable.) The final goal of this part of the problem is to derive a single differential equation for $\psi_1(r) = r\xi_r(r)$ describing the linear stability of the system. For simplicity of notation introduce the Doppler shifted frequency $\sigma = \omega - m\Omega$.

(c) Form an integral relation for $\sigma$ by multiplying the differential equation by $\psi_1^*$ and integrating over the volume of the plasma. At this point allow the profiles to be arbitrary, assuming only that $p(a) = 0$. Assume the modes of interest correspond to external modes: $\psi_1(a) \neq 0$.

(d) Estimate the eigenvalue $\sigma$ by substituting the equilibrium profiles given above and using a trial function "guess" given by

$$\psi_1(r) = \psi_a \left(\frac{r}{a}\right)^m.$$

(e) Based on the results, answer the following questions. Which if any $m$ modes are unstable? Is there a critical rotation speed for the onset of instability and if so what is its value?

# 13

# Magnetic fusion concepts

## 13.1 Introduction

The goal of Chapter 13 is to describe the various magnetic configurations currently under investigation as potential fusion reactors. As will become apparent there is a substantial number of concepts to discuss. To succeed, each of these concepts has to successfully overcome the problems not only of MHD equilibrium and stability ($p$), but also of transport ($\tau_E$) and heating ($T$). Even so, it still makes sense to introduce the concepts at this point in the book, immediately following MHD. The reason is that the underlying geometric features that distinguish each concept are primarily determined by MHD behavior. In contrast, transport is a far more difficult issue and significant progress has been made only for the tokamak configuration. With respect to heating, there are several techniques available providing a reasonable number of options. Because of this flexibility, heating can be accommodated in most fusion configurations, and thus is not a dominant driver of the geometry.

To motivate the discussion recall that the main objective of MHD is to discover magnetic geometries that are capable of stably confining sufficiently high plasma pressures to be of relevance to a fusion reactor. The leader for many years in terms of overall performance has been the tokamak which will therefore serve as the reference configuration against which all other concepts must be measured. Towards this goal, it is helpful to briefly and qualitatively discuss the pros and cons of the tokamak in order to understand exactly what problem each of the alternative concepts is hoped to solve.

The tokamak is an axisymmetric configuration with a large toroidal magnetic field and a significant DC toroidal current. Tokamaks have achieved stable operation at near reactor relevant pressures, confinement times, and temperatures. In other words in terms of physics performance the tokamak has met nearly all the requirements for a reactor and it is expected that a next generation experiment (e.g. ITER) will close the remaining gaps. What then are the outstanding problems facing the tokamak? There are two primary issues, the first involving plasma physics and the other involving technology. Each is described below.

The main outstanding plasma physics problem is related to the need for steady state operation. To appreciate the issue requires a somewhat lengthy sequence of reasoning. To begin note that the DC toroidal current required in the tokamak configuration is usually inductively generated by means of a transformer, a method that works well in present day pulsed devices but which clearly cannot work in a steady state device. Instead, some means of external non-inductive current drive is required. There are several methods available but they are not very efficient when converting power to current and would lead to an unacceptable power balance if the entire toroidal current had to be driven by such means. Fortunately, there is a naturally occurring transport driven current, known as the "bootstrap current," which can, in principle, provide most of the toroidal current without the need for external current drive. However, the amount of bootstrap current depends sensitively on the pressure profile. To achieve a large enough fraction of bootstrap current for a favorable power balance, one finds that the required plasma pressure invariably exceeds the MHD $\beta$ limit corresponding to a plasma configuration surrounded by a vacuum region extending to infinity. However, if a close fitting, perfectly conducting wall surrounds the plasma, there is a strong stabilizing effect so that the required plasma $\beta$ remains below the MHD $\beta$ limit.

One sees that there is a subtle but crucial issue in determining exactly what is meant by the MHD stability boundary. The difficulty involves the behavior of the first wall surrounding the plasma. While a perfectly conducting wall provides a strong stabilizing influence, this effect vanishes with a more realistic resistive wall. However, the resistive wall growth time (typically milliseconds) is much longer than for the vacuum case (typically microseconds), thereby opening up the possibility of feedback stabilization. The $\beta$ limits with a perfectly conducting wall and with a resistive wall are thus quite different and the one that is experimentally relevant depends quite strongly on one's belief in being able to feedback stabilize the resistive wall mode. The ability to stabilize the resistive wall mode is thus a crucial issue directly impacting whether or not a tokamak can achieve steady state operation with a sufficiently low current-drive power requirement to maintain a favorable overall power balance.

This then is one important area where new ideas and alternative concepts can improve on the performance of the standard tokamak: achieving steady state operation without the need for excessive current drive and without violating the appropriate MHD stability limit. One method to address this problem involves a new mode of tokamak operation known as "advanced tokamak (AT) operation." Here a combination of profile control and feedback stabilization is predicted to resolve the difficulties, but this still needs to be demonstrated experimentally and certainly adds to the physics complexity of the tokamak concept. Another approach that directly addresses the problem involves an alternative concept known as the stellarator, which is a 3-D, inherently steady state configuration with no need for non-inductive current drive. Its main disadvantage with respect to the tokamak is the need for technologically more complex and expensive magnets to generate the 3-D stellarator magnetic field.

The second important problem facing the tokamak is technological in nature. The large toroidal field in a tokamak needed to provide MHD stability requires a set of high-**B** toroidal field magnets, which must be superconducting for favorable steady state power balance. While such magnets are certainly feasible from an engineering point of view, they are expensive and add technological complexity to a tokamak reactor. It would clearly improve the attractiveness of a fusion reactor if the requirements on the toroidal field system were greatly reduced or even eliminated. While the stellarator makes this problem more difficult, alternative concepts substantially improve the situation. These concepts include the spherical tokamak, reversed field pinch (RFP), spheromak, field reversed configuration (FRC), and the LDX. However, the RFP, spheromak, and FRC concepts have noticeably poorer confinement properties and lower no-wall MHD $\beta$ limits than the tokamak. The spherical tokamak, RFP and spheromak also have more difficult current-drive problems. Even so, some very clever ideas have been introduced to alleviate these difficulties and are discussed later in the chapter. The LDX is a new concept that is just becoming operational so it is still too early to assess its performance.

Overall, one sees that tradeoffs are involved – each of the alternative concepts improves one aspect of the tokamak concept but suffers with respect to others. After many years of experimental research, the balancing of these tradeoffs has firmly placed the tokamak in the role of leader on the path to a fusion reactor. Time will tell if any of the alternatives can overtake the tokamak in terms of overall reactor desirability, which involves improvements in both plasma physics and fusion technology.

In presenting the material to substantiate these claims, one would ideally like to describe the tokamak first to set the stage for comparison with all other alternates. However, the tokamak is one of the more difficult concepts to analyze in terms of its MHD behavior. Consequently, for the sake of presentation the discussion begins with the simplest configurations and works it way towards the more complicated ones. The goal for each configuration is to develop an understanding of its MHD equilibrium and to then calculate the corresponding $\beta$ and pressure limits against MHD instabilities.

Based on this strategy the LDX and FRC are discussed first and the relevant results can be directly obtained from the $Z$-pinch analysis presented in the previous chapter. The remaining concepts are considerably more difficult to analyze. To derive the desired information a particularly simple model, known as the surface current model, is introduced. Although the profiles in the model are rather gross approximations to actual experimental profiles, the analysis can at least be carried to completion. Comparisons with advanced theoretical analysis as well as numerical MHD studies using more realistic profiles show good qualitative agreement and modest quantitative agreement with respect to MHD equilibrium and stability $\beta$ limits. By using the surface current model, MHD results are obtained for the reversed field pinch, spheromak, tokamak, elongated tokamak, and advanced tokamak (AT). The discussion of the stellarator, because of its complex 3-D equilibrium, is primarily descriptive, with reference made to appropriate numerical studies.

## 13.2 The levitated dipole (LDX)

### 13.2.1 Overview of the LDX

LDX is a relatively new fusion concept initially motivated by astrophysical observations, in particular the existence of a stable, long-lasting, plasma ring confined in the dipole magnetic field surrounding Jupiter. The goal of the LDX experiment, whose construction was completed at MIT in 2005, is to reproduce a similar, stable, long-lasting plasma, on a laboratory scale on earth.

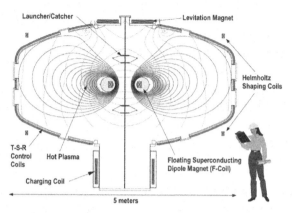

Figure 13.1 Conceptual drawing of the LDX experiment (courtesy of D. Garnier).

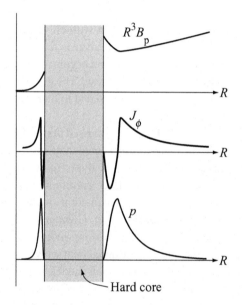

Figure 13.2 LDX profiles of $p$, $B_p$, and $J_\phi$ along the midplane $Z = 0$ (courtesy of L. Guazzotto).

The LDX magnetic configuration basically consists of a single ring current that produces a dipole magnetic field as illustrated in Fig. 13.1. The coil current and plasma current are purely toroidal, producing a purely poloidal magnetic field. The plasma forms a hollow toroidal shell surrounding the dipole coil. Typical profiles of the pressure $p$, poloidal magnetic field $B_p$, and current density $J_\phi$ along the mid plane $Z = 0$ are shown in Fig. 13.2. Note that while the direction of the current density must change sign in the plasma, there is always a net plasma current that flows in the same direction as the coil current in order to provide toroidal force balance.

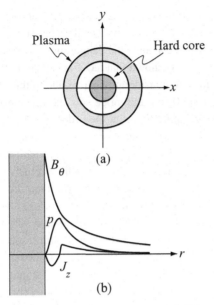

Figure 13.3 (a) An equivalent cylindrical LDX. (b) Typical radial profiles of $p$, $B_\theta$, and $J_z$.

A critical feature of LDX is the need to levitate the coil in the vacuum chamber. The reason is that if any type of mechanical supports are used, the plasma by topological necessity must intersect these supports at some location. This would cool the plasma and heat the supports, an unacceptable situation in an experiment, and certainly in a reactor. In practice, long-time levitation of the coil is possible if it is constructed of a superconducting material. This clearly adds to the technological complexity of the concept, but is nonetheless feasible, at least for laboratory-scale experiments.

The scientific goal of LDX is to produce a hot, well-confined plasma by means of microwave heating, whose frequency is chosen to resonate with the electron gyro frequency of the particles. One wants to learn how high a plasma pressure can be stably confined for a given input microwave power and what corresponding temperatures are achievable. The pressure limits are determined primarily by MHD considerations, while the achievable temperatures are largely determined by transport, particularly heat conduction and possibly convection. There are quantitative predictions for the pressure limits in LDX and these are the main focus of the present section. Transport is much more difficult to predict, and one will have to wait for actual experimental data to learn about this important property of the LDX concept.

Consider now the MHD pressure limits. Detailed analysis has shown that the essential physics of LDX is accurately described by a linear, cylindrical geometry. Toroidal effects make modest quantitative changes in the results, but are not necessary for a qualitative, or semi-quantitative, understanding of the MHD behavior. The analysis described in this section, therefore, makes use of a cylindrical model for LDX as illustrated in Fig. 13.3.

Observe that the non-trivial equilibrium field components are $p(r)$ and $B_\theta(r)$, similar to a pure linear Z-pinch. However, the central coil acts like a hard core imbedded in the Z-pinch, and is crucial for stabilizing LDX against the dangerous $m = 0$ and $m = 1$ modes.

In the analysis that follows it is shown that LDX is subject to three pressure limits (i.e., $\beta$ limits): one due to equilibrium considerations, one due to the $m = 1$ mode, and one due to the $m = 0$ mode. The $m = 0$ sausage mode sets the most stringent $\beta$ limit. Even so, the resulting critical $\beta$ is still quite high, on the order of $1/2$. Furthermore, this stability is achieved without the need for a perfectly conducting wall. A second important result obtained from the analysis is related to the value of the hard core coil current. Clearly, without this current, the configuration reduces to a pure Z-pinch, which is always unstable. The analysis predicts the minimum amount of coil current required to provide the necessary stabilization against the $m = 0$ and $m = 1$ modes.

Finally, although the study of the levitated dipole is at a very early stage, one can still assess several potential advantages and disadvantages with respect to the tokamak in terms of its ultimate viability as a fusion reactor. These are as follows.

In terms of advantages, the LDX possesses a very simple magnetic geometry, requiring only a single coil for plasma confinement. A single coil is highly desirable from the physics point of view as well as for simplicity in the design and operation of early stage experiments. Also advantageous is the fact that complete ideal MHD stability can be achieved at high $\beta$ without the need for a perfectly conducting wall. With respect to the tokamak, the absence of any toroidal field coils is an important technological advantage.

With respect to disadvantages, there are two worth noting. First, there is the obvious problem of levitating the superconducting coil, a problem that becomes increasingly more challenging technologically as the experiments grow in size, and the plasma environment becomes more hostile (i.e., increased density and temperature). While levitation itself is feasible, it is protection against failure that provides the challenges. Second, for the coil to remain superconducting, it must be shielded from fusion byproducts. The magnetic field should be able to shield charged particles, but not the large number of high-energy neutrons produced by the D–T reaction. Thus, to reduce the number of high-energy neutrons, a levitated dipole reactor must be based on one of the more difficult D–D or D–He$^3$ fusion reactions (i.e., the ones with a smaller cross section). As an example, for D–D the consequence is that at a typical reactor temperature of 40 keV the required performance parameter $p\tau_E$ is increased to $p\tau_E \approx 170$ atm s as compared to $p\tau_E \approx 8$ atm s for a D–T tokamak. While using D–D and D–He$^3$ as a fuel is technologically desirable in that relatively few neutrons are produced, the higher required $p\tau_E$ will likely be the dominant issue.

With the overview complete, the remainder of the section now addresses the MHD equilibrium and stability limits of the LDX concept.

### *13.2.2 LDX equilibrium*

The analysis begins with a calculation of equilibrium profiles in a cylindrically symmetric LDX geometry. A representative pressure profile is chosen, after which the poloidal magnetic field is determined from the 1-D radial pressure balance relation. Analytic expressions are obtained for the fields and it is shown that there is an equilibrium $\beta$ limit. Also derived is a general equilibrium condition that relates the plasma $\beta$ to the coil current and plasma current.

### The pressure profile

The first step is to specify the plasma pressure profile. A simple choice that accurately models the essential LDX physics is

$$p(r) = K \frac{r^2 - r_1^2}{(r^2 + r_2^2)^{\gamma+1}}, \tag{13.1}$$

which is the profile illustrated in Fig. 13.3. Note that $p(r)$ vanishes on the surface of the coil $r = r_1$. The pressure is hollow, first increasing and then gradually decreasing at large $r$: $p(r) \sim 1/r^{2\gamma}$ as $r \to \infty$. The radial scaling at large $r$ anticipates the dependence needed to stabilize the $m = 0$ mode in regions of low pressure. The parameter $K$ is a measure of the peak pressure, while the parameter $r_2$ is directly related to the location of the pressure maximum.

Even with this simple model the algebra quickly becomes quite complicated. To reduce the complexity, several assumptions are made that greatly simplify the analysis but still capture the essential physics. Specifically, the parameter $r_2$ is chosen as $r_2 = r_1$ and $\gamma$ is set to $\gamma = 2$ rather than $\gamma = 5/3$. As a further simplification a normalized radius $x$ is introduced through the definition $x = r^2/r_1^2$. Under these assumptions the pressure profile can be rewritten as

$$p(x) = 27 p_{\text{max}} \frac{x - 1}{(1 + x)^3}, \tag{13.2}$$

where $p_{\text{max}}$ is the maximum pressure, occurring at $r_{\text{max}}^2/r_1^2 = x_{\text{max}} = 2$.

### The global radial pressure balance relation

The next step in the equilibrium analysis is to calculate the magnetic field profile. As a prelude to this calculation it is first useful to derive a general relation for the LDX global equilibrium pressure balance, which motivates a practical definition of $\beta$. Following the discussion of radial pressure balance in Chapter 11, one starts with the equilibrium relation

$$p' + \frac{B_\theta}{\mu_0 r} (r B_\theta)' = 0. \tag{13.3}$$

Multiplying by $r^2$ and integrating over the entire plasma region yields

$$\int_{r_1}^{\infty} r^2 p' dr + \frac{1}{2\mu_0} \left. (r^2 B_\theta^2) \right|_{r_1}^{\infty} = 0, \tag{13.4}$$

which reduces to

$$\int_{r_1}^{\infty} p r \, dr = \frac{\mu_0}{16\pi^2} [(I_c + I_p)^2 - I_c^2]. \tag{13.5}$$

Here, $I_c$ is the hard core coil current and $I_p$ is the plasma current.

In analogy with the screw pinch pressure balance relation, it is useful to introduce the following definition of $\beta$:

$$\beta \equiv \frac{16\pi^2}{\mu_0(I_c + I_p)^2} \int_{r_1}^{\infty} p\, r\, dr. \tag{13.6}$$

Note that $\beta$ is normalized to the total current in the system $I_c + I_p$. The general LDX radial pressure balance expression, which relates $\beta$ to the coil current and plasma current, is obtained directly from Eq. (13.5):

$$\beta = 1 - i_c^2, \tag{13.7}$$

where

$$i_c = \frac{I_c}{I_c + I_p} \tag{13.8}$$

is proportional to the coil current. From the definition it is clear that $0 \le \beta \le 1$. Equation (13.7) is ultimately quite useful in understanding the stable operating regime of LDX.

### The poloidal magnetic field

The equilibrium solution is now completed by calculating the poloidal magnetic field from the integrated form of the local radial pressure balance relation:

$$\frac{1}{2\mu_0} \left( r^2 B_\theta^2 \right) \Big|_{r_1}^{r} = -\int_{r_1}^{r} r^2 p'\, dr. \tag{13.9}$$

After a slightly tedious calculation, one can evaluate the integral, obtaining

$$b_\theta^2(x) = 1 - 8\beta \frac{x^2}{(1+x)^3}, \tag{13.10}$$

where $b_\theta \propto r B_\theta$ is a normalized form of the poloidal magnetic field defined by

$$b_\theta(x) \equiv \frac{2\pi r B_\theta(r)}{\mu_0(I_c + I_p)} \tag{13.11}$$

and use has been made of the integral definition of $\beta$ given by Eq. (13.6) to express the pressure scaling constant $p_{max}$ in terms of $\beta$:

$$p_{max} = \frac{\mu_0(I_c + I_p)^2}{54\pi^2 r_1^2} \beta. \tag{13.12}$$

Equation (13.10) is the expression for $b_\theta^2$ required for the stability analysis.

### The equilibrium $\beta$ limit

An interesting feature of the $b_\theta$ solution is the existence of an equilibrium $\beta$ limit. This limit can be seen by plotting $b_\theta^2$ vs. $x$ for various $\beta$ as illustrated in Fig. 13.4. Observe that as $\beta$ increases, the diamagnetic dip in $b_\theta^2$ becomes deeper and deeper. Eventually a value of $\beta = \beta_{max}$ is reached at which the minimum crosses the axis. For values of $\beta > \beta_{max}$, $b_\theta^2$

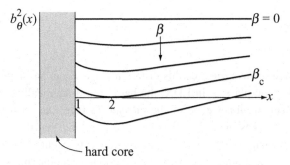

Figure 13.4 Curves of $b_\theta^2$ vs. $x$ for various $\beta$. The largest allowable value of $\beta$ occurs for the curve $\beta = \beta_c$, which is tangent to the axis at $x_{max} = 2$.

becomes negative, which is non-physical. The critical $\beta$ is easily found by simultaneously setting $b_\theta^2 = db_\theta^2/dx = 0$. A short calculation yields

$$\beta \le \beta_{max} = \frac{27}{32} \approx 0.84. \tag{13.13}$$

The limit is only a mild constraint. It arises because a minimum amount of coil current is needed to keep the pressure profile hollow – to push the plasma away from the surface of the coil at $r = r_1$. This value of coil current is determined from Eq. (13.7), the general equilibrium relation, and is given by

$$i_c \ge i_{min} = \sqrt{\frac{5}{32}} \approx 0.40. \tag{13.14}$$

At least 40% of the total current must flow in the hard core.

In summary, a simple equilibrium model has been introduced that accurately describes the basic features of the LDX configuration. The profiles exhibit an equilibrium $\beta$ limit which is weak and does not present a serious limitation on possible performance.

### 13.2.3 $m = 1$ stability

The stability of the levitated dipole is determined by the $m = 0$ and $m = 1$ local criteria derived for the $Z$-pinch. The fact that LDX has a hard core does not affect the general form of the stability criteria since the corresponding derivations only assume a cylindrical equilibrium with a purely poloidal magnetic field. Clearly, however, the quantitative prediction of marginal stability boundaries depends upon the actual shape of the profiles.

For LDX the parameters of the simple model equilibrium profile must be chosen such that both $m = 0$ and $m = 1$ are stable at every value of $r$ in the plasma. In practice, this leads to a $\beta$ limit for each mode. The analysis below presents a derivation of the $\beta$ limit for the $m = 1$ mode.

Recall that stability against the $m = 1$ mode requires that the $B_\theta$ profile satisfy

$$\frac{d}{dr}\left(r B_\theta^2\right) < 0 \tag{13.15}$$

or, equivalently

$$\frac{d}{dx}\left(\frac{b_\theta^2}{x^{1/2}}\right) < 0. \tag{13.16}$$

The simple Z-pinch always violates this criterion since $B_\theta$ is an increasing function of $r$ near the center of the plasma. In LDX, the hard core dramatically alters this situation since $B_\theta \sim 1/r$ near the inner edge of the plasma. In fact, one sees by inspection that for $r_1 < r < r_{\max}$ the pressure gradient is positive implying that the field-line curvature is favorable. Refer again to Fig. 13.3. Instability can only occur for $r > r_{\max}$, where the curvature becomes unfavorable. Intuitively, it is clear that it becomes more difficult to satisfy the stability criterion as $\beta$ increases, since the slope of $b_\theta^2$ becomes increasingly positive.

A quantitative prediction for the critical $\beta$ is obtained by substituting the equilibrium profile into Eq. (13.16). After a straightforward calculation, one finds that stability requires

$$\beta < \frac{1}{24}\frac{(x+1)^4}{x^2(x-1)}. \tag{13.17}$$

The function on the right hand side of Eq. (13.17) has a minimum and it is this minimum that sets the strictest limit on $\beta$. A short calculation shows that the minimum occurs at

$$x_{\min} = \frac{1}{2}(5 + \sqrt{17}) \approx 4.56. \tag{13.18}$$

The corresponding $\beta$ limit is given by

$$\beta \leq \beta_{\max} = \frac{51\sqrt{17} - 107}{192} \approx 0.54. \tag{13.19}$$

The $m = 1$ stability limit is considerably lower than the equilibrium $\beta$ limit, but is, nonetheless, still quite high in absolute value.

Lastly the critical $\beta$ is substituted into the general equilibrium pressure balance relation (i.e. Eq. (13.7)) to determine the minimum coil current to suppress the $m = 1$ mode. One finds

$$i_c \geq i_{\min} = (1 - \beta_{\max})^{1/2} = 0.68. \tag{13.20}$$

Nearly 70% of the total current must flow in the hard core.

### 13.2.4 $m = 0$ stability

A similar stability analysis holds for the $m = 0$ mode. In this case the condition for stability is

$$-\frac{r}{p}\frac{dp}{dr} < \frac{2\gamma B_\theta^2}{B_\theta^2 + \mu_0\gamma p}. \tag{13.21}$$

For $\gamma = 2$ the criterion can be rewritten in terms of the normalized variables as follows:

$$-\frac{x}{\hat{p}}\frac{d\hat{p}}{dx} < \frac{2b_\theta^2}{b_\theta^2 + x\hat{p}}, \tag{13.22}$$

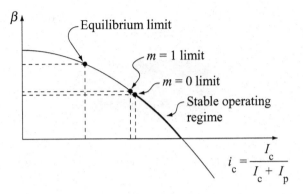

Figure 13.5 Plot of $\beta$ vs. $i_c$ as determined by the general equilibrium relation. Also shown are the equilibrium, $m = 1$, and $m = 0$ marginal stability limits.

where $\hat{p}(x)$ is the normalized pressure defined by

$$\hat{p}(x) = \frac{8\pi^2 r_1^2}{\mu_0 (I_c + I_p)^2} p(r) = 4\beta \frac{x-1}{(x+1)^3}. \tag{13.23}$$

The next step is to substitute the equilibrium profiles into Eq. (13.22). After a straight-forward calculation one obtains a stability condition on $\beta$ given by

$$\beta < \frac{(x+1)^2 (2x-1)}{4x^3}. \tag{13.24}$$

The strictest limit on $\beta$ again occurs in the region of unfavorable curvature, at large values of $x \rightarrow \infty$. The $m = 0$ $\beta$ limit thus has the value

$$\beta \leq \beta_{\max} = \frac{1}{2}, \tag{13.25}$$

with a corresponding minimum current

$$i_c \geq i_{\min} = \frac{1}{\sqrt{2}} \approx 0.707. \tag{13.26}$$

For the model profile chosen, the $m = 0$ mode has the lowest $\beta$ limit, slightly below that of the $m = 1$ mode. Still, the limit is quite substantial. Also more than 70% of the total current must flow in the hard core.

### 13.2.5 Summary of the levitated dipole

In summary, the levitated dipole is a new fusion configuration whose physics performance will be first tested in LDX. It has a simple closed field-line geometry produced by a levitated superconducting coil. The MHD equilibrium and stability $\beta$ limits are quite high without the need of a perfectly conducting wall, and are summarized in Fig. 13.5. This figure shows a curve of $\beta$ vs. the normalized coil current, with the stable region lying to the right of the critical $\beta$ points. The $m = 0$ mode sets the strictest stability limits, requiring that $\beta < 1/2$

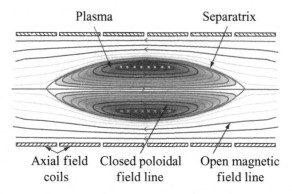

Figure 13.6  Schematic diagram of an FRC (courtesy of R. Milroy).

and $i_c > 1/\sqrt{2}$. The main long-range issues facing the concept are the technological complexities associated with the construction of the levitated coil and the need to use the more difficult-to-excite D–D or D–He$^3$ reaction. Also transport is still an unknown, but will be investigated as LDX data become available.

### 13.3  The field reversed configuration (FRC)

#### *13.3.1  Overview of the FRC*

The FRC is an ultra compact axisymmetric toroidal configuration in which the plasma is entirely confined by a poloidal magnetic field. There is no applied toroidal magnetic field. Hence, there is no need for a set of toroidal field coils. Also, there is no ohmic transformer passing through the center of the device. As such, the FRC is an inherently pulsed device, one that is quite simple from the point of view of its technological structure.

A schematic diagram of an FRC is shown in Fig. 13.6. The operation of the device is as follows. Initially a small axial $\theta$-pinch bias field is created in the vacuum chamber. Neutral gas is injected and a separate power supply pre-ionizes the gas, forming a plasma. At this point the main power supply is energized producing a large axial $\theta$-pinch magnetic field. The field is about a factor of 10 larger than the bias field and points in the opposite direction. The main field rises very rapidly causing the plasma to compress and heat. At some point, assuming the initial plasma has been properly prepared and the fields are appropriately programmed, the axial magnetic field lines tear and reconnect as shown in the diagram. The end result is an elongated configuration with closed flux surfaces contained within the separatrix flux surface.

Typical profiles for the pressure and field are shown in Fig. 13.7. When viewed in the context of the originating $\theta$-pinch fields one sees that $B_Z$ reverses sign at some point off axis. This is the origin of the name "field reversed configuration." Interestingly, although two $\theta$-pinch fields have been used to form the plasma, the final configuration actually has the topology of a flattened toroidal Z-pinch.

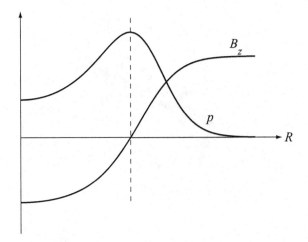

Figure 13.7 Typical midplane profiles $B_z(R, 0)$, $p(R, 0)$ for an FRC.

The Z-pinch topology allows one to determine the ideal MHD stability of the FRC from results already obtained for the pure Z-pinch. No new analysis is required. The reasoning is as follows. The configuration has unfavorable poloidal field-line curvature around each end of the plasma. In the flattened central portion of the plasma the poloidal field lines are essentially straight implying neutral field-line curvature. Also, there is no stabilizing toroidal magnetic field. The end result is that the FRC has regions of unfavorable or neutral curvature but no regions of favorable curvature. Therefore, according to ideal MHD theory, the FRC should always be strongly unstable. This conclusion is indeed borne out in numerical calculations.

Since the configuration is strongly MHD unstable, one may ask why there should be interest in the FRC for the fusion program. There are two main reasons. The first is related to experimental observations on FRC experiments. A number of such experiments have shown that FRC plasmas appear to be substantially more stable than predicted by ideal MHD. Although the reason for this favorable behavior is not completely understood, most researchers believe that plasma kinetic effects play an important role. These effects include the Hall term in the Ohm's law and the fact that the ion gyro radius is a substantial fraction of the minor plasma radius. Both effects modify the ideal MHD model, usually in a stabilizing manner. However, most researchers also believe that when one scales up an FRC to the much larger sizes required in a reactor, then kinetic effects become less important, implying that the plasma would become unstable to ideal MHD modes.

Motivated by the better-than-expected performance at small size and the worrisome prognosis at large size, researchers recognized that an FRC might serve as a good plasma source for a separate concept known as "magnetized target fusion" (MTF). MTF is the second basic reason for studying the FRC in the context of fusion energy. The MTF idea is to position a preheated plasma source with a high density ($10^{23}$ m$^{-3}$) and moderate temperature (300 eV) inside a hollow metal cylinder and then squeeze the cylinder with the

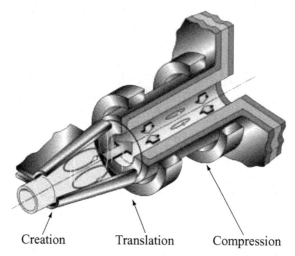

Creation        Translation        Compression

Figure 13.8 Schematic diagram of the three-step process involved in MTF: creation, translation, and compression (courtesy of R. Sieman).

target plasma inside to achieve extremely high particle pressure and magnetic field. The imploding plasma flow eventually stagnates and then re-expands. Near the stagnation point a large number of fusion reactions are predicted to occur because of the ultrahigh-intensity plasma, enough that the fusion energy released can exceed the input compression energy. A number of options have been considered for the role of providing the target plasma with the FRC perhaps having the most desirable features.

The remaining part of this section is devoted to a more detailed explanation of MTF using an FRC target plasma. Once the idea is understood, it will then be possible to discuss the advantages and disadvantages with respect to fusion energy.

### 13.3.2 The MTF concept

MTF is a three-step process as illustrated in Fig. 13.8: creation, translation, and compression.

#### Creation

The first step in the process is plasma "creation." Here, more or less standard FRC technology is utilized. The FRC is a good configuration for MTF because of its technological simplicity and the fact that there is no toroidal field within the plasma. The reason for the desirability of no toroidal field will become apparent shortly. The only constraint on the initially formed FRC is that it should satisfy the "smallness criterion" to be stable against kinetically modified MHD modes. This criterion, which is based primarily on empirical experimental data, requires that

$$S/\kappa < 3.5,$$ 

(13.27)

where $\kappa = Z_S/R_S$ represents the elongation of the separatrix and $S = R_S/\delta_i$ is the ratio of the radius of the separatrix to the so-called ion collisionless skin depth: $\delta_i = c/\omega_{pi}$.

For MTF applications the initial FRC plasma requires a high density $n_e \approx 10^{23}$ m$^{-3}$, a substantial elongation $\kappa \approx 10$, and a separatrix radius $R_S \approx 0.05$ m. With these parameters, Eq. (13.27) is marginally satisfied. The initial temperature and magnetic field required are about $T_e = T_i \approx 0.3$ keV and $B = 5$ T corresponding to $\beta \approx 1$ and a pressure of 100 atm. Note that the initial FRC density is about three orders of magnitude higher than typical densities for other fusion configurations. The one other point worth noting about FRC creation is the addition of a small set of coils at each end of the main $\theta$-pinch coils. These coils produce a small mirror field which causes the FRC plasma to form in the center of the $\theta$-pinch system during each pulse. In addition they produce a "guide field" in the translation chamber.

## Translation

The second step in the process corresponds to "translation." The goal here is to translate the plasma from the initial FRC creation chamber to the MTF compression chamber. This is accomplished by slightly decreasing the magnetic field at the right hand side of the main $\theta$-pinch coil system. Since the field is then slightly larger on the left hand side there is a small left-to-right imbalance of the magnetic pressure with the pressure being higher on the left. This imbalance propels the plasma to the right, along the guide field in the translation chamber, until it enters the compression chamber.

## Compression

The heart of the MTF concept is the compression chamber. The key component of the chamber is a thin metallic cylinder, sometimes called a liner, similar in size to a soda can. Once the FRC plasma drifts midway into the compression chamber, an enormous $Z$ directed current is driven along the cylinder. This produces a corresponding enormous $B_\phi$ magnetic field just outside the cylinder. The resulting magnetic pressure on the surface drives the cylinder radially inward at a very high velocity on the order of $v_R \approx 10^6$ m/s. This inward motion adiabatically compresses the FRC plasma to a very high temperature and density, sufficient to produce a large number of fusion reactions.

The compression process works as follows. During the relatively short time of the compression phase both the liner and the plasma behave like perfect conductors – their skin diffusion times are very long. Consequently, the magnetic flux in the narrow region between the plasma edge and the liner must be conserved. The rapid inward motion of the liner compresses this flux, thereby greatly increasing the value of the magnetic field on the plasma surface. The corresponding increased surface magnetic pressure then adiabatically compresses the FRC plasma (as well as the internal magnetic flux) into a volume with a very small radius and with a very high temperature and density.

At some point in time the liner motion finally stagnates because of the increased particle and internal magnetic field back pressures. If the system is properly designed, the stagnation

Table 13.1. *MTF parameters for the pre compression state and the maximum compression state at the liner stagnation point*

| FRC parameters | Pre compression | At stagnation |
|---|---|---|
| Density | $10^{23}$ m$^{-3}$ | $10^{25}$ m$^{-3}$ |
| Temperature ($T_e \approx T_i$) | 300 eV | 10 keV |
| Magnetic field | 5 T | 500 T |
| Plasma duration | 20 μs | 200 ns |
| Liner inner radius | 5 cm | 0.5 cm |

temperature should be on the order of 10 keV. This initiates the production of a large number of fusion reactions. These reactions continue for a period of time $\tau_D$ comparable to the dwell time of the liner, which is determined by the mass and energy of the liner as well as the final properties of the FRC plasma. Typical parameters for the final state of the compressed FRC plasma as determined by MTF reactor studies are presented in Table 13.1. Observe that the final densities and magnetic fields are huge, while the dwell time of the reactions is much, much shorter compared to typical values of $n$, $B$, $\tau_E$ in a tokamak.

In order to produce enough fusion reactions to be of interest for energy applications a substantial fraction of the fuel must be burned. In an optimized design this requires that $\tau_D \approx \tau_E$. If the energy confinement time is too short, $\tau_D \gg \tau_E$, the plasma cools before many fusion reactions can take place. On the other hand, if $\tau_D \ll \tau_E$, there is more than enough plasma confinement and the system can be reduced in size, thereby reducing the cost.

The qualitative picture just presented shows why the FRC, which has no internal toroidal magnetic field, is a good candidate for MTF. The reason is that if there was a toroidal magnetic field, some of the energy of the imploding liner would have to be allocated to compressing this field as well as the plasma pressure and poloidal magnetic field, since in ideal MHD the fields and plasma move together. This clearly reduces the efficiency of the implosion and higher initial liner kinetic energies would be required to obtain the same end conditions for the plasma as those for the case in which there is no toroidal field case. Too much of a reduction in efficiency would lead to an unfavorable power balance.

This line of reasoning raises another question. If the absence of a toroidal field improves efficiency, would it not be desirable to somehow also eliminate the poloidal field in the compression chamber? Hypothetically, about half of the liner energy is transferred to the plasma and the other half to the internal poloidal magnetic field. Eliminating the poloidal field would therefore lead to a doubling of the efficiency. The answer is emphatically "no!" The reason is associated with the difference in heat transport losses in a plasma with and without a magnetic field. It is shown in the next chapter that transport losses across a magnetic field are many orders of magnitude lower than along the field, or equivalently in the absence of a field. The poloidal magnetic field thus serves the critical purpose of providing thermal insulation between the plasma and the liner. Without the poloidal magnetic field the energy confinement time $\tau_E$ would be very short and there would be no chance of attaining a favorable power balance.

### 13.3.3 The FRC as a source of fusion energy

With the basic idea of MTF now established one can now assess its pros and cons with respect to fusion reactor viability. At the outset it should be noted that because of the wide difference in parameters between an MTF system and all other fusion concepts, the MTF is viewed as a "dark horse" candidate by most of the fusion community. The reason is that there will likely be relatively little exchange of detailed useful scientific knowledge since the regimes of operation are so different. Even so, the MTF concept offers a true alternative to conventional magnetic fusion concepts, and in this context is certainly worth examining as a potential source of fusion energy.

Consider now the potential advantages of the MTF concept. First, since the fusion reactivity scales as the square of the number density, the very high values of $n$ anticipated in MTF imply a very high fusion power density. Second, the high power density in turn implies a much smaller geometric scale for an MTF system. Since capital cost scales with size this is a substantial advantage. Third, although high-voltage, pulsed technology is required for MTF, this is, in general, still less complicated and less costly than, for instance, microwave power, laser power, or neutral beam power. All of these are potential advantages over the tokamak.

These advantages combine to provide the overall motivation for pursuing MTF research: the MTF concept has the potential to investigate certain aspects of ignition physics and to test scientific feasibility on experimental facilities that cost far less than conventional magnetic fusion approaches. An added attractive practical feature is that many of the pulsed power supplies needed for MTF research have already been built at the Los Alamos and Sandia National Laboratories under the auspices of US DoE's Defense Programs.

There are also a number of scientific and engineering problems that must be addressed by the MTF community. These include the following. First, a suitable initial FRC plasma must be created with the required pre-compression parameters. This requires an extrapolation from existing FRC plasma generation although there is an active experimental program underway and progress is being made.

Second, as the liner implodes there are several related plasma science–materials science problems that arise. Among these are the problems of: (1) mixing of the liner wall material and plasma during the implosion since the inner surface of the liner will be transformed into a liquid; (2) the plasma thermal transport during implosion, characterized by $\tau_E$, is unknown at present and a sufficiently high $\tau_E$ is critical for the success of MTF; and (3) it must be demonstrated that MHD stability persists during the entire implosion to prevent plasma from leaking out of the ends of the FRC.

Third, and probably the area that causes the most concern to a large majority of the traditional fusion community is the extrapolation of MTF technology to a power producing reactor. While many would agree that individual MTF implosions may be possible, a reactor requires a rapidly continuing sequence of such implosions, perhaps one every few seconds. This produces difficult constraints on the high-voltage technology as well as on the compression chamber technology, which must be fully evacuated of debris between each implosion. Also, since the liner is destroyed during every pulse, a continuing supply of new liners is required. The cost per liner must be kept to a sufficiently low value in order to achieve economic viability and is definitely a technological challenge.

Lastly, the liner itself faces difficult engineering challenges, among them the need to guarantee that the liner does not burst during implosion because of the enormous internal pressures resulting from joule heating. Many of these systems issues are close analogs to those facing inertial fusion energy.

Since MTF is at an early phase of development it probably makes sense to focus on whether or not a fully integrated system can be developed that leads to ignition during a single, isolated implosion. If this mission is successful it would then be sensible to more carefully consider the reactor technology challenges in order to determine the desirability of constructing new facilities.

### 13.3.4 Summary of the FRC

The FRC is a compact, elongated toroidal $Z$-pinch. Although predicted to be very unstable according to ideal MHD, small devices are experimentally observed to be unexpectedly stable, presumably because of additional plasma kinetic effects. This experimental observation, combined with absence of a toroidal field, suggests that that an FRC may be a good candidate for providing the target plasma in the MTF concept. In MTF a small, high-density FRC plasma is created and then injected into a compression chamber. Here a metallic liner compresses the plasma to very high densities and temperatures, leading to a large number of fusion reactions.

The underlying motivation for MTF is the potential to create fusion plasmas at a far reduced cost compared to other fusion concepts because of the small size and relatively lost-cost high-voltage technology. The primary concerns regarding fusion energy generation involve the ability to extrapolate the technology to a rapidly pulsed reactor environment.

## 13.4 The surface current model

### 13.4.1 Introduction

With the exceptions of LDX and the FRC, all of the other fusion concepts have a toroidal as well as a poloidal magnetic field. The inclusion of a toroidal field substantially increases the complexity of the analysis and, indeed, most of the literature in the field lies beyond the scope of the present book. To overcome this difficulty, a surprisingly simple model is introduced that greatly reduces the analysis and allows one to obtain the desired information concerning MHD $\beta$ limits. This model is known as the 'surface current model."

The surface current model is a simple MHD model that allows one to analytically calculate the pressure and current limits against external ballooning-kink modes in a wide variety of fusion configurations. Since these modes usually set the most stringent stability limits, this is precisely the information required to assess reactor viability with respect to macroscopic MHD behavior.

The idea behind the model is to consider a special equilibrium in which $\mathbf{J}$ consists solely of an idealized surface current. This current is assumed to flow on a single surface, separating the plasma from the vacuum region and results in an enormous simplification in both the equilibrium and stability analyses. Presented here is a formulation of the surface current model for subsequent use in determining pressure and current limits. The formulation is valid for 1-D and 2-D plasmas and thus covers most of the configurations. In order to keep the mathematics maximally tractable, the cross section of the plasma is assumed to be circular, although this is not essential to the model.

The end result of the formulation is a four-step procedure leading to a determination of MHD marginal stability limits. Qualitatively, these steps are as follows:

- Equilibrium: calculate a 2-D surface current equilibrium.
- Stability: solve for the perturbed magnetic field in the plasma region.
- Stability: solve for the perturbed magnetic field in the vacuum region.
- Stability: match across the plasma–vacuum interface using external mode boundary conditions to determine the conditions for marginal stability.

A description of each of these steps is presented below.

### 13.4.2 The 2-D surface current equilibrium

Consider a 2-D toroidal plasma with circular cross section as shown in Fig. 13.9. The task is to calculate the pressure and magnetic fields just inside and just outside the plasma–vacuum interface. An important simplification of the surface current model is that only the equilibrium fields just inside and just outside the surface are needed for the stability analysis.

The first quantity of interest is the pressure which is found as follows. By assumption, no current flows in the plasma: $\mathbf{J} = 0$. Therefore, in equilibrium $\mathbf{J} \times \mathbf{B} - \nabla p = 0$ reduces to $\nabla p = 0$. The solution to this equation is simply

$$p = \text{const.} \tag{13.28}$$

The pressure is constant everywhere within the plasma region. The second quantity of interest is the toroidal magnetic field in the plasma $B_\phi$. The assumption $\nabla \times \mathbf{B} = 0$ implies that $B_\phi$ can be written as

$$B_\phi(r, \theta) = B_i \frac{R_0}{R} = B_i \frac{R_0}{R_0 + r\cos\theta}, \tag{13.29}$$

where $B_i$ is the internal toroidal magnetic field at the major radius $R = R_0$.

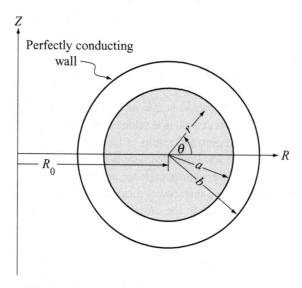

Figure 13.9 Two-dimensional geometry for the surface current model.

Next, note that in an axisymmetric geometry, the absence of any toroidal current flowing in the plasma requires that the poloidal magnetic field within the plasma be identically zero: $\mathbf{B}_p(r, \theta) = 0$. This completes the specification of the internal plasma fields.

Consider now the vacuum side of the interface. The pressure is clearly zero in a vacuum: $p = 0$. Next, in analogy with the plasma region, the toroidal field also satisfies the vacuum relation $\nabla \times \mathbf{B} = 0$ and thus can be expressed as

$$\hat{B}_\phi(r, \theta) = B_0 \frac{R_0}{R} = B_0 \frac{R_0}{R_0 + r \cos\theta}. \tag{13.30}$$

Here $B_0$ is the applied toroidal field at $R = R_0$. Lastly, note that in the region outside the surface current the poloidal field $\hat{\mathbf{B}}_p(r, \theta)$ is non-zero. It is quite difficult, in general, to calculate $\hat{\mathbf{B}}_p(r, \theta)$ everywhere in the vacuum region but the fields are required only on the plasma surface for the stability analysis. This limited information about the surface poloidal field can be easily obtained by means of the MHD pressure balance "jump condition" as follows.

Start with the exact non-linear momentum equation written in the following form:

$$\rho \frac{d\mathbf{v}}{dt} = \frac{1}{\mu_0}(\mathbf{B} \cdot \nabla)\mathbf{B} - \nabla\left(p + \frac{B^2}{2\mu_0}\right). \tag{13.31}$$

Integrate the equation an infinitesimal radial distance across the plasma–vacuum interface as shown in Fig. 13.10, keeping in mind that the values of the magnetic fields and pressure are allowed to jump across the surface. Their normal gradients thus produce a delta function at the interface. However, the pressure and fields are smooth within the surface on both the plasma side and the vacuum side. This implies that the $\mathbf{B} \cdot \nabla\mathbf{B}$ term integrates to zero across the interface. The reason is as follows. Although the magnetic field has a jump, the

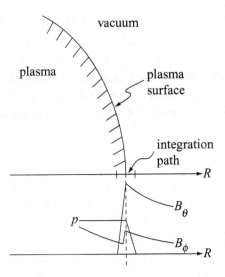

Figure 13.10 Integration across the plasma–vacuum interface.

operator $\mathbf{B} \cdot \nabla$ only involves surface derivatives since $\mathbf{B}$ lies in the surface; that is, there is no normal derivative since $\mathbf{n} \cdot \mathbf{B} = 0$ on the surface. Integration of the $\mathbf{B} \cdot \nabla \mathbf{B}$ term therefore involves integrating a step function across an infinitesimal distance and this produces a zero contribution. Similarly, the inertial term must integrate to zero across the interface. For the contribution to be finite, the velocity $\mathbf{v}$ would have to have a jump and this would imply an infinite acceleration (from the $\mathbf{v} \cdot \nabla \mathbf{v}$ term). This is unphysical and is not allowed to occur.

Based on this discussion it follows that the exact non-linear jump condition across the plasma–vacuum interface arises solely from the normal component of the gradient term and can be written as

$$\left( p + \frac{B^2}{2\mu_0} \right) \Bigg|_{S_p^-}^{S_p^+} = 0, \tag{13.32}$$

where $S_p$ is the exact plasma surface including both the equilibrium and stability contributions: $S_p = a + \xi_r(a, \theta, \phi)$.

The desired expression for the equilibrium vacuum poloidal magnetic field on the surface is now obtained by applying Eq. (13.32) across the equilibrium plasma–vacuum interface $S_p = a$. One finds

$$\hat{B}_\theta^2(a, \theta) = 2\mu_0 p + \frac{B_i^2 - B_0^2}{(1 + \varepsilon \cos \theta)^2}, \tag{13.33}$$

where $\hat{\mathbf{B}}_p = \hat{B}_\theta \mathbf{e}_\theta$ and $\varepsilon = a/R_0$ is the inverse aspect ratio of the plasma. In terms of the stability analysis, Eq. (13.33) is the only equilibrium information required. Although simple in appearance, it describes a wide range of fusion configurations, each distinguished by a different choice of the parameters $p$, $B_i$, $B_0$, and $\varepsilon$.

### 13.4.3 The perturbed magnetic field in the plasma

The equilibrium has now been defined. The next step is to carry out a linear MHD stability analysis. In terms of the general formulation, it is shown that the eigenfunction to be determined reduces solely to the normal component of plasma displacement evaluated on the plasma surface; that is, the eigenfunction is $\xi(\theta, \phi) \equiv \xi_r(a, \theta, \phi)$. Furthermore, in an axisymmetric system one can Fourier analyze with respect to $\phi$ and then treat each Fourier mode separately.

$$\xi(\theta, \phi) = \xi(\theta) \exp(-in\phi). \tag{13.34}$$

Here, $n$ is the toroidal wave number characterizing each Fourier mode. The unknown eigenvalue is the marginally stable (i.e., $\omega^2 = 0$) value of either the plasma $\beta$ or plasma $I$ depending upon application.

In the stability analysis the plan is to calculate the perturbed fields in the plasma region and the vacuum region, expressing each in terms of $\xi(\theta, \phi)$. The eigenvalue is then determined by the condition of matching these solutions across the plasma–vacuum interface using the linearized form of the pressure balance jump condition given by Eq. (13.32).

This subsection describes the formulation for determining the perturbed fields in the plasma region. The key feature in the formulation is that in the surface current model the calculation of $\xi$ can be reduced to the problem of solving Laplace's equation. The analysis proceeds as follows. For marginal stability the linearized momentum equation reduces to

$$B_\phi \mathbf{J}_1 \times \mathbf{e}_\phi - \nabla p_1 = 0. \tag{13.35}$$

The $\mathbf{e}_\phi$ component of this equation yields $(in/R) p_1 = 0$ or $p_1 = -\gamma p \nabla \cdot \boldsymbol{\xi} = 0$. For mode numbers $n \neq 0$ the perturbed pressure is zero and the eigenfunction is incompressible.

With $p_1 = 0$ the momentum equation reduces to $\mathbf{J}_1 \times \mathbf{e}_\phi = 0$, which has as its solution $\mathbf{J}_1 = \lambda_1(r, \theta) \exp(-in\phi)\mathbf{e}_\phi$. Here, $\lambda_1$ is an arbitrary scalar function. Next, note that the condition $\nabla \cdot \mathbf{J}_1 = 0$ reduces to $(in/R)\lambda_1 = 0$ or $\lambda_1 = 0$. The conclusion is that in the surface current model the perturbed current in the plasma must also vanish:

$$\mathbf{J}_1 = 0. \tag{13.36}$$

Therefore, since $\mathbf{B}_1$ is a vacuum magnetic field (i.e., $\nabla \times \mathbf{B}_1 = \nabla \cdot \mathbf{B}_1 = 0$) it can be written as

$$\mathbf{B}_1 = \nabla V_1, \tag{13.37}$$

with $V_1$ satisfying

$$\nabla^2 V_1 = 0. \tag{13.38}$$

The potential function $V_1$ is coupled to the surface displacement through the boundary conditions. These require: (1) regularity of $V_1$ within the plasma; and (2) that the normal component of perturbed magnetic field on the surface satisfy $\mathbf{n} \cdot \mathbf{B}_1|_a = \mathbf{n} \cdot \nabla \times (\boldsymbol{\xi} \times \mathbf{B})|_a$.

This last condition reduces to

$$\left.\frac{\partial V_1}{\partial r}\right|_a = -\left.\frac{in\,B_\phi}{R}\xi\right|_a = -\frac{in\,B_i}{R_0(1+\varepsilon\cos\theta)^2}\xi. \tag{13.39}$$

Equations (13.38) and (13.39) represent the desired formulation of the problem of calculating the perturbed fields in the plasma region. They will be solved under a variety of conditions as the applications progress. In all cases, once the equations are solved, the quantity needed for the pressure balance matching condition is $\mathbf{B}\cdot\mathbf{B}_1|_a$, which is easily expressed in terms of $V_1$ as follows:

$$\mathbf{B}\cdot\mathbf{B}_1|_a = -\left.\frac{in\,B_\phi}{R}V_1\right|_a = -\frac{in\,B_i}{R_0(1+\varepsilon\cos\theta)^2}V_1\bigg|_a. \tag{13.40}$$

This completes the stability formulation of the plasma region.

### 13.4.4 The perturbed magnetic field in the vacuum

The calculation of the perturbed fields in the vacuum region is very similar to that in the plasma. Since the region of interest is by definition a vacuum, one can immediately write the perturbed magnetic field in terms of a potential function

$$\hat{\mathbf{B}}_1 = \nabla\hat{V}_1, \tag{13.41}$$

with $\hat{V}_1$ satisfying

$$\nabla^2\hat{V}_1 = 0. \tag{13.42}$$

There are two boundary conditions on $\hat{V}_1$. The first assumes that a perfectly conducting, circular cross section, concentric wall of radius $b$ surrounds the plasma. Refer again to Fig. 13.9. On a perfectly conducting wall the normal component of magnetic field must vanish implying that the first boundary condition can be written as

$$\left.\frac{\partial\hat{V}_1}{\partial r}\right|_b = 0. \tag{13.43}$$

Note that the special limit $b/a \to \infty$ corresponds to the no-wall case. For simplicity, the case of a resistive wall is deferred until later.

The second boundary condition couples $\hat{V}_1$ to the surface perturbation. Although the final answer is relatively simple, a rigorous derivation requires a surprisingly large amount of algebra. One can, however, obtain the correct result intuitively as follows. Recall that in the plasma region the normal component of the perturbed magnetic field is related to the displacement vector at the boundary by $\mathbf{n}\cdot\mathbf{B}_1|_a = \mathbf{n}\cdot\nabla\times(\boldsymbol{\xi}\times\mathbf{B})|_a$. If one now imagines extending the plasma just slightly into the vacuum region, then by analogy the corresponding boundary condition for the perturbed vacuum magnetic fields is $\mathbf{n}\cdot\hat{\mathbf{B}}_1|_a = \mathbf{n}\cdot\nabla\times(\boldsymbol{\xi}\times\hat{\mathbf{B}})|_a$. Note the use of the vacuum rather than plasma magnetic fields. After a short calculation, the second boundary condition simplifies to

$$\left.\frac{\partial \hat{V}_1}{\partial r}\right|_a = \left.\left(-\frac{in\,\hat{B}_\phi}{R} + \frac{\hat{B}_\theta}{r}\frac{\partial}{\partial \theta}\right)\xi\right|_a = \left[-\frac{in\,B_0}{R_0(1 + \varepsilon \cos \theta)^2} + \frac{\hat{B}_\theta}{a}\frac{\partial}{\partial \theta}\right]\xi. \quad (13.44)$$

Equations (13.42)–(13.44) describe the stability problem in the vacuum region. They will be solved under a variety of conditions as the applications progress. Finally, as for the plasma region, once the vacuum solutions are obtained, the quantity required for the pressure balance matching condition is just $\hat{\mathbf{B}} \cdot \hat{\mathbf{B}}_1|_a$, which can be expressed in terms of $\hat{V}_1$ as follows:

$$\hat{\mathbf{B}} \cdot \hat{\mathbf{B}}_1|_a = \left.\left(-\frac{in\,\hat{B}_\phi}{R} + \frac{\hat{B}_\theta}{r}\frac{\partial}{\partial \theta}\right)\hat{V}_1\right|_a = \left.\left[-\frac{in\,B_0}{R_0(1 + \varepsilon \cos \theta)^2} + \frac{\hat{B}_\theta}{a}\frac{\partial}{\partial \theta}\right]\hat{V}_1\right|_a. \quad (13.45)$$

The formulation of the vacuum stability analysis is now complete.

### 13.4.5 The pressure balance matching condition

As the formulation now stands, the perturbed plasma and vacuum fields can, in principle, be calculated for any arbitrary choice of surface displacement $\xi(\theta, \phi)$. Stated differently, there is not yet sufficient information to determine the eigenfunction and eigenvalue. The additional information required to complete the formulation is the pressure balance matching condition across the plasma–vacuum interface.

The desired relation is obtained by linearizing the exact pressure balance jump condition given by Eq. (13.32). This is straightforward if one recalls that the surface appearing in Eq. (13.32) is the exact surface – the equilibrium plus perturbed surface: $S_P = a + \xi(\theta, \phi)$. A straightforward Taylor expansion yields

$$\left[p_1 + \frac{1}{\mu_0}\mathbf{B} \cdot \mathbf{B}_1 + \xi\frac{\partial}{\partial r}\left(p + \frac{B^2}{2\mu_0}\right)\right]_{a^-} = \left[\frac{1}{\mu_0}\hat{\mathbf{B}} \cdot \hat{\mathbf{B}}_1 + \xi\frac{\partial}{\partial r}\left(\frac{\hat{B}^2}{2\mu_0}\right)\right]_{a^+}. \quad (13.46)$$

This expression can be simplified by noting that $p_1 = 0$ and $p = \text{const.}$ Also, the radial derivative of $B_\phi$ at the surface can be easily evaluated as follows:

$$\left.\frac{\partial B_\phi}{\partial r}\right|_a = \left.\frac{\partial}{\partial r}\frac{B_i R_0}{(R_0 + r \cos \theta)}\right|_a = -\frac{B_i \cos \theta}{R_0(1 + \varepsilon \cos \theta)^2}. \quad (13.47)$$

A similar expression holds for $\hat{B}_\phi$. Lastly, $\partial \hat{B}_\theta/\partial r|_a$ is directly evaluated in terms of $\hat{B}_\theta(a, \theta)$ by using the condition $\nabla \times \hat{\mathbf{B}} = 0$ and noting that $\hat{B}_r(a, \theta) = 0$ (corresponding to $\mathbf{n} \cdot \hat{\mathbf{B}}(a, \theta) = 0$)

$$\left.\frac{\partial \hat{B}_\theta}{\partial r}\right|_a = -\frac{\hat{B}_\theta}{a}. \quad (13.48)$$

Combining these results leads to the desired, although somewhat complicated, form of the pressure balance jump condition:

$$\left[ \mathbf{B} \cdot \mathbf{B}_1 - \frac{B_i^2 \cos\theta}{R_0(1 + \varepsilon\cos\theta)^2}\xi \right]_{a^-} = \left[ \hat{\mathbf{B}} \cdot \hat{\mathbf{B}}_1 - \frac{B_0^2 \cos\theta}{R_0(1 + \varepsilon\cos\theta)^2}\xi - \frac{\hat{B}_\theta^2}{a}\xi \right]_{a^+}. \quad (13.49)$$

The mathematical formulation of the surface current stability problem is now complete. Admittedly, the formulation at this point is still conceptual in nature, with the actual path to determining the marginal pressure or current not as yet transparent. The best way to remedy the situation is actually to investigate the stability of a variety of magnetic fusion configurations. These applications demonstrate the practical utility of the surface current model to analytically predict MHD pressure and current stability limits.

### 13.4.6 Summary of the surface current model

For convenience the steps involved in solving the surface current model are summarized below.

*Equilibrium*

$$\hat{B}_\theta^2(a, \theta) = 2\mu_0 p + \frac{B_i^2 - B_0^2}{(1 + \varepsilon\cos\theta)^2}. \quad (13.50)$$

*Perturbed magnetic field in the plasma*

$$\nabla^2 V_1 = 0,$$
$$\left.\frac{\partial V_1}{\partial r}\right|_a = -\frac{in B_i}{R_0(1 + \varepsilon\cos\theta)^2}\xi, \quad (13.51)$$
$$\mathbf{B} \cdot \mathbf{B}_1|_a = -\frac{in B_i}{R_0(1 + \varepsilon\cos\theta)^2} V_1\bigg|_a.$$

*Perturbed magnetic field in the vacuum*

$$\nabla^2 \hat{V}_1 = 0,$$
$$\left.\frac{\partial \hat{V}_1}{\partial r}\right|_b = 0, \quad \left.\frac{\partial \hat{V}_1}{\partial r}\right|_a = \left[ -\frac{in B_0}{R_0(1 + \varepsilon\cos\theta)^2} + \frac{\hat{B}_\theta}{a}\frac{\partial}{\partial\theta} \right]\xi, \quad (13.52)$$
$$\hat{\mathbf{B}} \cdot \hat{\mathbf{B}}_1|_a = \left[ -\frac{in B_0}{R_0(1 + \varepsilon\cos\theta)^2} + \frac{\hat{B}_\theta}{a}\frac{\partial}{\partial\theta} \right]\hat{V}_1\bigg|_a.$$

*Pressure balance jump condition*

$$[\hat{\mathbf{B}} \cdot \hat{\mathbf{B}}_1 - \mathbf{B} \cdot \mathbf{B}_1]_a = \frac{(B_0^2 - B_i^2)\cos\theta}{R_0(1 + \varepsilon\cos\theta)^2}\xi + \frac{\hat{B}_\theta^2}{a}\xi. \quad (13.53)$$

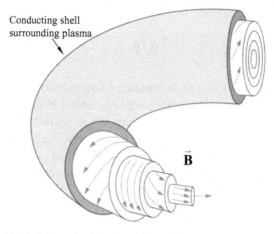

Figure 13.11 Schematic diagram of the RFP (courtesy of J. S. Sarff).

These are the equations that will be used to investigate the stability limits for the various fusion configurations.

## 13.5 The reversed field pinch (RFP)

### 13.5.1 Overview of the RFP

The RFP is the first configuration analyzed by means of the surface current model. The RFP is an axisymmetric toroidal configuration characterized by a large toroidal current, a moderate sized toroidal magnetic field, and a relatively high $\beta$. Currently there are several major RFPs in the world fusion program: (1) the Madison Symmetric Torus (MST) at the University of Wisconsin (USA), (2) the Reversed Field Experiment (RFX) operated by an Italian consortium consisting of the University of Padua, government agencies and businesses, and (3) the RFP at the AIST laboratory in Tsukuba, Japan. A smaller experiment, Extrap T2R, operates in Sweden.

A schematic diagram of an RFP is shown in Fig. 13.11. The device operates as follows. Initially a small toroidal bias field fills the vacuum chamber. A large toroidal current is then ramped up, and this compresses both the plasma and the toroidal bias field. In addition, the current raises the plasma temperature by means of ohmic heating. At the end of the current ramp, the toroidal and poloidal magnetic fields within the plasma are of comparable magnitude. However, since most of the toroidal magnetic flux has been trapped and compressed within the plasma, there remains only a small residual toroidal field at the plasma edge. Remarkably, in certain regimes of operation the edge toroidal magnetic field spontaneously reverses direction – hence the name "reversed" field pinch. This is usually the most desirable regime of operation.

Typical equilibrium profiles are illustrated in Fig. 13.12 for an equivalent cylindrical RFP. Observe the flat central pressure profile and the $B_\phi$ reversal at the plasma edge. Detailed

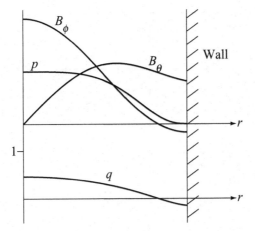

Figure 13.12 Typical radial profiles for an RFP (courtesy of J. S. Sarff).

analysis has shown that the essential physics of the RFP is accurately described in a linear geometry. The analysis in this section adopts the cylindrical approach, thereby substantially simplifying the mathematics.

In terms of a reactor vision the RFP has several advantages and disadvantages with respect to the tokamak. The advantages are as follows. First, only a very modest set of toroidal field coils is needed to provide the bias field. This ultimately leads to a compact, high-power-density plasma, which is clearly advantageous from an economics point of view. Second, the RFP is expected to achieve substantially higher values of $\beta$ than a tokamak. With a perfectly conducting wall, a maximum stable $\beta$ of about 0.5 is theoretically predicted, while experimental values on the order of 0.15–0.25 have already been achieved. Third, the high value of toroidal current may be sufficient, or at least nearly so, to ohmically heat the plasma to ignition. The simplicity, efficiency, and low cost of ohmic heating compared to the alternative of external microwave heating give the RFP another important advantage.

There are several disadvantages as well that must be considered. First, the low toroidal field at the plasma edge implies a low edge safety factor, which in turn leads to resistive MHD turbulence. This turbulence manifests itself in the form of strongly enhanced energy transport. The value of $\tau_E$ is usually much less for an RFP than a tokamak. Recent novel experimental techniques involving control of the current density profile have helped considerably, but the RFP still has a way to go to match the performance of a tokamak.

Second, since an ohmic transformer cannot operate in DC steady state, some type of external current drive is required. The amount of current drive is large since an RFP requires a large toroidal current. This adds to the expense and complexity of a reactor. For these reasons the current drive methods under consideration for a tokamak would probably not be viable for an RFP. However, there is another method known as "oscillating field current drive (OFCD)" that may be effective for an RFP but not for a tokamak. OFCD makes use of an AC driving voltage and thus can achieve sinusoidal steady state through an inductively driven transformer. This method is at an early stage of development in laboratory experiments.

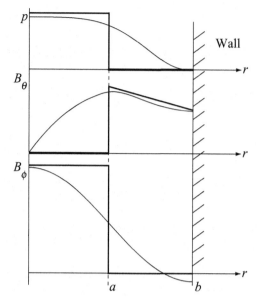

Figure 13.13 Surface current model approximating the RFP configuration.

Lastly, the high MHD stable $\beta$ in an RFP depends upon the presence of a perfectly conducting wall. If the wall has a finite conductivity, the plasma becomes subject to the resistive wall mode. The critical $\beta$ is reduced to zero and even at $\beta = 0$ the plasma still remains unstable to between 5–10 separate current-driven modes. The implication is that a relatively sophisticated feedback system is required.

The analysis presented in this section focuses on MHD $\beta$ limits, which are the primary motivation for the RFP magnetic configuration. Using the surface current model it is shown that the $m = 0$ and $m = 1$ modes lead to the strictest stability boundaries. The analysis is carried out first for a perfectly conducting wall and then for a resistive wall. For the perfectly conducting wall the surface current model predicts a critical $\beta$ of approximately 0.5. The resistive wall leads to a critical $\beta$ equal to zero with about six remaining current driven modes. The ultimate viability of an RFP as a fusion reactor is thus strongly connected to the likelihood of improving transport, developing OFCD to drive a large toroidal current, and being able to stabilize the resistive wall mode.

### 13.5.2 RFP surface current equilibrium

The first step in the analysis is to reduce the general 2-D surface current equilibrium relation to a 1-D form that accurately represents a cylindrical RFP. The transition from 2-D to 1-D is easily made by letting the inverse aspect ratio approach zero: $a/R_0 \equiv \varepsilon \to 0$. Next, the magnetic field and pressure parameters are chosen to approximate typical RFP profiles as illustrated in Fig. 13.13.

Note that the appropriate scaling for an RFP assumes that $p \sim B_i^2/2\mu_0 \sim \hat{B}_\theta^2/2\mu_0$. All the pressures are of comparable order. Also the external toroidal magnetic field is set to zero: $B_0 = 0$. Strictly speaking, the toroidal field in an RFP is small and reverses sign near the plasma edge: $B_0/B_i < 0$ and $|B_0/B_i| \ll 1$. This reversal is required to stabilize certain internal MHD modes driven by the pressure and current gradients that can appear when smooth profiles are used. However, these modes simply do not exist within the surface current model. Therefore, since the internal mode information is absent from the outset, nothing additional is lost by setting $B_0 = 0$. More importantly, the strictest stability limits result from the external kink modes, not the internal modes, and a more detailed analysis shows that these external kinks are only weakly dependent on the value of $B_0$ when $|B_0/B_i| \ll 1$. Based on the above discussion the surface current equilibrium relation for an RFP reduces to

$$p + \frac{B_i^2}{2\mu_0} = \frac{\hat{B}_\theta^2}{2\mu_0}. \tag{13.54}$$

Observe that in a cylindrical model $\hat{B}_\theta(a, \theta) = \text{const}$. The poloidal field is uniform around the plasma surface. Equation (13.54) implies that the sum of the plasma pressure and internal toroidal magnetic pressure is balanced by the tension in the poloidal magnetic field.

The form of Eq. (13.54) also suggests a useful definition of $\beta$ in an RFP, given by

$$\beta \equiv \frac{2\mu_0 p}{\hat{B}_\theta^2} = \frac{8\pi^2 a^2 p}{\mu_0 I^2}, \tag{13.55}$$

where $I$ is the toroidal plasma current. In an RFP, the total $\beta$ is equal to the poloidal $\beta$: $\beta = \beta_p$. With this definition it follows that $0 \leq \beta \leq 1$, indicating that there is no equilibrium $\beta$ limit.

Equations (13.54) and (13.55) represent the equilibrium information required for the stability analysis.

### 13.5.3 RFP surface current stability

This subsection applies the general surface current stability analysis to the RFP. The plasma and vacuum contributions are calculated and then substituted into the pressure balance matching condition. The end result is an explicit condition for marginal stability in an RFP.

#### The plasma contribution

The first step is to reduce the 2-D stability formulation for a torus to a 1-D formulation for a straight cylinder. This is easily done by letting $\varepsilon \to 0$, and defining the axial coordinate $z$ in terms of the toroidal coordinate $\phi$ as follows: $z = R_0 \phi$. Similarly, the relation between the toroidal mode number $n$ and the axial wave number $k$ can be expressed as $k = n/R_0$. Lastly, the straight cylindrical equilibrium possesses poloidal as well as axial symmetry.

Therefore, one can also Fourier analyze with respect to the poloidal angle. Under these transformations, the normal component of the surface displacement has the form $\xi(\theta, \phi) \rightarrow \xi_0 \exp[i(m\theta - kz)]$, where $\xi_0$ is a constant. Stability must now be tested for all $m$ and $k$.

Consider now the plasma contribution to the stability analysis. In accordance with the general formulation of the surface current stability problem, the plasma contribution is obtained by solving

$$\frac{1}{r}\frac{d}{dr}\left(r\frac{dV_1}{dr}\right) - \left(\frac{m^2}{r^2} + k^2\right)V_1 = 0, \tag{13.56}$$

subject to regularity within the plasma and

$$\left.\frac{dV_1}{dr}\right|_a = -ik B_i \xi_0. \tag{13.57}$$

The solution is easily obtained in terms of modified Bessel functions and can be written as

$$V_1(r) = -\frac{i B_i \xi_0}{I_m'(ka)} I_m(kr). \tag{13.58}$$

Here, $I_m'$ denotes the derivative with respect to the argument. Also, recall that the $I_m$ Bessel function behaves qualitatively like a non-oscillatory, growing exponential which is regular at the origin. The complementary modified Bessel function $K_m$ behaves qualitatively like a non-oscillatory decaying exponential which is singular as $r \rightarrow 0$.

It is now straightforward to calculate the quantity $\mathbf{B} \cdot \mathbf{B}_1|_a$ required for the pressure balance matching condition:

$$\mathbf{B} \cdot \mathbf{B}_1|_a = \mathbf{B} \cdot \nabla V_1|_a = -\frac{B_i^2}{a}\frac{\rho_a I_a}{I_a'}\xi_0, \tag{13.59}$$

where $\rho_a = ka$ is a normalized wave number and the following shorthand notation has been introduced: $I_a \equiv I_m(\rho_a)$ and $I_a' \equiv d I_m(\rho_a)/d\rho_a$.

### The vacuum contribution

The contribution of the vacuum to the stability analysis is derived in a similar manner. In this region the relevant equation to be solved is

$$\frac{1}{r}\frac{d}{dr}\left(r\frac{d\hat{V}_1}{dr}\right) - \left(\frac{m^2}{r^2} + k^2\right)\hat{V}_1 = 0, \tag{13.60}$$

subject to

$$\left.\frac{d\hat{V}_1}{dr}\right|_b = 0,$$

$$\left.\frac{d\hat{V}_1}{dr}\right|_a = \frac{im\hat{B}_\theta}{a}\xi_0. \tag{13.61}$$

The solution is a combination of $I_m$ and $K_m$ Bessel functions with the arbitrary multiplicative constants chosen to satisfy Eq. (13.61). A short calculation yields

$$\hat{V}_1(r) = \frac{im\hat{B}_\theta \xi_0}{ka} \left[ \frac{I_b' K_m(kr) - K_b' I_m(kr)}{I_b' K_a' - K_b' I_a'} \right]. \tag{13.62}$$

The quantity $\hat{\mathbf{B}} \cdot \hat{\mathbf{B}}_1|_a$ required for the matching condition is now easily calculated as follows:

$$\hat{\mathbf{B}} \cdot \hat{\mathbf{B}}_1|_a = \hat{\mathbf{B}} \cdot \nabla \hat{V}_1|_a = -\frac{m^2 \hat{B}_\theta^2}{a} \frac{K_a}{\rho_a K_a'} \Lambda_b \xi_0, \tag{13.63}$$

where $\Lambda_b$ represents the influence of the conducting wall and is given by

$$\Lambda_b = \frac{1 - (K_b' I_a)/(I_b' K_a)}{1 - (K_b' I_a')/(I_b' K_a')}. \tag{13.64}$$

### The pressure balance matching condition

The last step in the stability analysis is to substitute the plasma and vacuum contributions into the pressure balance matching condition, which in the cylindrical limit reduces to

$$[\hat{\mathbf{B}} \cdot \hat{\mathbf{B}}_1 - \mathbf{B} \cdot \mathbf{B}_1]_a - \frac{\hat{B}_\theta^2}{a} \xi_0 = 0. \tag{13.65}$$

After substituting, one obtains

$$\left[ \frac{m^2 \hat{B}_\theta^2}{a} \frac{K_a}{\rho_a K_a'} \Lambda_b - \frac{B_i^2}{a} \frac{\rho_a I_a}{I_a'} + \frac{\hat{B}_\theta^2}{a} \right] \xi_0 = 0. \tag{13.66}$$

Note that the amplitude $\xi_0$ is a common factor in all the terms. Therefore, for a non-trivial solution to exist the coefficient in the square bracket must vanish. This is the condition that determines the eigenvalue, which for the RFP can be considered to be the marginally stable value of $\beta$. Eliminating $B_i^2$ using the equilibrium relation leads to the desired form of the eigenvalue relation:

$$(1 - \beta) \frac{\rho_a I_a}{I_a'} - m^2 \frac{K_a}{\rho_a K_a'} \Lambda_b - 1 = 0. \tag{13.67}$$

Two general conclusions can be drawn from this relation before proceeding with a more detailed analysis. First, recall that $I_a$, $I_a'$, $K_a$ are each positive, while $K_a'$ is negative. Also it can be shown that the quantity $\Lambda_b$ is always positive. Consequently, the first two terms in Eq. (13.67), representing the plasma and vacuum contributions respectively, are positive. Since they arise from vacuum fields they are always stabilizing (i.e., their contributions are similar to the plasma wave solutions for the infinite, homogeneous, current-free plasma equilibrium). Second, the destabilizing "$-1$" term arises from the difference in surface gradients of the magnetic field, $\hat{B}_\theta^2/a$. This contribution corresponds to the unfavorable curvature associated with the poloidal magnetic field. Since the toroidal magnetic field is straight it makes no curvature contribution, one way or the other.

The last point to make is that while the eigenvalue condition is relatively simple, it implications are still not transparent because of the appearance of a large number of transcendental Bessel functions. In the following subsections simple approximations are made for the Bessel functions which lead to explicit limits on the marginally stable value of $\beta$.

### 13.5.4 The $m = 0$ mode

Consider now the application of the general eigenvalue relation (Eq. (13.67)) to the $m = 0$ mode, which has some similarities with the sausage instability in a pure $Z$-pinch. The critical difference is that an RFP has an internal axial magnetic field. Consequently, the sausage perturbation must compress both plasma and internal magnetic field in an RFP as opposed to only plasma in a $Z$-pinch. Compression of the axial magnetic field is an additional stabilizing effect and one might expect that a sufficiently large value of axial field $B_i$ would stabilize the $m = 0$ mode. This is indeed the case. Since increasing $B_i$ corresponds to decreasing $\beta$, stability against the $m = 0$ mode leads to a limit on the maximum allowable $\beta$.

An explicit expression for the critical $\beta$ is obtained by noting that the $I_m$ Bessel function for $m = 0$ behaves like $I_0(\rho) \approx 1 + \rho^2/4$ for $\rho \ll 1$ and $I_0(\rho) \approx \sqrt{1/2\pi\rho}\, \exp(\rho)$ for $\rho \gg 1$. It is thus reasonable to approximate

$$\frac{\rho_a I_a}{I_a'} \approx 2 + ka, \tag{13.68}$$

which matches at both large and small $\rho$ and is fairly accurate for intermediate values. With this approximation the eigenvalue condition for $m = 0$ and $ka > 0$ (the most unstable range of wave numbers) simplifies to

$$(1 - \beta)(2 + ka) - 1 = 0 \tag{13.69}$$

or, solving for the marginal $\beta$,

$$\beta \leq \beta_c(ka) \equiv \frac{1 + ka}{2 + ka}. \tag{13.70}$$

Clearly the strictest limit on $\beta$ occurs for $ka \to 0$ and is given by

$$\beta \leq \beta_{max} = \tfrac{1}{2}. \tag{13.71}$$

The conclusion is that a value of $B_i \geq \hat{B}_\theta/\sqrt{2}$ is required to stabilize the $m = 0$ mode. The corresponding maximum stable $\beta = \tfrac{1}{2}$ is less than the maximum equilibrium $\beta = 1$ but is nevertheless still quite high for fusion power purposes.

### 13.5.5 The $m = 1$ mode

For $m \neq 0$, the most dangerous mode in an RFP corresponds to $m = 1$. Qualitatively, the $m = 1$ mode in an RFP has similarities to the $m = 1$ mode in a pure $Z$-pinch. Again, however, the internal toroidal field is a stabilizing influence, since an $m = 1$ perturbation requires bending of the magnetic lines. Even so, the fact that the edge safety factor is small (i.e. $q(a) = 0$ for the present model) leads to a reduction of the stabilizing influence of the vacuum region. There is no toroidal field line bending in the vacuum region and instabilities persist unless $\beta$ is sufficiently small and the conducting wall is sufficiently close.

The analysis below examines the stability of the $m = 1$ mode, leading to a marginally stable $\beta$ that is a function of wall position. For a close fitting wall, high values of stable $\beta$ are possible. For a distant wall, the plasma is unstable even for $\beta = 0$. These results are quantified below by again approximating the Bessel functions and then solving for the marginal $\beta$.

The analysis begins by noting that $I_1(\rho) \approx \rho/2$ and $K_1(\rho) \approx 1/\rho$ for $\rho \ll 1$. For $\rho \gg 1$ the asymptotic forms are given by $I_1(\rho) \approx \sqrt{1/2\pi\rho} \exp(\rho)$ and $K_1(\rho) \approx \sqrt{\pi/2\rho} \exp(-\rho)$. These expansions can be used to approximate the various combinations of Bessel functions appearing in the eigenvalue equation by simple functions that, in analogy with $m = 0$, match at both small and large $\rho$ and are reasonably accurate at intermediate values. Specifically, for $m = 1$ and $ka > 0$ (the most unstable range of wave numbers) one can approximate

$$\frac{\rho_a I_a}{I_a'} \approx \frac{k^2 a^2}{1 + ka},$$
$$\frac{K_a}{\rho_a K_a'} \approx -\frac{1}{1 + ka}. \tag{13.72}$$

Similarly, the wall term simplifies to

$$\Lambda_b \approx \frac{1 + a^2/b^2}{1 - a^2/b^2}. \tag{13.73}$$

Note that for a finite wall position, $\Lambda_b > 1$. For a wall at "infinity", $\Lambda_\infty \to 1$.

With these approximations the $m = 1$ eigenvalue relation reduces to

$$(1 - \beta)\frac{k^2 a^2}{1 + ka} + \frac{\Lambda_b}{1 + ka} - 1 = 0 \tag{13.74}$$

or, solving for the marginal $\beta$,

$$\beta \le \beta_b(ka) \equiv \frac{k^2 a^2 - ka + \Lambda_b - 1}{k^2 a^2}. \tag{13.75}$$

The function of $ka$ on the right hand side has a minimum, which leads to the strictest limit on $\beta$. The minimizing wave number is easily found to be $ka = 2(\Lambda_b - 1)$. The corresponding range of stable $\beta$ is

$$\beta \le \beta_{max} = \frac{1}{8}\left(9 - \frac{b^2}{a^2}\right). \tag{13.76}$$

The quantity $\beta_{max}$ is plotted vs. $b/a$ in Fig. 13.14. Observe that for $\beta_{max}$ to be positive, as it must be for a physically realizable plasma, the perfectly conducting wall can be no further out than $b/a = 3$. For $b/a > 3$ the plasma is unstable even for $\beta = 0$. Also, if $b/a = \sqrt{5} \approx 2.24$, then $\beta_{max} = 1/2$ and the $m = 1$ and $m = 0$ stability boundaries coincide at this high value of $\beta$.

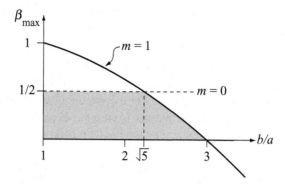

Figure 13.14 Maximum stable $\beta_{max}$ vs. $b/a$ for the $m = 1$ mode in an RFP. Also shown is the value of $\beta_{max}$ for the $m = 0$ mode. The shaded region is stable.

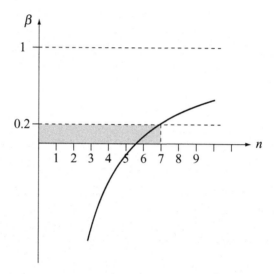

Figure 13.15 Marginal $\beta$ vs. toroidal mode number $n$ for $R_0/a = 5.5$ and $b/a \to \infty$. The integer numbers $n$ lying in the shaded region are unstable for $\beta = 0.2$.

The last point to discuss concerns the situation in which the wall is far from the plasma. In this limit the RFP is always unstable and the problems of interest are to determine precisely which, and how many, modes are unstable. To obtain the information one sets $\Lambda_\infty = 1$ and considers the equivalent torus with $k$ replaced by $k \to n/R_0$. The marginal stability relation reduces to

$$\beta \le \beta_\infty(n) \equiv 1 - \frac{R_0}{na}. \tag{13.77}$$

The marginal $\beta$ is plotted vs. mode number $n$ in Fig. 13.15 for the practical case $R_0/a = 5.5$. Observe that for $\beta = 0$ the plasma is unstable for $0 \le n \le 5$. As $\beta$ increases additional modes become unstable. For example, when $\beta = 0.2$ the $n = 6$ mode also becomes unstable.

The conclusion is that an RFP without a perfectly conducting wall is unstable to at least several different values of $n$ even at $\beta = 0$. The no-wall case is important to understand in connection with the resistive wall instability.

### 13.5.6 The resistive wall mode

The analysis just presented has shown that an RFP can be MHD stable at high values of $\beta$ if a perfectly conducting wall is present. Without a wall the plasma is unstable at $\beta = 0$. This subsection investigates the more realistic situation in which a wall is present but has a finite conductivity. It is shown that the resistive wall, marginal stability criterion exactly coincides with the no-wall stability boundary; that is, a resistive wall has no stabilizing effect on the marginal $\beta$. However, the growth rates are greatly reduced from their ideal MHD values to values associated with the resistive diffusion time of the wall. This is a crucial point since practical feedback systems are plausible when the required circuit response time is comparable to the wall diffusion time (typically many milliseconds). In practice, feedback is not plausible for modes growing on the ideal MHD time scale (typically several microseconds).

The analysis consists of recalculating the vacuum contribution including the presence of the resistive wall. The other contributions to the stability analysis remain unchanged. The reason is that attention is focused on very slowly growing modes so that the neglect of inertial effects (i.e., setting $\omega^2 = 0$ in the plasma) remains a valid assumption. The calculation involves solving for the perturbed vacuum magnetic fields on both sides of the resistive wall as well as within the resistive wall itself using the "thin wall" approximation. Appropriate matching conditions across the wall and across the plasma–vacuum interface ultimately allow one to calculate $\hat{\mathbf{B}} \cdot \hat{\mathbf{B}}_1|_a$, the quantity required for the stability analysis. The end result is an explicit formula for the growth rate of the resistive wall mode expressed in terms of the two ideal marginal stability $\beta$ limits, one with the wall and the other without the wall. Lastly, it is worth noting that while the calculation is carried out for the cylindrical surface current model, the results are general and valid for arbitrary profiles in multi-dimensional geometries.

### The vacuum and resistive wall magnetic fields

The analysis of the region exterior to the plasma in the presence of a resistive wall is actually a three-region problem as illustrated in Fig. 13.16. The magnetic fields in the inner and outer vacuum regions can again be written in terms of potential functions satisfying Laplace's equation. The corresponding solutions for the potentials are

$$\hat{V}_{\mathrm{I}} = [C_1 K_m(kr) + C_2 I_m(kr)] \exp[\omega_i t + i(m\theta - kz)],$$
$$\hat{V}_{\mathrm{II}} = C_3 K_m(kr) \exp[\omega_i t + i(m\theta - kz)].$$
(13.78)

Here, the $C_j$ are unknown coefficients to be determined from the yet to be defined matching conditions. The time dependence has been explicitly displayed by writing $\omega = i\omega_i$ in anticipation of the fact that the resistive wall instability will be shown to be a purely growing mode.

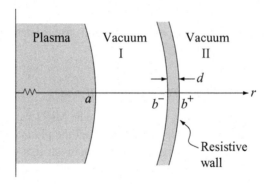

Figure 13.16 Geometry of the three-region problem for the resistive wall mode.

Also, note that $\hat{V}_{II}$ has only a $K_m(kr)$ contribution since regularity requires that $\hat{V}_{II} \to 0$ as $r \to \infty$. When the $C_j$ are finally determined, they are back substituted to evaluate the quantity

$$\hat{\mathbf{B}} \cdot \hat{\mathbf{B}}_1|_a = \hat{\mathbf{B}} \cdot \nabla \hat{V}_I|_a = \frac{im\hat{B}_\theta}{a}(C_1 K_a + C_2 I_a),\tag{13.79}$$

which is needed for the stability analysis.

The magnetic field within the resistive wall is determined from Faraday's law, after substituting $\hat{\mathbf{E}} = \eta\hat{\mathbf{J}} = (\eta/\mu_0)\nabla \times \hat{\mathbf{B}}_w$ with $\eta$ the resistivity of the wall. Using the fact that $\nabla \cdot \hat{\mathbf{B}}_w = 0$ one obtains the standard magnetic diffusion equation

$$\frac{\partial \hat{\mathbf{B}}_w}{\partial t} = \frac{\eta}{\mu_0}\nabla^2 \hat{\mathbf{B}}_w.\tag{13.80}$$

The solution to Eq. (13.80) is not very complicated but does require some slightly subtle analysis. The key point is to assume that $d \ll b$ allowing the introduction of the "thin wall" approximation by means of the following simplifications. (1) In this limit only the radial component $\hat{B}_{wr}$ is required. (2) It is convenient to introduce a local slablike radial coordinate $x$ as follows: $r = b + x$ with $0 \le x \le d$. (3) In the thin wall model there is rapid variation over the narrow distance in $x$. The variation over $\theta$ and $z$ is much slower. Therefore, it is valid to approximate $\nabla^2 \approx \partial^2/\partial x^2$. (4) Fourier analysis implies that $\hat{B}_{wr}$ can be written as $\hat{B}_{wr} = \hat{B}_{wr}(x)\exp[\omega_i t + i(m\theta - kz)]$. (5) The typical magnitude of $\omega_i$ corresponds to the inverse wall diffusion time: $\omega_i \sim \eta/\mu_0 bd$. This assumption is necessary in order to scale various terms appearing in the analysis. Under these conditions, Eq. (13.80) reduces to

$$\frac{\partial^2 \hat{B}_{wr}}{\partial x^2} - \frac{\mu_0\omega_i}{\eta}\hat{B}_{wr} = 0.\tag{13.81}$$

The solution of Eq. (13.81) is obtained by a simple iteration procedure based on the already introduced assumptions that $\partial^2/\partial x^2 \sim 1/d^2$ and $\mu_0\omega_i/\eta \sim 1/bd$. One expands $\hat{B}_{wr} = \hat{B}_{r0} + \hat{B}_{r1} + \cdots$, where $d/b \ll 1$ is the expansion parameter. The leading and first order equations are

$$\frac{\partial^2 \hat{B}_{r0}}{\partial x^2} = 0,$$

$$\frac{\partial^2 \hat{B}_{r1}}{\partial x^2} = \frac{\mu_0 \omega_i}{\eta} \hat{B}_{r0}. \tag{13.82}$$

The solution is

$$\hat{B}_{r0} = B_{r0},$$

$$\hat{B}_{r1} = B_{r0} \left( \frac{\mu_0 \omega_i}{2\eta} x^2 \right) + B_{r1} \left( \frac{x}{d} \right), \tag{13.83}$$

$$\hat{B}_{wr} \approx \hat{B}_{r0} + \hat{B}_{r1} = B_{r0} \left( 1 + \frac{\mu_0 \omega_i}{2\eta} x^2 \right) + B_{r1} \left( \frac{x}{d} \right),$$

where $B_{r0}$ and $B_{r1}$ are unknown constants of zeroth and first order respectively. Note that the $\omega_i$ term is also of first order.

The solutions in all three regions have now been specified in terms of the five unknown constants $C_1$, $C_2$, $C_3$, $B_{r0}$, $B_{r1}$. These constants are determined by appropriate matching conditions.

### The matching conditions

There are five matching conditions: one on the plasma surface and two on each side of the resistive wall. The wall conditions are reduced to a simple set of jump conditions relating the potential functions $\hat{V}_{\mathrm{I}}$ to $\hat{V}_{\mathrm{II}}$ at $r = b$.

Consider first the condition on the plasma surface. This condition remains unchanged from the perfectly conducting wall case:

$$\left. \frac{\partial \hat{V}_{\mathrm{I}}}{\partial r} \right|_a = k \left( C_1 K'_a + C_2 I'_a \right) = \frac{im \hat{B}_\theta}{a} \xi_0. \tag{13.84}$$

Next consider the wall. The jump conditions here are straightforward, although somewhat complicated in form. The key point is that the wall, while thin, still has a finite thickness. Therefore there are no surface currents on either surface of the wall. The implication is that both the normal and tangential components of the magnetic field in the wall are continuous with the vacuum fields on both wall surfaces. Continuity of the normal component of magnetic field requires

$$\left. \frac{\partial \hat{V}_{\mathrm{I}}}{\partial r} \right|_{b^-} = \hat{B}_{wr} \big|_{x=0} \qquad \left. \frac{\partial \hat{V}_{\mathrm{II}}}{\partial r} \right|_{b^+} = \hat{B}_{wr} \big|_{x=d}. \tag{13.85}$$

To evaluate continuity of the tangential magnetic field note that the condition $\nabla \cdot \mathbf{B} = 0$ in the wall implies that $i\mathbf{k} \cdot \hat{\mathbf{B}}_w = -\partial \hat{B}_{wr}/\partial x$. Similarly, in the vacuum regions $i\mathbf{k} \cdot \hat{\mathbf{B}}_1 = -k_0^2 \hat{V}$, where $k_0^2 = k^2 + m^2/b^2$. Thus continuity of the tangential fields requires

$$\hat{V}_{\mathrm{I}}\big|_{b-} = \frac{1}{k_0^2} \frac{\partial \hat{B}_{wr}}{\partial x}\bigg|_{x=0} \,, \qquad \hat{V}_{\mathrm{II}}\big|_{b+} = \frac{1}{k_0^2} \frac{\partial \hat{B}_{wr}}{\partial x}\bigg|_{x=d} \,. \tag{13.86}$$

After a short calculation the constants $\hat{B}_{r0}$ and $\hat{B}_{r1}$ can be eliminated resulting in the following set of jump conditions for the vacuum potential functions. Each condition contains the first non-vanishing contribution in the $d/b$ expansion:

$$\frac{\partial \hat{V}_{\mathrm{I}}}{\partial r}\bigg|_{b-} = \frac{\partial \hat{V}_{\mathrm{II}}}{\partial r}\bigg|_{b+} \,,$$

$$\hat{V}_{\mathrm{I}}\big|_{b-} = \hat{V}_{\mathrm{II}}\big|_{b+} - \left(\frac{\mu_0 \omega_i d}{\eta k_0^2}\right) \frac{\partial \hat{V}_{\mathrm{II}}}{\partial r}\bigg|_{b+} \,. \tag{13.87}$$

In terms of the coefficients $C_j$ these conditions can be written as:

$$C_1 K_b' + C_2 I_b' - C_3 K_b' = 0,$$

$$C_1 K_b + C_2 I_b - C_3 \left(K_b - \frac{\mu_0 \omega_i d k}{\eta k_0^2} K_b'\right) = 0. \tag{13.88}$$

Equations (13.84) and (13.88) are the desired relations, their derivations admittedly having required a somewhat complex analysis. These relations should be viewed as three linear algebraic equations for the three unknown coefficients $C_1$, $C_2$, $C_3$ in terms of the surface displacement $\xi_0$. After a somewhat tedious calculation one can find the coefficients and substitute them into Eq. (13.79) in order to obtain the resistive wall eigenvalue relation.

### The resistive wall stability boundary

Following the procedure described above one obtains an expression for the vacuum contribution to the pressure balance matching condition given by

$$\mathbf{B} \cdot \mathbf{B}_1\big|_a = -\frac{m^2 \hat{B}_\theta^2}{a} \frac{K_a}{\rho_a K_a'} \left[\frac{\Omega_i\left(1 - K_b' I_a/I_b' K_a\right) + 1}{\Omega_i\left(1 - K_b' I_a'/I_b' K_a'\right) + 1}\right] \xi_0$$

$$\approx \frac{\hat{B}_\theta^2}{a} \frac{1}{1 + ka} \left[\frac{\Omega_i(1 + a^2/b^2) + 1}{\Omega_i(1 - a^2/b^2) + 1}\right] \xi_0. \tag{13.89}$$

Here, $\Omega_i$ is a normalized form of the growth rate

$$\Omega_i = -\left(\frac{\rho_b^2 I_b' K_b'}{m^2 + \rho_b^2}\right) \frac{\mu_0 \omega_i b d}{\eta} \approx \left(\frac{1 + kb}{1 + k^2 b^2}\right) \frac{\mu_0 \omega_i b d}{2\eta} \tag{13.90}$$

and the simplified forms where the Bessel functions are approximated correspond to $m = 1$.

The next step is to substitute this expression into the pressure balance matching condition. The resulting eigenvalue equation is easily rearranged to give an explicit expression for the growth rate, which, for $m = 1$, can be written as

$$\omega_i \tau_w \approx -\left[\frac{2(1 + k^2 b^2)}{(1 + kb)(1 - a^2/b^2)}\right]\left(\frac{\beta - \beta_\infty}{\beta - \beta_b}\right), \tag{13.91}$$

where

$$\beta_\infty = 1 - \frac{1}{ka},$$

$$\beta_b = 1 - \frac{1}{ka} + \frac{\Lambda_b - 1}{k^2 a^2} = 1 - \frac{1}{ka} + \left(\frac{2}{b^2/a^2 - 1}\right)\frac{1}{k^2 a^2}, \tag{13.92}$$

and $\tau_w = \mu_0 bd/\eta$ is the wall diffusion time. This is the desired eigenvalue relation for the resistive wall mode.

In analyzing Eq. (13.91) observe first that the term in the square bracket is always positive. Second, recall that $\beta_b(ka)$ and $\beta_\infty(ka)$ are the marginal stability values of $\beta$ corresponding to a perfectly conducting wall at $r = b$ and $r = \infty$ respectively. Also, as $ka$ varies, both $\beta_b(ka)$ and $\beta_\infty(ka)$ can be either positive or negative but in all cases $\beta_b > \beta_\infty$. Lastly, keep in mind that the resistive wall mode is a new mode supported by the plasma in addition to the ideal modes.

With this as background there are three interesting regimes to consider. In the first case, assume a value of $k$ is chosen so that the plasma is stable to the ideal mode with the wall at infinity: $\beta < \beta_\infty < \beta_b$. The eigenvalue relation implies that $\omega_i \tau_w < 0$. The plasma is stable to both the ideal and resistive wall mode. Since a perfectly conducting wall is not required for ideal stability, making it a resistive wall does not alter the stability properties.

In the second regime, a perfectly conducting wall is present but it is not close enough to stabilize the ideal mode: $\beta > \beta_b > \beta_\infty$. Here too, $\omega_i \tau_w < 0$. The resistive wall mode is stable. However, the plasma remains unstable to the fast, potentially catastrophic ideal mode. If a perfectly conducting wall is not close enough to stabilize the ideal mode, making it resistive does not help.

The third, and most interesting, regime corresponds to the situation where a perfectly conducting wall is present and is close enough to stabilize the ideal mode. However, without a wall the plasma would be unstable to the ideal mode: $\beta_b > \beta > \beta_\infty$. In this regime $\omega_i \tau_w > 0$, implying that the resistive wall mode is unstable. The stabilizing eddy currents that develop in a perfectly conducting wall diffuse away in a wall diffusion time when the wall is resistive. On this time scale the effect of the resistive wall vanishes and the stability boundary reverts back to the no-wall case.

These results are illustrated in Fig. 13.17 for an RFP with $R_0/a = 5.5$ and $b/a = \sqrt{5}$. The figure shows the resistive wall growth rate $\omega_i \tau_w$ vs. $\beta$ for the $n = 3$ mode. Note that the region to the right of $\beta = \beta_b \approx 0.85$ is unstable to the ideal mode. The region lying to the left of $\beta = \beta_\infty \approx -0.83$ is stable to both the ideal and resistive wall modes but is clearly inaccessible physically since $\beta$ cannot be negative. The intermediate region is stable to the ideal mode but unstable to the resistive wall mode. As previously stated, the mode is unstable even when $\beta = 0$.

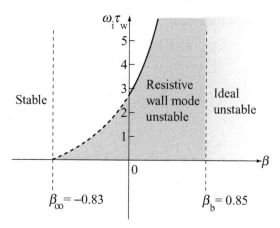

Figure 13.17 The resistive wall growth rate vs. $\beta$ for $R_0/a = 5.5$, $b/a = \sqrt{5}$, and $n = 3$.

### 13.5.7 Summary

The RFP is a relatively large aspect ratio, axisymmetric, toroidal configuration. In terms of MHD its stability is greatly improved over that of a pure $Z$-pinch because of the presence of an internal toroidal magnetic field. However, since the edge safety factor is small it is still susceptible to a number of external kink modes, even at $\beta = 0$, if there is no perfectly conducting wall surrounding the plasma. With a reasonably close fitting, perfectly conducting wall, the RFP is stable to the most severe $m = 0$ and $m = 1$ modes at quite high values of $\beta \sim 1/2$. For a realistic resistive wall the unstable modes return, although with a much slower growth. Ultimately some form of feedback stabilization system will be required.

Even so, transport still remains as a major priority for RFP research and this is the main focus of the program on the MST device at the University of Wisconsin. Early experiments had loss rates nearly two orders of magnitude faster than a comparably sized tokamak. In these experiments thick copper walls were used that effectively acted liked a perfectly conducting wall on the time scale of the experiment. Therefore, the resistive wall mode was not a dominant experimental effect. However, several innovative experimental ideas have helped to dramatically improve the transport situation by more than an order of magnitude, indeed a very favorable result.

As the transport situation hopefully continues to improve, accompanied by longer experimental pulse lengths, the resistive wall MHD modes will become more important. Recognition of this has led other RFP groups (e.g. those working on RFX in Italy and Extrap T2R in Sweden) to focus their research programs on developing methods for active control of MHD behavior, particularly the resistive wall mode.

Several of the main plasma physics problems faced by the RFP are under investigation by the worldwide RFP community. Time will tell if the overall physics performance of an RFP can be made to match that of a tokamak.

## 13.6 The spheromak

### 13.6.1 Overview of the spheromak

The spheromak is a compact, axisymmetric, toroidal configuration. It has comparably sized toroidal and poloidal magnetic fields and should potentially be capable of achieving high pressures although experiments to date have been only operated at relatively low $\beta$. The unique features of the spheromak are the absence of both toroidal field coils and an ohmic transformer; there are no coils on the inside of the torus. The vacuum chamber thus has a spherical rather than toroidal topology, suggesting the name *spher*omak.

Without toroidal field coils and an ohmic transformer one can ask how the spheromak generates both a toroidal and poloidal magnetic field. The answer involves an innovative procedure for driving steady state currents coupled with an externally applied poloidal magnetic field. A simple schematic diagram of a spheromak is shown in Fig. 13.18. Also shown is a practical realization of the concept, the Sustained Spheromak Physics Experiment (SSPX) at Lawrence Livermore National Laboratory.

The operation of the device is as follows. First, the external bias coil is energized on a slow time scale, producing a poloidal magnetic field which intersects the toroidally symmetric gap in the vacuum chamber. A small amount of neutral gas is then injected into the chamber. The plasma phase begins with the application of a high voltage across the gap, often produced by a capacitor bank. The plasma breaks down and initially a current flows along the magnetic field lines connecting the two electrodes. As the current increases it eventually reaches a critical value above which the poloidal magnetic field lines break and reconnect forming closed flux surfaces. The current inside the closed surfaces produces ohmic heating, thereby forming the spheromak plasma.

The mechanism of field-line breaking and reconnection is a complex physical process. The standard view in the spheromak community, the one focused on in this section, is that the process involves small-scale, symmetry breaking, resistive MHD turbulence. This turbulence persists even after the spheromak plasma is formed, causing it to non-linearly relax to a lowest-energy state, a process often called "self-organization" although "turbulent relaxation" might be more appropriate. A more recent idea is that larger-scale symmetry breaking modes are involved in the reconnection process leading to the creation of stochastic magnetic field lines. This effect may be dominant during spheromak creation but can then be made to diminish after the plasma has formed by lowering the gap current. With either explanation there is considerable enhanced transport. Further studies are needed to sort out this complex behavior.

Interestingly, although the externally driven current between the electrodes flows in the poloidal direction the relaxation process within the spheromak plasma induces a toroidal current as well. From a plasma physics point of view, the analysis of the relaxation process is another fascinating topic. However, because of its complexity, it too is beyond the scope of the present book and interested readers should refer to *Spheromaks* by P. M. Bellan 2000. (London: Imperial College Press). For present purposes it suffices to assume that

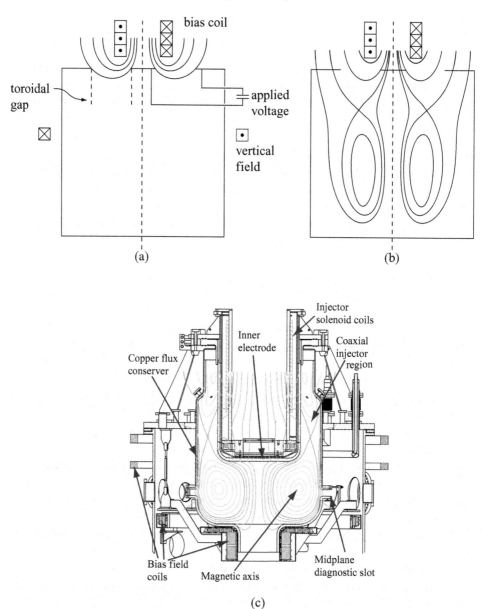

Figure 13.18 Conceptual diagram of a spheromak: (a) configuration with a low poloidal current; (b) configuration with a high poloidal current; (c) the SSPX experiment at Livermore National Laboratory. (Courtesy of E. B. Hooper.)

a spheromak plasma has been generated and the task is to understand its behavior in the context of MHD equilibrium and stability and ultimately as a potential reactor.

In terms of MHD equilibrium, present spheromaks are held in toroidal force balance by means of eddy currents flowing in the close fitting conducting shell that serves as the vacuum chamber. Most experiments, because of power supply constraints, do not operate for pulse lengths long compared to the wall diffusion time. In other words, the wall acts like a perfect conductor on the time scale of the experiment. For longer pulse experiments, external vertical field coils need to be added. Ultimately, true steady state operation is possible solely by the application of three DC sources: (1) the current in the external bias coil, (2) the voltage across the electrodes, and (3) the current in the vertical field coils. Also, in present experiments no external heating supplies are utilized and the achievable values of $\beta$ are limited by the amount of ohmic heating current available and the transport losses.

With respect to stability, the safety factor decreases weakly with minor radius and typically has a value too low to provide stability without a perfectly conducting wall. The most dangerous instability normally corresponds to $n = 1$ with $m = 1$ being the dominant poloidal harmonic. This is an external kink mode best described as a tilting motion of the plasma. Specifically, without a conducting wall the plasma would be held in equilibrium by a set of vertical field coils whose current flows in the opposite direction to the plasma current. This opposing current orientation causes the ring of spheromak current to try to flip $180°$ to an upside down position. For short times the flipping motion has the appearance of a tilt. In practice, because of the relatively short experimental pulse lengths the vacuum chamber behaves like a perfectly conducting wall and the instability is not observed. As experimental pulse lengths increase, the instability is expected to reappear as a resistive wall mode and some form of feedback stabilization will be required.

Consider next the potential advantages and disadvantages of the spheromak with respect to the tokamak as a fusion reactor. On the positive side, a spherical configuration with no central hardware is technologically simpler than a torus. This should lead to a more compact device with a lower capital cost, clearly an important advantage. A second positive point has to do with the current flowing in the plasma. A spheromak reactor is projected to have a toroidal current of 30 MA and a poloidal current of 100 MA. The total current is indeed very much larger, by more than an order of magnitude, than for a tokamak. This large current should, in principle, be able to ohmically heat the plasma to ignition without the need for expensive auxiliary microwave or neutral beam power. A related advantage is that no expensive, sophisticated auxiliary power supplies are needed for current drive to maintain steady state operation. The spheromak, as previously stated, is potentially capable of steady state operation using only three, relatively low-technology, DC power supplies.

There are also several disadvantages to consider. First, the turbulence associated with the relaxation process also produces enhanced transport: the values of $\tau_E$ thus far achieved experimentally are noticeably less than for the mainline tokamak. Since the relaxation process is an inherent component of the spheromak, it is not clear how much the resulting enhanced transport can be ultimately reduced. This is a major topic of current research. Second, the resistive wall tilt instability, caused by the relatively low safety factor, is present even at zero $\beta$. The implication is that some form of feedback stabilization is required, adding to the technological complexity, particularly in a very compact geometry. Lastly, while it is advantageous that steady state current drive can be provided by low-technology DC power supplies, the question of efficiency remains an important issue. How many watts of power are required to drive 1 A of current? Presently, the relaxation generated current drive is very inefficient and would not allow a favorable power balance if directly extrapolated to a reactor. However, in fairness, improving the current drive efficiency has not been a major topic in spheromak research. This will become a more important issue if the transport problem can be alleviated.

The analysis presented in this section focuses on spheromak MHD equilibrium and stability in the context of the surface current model. Although transport is currently the most critical issue facing the spheromak experimental program, the motivation for the spheromak geometry is primarily driven by MHD considerations and understanding this motivation is the primary goal of the section. The main aim of the analysis is to determine a relationship between the maximum stable $\beta$ against the tilt mode and the radius of the perfectly conducting wall. The results show that for reasonable values of aspect ratio and wall radius quite high values of stable $\beta$ can be achieved: $\beta \sim 0.5$. When the wall has a finite conductivity, the instability reappears as a resistive wall mode with typically only a single mode, the $n = 1$ mode, becoming unstable. Some form of feedback is required to stabilize this mode.

### 13.6.2 Spheromak surface current equilibrium

The analysis begins with a reduction of the general 2-D surface current equilibrium relation to a simpler form applicable to the spheromak. The parameters of the surface current model are chosen by examining a set of numerically calculated midplane profiles as illustrated in Fig. 13.19. Also superimposed is the surface current approximation. First, note that for a spheromak the proper scaling assumes that $p \sim B_0^2/2\mu_0 \sim \hat{B}_\theta^2/2\mu_0$. All the pressures are of comparable order.

Second, observe that it is reasonable to approximate the cross section of the spheromak as a circle. Elongation and shaping of the cross section are not essential features of the configuration.

Lastly, an examination of the numerically computed spheromak profiles shows that toroidal effects should be important because of the compact, tight aspect ratio geometry. However, including toroidal effects in the tilt stability analysis leads to a very complicated calculation that requires large amounts of mathematics. To avoid this difficulty one again introduces the large aspect ratio expansion: $\varepsilon \ll 1$. No essential physics is lost by utilizing this rather poor approximation although the results must be considered only qualitatively correct. The simplification that follows is that the combination of circular cross section and $\varepsilon \ll 1$ reduces the toroidal stability problem to an equivalent straight cylindrical problem, similar to that of the RFP except that $B_0 \neq 0$.

After combining the assumptions made above, one obtains the following reduced equilibrium pressure balance relation for the surface current model of a "straight cylindrical"

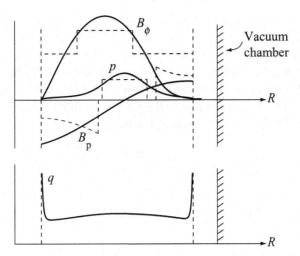

Figure 13.19 Typical midplane profiles as a function of $R$ at $Z = 0$ for a spheromak. (Courtesy of E. B. Hooper.)

spheromak:

$$\hat{B}_\theta^2(a) = 2\mu_0 p + B_i^2 - B_0^2. \tag{13.93}$$

For the stability analysis it is necessary to define $\beta$ and $q$. In the surface current model an appropriate definition of $\beta$ for a spheromak is

$$\beta = \frac{2\mu_0 p}{B_0^2 + \hat{B}_\theta^2}. \tag{13.94}$$

Similarly, the definition of the safety factor implies that

$$q = \frac{\varepsilon B_0}{\hat{B}_\theta}. \tag{13.95}$$

In a spheromak the safety factor is usually comparable to, but somewhat smaller than, unity. A typical value is $q \approx 0.7$. If one thus assumes that the appropriate ordering for the safety factor is $q \sim 1$ and stretches the limit of the aspect ratio expansion so that $\varepsilon \approx 0.6$, it follows that

$$\beta = \frac{2\mu_0 p}{B_0^2(1 + \varepsilon^2/q^2)}. \tag{13.96}$$

This completes the specification of the equilibrium.

### 13.6.3 *The* **m** $= 1$ *tilt instability*

The most important external MHD instability affecting the spheromak is the $m = 1, n = 1$ tilt mode. The basic features of the instability can be determined from an analysis of the large-aspect-ratio spheromak model. The details are almost identical to those presented for the RFP and there is no need to redo the analysis here. There are, however, several minor differences discussed below. Also, one should keep in mind that the goal of the calculation is to derive a marginal stability relation of the form $\beta$ vs. $b/a$ for the tilt mode.

The main difference between the spheromak and RFP analysis is the boundary condition on the perturbed vacuum magnetic field. For the RFP this condition has been derived assuming that $B_0 = 0$. For the spheromak $B_0 \neq 0$ and the RFP boundary condition on the perturbed vacuum potential must be modified as follows:

$$\left.\frac{\partial \hat{V}_1}{\partial r}\right|_a = \frac{i m \hat{B}_\theta}{a}\xi_0 \to \frac{i(m\hat{B}_\theta - ka B_0)}{a}\xi_0. \tag{13.97}$$

Using this replacement one can easily show that for the RFP pressure balance matching condition to be applicable to a spheromak only one modification is required. The new spheromak matching condition is

$$\left[\frac{(m\hat{B}_\theta - \rho_a B_0)^2}{a}\frac{K_a}{\rho_a K_a'}\Lambda_b - \frac{B_i^2}{a}\frac{\rho_a I_a}{I_a'} + \frac{\hat{B}_\theta^2}{a}\right]\xi_0 = 0. \tag{13.98}$$

The only change is in the first term, where $m\hat{B}_\theta$ has been replaced by $m\hat{B}_\theta - ka B_0$.

To apply this eigenvalue relation to the spheromak several simplifications are necessary. First, the tilt instability requires setting $m = 1$. Second the parameter $\rho_a = ka = n\varepsilon$. For the tilt instability one must set $n = 1$, implying that $\rho_a = \varepsilon$. Third, in the context of the large-aspect-ratio expansion one can approximate the Bessel function expressions by $\rho_a I_a/I_a' \approx \varepsilon^2$ and $K_a/\rho_a K_a' \approx -1$. Similarly, the wall stabilization term reduces to $\Lambda_b = (w^2 + 1)/(w^2 - 1)$, where $w = b/a$ is the ratio of wall radius to plasma radius. Lastly, the quantity $B_i^2$ is eliminated by means of the equilibrium relation: $B_i^2 = B_0^2 + \hat{B}_\theta^2(a) - 2\mu_0 p = (1 - \beta)(1 + \varepsilon^2/q^2)B_0^2$. Combining these simplifications leads to the following expression for the marginally stable $\beta$ against the tilt instability:

$$\beta \leq \beta_{max} \equiv \frac{[2(1-q) - \varepsilon^2] - w^2\,[2(1-q) - \varepsilon^2]}{(q^2 + \varepsilon^2)\,(w^2 - 1)}. \tag{13.99}$$

This expression is illustrated in Fig. 13.20 for the case $q = 0.7$, $\varepsilon = 0.6$. Observe that when the wall is far enough away to satisfy $w > [2(1-q) - \varepsilon^2]^{1/2}/[2(1-q)q - \varepsilon^2]^{1/2} \approx 2$ the spheromak is unstable to the tilt instability even for $\beta = 0$. On the other hand for a very close wall satisfying the condition $w < [1 + (1-q)^2]^{1/2}/[1 - (1-q)^2]^{1/2} \approx 1.1$, the spheromak is stable for $\beta = 1$.

Clearly within the context of the surface current model the critical $\beta$ is a strong function of wall position. More exact numerical studies including toroidal effects also show that the wall must be quite close to suppress the tilt instability. The precise results depend upon profiles and the shape of the conducting wall so there is no simple universal relation that determines the critical wall radius to stabilize a specified $\beta$. Without a wall, however, there is agreement that an isolated spheromak plasma would always be unstable, even at $\beta = 0$, to the tilt instability. For discharges whose duration is long compared to the wall diffusion time, the stabilizing effect of the wall vanishes and the tilt instability reappears in the form

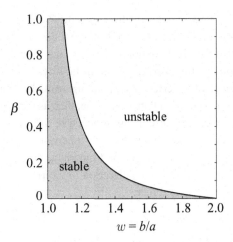

Figure 13.20 Critical $\beta$ for tilt stability as a function of wall radius $w = b/a$ for $q = 0.7$.

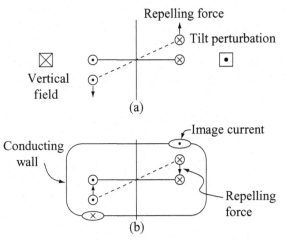

Figure 13.21 Physical mechanism of the tilt instability: (a) without a conducting wall, and (b) with a conducting wall.

of a resistive wall mode. As for the RFP some form of feedback stabilization will be required for long-pulse operation.

The last topic in this subsection is a brief qualitative description of the basic mechanism driving the tilt instability. A simple wire model provides the necessary explanation. Consider a spheromak held in equilibrium by a vertical field coil as shown in Fig. 13.21. Assume for now that there is no perfectly conducting shell. The spheromak can be approximated by a circular loop of wire carrying a current $I$. The vertical field coil must carry a current $I_{\text{vert}}$ in the opposite direction in order to provide an inward, repelling force along $R$ to hold the plasma in toroidal force balance. Now assume the plasma undergoes a small tilting motion as shown in the figure. A small force develops on the spheromak current in the vertical $Z$

direction. Since $I$ and $I_{vert}$ flow in opposite directions the $Z$ force is upward (i.e., repelling) on the right hand portion of the spheromak current and downward (i.e., repelling) on the left hand portion. In other words the induced force points in the direction to amplify the initial tilting perturbation and this corresponds to instability.

Consider now the effect of a perfectly conducting wall. If such a wall surrounds the plasma, particularly with a somewhat flattened shape, then as the spheromak plasma undergoes a tilting perturbation image currents are induced in the wall as shown in the figure. These currents flow in the opposite direction to the spheromak current, and thus produce a repelling force. However, because the location of the image currents rotates with the tilting motion, the repelling force is actually in a direction to return the spheromak current to its original position. This represents stability.

### 13.6.4 Summary of the spheromak

The spheromak is a compact axisymmetric configuration with the unique properties of not requiring toroidal field magnets or an ohmic transformer. Comparably sized poloidal and toroidal magnetic fields are generated by means of an externally provided poloidal current and poloidal bias field, both in principle driven by DC sources. The plasma itself is formed by a complex non-linear relaxation process. Although the geometry of the spheromak is quite compact compared to an RFP, the basic physics issues are nonetheless somewhat similar.

The dominant experimental issue currently facing the spheromak concept is the understanding and reduction of anomalous energy transport arising from the relaxation process. The goal is to achieve sustained operation at reasonably high temperatures ($T_e \sim T_i \sim$ several hundred electronvolts). If sustained, near steady state operation were achieved, one would expect that the relatively low edge safety factor would cause the spheromak to become unstable to the resistive wall tilt mode. If this mode could be stabilized by feedback, then quite high $\beta$ should be achievable. Compared to that of the tokamak, spheromak research is still at a relatively early stage. Consequently, the fusion community will have to wait to see whether or not the spheromak can overcome the plasma physics issues inherent in the concept.

## 13.7 The tokamak

### 13.7.1 Overview of the tokamak

The tokamak is an axisymmetric torus with a large toroidal magnetic field, a moderate plasma pressure and a relatively small toroidal current. It is presently the leading candidate to become the world's first fusion reactor, a status earned by virtue of its excellent physics performance. Specifically, the achieved values of $p\tau_E$ at high $T$ in a tokamak exceed those of any of the other concepts. Because of its performance, there is a large number of major tokamak experimental facilities operating or being constructed in the international fusion program.

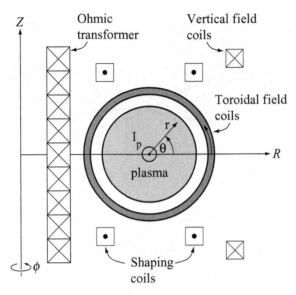

Figure 13.22 Schematic diagram of a tokamak.

A schematic diagram of a tokamak is shown in Fig. 13.22. Observe that there are four basic magnet systems in the tokamak: (1) the toroidal field coils, which produce the large toroidal field; (2) the ohmic transformer, which induces the toroidal plasma current required for equilibrium and ohmic heating; (3) the vertical field system, which is required for toroidal force balance; (4) shaping coils, which produce a non-circular cross section to improve MHD stability limits and alleviate plasma–wall impurity problems.

Typical operation of a tokamak discharge starts with the establishment of a large, steady, toroidal, magnetic field. Next, neutral gas is injected into the vacuum chamber and often pre-ionized. The transformer induced toroidal current is then ramped up to its maximum value and maintained for the "flat top" portion of the pulse. During flat top operation external heating power in the form of RF or neutral beams is applied to the plasma. The magnitude of the external power is usually substantially greater than that of the ohmic power. Most of the interesting experimental plasma physics occurs during the flat top period.

The characteristic equilibrium profiles of a tokamak during flat top operation are illustrated in Fig. 13.23. Note that the toroidal magnetic field has a slight diamagnetic dip, which is responsible for holding the plasma in radial pressure balance. A crucial feature is the behavior of the safety factor $q(r) \approx r B_\phi / R_0 B_\theta$. For a tokamak $q(r)$ is an increasing function of radius and, most importantly, is always large: $q(r) > 1$ over almost the entire plasma, a consequence of the large toroidal magnetic field.

In terms of reactor desirability, the tokamak has a number of advantages and a few problems. The main advantages are associated with good physics performance. The large toroidal field and correspondingly large edge safety factor lead to finite values of MHD stable $\beta$ without a conducting wall and to reasonably high experimental values of the energy

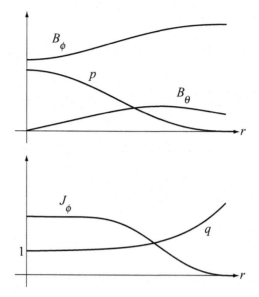

Figure 13.23 Typical profiles in a tokamak in the large-aspect-ratio limit $R_0/a \to \infty$.

confinement time $\tau_E$. Good confinement allows the plasma to heat up to high temperatures using only a moderate amount of external heating. In addition the resulting values of $\beta$ lie in the regime of reactor interest. As an example, consider a high performance D–T shot on the Joint European Torus (JET) located near Oxford in the UK. At $B_\phi \sim 3.6$ T and external heating power $P_h = 25$ MW, JET has achieved the following performance for a pulse lasting several seconds: $n(0) \approx 0.4 \times 10^{20}$ m$^{-3}$, $T_i(0) = 28$ keV, $T_e(0) = 14$ keV, $\langle \beta \rangle \approx \beta(0)/3 \approx 0.018$, and $\tau_E \approx 0.9$ s corresponding to $\langle p \rangle \tau_E \sim 0.84$ atm s. For comparison recall that a value of $p\tau_E \sim 8$ atm s is required for ignition. Clearly, existing tokamak experimental performance is starting to approach the regime of reactor interest.

There are several problems facing the tokamak. First, the need for a large toroidal magnetic field increases the technological complexity and cost of the reactor. Most of the alternative concepts, with the exception of the stellarator, have been designed to alleviate this problem by utilizing a small external toroidal magnetic field, which in turn leads to a small edge safety factor. Philosophically, these concepts are trading off more difficult plasma physics for simpler reactor technology. Tokamak reactor designs have shown that high toroidal magnetic fields are certainly achievable from a practical engineering point of view – it is just that it would be technologically simpler and economically less expensive if such a large field were not required.

The second main issue arises because a reactor will almost certainly need to operate as a steady state device. This requirement is incompatible with an ohmic transformer, which cannot inductively drive a DC current for an indefinite period of time. Some form of external current drive is required. In general, the methods of external current drive involve costly, high-technology power sources, such as microwaves or neutral beams. Furthermore, current-drive efficiency is not very high – the number of input watts per output ampere is large. The net result is that if too much current needs to be driven, power balance becomes unfavorable and the economics become unattractive.

Fortunately, in a tokamak there is a transport-driven toroidal current, known as the "bootstrap current." This current arises naturally, and does not require any external sources. Depending upon the pressure and magnetic field profiles, the bootstrap current can provide between virtually none and 95% of the total current. The present consensus in the fusion community is that 75% or more of the current needs to be provided by the bootstrap effect for an attractive reactor, implying that 25% or less must be provided by external current drive. A further complication is that achieving a high bootstrap fraction requires a high $\beta$, whose value invariably lies above the no-wall $\beta$ limit. Thus, high bootstrap tokamaks will likely need to stabilize the resistive wall mode.

The second problem facing the tokamak can be summarised as follows. A successful tokamak must achieve a high-$\beta$, high-bootstrap-fraction plasma that can be sustained in steady state with only a small amount of external current drive and no ohmic transformer. This is one of the main plasma physics missions of future research. The current drive and bootstrap current issues are discussed in more detail in Chapters 15 and 14 respectively.

Finally, the importance of satisfying the MHD $\beta$ limit against external ballooning-kink modes must be re-emphasized. When this criterion is violated experimentally there is almost always a catastrophic collapse of the plasma pressure and current. Such events are appropriately called "major disruptions." They must be avoided in large experiments and reactors since the large transient forces developed can cause actual physical damage to the structure. The conclusion is that it is important to accurately know the MHD $\beta$ stability limit against ballooning-kink modes and to learn how to operate as close to the limit as possible to maximize performance.

The topics discussed in this section are focused on the MHD behavior of the tokamak, including several closely related variations. As with most other concepts, MHD considerations are the primary motivation for the basic magnetic field configuration. The discussion begins with an analysis of the large-aspect-ratio, circular cross section tokamak. In spite of the large aspect ratio, it is essential to maintain toroidal effects in both the equilibrium and stability calculations.

It is shown that within the context of the surface current model there are three important $\beta$ and $I$ limits. First, there is an equilibrium limit that sets the maximum value of $\beta/I^2$. Second, as $\beta \rightarrow 0$ there is a stability limit that sets the maximum value of $I$ due to low $n$ number kink modes. Third, there is a stability limit due to low-$n$ ballooning-kink modes that sets a limit on the maximum allowable $\beta$ at a corresponding optimized value of $I$. The strictest $\beta$ limit results from the ballooning-kink mode. The stability limits on $\beta$ and $I$ are initially derived assuming that no perfectly conducting wall is present.

It is also worth noting that there is an internal instability that affects the operation of a tokamak, corresponding to an $m = 1, n = 1$ mode. This instability is localized near the magnetic axis of the plasma, and generates internal relaxation oscillations known as "sawteeth." However, from the present perspective this mode is not overly dangerous, although it does set an operational limit $q(0) > 1$. The physics of the sawtooth oscillation is not covered by the surface current model and is mentioned here primarily for completeness.

Consider now variations on the basic tokamak. The first variation discussed involves elongation of the cross section. It has been known for many years that a combination of elongation and triangularity of the plasma cross section increases the $\beta$ limit against low-$n$ ballooning-kink modes. There is, however, a practical limit to the maximum achievable elongation due to the excitation of an $n = 0$ axisymmetric mode. Typically, it is very difficult to achieve elongations greater than a factor of height/width $\sim 2$. A simple qualitative picture of the effects of elongation on tokamak stability is presented.

The next variation discussed concerns operation in the "advanced tokamak" (AT) regime. It is AT operation that is intended to address the main plasma physics issue discussed above. In AT operation the profiles are externally adjusted to produce a hollow toroidal current density and a high $\beta_p$, the right combination to generate the large bootstrap current essential for minimizing the current-drive requirements. Profile control is accomplished experimentally by means of programming the time dependence of the plasma current and the radial and time dependence of the external heating sources. The MHD issue that arises here is that for a large bootstrap current, the ballooning-kink $\beta$ limit without a perfectly conducting wall is invariably violated and the resulting resistive wall mode must be feedback stabilized. A calculation is presented that determines exactly how close the wall must be to make the transition from an ideal mode to a resistive wall mode in order for feedback stabilization to be a practical possibility.

The last tokamak variation considered is the spherical tokamak, which is actually an ultra tight aspect ratio tokamak. Recall that topologically, the aspect ratio must satisfy $R_0/a \geq 1$. A standard tokamak has $R_0/a \approx 3$, while for a spherical tokamak $R_0/a \approx 1.2$. The tight aspect ratio is motivated by the theoretical result (derived shortly) that the maximum stable $\beta$ against ballooning-kink modes in a tokamak scales as $\beta \sim a/R_0$. Thus, a tight aspect ratio leads to a high $\beta$. Compared to a standard tokamak, a spherical tokamak extrapolates to a more compact, potentially lower-cost reactor. This gain must be balanced against increased technological problems with the compact central core and the need to maintain a very large fraction of the toroidal current by the bootstrap effect. Most of the spherical tokamak discussion is focused on the $\beta$ limit and its impact on fusion power density.

Finally, the simple conceptual reactor design presented in Chapter 5 is revisited. The goal is to quantitatively investigate one of the assumptions made in the design. Specifically, it was stated that when MHD $\beta$ limits are taken into account, the optimum reactor design should operate at the maximum allowable field at the inside of the toroidal field coils, even though a large field increases the costs. This point is addressed using the explicit MHD $\beta$ limits derived for tokamaks.

In summary, there is a consensus among fusion researchers that tokamak physics parameters will ultimately meet the requirements of a fusion reactor, although probably not by a large margin. The key point to keep in mind is that these parameters need to be maintained in steady state. This requires the achievement of a high bootstrap fraction with either no conducting wall or, more likely, feedback stabilization of the resistive wall mode. The present

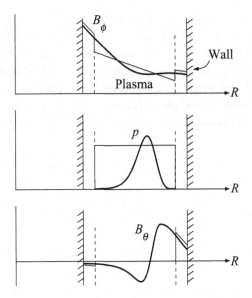

Figure 13.24 Typical midplane profiles (smooth curves) for the TFTR experiment. Superimposed is the surface current model approximation (box-like curves).

chapter focuses on one part of the problem – the determination of the MHD $\beta$ limits. In future chapters the $\beta$ limits are coupled to the generation of the bootstrap current (Chapter 14) and the power requirements for current drive (Chapter 15).

### 13.7.2 The circular cross section tokamak – equilibrium

The equilibrium analysis of the circular cross section tokamak involves the simplification of the exact pressure balance relation using a slightly subtle aspect ratio expansion. The result is a simple expression for the poloidal magnetic field around the plasma surface. It is shown that an equilibrium limit exists for the maximum achievable value of $\beta/I^2$. This limit arises because of the need to increase the external vertical field continually to provide toroidal force balance as $\beta$ increases assuming that $I$ remains fixed. The current $I$ is assumed fixed in order to maintain high values of the safety factor; that is, supporting higher pressures in toroidal force balance by increasing the plasma current while holding the vertical field fixed would reduce the value of the safety factor. Ultimately, when the external vertical field becomes large enough a separatrix moves onto the inner surface of the plasma and this represents the equilibrium limit.

### The aspect ratio expansion

To motivate the choice of parameters for the surface current equilibrium, examine Fig 13.24, which illustrates the typical magnetic field and pressure profiles for the large, circular cross section Tokamak Fusion Test Reactor (TFTR). This device operated at the Princeton Plasma Physics Laboratory in the USA from 1982 to 1997 and used D–T plasmas during the last several years. Superimposed on the TFTR profiles is the surface current approximation. The task now is to obtain a simplified expression for the general surface current pressure balance relation by ordering the various parameters with respect to the inverse aspect ratio $\varepsilon \equiv a/R_0$, assuming a large aspect ratio expansion $\varepsilon \ll 1$. The following simple physical arguments suggest the appropriate ordering scheme for the tokamak.

The key to the tokamak aspect ratio expansion is the assumption that the safety factor must be treated as a finite quantity in order to achieve good MHD stability. Specifically, it is assumed that

$$q \sim 1. \tag{13.100}$$

Since $q \approx r B_0 / R_0 B_\theta$, this implies that the appropriate scaling for the poloidal magnetic field is

$$\frac{B_\theta}{B_\phi} \sim \varepsilon. \tag{13.101}$$

The poloidal field is small compared to the toroidal field.

Next, consider radial pressure balance. Recall that a small poloidal field implies that the total $\beta$ is approximately equal to the $\beta_t$:

$$\beta \approx \frac{2\mu_0 p}{B_\phi^2 + B_\theta^2} \approx \frac{2\mu_0 p}{B_\phi^2} \approx \beta_t. \tag{13.102}$$

Note that if the toroidal field is a pure vacuum field (i.e., $B_i = B_0$), then radial pressure balance is provided solely by the poloidal magnetic field and the amount of plasma that can be confined is indeed small: $p \approx B_\theta^2 / 2\mu_0$, implying that $\beta_t \sim \varepsilon^2$ and $\beta_p \sim 1$. This is not the most interesting regime for a tokamak.

Higher pressures can be confined by a small diamagnetic dip in the toroidal magnetic field. If one assumes that $B_0 - B_i \sim \varepsilon B_0$, then for radial pressure balance $p \approx (B_0 - B_i) B_0 / \mu_0$ leading to $\beta_t \sim \varepsilon$ and $\beta_p \sim 1/\varepsilon$. This is the interesting regime for a tokamak. One cannot make the diamagnetic dip any deeper than order $\varepsilon$ because, as is shown shortly, toroidal force balance cannot be achieved if $q$ is constrained to remain of order unity. Experimentally, the diamagnetic dip is created by the application of an external heating power greater in magnitude than the ohmic heating power.

To summarize, the appropriate aspect ratio expansion for a tokamak is defined by

$$q \sim B_\theta / \varepsilon B_\phi \sim (1 - B_i / B_0) / \varepsilon \sim \beta_t / \varepsilon \sim \varepsilon \beta_p \sim 1. \tag{13.103}$$

This ordering scheme can now be introduced into the surface current pressure balance relation to obtain a simple expression for the poloidal magnetic field.

### The surface current pressure balance relation

The general surface current pressure balance relation

$$\hat{B}_p^2(\theta) = 2\mu_0 p + \frac{B_i^2 - B_0^2}{(1 + \varepsilon \cos \theta)^2} \tag{13.104}$$

is simplified by Taylor expanding the denominator of the toroidal field term and introducing a new constant $\lambda$ to replace $B_i$ as follows:

$$\frac{B_i^2}{B_0^2} \equiv 1 - \beta + \varepsilon^2 \lambda. \tag{13.105}$$

Here,

$$\beta \equiv \frac{2\mu_0 p}{B_0^2} \tag{13.106}$$

and to be consistent with the aspect ratio expansion $\lambda \sim 1$. A short calculation yields

$$\left( \frac{\hat{B}_\theta}{\varepsilon B_0} \right)^2 = \lambda + 2\frac{\beta}{\varepsilon} \cos \theta. \tag{13.107}$$

It is now convenient to replace $\lambda$ with yet another equivalent constant $k$ through the definition

$$\lambda \equiv \frac{2\beta}{\varepsilon k^2}(2 - k^2). \tag{13.108}$$

Equation (13.107) reduces to

$$\left( \frac{\hat{B}_\theta}{\varepsilon B_0} \right)^2 = \frac{4\beta}{\varepsilon k^2} \left( 1 - k^2 \sin^2 \frac{\theta}{2} \right). \tag{13.109}$$

Although Eq. (13.109) is relatively simple in form it is not completely satisfactory in the sense that $\hat{B}_\theta$ is expressed in terms of the constant $k$, which at this point has no simple physical interpretation. It would be far more desirable to express $\hat{B}_\theta$ in terms of $\beta$ and $I$ rather than $\beta$ and $k$. This is the last step in the equilibrium derivation.

The current $I$ can be introduced through the definition of the safety factor, although there is an important subtlety that arises in a tokamak. Recall that in a straight cylinder $q(a) = a B_0/R_0 B_\theta(a) \sim 1/I$. In a toroidal tokamak $q$ and $I$ are inversely related only at low $\beta/\varepsilon$. At high $\beta/\varepsilon$ this relationship breaks down. To sort out the issues one can calculate $q = q(k, \beta)$ directly from the definition of $q$ and then compare it with a new quantity $q_* = q_*(k, \beta)$, which by its definition is always inversely proportional to $I$. Specifically, from the definition of the safety factor it follows that

$$q(a) = \frac{1}{2\pi} \int_0^{2\pi} \left( \frac{r B_\phi}{R B_\theta} \right) \Bigg|_a d\theta \approx \frac{\varepsilon B_0}{2\pi} \int_0^{2\pi} \frac{d\theta}{\hat{B}_\theta} = \frac{kK(k)}{\pi} \frac{1}{(\beta/\varepsilon)^{1/2}}, \tag{13.110}$$

where $K(k)$ is the complete elliptic integral of the first kind. Next, the quantity $q_*(k, \beta)$ is defined so that it coincides with $q(a)$ at small $k$ but is forced to be inversely proportional to $I$ at all $k$:

$$q_* \equiv \frac{2\pi a^2 B_0}{\mu_0 R_0 I} = \varepsilon B_0 \left[ \frac{1}{2\pi} \int_0^{2\pi} \hat{B}_\theta d\theta \right]^{-1} = \frac{\pi k}{4E(k)} \frac{1}{(\beta/\varepsilon)^{1/2}}. \tag{13.111}$$

Here $E(k)$ is the complete elliptic integral of the second kind.

Note that at small $k$, $K \approx E \approx \pi/2$ and $q(a) \approx q_* \approx k/[2\,(\beta/\varepsilon)^{1/2}]$. However, as $k \to 1$, $K \approx \ln[4/(1-k^2)^{1/2}]$ and $E \approx 1$. In other words $q(a)$ diverges while $q_*$ remains finite. The significance of the divergence in $q(a)$ is discussed shortly. For present purposes, it suffices to say that in terms of the stability analysis that follows it is $q_*$ rather than $q(a)$ that is the critical parameter. Stability limits are more closely connected to the value of the current than the average pitch angle of the magnetic field lines. Finally, using the limiting values of $E(k)$ one can approximate

$$E(k) \approx \left[ k^2 + \frac{\pi^2}{4}(1 - k^2) \right]^{1/2}. \tag{13.112}$$

Using this approximation to eliminate $k$ leads to an expression for the poloidal magnetic field in terms of $\beta/\varepsilon$ and $q_*$ given by

$$\left( \frac{\hat{B}_\theta}{\varepsilon B_0} \right)^2 \approx \frac{1}{q_*^2} \left( 1 - \hat{\beta} + \frac{\pi^2}{4}\hat{\beta}\cos^2\frac{\theta}{2} \right), \tag{13.113}$$

where

$$\hat{\beta} \equiv \left( \frac{16}{\pi^2} \right) \left( \frac{\beta q_*^2}{\varepsilon} \right). \tag{13.114}$$

This is the desired relation.

### The equilibrium $\beta$ limit

An interesting feature of the solution is the existence of an equilibrium $\beta$ limit when the current (i.e., $q_*$) is held fixed. This limit arises mathematically because $\hat{B}_\theta$ must always be a real quantity. From Eq. (13.109) it follows that the angle $\theta = \pi$, corresponding to the inside of the torus, sets the strictest limit on the reality of $\hat{B}_\theta$, requiring that $k^2 \leq 1$. Equivalently, the corresponding limit from Eq. (13.113) requires that $\hat{\beta} \leq 1$. This translates into the following equilibrium $\beta$ limit:

$$\frac{\beta q_*^2}{\varepsilon} \leq \frac{\pi^2}{16} \approx 0.62, \tag{13.115}$$

which for a fixed geometry is actually a limit on $\beta/I^2$.

The physical origin of the equilibrium limit can be understood by examining Fig. 13.25, which shows schematically how toroidal force balance is achieved by means of an externally applied vertical field. Note that on the inside of the torus the vertical field and the $B_\theta$ contribution from the plasma current point in opposite directions. Clearly at some point outside the plasma these contributions exactly cancel; the net poloidal field is zero. The point where $\hat{B}_p = 0$ is referred to as the "X-point" of the separatrix flux surface.

Assume now that $\beta$ is increased while holding the plasma current fixed. In terms of Fig. 13.25 this implies that the $B_\theta$ contribution from the plasma remains fixed in amplitude. Therefore, to maintain toroidal force balance the vertical field must be increased. The result is that the X-point moves closer to the plasma surface. Eventually, at sufficiently high $\beta$

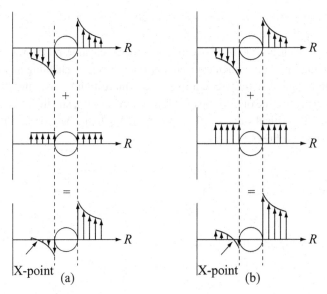

Figure 13.25 Physical origin of the equilibrium limit showing the location of the X-point for: (a) a low-$\beta$ case, and (b) a high-$\beta$ case. In both cases the plasma current is held fixed. In the high-$\beta$ case the vertical field is increased to provide toroidal force balance.

the X-point moves onto the plasma surface. When this occurs the plasma has reached its equilibrium $\beta$ limit. The value of $\beta$ cannot be increased any further with a circular cross section plasma. Attempts to increase $\beta$ further require cutting off and eliminating the inside portion of the plasma.

Lastly, note that when the X-point moves onto the plasma surface $\hat{B}_\theta = 0$ at $\theta = \pi$. This is the reason that $q(a) \to \infty$ at the equilibrium limit. However, even though $\hat{B}_\theta(a, \pi) = 0$, the plasma current remains finite at its fixed value. This is the explanation for the divergence in values between $q(a)$ and $q_*$ at high $\beta/\varepsilon$.

The surface current equilibrium model for the tokamak is now complete. The next topic is the corresponding MHD stability analysis.

### 13.7.3 The circular cross section tokamak – stability

The surface current stability analysis is now applied to the circular tokamak. The problem is more complicated than for the cylinder because toroidal effects couple all the poloidal harmonics of the perturbation. In other words the appropriate Fourier analysis $\xi(\theta, \phi)$ with respect to the toroidal angle $\phi$ and the poloidal angle $\theta$ is

$$\xi(\theta, \phi) = e^{-in\phi} \sum_{-\infty}^{\infty} \xi_m e^{im\theta}. \tag{13.116}$$

The result is that the final eigenvalue relation consists of an infinite set of coupled algebraic equations in the unknown Fourier amplitudes $\xi_m$. The condition for marginal stability is obtained by setting the determinant of this set of equations to zero. Also, the most unstable toroidal mode numbers are shown to correspond to $n \sim 1$.

The approach taken here is to derive the general eigenvalue relation arising from the infinite set of equations and to then truncate the Fourier series, enabling an analytic determination of the marginal stability condition in two interesting limits – the $\beta/\varepsilon \to 0$ limit, which yields a critical value of $q_*$ corresponding to kink modes, and the $\hat{\beta} \to 1$ equilibrium limit, which yields a critical value for $\beta/\varepsilon$ corresponding to the ballooning-kink mode. The analysis proceeds as follows.

### The plasma contribution

The plasma contribution to the stability analysis is found by solving

$$\nabla^2 V_1 = 0. \tag{13.117}$$

For $n \sim 1$ the large aspect ratio expansion implies that

$$\nabla^2 V_1 \approx \nabla_p^2 V_1 = \frac{1}{r}\frac{\partial}{\partial r}\left(r\frac{\partial V_1}{\partial r}\right) + \frac{1}{r^2}\frac{\partial^2 V_1}{\partial \theta^2} = 0. \tag{13.118}$$

Equation (13.118) must be solved subject to regularity in the plasma and, to leading order in $\varepsilon$,

$$\left.\frac{\partial V_1}{\partial r}\right|_a = -\frac{in B_0}{R_0}\xi(\theta, \phi). \tag{13.119}$$

The solution is easily found, and is

$$V_1 = e^{-in\phi}\sum_{m\neq 0} A_m \left(\frac{r}{a}\right)^{|m|} e^{im\theta}, \tag{13.120}$$

where

$$A_m = -\frac{in\varepsilon B_0}{|m|}\xi_m. \tag{13.121}$$

Note that the $m = 0$ term is eliminated from the sum. This is a consequence of the fact that in the large aspect ratio expansion the condition $\nabla \cdot \boldsymbol{\xi} = 0$ implies that $\xi_0 = 0$. The quantity $\mathbf{B} \cdot \mathbf{B}_1|_a$ required for the stability analysis is now easily calculated:

$$\mathbf{B} \cdot \mathbf{B}_1|_a = \mathbf{B} \cdot \nabla V_1|_a = -\frac{a B_0^2}{R_0^2}e^{-in\phi}\sum_{m\neq 0}\frac{n^2}{|m|}\xi_m e^{im\theta}. \tag{13.122}$$

This is the desired expression.

### The vacuum contribution

The vacuum contribution follows in an analogous manner. In the large aspect ratio limit the vacuum potential satisfies

$$\nabla^2 \hat{V}_1 \approx \nabla_p^2 \hat{V}_1 = \frac{1}{r}\frac{\partial}{\partial r}\left(r\frac{\partial \hat{V}_1}{\partial r}\right) + \frac{1}{r^2}\frac{\partial^2 \hat{V}_1}{\partial \theta^2} = 0. \tag{13.123}$$

The boundary conditions are slightly more complicated. If the wall is far from the plasma, then $V_1$ must be regular for $a < r < \infty$. At the plasma–vacuum interface the boundary condition can be written as

$$\left.\frac{\partial \hat{V}_1}{\partial r}\right|_a = \mathbf{e}_r \cdot \nabla \times (\boldsymbol{\xi} \times \hat{\mathbf{B}})\Big|_a = -\frac{i n B_0}{R_0}\xi + \frac{1}{a}\frac{\partial}{\partial \theta}(\hat{B}_\theta \xi). \tag{13.124}$$

The solution can be written as

$$\hat{V}_1 = e^{-in\phi}\sum_{m \neq 0}\hat{A}_m\left(\frac{a}{r}\right)^{|m|}e^{im\theta}, \tag{13.125}$$

where, after a short calculation it can be shown that

$$\hat{A}_m = -\frac{i\varepsilon B_0}{|m|}\sum_{p \neq 0}G_{mp}\xi_p \tag{13.126}$$

and the matrix element $G_{mp}$ is given by

$$G_{mp} = \frac{1}{2\pi}\int_0^{2\pi}\left(\frac{m\hat{B}_\theta}{\varepsilon B_0} - n\right)\cos\left[(m - p)\theta\right]d\theta. \tag{13.127}$$

In practice $G_{mp}$ is evaluated numerically, although for several interesting special cases it can be evaluated analytically. The fact that $G_{mp}$ is not a diagonal matrix explicitly shows the coupling of different poloidal harmonics.

The last step in the vacuum analysis is to calculate the quantity $\hat{\mathbf{B}} \cdot \hat{\mathbf{B}}_1|_a$ required for the stability analysis. After another short calculation one finds

$$\hat{\mathbf{B}} \cdot \hat{\mathbf{B}}_1|_a = \hat{\mathbf{B}} \cdot \nabla\hat{V}_1|_a = \frac{aB_0^2}{R_0^2}e^{-in\phi}\sum_{p \neq 0}\sum_{l \neq 0}\frac{G_{lm}G_{lp}}{|l|}\xi_p e^{im\theta}. \tag{13.128}$$

The double sum will be evaluated analytically for the two special cases of interest.

### The pressure balance matching condition

The last step in the stability analysis is to substitute the plasma and vacuum contributions into the pressure balance matching condition. For a large aspect ratio, circular cross section tokamak this condition reduces to

$$[\hat{\mathbf{B}} \cdot \hat{\mathbf{B}}_1 - \mathbf{B} \cdot \mathbf{B}_1]_a = \frac{\hat{B}_\theta^2}{a}\xi + \frac{\beta B_0^2}{R_0}\xi\cos\theta. \tag{13.129}$$

The first term on the right hand side represents the destabilizing effect of the poloidal field curvature. Although the term varies with the angle $\theta$, it is always positive, corresponding to unfavorable curvature. The second term on the right hand side represents the effect of the toroidal field-line curvature. This term is positive on the outside of the torus ($\theta = 0$) and negative on the inside of the torus ($\theta = \pi$). On the outside of the torus the field lines curve towards the plasma indicating unfavorable curvature. Conversely, on the inside of the torus the field lines bend away from the plasma indicating favorable curvature. This oscillation in curvature is the primary reason why the most dangerous perturbations in a tokamak have a strong component of ballooning, with the displacement enhanced on the outside of the torus and reduced on the inside.

The evaluation of the right hand side of Eq. (13.129) is straightforward. Substitution of the equilibrium fields shows that

$$
\begin{aligned}
\frac{\hat{B}_\theta^2}{a}\xi &= \frac{aB_0^2}{R_0^2}e^{-in\phi}\sum_{m\neq0}\frac{1}{q_*^2}\left(1-\hat{\beta}+\frac{\pi^2}{4}\hat{\beta}\cos^2\frac{\theta}{2}\right)\xi_m e^{im\theta}, \\
\frac{\beta B_0^2}{R_0}\xi\cos\theta &= \frac{aB_0^2}{R_0^2}e^{-in\phi}\sum_{m\neq0}\left(\frac{\beta}{\varepsilon}\xi_m\cos\theta\right)e^{im\theta},
\end{aligned}
\tag{13.130}
$$

where, as before, $\hat{\beta} = (16/\pi^2)(\beta q_*^2/\varepsilon)$.

The final form of the pressure balance matching condition can now be deduced. After a short calculation one obtains

$$
\sum_{p\neq0}W_{mp}\xi_p e^{im\theta} = 0.
\tag{13.131}
$$

Here, the matrix element $W_{mp} = W_{pm}$ is given by

$$
W_{mp} = \left\{\frac{n^2}{|m|}-\left[\frac{1}{q_*^2}+\left(2-\frac{16}{\pi^2}\right)\frac{\beta}{\varepsilon}\right]\right\}\delta_{m-p}
$$
$$
-\frac{3}{2}\frac{\beta}{\varepsilon}(\delta_{m-p-1}+\delta_{m-p+1})+\sum_{l\neq0}\frac{G_{lm}G_{lp}}{|l|}
\tag{13.132}
$$

and $\delta_k$ is the familiar Kronecker delta function: $\delta_0 = 1$, $\delta_{k\neq0} = 0$. Since each poloidal harmonic $m$ is independent, Eq. (13.131) actually represents a set of equations; that is, the coefficient of each harmonic must separately vanish. This leads to a set of linear, homogeneous, algebraic equations for the unknown $\xi_m$:

$$
\sum_{p\neq0}W_{mp}\xi_p = 0 \qquad \text{(for each } m \neq 0\text{)}.
\tag{13.133}
$$

Equation (13.133) is the desired expression for the eigenvalue condition. For a non-trivial solution to exist the determinant of $\overset{\leftrightarrow}{\mathbf{W}}$ must vanish. Setting $\det\overset{\leftrightarrow}{\mathbf{W}} = 0$ then leads to the condition for marginal stability.

### The $\beta/\varepsilon \to 0$ kink instability

The first application of the general eigenvalue relation concerns $\beta/\varepsilon \to 0$ kink instabilities. It is shown that in the low-$\beta$ limit there is a marginal value of current (i.e., $q_*$). If the current is too high, the plasma is unstable. Conversely, when the current is sufficiently small the plasma is stable to the kink mode.

The mathematics of the kink mode is greatly simplified because the matrix $\overset{\leftrightarrow}{\mathbf{W}}$ becomes diagonal as $\beta/\varepsilon \to 0$. Thus, in this limit the poloidal harmonics decouple and each $m$ value can be treated separately. The reason for the decoupling is associated with the fact that in the $\beta/\varepsilon \to 0$ limit, the poloidal field reduces to

$$\frac{\hat{B}_\theta}{\varepsilon B_0} = \frac{1}{q_*} = \text{const.} \tag{13.134}$$

There is no angular coupling. This can be explicitly demonstrated by evaluating the vacuum matrix element $G_{lm}$ which reduces to

$$G_{lm} = \left(\frac{l}{q_*} - n\right)\delta_{l-m}. \tag{13.135}$$

After substituting Eq. (13.135) into Eq.(13.132) one obtains the following simple expression for the matrix element $W_{mp}$:

$$
\begin{aligned}
W_{mp} &= \left[\frac{n^2}{|m|} - \frac{1}{q_*^2} + \frac{1}{|m|}\left(\frac{m}{q_*} - n\right)^2\right]\delta_{m-p} \\
&= \frac{2}{|m|\,q_*^2}\left[\left(nq_* - \frac{m}{2}\right)^2 + \frac{|m|\,(|m| - 2)}{4}\right]\delta_{m-p}.
\end{aligned}
\tag{13.136}
$$

Stability requires that each matrix element $W_{mp} \geq 0$. From the second form in Eq. (13.136) it immediately follows that for any $m \geq 2$ the plasma is stable to the kink mode. The only possibility for an instability occurs when $m = 1$. In this case a simple calculation shows that an instability exists if $0 < nq_* < 1$. The strictest limit on $q_*$ occurs for $n = 1$ leading to the stability condition

$$q_* > 1 \qquad \text{(for stability)}. \tag{13.137}$$

This stability boundary is known as the Kruskal–Shafranov limit. It can be rewritten as a limit on the maximum allowable current

$$I \leq I_{\max} \equiv \frac{2\pi a^2 B_0}{\mu_0 R_0}. \tag{13.138}$$

For the simple reactor designed in Chapter 5 one finds that $I_{\max} \approx 19\,\text{MA}$. More realistic numerical calculations using smooth profiles show that the surface current model is somewhat optimistic. Depending upon profiles, the more accurate value of $q_*$ typically lies in the range $2 < q_* < 3$, which reduces the maximum current to $6.3\,\text{MA} < I_{\max} < 9.4\,\text{MA}$. It should be noted that $I_{\max}$ can be increased substantially from these values if the tokamak has a non-circular cross section. This point is discussed in the next subsection.

The conclusion from the analysis is that for a given geometry the toroidal current cannot be made too large with respect to the toroidal field or else a current-driven instability will be excited. Stated differently, a sufficiently large toroidal magnetic field is required to stiffen the tokamak against the $m = 1$ mode, which is always unstable in a pure $Z$-pinch.

### The $\beta q_*^2/\varepsilon \to \pi^2/16$ ballooning-kink instability

Perhaps the most important application of the general pressure balance matching condition involves the calculation of the $\beta$ limit against ballooning-kink modes. Within the context of the surface current model, the $\beta$ limit due to such instabilities can be calculated by assuming that the plasma is operating at some point along the maximum equilibrium $\beta$ curve defined by $\beta q_*^2/\varepsilon = \pi^2/16$. The equilibrium $\beta$ limit is actually a limit on $\beta/I^2 \sim \beta q_*^2/\varepsilon$. Application of the pressure balance matching condition determines the highest current or equivalently the lowest $q_*$ that is stable. Once this critical value of $q_* = q_{*\text{crit}}$ is determined the maximum stable $\beta$ is simply determined from the relation $(\beta/\varepsilon)_{\max} = \pi^2/16\, q_{*\text{crit}}^2$.

Although the calculation of the $\beta$ limit is straightforward, the details are rather tedious. Consequently, to maintain cohesiveness only the main points are highlighted. In this connection there are two important simplifications in the analysis that should be noted. First, although the determinant of $\overset{\leftrightarrow}{\mathbf{W}}$ has been evaluated numerically maintaining a large number of poloidal harmonics, it suffices for present purposes to truncate the matrix keeping only three terms: $m = 1, 2, 3$. The results obtained are quite accurate compared to the numerical results. The second simplification is a result of the fact that at the equilibrium $\beta$ limit, the poloidal magnetic field reduces to

$$\frac{\hat{B}_\theta}{\varepsilon B_0} = \frac{\pi}{2q_*}|\cos(\theta/2)|. \tag{13.139}$$

This simple form allows an analytic evaluation of the matrix element $G_{lm}$. A short calculation yields

$$G_{lm} = -n\delta_{l-m} - (-1)^{l-m}\frac{1}{q_*}\frac{l}{4(l-m)^2 - 1}. \tag{13.140}$$

The marginal stability criterion is obtained by substituting these simplifications into the $3 \times 3$ truncated form of the pressure balance matching condition, which can be written as

$$\begin{bmatrix} W_{11} & W_{12} & W_{13} \\ W_{12} & W_{22} & W_{23} \\ W_{13} & W_{23} & W_{33} \end{bmatrix} \cdot \begin{bmatrix} \xi_1 \\ \xi_2 \\ \xi_3 \end{bmatrix} = 0. \tag{13.141}$$

Here use has been made of the symmetry relation $W_{mp} = W_{pm}$. The matrix elements are determined directly from Eq. (13.132). A short calculation yields the following admittedly unintuitive and uninspiring expressions for the $W_{mp}$

$$W_{11}/n^2 = 2 - 2\left(\frac{1}{nq_*}\right) - \left[\frac{\pi^2}{8} - M_{11}\right]\left(\frac{1}{n^2 q_*^2}\right),$$

$$W_{22}/n^2 = 1 - 2\left(\frac{1}{nq_*}\right) - \left[\frac{\pi^2}{8} - M_{22}\right]\left(\frac{1}{n^2 q_*^2}\right),$$

$$W_{33}/n^2 = \frac{2}{3} - 2\left(\frac{1}{nq_*}\right) - \left[\frac{\pi^2}{8} - M_{33}\right]\left(\frac{1}{n^2 q_*^2}\right),$$

$$W_{12}/n^2 = -\frac{2}{3}\left(\frac{1}{nq_*}\right) - \left[\frac{3\pi^2}{32} + M_{12}\right]\left(\frac{1}{n^2 q_*^2}\right), \tag{13.142}$$

$$W_{23}/n^2 = -\frac{2}{3}\left(\frac{1}{nq_*}\right) - \left[\frac{3\pi^2}{32} + M_{23}\right]\left(\frac{1}{n^2 q_*^2}\right),$$

$$W_{13}/n^2 = \frac{2}{15}\left(\frac{1}{nq_*}\right) + M_{13}\left(\frac{1}{n^2 q_*^2}\right),$$

where the numerical coefficients $M_{mp}$ are given by

$$M_{mp} = \sum_{-\infty}^{\infty} \frac{|l|}{[4(l-m)^2 - 1][4(l-p)^2 - 1]}. \tag{13.143}$$

The main point to note about the $W_{mp}$ elements is their form. By focusing on equilibria that are constrained to operate along the equilibrium $\beta$ limit curve, each of the $W_{mp}$ can be written as a polynomial function of only the single quantity $1/nq_*$: $W_{mp} = A_{mp} + B_{mp}/nq_* + C_{mp}/n^2 q_*^2$. Therefore once the coefficients $M_{mp}$ are evaluated it is a simple matter to calculate the determinant of the truncated matrix $\overset{\leftrightarrow}{\mathbf{W}}$, which has the form of a sixth order polynomial in $1/nq_*$. The $M_{mp}$ are easily evaluated numerically, although after a rather lengthy calculation they can also be evaluated analytically: $M_{11} = 5/4, M_{22} = 89/36$, $M_{33} = 1111/300, M_{12} = -11/12, M_{23} = -277/180$, and $M_{13} = 1/180$.

The roots of the $1/nq_*$ polynomial are easily found numerically. The most restrictive root in terms of stability is found to be $nq_{*\text{crit}} = 1.67$. A more accurate numerical calculation using 20 poloidal harmonics yields a slightly higher value, $nq_{*\text{crit}} = 1.71$, and this value will be used hereafter. Also, since $q_* \geq q_{*\text{crit}}$ for stability it is clear that $n = 1$ is the most unstable mode.

The final result is that the lowest $q_*$ and corresponding highest $\beta$ that are stable against the ballooning-kink mode in a circular cross section tokamak are given by

$$q_* \geq q_{*\text{crit}} = 1.71,$$
$$\beta \leq \beta_{\text{crit}} = \left(\frac{\pi}{4q_{*\text{crit}}}\right)^2 \varepsilon = 0.21\frac{a}{R_0}. \tag{13.144}$$

Equations (13.144) can be interpreted as follows. First, observe that for the simple reactor design characterized by an aspect ratio $R_0/a = 2.5$, the value of $\beta_{\text{crit}}$ is 0.084, which is in the desired range. However, as is shown shortly, the surface current model predictions are somewhat optimistic when compared to numerical calculations using realistic diffuse profiles. Next, note that the scaling $\beta_{\text{crit}} \sim a/R_0$ suggests that a tight aspect ratio tokamak is desirable, at least from the MHD point of view, although as discussed in the context of the spherical torus there are limits as to how tight the aspect ratio can be made. Lastly, note that the lowest value of stable $q_*$ has been raised from $q_* = 1$ for $\beta/\varepsilon \to 0$ to $q_* = 1.71$ for $\beta/\varepsilon = 1$. The maximum allowable plasma current has decreased by nearly a factor of 2.

The overall picture of stability can be summarized by plotting $\beta$ vs. $I$ as illustrated in Fig. 13.26, which shows the curves for the equilibrium $\beta$ limit and the full stability curve as obtained numerically. The analytic stability limits correspond to the end values on the stability curve. Observe that there is a stable operating regime for sufficiently low $\beta$ and sufficiently low $I$.

The last point to consider involves the shape of the eigenfunction. The task here is to verify the assertion that the most unstable mode has the form of a ballooning-kink perturbation of the plasma surface. This is easily demonstrated by back-solving Eq. (13.141) to determine the relative amplitudes of the $\xi_m$. A short calculation using parameters corresponding to the

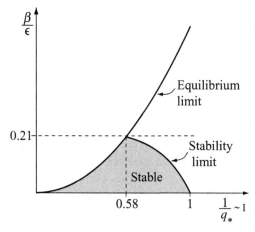

Figure 13.26 Overall stability diagram for the surface current model of a circular tokamak showing the equilibrium limit and the numerically computed stability limit (D'Ippolito *et al.* (1976). *Plasma Physics and Controlled Nuclear Fusion Research*, **1**, 523).

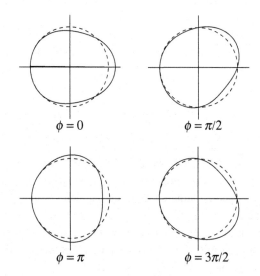

Figure 13.27 Perturbed plasma cross section at four different toroidal angles.

point of marginal stability shows that the perturbation is dominated by the $m = 2$ poloidal harmonic and that the relative amplitudes of the side bands are given by

$$\xi_1/\xi_2 \approx \xi_3/\xi_2 \approx 0.43. \tag{13.145}$$

The shape of the eigenfunction is illustrated in Fig. 13.27. The figure shows the overlaying of the unperturbed surface $r = a$, on the perturbed surface $r = a + \xi_1 \cos(\theta - \phi) + \xi_2 \cos(2\theta - \phi) + \xi_3 \cos(3\theta - \phi)$ at four different toroidal angles, assuming $\xi_2/a = 0.09$. Observe that the surface distortion rotates with the toroidal angle demonstrating the kink nature of the instability. In addition, the perturbation is always much larger on the outside of the torus than the inside: $r(\theta = 0) - a = 1.86 \, \xi_2 \cos\phi$, while $r(\theta = \pi) - a = 0.14 \, \xi_2 \cos\phi$. This demonstrates the tendency of the perturbation to balloon in the region of unfavorable curvature.

The analysis of the MHD stability of a large aspect ratio, circular cross section tokamak using the surface current model is now complete. For practical experimental applications it is of interest to compare the predicted $\beta/\varepsilon$ limit with that obtained from more detailed numerical results using realistic diffuse profiles.

### The Troyon $\beta$ limit

Many numerical studies have been carried out to determine the overall $\beta$ limit against ideal MHD instabilities. One of the most widely used results is due to Troyon and coworkers and is usually referred to as the "Troyon limit." These studies require stability against all ideal MHD modes: external kink modes and ballooning-kink modes, internal modes, localized modes, etc. although usually the external ballooning-kink mode is the most restrictive. There is no perfectly conducting wall to provide additional stability.

The aim of the stability analysis is to determine the highest possible stable $\beta$ for realistic geometries and profiles. Towards this goal Troyon and coworkers carry out an optimization procedure by varying the shape of the plasma cross section as well as the pressure and current profiles over a reasonably wide range of realistic situations.

Numerical studies show that elongated, outward pointing triangular cross sections are most favorable for stability. Broad pressure profiles and an internally flat safety factor are also desirable. After a rather extensive set of numerical studies Troyon and coworkers obtained a remarkably simple empirical relation for the optimized $\beta$ limit:

$$\beta \le \beta_{\mathrm{crit}} \equiv \beta_{\mathrm{N}} \frac{I}{a \, B_0}. \tag{13.146}$$

Here $\beta_{\mathrm{N}}$ is a numerically determined coefficient that has the value $\beta_{\mathrm{N}} = 0.028$ when $I$ is measured in megaamperes. Often in the literature, $\beta$ is measured as a percentage in which case $\beta_{\mathrm{N}} = 2.8$. However, for consistency the decimal notation is used here and in following sections.

Note that while the highest values of $I/a B_0$ occur for elongated, triangular cross sections, the $\beta$ limit given by Eq. (13.146) is valid for any given cross section as long as it is held fixed during the optimization over the current and pressure profiles. Therefore, in order to compare with the surface current model, one can rewrite Eq. (13.146) in the following more convenient form corresponding to a circular cross section tokamak:

$$\beta \le \beta_{\mathrm{crit}} \equiv 0.14 \frac{\varepsilon}{q_*}. \tag{13.147}$$

This expression is illustrated in Fig. 13.28, where it is superimposed on the surface current stability diagram. Again, the surface current model is shown to be somewhat optimistic. In terms of determining the highest overall stable $\beta$ the Troyon limit is not completely satisfactory in that the minimum value of stable $q_*$ is not specified. In general, the minimum

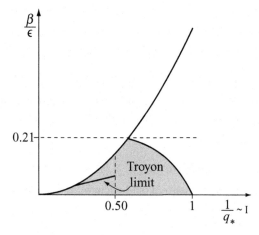

Figure 13.28 The Troyon limit superimposed on the surface current stability diagram.

$q_*$ is a weak function of geometry and profiles and for a circular cross section is found to be approximately $q_* \approx 2$. Using this admittedly slightly ambiguous value one finds that the maximum stable $\beta$ in a circular cross section tokamak is given by

$$\beta \leq \beta_{max} \equiv 0.07 \frac{a}{R_0}. \tag{13.148}$$

Observe that the Troyon limit and the surface current model yield the same scaling of $\beta$ with aspect ratio: $\beta \sim a/R_0$. However, the numerical coefficient is substantially reduced, by about a factor of 3. The Troyon limit thus predicts a maximum $\beta$ of about 0.028 for $R_0/a = 2.5$. This value is lower than would be desired in a reactor and has motivated fusion researchers to find new ideas to raise the $\beta$ limit in a tokamak. These ideas are discussed in the following subsections.

### 13.7.4 The non-circular cross section tokamak

One idea to improve the $\beta$ limit in a tokamak is to allow the plasma to have a non-circular cross section. Both theoretical and experimental studies have shown that stability is optimized for a cross section possessing a combination of elongation and outward pointing triangularity. Virtually all modern tokamaks make use of this improvement. One prominent example is JET located near Oxford in the UK. A drawing of JET exhibiting the non-circularity is shown in Fig. 13.29.

In this subsection the effects of non-circularity on MHD stability are discussed. There are two important topics. The first involves the $\beta$ limit due to the $n = 1$ ballooning-kink mode and is addressed by means of the surface current model. It is shown that in the regime of interest the critical $\beta$ increases linearly with elongation, indeed a very encouraging result. There is, however, a practical limit to the maximum achievable elongation and this is the second topic discussed. It is shown that $n = 0$ axisymmetric perturbations (i.e., a

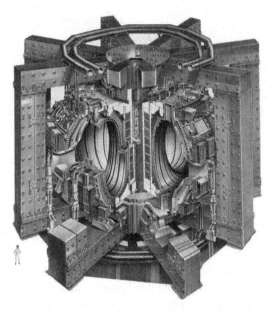

Figure 13.29 Drawing of JET showing a non-circular cross section with elongation and outward pointing triangularity (courtesy of EFDA-JET).

uniform vertical displacement) drive instabilities in elongated tokamaks. These instabilities have the form of resistive wall modes and practical feedback systems are only possible for elongations of less than a factor of about 1.8. The $n = 0$ mode is investigated by means of a simple wire model for the plasma.

The end result is that in experiments elongated tokamaks have achieved reasonably high values of stable $\beta$, close to that required in a reactor. The one remaining limitation is that the pressure and current profiles, while stable, produce only a moderate fraction of bootstrap current, not as high as one would like in a reactor. Other ideas have been suggested to improve upon this situation, but for the moment attention is focused solely on the effects of elongation on MHD stability.

### *The $n = 0$ axisymmetric instability*

The $n = 0$ instability has been well known in tokamak experiments for many years and feedback stabilization of this mode is now almost a routine task. The essential physics of the mode does not require toroidicity. Nor does it require a finite plasma pressure. It occurs in straight, low-$\beta$ tokamaks with elongated cross sections. Remarkably the key features of the mode are easily established by modeling the plasma as a simple wire carrying a current $I$. The problem is thereby reduced to one of pure magnetostatics.

The simplest explanation for the $n = 0$ mode can be ascertained by examining Fig. 13.30. Assume the plasma wire is held in equilibrium by two equally spaced wires as shown in the diagram. To elongate the plasma vertically the currents must flow in the same direction as

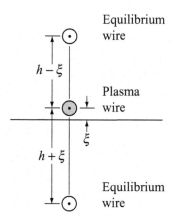

Figure 13.30 Simple wire model showing the $n = 0$ vertical instability.

the plasma current. The resulting attractive force due to each equilibrium wire "pulls" on the top and bottom of the plasma causing it to become elongated.

Now assume the plasma wire undergoes a small vertical displacement $\xi$ towards the upper wire. Recall that the force of the upper wire on the plasma wire is inversely proportional to the distance separating them. Thus, the upward pulling force is increased since the wires are slightly closer to each other. Similarly the pulling force of the lower wire is decreased since it is now slightly further away from the plasma. The conclusion is that there is a net force acting on the wire pointing in the upward direction. The direction of this force is to move the plasma wire even further away from its equilibrium position. This clearly corresponds to an unstable situation.

The behavior of the instability can be quantified by means of a slightly more realistic wire model of the plasma and the equilibrium coils.

*A more realistic n = 0 wire model*

Consider the $n = 0$ wire model illustrated in Fig. 13.31. Here, the plasma is still modeled by a single wire carrying a current $I$. It is held in equilibrium by four equally spaced "equilibrium" wires, each separated by a distance $c$ from the plasma wire. To model the actual situation of radial pressure balance each equilibrium wire carries a current flowing in the opposite direction to the plasma current. The resulting repelling forces are the "pushing" forces required for pressure balance. Elongation is achieved by assuming that $I_y < I_x$. The vertical wires push less than the horizontal wires, thereby producing an elongated plasma.

A key point in the analysis is to determine the specific relationship between $I_y$ and $I_x$ that produces a given desired elongation. This relationship is determined as follows. Assume the horizontal half-width of the equivalent plasma is equal to $s_x/2 = a$. Similarly the half-height of the equivalent plasma is assumed to be $s_y/2 = \kappa a$, where $\kappa = s_y/s_x$ is defined as the elongation. The relation between $I_y$ and $I_x$ is determined by requiring that both the horizontal and vertical boundary points lie on the same flux surface. In terms of the vector

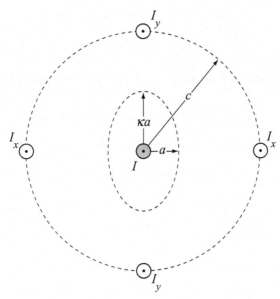

Figure 13.31 Wire model for the analysis of the $n = 0$ vertical instability.

potential this corresponds to $A_z(\kappa a, 0) = A_z(0, a)$. To evaluate the condition one makes use of the well-known result from magnetostatics that relates the vector potential (i.e., the flux) to the current flowing in a wire: $A_{z,j} = (\mu_0 I_j / 2\pi) \ln r_j$, where $r_j$ is radial distance measured from the center of the $j$th wire. A short calculation, summing over the plasma and equilibrium wires, leads to the desired relation between $I_y$ and $I_x$:

$$I_y \ln \left( \frac{c^2 - \kappa^2 a^2}{c^2 + a^2} \right) + I_x \ln \left( \frac{c^2 + \kappa^2 a^2}{c^2 - a^2} \right) - I \ln \kappa = 0. \tag{13.149}$$

This relation can be simplified by taking the interesting limit where the equilibrium wires are far from the plasma: $c \gg a$. Equation (13.149) reduces to

$$I_x - I_y = \frac{c^2}{a^2} \frac{\ln \kappa}{1 + \kappa^2} I. \tag{13.150}$$

The next step is to carry out the $n = 0$ stability analysis. As before assume the plasma undergoes a small positive vertical displacement $\xi$ keeping all wire currents, including the plasma current, fixed. The stability analysis requires a calculation of the net perturbed vertical force acting on the plasma wire due to all the equilibrium wires. If this force is positive, it reinforces the displacement and the plasma is unstable. If the force is negative, it returns the plasma to its equilibrium position, corresponding to stability. One now makes use of the well-known result from magnetostatics that the force acting on a current $I_j$ due to another current $I_i$ is given by $\mathbf{F}_{ij} = -(\mu_0 L I_i I_j / 2\pi r_{ij}) \mathbf{e}_{ij}$, where $L$ is the length of the wire and $\mathbf{e}_{ij}$ is a unit vector pointing along the line connecting wire $i$ to wire $j$. Note that for opposing current directions $\mathbf{F}_{ij}$ corresponds to a repelling force. The resulting vertical force acting on the plasma wire is easily calculated and is given by

$$F_y = \frac{\mu_0 L I}{2\pi} \left( -\frac{I_y}{c - \xi} + \frac{I_y}{c + \xi} + 2\frac{\xi I_x}{(c^2 + \xi^2)^{1/2}} \right). \tag{13.151}$$

Linearizing leads to the following expression for the perturbed vertical force:

$$\delta F_y = \frac{\mu_0 L I^2}{\pi} \left( \frac{I_x - I_y}{I} \right) \frac{\xi}{c^2}. \tag{13.152}$$

The condition for stability, $\delta F_y < 0$, simplifies to

$$\frac{I_x - I_y}{I} < 0 \qquad \text{(for stability).} \tag{13.153}$$

The final result is obtained by substituting the equilibrium relation given by Eq. (13.150). This yields a stability condition expressed in terms of the elongation:

$$\frac{\ln \kappa}{1 + \kappa^2} < 0 \qquad \text{(for stability).} \tag{13.154}$$

As anticipated, any vertically elongated plasma, $\kappa > 1$, is unstable. Flattened plasmas, $\kappa < 1$, are stable against a vertical displacement. However, by symmetry, such plasmas would be unstable against a horizontal displacement.

How then can one produce an elongated plasma? The answer is closely tied to the critical assumption made in the analysis that all currents remain fixed as the plasma is perturbed from its equilibrium position. This constraint is altered if one inserts a perfectly conducting wall between the plasma current and the equilibrium currents. As the plasma is displaced, conservation of flux induces eddy currents in the wall, which are in the direction to produce stability. The quantitative effect of a perfectly conducting wall on $n = 0$ stability is the next topic for discussion.

### *Stabilization of the $n = 0$ mode by a conducting wall*

A simple way to analyze the effects of a perfectly conducting wall is illustrated in Fig. 13.32. Here, a circular, perfectly conducting wall of radius $\kappa b$ is inserted between the plasma and the equilibrium wires. The factor $\kappa$ is a mathematical convenience chosen so that the limit of the wall moving onto the plasma surface corresponds to $b/a \to 1$. The goal of the analysis is to derive the marginal stability boundary determining the maximum achievable $\kappa$ as a function of $b/a$.

In formulating the problem, a useful way to think about the conducting wall is as follows. Assume that initially the wall has a large but finite conductivity. After a long enough period of time the fields due to the equilibrium wires completely diffuse through the wall, achieving the same values they had without the wall. At this point in time assume the wall becomes superconducting, after which the plasma wire is shifted vertically upwards by an amount $\xi$. The fields due to the equilibrium wires remain unchanged since they are locked in at their no-wall values by the superconductivity. In other words they produce the same destabilizing force on the plasma as without the wall. However, the motion of the plasma induces eddy

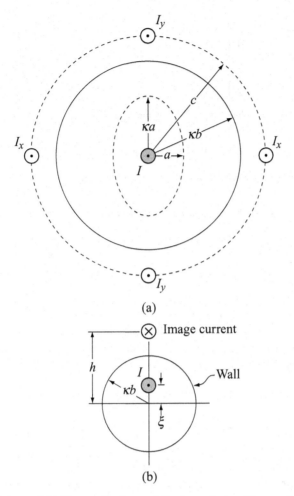

(a)

(b)

Figure 13.32 (a) Wire model with a conducting wall for the analysis of the $n = 0$ vertical instability. (b) Equivalent image current for a vertically shifted plasma.

currents in the wall and these currents produce an additional force on the plasma wire, which is in the stabilizing direction.

The size of the stabilizing force can be easily determined using the method of images. It is well known from the theory of magnetostatics that the effects of the eddy currents can be calculated by replacing the wall with an image current located directly above the wall as shown in Fig. 13.32. The value of the image current must be chosen so that $I' = -I$ and its distance from the origin must be set to $h = \kappa^2 b^2/\xi$. With this choice of parameters the surface $r = \kappa b$ remains a flux surface (i.e., $A_z(x, y) = \text{const.}$ for $x = \kappa b \cos\theta$ and $y = \kappa b \sin\theta$).

It is now straightforward to generalize the stability analysis to include the effect of the image current. Following the procedure associated with Eq. (13.151) one can easily show

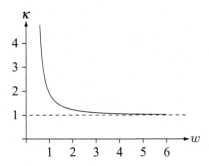

Figure 13.33 Marginally stable elongation $\kappa$ vs. wall radius $w = b/a$ for the wire model.

that the generalized form of the perturbed vertical force on the plasma wire is

$$\delta F_y = \frac{\mu_0 L I^2}{2\pi} \left[ 2 \left( \frac{I_x - I_y}{I} \right) \frac{\xi}{c^2} + \left( \frac{I'}{I} \right) \frac{1}{h} \right]. \tag{13.155}$$

The stability condition thus reduces to

$$2 \left( \frac{I_x - I_y}{I} \right) \frac{\xi}{c^2} + \left( \frac{I'}{I} \right) \frac{1}{h} < 0 \qquad \text{(for stability)}. \tag{13.156}$$

After substituting for $I_x - I_y$, $I'$, and $h$ one obtains the marginal stability boundary relating the maximum achievable $\kappa$ to the wall position $w \equiv b/a$:

$$\frac{2\kappa^2}{1 + \kappa^2} \ln \kappa \le \frac{1}{w^2}. \tag{13.157}$$

This relation is plotted in Fig. 13.33. Observe that the behavior is consistent with intuition in the limit where the wall is far from the plasma: $w \to \infty$. Here, the maximum stable elongation reduces to that of a circle: $\kappa \to 1$. When the wall moves onto the plasma surface corresponding to $w \to 1$, there is strong wall stabilization: $\kappa \approx 1.9$. However, the stabilization is not infinite ($\kappa \to \infty$) as one might expect. The reason is that the wire model stability analysis conserves the flux between the center of the wire ($r = 0$) and the wall ($r = \kappa b$) and not the plasma surface ($x = a \cos \theta$, $y = \kappa a \sin \theta$) and the wall. In other words, for the wire model the wall moving onto the plasma actually corresponds to $w \to 0$ and not $w \to 1$.

In any event for more practical elongations the error, as compared to realistic numerical calculations, is not very large. As an example, an elongation of $\kappa = 1.6$ requires the wall to be closer than $w < 1.2$ for the plasma to be stable. This is slightly pessimistic with respect to numerical studies, which typically conclude that for a plasma with an aspect ratio $R_0/a \approx 3$, an elongation of $\kappa \approx 1.8$ requires $w < 1.2 - 1.4$, depending upon details of the profiles and cross section.

Finally, as previously discussed, a real wall has a finite conductivity. The implication is that the vertical displacement instability converts into an $n = 0$ resistive wall mode, which requires feedback for stabilization. All modern tokamak experiments have a close fitting conducting shell and an $n = 0$ feedback system, which in practice have together been capable of producing elongations up to about $\kappa \approx 1.8$. Therefore, for future reference the value $\kappa = 1.8$ is used as the maximum allowable elongation.

### The $n = 1$ ballooning-kink instability

The $n = 1$ stability analysis of elongated tokamaks has been carried out using the surface current model. The results are qualitatively similar to that of the circular cross section tokamak. There is an equilibrium $\beta/I^2$ limit, a low-$\beta$ kink limit on the maximum $I$, and a ballooning-kink limit that sets the maximum $\beta/\varepsilon$. Quantitatively, all the scaling relations with respect to inverse aspect ratio are preserved. The main difference is that the numerical coefficients defining each of the limits now become functions of elongation. Overall, one of the main conclusions is that the $\beta$ limit associated with the ballooning-kink instability increases linearly with elongation, a very favorable scaling. Ultimately the highest stable $\beta$ is achieved at the maximum elongation set by the $n = 0$ vertical instability.

The mathematical details of the analysis are very similar to the circular case except that the cross section is assumed to be elliptical with an elongation $\kappa$. However, even for this simple cross section the results must be obtained numerically. In addition to the $\cos\theta$ coupling associated with toroidicity, there is $\cos 2\theta$ coupling associated with ellipticity. While the analysis again leads to a matrix formulation of the stability problem, no simple approximations can be made for finite ellipticity that yield the desired scaling relations for $\beta$ and $q_*$ as a function of $\kappa$.

For these reasons it suffices to give simple approximate relations for the critical $\beta_{\text{crit}}$ and $q_{*\text{crit}}$ as determined from the surface current numerical results. As stated, these quantities scale favorably as elongation is increased. The physical explanation for this is somewhat complicated, although within the context of the surface current model there is one effect that is easy to understand. For a given minor radius the unfavorable poloidal field curvature in an elongated tokamak is comparable to that of a circular tokamak around the top and bottom portions of the cross section: $\kappa_r^{(\text{pol})} \approx -B_\theta^2/aB_\phi^2$. However, the sides of the cross section near the midplane are nearly straight vertical lines implying very weak poloidal field curvature in this region: $\kappa_r^{(\text{pol})} \approx 0$. Thus, one sees intuitively that the neutral stability associated with the vertical sides should not affect the overall stability very much although the greater cross sectional area should allow a higher current to flow. This should improve both the equilibrium and ballooning-kink $\beta$ limits which depend on the maximum allowable current.

With this as introduction, consider now the three limits of interest: (1) the equilibrium $\beta$ limit, (2) the low-$\beta$ kink stability limit, and (3) the ballooning-kink stability $\beta$ limit. To begin it is necessary to define the critical plasma parameters in an elongated elliptical tokamak. The plasma $\beta$ has the same definition as for the circular cross section: $\beta \approx \beta_t \equiv 2\mu_0 p/B_0^2$. Similarly the inverse aspect ratio is defined in terms of the horizontal minor radius: $\varepsilon \equiv a/R_0$. The kink safety factor $q_*$ is defined such that it coincides with the actual safety factor $q$ in the limit of low $\beta$. The surface current analysis shows after considerable algebra that the appropriate definition is

$$q_* = \frac{2\pi a^2 \kappa B_0}{\mu_0 R_0 I} G(\kappa),$$
(13.158)

where

$$G(\kappa) = \frac{4\kappa E^2(\alpha)}{\pi^2} \approx \frac{1}{\kappa}\left[1 + \frac{4}{\pi^2}(\kappa^2 - 1)\right]. \tag{13.159}$$

Here, $E(\alpha)$ is the elliptic integral and $\alpha^2 = 1 - 1/\kappa^2$. The approximate formula uses a simple curve fit for the elliptic integral. The key point to note is that over the interesting regime of elongations, $1 < \kappa < 2$, the value of $G(\kappa)$ is very nearly a constant: $G(\kappa) \approx 1$. With this approximation the kink safety factor reduces to

$$q_* = \frac{2\pi a^2 \kappa B_0}{\mu_0 R_0 I}. \tag{13.160}$$

It is identical to the circular definition except for the additional factor of $\kappa$ in the numerator. Hereafter, Eq. (13.160) is used as the definition of $q_*$.

The first result of interest corresponds to the equilibrium $\beta$ limit. As for the circular plasma, the elliptical plasma exhibits an equilibrium $\beta$ limit when $\beta$ becomes sufficiently large that the X-point of the vacuum separatrix moves onto the plasma surface. The surface current analysis shows that to a good approximation the equilibrium $\beta$ limit in the regime $1 < \kappa < 2$ is given by

$$\frac{\beta q_*^2}{\varepsilon} \leq \frac{\pi^2}{16}\kappa. \tag{13.161}$$

To the extent that the same value of $q_*$ can be maintained for the ellipse as for the circle, it follows that the equilibrium $\beta$ limit scales linearly with elongation. This is the first example of the favorable scaling with elongation.

The next limit involves low-$\beta$ kink modes. Recall that in the circular case the low-$\beta$ limit caused the stability matrix $\overset{\leftrightarrow}{\mathbf{W}}$ to become diagonal. Stability was then easily determined by examining the sign of each diagonal matrix element separately. For the elliptical case the matrix does not become diagonal at low $\beta$ because of the $\cos 2\theta$ coupling. The coupled harmonic stability problem has to be solved numerically although the final result, in analogy with the circle, still consists of a limit on the minimum stable value of $q_*$. These results show that in the regime $1 < \kappa < 2$ the kink stability limit is approximately given by

$$q_* \geq \frac{1 + \kappa}{2}. \tag{13.162}$$

Note that $q_{*\text{crit}}$ increases with $\kappa$ implying that at low $\beta$ more "safety" is required as the plasma becomes elongated. This would seem to be an unfavorable scaling, contradicting the intuition about elongation leading to larger allowable currents. However, if one evaluates the current using the definition of $q_*$ one finds

$$I \leq I_{\text{max}} = \frac{2\pi a^2 B_0}{\mu_0 R_0}\frac{2\kappa}{1 + \kappa} \sim \frac{2\kappa}{1 + \kappa}. \tag{13.163}$$

The conclusion is that the maximum $I$ increases with $\kappa$ (e.g., $I$ increases by a factor of 4/3 as $\kappa$ increases from 1 to 2) although the effect is not very large. The explanation for this behavior is related to the fact that at low $\beta$ the perturbations can localize in the bad curvature regions at the top and bottom of the ellipse without paying much of a penalty to the pressure-driven ballooning effect. Stated differently, the pressure-driven ballooning effect is destabilizing when the perturbation is localized on the outside of the plasma. It becomes slightly stabilizing for perturbations localized on the top and bottom of the plasma. However, this stabilization effect is small in the low-$\beta$ limit.

The last limit of interest involves the ballooning-kink mode. Here, as in the circular case, one assumes that the tokamak is operating at some point along the equilibrium $\beta$ limit curve and then (numerically) determines the lowest value of $q_*$ for stability. This value of $q_*$ is then substituted into the expression for the equilibrium $\beta$ limit leading to the maximum value of stable $\beta$.

Interestingly, the numerical results show that as the elongation increases, $q_{*\mathrm{crit}}$ remains virtually unchanged from its circular value; that is, the minimum $q_*$ is independent of $\kappa$ in the high $\beta$ regime: $q_{*\mathrm{crit}} \approx 1.7$. Physically, at higher values of $\beta$ the perturbation cannot easily localize around the top and bottom of the cross section because of the pressure-driven ballooning effect, which causes localization on the outside of the plasma. Therefore, in agreement with the intuitive argument previously presented, substantially larger currents are indeed possible as $\kappa$ increases. This increase is balanced by the factor of $\kappa$ in the definition of $q_*$ leading to the conclusion that $q_{*\mathrm{crit}}$ is independent of $\kappa$.

Using the value of $q_{*\mathrm{crit}} = 1.7$ in the expression for the equilibrium $\beta$ limit leads to the following limit on $\beta$ due to ballooning-kink modes:

$$\beta \le \beta_{\max} = 0.21 \frac{a}{R_0} \kappa. \tag{13.164}$$

As an example, if one now sets $\kappa = 1.8$, corresponding to the maximum elongation allowed by the $n = 0$ mode, then for a tokamak with $R_0/a = 2.5$, $\beta_{\mathrm{crit}}$ becomes 0.15, a value more than sufficient for a reactor. However, as for the circular case the more realistic numerical results corresponding to the Troyon limit are not this favorable, although they do improve as $\kappa$ is increased.

To see the effects of elongation on the Troyon limit recall the following remarkable feature of the numerical results. The Troyon limit

$$\beta \le \beta_{\mathrm{crit}} = \beta_N \frac{I}{a B_0} = 0.028 \frac{I}{a B_0} \tag{13.165}$$

was obtained by optimizing over cross section as well as profiles. The expression given by Eq. (13.165) is thus valid for any elongation in the range studied. Note that there is no explicit dependence on $\kappa$. There is only an implicit improvement with elongation to the extent that as $\kappa$ increases, the maximum $I$ also increases.

The effects of elongation can be made explicit by rewriting Eq. (13.165) in a form more readily comparable with the surface current results:

$$\beta \le \beta_{\mathrm{crit}} = 0.14 \frac{\varepsilon \kappa}{q_*}. \tag{13.166}$$

If it is again assumed that the minimum $q_*$ is independent of elongation for diffuse profiles, then setting $q_* \approx 2$ leads to the Troyon limit on $\beta$ given by

$$\beta \le \beta_{\max} = 0.07 \frac{a}{R_0} \kappa. \tag{13.167}$$

Observe that the Troyon limit and the surface current model predict the same scaling of $\beta_{max}$ with $\varepsilon$ and $\kappa$. Only the numerical coefficient is different. For $R_0/a = 2.5$ and $\kappa = 1.8$ the Troyon limit has the value $\beta_{max} = 0.05$. This value is close to but slightly below the value $\beta \approx 0.08$ derived for the simple reactor design. However, the simple reactor design does not include the effect of elongation and is clearly based on a very highly simplified analysis. There is general consensus in the tokamak community that values of $\beta \sim 0.05$ are probably adequate for a reactor. This is encouraging since such values have been obtained experimentally.

The one related MHD problem remaining for the tokamak is associated with current drive and the bootstrap current. The stable profiles obtained by Troyon and coworkers do not produce as much natural bootstrap current as desired, thereby putting strong requirements on the current-drive system. These requirements can be relaxed by means of another modification to the tokamak, known as "advanced operation." This is the next topic for discussion.

### 13.7.5 The advanced tokamak (AT)

The AT actually refers to a special mode of operation of a tokamak experiment. The goal is to create pressure and current profiles that produce a high fraction of bootstrap current, thereby reducing the requirements on the current drive system.

As will be shown in Chapter 14, the most effective profiles for generating a high bootstrap fraction tend to have a hollow current profile, which leads to a qualitative change in the shape of the $q$ profile. Rather than being a monotonically increasing function of radius, the $q$ profile first decreases and then increases. The safety factor has an off-axis minimum. A typical example from the DIII-D tokamak at General Atomics in San Diego is shown in Fig. 13.34. These profiles are experimentally generated by a combination of current programming and profile control utilizing localized auxiliary heating. They are often referred to as "reversed shear profiles" because the shear, defined as $rq'/q$, reverses sign at the $q$ minimum.

Qualitatively, reversed shear profiles help the current-drive problem in two ways. First, since $q_{min}(r_{min}) > 2$ for the optimum profiles, the total current flowing in the tokamak is less than for typical tokamak operation where $q_{min}(0) \approx 1$. Therefore, there is less total current to drive. Second, a hollow current profile closely overlaps with the naturally forming bootstrap current profile. One does not have to drive large amounts of current on-axis, since the reverse shear and natural bootstrap current profiles are both hollow.

These advantages are partially offset by the introduction of a new MHD stability problem. Obtaining the high bootstrap fractions demanded by reactor economics (i.e., $f_B \sim 0.7{-}0.9$) requires both high total $\beta$ and high poloidal $\beta$. However, the lower currents, which are good for current drive, lead to lower MHD stability limits against the $n = 1$ ballooning-kink mode. Recall that from the Troyon limit $\beta \leq \beta_N(I/aB) \sim I$. Typically values of $\beta_N \sim 0.04{-}0.05$ are required to achieve the necessary bootstrap fractions in a reactor. However, this value is nearly twice the value obtained by Troyon: $\beta_N \sim 0.028$. The implication is that a perfectly

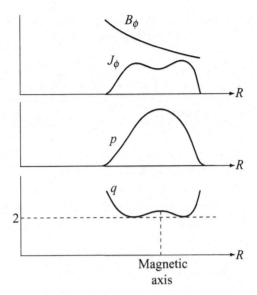

Figure 13.34 Typical midplane profiles for the DIII-D tokamak showing the off-axis minimum in $q(\psi)$ (i.e., reversed shear) and the hollow $J_\phi$ profile (Strait *et al.* (1995). *Physical Review Letters*, **75**, 4421).

conducting wall must be utilized to increase the $\beta$ limit. (Troyon's studies assumed there was no conducting wall surrounding the plasma.)

The purpose of the present subsection is to determine how close the perfectly conducting wall must be to achieve the necessary gains in the $\beta$ limit. The answer is easily obtained from the surface current model and the results are in good agreement with those obtained from more realistic numerical studies. The results show that the required wall radius is typically $b/a \sim 1.4$, which is similar to that required to stabilize the $n = 0$ vertical instability in elongated tokamaks. Also, keep in mind that when the wall has a finite conductivity, the instability is transformed into a resistive wall mode requiring feedback stabilization. This technique is not as well established as for the $n = 0$ mode and is an active area of current research.

### The effect of a wall on the kink current limit

The effect of a perfectly conducting wall on the $n = 1$ mode in a tokamak can be determined from the surface current model. For simplicity the results are calculated using a circular cross section plasma. The analysis only requires modifying the vacuum contribution to the pressure balance relation to include the effects of the wall. These have already been calculated for the RFP and in the appropriate long wavelength limit all one needs to do is to replace

$$\sum_{l\neq 0} \frac{G_{lm}G_{lp}}{|l|} \quad \rightarrow \quad \sum_{l\neq 0} \frac{G_{lm}G_{lp}}{|l|} \left[ \frac{1 + (a/b)^{2|l|}}{1 - (a/b)^{2|l|}} \right] \tag{13.168}$$

in the tokamak pressure balance relation.

Consider now the low-$\beta$ kink limit. As in the no-wall case the stability matrix becomes diagonal with only the $m = 1$ mode possibly unstable. The modified stability condition for $m = 1$, $n = 1$ becomes

$$
\begin{aligned}
W_{11} &= 1 - \frac{1}{q_*^2} + \left(1 - \frac{1}{q_*}\right)^2 \left(\frac{w^2 + 1}{w^2 - 1}\right) > 0, \\
&= \frac{2w^2}{(w^2 - 1)q_*^2} (q_* - 1)(q_* - 1/w^2) > 0,
\end{aligned}
\tag{13.169}
$$

where $w = b/a > 1$. Observe that one still requires $q_* > 1$ for stability. The conclusion is that a perfectly conducting wall does not have any effect on the value of $I_{max}$ for stability against low-$\beta$ kink modes. Note that the theoretically stable region at low $q_*$ (i.e., $0 < q_* < 1/w^2$) is inaccessible in a tokamak since to enter this parameter regime requires passing through the violently unstable region $1/w^2 < q_* < 1$ as the current rises or equivalently as $q_*$ decreases.

### The effect of a wall on the ballooning-kink $\beta$ limit

The ballooning-kink instability is the most dangerous mode to overcome in AT operation. The critical issue is to determine how close to place the perfectly conducting wall in order to raise the maximum stable $\beta$ a specified amount above the no-wall limit.

The information is derived from the surface current model by again constraining the plasma to operate along the equilibrium limit curve $\beta q_*^2/\varepsilon = \pi^2/16$ (which is unaffected by the wall) and then determining the critical $q_*$ for stability. In contrast to the low-$\beta$ regime, $q_{*crit}$ for stability does depend upon wall position in the high-$\beta$ regime. Once $q_{*crit}$ is determined, the corresponding maximum $\beta$ is found by substituting back into the equilibrium limit relation.

The desired stability relation $\beta_{max}/\varepsilon$ vs. $b/a$ is easily obtained numerically by setting the determinant of the $\overleftrightarrow{W}$ matrix to zero using the wall corrected vacuum contribution. The results are illustrated in Fig. 13.35. Observe that the wall has relatively little effect until $b/a \approx 2$. The reason is that the dominant harmonic in the mode corresponds to $m = 2$ for which the wall effect enters only weakly as $(a/b)^4$.

Using Fig. 13.35 one can now address the important question of how close the wall must be to achieve a high fraction of bootstrap current. To formulate the question, recall that reactor studies have shown that in the context of the Troyon limit, the critical $\beta_N$ must be raised from $\beta_N = 0.028$ to $\beta_N \approx 0.045$ to obtain the necessary high bootstrap fraction. This corresponds to a $\beta_N$ ratio of $0.045/0.028 = 1.6$. Under the assumption that for diffuse plasma models the stabilizing effect of a conducting wall is similar to that for the surface current plasma model, one therefore needs to know the critical wall position corresponding to a gain in the surface current $\beta_{max}$ by this same factor of 1.6. In other words for a given aspect ratio the critical $\beta/\varepsilon$ must increase from $\beta/\varepsilon = 0.21$ to $\beta/\varepsilon = 0.34$. From Fig. 13.35 it follows that the critical wall position is $b/a \approx 1.4$. This value is slightly optimistic in comparison to diffuse profile numerical stability studies. It is comparable in value to the wall position needed for $n = 0$ stability in elongated plasmas.

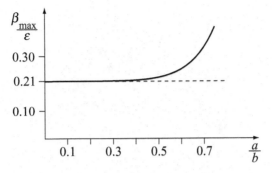

Figure 13.35 Curve of $\beta_{max}/\varepsilon$ vs. $a/b$ showing how the $\beta$ limit increases as the wall is moved closer to the plasma.

To summarize, AT operation using reversed shear profiles can lead to high fractions of bootstrap current. However, the high $\beta$ required for this regime causes the plasma to exceed the no-wall limit. A conducting wall is required typically with $b/a < 1.4$. Since a real wall has finite conductivity, the instability is converted in a resistive wall mode requiring feedback or rotational stabilization. At present the problem of simultaneously achieving high bootstrap fraction and high stable $\beta$ is a major area of tokamak research.

### 13.7.6 The spherical tokamak (ST)

The ST is the last tokamak configuration to be discussed. The new feature in this configuration is an ultratight aspect ratio, typically on the order of $R_0/a \sim 1.2 - 1.4$. A number of spherical tokamak experiments have been built with the largest two being the Mega-Ampere Spherical Tokamak (MAST) at the Culham Laboratory in the UK and the National Spherical Torus Experiment (NSTX) at the Princeton Plasma Physics Laboratory in the USA. A drawing of NSTX is shown in Fig. 13.36.

One of the main motivations for the ST configuration is to exploit the MHD $\beta$ limit scaling $\beta \sim \varepsilon$. Clearly, as the aspect ratio becomes tighter, $\beta_{crit}$ should increase and, indeed, experiments have demonstrated this behavior. The hope is that higher stable $\beta$s will add to the attractiveness of a fusion reactor by allowing the use of lower toroidal magnetic fields or, alternatively, a smaller, more compact geometry. Either option would lead to a lower cost.

More detailed analysis has shown that there are both pros and cons with respect to the desirability of the ST as a reactor. In fact, as demonstrated in this subsection, when one balances the tradeoffs compared to the standard tokamak it is not clear that an ST would lead to a noticeably more attractive reactor, even if the high-$\beta$ MHD physics predictions are validated. However, there is an alternative application of the ST for which it may well be better suited – a volume neutron source.

The basis for these statements is as follows. The gains in $\beta$ and compactness must be balanced against several problems that are more serious than for the standard tokamak.

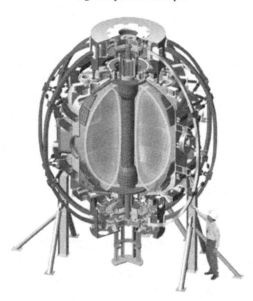

Figure 13.36 Drawing of the NSTX experiment at the Princeton Plasma Physics Laboratory (courtesy of M. Peng).

First, recall that a fusion reactor requires high plasma pressure which is related to, but is not the same, as $\beta$. Specifically, the constraint of a limiting toroidal $B_{max}$ on the inner leg of the toroidal field magnet leads to a much larger reduction in $B_\phi$ at the plasma center where $\beta$ is defined because of the stronger $1/R$ effect at tight aspect ratio (i.e., $R_0/a \to 1$). The end result is that the maximum achievable pressures are typically less in an ST than in a standard tokamak. This point is discussed in more detail shortly.

Second, to achieve a very tight aspect ratio the blanket and almost all the shield must be removed from the inboard side of the plasma. One consequence is that the central leg of the toroidal field magnet must be copper since there is no longer adequate shielding for a superconducting magnet. A copper central leg dissipates a substantial amount of ohmic power that must be accounted for in the overall power balance. Detailed reactor studies have shown that to achieve a favorable power balance $B_{max}$ is constrained to be less than 7.5 T, a considerably lower value than in a standard tokamak using superconducting magnets. This constraint also leads to a lower maximum achievable pressure in an ST.

Third, to achieve a safety factor in a tight aspect ratio device similar to that in a standard tokamak requires a much larger toroidal current. Therefore, the issue of current drive in an ST is more difficult and in fact bootstrap fractions well above 0.9 are required. As in the AT this causes the MHD no-wall $\beta$ limit to be violated, in spite of the higher $\beta$s achievable because of tight aspect ratio. The ST must then also be able to stabilize the resulting resistive wall mode.

The conclusion from this discussion is that an ST reactor does not provide the obvious net gains originally anticipated over a standard tokamak in terms of reactor attractiveness. There is, however, another fusion application for which the ST may be more desirable – a volume neutron source. A source of 14 MeV neutrons would play an important role in the testing and development of new materials for a fusion reactor. Durability and environmental impact (particularly with respect to levels of radioactivity) are clearly important issues when considering how attractive a material is for use in a fusion reactor. Currently there is no such source of 14 MeV neutrons. A key feature of such a neutron source is that its main goal is the production of a high-intensity neutron flux in a relatively small volume to test small samples of fusion materials. It does not need to achieve a favorable power balance to be successful. Once the power balance constraint is relaxed, the ST, because of its compact size and use of copper magnets, may be well suited for the role of a relatively low cost 14 MeV neutron source.

With this as an introduction, the remainder of the subsection is focused on the MHD behavior of the ST. Of particular interest is a determination of the $\beta$ limit in an ST to see whether the favorable aspect ratio scaling persists as $\varepsilon \to 1$. Also presented is a comparison of the $\beta$–pressure relationship between an ST and a standard tokamak including the effect of a magnetic field limit on the inner leg of the toroidal field magnet. Although the analysis could, in principle, be carried out using the surface current model, no such results have yet been obtained in the parameter range of interest. Instead, the discussion is based on some numerical studies similar in flavor to Troyon's early studies, but generalized to include the very tight aspect ratio of the spherical torus.

### MHD $\beta$ limit in a spherical torus

The original studies of Troyon and coworkers involved a numerical optimization over profiles and plasma shape to determine the maximum stable $\beta$ against all MHD modes with the $n = 1$ external ballooning-kink instability usually setting the strictest limit. These studies covered a reasonably wide range of configurations, although not the tight aspect ratio ST, since the ST was not yet under consideration at the time.

The construction of MAST and NSTX, however, has motivated additional numerical studies to extend the results of Troyon and coworkers to the regime of very tight aspect ratio. One set of studies has been carried out at the Princeton Plasma Physics Laboratory (Menard and coworkers). With the introduction of several improved definitions of the critical plasma parameters, the Princeton group has shown that the simple dependence of $\beta_{crit}$ on aspect ratio ($\beta_{crit} \sim \varepsilon$) persists even for very tight aspect ratios. They have also taken into account the $q_*$ dependence associated with kink modes. By combining these results one can determine the optimum $q_*$ and corresponding $\beta_{max}$ as a function of aspect ratio and elongation. The details of this analysis are as follows.

The first step is to introduce the improved definitions of the critical plasma parameters, in this case $\beta$ and $q_*$. The existing definition of $\beta$ is replaced by

$$\beta \equiv \frac{2\mu_0 \langle p \rangle}{B_0^2} \to \frac{2\mu_0 \langle p \rangle}{\langle B^2 \rangle} \tag{13.170}$$

as first suggested by Troyon and coworkers. Here, $\langle \ \rangle$ denotes the volume averaged value. Observe that in the new definition the vacuum magnetic energy on-axis, $B_0^2$, is replaced with the volume average of the total magnetic energy, $\langle B^2 \rangle = \langle B_\phi^2 + B_p^2 \rangle$. In the large aspect ratio limit $\varepsilon \to 0$ these definitions coincide. Next, a new definition of the kink safety factor is introduced with a slightly different dependence on elongation:

$$q_* \equiv \frac{2\pi B_0 a^2}{\mu_0 R_0 I} \kappa \to \frac{2\pi B_0 a^2}{\mu_0 R_0 I} \left( \frac{1+\kappa^2}{2} \right). \tag{13.171}$$

These definitions coincide for $\kappa = 1$ and differ only by the ratio $5/4$ for $\kappa = 2$.

Using these definitions the Princeton group carried out an extensive numerical study to determine the maximum stable $\beta$ over a wide range of cross sections, profiles, safety factors, and aspect ratios, including the tight aspect ratios associated with the ST configuration. The data from their numerical results can be summarized (using the new definitions) by a simple analytic fit given by

$$\beta \le \beta_{\text{crit}} \equiv \langle \beta_N \rangle \frac{I}{a B_0} = 5 \langle \beta_N \rangle \left( \frac{1+\kappa^2}{2} \right) \frac{\varepsilon}{q_*}, \tag{13.172}$$

where $\langle \beta_N \rangle$ is a new normalizing coefficient which remarkably turns out to be independent of both aspect ratio and elongation. It is a function only of $q_*$.

Observe that the linear scaling of $\beta_{\text{crit}}$ with $\varepsilon$ persists even into the regime of the spherical torus. Also, since $\langle \beta_N \rangle$ is independent of $\kappa$, the advantages of elongation also persist for tight aspect ratio: $\beta_{\text{crit}} \sim 1 + \kappa^2$ increases with elongation. With respect to the $q_*$ dependence of $\langle \beta_N \rangle$ set by the kink instability, it is found numerically that for $q_* > 2$ the value of $\langle \beta_N \rangle$ is approximately a constant: $\langle \beta_N \rangle \approx 0.03$. For lower $q_*$ the value of $\langle \beta_N \rangle$ decreases, eventually reaching zero when $q_* \approx 1$. The $q_*$ dependence can be modeled analytically by the simple expression

$$\langle \beta_N \rangle \approx 0.03 \frac{(q_* - 1)}{[(3/4)^4 + (q_* - 1)^4]^{1/4}}. \tag{13.173}$$

The overall $q_*$ dependence of $\beta_{\text{crit}}$ is determined by the ratio $\langle \beta_N \rangle / q_*$. It can easily be shown that this function has a maximum at $q_* = 1 + (3/4)^{4/5} \approx 1.8$ and it is at this value that $\beta_{\text{crit}}$ reaches its maximum value $\beta_{\text{max}}$. Specifically, the $\beta$ limit in either a standard tokamak or a spherical torus is given by

$$\beta \le \beta_{\text{max}} \equiv 0.072 \left( \frac{1+\kappa^2}{2} \right) \varepsilon. \tag{13.174}$$

A spherical torus with an elongation $\kappa = 2$ and an inverse aspect ratio $\varepsilon = 1/1.25 = 0.8$ has an impressive maximum $\beta$ of $\beta_{\text{max}} = 0.14$, thereby validating the original ST motivation to increase the value of stable $\beta$.

### Relation between $\beta$ and pressure in tokamaks

The analysis just presented shows that the definition of $\beta$ using $\langle B^2 \rangle$ is convenient theoretically since the inverse aspect ratio scaling is simple, even for very tight aspect ratios: $\beta \sim \varepsilon$ as $\varepsilon \to 1$. A more convenient experimental definition replaces $\langle B^2 \rangle$ with $B_0^2$, a much simpler quantity to measure. Using the $B_0^2$ definition leads to a more complicated $\varepsilon$ dependence

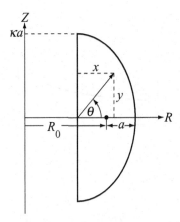

Figure 13.37 Non-circular half-ellipse model for the ST plasma.

of $\beta_{max}$, characterized by a $\beta$ limit that is substantially higher than with the $\langle B^2 \rangle$ definition. One can then ask what is the best definition of $\beta$?

The answer to this question can be found by returning to the requirements of a fusion reactor. The critical point to recall is that the fusion power density is approximately proportional to the square of the plasma pressure:

$$S_f = \frac{E_f}{4} n^2 \langle \sigma v \rangle = \frac{E_f}{16} \frac{\langle \sigma v \rangle}{T^2} p^2 \approx \frac{E_f}{16} \left[ \frac{\langle \sigma v(T_f) \rangle}{T_f^2} \right] p^2, \tag{13.175}$$

where $T_f \approx 15$ keV. Therefore, an economically attractive fusion reactor is one that stably confines a high value of plasma pressure. On this basis the best definition of $\beta$ is one that gives an accurate measure of the pressure. Since $p \sim \beta B^2$ with $B$ limited by $B \leq B_{max}$ on the inside of the toroidal field coil, one can introduce the definition

$$\beta_E \equiv \frac{2\mu_0 \langle p \rangle}{B_{max}^2}. \tag{13.176}$$

Here, $\beta_E$ is known as the "engineering-beta" and represents the "best" definition in the context of a fusion reactor.

The task now is to relate $\langle B^2 \rangle$ to $B_{max}^2$ and to calculate and compare the highest stable pressures achievable in an ST with those of a standard tokamak. This task is carried out in two steps. First, $\langle B^2 \rangle$ is related to $B_0^2$. Second, $B_0^2$ is related to $B_{max}^2$.

In general, the relation between $\langle B^2 \rangle$ and $B_0^2$ requires detailed numerical calculations. However, a reasonably accurate analytic approximation can be obtained by making several simplifying assumptions. Consider first the $\langle B_\phi^2 \rangle$ contribution to $\langle B^2 \rangle$. Assume that in the plasma $B_\phi$ is equal to its vacuum value: $B_\phi \approx B_0 (R_0/R)$. Neglecting the plasma diamagnetic contribution slightly overestimates $\langle B_\phi^2 \rangle$, but this is not a major effect. More important is the fact that typical plasma cross sections are elongated with outward pointing triangularity. The shaping is important because the higher fields on the inside of the torus are more heavily weighted than for a circle because of the relatively larger inner volume. A simple model for the cross section is a half-ellipse as illustrated in Fig. 13.37. Under these

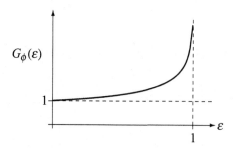

Figure 13.38 The function $G_\phi(\varepsilon) \equiv \langle B_\phi^2 \rangle / B_0^2$ vs. $\varepsilon$.

assumptions the relation between $\langle B_\phi^2 \rangle$ and $B_0^2$ is given by

$$\langle B_\phi^2 \rangle = B_0^2 \int (R_0/R)^2 \, d\mathbf{r} \Big/ \int d\mathbf{r}. \tag{13.177}$$

These integrals can be evaluated in a straightforward way by introducing new coordinates $\rho, \theta$ as follows: $R = R_0 - a + 2a\rho \cos\theta$ and $Z = \kappa a\rho \sin\theta$ with $0 \le \rho \le 1, -\pi/2 \le \theta \le \pi/2$. A short calculation yields a somewhat complicated expression that can be written as

$$\frac{\langle B_\phi^2 \rangle}{B_0^2} = G_\phi(\varepsilon)$$

$$= \begin{cases} \dfrac{1}{\varepsilon\delta} \left[ \dfrac{1 - 2\delta/\pi - (4/\pi)\sqrt{1-\delta^2}\tan^{-1}\sqrt{(1-\delta)/(1+\delta)}}{1 - (1 - 8/3\pi)\varepsilon} \right] & \varepsilon < 1/3, \\[2em] \dfrac{1}{\varepsilon\delta} \left[ \dfrac{1 - 2\delta/\pi + (2/\pi)\sqrt{\delta^2-1}\ln(\delta + \sqrt{\delta^2-1})}{1 - (1 - 8/3\pi)\varepsilon} \right] & \varepsilon > 1/3, \end{cases} \tag{13.178}$$

where $\delta = 2\varepsilon/(1-\varepsilon)$. The function is plotted in Fig. 13.38. Note that $\langle B_\phi^2 \rangle / B_0^2$ is a monotonically increasing function of $\varepsilon$. As $\varepsilon \to 0$, one sees that $\langle B_\phi^2 \rangle / B_0^2 \to 1$ as expected. For a typical tokamak with $\varepsilon = 1/3$ the ratio increases to $\langle B_\phi^2 \rangle / B_0^2 = 1.15$. There is only a small difference in values. For an ST with $\varepsilon = 1/1.25 = 0.8$ the ratio is $\langle B_\phi^2 \rangle / B_0^2 = 1.76$. The difference is now substantial.

Consider next the $\langle B_p^2 \rangle$ contribution to $\langle B^2 \rangle$. This is the smaller contribution, even at tight aspect ratio. Therefore, to simplify the analysis one can use the large aspect ratio approximation for the poloidal field. A simple model in which the current density increases with major radius, as it does for realistic profiles, is given by the following form for the vector potential:

$$A_\phi = C \left( \frac{x^2}{4a^2} + \frac{y^2}{\kappa^2 a^2} - 1 \right) x, \tag{13.179}$$

where $R = R_0 - a + x$ and $Z = y$. The constant $C$ is easily expressible in terms of the plasma current by calculating $B_x, B_y$ and then evaluating the line integral of $\mathbf{B} \cdot d\mathbf{l}$ around the plasma surface. One finds

$$\mu_0 I = 4aC \left( \frac{\kappa^2 + 4/3}{\kappa} \right). \tag{13.180}$$

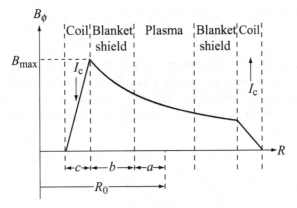

Figure 13.39 Geometry of a tokamak reactor showing the location of $B_{max}$. Note that $B_\phi$ vanishes outside of the toroidal field coil.

It is now a straightforward matter to evaluate the magnetic field and to then calculate $\langle B_p^2 \rangle$. A short calculation yields

$$\frac{\langle B_p^2 \rangle}{B_0^2} = G_p(\varepsilon) = \frac{\pi^2}{32} \frac{(\kappa^2 + 1)^2}{(\kappa^2 + 4/3)} \frac{\varepsilon^2}{q_*^2}. \tag{13.181}$$

Note that for $\kappa = 2$ and $q_* = 1.8$, $G_p = 0.05$ for a standard tokamak with $\varepsilon = 1/3$ and $G_p = 0.29$ for an ST with $\varepsilon = 1/1.25 = 0.8$. The values are finite but small compared to the toroidal field contribution.

Combining these results leads to a simple relation between $\langle B^2 \rangle$ and $B_0^2$ given by

$$\langle B^2 \rangle / B_0^2 \equiv G(\varepsilon) = G_\phi(\varepsilon) + G_p(\varepsilon). \tag{13.182}$$

For the standard values $\kappa = 2$ and $q_* = 1.8$, $G = 1.2$ for $\varepsilon = 1/3$ and $G = 2.05$ for $\varepsilon = 0.8$. Note that for the ST example, the predicted value of maximum stable $\beta$ varies by more than a factor of 2 depending whether $\langle B^2 \rangle$ or $B_0^2$ is used in the definition with the $B_0^2$ choice corresponding to the higher value. Clearly, for an ST the precise definition of $\beta$ is an important issue.

When considering a fusion reactor it is $\beta_E$ that is the appropriate definition. To compute $\beta_E$ one needs to complete the second step in the derivation, which is to determine a relation between $B_0$ and $B_{max}$. This is easily accomplished by re-examining the geometry of the inboard side of a tokamak reactor as illustrated in Fig. 13.39. Since the toroidal magnetic field $B_\phi \sim 1/R$, it immediately follows that the required relation is

$$B_0 = (1 - \varepsilon_b - \varepsilon) B_{max}, \tag{13.183}$$

where $\varepsilon_b = b_i / R_0$ and $b_i$ is the inboard thickness of the blanket-and-shield.

Collecting the results from this section leads to the desired expression for the $\beta_E$ for the optimized case $q_* = 1.8$:

$$(\beta_E)_{max} = 0.036(1 + \kappa^2)\varepsilon(1 - \varepsilon_b - \varepsilon)^2 G(\varepsilon). \tag{13.184}$$

The more practical parameter, the maximum stable pressure, is given by

$$\langle p \rangle = 0.14(1 + \kappa^2)\varepsilon(1 - \varepsilon_b - \varepsilon)^2 G(\varepsilon)B_{max}^2 \qquad \text{atm.} \qquad (13.185)$$

One is finally in a position to compare the maximum achievable pressure (without a conducting wall) in an ST and a standard tokamak. The standard tokamak reactor has $\kappa = 2$, $\varepsilon = 2/5 = 0.4$, and superconducting toroidal field magnets with a maximum allowable field of $B_{max} = 13$ T. An inboard blanket thickness of $b_i = 1.2$ m corresponds to $\varepsilon_b = 1.2/5 = 0.24$. Using these values leads to

$$\langle p \rangle_{max} = 7.7 \text{ atm.} \qquad (13.186)$$

This is about the value necessary for a reactor, but, the corresponding profiles do not produce quite enough bootstrap current.

Consider now a hypothetical ST fusion reactor with $\varepsilon = 0.8$, $\kappa = 2$, and $q_* = 1.8$ as typical values. Also, the thickness of the inboard blanket can be neglected: $\varepsilon_b = 0$. The key point is that the maximum toroidal field is set to $B_{max} \approx 7.5$ T. This is due to the need to keep the ohmic dissipation low enough to achieve a favorable power balance. Current experiments operate well below this limit. Using these parameters yields

$$\langle p \rangle_{max} = 2.6 \text{ atm.} \qquad (13.187)$$

In spite of the higher achievable values of $\beta$ in an ST the maximum pressure is substantially lower than in a standard tokamak due to a combination of the strong $1/R$ effect at tight aspect ratio and the lower limit on $B_{max}$.

A far more comprehensive and self-consistent ST reactor design has been developed by the ARIES group at the University of California, San Diego. In their design plasma performance is improved by a combination of increasing the elongation to $\kappa = 3.4$ and approximately doubling the value of $\langle \beta_N \rangle$. With these parameters the $n = 0$ vertical insta-bility becomes more difficult to stabilize and the $n = 1$ ballooning-kink resistive wall mode is excited requiring feedback stabilization. Also the plasma current is about 30 MA (i.e., $q_* = 2.8$) requiring a bootstrap fraction of $f_B = 0.96$ to minimize the current-drive require-ments, particularly near the center of the plasma. Interestingly, the ARIES study shows that there is an optimum inverse aspect ratio given by $\varepsilon \approx (1.7)^{-1} = 0.59$. This optimum arises from the competition to increase $\langle p \rangle$ by increasing $\varepsilon$ for stability versus decreasing $\varepsilon$ because of the $1 - \varepsilon$ reduction in $B_0$ due to the $1/R$ effect. The conclusion is that the opti-mized "ST reactor" aspect ratio actually lies between a standard tokamak and current ST experiments.

Lastly, it is worth noting that according to the ARIES study the projected cost of electricity (COE) for an ST reactor is about the same as for a conventional tokamak reactor. The ST, however, requires more aggressive assumptions with respect to improved plasma physics performance as well as having to deal with the complex technological issues associated with the central core. On this basis one can conclude that the benefits of the ST associated with the achievement of very high $\beta$ do not lead to an obviously more attractive fusion reactor as compared to the standard tokamak.

### The ST as a neutron source

One application in which the ST configuration may have an advantage is as a volume neutron source. The purpose of such a facility is to produce a high-intensity 14.1 MeV neutron flux for testing advanced fusion materials. The volume need not be as large as a reactor since materials can be tested in small samples. The critical issue here is not a favorable overall power balance but is instead the production of a high neutron flux.

The advantage can be explained as follows. As for a reactor, using an ST as a neutron source eliminates the need for the blanket-and-shield on the inboard side of the torus. Also, the aspect ratio is chosen to optimize neutron production, again resulting in an intermediate value of $\varepsilon$. This combination of features leads to a relatively compact design corresponding to a low cost. Countering these gains is the limit of $B_{max}$ on the inside of the toroidal field magnet. The issue here is not primarily power balance but perhaps more importantly, the need to keep the magnitude of the ohmic heating power sufficiently low (i.e., less than 250 MW) that the operating costs stay within reason.

The net result is that an ST potentially produces a less expensive neutron source than a standard tokamak for low-to-moderate neutron fluxes. For higher neutron fluxes, the standard tokamak is more economical, although many more neutrons may be produced than are needed. These points are demonstrated in the analysis below. Two important assumptions are that both the ST and standard tokamak are assumed to operate with moderate elongations ($\kappa \approx 2$) and at values of $\beta$ below the no-wall $\beta_{max}$ limit. The motivation for these assumptions is that a neutron source will be needed well before an actual reactor is built and it is, therefore, prudent to develop designs that do not require aggressive physics improvements above current performance levels.

The analysis consists of calculating and comparing the neutron flux from an ST and a standard tokamak. In both cases the minor radius $a$ is chosen to maximize the neutron flux $P_W$ for a given value of major radius $R_0$. The optimized $P_W$ can then be plotted as a function of $R_0$. Next, the volume $V$ of the blanket, shield, and coils is easily evaluated and plotted as a function of $P_W$. The quantity $V$ serves as a measure of the capital cost. An examination of this curve shows qualitatively in which regions the ST is a more economical neutron source than a standard tokamak.

To begin, consider the evaluation of the neutron flux. The geometry of the neutron source is illustrated in Fig. 13.40. Following the discussion in Chapter 5, one can express the average neutron power flux passing through the first wall as

$$P_W = \frac{E_n n^2 \langle \sigma v \rangle}{4} \frac{V_p}{A_p} \approx \left[ \frac{E_n}{16} \frac{\langle \sigma v (T_f) \rangle}{T_f^2} \right] \frac{V_p}{A_p} p^2 \qquad \text{W/m}^2. \qquad (13.188)$$

Here, $T_f = 15 \text{ keV}$, $E_n = 14.1 \text{ MeV}$, and $V_p$, $A_p$ are the plasma volume and surface area respectively. Equation (13.188) states that the neutron power flux passing through the first wall is equal to the neutron energy produced per second per cubic meter multiplied by the plasma volume and divided by the surface area of the plasma. For both the ST and the

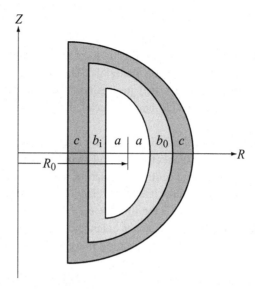

Figure 13.40 Geometry of a fusion neutron source.

standard tokamak, the plasma volume and surface area are given by

$$V_p = \pi \kappa a^2 R_0 [\pi - (\pi - 8/3)\varepsilon],$$
$$A_p \approx 2\pi \kappa a R_0 [\pi + 2 - (\pi - 2)\varepsilon]. \tag{13.189}$$

The second formula is a reasonably good approximation over the regime of interest. The pressure is now set to its maximum no-wall value given by Eq. (13.185). After a short calculation one obtains the following expression for the neutron power flux, valid for either configuration:

$$P_W = 4.29 \times 10^{-4} R_0 B_{max}^2 (1 + \kappa^2)^2 H(\varepsilon) \qquad \text{MW/m}^2, \tag{13.190}$$

where

$$H(\varepsilon) = \frac{1 - 0.15\varepsilon}{1 - 0.22\varepsilon} G^2(\varepsilon)\varepsilon^3 (1 - \varepsilon_b - \varepsilon)^4. \tag{13.191}$$

The function $H(\varepsilon)$ has a maximum and it is at this value that the neutron power flux is maximized. The optimizing $\varepsilon$ is plotted as a function of $\varepsilon_b$ in Fig. 13.41. Observe that for an ST ($\varepsilon_b = 0$) the optimum aspect ratio is about $R_0/a \approx 2$. For standard tokamaks the optimum aspect ratio is larger: $R_0/a > 2$. More realistic calculations lead to slightly smaller values of aspect ratio but for present purposes the simple estimate given above suffices.

On substituting the optimum $\varepsilon$ into Eq. (13.190), one obtains a relationship between the neutron power flux and the major radius. This relationship is illustrated in Fig. 13.42 for an ST and a standard tokamak assuming $\kappa = 2$ for both configurations; also, $B_{max} = 7.5$ T for the ST while $B_{max} = 13$ T for the standard tokamak. For the standard tokamak, $b_i = 1.2$ m. Note that the ST produces a larger neutron power flux for major radii of less than about 4.7 m.

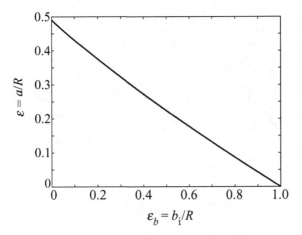

Figure 13.41 The optimizing aspect ratio $\varepsilon$ as a function of $\varepsilon_b = b_i/R_0$.

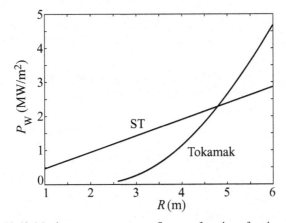

Figure 13.42 Maximum neutron power flux as a function of major radius.

Above this value the standard tokamak produces the larger neutron flux. At the transition point $P_W \approx 2.3$ MW/m$^2$.

The last point of interest is to calculate the volume $V$ of the blanket, shield, and coil for both configurations to obtain some insight into the relative capital costs. After some straightforward geometry one obtains an expression for $V$ that can be written as

$$
\begin{aligned}
V &= \pi^2\{wh\bar{R} - 2\kappa a^2 [R_0 - (1 - 8/3\pi)a]\}, \\
w &= a + c + (b_i + b_o)/2, \\
h &= \kappa a + b_o + c, \\
\bar{R} &= R_0 - a - b_i - c + (8/3\pi)w,
\end{aligned}
\tag{13.192}
$$

where $b_i$ and $b_o$ are the inboard and outboard blanket-and-shield thicknesses respectively. The curves of volume versus neutron power flux are plotted in Fig. 13.43. As anticipated the

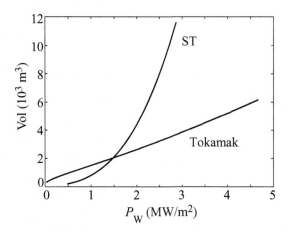

Figure 13.43 Volume of the blanket-and-shield and coils as a function of neutron power flux in a fusion neutron source.

ST has a smaller volume than the standard tokamak at low-to-moderate values of neutron power flux, implying a lower cost.

The simplified analysis just presented provides a qualitative insight into the performance of an ST relative to a standard tokamak. The conclusion is that the ST geometry has a potential advantage over the standard tokamak for application as a neutron source for low-to-moderate values of neutron flux.

### 13.7.7 Summary of the tokamak

The tokamak is currently the leading contender to become the first fusion reactor, largely because of its high-quality physics performance. Tokamaks with elongated, non-circular cross sections are capable of stably confining reactor grade plasma pressures without the need of a perfectly conducting wall. The biggest physics drawback is that the corresponding pressure profiles do not produce quite enough bootstrap current to keep the current-drive requirements at a sufficiently low level to satisfy reactor economic constraints.

The regime of AT operation, characterized by hollow current and hollow safety factor profiles, produces a much higher fraction of bootstrap current, thereby resolving the difficulty. However, to achieve this performance the Troyon limit is violated, usually leading to the excitation of an $n = 1$ ballooning-kink resistive wall mode. The consequence is that an $n = 1$ feedback system is necessary. Still, at present, the AT tokamak approach seems to be the most promising path towards a fusion reactor.

Lastly, a more recent configuration, known as the ST has been introduced as a possible contender for a fusion reactor. The ST is essentially a very tight aspect ratio tokamak designed to take advantage of the MHD stability scaling $\beta \sim \varepsilon$ by letting $\varepsilon \to 1$. More detailed analysis has shown that while higher $\beta$ values are achievable, the corresponding plasma pressures are usually lower than in a standard tokamak. The conclusion is that the

ST does not offer any substantial advantages over the standard tokamak as a reactor but, instead, may be more attractive as a volume neutron source for low-to-moderate neutron power fluxes.

## 13.8 The stellarator

### *13.8.1 Overview of the stellarator*

The stellarator is the last major fusion concept to be discussed. It is an inherently 3-D configuration, best described as a helically symmetric system bent into a torus. There are many different ways to construct and arrange coils to produce stellarator magnetic fields. In the end, however, three types of field are required: (1) a relatively large axisymmetric toroidal field $B_\phi(r, \theta)$, (2) a moderately sized helical field characterized by $l\theta - n\phi$ symmetry, almost always with a sum of multiple field harmonics with different $l$ and $n$, and (3) a small axisymmetric vertical field $B_V(r, \theta)$. There is no ohmically or externally driven toroidal current and this is a crucial feature of a stellarator. In other words, the only net toroidal current that possibly flows in a stellarator is due to the naturally occurring bootstrap effect.

From a physics performance point of view stellarators are most similar to the tokamak. The relatively large toroidal field is effective in stably confining modest amounts of plasma pressure corresponding to $\beta \sim 0.05$. Stellarators do this without the need for a perfectly conducting wall – one does not, therefore, expect the resistive wall mode to be important.

The combination of helical fields and vertical fields in a torus can have favorable average curvature and produces safety factors above unity. These fields are also crucial for single-particle confinement. Specifically, because of the 3-D nature of the configuration, a poorly designed stellarator magnetic field can lead to a rapid loss of particles by purely classical mechanisms, losses large enough to even dominate micro-turbulence-driven anomalous transport. These rapid classical losses can occur because a poorly designed stellarator field generates loss cones and large particle excursions off the flux surfaces. Over many years of study, stellarator researchers have learned how to minimize these geometrically driven losses so they do not dominate transport.

In terms of physics design strategy, there are two basic approaches that have been used to optimize the multiple degrees of freedom characterizing the stellarator configuration. One approach focuses on optimizing MHD equilibrium and stability performance. The second approach focuses on optimizing single-particle confinement. Furthermore, because of the difficulty of analyzing 3-D configurations it has been only since the late 1990s that theoretical and computational tools have become available that allow a nearly complete optimization to be carried out using either approach. The net result of increased understanding of 3-D stellarator physics is that substantial progress has been made in improving the performance of stellarator experiments. The maximum experimental values of $\beta$ and $\tau_E$ are now approaching and are sometimes comparable to those of similar size tokamaks. This is indeed an excellent achievement.

Consider now the main pros and cons of a stellarator with respect to a fusion reactor. It offers two major improvements over the standard tokamak. First, it is inherently steady state since no ohmic current or current drive is required. Second, with little or no net toroidal current flowing, the likelihood of exciting major disruptions is substantially reduced. Both of these advantages are clearly highly desirable.

The stellarator has one main disadvantage compared to the tokamak. As described shortly, the coil system needed to generate a stellarator magnetic field is substantially more complicated technologically than the tokamak coil system. This leads to an increased cost and possibly also to stricter limitations on the maximum achievable magnetic field. Ultimately this disadvantage must be weighed against the difficulty the tokamak faces in overcoming its current drive and possible disruption problems by means of the advanced operation mode (i.e., AT operation).

The discussion presented in this subsection is largely motivated by the recognition that the analysis of the stellarator is quite complicated mathematically because of the 3-D nature of the geometry. The analysis is beyond the scope of this book and will not be discussed further. Instead the three largest stellarator projects currently under study are described. These are: (1) the Large Helical Device (LHD) at the National Institute for Fusion Science in Japan, (2) the Wendelstein 7-X (W7-X) stellarator at the Max-Planck Institute for Plasma Physics in Germany, and (3) the National Compact Stellarator Experiment (NCSX) at the Princeton Plasma Physics Laboratory in the USA. Of these only the LHD is completed and operational.

The goal is to understand each of these devices and to learn the motivation and philosophy behind their designs, which as it turns out are quite different. When necessary, experimental results or numerically obtained theoretical predictions will simply be quoted.

### 13.8.2 *The Large Helical Device (LHD)*

The LHD is a billion dollar class stellarator experiment built and operating in Japan. The LHD magnet configuration corresponds to a special form of stellarator known as the "heliotron." Specifically, the magnet system consists of two continuous, intertwined helical–toroidal coils plus a set of axisymmetric vertical field coils as illustrated in Fig. 13.44. The helical–toroidal coils provide the toroidal field and the helical field. They also produce a substantial vertical field which is largely canceled out by the external vertical field coils, leaving a small net, adjustable vertical field to optimize physics performance.

The helical–toroidal coils are superconducting and because of topological constraints had to be wound in place. There is wide consensus that the design and construction of the LHD helical–toroidal coils represent a remarkable engineering achievement. Although the continuous coils work well in LHD there is also consensus that as one extrapolates stellarator designs into the reactor regime, such coils do not correspond to the optimum design approach. The need to wind the coils in place combined with the difficulties associated with maintenance and the high cost of coil replacement indicate that an alternative approach would be desirable. This alternative approach involves the use of modular coils and is discussed shortly in the context of W7-X and NCSX.

Figure 13.44  Schematic diagram of the LHD in Japan (courtesy of A. Komori).

Consider now the physics motivation behind LHD. At the time the LHD was designed, many of the advanced 3-D computational tools now available did not exist. In addition, some of the subtleties connected with single-particle confinement were not full appreciated. Using the best information available at the time, the LHD design team chose to optimize the configuration so as to maximize performance with respect to MHD equilibrium and stability as opposed to single-particle confinement. A further complication was that 3-D macroscopic MHD codes, capable of testing stability against the dangerous external kink modes had not yet been developed to the point where they could be used in a routine fashion as a design tool. Accuracy and run-time were major issues. The design team instead chose to focus much of the geometric optimization on maximizing $\beta$ against a class of localized, internal MHD modes (i.e., essentially interchange modes) that could be more easily tested numerically. It was on this basis that the coil design, diagnostic location, and "standard" operation of LHD were determined.

Now that the device is in operation one can ask "how well does LHD work?" The early operation of the device produced acceptable performance but not as good as originally anticipated. Improvements in the understanding of stellarator physics during the design and construction phase suggested that MHD stability against localized modes was probably not the most critical issue. These modes either were not excited or, if they were, were saturated at a low, relatively harmless level. Instead, poor single-particle confinement was likely a more serious issue. Fortunately, relatively small changes in the operation of the device led to substantially improved performance. Specifically, the coil currents were reprogrammed to shift the plasma further inward than in the original mode of operation, and this shift greatly increased the quality of the single-particle confinement.

The optimum performance of LHD (as of 2004) is summarized in Table 13.2 in which are listed the parameters for several different type of optimized discharges, one type to

Table 13.2. *Experimental parameters for various LHD discharges optimized with respect to electron temperature, ion temperature, and energy confinement time. Also listed are the maximum achieved values of $p_i \tau_E$, $\beta$, and $n_e$*

|                          | $T_e(\text{keV})$ | $T_i(\text{keV})$ | $\tau_E(s)$ | $P_{\text{aux}}(\text{MW})$ | $n_e(\text{m}^{-3})$ |
|--------------------------|--------------------|--------------------|--------------|------------------------------|----------------------|
| $T_e$ maximum            | 10.0               | 2.0                | 0.06         | 1.2                          | $5.0 \times 10^{18}$ |
| $T_i$ maximum            | 4.2                | 7.0                | 0.06         | 3.1                          | $2.9 \times 10^{18}$ |
| $\tau_E$ maximum         | 1.3                | 1.3                | 0.36         | 1.5                          | $4.8 \times 10^{19}$ |
| $p_i \tau_E$ maximum     | 0.035 atm s        |                    |              |                              |                      |
| $\beta$ maximum          | 0.032 at $B_\phi = 0.5$ T |             |              |                              |                      |
| $n_e$ maximum            | $1.6 \times 10^{20}$ m$^{-3}$ |          |              |                              |                      |

maximize $T_e$, one type to maximize $T_i$, and one type to maximize $\tau_E$. Also listed are the maximum achieved values of $p_i \tau_E$, $\beta$, and $n_e$. These values are comparable to, or perhaps slightly lower than, those obtained in similar sized tokamaks. The inherent steady state nature of the stellarator has also been demonstrated. The combination of superconducting coils plus the lack of need for current drive has allowed LHD to operate discharges for several hundred seconds.

Future research will involve various upgrades to the device, primarily in the auxiliary heating supplies, to allow longer discharges at higher densities and temperatures. With reasonably good single-particle confinement the physics objectives will focus on maximizing the energy confinement time in the presence of micro-turbulence-driven anomalous transport and determining the maximum achievable macroscopically stable MHD $\beta$.

### 13.8.3 Guiding center particle orbits in a stellarator

The other two stellarators, W7-X and NCSX, have not yet been fully constructed (as of 2004). The W7-X is currently in the midst of construction, while the NCSX is nearing completion of its engineering design, and construction should start shortly. The primary physics goal of each device is to maximize performance by focusing on the problem of single-particle confinement with MHD equilibrium and stability being the secondary goal. However, the strategies chosen to achieve these goals are quite different.

To appreciate the motivation behind each device one has to understand the behavior of the guiding center orbits in a 3-D geometry. This requires a somewhat complicated and abstract analysis, but is nonetheless essential if one wants to understand the stellarator. A summary of the essential features of guiding center behavior in 3-D is presented in this subsection. The analysis is presented in Appendix C for interested readers.

The basic goal of the analysis is to demonstrate that the guiding center behavior of the particle orbits on any given flux surface depends only on the magnitude of the magnetic field ($B = |\mathbf{B}|$) and not on its vector nature. This is by no means an obvious result, but once established, it sets the stage for inventing various ways to optimize single-particle confinement in a stellarator. Two of these are demonstrated by W7-X and NCSX and are discussed in the following subsections.

Two steps are involved in understanding guiding center behavior in three dimensions. First, one must introduce an appropriate set of coordinates that automatically takes into account several of the constraints imposed by the MHD equilibrium equation. Second, these coordinates are then substituted into the general expression for the guiding center trajectories leading to the desired conclusion that the orbits on any flux surface depend only on $B$. These steps are outlined below.

### Stellarator coordinates

The basic problem being addressed can be understood by recalling the equations describing the guiding center motion of a particle in a general 3-D geometry as derived in Chapter 8. For steady state equilibrium, the guiding center trajectory can be written as $\dot{\mathbf{r}} = \dot{\mathbf{r}}_g(t) + v_\parallel(t)\mathbf{b}$, where $\dot{\mathbf{r}}_g(t)$ represents the perpendicular drifts (i.e., the grad-$B$ drift and the curvature drift) while $v_\parallel(t)$ represents the parallel motion). The equations determining $\mathbf{r}_g(t)$ and the position $l(t)$ along a field line (for a positively charged particle) are repeated here for convenience:[1]

$$\frac{d\mathbf{r}_g}{dt} = \frac{v_\perp^2}{2\omega_c}\frac{\mathbf{B} \times \nabla B}{B^2} + \frac{v_\parallel^2}{\omega_c}\frac{\mathbf{R}_c \times \mathbf{B}}{R_c^2 B};$$

$$\frac{dl}{dt} = v_\parallel.$$

(13.193)

At first glance the appearance of the vectors $\mathbf{B}$ and $\mathbf{R}_c$ would seem to imply that the guiding center trajectories depend upon the vector nature of the magnetic field. This need not necessarily be so. A simple example that illustrates the point is the motion of a charged particle in a magnetic field $\mathbf{B} = B_y\mathbf{e_y} + B_z\mathbf{e_z}$, where $B_y$, $B_z$ are constants. The equations describing the trajectory are

$$m\dot{v}_x = e(v_y B_z - v_z B_y),$$
$$m\dot{v}_y = -ev_x B_z,$$
$$m\dot{v}_z = ev_x B_y.$$

(13.194)

Superficially, the solution to the problem might appear to depend upon the vector nature of the magnetic field since $B_y$, $B_z$ enter the equations of motion separately. However, the introduction of a set of "clever" new variables $v_n = v_x, v_t = (v_y B_z - v_z B_y)/B$, and $v_\parallel = (v_y B_y + v_z B_z)/B$ with $B = (B_y^2 + B_z^2)^{1/2}$ shows that the equations reduce to

$$m\dot{v}_n = eBv_t,$$
$$m\dot{v}_t = -eBv_n,$$
$$m\dot{v}_\parallel = 0.$$

(13.195)

---

[1] There is actually a first order correction to the equation for $l(t)$ that has not been calculated in Chapter 8 but which is needed for a complete treatment of guiding center motion in three dimensions. However, this correction is not required for the limited goals of the present analysis.

Observe that in the new coordinates the trajectory depends only upon the single quantity $B$ and not the separate vector components $B_y$, $B_z$. The corresponding analysis for a general 3-D field is obviously more complicated but is philosophically the same.

The 3-D analysis begins with the introduction of an appropriate set of "clever" variables. As a way to begin to think about the problem, imagine that a numerical solution has been found to the MHD equations that yields $\mathbf{B} = \mathbf{B}(r, \theta, \phi)$ and $p = p(r, \theta, \phi)$. In principle, the guiding center orbits can be calculated by relating $\mathbf{r}_g$ and $l$ to $r$, $\theta$, $\phi$ as follows. One equates the two equivalent expressions for $\dot{\mathbf{r}}$: $\dot{\mathbf{r}} = \dot{\mathbf{r}}_g + \dot{l}\mathbf{b} = \dot{r}\,\mathbf{e}_r + r\dot{\theta}\,\mathbf{e}_\theta + (R_0 + r\cos\theta)\,\dot{\phi}\mathbf{e}_\phi$ and then solves for $r(t)$, $\theta(t)$, and $\phi(t)$ by substituting into Eq. (13.193). However, while using $r, \theta, \phi$ is an intuitively obvious way to define the location of the guiding center, these coordinates do not have the property of showing that the orbits depend only upon $B$.

The desired coordinates are more abstract and are defined as follows. In general, one needs three coordinates to specify the location of the guiding center and rather than use $r, \theta, \phi$ an equivalent set of "flux" coordinates is introduced, denoted by $\psi, \chi, \zeta$. Here, within a factor of $2\pi$, $\psi$ represents the poloidal flux contained within a $p = \text{const.}$ contour. It is essentially a radial-like coordinate that specifies the flux surface upon which the guiding center is located. Note that, as for tokamaks, it is more convenient to use $\psi$ rather than $p$ to label the flux surfaces. However, since $\mathbf{B} \cdot \nabla p = 0$ implies that one can always write $p = p(\psi)$ with $\mathbf{B} \cdot \nabla\psi = 0$, $p$ and $\psi$ are equivalent and equally valid labels. Next, the quantity $\chi$ is a poloidal like angle whose value increases by $2\pi$ every time $\theta$ increases by $2\pi$. In general, it also has superimposed periodic modulations in both the toroidal and poloidal angles. Similarly, $\zeta$ is a toroidal-like angle that increases by $2\pi$ every time $\phi$ increases by $2\pi$. It too has periodic toroidal and poloidal modulations. In summary, the general relations between the abstract $\psi, \chi, \zeta$ coordinates and the more familiar $r, \theta, \phi$ coordinates can be written as

$$\psi(r, \theta, \phi) = \psi_0(r) + \sum_{l,n} \psi_{ln}(r)e^{i(l\theta+n\phi)},$$

$$\chi(r, \theta, \phi) = \theta + \sum_{l,n} \theta_{ln}(r)e^{i(l\theta+n\phi)}, \tag{13.196}$$

$$\zeta(r, \theta, \phi) = -\phi + \sum_{l,n} \phi_{ln}(r)e^{i(l\theta+n\phi)}.$$

(The minus sign in $\phi$ is used to make $\psi, \chi, \zeta$ a right handed system.) It is clear that either set of coordinates represents an equally valid description of the location of the guiding center.

At this point the $\psi_{ln}, \theta_{ln}, \phi_{ln}$ coefficients are arbitrary. A particularly clever choice for these coefficients has been given by Boozer. He has shown that a set of $\psi_{ln}, \theta_{ln}, \phi_{ln}$ can always be found that allows the magnetic field to be expressed in two alternative but equivalent forms as follows:

$$\mathbf{B} = \nabla\zeta \times \nabla\psi + q(\psi)\nabla\psi \times \nabla\chi,$$

$$\mathbf{B} = i_t(\psi)\nabla\chi + i_p(\psi)\nabla\zeta + \tilde{i}(\psi, \chi, \phi)\nabla\psi. \tag{13.197}$$

In these expressions $\psi = \Psi_p/2\pi$, where $\Psi_p$ is the poloidal flux. The quantity $q(\psi)$ is the safety factor. Also, $i_t(\psi) = (\mu_0/2\pi)I_t(\psi)$ and $i_p(\psi) = (\mu_0/2\pi)I_p(\psi)$, where $I_t(\psi)$ is the toroidal plasma current and $I_p(\psi)$ is the total poloidal current due to the coil and the plasma. Lastly $\tilde{i}(\psi, \chi, \phi)$ is an arbitrary function with zero average value over any period of $2\pi$ in either $\chi$ or $\zeta$. The proof that two such equivalent forms exist for the magnetic field is not at all obvious and requires some advanced analysis. The details are presented in Appendix C for interested readers. Coordinates with the dual representations given by Eq. (13.197) are very important for understanding stellarators and are known as "Boozer coordinates."

Assume now that a set of Boozer coordinates exists. One can then ask what the value is of such a dual representation. The answer is found by forming the dot product of one form with the other form. A short calculation yields

$$\nabla\psi \cdot (\nabla\chi \times \nabla\zeta) = \frac{B^2}{i_t + qi_p} = f(\psi)B^2. \tag{13.198}$$

The quantity $\nabla\psi \cdot (\nabla\chi \times \nabla\zeta)$ represents the Jacobian $J$ of the transformation from the $\psi, \chi, \zeta$ coordinates to the $r, \theta, \phi$ coordinates. Therefore, the advantage of the Boozer coordinates is that the Jacobian depends only on $\psi$ and $B = B(\psi, \chi, \zeta)$. It does *not* depend upon the vector nature of the magnetic field on any given flux surface since there is no separate appearance of any of the individual vector components. This is the critical property responsible for simplifying the guiding center particle trajectories as is next shown.

### The guiding center orbits in Boozer coordinates

The goal now is to determine how the $\psi, \chi, \zeta$ coordinates of a given particle evolve in time as the particle moves along its guiding center trajectory. The basic equations of motion are

$$\frac{d\psi}{dt} = \frac{\partial\psi}{\partial t} + \frac{d\mathbf{r}}{dt} \cdot \nabla\psi,$$

$$\frac{d\chi}{dt} = \frac{\partial\chi}{\partial t} + \frac{d\mathbf{r}}{dt} \cdot \nabla\chi, \tag{13.199}$$

$$\frac{d\zeta}{dt} = \frac{\partial\zeta}{\partial t} + \frac{d\mathbf{r}}{dt} \cdot \nabla\zeta.$$

In steady state $\partial/\partial t = 0$. Furthermore, to move with the particle along its guiding center trajectory one must choose the directional derivative such that $\dot{\mathbf{r}} = \dot{\mathbf{r}}_g + v_\parallel \mathbf{b}$. The guiding center trajectory equations thus reduce to

$$\frac{d\psi}{dt} = (\dot{\mathbf{r}}_g + v_\parallel \mathbf{b}) \cdot \nabla\psi,$$

$$\frac{d\chi}{dt} = (\dot{\mathbf{r}}_g + v_\parallel \mathbf{b}) \cdot \nabla\chi, \tag{13.200}$$

$$\frac{d\zeta}{dt} = (\dot{\mathbf{r}}_g + v_\parallel \mathbf{b}) \cdot \nabla\zeta.$$

The next point to note is that of these three equations the one determining $d\psi/dt$ is the most important. This equation describes how the guiding center motion of a particle can result in a drift across flux surfaces. Such motion, if unidirectional, can lead to a direct drift of particles out of the plasma to the wall (i.e., as in a loss cone). In other cases it can lead to a rapid loss of particles by means of enhanced diffusion (i.e., neoclassical transport) as discussed in the next chapter. The changes in the $\chi$ and $\zeta$ positions of the guiding center are not as important since these represent motion within the surface and are not directly responsible for radial loss of particles. Based on this discussion and the fact that $\mathbf{B} \cdot \nabla \psi = 0$, it follows that the critical equation to examine is

$$\frac{d\psi}{dt} = \dot{\mathbf{r}}_g \cdot \nabla \psi, \tag{13.201}$$

where $\dot{\mathbf{r}}_g$ is given by Eq. (13.193).

A straightforward calculation given in Appendix C shows that Eq. (13.201) can be written as

$$\frac{d\psi}{dt} = \left( \frac{2E - \mu B}{eB} \right) \left[ \frac{i_t}{i_t + qi_p} \frac{\partial B}{\partial \zeta} - \frac{i_p}{i_t + qi_p} \frac{\partial B}{\partial \chi} \right]. \tag{13.202}$$

Here, $E = (m/2)(v_\perp^2 + v_\parallel^2)$ and $\mu = mv_\perp^2/2B$ are both constants of the motion; that is, they are both constant with respect to the time derivative in Eq. (13.202).

The crucial feature of the equation for $d\psi/dt$ is that the right hand side depends only on the flux $\psi$ and the magnitude of the magnetic field $B$. There is no dependence on the vector nature of the magnetic field. A similar conclusion holds for the $d\chi/dt$ and $d\zeta/dt$ equations. It is this feature that will be exploited to optimize the design of the next generation of stellarator experiments.

Optimized stellarators are based on the idea of "quasi-symmetry," a property that provides inherently good single-particle confinement. Quasi-symmetry can be described qualitatively as follows. A stellarator magnetic field is, in general, a complicated 3-D vector function of space: $\mathbf{B} = \mathbf{B}(\psi, \chi, \zeta)$. By cleverly choosing the amplitudes and phases of the separate helical harmonics contributing to the total field, it is possible to create a configuration where the vector components of $\mathbf{B}$ combine in such a way that $B = |\mathbf{B}(\psi, \chi, \zeta)|$ is approximately a 2-D function. A configuration with this approximate 2-D property is said to possess quasi-symmetry. The three possibilities are quasi-toroidal symmetry $[B \approx B(\psi, \chi)]$, quasi-poloidal symmetry $[B \approx B(\psi, \zeta)]$, and quasi-helical symmetry $[B \approx B(\psi, l\chi - n\zeta)]$. The reason why quasi-symmetry helps single-particle confinement is that such configurations have two conserved constants of the motion, the energy and a canonical momentum associated with the symmetry direction. It is the existence of the conserved canonical momentum that can be shown to lead to confined orbits. In a fully 3-D configuration, no conserved canonical momentum exists and the orbits are not guaranteed to remain confined.

In practice, optimized quasi-symmetric designs require that two simultaneous conditions be satisfied. The first requires that the magnetic configuration possess quasi-symmetry as closely as possible so that the particles do not drift far off a flux surface and that loss cones do not exist. The second requires that when $B(\psi, \chi, \zeta)$ is transformed back into the practical $r, \theta, \phi$ geometric space, the resulting magnetic field must be physically realizable and able to be generated by a technologically credible set of magnets. The different strategies used in the design of W7-X and NCSX are the subject of discussion in the next two subsections.

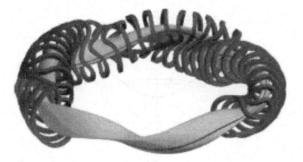

Figure 13.45 Schematic diagram of the W7-X in Germany.

### 13.8.4 The Wendelstein 7-X (W7-X)

The W7-X is a billion dollar class device that is being built at the Max-Planck Institute for Plasma Physics in Germany. It utilizes modular superconducting coils and therefore should be capable of long-pulse operation. A schematic diagram of W7-X is shown in Fig. 13.45. Observe the technological complexity of each modular coil.

The design of the W7-X is based on the concept of quasi-poloidal symmetry ($B = B(\psi, \zeta)$), first demonstrated in the favorable experimental results from the smaller Wendelstein Advanced Stellarator (W-AS). Qualitatively, quasi-poloidal symmetry has some similarities to a simple straight mirror machine in which $B = B(r, z)$. There is no variation with respect to the poloidal angle and there is no current flowing parallel to the magnetic field. Also particles are trapped between the maxima of the magnetic field. The end loss problem is resolved by connecting a series of linked mirror machines bent into the shape of a torus. Thus, particles leaving the end of one mirror cell simply flow into the adjacent mirror cell.

The actual W7-X configuration is, however, considerably more complicated than this simple picture. The field must be designed to provide toroidal force balance without any parallel current. Also the details of the cross sectional shaping are very important since a simple mirror machine can be shown to possess average unfavorable curvature (see Section 12.3.2), leading to MHD instability. Therefore the motivation for the W7-X design is to create a quasi-poloidal configuration without parallel currents and possessing average favorable curvature.

Consider now in more detail the reason for the no-parallel current goal. The purpose of minimizing the parallel current is to improve the robustness of single-particle confinement as the plasma pressure varies. In general, the parallel current in a stellarator comprises two main contributions as can be seen in the following easily derived expression:

$$\mu_0 \frac{\mathbf{J} \cdot \mathbf{B}}{B^2} = \mu_0 \frac{J_\parallel}{B} = \frac{1}{i_t + q\,i_p} \left[ \left( i_p \frac{di_t}{d\psi} - i_t \frac{di_p}{d\psi} \right) + \left( i_t \frac{\partial \tilde{i}}{\partial \zeta} - i_p \frac{\partial \tilde{i}}{\partial \chi} \right) \right]. \quad (13.203)$$

The first term contains the net toroidal current arising from the bootstrap effect and, as will be shown in Chapter 14, is proportional to $dp/d\psi$. The second term contains a dipole-like

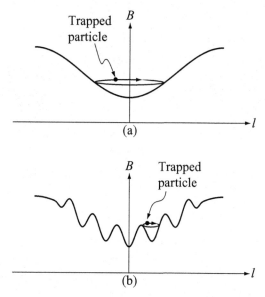

Figure 13.46 (a) Curve of $B$ vs. $l$ for: (a) an axisymmetric tokamak, and (b) a 3-D stellarator. Note the possibility of multiple minima in the stellarator.

contribution that can be shown to be proportional to $(dp/d\psi)\cos\theta$. This contribution is generated by the shift of the plasma in response to toroidal force balance and is usually referred to as the Pfirsch–Schluter current. By minimizing $J_\parallel$ in the design, one sees that the sensitivity of the total field (applied plus induced) against changes in the pressure profile is minimized. In other words, if the device has good single-particle confinement at low $\beta$, this behavior should persist as $\beta$ increases.

Next, consider the principle of quasi-poloidal symmetry in the context of Boozer coordinates to show in detail how this helps to minimize the particle drifts off the flux surface. At the simplest level, an examination of Eq. (13.202) suggests that choosing $B = B(\psi)$ would be ideal. The configuration would behave like a straight 1-D cylinder and no particles would drift off the surface. Unfortunately, this is not physically achievable in a toroidal configuration. At least a 2-D configuration is required and the W7-X team has focused on the quasi-poloidal concept $B = B(\psi, \zeta)$. Even so, an exact quasi-poloidal configuration is not geometrically possible. Consequently, in keeping with the basic philosophy of single-particle confinement, the W7-X team focused on maximizing quasi-poloidal symmetry for the most dangerous class of particles, those likely to have the largest excursions off the flux surface. These turn out to be trapped particles. The team then designed W7-X to minimize the loss of such particles.

Trapped particles occur because $\nabla_\parallel B \neq 0$ in a toroidal geometry. A typical profile of $B$ vs. $l$ (where $l$ is distance along a field line) is depicted in Fig. 13.46. Figure 13.46(a) shows the profile for an axisymmetric tokamak and Fig. 13.46(b) that for a stellarator or a

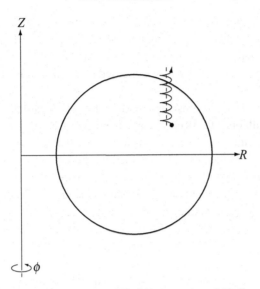

Figure 13.47 A deeply trapped particle whose $\nabla B$ drift causes it to drift directly out of the plasma.

tokamak with a toroidal ripple (due to the use of a finite number of spaced toroidal field coils). Particles with a small $v_\parallel$ are trapped in regions where $B$ has a minimum.

To understand the behavior of trapped particles it is necessary to analyze the behavior of $B(\psi, \chi, \zeta)$ near a minimum. A simple calculation shows that such minima occur whenever

$$\nabla_\parallel B = \mathbf{b} \cdot \nabla B = \frac{J}{B} \left( \frac{\partial B}{\partial \chi} + q \frac{\partial B}{\partial \zeta} \right) = 0. \qquad (13.204)$$

Consider next a deeply trapped particle, one for which $v_\parallel \approx 0$. In a torus such particles do not, in general, stay confined within a narrow region about a flux surface. The $\nabla B$ drift, which has a purely vertical component because of the $1/R$ dependence of $B_\phi$, often cannot be cancelled out on the top and bottom of the plasma since the particle samples only a small segment of the field line.

The geometric variation of $B$ in the vicinity of $B_{\max}$ determines whether or not a particle drifts off the flux surface. This geometric variation is proportional to the component of $\nabla B$ that lies in the flux surface but is perpendicular to the field lines:

$$\frac{d\psi}{dt} \sim \nabla_\perp B = \frac{1}{|\nabla \psi|} (\nabla \psi \times \mathbf{b}) \cdot \nabla B = -\frac{J}{B |\nabla \psi|} \left( i_p \frac{\partial B}{\partial \chi} - i_t \frac{\partial B}{\partial \zeta} \right). \qquad (13.205)$$

A deeply trapped particle for which $\nabla_\perp B \neq 0$ can drift out of the plasma without encircling the origin as shown in Fig. 13.47. Even if the particle is not lost, its excursion off the flux surface is likely to be quite large, leading to large diffusive losses. Although the fraction of rapidly lost trapped particles at any instant of time may be small, their population is steadily replenished by velocity space scattering resulting from Coulomb collisions among untrapped particles. In other words, the trapped particles often represent the fastest particle and energy loss channels in the plasma.

The goal of the design for the W7-X is to minimize the $\nabla_\perp B$ drift at the minima where $\nabla_\parallel B = 0$. To the extent that this is possible, deeply trapped particles do not leave the flux surface and remain well confined. In other words, in the design of W7-X the quasi-poloidal symmetry property $\partial B / \partial \chi = 0$ is focused in the regions of $B$ field minima where $\partial B / \partial \zeta = 0$.

The W7-X design takes into consideration the orbits of all particles, not only the deeply trapped particles. Still, once the deeply trapped particles are well confined, the off-surface excursions of the remaining particles are also tolerably small. The W7-X geometry is therefore optimized to achieve quasi-poloidal symmetry at the field minima while simultaneously minimizing $J_\parallel$. Extensive studies show that the optimized design occurs for a relatively large aspect ratio ($R_0/a \approx 10.6$). Qualitatively, the simultaneous vanishing of $\nabla_\perp B$ and $\nabla_\parallel B$ at certain locations along the field lines implies that the magnetic field is locally "straight" at these locations. Thus, from the point of view of single-particle confinement the W7-X can be viewed as a series of toroidally linked mirror machines, with each mirror "cylindrically symmetric" near its field minimum.

When completed and operational the W7-X will test important principles of plasma confinement in 3-D geometries. The W7-X team anticipates a successful experimental program based on great improvements in 3-D analysis and the experimental results of W-AS.

### 13.8.5 The National Compact Stellarator Experiment (NCSX)

The NCSX represents a different approach to achieving high-quality single-particle confinement in a 3-D geometry. Its design is based on the concept of quasi-toroidal symmetry which is described below. The experiment itself is just completing its design phase and construction is expected to begin shortly at the Princeton Plasma Physics Laboratory in the USA. It is an innovative experiment that uses modular copper coils in a compact geometry. A drawing of the NCSX is shown in Fig. 13.48. As with the W7-X, the modular coils are quite complex technologically. The experiment is estimated to cost on the order of $100M.

From the MHD point of view the NCSX can be viewed as a hybrid stellarator–tokamak. Unlike the W7-X, the NCSX configuration allows, and in fact requires, a substantial net toroidal current in the plasma to produce some of the favorable MHD, transport, and technological properties of a tokamak. This toroidal current is generated by the natural bootstrap effect so that no external current drive is required. Typically the rotational transform due to the net current is comparable to but less than the transform due to the helical fields in order that the configuration not be overly sensitive the pressure profile (which determines the profile of the bootstrap current). Most importantly, the toroidal current also ultimately leads to a more compact design with $R_0/a \sim 4$. A compact design is expected to result in a smaller unit size (i.e., lower total power output) when extrapolating into the reactor regime, implying a lower total capital cost (although not necessarily a lower capital cost per watt).

Figure 13.48 Schematic diagram of the NCSX in the USA (courtesy of A. R. DeMeo).

Compactness is usually considered to be a more important issue in the USA than in Europe or Japan.

The key concept behind the NCSX design is quasi-toroidal symmetry, a concept that provides inherently good single-particle confinement. Qualitatively, quasi-toroidal symmetry ($B = B(\psi, \chi)$) has some similarities to an axisymmetric tokamak ($B = B(r, \theta)$). The particles behave as if they are in an axisymmetric torus with the vertical $\nabla B$ drift canceling on the top and bottom of the cross section. Also a substantial toroidal current flows in the plasma and good MHD stability is possible at finite $\beta$ without a conducting wall.

One can next ask in more detail how quasi-axisymmetry leads to good single-particle confinement. Focus again on the trapped particles, which are the ones most susceptible to rapid loss out of the plasma. For comparison recall that in W7-X the idea is to design $B$ such that deeply trapped particles do not drift off the flux surface; $\nabla_\perp B = 0$ when $\nabla_\parallel B = 0$. If $\nabla_\perp B \neq 0$, particles could be trapped in local minima of $B$ at any poloidal location and drift directly out of the plasma, since the vertical $\nabla B$ drift would not, in general, cancel on the top and bottom of the plasma.

Similarly, the NCSX with quasi-toroidal symmetry does not allow deeply trapped particles to drift off the flux surfaces. This follows directly from the Boozer representation of the guiding center orbits. Quasi-toroidal symmetry is equivalent to the requirement $\partial B/\partial \zeta = 0$ everywhere. Thus, the minimum point in $B$, where the deeply trapped particles are located, is defined by the angle $\chi$ for which

$$\nabla_\parallel B = \frac{J}{B}\frac{\partial B}{\partial \chi} = 0. \tag{13.206}$$

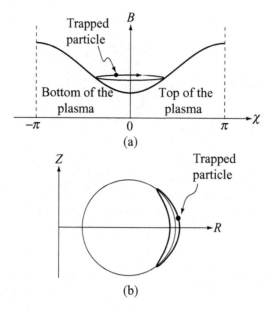

Figure 13.49 (a) Trapped particles in a quasi-axisymmetric stellarator; (b) trapped particles spend equal times on the top and bottom of the plasma, thus canceling out the vertical $\nabla B$ drift.

With quasi-toroidal symmetry Eq. (13.206) also implies that the perpendicular drift off the surface is zero at this point:

$$\nabla_{\perp} B = \frac{J i_p}{B |\nabla \psi|} \frac{\partial B}{\partial \chi} = 0. \tag{13.207}$$

The deeply trapped particles cannot be directly lost in a quasi-toroidal configuration.

Interestingly, quasi-toroidal symmetry also guarantees that for even moderately or weakly trapped particles, the corresponding vertical drifts always cancel on the top and bottom of the plasma. This point is illustrated in Fig. 13.49, which plots the projection of $B$ vs. $\chi$ in a poloidal plane. The key point is that with tokamak-like symmetry there is only one minimum in $B$ and this occurs on the outside of the torus (i.e., $\chi = \theta = 0$). Therefore all trapped particle orbits have up–down symmetry as shown in Fig. 13.49, resulting in the exact cancellation of the vertical $\nabla B$ drift on the top and bottom of the plasma. The conclusion is that quasi-axisymmetry does not allow trapped particles to be directly lost from the plasma.

Observe that, while there are similarities between quasi-poloidal symmetry and quasi-toroidal symmetry with respect to trapped particle confinement, there is also one important difference. Quasi-poloidal systems can be designed with little or no parallel current. In a quasi-toroidal system there will always be a net parallel current. Therefore, the contrasting goals of eliminating or exploiting this parallel current lead to very different looking W7-X and NCSX experiments.

Lastly, recall that, even without the direct loss of trapped particles, it was previously stated that excursions off the flux surface by moderately or weakly trapped particles lead to a more rapid loss of particles by diffusive processes than occurs for untrapped particles. This is indeed the case for any quasi-toroidal configuration including the perfectly axisymmetric tokamak. This enhanced loss is known as "neoclassical" transport and is discussed in Chapter 14. Still, from the overall requirements of confinement in a fusion reactor, it turns out that the neoclassical losses are quite tolerable and, in fact, are usually dominated by anomalous micro-turbulence-driven transport. The overall conclusion is that the quasi-poloidal linked mirror idea in W7-X and the quasi-toroidal tokamak-like idea in NCSX are two ways to prevent an unacceptably large loss of plasma and energy from a 3-D stellarator.

When NCSX is operational it will provide an important test of the principle of quasi-toroidal symmetry. The combination of extensive 3-D analysis and the many tokamak-like features inherent in the configuration has generated confidence for a successful experimental program in the NCSX team.

### 13.8.6 Summary

The stellarator is a 3-D magnetic configuration. All versions of the concept have the major advantage of being inherently steady state devices. No current drive is required to sustain the discharge. In general, the net toroidal current is either modest or zero, implying that major disruptions due to MHD instabilities should be less of an issue than in a tokamak.

Currently there are three major stellarator experiments: the LHD in Japan, the W7-X in Germany and the NCSX in the USA. The LHD has been constructed and is fully operational. It uses continuously wound superconducting coils and its performance to date has been comparable to, or perhaps slightly below, that of similar size tokamaks. The W-7X is currently under construction. It is a large superconducting device using modular coils. A major goal of W-7X is to test the principles of quasi-poloidal linked mirror symmetry, which provides good single-particle confinement with virtually no net bootstrap-driven parallel current or dipole-like Pfirsch–Schluter current. The NCSX is just completing its design phase and construction should begin shortly. It uses modular copper coils and has a compact design. The NCSX is a stellarator–tokamak hybrid with a substantial toroidal bootstrap current. One of its primary goals is to obtain good particle and energy confinement by using the principle of quasi-toroidal tokamak-like symmetry.

In terms of plasma physics performance the stellarator is the main competition for the tokamak. Ultimately, the hoped-for physics advantages must be weighed against the increased complexity and cost required to construct the modular superconducting coils required in large-scale experiments and reactors.

## 13.9 Revisiting the simple fusion reactor

### 13.9.1 Goal of the analysis

In this section the simple reactor design presented in Chapter 5 is revisited. Recall that the goal of the simple design was to determine the geometric and plasma physics parameters of a fusion reactor such that the combined volume of the blanket-and-shield and coil system was minimized. The assumption was that the capital cost of the reactor was approximately proportional to the volume of material used to construct the reactor with the blanket-and-shield and coils being the dominant components. The minimization was carried out subject to the constraints of a fixed electrical output power $P_E = 1000$ MW, a maximum allowable neutron wall loading $P_W = 4$ MW/m$^2$, a maximum thermal to electrical conversion efficiency $\eta_t = 0.4$, and a minimum blanket-and-shield thickness $b = 1.2$ m set by nuclear physics cross sections.

As the analysis progressed there came a point where one had to decide what value of magnetic field at the coil, $B_c$, should be used for the reactor. The assumption was made that whatever value of $B_c$ was chosen, the magnet would be designed to operate at the maximum allowable stress level $\sigma_{max} = 300$ MPa $\approx 3000$ atm in order to minimize the coil thickness $c = c(B_c)$ and hence the corresponding coil volume. It was pointed out that $c(B_c)$ is a monotonically increasing function of $B_c$. Therefore, operating at the maximum allowable magnetic field $B_c = B_{max} = 13$ T *maximized* the coil thickness (i.e., max $[c(B_c)] = c(B_{max})$) but led to the *lowest* possible value of the required plasma $\beta$. Choosing a lower value of $B_c$ would reduce the coil thickness but would require a higher plasma $\beta$. It was asserted, but not proved, that the maximum $\beta$ allowed by MHD stability considerations would be sufficiently low that in the tradeoff between coil thickness and allowable $\beta$ one would be forced to set $B_c = B_{max}$ in order to achieve power balance without violating the MHD stability limit: $\beta$ (required) $< \beta$ (MHD). This assertion can now be tested for a tokamak reactor by using the MHD stability limits derived in Section 13.7. The main goal of the present section is to show that the assertion is indeed correct.

### 13.9.2 Reactor analysis

The analysis begins by first reviewing the basic relations involved in the design of the reactor. The main difference is that instead of setting $B_c = B_{max}$, one lets $B_c$ remain a variable parameter. Next, the volume per watt of the reactor ($V_I/P_E$) is calculated as a function of $B_c$. The point is to show that, as expected, $V_I/P_E$ increases as $B_c$ (and hence $c$) increases. The last step is to calculate the value of $\beta$ required for power balance ($\beta_{PB}$) as a function of $B_c$ and to compare it with the MHD $\beta$ limit for a tokamak ($\beta_{MHD}$). This comparison confirms the original assertion that one should set $B_c = B_{max}$ in a tokamak reactor.

Consider now the basic relations involved in the design of the reactor. Recall that the critical parameter to be minimized is the volume of the blanket-and-shield and coil system per watt of produced electricity:

$$\frac{V_I}{P_E} = \frac{2\pi^2 R_0[(a+b+c)^2 - a^2]}{P_E}. \tag{13.208}$$

The radius $R_0$ is eliminated by using the relationship between wall loading and the total electrical power output.

$$R_0 = \left(\frac{1}{4\pi^2\eta_t} \frac{E_n}{E_\alpha + E_n + E_{Li}}\right) \frac{P_E}{a P_W} = 0.04 \frac{P_E}{a P_W} = \frac{10}{a} \text{ m}. \tag{13.209}$$

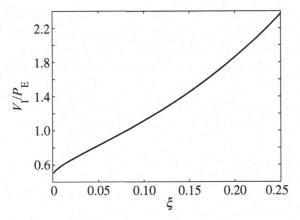

Figure 13.50 The cost function $V_I/P_E$ as a function of $\xi \propto B_c^2$.

For simplicity, here and below the specified values of the quantities $P_E$, $P_W$, and $\eta_t$ are substituted into all relations.

The coil thickness $c$ is eliminated in terms of $a, b,$ and $B_c$ using the assumption that $c$ is minimized by operating at the maximum allowable stress level of the magnet $\sigma_{max} = 300$ MPa:

$$c = \frac{2\xi}{1 - \xi}(a + b),\qquad (13.210)$$

where $\xi = B_c^2/4\mu_0\sigma_{max} = 6.63 \times 10^{-4}B_c^2$ is a dimensionless parameter. In the present analysis $\xi$ is treated as a variable parameter rather than setting $\xi = \xi_{max}$ corresponding to $B_c = B_{max}$.

Substituting these relations into Eq. (13.208) leads to a reduced expression for $V_I/P_E$:

$$\frac{V_I}{P_E} = \frac{0.2}{(1 - \xi)^2}\left[(4\xi)a + (1 + \xi)^2\frac{b^2}{a} + 2(1 + \xi)^2 b\right] \quad \text{m}^3/\text{MW} \qquad (13.211)$$

This quantity, which is proportional to the cost per watt, has a minimum with respect to $a$ which is found to be

$$a = \frac{1 + \xi}{2\xi^{1/2}}b = 0.6\frac{1 + \xi}{\xi^{1/2}}\ \text{m} \qquad (13.212)$$

for $b = 1.2$ m. The corresponding value of $V_I/P_E$ is

$$\frac{V_I}{P_E} = 0.47\frac{1 + \xi}{(1 - \xi^{1/2})^2}\ \text{m}^3/\text{MW} \qquad (13.213)$$

The cost function $V_I/P_E$ is plotted as a function of $\xi \sim B_c^2$ in Fig. 13.50. As expected $V_I/P_E$ increases monotonically as $B_c$ increases: larger $B_c$ requires larger $c$, implying a larger coil volume and hence a larger cost. From an economic point of view one clearly wants to operate at as low a value of $\xi$ as possible.

The lowest allowable value of $\xi$ must, however, be compatible with MHD stability constraints on $\beta$. This point is demonstrated by calculating the required value of $\beta$ for power balance as a function of $\xi$. The first step in the demonstration is to calculate the plasma pressure required to produce the thermal fusion power corresponding to the specified electrical power output:

$$p = \left( \frac{16}{E_\alpha + E_n + E_{Li}} \frac{P_E}{\eta_t V_P} \frac{T^2}{\langle \sigma v \rangle} \right)^{1/2} = \frac{10.4}{a^{1/2}} \text{ atm,} \qquad (13.214)$$

where $T = 15$ keV and $V_p$ is the volume of the plasma: $V_p = 2\pi^2 R_0 a^2 = 20\pi^2 a$. Also, $a = a(\xi)$ as given by Eq. (13.212). The value of $\beta$ required for power balance is now easily evaluated using the definition $\beta = 2\mu_0 p / B_0^2$ and the relation between $B_0$ and $B_c$ arising from the familiar $1/R$ effect:

$$\beta_{PB} = \frac{1.7 \times 10^{-3}}{\xi a^{1/2}(1 - 0.1a^2 - 0.12a)^2}. \qquad (13.215)$$

The quantity $\beta_{PB}$ is plotted as a function of $\xi$ in Fig. 13.51. Note, as intuitively expected, the required $\beta_{PB}$ increases as $\xi \sim B_c^2$ decreases. This curve must now be compared with the maximum allowable $\beta$ set by the MHD marginal stability boundary. For a tokamak the stability boundary is given by the Troyon limit. Strictly speaking one should use the Troyon limit corresponding to a circular cross section plasma as this is the geometry of the simple reactor. However, a Tokamak reactor will almost certainly have an elongation on the order of $\kappa \approx 2$ to improve the MHD $\beta$ limit. Thus, while not completely self-consistent, the higher value of the $\beta$ limit is used in the comparison. The end result is that MHD stability limit is taken as

$$\beta \le \beta_{MHD} \equiv 0.072 \left( \frac{1 + \kappa^2}{2} \right) \frac{a}{R_0} = 0.018 \, a^2. \qquad (13.216)$$

Here, for simplicity the difference between $\langle B^2 \rangle$ and $B_0^2$ used in the definition of the Troyon $\beta$ has been ignored.

The curve of $\beta_{MHD}$ vs. $\xi$ is also illustrated in Fig. 13.51. Stability requires operation below this curve. Observe that $\beta_{MHD}$ has a much weaker dependence on $\xi$ than $\beta_{PB}$. The interpretation of Fig. 13.51 is as follows. The key point to keep in mind is that the cost per watt of the reactor, as modeled by $V_I / P_E$, is minimized by operating at as small a value of $\xi$ as possible. Now assume that a relatively large value of $\xi$ is chosen such that $\beta_{PB} < \beta_{MHD}$. In this regime there is a substantial safety margin between the required $\beta$ and the allowable $\beta$. The implication is that there is no need to operate at such a high value of magnetic field. By lowering the field, operation shifts to the left to a smaller value of $\xi$ corresponding to a lower cost per watt.

On the other hand, one cannot shift too far to the left. In the regime of small $\xi$ the required $\beta$ substantially exceeds the value allowed by MHD and is therefore inaccessible. The furthest that one can shift to the left is the point where the curves intersect. This defines

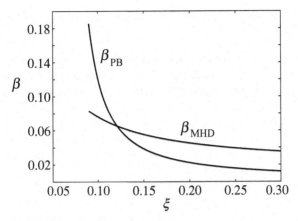

Figure 13.51 Comparison of the required value of $\beta$ for power balance ($\beta_{PB}$) and the maximum allowable $\beta$ for MHD stability ($\beta_{MHD}$) as a function of $\xi \propto B_c^2$.

the optimum value of magnetic field to be used in the reactor. However, one must now verify that the optimum magnetic field is consistent with the maximum value allowed by technology. For $B_c \leq B_{max} = 13$ T this corresponds to $\xi \leq \xi_{max} = 0.11$. One sees that, perhaps slightly fortuitously, for the tokamak the optimum magnetic field is approximately equal to the maximum technologically allowable magnetic field. Using a substantially lower field would cause the reactor to operate above the MHD stability limit, which would be unacceptable. Returning to the original goal of this section, the discussion just presented has indeed confirmed the original assertion used in the simple reactor design that the reactor should be operated at the largest allowable magnetic field.

There are several additional points that should be made. First, alternative concepts which have a much higher MHD $\beta$ limit (assuming that resistive wall modes can be stabilized) cause the intersection of the $\beta_{PB}$ and $\beta_{MHD}$ curves to shift further to the left. This reduces the cost per watt and is one primary motivation for studying such configurations.

Second, assume that technological advances lead to increased values of $B_{max}$ in super-conducting coils. How can one take advantage of this improvement? The answer is that for a given $\beta$ limit, a higher magnetic field allows a higher pressure. This, in turn, leads to a smaller plasma volume to produce the same total thermal fusion power which ultimately leads to a lower cost per watt. However, the same fusion neutron energy passing through the first wall of a smaller plasma chamber results in a larger wall loading. Therefore, to take full advantage of improved magnet technology, there must also be an accompanying improvement in first wall material to allow a higher wall loading.

## 13.10  Overall summary

This chapter has described the major configurations currently under consideration as potential fusion reactors. It has presented a derivation of the critical MHD properties of each concept including equilibrium and stability $\beta$ limits. A summary of the results is given below.

**The tokamak**   The tokamak is the leading contender to become the first fusion power reactor. It has good MHD, transport, and heating, probably sufficient for a reactor. Its biggest problems are the need to achieve steady state operation by means of current drive and the requirement of large, high-field superconducting magnets.

**The AT**   The AT employs a mode of operation in which profiles are adjusted to produce a large bootstrap current. This reduces the requirements on the current-drive system, allowing a favorable overall power balance. However, in this regime the plasma is invariably unstable to the resistive wall mode, which must then be feedback stabilized.

**The ST**   The ST is a tokamak with an ultra tight aspect ratio. Since the MHD stability limit scales as $\beta \sim \varepsilon$, the ST is capable of achieving high $\beta$. Even so, technological limitations on the copper central core lead to the conclusion that the anticipated advantages of the ST over the standard tokamak for use as a reactor are modest at best. Even so, the ST may have a decided advantage as a volume neutron source to test advanced fusion materials.

**The stellarator**   The stellarator is currently the main competition for the tokamak. It is a helical–toroidal device whose geometry is inherently 3-D, which is capable of inherently steady state operation, thereby eliminating the current-drive problem faced by the tokamak. The 3-D nature of the configuration can result in very poor single-particle confinement unless particular care is taken in the design of the magnetic field. Two ideas that promise high performance are linked-mirror symmetry and quasi-axisymmetry. The biggest problems facing the stellarator are the increased technological complexity and cost involved in designing and constructing the modular superconducting coils.

**The RFP**   The RFP is a relatively large aspect ratio, axisymmetric torus with a very modest toroidal field requirement. This gives it a technological advantage over the tokamak. However, the low toroidal field leads to a low edge safety factor. This results in the potential excitation of multiple resistive wall modes, which ultimately need to be feedback stabilized. In addition, the low safety factor results in resistive MHD turbulence, which increases the transport losses. Lastly, the toroidal current in an RFP is large implying the need for a large amount of non-inductive current drive. In spite of these problems, researchers continue to improve RFP performance by means of some very innovative experimental ideas.

**The spheromak**   The spheromak is a relatively tight aspect ratio device with no central ohmic transformer and no toroidal field system. The plasma is produced by a complicated relaxation process, and is potentially capable of steady state operation using three DC power sources. These features represent important technological improvements over the tokamak. However, like the RFP, the spheromak has a low edge safety factor leading to relatively poor transport and the presence of a resistive wall tilt mode. Also, at present, a large amount of power is required to drive the plasma current. The spheromak is at an early stage of experimentation and the future will tell how well it performs.

**The FRC**   The FRC is essentially an elongated, toroidal $Z$-pinch. It is potentially very unstable to ideal MHD modes. However, experiments with small devices where kinetic modifications to MHD behavior become important have demonstrated much better stability than expected. This has led researchers to consider the FRC as a prime candidate for the plasma source in the MTF concept. In MTF a small FRC plasma is formed and then compressed by a rapidly imploding cylindrical shell.

Although extrapolations to reactor scale are problematic, MTF could offer a quick, inexpensive way to create a burning plasma. The MTF is also at an early stage of development.

**The LDX** The LDX configuration basically consists of a single levitated coil carrying a toroidal current. From an MHD point of view LDX is similar to a hard core $Z$-pinch which has favorable stability properties due to the effect of plasma compressibility. The simplicity of the coil configuration gives it an important advantage over the tokamak. However, the need to levitate the coil clearly adds to the technological complexity. One problem is that in order to protect the levitated superconducting coil from fusion neutrons, the LDX concept will not be able to use D–T as a fuel. Using D–D as a fuel raises the required $p\tau_E$ by more than an order of magnitude. The LDX is just becoming operational so one will have to wait to see how well it performs.

The overall conclusion from this chapter is that fusion researchers have been very creative, inventing a large number of configurations for use in a fusion reactor. All of these concepts involve various tradeoffs in the plasma physics and the fusion engineering. At present, the tokamak is the leader. It remains to be seen whether any of the alternatives can overtake the tokamak and assume this role.

## Bibliography

Chapter 13 describes a wide variety of magnetic fusion concepts, primarily from the point of view of MHD. The references listed below describe some of the more advanced theoretical MHD equilibrium and stability limits. Also listed are specific references for each of the concepts discussed in the chapter.

### *Advanced MHD theory*

Bateman, G. (1978). *MHD Instabilities*. Cambridge, MA: MIT Press.
Bernstein, I. B. *et al.* (1958). An energy principle for hydromagnetic stability problems. *Proceedings of the Royal Society*, **A223**, 17.
Connor, J. W., Hastie, R. J., and Taylor, J. B. (1978). Shear, periodicity, and plasma ballooning modes. *Physical Review Letters*, **40**, 396.
Coppi, B. (1977). Topology of ballooning modes. *Physical Review Letters*, **39**, 939.
Freidberg, J. P. (1987). *Ideal Magnetohydrodynamics*. New York: Plenum Press.
Goedbloed, H. and Poedts, S. (2004). *Principles of Magnetohydrodynamics*. Cambridge, England: Cambridge University Press.
Laval, G., Pellat, R., and Soule, J. S. (1974). Hydromagnetic stability of a current-carrying pinch with noncircular cross section. *Physics of Fluids*, **17**, 835.
Menard, J. E., Bell, M. G., *et al.* (2004). Aspect ratio scaling of ideal no-wall stability limits in high bootstrap fraction tokamak plasmas. *Physics of Plasmas*, **11**, 639.
Sykes, A. and Wesson, J. A. (1974). Two-dimensional calculation of tokamak stability. *Nuclear Fusion*, **14**, 645.
Troyon, F. (1984). MHD limits to plasma confinement. *Plasma Physics and Controlled Fusion*, **26** (1A), 209.
Wesson, J. A. (1978). Hydromagnetic stability of tokamaks. *Nuclear Fusion*, **18**, 87.
Wesson, J. (2004). *Tokamaks*, third edn. Oxford: Oxford University Press.

### *LDX*

LDX Group (1998). Levitated Dipole Whitepaper. *Innovative Confinement Concepts Workshop*, Princeton Plasma Physics Laboratory, Princeton, New Jersey.


### *FRC-MTF*

Barnes, D. C. (1997). Scaling relations for high gain, magnetized target fusion systems. *Comments on Plasma Physicss and Controlled Fusion*, **18**, 17.
Barnes, D. C. *et al.* (2002). Field-Reversed Configuration (FRC) equilibrium and Stability. 19th *IAEA Fusion Energy Conference*, Lyon, France. Paper TH/4–5. Vienna: IAEA.
FRC community (1998). FRC Development Whitepaper. *Innovative Confinement Concepts Workshop*. Princeton: New Jersey Princeton Plasma Physics Laboratory.
Siemon, R. E., Lindemuth, I. R., and Schoenberg, K. F. (1999). Why magnetized target fusion offers a low-cost development path for fusion energy. *Comments on Plasma Physics and Controlled Fusion*, **18**, 363.


### *RFP*

Bodin, H. A. B. and Newton, A. A. (1980). Reversed field pinch research. *Nuclear Fusion*, **20**, 1255.
Miyamoto, K. (2001). *Fundamentals of Plasma Physics and Controlled Fusion*, revised edn. Toki City: National Institute for Fusion Science.
RFP Research Community (1998). The Reversed Field Pinch Whitepaper. *Innovative Confinement Concepts Workshop*. Princeton: New Jersey Princeton Plasma Physics Laboratory.


### *Spheromak*

Bellan, P. M. (2000). *Spheromaks*. London, England: Imperial College Press.
Hooper, E. B., Pearlstein, L. D., and Ryutov, D. D. (1998). The spheromak path to fusion energy. *Innovative Confinement Concepts Workshop*. Princeton, New Jersey: Princeton Plasma Physics Laboratory.
Hooper, E. B. (1999). Spheromak overview. *Fusion Summer Study*. Colorado: Snowmass.


### *Tokamak*

Wesson, J. (2004). *Tokamaks*, third edn. Oxford: Oxford University Press.


### *Spherical tokamak*

Peng, M. (1998). The spherical torus pathway to fusion power. *Innovative Confinement Concepts Workshop*. Princeton: New Jersey Princeton Plasma Physics Laboratory.
Spherical Torus White Paper (1999). US Spherical Torus Fusion Energy Science Research. *Fusion Summer Study*, Snowmass, Colorado.

## Stellarator

Boozer, A. H. (1982). Establishment of magnetic coordinates for a given magnetic field. *Physics of Fluids*, **25**, 520.

Boozer, A. H. (2004). Physics of magnetically confined plasmas. *Reviews of Modern Physics*, **76**, 1071.

Miyamoto, K. (2001). *Fundamentals of Plasma Physics and Controlled Fusion*, revised edn. Toki City: National Institute for Fusion Science.

National Stellarator Program Planning Committee (1998). US Stellarator program plan. *Innovative Confinement Concepts Workshop*. Princeton: New Jersey, Princeton Plasma Physics Laboratory.

## Problems

13.1 Calculate the three $\beta$ limits (i.e., equilibrium, $m = 1$, $m = 0$) for LDX assuming the pressure profile is given by

$$p(r) = K \frac{r^{2(\gamma-1)} - r_1^{2(\gamma-1)}}{\left[r^{2(\gamma-1)} + r_2^{2(\gamma-1)}\right]^{(2\gamma-1)/(\gamma-1)}}.$$

Assume $\gamma = 5/3$ and plot the $\beta$ limits as a function of $r_1/r_2$.

13.2 The following simple model can be used to describe the evolution of an RFP. Assume a straight cylindrical plasma is contained within a perfectly conducting wall of radius $b = 0.4$ m. Initially the chamber is filled with pre-ionized plasma characterized by $n = 10^{20}$ m$^{-3}$, $T_e = T_i = 10$ eV, and a toroidal bias field $B_z = 0.05$ T. At $t = 0$ a toroidal current flows on the plasma surface. As time evolves, $I$ increases slowly in amplitude, but always remains a surface current.
   (a) Using the conservation of mass, flux, energy, and momentum, calculate the toroidal current $I$ required to compress the plasma to a radius $r = 0.16$ m. Assume the evolution is adiabatic with $\gamma = 5/3$.
   (b) Calculate the final value of $\beta_p$.
   (c) Calculate the final value of $B_z(0)$.

13.3 High-performance operation of the high-field Alcator C-Mod tokamak is characterized by $\bar{n} = 2 \times 10^{20}$ m$^{-3}$, $\overline{T}_e = \overline{T}_i = 3$ keV and $B_\phi = 8$ T. Assume the density profile is uniform and the temperature profiles have the form $T = 2\overline{T}(1 - r^2/a^2)$. For simplicity assume a circular cylindrical geometry.
   (a) Calculate the average particle pressure in atmospheres.
   (b) Calculate the value of $\bar{\beta}$.
   (c) Assume $q(0) = 1$, $q(a) = 3$, $R_0 = 0.68$ m. The equivalent circular minor radius is $a = 0.3$ m. If the toroidal current density is of the form $J_z = J_0(1 - r^2/a^2)^\nu$ calculate the value of $\nu$ and the total toroidal current $I$ flowing in the plasma.
   (d) Calculate the ratio of the poloidal to toroidal magnetic field at $r = a$.

13.4 A superconducting tokamak reactor is to be designed with the following properties: (1) $P_W = 4$ MW/m$^2$ (maximum wall loading), (2) $B_{max} = 13$ T (maximum TF field at the coil), (3) $b = 1.2$ m (blanket-and-shield thickness), and $\overline{T}_e = \overline{T}_i = 15$ keV. Assume the reactor is circular in cross section. To make efficient use of the blanket the minor radius is chosen as $a = (3/2)b$. Show that to achieve the desired wall loading there is an optimum value of $\varepsilon = a/R_0$ that minimizes the required value of $\bar\beta/\varepsilon$. Calculate the optimum $\varepsilon$ and the minimum $\bar\beta/\varepsilon$.

13.5 A simple model for the ohmic heating transformer circuit of a tokamak is given by

$$V = R_1 I_1 + L_1 \dot I_1 - M \dot I_2 \qquad \text{primary circuit (power supply)},$$
$$0 = R_2 I_2 + L_2 \dot I_2 - M \dot I_1 \qquad \text{secondary circuit (plasma)}.$$

For simplicity neglect $R_1$ and assume $V = \text{const.}$
(a) Derive an expression for the time required for the plasma current $I_2$ to reach 95% of its final value at $t \to \infty$ assuming $I_2(0) = 0$.
(b) In (a) the plasma resistance $R_2$ is assumed constant during the entire evolution. In reality $R_2$ is large initially when the plasma is cold and the current is low. As time progresses $R_2$ decreases as a result of increased plasma heating. This effect can be modeled by assuming that $R_2 = R_2(I_2) = R_f(I_f/I_2)^{1/2}$, where the subscript f denotes the values at $t \to \infty$. Using this model derive an expression for the time required for $I_2$ to reach 95% of its final value. Compare this result with that obtained in part (a).

13.6 A cylindrical tokamak with minor radius $a$ has a safety factor profile given by $q(r) = q_0(1 + 2r^2/a^2)$. Find the average radius of the plasma, defined as $\langle a^2 \rangle \equiv I/\pi J_\phi(0)$, if $a = 0.2$ m, $R_0 = 0.8$ m.

13.7 It is often said that a tokamak with elliptic flux surfaces near the magnetic axis is capable of sustaining a higher current density on axis in the presence of sawtooth oscillations. This statement is investigated as follows. Assume a "straight" elliptically shaped tokamak of length $2\pi R_0$. The vector potential for the poloidal magnetic field near the axis is given by
$$A(x, y) \approx A_{xx}(0, 0)(x^2/2) + A_{yy}(0, 0)(y^2/2).$$ Now, define the flux surface elongation as $\kappa = (A_{xx}/A_{yy})^{1/2}$ and the current density on axis as $\mu_0 J_0 = -\nabla^2 A = -(A_{xx} + A_{yy})$.
(a) Calculate the safety factor on-axis and show that it can be written as $q_0 = K/(A_{xx}A_{yy})^{1/2}$. Find $K$.
(b) Show that the current density on-axis can be written as $J_0 = 2B_0 f(\kappa)/\mu_0 R_0 q_0$. Find $f(\kappa)$ and determine whether higher $\kappa$ allows higher $J_0$ when $q_0 = 1$.

13.8 This problem provides an estimate of how much paramagnetism an ohmic discharge develops as the toroidal current is increased from zero to a final value $I$. To make the analysis simple assume that as the current rises, the plasma pressure remains negligible. Thus at any instant during the evolution the 1-D radial pressure balance relation can be considered force free: $\mu_0 \mathbf{J} = \alpha \mathbf{B}$. As a further simplification assume that during the evolution $\alpha(r, t) \approx \alpha(t)$ is independent of radius and slowly varying in time.
(a) Calculate $B_z(r, t)$ and $B_\theta(r, t)$ for arbitrary $\alpha(t)$. Assume the edge of the plasma is at $r = a$ and that $B_z(a, t) = B_0$.
(b) Calculate the diamagnetism $\Delta B_z/B_0$, where $\Delta B_z$ is the change in $B_z(0, t)$ from the final to initial state. Derive an analytic expression for $\Delta B_z/B_0$ as a function of $\varepsilon$ and $q_a$. Evaluate $\Delta B_z/B_0$ for $\varepsilon = 1/4$ and $q_a = 3$.

13.9 The purpose of this problem is to calculate the duration of the flat top portion of a transformer driven tokamak pulse. Consider the device illustrated in Fig. 13.52.

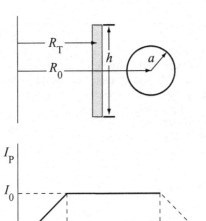

Figure 13.52

(a) The transformer operates in a "double swing" mode. That is, at time $t = 0$ there is a negative DC current in the transformer producing a magnetic field in the coil equal to $B(0) = -8$ T. A voltage is then applied to the transformer coil that swings the current through zero to a maximum value at $t = t_2$ corresponding to a field in the coil equal to $B(t_2) = +8$ T. The various dimensions of the system are as follows: $a = 0.2$ m, $R_0 = 0.8$m, $R_T = 0.5$ m, $h = 1$ m. How many volt seconds are available from the transformer from $t = 0$ to $t = t_2$?

(b) The next step involves the steady state power balance between ohmic heating and thermal conduction losses. All other sources and sinks are negligible. Calculate the plasma resistance $R_p$ if the energy confinement time $\tau_E = 0.05$ s, the plasma density $n = 2 \times 10^{20}$ m$^{-3}$, and the plasma resistivity $\eta = 2.5 \times 10^{-8}/T_k^{3/2}$ $\Omega$ m. Your answer should be of the form $R_p = R_p(I)$, where $I$ is the toroidal plasma current. Note that $R_p$ is a non-linear resistance.

(c) The self-inductance of the transformer and plasma are $L_T = 1.0 \times 10^{-6}$ H and $L_p = 2 \times 10^{-6}$ H respectively. Their mutual inductance is $M = 0.8 \times 10^{-6}$ H. Write down a set of coupled circuit equations describing the primary and secondary of the transformer. Include the plasma resistance calculated in part (b) but neglect the resistance in the primary.

(d) Assume the desired current in the plasma has the form shown in Fig. 13.52, where the flat-top current $I_0 = 1$ MA. Set the value of $t_1 = 1$ s in order not to ramp the current up too fast. Calculate the flat-top time $\Delta t = t_2 - t_1$ such that at $t = t_2$ the total volt seconds, as calculated in part (a), have been consumed. To do this part of the problem you will have to calculate the initial current in the primary so that at $t = 0$, the field in the coil is $B(0) = -8$ T.

13.10 Analytic theory shows that the MHD stability limit against ballooning modes in a large aspect ratio, circular cross section torus is approximately given by $\alpha \leq 0.6s$, where $s(r) = rq'/q$ is the shear and $\alpha(r) = -q^2 R_0 \beta'$ is a normalized form of the gradient in toroidal $\beta$. This is a local inequality that must be satisfied at every radius in the plasma.

(a) To determine the maximum stable $\beta$ against ballooning modes assume that the stability condition is marginally satisfied at every value of radius; that is, $\alpha(r) = 0.6s(r)$ for $0 < r < a$. If the safety factor is given by $q(r) = q_0/(1 - kr^3/a^3)^{1/2}$ with $k$ determined by the condition $q_a = q_0/(1 - k)^{1/2}$, find the maximum possible volume averaged $\beta$ as a function of $a$, $R_0$, $q_0$, $q_a$.

(b) What is the numerical value when $a = 2$ m, $R_0 = 6$ m, $q_0 = 1$, $q_a = 2.5$? Compare your results with the Troyon limit.

# 14

# Transport

## 14.1 Introduction

For the plasma in a fusion reactor to be self-sustaining at $T = 15$ keV the alpha particle heating must balance the losses due to thermal conduction as described by the familiar condition

$$p\tau_E = \frac{24}{E_\alpha} \frac{T^2}{\langle \sigma v \rangle}. \tag{14.1}$$

Recall now that in a reactor the pressure is set by the power density required to achieve a desired output power, while the temperature is determined by minimizing the ratio $T^2/\langle \sigma v \rangle$. Therefore, achieving self-sustained ignited operation leads to a requirement on the value of the energy confinement time $\tau_E$.

Understanding and controlling energy confinement is the domain of transport theory, and is the main objective of Chapter 14. In a plasma there are three important types of transport: heat conduction, particle diffusion, and magnetic field diffusion. Of these, heat conduction is the most serious loss mechanism and consequently is the main focus of the discussion. Also, most of the analysis involves the tokamak since it is for this configuration that most of the theory is formulated and most of the data have been collected and analyzed.

Most fusion researchers would agree that understanding heat conduction has been the most difficult challenge on the path to a reactor. The reason is that transport in a plasma is almost always dominated, not by Coulomb collisions, but by plasma turbulence driven by micro-instabilities. Understanding the resulting anomalous transport requires sophisticated kinetic models and non-linear, multi-dimensional numerical simulations.

After many years of research great strides have been made towards determining a first principles theory of anomalous heat conduction. Even so, there is still a long way to go before the theory can be used as a reliable design tool for new devices. By and large, new designs are based on empirical scaling relations derived from an extensive database of experimental measurements. These empirical relations predict that the tokamak has a reasonably good chance of achieving self-sustained ignited operation, although the safety margins are not large. Also, the predictions should be treated with caution as they invariably involve extrapolations beyond the experimental regimes represented in the database.

To introduce some order into the understanding of the plasma transport problem, Chapter 14 starts with a simple model, classical transport in a 1-D cylinder, and adds increasing levels of complexity, finally arriving at transport in a tokamak reactor. The material is organized as follows.

First, the fluid equations describing heat, particle, and magnetic field diffusion in a 1-D cylinder are derived using the low-$\beta$ tokamak expansion. The analysis shows how both particle and magnetic field diffusion arise naturally from the resistive MHD model. The dominant thermal diffusion coefficient is, however, not derivable from this model. It is instead derived independently by a simple heuristic calculation based on the "random walk" model. The analysis shows that heat transport is much larger than particle transport and is dominated by the ions. The end result of this discussion is a set of well-posed, 1-D transport equations.

Second, even though experimentally observed diffusion coefficients are much larger that the simple 1-D values derived above, it is a worthwhile preliminary goal to learn how to mathematically formulate and solve transport equations assuming the diffusion coefficients are known. Towards this goal several simple applications of the 1-D transport model are investigated, including temperature equilibration, off-axis heating, and ohmically heating to ignition.

The third topic involves neoclassical transport theory, which corresponds to classical Coulomb transport in a toroidal geometry. Although one might initially expect toroidicity to simply add small $a/R_0$ corrections to the cylindrical results, the actual results are much different. Toroidal effects typically lead to an increase of nearly two orders of magnitude in the ion thermal diffusivity, a consequence of the effects of guiding center particle drifts in toroidal geometry, particularly as they affect trapped particles. Even with this large increase the neoclassical ion thermal diffusivity is still noticeably smaller than experimental observations. The final neoclassical topic is a simple derivation of the bootstrap current $J_B$. Recall, that $J_B$ is a natural, transport-induced, toroidal plasma current essential for the AT mode of tokamak operation, the purpose of which is to reduce the requirements on the external current drive system.

The next topic in the logical progression should involve a discussion of micro-instability driven anomalous transport. This is, however, beyond the scope of the present book. Instead, attention is focused on the macroscopic consequences of anomalous transport which appear as empirical scaling relations for $\tau_E$. It is shown how the generalized 1-D transport equations are simplified to a 0-D form in which the thermal transport is modeled in terms of $\tau_E$. A description is presented showing how $\tau_E$ is calculated in practice and expressions are given for two typical modes of tokamak operation: the L-mode (for low confinement) and the H-mode (for high confinement). It is worth emphasizing that $\tau_E$ models the global thermal transport of the plasma core. Even so, there are several important transport phenomena associated with the plasma edge that directly and indirectly affect core transport. Here too the treatment is primarily empirical. A brief description is given of these transport-related edge phenomena.

Lastly, these results are combined to investigate several practical transport applications related to fusion power production. First, a simple optimized tokamak ignition experiment is designed. The resulting parameters are very similar to those of ITER. Also, the analysis shows that the size of the ignition experiment is comparable to a full scale reactor. Next, the questions of thermal stability and the minimum auxiliary power required for ignition are re-examined in the context of the empirical scaling relations. The results are noticeably more favorable than those derived in Chapter 4. Finally, the fraction of the total current carried by the bootstrap current is calculated for standard and AT tokamaks. It is shown that this fraction is small for standard operation. The higher required values of bootstrap fraction are potentially achievable for AT operation but will likely require operation in a regime where the resistive wall mode is excited, thus necessitating the need for feedback stabilization.

## 14.2 Transport in a 1-D cylindrical plasma

### 14.2.1 Fluid model

#### The starting equations

In this subsection a derivation is presented of the fluid equations describing the transport of mass, energy, and magnetic flux in a plasma. The transport of momentum associated with viscosity is neglected as this is usually not a dominant effect. For simplicity, the analysis is carried out in a 1-D cylindrical geometry. Even so, because of the large number of physical variables involved, the starting model is quite complicated and further approximations must be made to reduce the equations to a tractable form. The end goal is to obtain a set of diffusion-like equations of the form

$$\frac{\partial Q}{\partial t} = \frac{1}{r}\frac{\partial}{\partial r}\left(rD\frac{\partial Q}{\partial r}\right) + S(Q, r, t) \tag{14.2}$$

for each physical variable $Q$ and to identify the corresponding diffusion coefficient $D$ and source and sink terms contained in $S$.

The transport behavior is described by the single-fluid resistive MHD model with several modifications and caveats as described below.

(a) Since the characteristic time scale for transport is long compared to the ideal MHD time scale, it is a good approximation to neglect the inertial terms in the MHD momentum equation. Thus, as the system slowly evolves in time, the plasma passes through a continuing sequence of quasi-static MHD equilibria, each satisfying $\mathbf{J} \times \mathbf{B} = \nabla p$.

(b) Ohm's law is modified by separating the resistivity into perpendicular and parallel components: $\eta \mathbf{J} \rightarrow \eta_\perp \mathbf{J}_\perp + \eta_\parallel \mathbf{J}_\parallel$. This decomposition allows one to distinguish particle diffusion (related to $\eta_\perp$) from magnetic field diffusion (related to $\eta_\parallel$). Classically, it turns out that $\eta_\perp \approx 2\eta_\parallel$, so there is not much difference. However, in real experiments the particle diffusion is highly anomalous which can be approximately modeled by a strongly enhanced $\eta_\perp \gg \eta_\parallel$.

(c) The energy equation is generalized from the simple adiabatic form used in ideal MHD to the more general form familiar in fluid dynamics, and described in Section 4.2. The general form adds thermal conduction as well as sources and sinks to the adiabatic convection and compression effects. A single energy equation is used for the plasma based on the reasonable assumption that in a fusion plasma $T_e \approx T_i \equiv T$. The issue of temperature equilibration is discussed later in the section.

These modifications are substituted into the 1-D cylindrical MHD equations. The non-trivial physical variables are all functions of $(r, t)$ and correspond to $n$, $T$, $\mathbf{v} = v\mathbf{e}_r$, $\mathbf{B} = B_\theta \mathbf{e}_\theta + B_z \mathbf{e}_z$, $\mathbf{E} = E_\theta \mathbf{e}_\theta + E_z \mathbf{e}_z$. The starting equations describing the transport model can now be written as:

$$\frac{\partial n}{\partial t} + \frac{1}{r}\frac{\partial}{\partial r}(rnv) = 0 \qquad \text{mass;}$$

$$\frac{\partial}{\partial r}\left(p + \frac{B_z^2}{2\mu_0}\right) + \frac{B_\theta}{\mu_0 r}\frac{\partial}{\partial r}(rB_\theta) = 0 \qquad \text{momentum;}$$

$$\mathbf{E} + \mathbf{v} \times \mathbf{B} = \eta_\perp \mathbf{J}_\perp + \eta_\parallel \frac{J_\parallel}{B}\mathbf{B} \qquad \text{Ohm's law;}$$

$$3n\left(\frac{\partial T}{\partial t} + v\frac{\partial T}{\partial r}\right) + \frac{2nT}{r}\frac{\partial}{\partial r}(rv) = -\nabla \cdot \mathbf{q} + S \qquad \text{energy;} \qquad (14.3)$$

$$\partial B_\theta / \partial t = \partial E_z / \partial r \qquad \text{Maxwell;}$$

$$\frac{\partial B_z}{\partial t} = -\frac{1}{r}\frac{\partial}{\partial r}(rE_\theta) \qquad \text{Maxwell;}$$

$$\mu_0 J_\theta = -\partial B_z / \partial r \qquad \text{Maxwell;}$$

$$\mu_0 J_z = \frac{1}{r}\frac{\partial}{\partial r}(rB_\theta) \qquad \text{Maxwell.}$$

In these equations $p = 2nT$ and the currents $\mathbf{J}_\perp$, $J_\parallel$ appearing in Ohm's law are given by[1]

$$\mathbf{J}_\perp = \frac{J_\theta B_z - J_z B_\theta}{B^2}(B_z \mathbf{e}_\theta - B_\theta \mathbf{e}_z) = \frac{1}{B^2}\frac{\partial p}{\partial r}(B_z \mathbf{e}_\theta - B_\theta \mathbf{e}_z),$$

$$J_\parallel = \frac{J_\theta B_\theta + J_z B_z}{B}. \qquad (14.4)$$

The source and sink term $S$ in the energy equation consists of ohmic heating, external heating, fusion alpha particle heating and radiation losses. Finally, the heat flux vector is expressed in terms of thermal diffusivity in the usual manner

$$\mathbf{q} = -n\chi\frac{\partial T}{\partial r}\mathbf{e}_r. \qquad (14.5)$$

At this point the diffusivity $\chi$ is unspecified. Its value will be derived and discussed in future subsections. For the moment, readers should just assume it is a known quantity.

The starting equations are now specified and one can see they are quite complicated mathematically. The reason is as follows. Observe that there are four time evolution equations for the four quantities $n$, $T$, $B_\theta$, $B_z$. However, these quantities are also coupled through a fifth equation, the quasi-static pressure balance relation. The number of equations equals the number of unknowns because of the fifth unknown quantity $v$ which appears in the equations but whose behavior is not determined by a time evolution equation. In general, it is not easy to eliminate $v$ to obtain a closed set of transport equations. One approach to overcome this difficulty is to introduce the low-$\beta$ tokamak expansion into the model, resulting in an explicit determination of $v$. This is the next task.

---

[1] There is actually an additional term, not previously calculated, in Ohm's law known as the "thermo-electric effect," but this term does not have a dominant effect and is neglected for simplicity.

### Reduction of the model

The first step in the reduction of the model is to eliminate the electric field in Faraday's law by means of Ohm's law. A short calculation leads to the following time evolution equations for $B_\theta$, $B_z$:

$$\frac{\partial B_\theta}{\partial t} + \frac{\partial}{\partial r}(B_\theta v) = -\frac{\partial}{\partial r}\left(\frac{\eta_\perp B_\theta}{B^2}\frac{\partial p}{\partial r} - \eta_\parallel \frac{B_z}{B}J_\parallel\right);$$

$$\frac{\partial B_z}{\partial t} + \frac{1}{r}\frac{\partial}{\partial r}(r B_z v) = -\frac{1}{r}\frac{\partial}{\partial r}\left[r\left(\frac{\eta_\perp B_z}{B^2}\frac{\partial p}{\partial r} + \eta_\parallel \frac{B_\theta}{B}J_\parallel\right)\right].$$

$$(14.6)$$

The next step is to introduce the low-$\beta$ tokamak expansion, which basically assumes that the dominant component of magnetic field points in the axial direction and is independent of $r$ and $t$. Specifically, one writes $B_z(r, t) = B_0 + \delta B_z(r, t)$, where $B_0 = $ const. and $\delta B_z \ll B_0$. The ordering for the other quantities is given by

$$2\mu_0 p/B_0^2 \sim B_\theta^2/B_0^2 \sim \delta B_z/B_0 \ll 1. \qquad (14.7)$$

The simplification that arises is seen by examining the left hand side of the $B_z$ evolution equation and making use of the assumption $\delta B_z \ll B_0$:

$$\frac{\partial B_z}{\partial t} + \frac{1}{r}\frac{\partial}{\partial r}(r B_z v) \approx \left(\frac{\partial}{\partial t} + v\frac{\partial}{\partial r}\right)\delta B_z + \frac{B_0}{r}\frac{\partial}{\partial r}(rv) \approx \frac{B_0}{r}\frac{\partial}{\partial r}(rv). \qquad (14.8)$$

Using this approximation allows one to integrate the $B_z$ evolution equation with respect to $r$ obtaining an explicit expression for $v$:

$$v \approx -\frac{\eta_\parallel}{\mu_0 B_0^2}\frac{B_\theta}{r}\frac{\partial}{\partial r}(r B_\theta) - \frac{\eta_\perp}{B_0^2}\frac{\partial p}{\partial r}. \qquad (14.9)$$

Here use has been made of the ordering approximation $J_\parallel \approx J_z$. The expression for $v$ is now substituted into the time evolution equations for $n$, $T$, $B_\theta$, leading to a set of simplified transport equations that can be written as

$$\frac{\partial n}{\partial t} = \frac{1}{r}\frac{\partial}{\partial r}\left[r D_n\left(\frac{\partial n}{\partial r} + \frac{n}{T}\frac{\partial T}{\partial r} + \frac{2\eta_\parallel}{\beta_p \eta_\perp}\frac{n}{r B_\theta}\frac{\partial r B_\theta}{\partial r}\right)\right],$$

$$3n\frac{\partial T}{\partial t} = \frac{1}{r}\frac{\partial}{\partial r}\left(rn\chi\frac{\partial T}{\partial r}\right) + S,$$

$$\frac{\partial r B_\theta}{\partial t} = r\frac{\partial}{\partial r}\left(\frac{D_B}{r}\frac{\partial r B_\theta}{\partial r}\right).$$

$$(14.10)$$

Here, $\beta_p = 4\mu_0 n T/B_\theta^2 \sim 1$ and the magnetic field and particle diffusion coefficients $D_B$, $D_n$ are given by

$$D_B = \frac{\eta_\parallel}{\mu_0},$$

$$D_n = \frac{2n T \eta_\perp}{B_0^2}. \qquad (14.11)$$

Also, in the energy equation use has been made of the fact that in classical transport theory $\chi \gg D_n$ so that the convection and compression terms are small. The condition $\chi \gg D_n$ is proven in the next subsection.

These equations represent the desired model for classical transport in a 1-D cylinder. They have the form of three coupled non-linear partial differential equations whose form is similar to the generic transport equation given by Eq. (14.2). Note that, in general, the transport coefficients are not constants but are functions of $n$, $T$, $B_0$. Aside from several minor modifications due to the cylindrical geometry, the main difference from the generic form is in the density equation. This equation shows that the density evolution is also coupled to the temperature and magnetic field gradients. Lastly, observe that the quantity $\delta B_z$ does not appear in any of these equations. The quantity $\delta B_z$ is obtained from the pressure balance relation once the other quantities have been determined.

Although substantially simpler than the starting equations, the reduced transport model is still quite difficult to solve. Several special examples are discussed shortly. First, however, attention is focused on the problem of calculating $\chi$ and comparing it to the other transport coefficients.

### 14.2.2  Calculating transport coefficients from the random walk model

#### Introduction

The reduction of the fluid model has shown that particle diffusion and magnetic field diffusion arise from the presence of resistivity, which in turn arises from net momentum exchange Coulomb collisions. The corresponding Coulomb collision analysis, presented in Chapter 9, does not, however, lead to thermal diffusion, viscosity, or the different values of $\eta_\parallel$ and $\eta_\perp$. The reason is that the distribution functions used in the derivations are either pure Maxwellians, or else Maxwellians with slight shifts in average velocity and slightly different temperatures. The "missing" phenomena above arise from non-Maxwellian modifications to the distribution function not included in the analysis. To calculate these modifications requires the solution of a more basic kinetic model that directly determines the distribution functions $f_{e,i}(\mathbf{r}, \mathbf{v}, t)$. These are more complex calculations and are beyond the scope of the present book.

Instead, the approach taken here is to derive the transport coefficients using a much simpler model known as the "random walk model." The end result is a set of expressions for the electron and ion thermal conductivities $\chi_e$, $\chi_i$. It also reproduces the particle diffusion coefficient $D_n$ obtained from the fluid theory given by Eq. (14.11). The random walk model contains the essential physics of the diffusion process, although a more sophisticated kinetic theory is required to accurately calculate the numerical multipliers for each transport coefficient.

#### The random walk model

The idea behind the random walk model is to show how a particle diffuses away from its initial position as a result of a series of random collisions. The diffusion process is

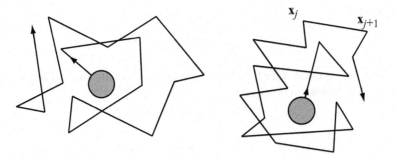

Figure 14.1 Trajectory of two particles undergoing random collisions. Note that on average $\Delta \mathbf{x} = 0$, while $(\Delta \mathbf{x})^2 \neq 0$.

characterized by a "diffusion coefficient $D$" which depends on the mean time and the average distance traveled between collisions.

To begin, consider the motion of a particle undergoing random collisions as shown in Fig. 14.1. Observe that the particle moves with its smooth, long-range velocity until it experiences a collision after which there is an abrupt change in the direction of motion. Each collisional change in direction is assumed to be random. Now, define $\Delta \mathbf{x}_j = \mathbf{x}_{j+1} - \mathbf{x}_j$, representing the difference in the position of the particle just after collision $j$ and just before collision $j + 1$. The total change in the position of a particle after $N$ collisions is given by

$$\Delta \mathbf{x} = \Delta \mathbf{x}_1 + \Delta \mathbf{x}_2 + \Delta \mathbf{x}_3 + \cdots + \Delta \mathbf{x}_N = \sum_j \Delta \mathbf{x}_j. \tag{14.12}$$

If the collisions are purely random, then one expects that for an ensemble of particles

$$\langle \Delta \mathbf{x} \rangle = 0. \tag{14.13}$$

A particle is just as likely to be above as below or to the left or right of its initial position.

On the other hand the mean square distance from its starting point is not zero. This follows by noting that the mean square distance is defined as

$$(\Delta x)^2 = \sum_{i,j} \Delta \mathbf{x}_i \cdot \Delta \mathbf{x}_j. \tag{14.14}$$

The terms in the sum with $i \neq j$ average to zero because of the random nature of the collisions. However, the contributions for $i = j$ do not cancel and the ensemble average reduces to

$$\langle (\Delta x)^2 \rangle = \sum_j (\Delta x_j)^2. \tag{14.15}$$

One now defines $(\Delta l)^2 \equiv [(\Delta x_1)^2 + (\Delta x_2)^2 + \cdots + (\Delta x_N)^2]/N$ as the magnitude of the average step size between collisions. Thus, after $N$ collisions a typical particle has diffused a mean square distance

$$\langle (\Delta x)^2 \rangle = \sum_j (\Delta x_j)^2 = N(\Delta l)^2. \qquad (14.16)$$

Next, assume that the average time between collisions is defined as $\tau$. Then, the time $\Delta t$ required for $N$ collisions to occur is just

$$\Delta t = N\tau. \qquad (14.17)$$

Eliminating $N$ then leads to the following relation between $\Delta x$ and $\Delta t$:

$$(\Delta x)^2 = D\Delta t, \qquad (14.18)$$

where

$$D = \frac{(\Delta l)^2}{\tau} \qquad (14.19)$$

is defined as the diffusion coefficient. Observe that the mean square distance traveled by the particle as a result of random collisions scales as $\Delta x \sim (\Delta t)^{1/2}$. This should be contrasted with collisionless directed motion in which case $\Delta x \sim \Delta t$. The square root dependence associated with collisional diffusion leads to a much slower motion of the particle because of the frequent random changes in direction.

The conclusion from this analysis is that in order to calculate the diffusion coefficient resulting from a series of random collisions one needs to know the average step size between collisions $\Delta l$ and the mean time between collisions $\tau$. Equation (14.19) then gives the diffusion coefficient. This formulation is now applied to derive the particle and energy diffusion coefficients in a magnetized cylindrical plasma.

### 14.2.3 Particle diffusion in a magnetized plasma

The perpendicular particle diffusion coefficient in a magnetized plasma can be evaluated in a reasonably straightforward manner as described above, although two modifications must be made in the analysis. First, perpendicular to the field the orbits between collisions are not straight lines, but are instead circular gyro orbits. Second, for like particle collisions each particle may have its orbit changed by a comparable amount after each collision. Thus, the average step size must include the motion of both particles during each collision in contrast to the single-particle analysis described in the previous subsection.

Based on these modifications, one might make the following simple estimate of the perpendicular particle diffusion coefficient. Consider the two-particle collision illustrated in Fig. 14.2. Assume both particles are ions. Before the collision particle 1 has a circular orbit with gyro radius $r_{Li}$. After the collision the particle has been scattered in a random direction and once again assumes its gyro motion. On average the guiding center of the particle has shifted by a distance comparable to a gyro radius. A similar shift takes place for

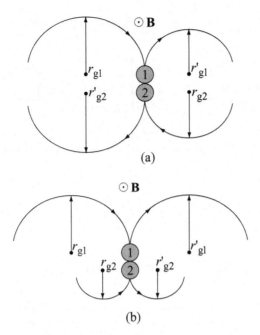

Figure 14.2 (a) Like particle collision in a magnetic field, (b) unlike particle collision in a magnetic field. Note that there is no shift in the center of mass for the like particles before and after the collision.

particle 2. Consequently one might deduce that the average step size between collisions is just $\Delta l \approx r_{Li}$. Combining this estimate with the fact that the mean time between collisions is $\tau_{ii} = (\nu_{ii})^{-1}$ leads to the conclusion that the particle diffusion coefficient for ions is given by $D_i \approx r_{Li}^2/\tau_{ii}$. This conclusion is *wrong*!

As appealing as the simple physical picture may be, the analysis presented below shows that like particle collisions do not lead to particle diffusion. It is only unlike collisions that lead to particle transport and in fact there is only a single diffusion coefficient, implying that both electrons and ions diffuse at the same rate.

The crucial step in the analysis is the definition of the step size between collisions involving two equal mass particles. The appropriate definition that makes physical sense is to define the step size as the difference in the location of the center of mass of the two particles before and after the collision. If there is a diffusion of the two-particle center of mass, both particles will ultimately be lost through a sequence of collisions. If not, the particles do not escape.

The analysis below calculates the step size based on the center of mass definition for two like particles colliding at an arbitrary angle with respect to one another and then having a random scattering collision. The resulting change in the center of mass is then averaged over all collisions showing that like particles do not lead to particle diffusion. The calculation is then repeated for unlike particle collisions and it is here that particle diffusion arises.

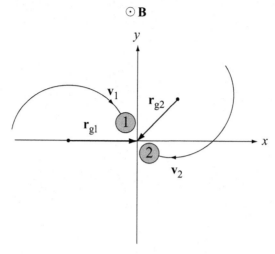

Figure 14.3 Geometry of an ion–ion collision in the laboratory reference frame.

The analysis itself is slightly simplified by assuming that all motion is 2-D in the plane perpendicular to the magnetic field. In other words, particles are not allowed to scatter in such a way that their parallel velocity is altered. This assumption is not essential to the final result but eliminates substantial amounts of unnecessary, complicating algebra.

### Like particle analysis

Consider now two colliding ions as illustrated in Fig. 14.3. The first task is to calculate the location of the guiding center of each particle assuming the coordinate system has been chosen such that the actual collision takes place at the origin $x = 0$, $y = 0$. Recall from Chapter 8 that the orbit of the ion is given by

$$
\begin{aligned}
v_x &= v_\perp \cos(\omega_c t - \phi), \\
v_y &= -v_\perp \sin(\omega_c t - \phi), \\
x &= x_g + r_L \sin(\omega_c t - \phi), \\
y &= y_g + r_L \cos(\omega_c t - \phi).
\end{aligned}
\tag{14.20}
$$

Here, $\omega_c = eB_0/m$ and $\mathbf{B} = B_0\mathbf{e}_z$. (An identical analysis follows for electron–electron collisions by simply replacing $\omega_c \to -\omega_c$.) If the collision takes place at the origin, then the location of the guiding center for each particle at the point of impact is given by

$$
\begin{aligned}
\mathbf{r}_{g1} &= x_{g1}\mathbf{e}_x + y_{g1}\mathbf{e}_y = \frac{\mathbf{v}_1 \times \mathbf{e}_z}{\omega_c}, \\
\mathbf{r}_{g2} &= x_{g2}\mathbf{e}_x + y_{g2}\mathbf{e}_y = \frac{\mathbf{v}_2 \times \mathbf{e}_z}{\omega_c},
\end{aligned}
\tag{14.21}
$$

where $\mathbf{v}_1$ and $\mathbf{v}_2$ are the perpendicular velocities of the particles.

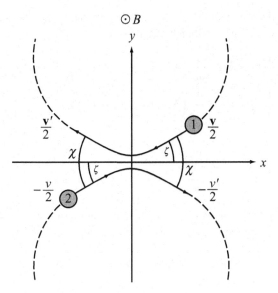

Figure 14.4 Geometry of an ion–ion collision in the center of mass frame. Here $\zeta$ is the angle between the relative velocity and the laboratory coordinate system, while $\chi$ is the random scattering angle.

The next step is to calculate the guiding centers of the particles just after a random Coulomb collision. This is most easily done in the center of mass reference frame, which for like particles is defined by the transformation

$$\mathbf{v}_1 = \frac{\mathbf{v}_1 + \mathbf{v}_2}{2} + \frac{\mathbf{v}_1 - \mathbf{v}_2}{2} = \mathbf{V} + \frac{1}{2}\mathbf{v},$$
$$\mathbf{v}_2 = \frac{\mathbf{v}_1 + \mathbf{v}_2}{2} - \frac{\mathbf{v}_1 - \mathbf{v}_2}{2} = \mathbf{V} - \frac{1}{2}\mathbf{v}.$$
(14.22)

In the center of mass frame moving with $\mathbf{V}$ the collision has the form illustrated in Fig. 14.4. Here $\mathbf{e}_\zeta = \mathbf{e}_x \cos \zeta + \mathbf{e}_y \sin \zeta$, where $\zeta$ is an arbitrary angle between the direction of the relative velocity vector and the laboratory coordinate system. The collision scatters the particle by a random angle $\chi$. Note that conservation of momentum and energy before and after the collision requires that each particle be scattered by the same angle $\chi$ and that $v'^2 = v^2$. Here and below unprimed and primed quantities refer to values before and after the collision respectively. Thus after the collision the particle velocities, transformed back into the laboratory frame are given by

$$\mathbf{v}_1' = \mathbf{V} + \frac{1}{2}(\mathbf{v} \cos \chi + \mathbf{e}_z \times \mathbf{v} \sin \chi),$$
$$\mathbf{v}_2' = \mathbf{V} - \frac{1}{2}(\mathbf{v} \cos \chi + \mathbf{e}_z \times \mathbf{v} \sin \chi).$$
(14.23)

The guiding center locations of the particles in the laboratory frame, expressed in terms of the center of mass velocities can then be written as

$$
\begin{aligned}
\mathbf{r}_{g1} &= \frac{1}{\omega_c}\mathbf{v}_1 \times \mathbf{e}_z = \frac{1}{\omega_c}\left(\mathbf{V} \times \mathbf{e}_z + \frac{1}{2}\mathbf{v} \times \mathbf{e}_z\right), \\
\mathbf{r}_{g2} &= \frac{1}{\omega_c}\mathbf{v}_2 \times \mathbf{e}_z = \frac{1}{\omega_c}\left(\mathbf{V} \times \mathbf{e}_z - \frac{1}{2}\mathbf{v} \times \mathbf{e}_z\right), \\
\mathbf{r}'_{g1} &= \frac{1}{\omega_c}\mathbf{v}'_1 \times \mathbf{e}_z = \frac{1}{\omega_c}\left(\mathbf{V} \times \mathbf{e}_z + \frac{1}{2}\mathbf{v} \times \mathbf{e}_z \cos\chi + \frac{1}{2}\mathbf{v}\sin\chi\right), \\
\mathbf{r}'_{g2} &= \frac{1}{\omega_c}\mathbf{v}'_2 \times \mathbf{e}_z = \frac{1}{\omega_c}\left(\mathbf{V} \times \mathbf{e}_z - \frac{1}{2}\mathbf{v} \times \mathbf{e}_z \cos\chi - \frac{1}{2}\mathbf{v}\sin\chi\right).
\end{aligned}
$$

(14.24)

One is now in a position to evaluate the center of mass of the two particles before and after the collision. Using the standard definition leads to the following expressions:

$$
\begin{aligned}
\mathbf{r}_{cm} &= \frac{1}{2}(\mathbf{r}_{g1} + \mathbf{r}_{g2}) = \frac{2}{\omega_c}\mathbf{V} \times \mathbf{e}_z, \\
\mathbf{r}'_{cm} &= \frac{1}{2}(\mathbf{r}'_{g1} + \mathbf{r}'_{g2}) = \frac{2}{\omega_c}\mathbf{V} \times \mathbf{e}_z.
\end{aligned}
$$

(14.25)

Note the cancellation of all the relative velocity terms.

The last step is to evaluate $\Delta\mathbf{r}$ for each collision and then average over all collisions. The quantity $\Delta\mathbf{r}$ is defined as the difference in the centers of mass of the two particles before and after the collision and is therefore given by

$$
\Delta\mathbf{r} = \mathbf{r}_{cm} - \mathbf{r}'_{cm} = 0.
$$

(14.26)

Remarkably, the shift in the center of mass for like particle collisions is identically zero for each and every collision. It is on this basis that the random walk model predicts that like particle Coulomb collisions produce no particle diffusion.

### Unlike particle analysis

A similar analysis can be carried out for unlike particle collisions. Two modifications are necessary. First separate masses must be introduced for each species. Second, the effects of the opposite charges must be taken into account. Steps analogous to the like particle analysis are now outlined below.

First, the transformation of the center of mass coordinates is defined by

$$
\begin{aligned}
\mathbf{v}_i &= \frac{m_i\mathbf{v}_i + m_e\mathbf{v}_e}{m_i + m_e} - \frac{m_e}{m_i + m_e}(\mathbf{v}_e - \mathbf{v}_i) = \mathbf{V} - \frac{m_e}{m_i + m_e}\mathbf{v}, \\
\mathbf{v}_e &= \frac{m_i\mathbf{v}_i + m_e\mathbf{v}_e}{m_i + m_e} + \frac{m_i}{m_i + m_e}(\mathbf{v}_e - \mathbf{v}_i) = \mathbf{V} + \frac{m_i}{m_i + m_e}\mathbf{v}.
\end{aligned}
$$

(14.27)

An illustration of the collision in the center of mass frame is given in Fig. 14.5. Note that conservation of momentum and energy again requires that the scattering angle $\chi$ be the

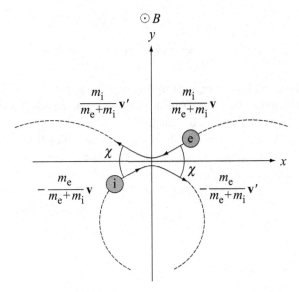

Figure 14.5 Geometry of an electron–ion collision in the center of mass frame. Note the opposite direction of rotation for the electron and ion.

same for both particles and that $v' = v$. Therefore the particle velocities after the collision are

$$\mathbf{v}'_{\mathrm{i}} = \mathbf{V} - \frac{m_{\mathrm{e}}}{m_{\mathrm{i}} + m_{\mathrm{e}}} (\mathbf{v} \cos \chi + \mathbf{e}_z \times \mathbf{v} \sin \chi),$$

$$\mathbf{v}'_{\mathrm{e}} = \mathbf{V} + \frac{m_{\mathrm{i}}}{m_{\mathrm{i}} + m_{\mathrm{e}}} (\mathbf{v} \cos \chi + \mathbf{e}_z \times \mathbf{v} \sin \chi). \tag{14.28}$$

The corresponding guiding center of mass locations can be written as

$$\mathbf{r}_{\mathrm{gi}} = \frac{1}{\omega_{\mathrm{ci}}} \left( \mathbf{V} \times \mathbf{e}_z - \frac{m_{\mathrm{e}}}{m_{\mathrm{i}} + m_{\mathrm{e}}} \mathbf{v} \times \mathbf{e}_z \right),$$

$$\mathbf{r}_{\mathrm{ge}} = -\frac{1}{\omega_{\mathrm{ce}}} \left( \mathbf{V} \times \mathbf{e}_z + \frac{m_{\mathrm{i}}}{m_{\mathrm{i}} + m_{\mathrm{e}}} \mathbf{v} \times \mathbf{e}_z \right),$$

$$\mathbf{r}'_{\mathrm{gi}} = \frac{1}{\omega_{\mathrm{ci}}} \left[ \mathbf{V} \times \mathbf{e}_z - \frac{m_{\mathrm{e}}}{m_{\mathrm{i}} + m_{\mathrm{e}}} (\mathbf{v} \times \mathbf{e}_z \cos \chi + \mathbf{v} \sin \chi) \right],$$

$$\mathbf{r}'_{\mathrm{ge}} = -\frac{1}{\omega_{\mathrm{ce}}} \left[ \mathbf{V} \times \mathbf{e}_z + \frac{m_{\mathrm{i}}}{m_{\mathrm{i}} + m_{\mathrm{e}}} (\mathbf{v} \times \mathbf{e}_z \cos \chi + \mathbf{v} \sin \chi) \right]. \tag{14.29}$$

Here $\omega_{\mathrm{c}j} = |e| B_0 / m_j$ and the sign of the charges has been taken into account.

One is again in a position to calculate $\Delta \mathbf{r}$ as the difference in the centers of mass before and after the collision:

$$\Delta \mathbf{r} = \frac{m_{\mathrm{i}} \mathbf{r}_{\mathrm{gi}} + m_{\mathrm{e}} \mathbf{r}_{\mathrm{ge}}}{m_{\mathrm{i}} + m_{\mathrm{e}}} - \frac{m_{\mathrm{i}} \mathbf{r}'_{\mathrm{gi}} + m_{\mathrm{e}} \mathbf{r}'_{\mathrm{ge}}}{m_{\mathrm{i}} + m_{\mathrm{e}}}. \tag{14.30}$$

A short calculation yields

$$\Delta \mathbf{r} = -\frac{1}{\omega_{cr}} [\mathbf{v} \times \mathbf{e}_z (1 - \cos \chi) - \mathbf{v} \sin \chi], \tag{14.31}$$

where $\omega_{cr} = eB_0/m_r$ and $m_r = m_e m_i/(m_i + m_e)$ is the reduced mass. Observe that for unlike particle collisions $\Delta \mathbf{r}$ does not vanish, even for equal masses. The non-vanishing is associated with the opposite sign of the charges.

The last step in the analysis requires the averaging over all collisions and all scattering angles assuming equal probabilities for all angles. Clearly the averaging over collisions (i.e., the averaging over the initial angle $\zeta$) leads to the conclusion that $\langle \Delta \mathbf{r} \rangle = 0$ as one would expect. However, the step size for diffusion involves the mean square average defined by

$$(\Delta l)^2 = \frac{1}{4\pi^2} \int_0^{2\pi} d\zeta \int_0^{2\pi} d\chi \, (\Delta \mathbf{r})^2. \tag{14.32}$$

A straightforward calculation yields

$$(\Delta l)^2 = 2 \frac{v^2}{\omega_{cr}^2} \approx 2 \frac{v^2}{\omega_{ce}^2}. \tag{14.33}$$

The final form is obtained by noting that for Maxwellian distribution functions in velocity $v = |\mathbf{v}_e - \mathbf{v}_i| \approx |\mathbf{v}_e| \sim (2T_e/m_e)^{1/2}$:

$$(\Delta l)^2 = 4 \frac{m_e T_e}{e^2 B_0^2}. \tag{14.34}$$

The particle diffusion coefficient $D_n$ is now easily evaluated by recognizing that the mean time between collisions is just $\bar{\tau}_{ei} = (\bar{\nu}_{ei})^{-1}$, where $\bar{\nu}_{ei}$ is the momentum exchange collision frequency given by Eq. (9.110). One obtains

$$D_n = \frac{(\Delta l)^2}{\bar{\tau}_{ei}} = 4 \frac{\bar{\nu}_{ei} m_e T_e}{e^2 B_0^2} \sim \frac{r_{Le}^2}{\bar{\tau}_{ei}}. \tag{14.35}$$

This is the desired result. The value of $D_n$ by definition includes the effects of both electrons and ions so that both species diffuse at the same rate, a phenomenon known as ambipolar diffusion. Physically the species must diffuse together since if one species were depleted faster than the other a large charge imbalance would occur. This charge imbalance would induce an electric field whose direction is such as to attract the species back to each other causing them to leave at the same rate.

A further interesting point is that the value of $D_n$ is smaller by a factor of $(m_e/m_i)^{1/2}$ than the original incorrect estimate $D_n \sim r_{Li}^2/\tau_{ii}$. Particle diffusion occurs on a slower time scale; that is, ions might originally be expected to diffuse faster because their larger gyro radius produces a larger step size after each collision. However, the fact that the center of mass in ion–ion collisions is invariant before and after a collision negates this original expectation.

## Comparison with the fluid model and numerical values

To test the validity of the random walk model it is useful to compare the value of $D_n$ just calculated in Eq. (14.35) with the value obtained from the fluid model and repeated here for convenience

$$D_n = \frac{2nT\eta_\perp}{B_0^2}. \qquad (14.36)$$

If one now recalls that $\eta_\perp = m_e \bar{\nu}_{ei}/ne^2$ it follows that the value of $D_n$ for the fluid model can be written as

$$D_n = 2\frac{\bar{\nu}_{ei} m_e T}{e^2 B_0^2}. \qquad (14.37)$$

The fluid model differs from the random walk model by an unimportant numerical factor of "2". The implication is that the fluid model is self-consistent in that it automatically takes into account the fact that particle diffusion is ambipolar.

Finally, it is useful as a point of reference to substitute numerical values into $D_n$ so that the scaling with respect to density, temperature, and magnetic field becomes apparent. An accurate calculation, including kinetic effects, has been carried out in a classic paper by Braginskii. His result can be written as

$$D_n = 2.0 \times 10^{-3} \frac{n_{20}}{B_0^2 T_k^{1/2}} \quad \text{m}^2/\text{s}. \qquad (14.38)$$

The fluid model value of $D_n$ in Eq. (14.37) yields the same numerical coefficient as that found by Braginskii. Observe that $D_n$ decreases with $T$ (fewer collisions), increases with $n$ (more collisions), and decreases with $B$ (smaller gyro radius).

For the simple fusion reactor with $T_k = 15$, $n_{20} = 1.5$, $B_0 = 4.7$, then $D_n$ is $3.5 \times 10^{-5}$ m$^2$/s. This value is enormously optimistic, by nearly five orders of magnitude, compared to typical experimentally measured values in a tokamak: $D_n \sim 1$ m$^2$/s. Part of the difference is associated with toroidal effects (neoclassical transport) but most is due to plasma-driven micro-turbulence. In any event, the calculation shows how to apply the random walk model and sets a reference value for purely classical transport in a straight cylinder.

### 14.2.4 Thermal conductivity of a magnetized plasma

The particle diffusion analysis just presented gives one confidence that the random walk model, despite its simplicity, is capable of reliably predicting transport coefficients. The task now is to utilize the random walk model to predict the thermal diffusivities, the last remaining unknown transport coefficients required to close the set of self-consistent fluid equations. The same simple model will be used for thermal diffusion as for particle diffusion; that is, the step size is calculated for a 2-D collision model where particles make circular gyro orbits and scatter in random directions after each collision.

Based on the similarities with the previous analysis one might initially think that like particle collisions do not lead to thermal diffusion. The same cancellation in centers of mass will occur, implying no like particle thermal diffusion. This conclusion is also *wrong*!

The reason is as follows. The correct way to calculate particle diffusion is by defining the step size in terms of the change in the two-particle center of mass before and after the collision. For thermal diffusion, however, one must define the step size in terms of the change in the two-particle "center of energy" (i.e., the energy centroid) before and after the collision. Thus, essentially all of the analysis for like particle diffusion is still valid, the one critical difference being the definition of $\Delta \mathbf{r}$, which for thermal diffusion is defined by

$$\Delta \mathbf{r} = \mathbf{r}_{cE} - \mathbf{r}'_{cE}, \qquad (14.39)$$

where

$$\begin{aligned}
\mathbf{r}_{cE} &= \frac{v_1^2 \mathbf{r}_{g1} + v_2^2 \mathbf{r}_{g2}}{v_1^2 + v_2^2}, \\
\mathbf{r}'_{cE} &= \frac{v_1'^2 \mathbf{r}'_{g1} + v_2'^2 \mathbf{r}'_{g2}}{v_1'^2 + v_2'^2}.
\end{aligned} \qquad (14.40)$$

After a straightforward but slightly tedious calculation one can substitute for all quantities in $(\Delta l)^2$, expressed in terms of the center of mass velocity variables, and then carry out the averaging over angles. One obtains

$$(\Delta l)^2 = \frac{2v^4 V^2}{\omega_c^2 (4V^2 + v^2)^2}. \qquad (14.41)$$

The last step is to average over $v_1$, $v_2$ or equivalently $v$, $V$. A reasonable estimate for the averages is given by

$$\begin{aligned}
v^2 &= v_1^2 + v_2^2 - 2\mathbf{v}_1 \cdot \mathbf{v}_2 \sim 2v_T^2, \\
V^2 &= \frac{1}{4}(v_1^2 + v_2^2 + 2\mathbf{v}_1 \cdot \mathbf{v}_2) \sim \frac{v_T^2}{2},
\end{aligned} \qquad (14.42)$$

where it is assumed that the $\mathbf{v}_1 \cdot \mathbf{v}_2$ terms average to zero. This estimate yields the following expression for the mean square step size:

$$(\Delta l)^2 = \frac{1}{4} \frac{v_T^2}{\omega_c^2}. \qquad (14.43)$$

The thermal diffusivities can now easily be evaluated by noting that the mean time between collisions is $\tau_{ii} = (\nu_{ii})^{-1}$ for ions and $\tau_{ee} = (\nu_{ee})^{-1}$ for electrons. Since $\tau_{jj} = \tau_{jj}(v)$ is a function of velocity one can again approximately average over $v$ by defining $\overline{\tau}_{jj} = \tau_{jj}(v_T)$. The random walk model thus predicts the following values for the ion and electron thermal diffusivities:

$$\begin{aligned}
\chi_i &= \frac{1}{4} \frac{v_{Ti}^2}{\omega_{ci}^2 \overline{\tau}_{ii}} \sim \frac{r_{Li}^2}{\overline{\tau}_{ii}}, \\
\chi_e &= \frac{1}{4} \frac{v_{Te}^2}{\omega_{ce}^2 \overline{\tau}_{ee}} \sim \frac{r_{Le}^2}{\overline{\tau}_{ee}}.
\end{aligned} \qquad (14.44)$$

Observe that the electron thermal diffusivity is comparable to the particle diffusion coefficient: $\chi_e \sim D_n$. The ion thermal diffusivity is larger by the square root of the mass ratio: $\chi_i \sim (m_i/m_e)^{1/2} \chi_e$. It is shown in the next subsection that the single thermal diffusivity appearing in the transport equations (Eqs. (14.10)) is actually given by $\chi = \chi_i + \chi_e \approx \chi_i$. The remarkable cancellation that occurs in the value of $\Delta \mathbf{r}$ for particle diffusion does not occur for thermal diffusion. This is the underlying reason why collisional thermal diffusion is so much larger than particle diffusion.

The numerical values of the thermal diffusivities (for a 50%–50% D–T plasma), using Braginskii's more accurate coefficients are given by

$$\chi_i = 0.10 \frac{n_{20}}{B_0^2 T_k^{1/2}} \quad \text{m}^2/\text{s},$$

$$\chi_e = 4.8 \times 10^{-3} \frac{n_{20}}{B_0^2 T_k^{1/2}} \quad \text{m}^2/\text{s}.$$

(14.45)

For the dominant ion diffusivity the numerical coefficient is approximately a factor of 2.4 larger than that obtained by the simple random walk model. Using parameter values for the simple test reactor yields an ion thermal diffusivity $\chi_i = 1.8 \times 10^{-3}$ m$^2$/s. This value is also highly optimistic by about three orders of magnitude from typical experimentally measured values $\chi_i \sim 1$ m$^2$/s.

### 14.2.5 Summary

The formulation of classical transport in a 1-D cylinder is now complete. The model, which makes use of the tokamak expansion, is described by Eq. (14.10). It consists of a closed set of coupled time evolution equations for the density, temperature, and poloidal magnetic field. The particle and magnetic diffusion coefficients follow directly from the fluid equations and result from electron–ion momentum exchange collisions. The expressions for $D_n$ and $D_B$ are given by Eqs. (14.11). Heat transport is dominated by the ions. Here the dominant mechanism is ion–ion collisions. A simple estimate of $\chi_i$ is given by Eqs. (14.44) by means of the random walk model. The predicted values of classical particle and heat transport are both highly optimistic with respect to typical experimentally measured values. They nevertheless serve as a useful point of reference.

### 14.3 Solving the transport equations

The transport model just derived describes classical transport in a simplified 1-D cylinder in the context of the tokamak expansion. In this section several specific problems described by the model are solved analytically. The purpose is two-fold. First, it is instructive to see in detail how to approach and cast each problem into a form amenable to solution. Many of the same ideas apply to more general toroidal calculations where the solutions must be obtained numerically. Second, the problems addressed are those for which the answers are not immediately obvious. For example, while it is obvious from dimensional analysis that the characteristic relaxation time for any diffusive process scales as $\tau \approx a^2/D$, there are more subtle questions that often need to be answered.

In this section three such problems are discussed. The first problem corresponds to temperature equilibration. In all the analysis thus far presented it has been assumed that $T_e \approx T_i = T$. The goal here is to derive a quantitative criterion for this condition to be satisfied. The answer is not immediately obvious because of the large difference in magnitudes between $\chi_e$ and $\chi_i$.

The second problem is concerned with the effects that the external heating deposition profile has on the central plasma temperature. For instance, does a highly peaked off-axis heating source lead to a temperature profile with a corresponding off-axis peak?

The third problem involves a solution of the steady state 1-D model assuming the heating source corresponds solely to ohmic heating. The issue here is to determine whether or not it is feasible to ohmically heat to ignition, thereby eliminating the need for external auxiliary heating sources. This would clearly be a highly desirable situation.

### 14.3.1 Temperature equilibration

As a model problem to investigate temperature equilibration, consider the two-fluid steady state energy equations:

$$\frac{1}{r}\frac{\partial}{\partial r}\left(rn\chi_e\frac{\partial T_e}{\partial r}\right) + S_e + \frac{3}{2}\frac{n(T_i - T_e)}{\overline{\tau}_{eq}} = 0,$$

$$\frac{1}{r}\frac{\partial}{\partial r}\left(rn\chi_i\frac{\partial T_i}{\partial r}\right) - \frac{3}{2}\frac{n(T_i - T_e)}{\overline{\tau}_{eq}} = 0. \tag{14.46}$$

The model corresponds to the situation in which the electrons are heated by a source $S_e$. The resulting energy gain is balanced by a combination of electron thermal conduction losses and collisional energy exchange to the ions. The ions have no external heating source. They are heated by energy exchange from the electrons and lose energy by ion thermal conduction. For simplicity the effects of compression and convection are neglected as they do not dominate the behavior. In order to obtain an analytic solution the coefficients $\chi_e$, $\chi_i$, $\overline{\tau}_{eq}$ are treated as constants with $\chi_e$, $\chi_i$ allowed to have anomalously high values if necessary. Similarly, for simplicity it is also assumed that the density profile $n(r)$ and heating deposition profile $S_e(r)$ are both constants.

The goal of this reduced problem is to calculate the equilibrium electron and ion temperature profiles and then to determine the conditions under which temperature equilibration is a good approximation. Specifically, one wants to determine the condition under which

$$R \equiv \frac{T_e(0) - T_i(0)}{T_e(0) + T_i(0)} \ll 1. \tag{14.47}$$

The solution is obtained in two steps. First, the equations are added together in order to annihilate the energy exchange terms:

$$\frac{1}{r}\frac{\partial}{\partial r}\left[r\frac{\partial}{\partial r}(n\chi_e T_e + n\chi_i T_i)\right] = -S_e. \tag{14.48}$$

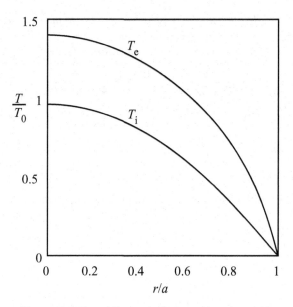

Figure 14.6 $T_e$ and $T_i$ vs. $r/a$ for $\alpha = 10$, $\chi_i/\chi_e = 10$.

For boundary conditions one requires regularity at the origin ($T_e'(0) = T_i'(0) = 0$) and a perfect sink condition at $r = a$ ($T_e(a) = T_i(a) = 0$). The solution to Eq. (14.48) is easily found and is

$$n\,(\chi_e T_e + \chi_i T_i) = \frac{S_e}{4}(a^2 - r^2).\tag{14.49}$$

The second step is to solve for $T_e$ and substitute the solution into the ion equation. A short calculation yields

$$\frac{1}{x}\frac{\partial}{\partial x}\left(x\frac{\partial T_i}{\partial x}\right) - \alpha^2 T_i = \alpha^2 T_0(1 - x^2),\tag{14.50}$$

where $x = r/a$ and

$$\alpha^2 = \frac{3}{2}\left(\frac{\chi_e + \chi_i}{\chi_e \chi_i}\right)\frac{a^2}{\tau_{eq}},$$

$$T_0 = \frac{1}{4}\frac{S_e a^2}{n(\chi_e + \chi_i)}.\tag{14.51}$$

The solution for $T_i$ (and then $T_e$) satisfying the boundary conditions can be written in terms of modified Bessel functions as follows:

$$\frac{T_i}{T_0} = 1 - x^2 - \frac{4}{\alpha^2}\left[1 - \frac{I_0(\alpha x)}{I_0(\alpha)}\right],$$

$$\frac{T_e}{T_0} = 1 - x^2 + \frac{4}{\alpha^2}\frac{\chi_i}{\chi_e}\left[1 - \frac{I_0(\alpha x)}{I_0(\alpha)}\right].\tag{14.52}$$

These profiles are illustrated in Fig. 14.6 for the case $\chi_i/\chi_e = 10$ and $\alpha = 10$.

Table 14.1. *Relationship between $\alpha^2$ and $\chi_i/\chi_e$ to insure that $R \ll 1$*

| Regime of $\chi_i/\chi_e$ | $R = R(F)$ | $R = R(\alpha, \chi_i/\chi_e)$ | Condition for $R \ll 1$ |
|---|---|---|---|
| $\chi_i/\chi_e \gg 1$ | $R \approx \dfrac{F}{F + 2\chi_e/\chi_i}$ | $R \approx \dfrac{4}{4 + 2\alpha^2 \chi_e/\chi_i}$ | $\alpha^2 \gg 2\chi_i/\chi_e$ |
| $\chi_i/\chi_e = 1$ | $R = F$ | $R \approx \dfrac{4}{4 + \alpha^2}$ | $\alpha^2 \gg 4$ |
| $\chi_i/\chi_e \ll 1$ | $R \approx \dfrac{F}{2 - F}$ | $R \approx \dfrac{2}{2 + \alpha^2}$ | $\alpha^2 \gg 2$ |

One is now in a position to calculate the equilibration parameter $R$. Substituting into Eq. (14.47) yields

$$R(\alpha, \chi_i/\chi_e) = \frac{(\chi_i/\chi_e + 1)\, F(\alpha)}{2 + (\chi_i/\chi_e - 1)F(\alpha)}, \tag{14.53}$$

where

$$F(\alpha) = \frac{4}{\alpha^2}\left[1 - \frac{1}{I_0(\alpha)}\right] \approx \frac{4}{4 + \alpha^2}. \tag{14.54}$$

The last form is an approximation that matches the behavior at both small and large $\alpha$. Note that $F(\alpha)$ is a decreasing function and that $F \ll 1$ when $\alpha \gg 2$.

The condition for good equilibration can be determined by examining Eq. (14.53) for different values of the ratio $\chi_i/\chi_e$. Specifically the condition $R \ll 1$ sets a requirement on $\alpha^2$ as shown in Table 14.1. From Table 14.1 it follows that a simple form for the condition on $\alpha^2$, valid for all values of $\chi_i/\chi_e$, can be written as

$$\alpha^2 \gg 2\frac{\chi_i + \chi_e}{\chi_e}. \tag{14.55}$$

In unnormalized units, Eq. (14.55) reduces to

$$\overline{\tau}_{eq} \ll \frac{3}{4}\frac{a^2}{\chi_i}. \tag{14.56}$$

Physically, good temperature equilibration occurs if the equilibration time $\overline{\tau}_{eq}$ is much shorter than the ion energy confinement time $a^2/\chi_i$. The electron energy confinement time $a^2/\chi_e$ has a strong influence on the final central temperature. However, regardless of the central temperature, the electrons and ions will equilibrate as long as the ions do not lose the energy transferred from the electrons too rapidly by ion thermal conduction.

One can now ask whether or not Eq. (14.56) is satisfied in most plasma experiments. The answer for classical diffusion is obtained by noting that $\overline{\tau}_{eq} \sim (m_i/m_e)\overline{\tau}_{ei}$, $\chi_i \sim r_{Li}^2/\overline{\tau}_{ii}$, and $\overline{\tau}_{ii} \sim (m_i/m_e)^{1/2}\overline{\tau}_{ei}$. The equilibration condition reduces to

$$\frac{a^2}{r_{Li}^2} \gg \left(\frac{m_i}{m_e}\right)^{1/2}, \tag{14.57}$$

which is easily satisfied in most experiments. However, when $\chi_i$ is anomalous, the condition is more difficult to satisfy. In this case, using the numerical value of $\overline{\tau}_{eq}$ from Eq. (9.119), one can rewrite the equilibration condition in the following form:

$$n_{20} \gg 0.017\frac{T_k^{3/2}\chi_i}{a^2} = 0.25\chi_i,$$ (14.58)

where the last value corresponds to the simple fusion reactor ($a = 2$, $T_k = 15$). For the reactor, the value $n_{20} = 1.5$ is sufficiently large that the condition would be reasonably well satisfied assuming that $\chi_i \sim 1\,\mathrm{m}^2/\mathrm{s}$, a typical experimental value in present tokamaks. Interestingly, most present tokamaks usually operate at somewhat lower densities ($n_{20} \sim 0.5$) so that the equilibration condition is only marginally satisfied.

The final point to discuss is the derivation of a single energy equation in the limit where there is good equilibration. Mathematically, good equilibration implies that $\overline{\tau}_{eq} \to 0$ and $T_i \to T_e$. The energy equilibration term in each equation therefore becomes indeterminate ($(T_e - T_i)/\overline{\tau}_{eq} \to 0/0$). The difficulty is resolved by adding the energy equations together to exactly annihilate the equilibration terms and to then set $T_e = T_i = T$. This leads to a single energy equation given by

$$\frac{1}{r}\frac{\partial}{\partial r}\left[rn(\chi_e + \chi_i)\frac{\partial T}{\partial r}\right] + S_e = 0.$$ (14.59)

One sees that the final equation balances heat conduction losses against the heating source term. The thermal diffusivity is just the sum of the separate components ($\chi = \chi_e + \chi_i \approx \chi_i$) and is dominated by the largest contribution, usually due to the ions.

The analysis just presented provides a good justification for considering a single energy equation when investigating the performance of fusion grade plasmas.

### 14.3.2 Effect of the heating profile on the central temperature

Next, the effect of the external source heating profile on the peak temperature and temperature profile is investigated. Of particular interest is the question of whether or not a highly localized off-axis heating source results in a corresponding peaked off-axis temperature profile. To answer this question two simple problems are considered. First the temperature profile is calculated for a heating source that is uniform in space. Second, the calculation is repeated assuming the same amount of total power is deposited off-axis in a highly localized region of space, modeled mathematically by a delta function. A comparison of the two solutions provides the answer to the question.

As a simple model consider a well-equilibrated plasma described by the following steady state energy equation:

$$\frac{1}{r}\frac{\partial}{\partial r}\left(rn\chi\frac{\partial T}{\partial r}\right) = -S(r).$$ (14.60)

As in the previous equilibrium problem convection and compression are neglected and $n$, $\chi$ are assumed to be constants.

First, assume that a total power $P_h$ is absorbed uniformly over the plasma cross section. If the volume of the plasma is denoted by $V = 2\pi^2 a^2 R_0$, then for this case $S(r) = P_h/2\pi^2 a^2 R_0 = \text{const}$. The boundary conditions again require regularity at the origin and a perfect heat sink at $r = a : T'(0) = 0$ and $T(a) = 0$. The solution to the transport equation is easily found and is given by

$$T = T_0 \left( 1 - \frac{r^2}{a^2} \right),$$ (14.61)

where

$$T_0 = \frac{P_h}{8\pi^2 n \chi R_0}.$$ (14.62)

Observe that the temperature profile decreases parabolically with radius. It is also of interest to evaluate the peaking factor, defined as the ratio of the peak temperature $T_0$ to the average temperature $\overline{T}$. Here,

$$\overline{T} = \frac{2}{a^2} \int_0^a T(r) r \, dr.$$ (14.63)

For the case of a constant heating profile the peaking factor has the value

$$T_0/\overline{T} = 2.$$ (14.64)

The calculation just presented serves as the reference case. The next step is to redo the calculation assuming a highly localized off-axis source, modeled by a delta function as follows:

$$S(r) = K \, \delta(r - \alpha a) = \frac{P_h}{4\pi^2 R_0 a \alpha} \, \delta(r - \alpha a).$$ (14.65)

Note that the heating source peaks at $r = \alpha a$ with $0 < \alpha < 1$. Also, the coefficient multiplying the delta function has been chosen so that the total power absorbed by the plasma is again equal to $P_h$.

The temperature is found by solving separately in the regions on either side of the delta function and then matching across the surface $r = \alpha a$. For $0 \le r \le \alpha a^-$ the solution that is regular at the origin is given by

$$T = C_1,$$ (14.66)

where $C_1$ is an as yet undetermined coefficient. For $\alpha a^+ \le r \le a$ the solution satisfying the sink condition at $r = a$ can be written as

$$T = C_2 \ln (a/r).$$ (14.67)

Here, $C_2$ is also an undetermined coefficient.

Next, there are two matching conditions across $r = \alpha a$ that must be satisfied. First, the temperature must be continuous, implying that $C_2 \ln (1/\alpha) = C_1$. Second, integrating across

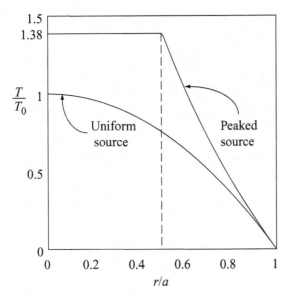

Figure 14.7 Temperature profiles for a uniform and a peaked heating source.

the delta function yields a jump condition on the heat fluxes which is given by

$$[rn\chi \,\partial T/\partial r]_{\alpha a^-}^{\alpha a^+} = -\int_{\alpha a^-}^{\alpha a^+} K\delta\,(r - \alpha a)\,r\,dr = -P_h/4\pi^2 R_0. \tag{14.68}$$

Using the profiles in each region one can easily evaluate the temperature derivatives leading to

$$C_1 = \frac{P_h \ln\,(1/\alpha)}{4\pi^2 n\chi R_0} = 2T_0 \ln\,(1/\alpha), \tag{14.69}$$

where $T_0$ has been defined in Eq. (14.62).

The resulting temperature profiles are thus given by

$$T = 2T_0 \ln\,(1/\alpha) \qquad 0 \le r \le \alpha a^-,$$
$$T = 2T_0 \ln\,(a/r) \qquad \alpha a^+ \le r \le a. \tag{14.70}$$

The solution is plotted in Fig. 14.7 for the case of $\alpha = 0.5$. Observe that even with a very peaked heating profile the temperature itself does not peak. Physically, the reason is as follows. Initially, the heating profile does indeed produce a peaked off-axis temperature profile. The heat then starts to diffuse in both directions away from the source. At the edge of the plasma the heat energy is absorbed because of the sink boundary condition. In the center, however, there is no sink, and the heat accumulates. In steady state a balance is reached where there is no net flow of energy in either direction in the central region, corresponding to a uniform temperature.

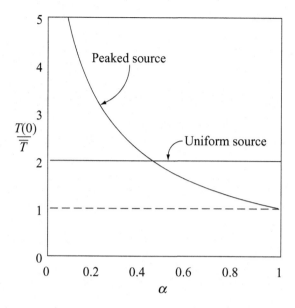

Figure 14.8 Peaking factor vs. normalized localization radius.

The last point of interest is to compare the temperature peaking factor between the two cases. For the localized heating source the peaking factor is easily evaluated, and can be expressed as

$$\frac{T(0)}{\overline{T}} = \frac{2\ln(1/\alpha)}{1-\alpha^2}.$$ (14.71)

The results are plotted in Fig. 14.8 as a function of $\alpha$. Note that the difference in peaking factors between the two cases is relatively modest in comparison to the dramatically different heating profiles.

The main conclusion from this problem is that while the average plasma temperature depends directly upon the total heating power supplied (i.e., $\overline{T} \sim P_h$), the actual temperature profile is relatively insensitive to the heating profile (except for the special limit $\alpha \to 0$).

### 14.3.3 Ohmic heating to ignition

The last problem discussed raises the question of whether or not it is possible to ohmically heat a plasma to ignition without the need of external power sources. This would presumably be highly desirable from a fusion engineering point of view. The problem is addressed in the present subsection within the context of classical transport, which is highly optimistic compared to actual experimental performance. Even so, this serves as a good point of reference and demonstrates the process of formulating and finding an approximate solution to the transport equations.

The goal of the calculation is to determine how much ohmic power is required to heat a plasma to a temperature of 7 keV, which is about half way to ignition. The calculation neglects alpha particle heating, which is a good approximation for low temperatures, although once the 7 keV point is reached, the alpha particle heating takes over and then the ohmic power can be neglected. The ohmic solution also allows one to calculate other plasma parameters, including the plasma current $I$, the plasma pressure $p$, the plasma beta $\beta$, the safety factor $q$, and the energy confinement time $\tau_E$, to make sure these values are compatible with engineering and plasma physics constraints. The analysis shows that ohmic heating to ignition is possible under the assumption of classical transport, although further thought reveals that this is a mixed blessing.

## The model

The ohmic heating model represents a subset of the complete 1-D cylindrical transport equations derived in the previous section. For simplicity one again assumes steady state and neglects compression and convection. Furthermore, to eliminate the need for solving a set of coupled differential equations the density is assumed to be uniform: $n = \text{const}$. This experimentally plausible assumption simplifies the analysis by removing the density equation from the calculation.[2] The ohmic heating model reduces to

$$r \frac{\partial}{\partial r} \left( \frac{D_B}{r} \frac{\partial r B_\theta}{\partial r} \right) = 0,$$

$$\frac{1}{r} \frac{\partial}{\partial r} \left( r n \chi \frac{\partial T}{\partial r} \right) + \eta_\parallel J^2 = 0, \qquad (14.72)$$

$$\mu_0 J = \frac{1}{r} \frac{\partial r B_\theta}{\partial r}.$$

Note that the heating source corresponds to ohmic power dissipation.

The task now is to simplify these equations in order to obtain a single, second order, differential equation for the temperature. The first step is to integrate the magnetic field equation, recalling that $D_B = \eta_\parallel / \mu_0 \sim 1/T^{3/2}$. This yields

$$D_B J = \text{const.} \qquad (14.73)$$

or

$$J = K \frac{\mu_0}{\eta_\parallel} = J_0 \left( \frac{T}{T_0} \right)^{3/2}. \qquad (14.74)$$

Here, $J_0$, $T_0$ are constants representing the on-axis values. They are presently unknown but will ultimately be related to the plasma current $I$ and the desired average temperature $\overline{T}_k = 7 \, \text{keV}$.

---

[2] The density can be included by a more complicated analysis, although considerable care must be taken with the boundary condition and/or the introduction of more realistic modifications to the physical model in the region near $r = a$. This is because indeterminate ratios appear in $\chi$ when the ideal sink boundary condition is used.

The next step is to substitute the expression for $J$ into the energy equation using the classical value of $\chi \approx \chi_i$ given by Eqs. (14.45). A short calculation leads to a single differential equation for the normalized temperature $U = T/T_0$. This equation can be written as

$$\frac{1}{x}\frac{\partial}{\partial x}\left(\frac{x}{U^{1/2}}\frac{\partial U}{\partial x}\right) + 2\alpha\, U^{3/2} = 0, \tag{14.75}$$

where $x = r/a$ and $\alpha$ is a dimensionless parameter defined (in practical units) by

$$\alpha = 10.3\left(\frac{B_0 J_{M0} a}{\bar{n}_{20} T_{k0}}\right)^2. \tag{14.76}$$

Here, $T_{k0} = T_0$ (keV) and $J_{M0} = J_0$ (MA/m$^2$).

The last step in the simplification of the model is to introduce a new dependent variable $V = U^{1/2}$. Equation (14.75) reduces to

$$\frac{1}{x}\frac{\partial}{\partial x}\left(x\frac{\partial V}{\partial x}\right) + \alpha\, V^3 = 0. \tag{14.77}$$

The boundary conditions again require regularity at the origin and a sink condition at the plasma edge: $V'(0) = 0$, $V(1) = 0$. The normalizing condition $V(0) = 1$ is an additional constraint which can only be satisfied by choosing an appropriate value for $\alpha$. In fact, the next step in the analysis is to determine an approximate solution for $V(x)$ and the corresponding value for $\alpha$. Once $V(x)$ and $\alpha$ are known all the desired physical properties of the ohmically heated plasma can easily be evaluated.

### Approximate solution to the problem

Equation (14.77) is a non-linear differential equation for which there is no simple closed form solution. While it can easily be solved numerically, this is not optimal in terms of developing physical insight. A different approach is used here to obtain an approximate form of the solution. The method is based on two points of mathematical insight. First, the solution for $V(x)$ is qualitatively simple. It starts out with a value of unity at the origin and then monotonically decreases to zero at the edge of the plasma. Second, all the physical quantities are calculated by integrating various functions of $V(x)$ over the plasma volume. Since integrated values are involved, the results are not very sensitive to the precise details of the $V(x)$ profile.

This mathematical insight suggests that a moment approach would be sufficiently accurate for present purposes. Specifically, the approximation is made that the $V(x)$ profile is of the form

$$V(x) = (1 - x^2)^\nu, \tag{14.78}$$

where $\nu$ is an as yet undetermined parameter. Clearly, $V(x)$ has the correct qualitative behavior of the exact solution.

The values of $\nu$ and $\alpha$ are determined by requiring that two low-order moments of the differential equation be exactly satisfied. These moments are defined by

$$\int_0^1 V \left[ \frac{1}{x} \frac{\partial}{\partial x} \left( x \frac{\partial V}{\partial x} \right) + \alpha V^3 \right] x \, dx = 0,$$

$$\int_0^1 x^2 V \left[ \frac{1}{x} \frac{\partial}{\partial x} \left( x \frac{\partial V}{\partial x} \right) + \alpha V^3 \right] x \, dx = 0.$$

(14.79)

Although the choice of which moment equations to use is clearly not unique, using low-order moments captures the main macroscopic features of the exact solution.

Substituting the approximate $V(x)$ into Eqs. (14.79) leads to two coupled algebraic equations for the unknowns $\nu$ and $\alpha$. The details are straightforward but slightly lengthy. Once obtained, the resulting algebraic equations can easily be solved analytically yielding the values

$$\nu = 2, \ \alpha = 12. \tag{14.80}$$

These values are used next to derive the desired physical properties of the ohmically heated plasma.

### Physical properties of the solution

First, the approximate temperature profile is considered. From the value $\nu = 2$ and the relation $U = V^2$ it follows that

$$T_k(r) = T_{k0}(1 - r^2/a^2)^4. \tag{14.81}$$

Note that this is a highly peaked profile, a consequence of the fact that near the edge of the plasma $\chi \sim T^{-1/2} \rightarrow \infty$. A high thermal diffusivity tends to flatten the temperature profile at the edge, thus making the central profile more peaked. A measure of the peaking is given by the peaking factor, which is the ratio of the peak to average temperature, determined as follows:

$$\overline{T}_k = 2 \int_0^1 T_k x \, dx = T_{k0}/5. \tag{14.82}$$

Thus the peaking factor is $T_{k0}/\overline{T}_k = 5$. For future numerical values $\overline{T}_k$ is set to $\overline{T}_k = 7$ keV, the target temperature for ohmic heating to ignition.

Second, the current density profile is examined. This is slightly more peaked than the temperature profile:

$$J_M(r) = J_{M0}(1 - r^2/a^2)^6. \tag{14.83}$$

The constant $J_{M0}$ can be expressed in terms of the total plasma current from the definition

$$I_M = 2\pi a^2 \int_0^1 J_M x \, dx = \pi a^2 J_{M0}/7. \tag{14.84}$$

Thus, $J_{M0} = 7 I_M/\pi a^2$.

These results, combined with the value $\alpha = 12$ lead to a relationship between the required plasma current and the desired ohmic heating temperature which can be written as

$$I_M = \left(\frac{\alpha}{2.05}\right)^{1/2} \frac{a\overline{n}_{20}\overline{T}_k}{B_0} = 2.4\frac{a\overline{n}_{20}\overline{T}_k}{B_0} = 10.7 \text{ MA.} \tag{14.85}$$

Note that the current scales linearly with the temperature for classical transport. The final numerical value corresponds to the simple reactor design parameters: $n_{20} = 1.5$, $a = 2$, $R_0 = 5$, $B_0 = 4.7$. One can now ask whether the resulting value of $q_*$ is high enough to suppress MHD instabilities. The value of $q_*$ (assuming an elongation $\kappa = 2$) is

$$q_* = \frac{2\pi a^2 \kappa B_0}{\mu_0 R_0 I} = 5\frac{a^2 \kappa B_0}{R_0 I_M} = 3.5. \tag{14.86}$$

This value is safely above the limit for exciting ballooning-kink modes. In other words the ohmic current required to reach a temperature of 7 keV in a tokamak with classical transport is well below the instability threshold, indeed a favorable conclusion.

Consider next the pressure and $\beta$. Their average values are easily calculated as follows:

$$\overline{p} = 2\overline{n}\overline{T} = 0.32\overline{n}_{20}\overline{T}_k = 3.4 \text{ atm,}$$
$$\overline{\beta} = 2\mu_0\overline{p}/B_0^2 = 0.25\overline{p}_a/B_0^2 = 0.038. \tag{14.87}$$

These are just the values that one would expect at 7 keV, about one half the way to a steady state fully ignited plasma at 15 keV. Here too, the $\overline{\beta}$ is below the threshold for MHD instability.

One is now in a position to calculate the total ohmic power required to reach 7 keV and compare it with the nominal electric power out of the reactor, 1000 MW. The ohmic power is defined as

$$P_\Omega = \int \eta_\parallel J^2 d\mathbf{r}. \tag{14.88}$$

A short calculation using the relation $\eta_\parallel = 3.3 \times 10^{-8}/T_k^{3/2}$ $\Omega\,$m leads to

$$P_\Omega = 4.1 \times 10^{-2}\frac{R_0 I_M^2}{a^2 \overline{T}_k^{3/2}} = 0.32 \text{ MW.} \tag{14.89}$$

The ohmic power required is very modest compared to the electric power output. This is a consequence of the good confinement associated with classical transport. If the plasma is good at confining its thermal energy (i.e., $\chi$ is small), then only a small amount of ohmic power is required to heat the plasma to a high temperature. Clearly the low requirement on $P_\Omega$ is a very favorable result.

The last parameter of interest, which quantifies "good confinement," is the energy confinement time. The quantity $\tau_E$ is defined by integrating the starting energy balance equation (Eq. (14.72)) over the plasma volume:

$$P_\Omega = -4\pi^2 R_0 a n \chi \left.\frac{\partial T}{\partial r}\right|_a \equiv \frac{3}{2}\frac{\int p\, d\mathbf{r}}{\tau_E}. \tag{14.90}$$

Observe that the energy confinement time is defined in terms of the heat flux at the plasma edge. If one knew the exact solution for the profiles, a simple evaluation of the edge derivative would yield the desired value of $\tau_E$. However, with the approximate profiles used in the calculation this is a very poor way to evaluate $\tau_E$, leading in fact to the result $\tau_E = \infty$. A much better way to use the approximate equilibrium solution is through less sensitive integral relations. A good estimate of the energy confinement time is thus obtained from Eq. (14.90) as follows:

$$\tau_E = \frac{3}{2}\frac{\overline{p}V}{P_\Omega} = 23\frac{a^4\overline{n}_{20}\overline{T}_k^{5/2}}{I_M^2} = 6.3 \times 10^2 \text{ s}. \qquad (14.91)$$

Classical confinement predicts an energy confinement time of over 600 s, which is very optimistic compared to present day experiments.

Finally, in practice $\tau_E$ is often used to predict the temperature of an experiment once the density, current, toroidal field, and geometry are specified. For this alternative application of transport theory, the temperature should be eliminated from $\tau_E$, in the present case by means of Eq. (14.85), leading to an expression solely in terms of parameters directly under experimental control. For classical confinement the new form of $\tau_E$ can be written as

$$\tau_E = 2.6\frac{a^{3/2}B_0^{5/2}I_M^{1/2}}{\overline{n}_{20}^{3/2}} \text{ s}. \qquad (14.92)$$

This form of the energy confinement time will be useful when comparing with the actual empirical value determined from experimental measurements.

### *Irony – too much confinement can be a disadvantage*

The analysis thus far seems to indicate that the combination of classical confinement and ohmic heating would be highly desirable in a fusion reactor. However, further thought shows that too much confinement is actually a disadvantage.

There are several ways to understand the issues. The basic problem arises from the fact that too much confinement leads to very high values of $p\tau_E$. For the case of classical confinement described above, the value of $p\tau_E$ at 7 keV is $2.1 \times 10^3$ atm s. This is more than a factor of 200 larger than that required to maintain the plasma in steady state equilibrium characterized by alpha power balancing thermal conduction losses. Specifically, the alpha particles are producing much more heat than is lost by thermal conduction. Should this situation arise, the plasma would continue to heat to much higher temperatures thereby increasing the plasma pressure and power density. Very quickly the critical plasma $\beta$ for stability as well as the maximum allowable neutron wall loading would be violated.

Another approach might be to lower the number density such that $p\tau_E$ is reduced to the value necessary for steady state ignition: $p\tau_E \approx 8.3$ atm s. The difficulty with this strategy is that the lower number density corresponds to a lower pressure, which in turns leads to a lower power density. When the power density is greatly reduced a much larger volume of plasma is required to produce the same total required power output. A larger reactor leads to a higher capital cost per watt which is clearly undesirable.

Yet another idea is to use good confinement to reduce the size of the plasma. Since $\tau_E \sim a^2/\chi$ one can reduce the size of the plasma until $\tau_E$ becomes short enough that steady state power balance is achieved. However, if the pressure and corresponding power density remain unchanged (i.e., $\overline{p} \sim 7$ atm), then the total power output is reduced because of the smaller plasma volume. Even so, the reactor volume remains large since the blanket-and-shield thickness must still be $b = 1.2$ m, a value determined by nuclear physics, not plasma physics. The net result is an inefficient use of the blanket-and-shield which raises the capital cost per watt. Raising the plasma pressure also does not help as this again increases $\beta$ above the MHD stability limit and causes the neutron flux to exceed the wall loading limit.

The one exception to these arguments involves the use of advanced fuels such as D–D. Here, the fusion cross section is much smaller than for D–T implying that a much larger value of $p\tau_E$ is required for plasma ignition.

In any event, the discussion suggests that there is no clear way to exploit the achievement of a very long energy confinement time in a D–T fusion reactor. While this is a valid conclusion one should remember that a large part of current fusion research is aimed at improving energy confinement. There is no contradiction here since the present experimentally achievable energy confinement times are still somewhat below that which is required in a reactor. However, once the reactor relevant energy confinement time is achieved there is little reason for further substantial enhancements in $\tau_E$.

### 14.3.4 Summary

Several energy diffusion problems have been solved for the 1-D cylindrical model within the context of classical transport. These include the problems of temperature equilibration, heating profile effects, and the question of ohmically heating to ignition. All the problems are focused on the energy transport equation as energy losses are the dominant loss mechanism in fusion grade plasmas.

It has been shown that: (1) energy equilibration requires that the ion energy confinement time be long compared to the energy equilibration time; (2) the temperature profile is only weakly dependent on the heating profile; and (3) one can easily ohmically heat to ignition with classical transport, although too much confinement is actually a disadvantage from the reactor point of view.

The idealized classical transport model is highly optimistic with respect to experimental values of $\chi_i$ but nonetheless serves as a useful point of reference. In the following sections more realism is added to the models to bring the theory and experimental data into closer agreement.

## 14.4 Neoclassical transport

### 14.4.1 Introduction

Neoclassical transport is classical transport including the effects of toroidal geometry. The transport is still driven purely by Coulomb collisions – no anomalous transport due to plasma microscopic instabilities is included. The development of neoclassical transport theory follows from a beautiful and sophisticated analysis of plasma kinetic theory. The final model is quite complete. It contains a two-fluid description including resistivity, viscosity, and thermal conduction. The model is also valid for arbitrary regimes of collisionality.

This section contains a derivation of several of the key features of neoclassical transport most relevant to a fusion reactor. These include the particle and thermal diffusion coefficients and the generation of the bootstrap current. The analysis avoids the need to solve complex kinetic equations by instead making use of guiding center theory and the random walk model. Also, use is made of the large aspect ratio, circular cross section, and low collisionality approximations.

At first glance one might expect that in the large aspect ratio limit, toroidal effects simply produce uninteresting $r/R_0$ corrections to the cylindrical results. This is an incorrect conclusion. Perhaps surprisingly, neoclassical effects actually produce increases in the plasma transport coefficients by nearly two orders of magnitude. Qualitatively, the reason is as follows. In a cylindrical system particles are confined to within a gyro radius of a flux surface. Consequently, the corresponding step size associated with Coulomb collisions is also on the order of a gyro radius. In a torus, however, particles drift off the flux surface because of the toroidally induced $\nabla B$ and curvature drifts. The radial excursions due to these drifts can be much larger than a gyro radius, leading to an increase in the collisional step size, and a corresponding increase in the transport coefficients. A further surprise is that the transport that arises from the small population of trapped particles actually dominates the transport resulting from the majority of passing (i.e., untrapped) particles.

The discussion below presents a random walk derivation of particle and heat transport, first due to passing particles and then due to trapped particles. Lastly a derivation is presented of the very important bootstrap current which has no simple analog in a cylindrical system. This current, which is essential to minimize the current drive requirements in a tokamak, is closely related to the magnetization current discussed in Chapter 11.

### 14.4.2 Neoclassical transport due to passing particles

The neoclassical transport associated with passing particles in a toroidal geometry is calculated by means of the random walk model. The critical task is to evaluate the average step size associated with the guiding center drifts off the flux surface.

Towards this goal consider a large aspect ratio, circular cross section tokamak as shown in Fig. 14.9, which depicts a flux surface and the guiding center orbits of two typical passing particles. Recall that in a tokamak $B \approx B_\phi(R)$, implying that the $\nabla B$ and curvature drifts are in the $\mathbf{e}_z$ direction; that is, the guiding center drift for an ion is always in the positive, upward direction. Now, note that if the pitch angle of the magnetic field is for instance positive, a particle with a positive $v_\parallel$ starting out at $\theta = 0$ traces out a circular-like orbit, whose radius is slightly larger than the radius of its starting flux surface. Similarly, in this same magnetic field, a particle with a negative $v_\parallel$ traces out a slightly smaller circle. As an aside, note that these results, including closure of the orbit, can be derived rigorously using the conservation of canonical toroidal momentum.

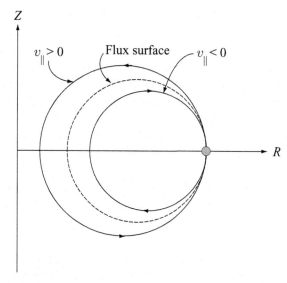

Figure 14.9 Poloidal projections of the guiding center orbits for two passing particles, one with $v_\parallel > 0$, the other with $v_\parallel < 0$.

In terms of transport, if a particle undergoes a $90°$ momentum collision changing $v_\parallel \to -v_\parallel$, then its orbit jumps from one flux surface to another. The average radii of these surfaces differ by an amount comparable to radial excursion off the surface: $\Delta r = r_i - r_f$. Clearly $\Delta r$ represents the appropriate step size for radial transport in a toroidal geometry.

The value of $(\Delta l)^2 = \langle (\Delta r)^2 \rangle$ is calculated by the following steps. First, one needs to determine the time $\tau_{1/2}$ it takes for a particle to complete a half-transit around the poloidal cross section. Second, during $\tau_{1/2}$ the particle drifts off the surface with a guiding center drift velocity $v_D$ due to the $\nabla B$ and curvature drifts. The corresponding radial excursion is then of the order $\Delta r \sim v_D \tau_{1/2}$. Third, the desired step size is then calculated by averaging over all collisions. The details of the calculation are given below.

### The half-transit time

The time it takes a particle to make a half-transit is given by $\tau_{1/2} = l/v_\parallel$, where $l$ is the total parallel distance traveled by the particle as it covers a poloidal distance $l_p = \pi r$. For a large aspect ratio tokamak

$$dl/B \approx dl/B_0 \approx dl_p/B_\theta. \tag{14.93}$$

Therefore,

$$l(r) \approx \frac{B_0}{B_\theta(r)} l_p = \frac{B_0}{B_\theta(r)} \pi r = \pi R_0 q(r). \tag{14.94}$$

The corresponding half-transit time is then given by

$$\tau_{1/2}(r) = l/v_\parallel = \pi R_0 q/v_\parallel. \tag{14.95}$$

## The radial excursion

The next step is to calculate the particle excursion off the flux surface. To begin, note that the combined $\nabla B$ and curvature drifts in the dominant toroidal magnetic field of the tokamak, which is nearly a vacuum field, can be written as

$$\mathbf{v}_D = \frac{m_i}{e}\left(v_\parallel^2 + \frac{v_\perp^2}{2}\right)\frac{\mathbf{R}_c \times \mathbf{B}}{R_c^2 B} \approx \frac{1}{\omega_{ci} R_0}\left(v_\parallel^2 + \frac{v_\perp^2}{2}\right)(\mathbf{e}_r \sin\theta + \mathbf{e}_\theta \cos\theta). \tag{14.96}$$

For simplicity the particle under consideration is a positive ion. The trajectory of the particle, assuming it starts at some arbitrary initial angle $\theta = \theta_0$ (i.e., it does not have to start at $\theta = 0$), is given by $\theta(t) = \omega_T t + \theta_0$, where $\omega_T = \pi/\tau_{1/2}$ is the full-transit frequency. The corresponding component of radial velocity thus has the form

$$v_{Dr}(t) \approx |v_{Dr}| \sin(\omega_T t + \theta_0), \tag{14.97}$$

where

$$|v_{Dr}| = \frac{1}{\omega_{ci} R_0}\left(v_\parallel^2 + \frac{v_\perp^2}{2}\right). \tag{14.98}$$

As expected, the velocity oscillates in sign, half the time moving towards the plasma axis and half the time moving away.

The radial position of the guiding center is now easily obtained by integrating $\dot{r} = v_{Dr}$, assuming the particle starts off at a radius $r = r_0$. Under the assumption that $v_\parallel$ and $v_\perp$ do not change very much during the orbit of a passing particle one obtains

$$r(t) \approx r_0 - \frac{|v_{Dr}|}{\omega_T}[\cos(\omega_T t + \theta_0) - \cos(\theta_0)]. \tag{14.99}$$

Equation (14.99) shows that the radial position of the particle oscillates in time about a mean value, corresponding to the radius of the flux surface $r_i$ to which the particle's average guiding center is attached. This radius is clearly given by

$$r_i = r_0 + \frac{|v_{Dr}|}{\omega_T}\cos(\theta_0). \tag{14.100}$$

## The step size

The step size can finally be calculated by assuming that the actual Coulomb collision takes place at the point $r = r_0$, $\theta = \theta_0$. For simplicity assume the particle undergoes a typical momentum collision that scatters its velocity from an initial value $\mathbf{v}_i = v_\parallel \mathbf{b} + \mathbf{v}_\perp$ to a final value $\mathbf{v}_f = -v_\parallel \mathbf{b} + \mathbf{v}_\perp$. In other words the collision reverses the sign of $v_\parallel$. Since $v_\parallel \sim |\mathbf{v}_\perp|$ for a passing particle, this corresponds to a $90°$ momentum collision as illustrated in Fig. 14.10.

The radius of the final flux surface is easily obtained by setting $v_\parallel \to -v_\parallel$ in Eq. (14.100), which is equivalent to setting $\omega_T \to -\omega_T$. This yields

$$r_f = r_0 - \frac{|v_{Dr}|}{\omega_T}\cos(\theta_0). \tag{14.101}$$

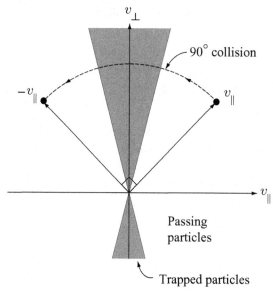

Figure 14.10 Velocity space showing a typical particle with $v_\parallel \sim v_\perp$ undergoing a 90° collision reversing the sign of $v_\parallel$.

The step size $\Delta r$ is defined as the difference in radii of the flux surfaces before and after the collision:

$$\Delta r = r_i - r_f = 2\frac{|v_{\mathrm{Dr}}|}{\omega_T}\cos\theta_0. \tag{14.102}$$

One next has to average over all collisions, which is equivalent to averaging over all (equally likely) starting positions $\theta_0$ and velocities. As expected $\langle\Delta r\rangle = 0$ when averaging over $v_\parallel$ because of the odd symmetry in $\omega_T$. However, the mean square step size is not zero and is given by

$$(\Delta l)^2 = \langle(\Delta r)^2\rangle = 4\frac{|v_{\mathrm{Dr}}|^2}{\omega_T^2}\langle\cos^2\theta_0\rangle = 2\frac{|v_{\mathrm{Dr}}|^2}{\omega_T^2}. \tag{14.103}$$

Substituting for $v_{\mathrm{Dr}}$ and $\omega_T$ leads to

$$(\Delta l)^2 = 2\frac{q^2}{\omega_{ci}^2}\frac{(v_\parallel^2 + v_\perp^2/2)^2}{v_\parallel^2} \approx 4\frac{q^2 v_{Ti}^2}{\omega_{ci}^2} \sim q^2 r_{Li}^2, \tag{14.104}$$

where the average over velocities has been taken by approximating $v_\perp^2/2 \approx v_\parallel^2 \sim v_{Ti}^2/2$. Note that the mean square step size is a factor $q^2$ larger than for classical transport. A similar expression holds for electrons.

### The transport coefficients

The transport coefficients in the random walk model are given by $(\Delta l)^2/\tau$, where $\tau$ is the mean time between collisions. For passing particles $\tau$ corresponds to the 90° momentum collision time. Consider first particle diffusion. As in the case of classical diffusion, a more detailed derivation of the step size shows that ambipolarity still holds. Like particle collisions do not produce density diffusion in neoclassical theory. Particle diffusion results from electron–ion momentum exchange collisions. The random walk neoclassical particle diffusion coefficient due to passing particles is thus given by

$$D_n^{(NC)} = \frac{(\Delta l)_e^2}{\overline{\tau}_{ei}} = 4q^2 \left( \frac{2m_e T_e}{e^2 B_0^2 \overline{\tau}_{ei}} \right) = 4q^2 D_n^{(CL)}. \tag{14.105}$$

Note that the neoclassical coefficient is a factor $4q^2 \sim 30$ larger than the classical value.

A similar conclusion holds for the thermal diffusivities which are again driven by like particle collisions:

$$\chi_e^{(NC)} \sim q^2 \chi_e^{(CL)} \sim q^2 \frac{r_{Le}^2}{\overline{\tau}_{ee}},$$

$$\chi_i^{(NC)} \sim q^2 \chi_i^{(CL)} \sim q^2 \frac{r_{Li}^2}{\overline{\tau}_{ii}}. \tag{14.106}$$

Numerical values indicate that the ion thermal diffusivity due to passing particles is approximately equal to $\chi_i^{(NC)} \approx 1.6 \times 10^{-2}$ m$^2$/s, which is still a factor of about 60 smaller than that observed in experiments. However, the transport due to passing particles is not the dominant loss mechanism in toroidal geometry. It is instead the transport that arises from the small population of trapped particles that dominates particle and heat loss and this is the next topic.

### 14.4.3 Neoclassical transport due to trapped particles

The neoclassical losses resulting from trapped particles are also calculated using the random walk model. Before proceeding with the analysis one can ask why there are trapped particles and why should their neoclassical losses dominate Coulomb transport? Qualitatively, the answers are as follows. Trapped particles exist because $B \approx B_\phi \approx B_0 (R_0/R)$ in a tokamak. Thus the magnetic field strength is weak on the outside of the torus and strong on the inside. Therefore, particles starting on the outside of the torus with a small ratio of $v_\parallel/v_\perp$ are mirror reflected as their parallel motion winds them towards the inside of the torus into a region of higher field. The particles are "trapped" on the outside of the torus.

There are several reasons why trapped particle transport is large. One main reason results from the fact that their parallel velocity is small. It takes a longer time for a trapped particle to complete one full cycle of its mirror motion than for a typical passing particle to make one full transit around the poloidal cross section. Since the trapped particle mirror period is longer, there is more time for particles to drift off the flux surfaces because of the $\nabla B$ and curvature drifts and this increases the step size.

The analysis of trapped particle transport is similar to that for the passing particles although three modifications must be made in the random walk model. First, only a small fraction of the plasma particles is trapped (in the large aspect ratio limit) and one needs to

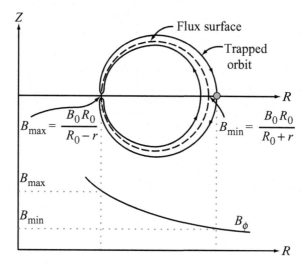

Figure 14.11 Geometry showing a trapped particle orbit and the maximum and minimum $B \approx B_\phi \approx B_0(R_0/R)$.

know this fraction. Second, the radial step size must be recalculated taking into account the typical ratio of $v_\parallel/v_\perp$ for trapped particles. Third, the mean time between collisions is modified because trapped particles have to scatter over a smaller angle (much less than $90°$) in order to move "one step." The details of the analysis are described below.

### The fraction of trapped particles

The fraction of trapped particles can easily be calculated by examining Fig. 14.11. One sees that the minimum (at the outside) and maximum (at the inside) magnetic field strengths are given by

$$
\begin{aligned}
B_{\min} &= B_0 \frac{R_0}{R_0 + r}, \\
B_{\max} &= B_0 \frac{R_0}{R_0 - r}.
\end{aligned}
\tag{14.107}
$$

Consider now a particle starting off on the outside of the torus ($\theta = 0$) with a velocity $\mathbf{v} = v_\parallel \mathbf{b} + \mathbf{v}_\perp$. Using the conservation of energy and magnetic moment as discussed in Chapter 8, it follows that the condition for particles to be trapped can be written as

$$
\frac{v_\parallel^2}{v^2} < 1 - \frac{B_{\min}}{B_{\max}} = 1 - \frac{R_0 - r}{R_0 + r} \approx 2\frac{r}{R_0},
\tag{14.108}
$$

where $v^2 = v_\parallel^2 + v_\perp^2$ and the last form follows from the large aspect ratio assumption $r/R_0 \ll 1$.

The boundary between trapped and untrapped particles is shown in velocity phase space in Fig. 14.12. Note that the critical angle $\theta_c$ is defined by $\cos\theta_c = v_\parallel/v \approx (2r/R_0)^{1/2}$. The

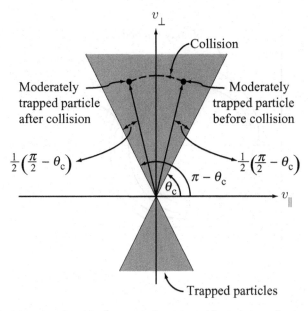

Figure 14.12 Velocity space showing the trapped–untrapped boundary and a moderately trapped particle undergoing a collision that reverses the sign of $v_\parallel$.

fraction $f$ of trapped particles can now be calculated by integrating over the trapped portion of the distribution function. For a Maxwellian distribution function $F_M(v)$ one finds

$$f = \frac{1}{n} \int_{\theta_c}^{\pi-\theta_c} \sin\theta \, d\theta \int_0^{2\pi} d\phi \int_0^\infty F_M(v) v^2 \, dv = \cos\theta_c \approx \left(\frac{2r}{R_0}\right)^{1/2}. \qquad (14.109)$$

In terms of the inverse aspect ratio $\varepsilon = a/R_0$, Eq. (14.109) implies that $f \sim \varepsilon^{1/2}$. In the limit of large aspect ratio only a small fraction of the particles are trapped, although for practical cases $f$ can easily exceed the value $f > \frac{1}{2}$. Even so, maintaining the expansion $\varepsilon \ll 1$ is still very useful for understanding the physics.

### The bounce frequency

The distance that a particle drifts off the flux surface is proportional to the time it takes for its guiding center to complete one full mirror trapping period. This time can be calculated by examining Fig. 14.13, which shows the orbits of a strongly trapped particle, a moderately trapped particle, and a weakly trapped particle. The average behavior of the trapped particles is approximately equal to that of the moderately trapped particles. Attention is therefore focused on this class of particles. Note that because of the shape of the guiding center trajectories, the trapped particle orbits are almost always referred to in the literature as "banana" orbits. Also, the trapping period is referred to as the "bounce" period.

Similarly to the passing particles, the trapped particles drift monotonically away from the surface for one half of a bounce period and then return to the surface during the second

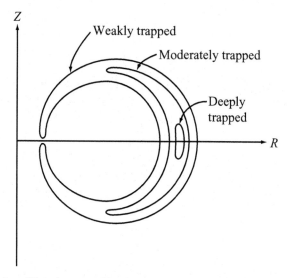

Figure 14.13 Drawing of three banana orbits corresponding to a deeply trapped particle, a moderately trapped particle, and a weakly trapped particle.

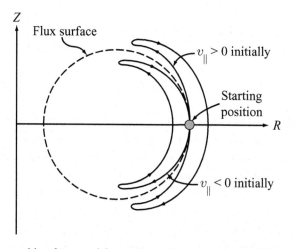

Figure 14.14 Banana orbits of two particles starting at the same point but with equal and opposite $v_\parallel$. Note that the $v_\parallel > 0$ particle drifts outward, while the $v_\parallel < 0$ particle drifts inward.

half. The banana orbits of two particles with equal but opposite initial $v_\parallel$ are illustrated in Fig. 14.14. Observe that when $v_\parallel > 0$ the banana orbit shifts outward, while the opposite is true for $v_\parallel < 0$.

Now, the half-bounce period is the time it takes a moderately trapped particle to move a poloidal distance equal to one half the circumference of the flux surface: $l_p = \pi r$. As for the passing particles, the actual distance $l$ that the trapped particle moves parallel to the field is given by $l \approx (B_0/B_\theta) l_p$. Therefore, the half-bounce period can be written as

$$\tau_{1/2} \approx \frac{l}{\bar{v}_\parallel} \approx \frac{2l}{v_\parallel} \approx \frac{2\pi R_0 q}{v_\parallel}, \tag{14.110}$$

where $\bar{v}_\parallel \approx v_\parallel/2$ is the average parallel velocity of the particle during its bounce motion and $v_\parallel$ is the outboard parallel velocity at $\theta = 0$. The full cycle bounce frequency is defined as $\omega_B = \pi/\tau_{1/2}$ and is given by

$$\omega_B = \frac{v_\parallel}{2 R_0 q}. \tag{14.111}$$

This is the quantity required to calculate the step size.

### The step size

The step size $\Delta r$ is defined as the distance between banana centers for particles with equal but opposite $v_\parallel$. The calculation is nearly identical to that of the passing particles. Consider a particle starting at a position $\theta = \theta_0$, $r = r_0$, where for moderate trapping $-\pi/2 < \theta_0 < \pi/2$. The radial component of the guiding center drift velocity can be expressed as

$$v_{Dr}(t) \approx |v_{Dr}| \sin(\omega_B t + \theta_0). \tag{14.112}$$

Here, if one takes into account that $v_\parallel \ll v_\perp \approx v$ for trapped particles, it then follows that $|v_{Dr}|$ is dominated by the $\nabla B$ drift and is given by

$$|v_{Dr}| = \frac{m_i}{e} \left( v_\parallel^2 + \frac{v_\perp^2}{2} \right) \left| \frac{\mathbf{R}_c \times \mathbf{B}}{R_c^2 B} \right| \approx \frac{v^2}{2\omega_{ci} R_0}. \tag{14.113}$$

The corresponding radial position of the guiding center is again obtained by integrating $\dot{r} = v_{Dr}$ leading to

$$r(t) \approx r_0 - \frac{|v_{Dr}|}{\omega_B} [\cos(\omega_B t + \theta_0) - \cos(\theta_0)]. \tag{14.114}$$

For a particle with $v_\parallel > 0$ the radius $r_B^+$ of the center of the banana orbit is given by

$$r_B^+ = \frac{1}{2}(r_{max} + r_{min}) = r_0 + \frac{|v_{Dr}|}{\omega_B} \cos\theta_0. \tag{14.115}$$

Assume now that a collision takes place at $r = r_0$, $\theta = \theta_0$ that switches the sign of $v_\parallel$. After the collision, the particle begins a new banana orbit with $v_\parallel \to -v_\parallel$, equivalent to $\omega_B \to -\omega_B$. The radius $r_B^-$ of the new banana orbit is then

$$r_B^- = \frac{1}{2}(r_{max} + r_{min}) = r_0 - \frac{|v_{Dr}|}{\omega_B} \cos\theta_0. \tag{14.116}$$

One can now easily calculate the step size defined as

$$\Delta r = r_B^+ - r_B^- = 2\frac{|v_{Dr}|}{\omega_B} \cos\theta_0. \tag{14.117}$$

The mean square value required for the random walk model is given by

$$(\Delta l)^2 = \langle (\Delta r)^2 \rangle = 4 \frac{|v_{Dr}|^2}{\omega_B^2} \langle \cos^2 \theta_0 \rangle = 2 \frac{q^2 v^4}{\omega_{ci}^2 v_\parallel^2}. \tag{14.118}$$

The average over velocities is carried out by the following approximations. For the total particle energy $v^2 \sim 3T/m = (3/2) v_{Ti}^2$, while for the parallel energy of a moderately trapped particle $v_\parallel^2 \approx (r/R_0) v^2 \sim (r/R_0)(3v_{Ti}^2/2)$. Substituting these approximations leads to

$$(\Delta l)^2 \approx 3 \left( q^2 \frac{R_0}{r} \right) \frac{v_{Ti}^2}{\omega_{ci}^2} \sim \left( q^2 \frac{R_0}{r} \right) r_{Li}^2. \tag{14.119}$$

Observe that the mean step size for trapped particles scales as $(q^2 r_{Li}^2/\varepsilon)^{1/2}$. It is thus larger by a factor of $1/\varepsilon$ than for passing particles and about a factor of 50–100 larger than for classical transport.

### The effective collision frequency

The last quantity needed for the random walk model is the mean time between collisions. For trapped particles this time is considerably shorter than the 90° momentum collision time. The reason is that trapped particles are characterized by a small $v_\parallel$ and consequently such particles need to scatter over a much smaller angle to become de-trapped.

To be specific, refer back to Fig. 14.12 and note that a moderately trapped particle has an initial pitch angle $\theta_i \approx \theta_c + (1/2)(\pi/2 - \theta_c) = \pi/4 + \theta_c/2$. A Coulomb collision that produces a mean square step size $\langle (\Delta r)^2 \rangle$ requires that $v_\parallel$ change sign. A typical scattering collision thus leaves the particle with a pitch angle $\theta_f = \pi - \theta_c - (1/2)(\pi/2 - \theta_c) = 3\pi/4 - \theta_c/2$. The change in pitch angle is given by $\Delta\theta = \theta_f - \theta_i = \pi/2 - \theta_c$.

The mean time between collisions is known as the "effective collision time" and is determined by recalling that angular diffusion in velocity space is the result of many small-angle collisions. The diffusive nature of the process implies that the mean square value of $\theta - \theta_i$ can be written as

$$\langle (\theta - \theta_i)^2 \rangle = D_\theta t, \tag{14.120}$$

where $D_\theta$ is determined in terms of the 90° collision time $\tau_{90}$ by setting $\theta - \theta_i = \pi/2$ and $t = \tau_{90}$. Here $\tau_{90}$ is equal to $\overline{\tau}_{ei}$, $\overline{\tau}_{ii}$, or $\overline{\tau}_{ee}$ depending upon the collisions under consideration. A simple calculation leads to $D_\theta = \pi^2/(4\tau_{90})$ and

$$\langle (\theta - \theta_i)^2 \rangle = \frac{\pi^2}{4} \left( \frac{t}{\tau_{90}} \right). \tag{14.121}$$

The calculation of the effective collision time is now completed by setting $t = \tau_{eff}$ and $\theta = \theta_f$. One obtains

$$\tau_{eff} = \frac{8}{\pi^2} \left( \frac{r}{R_0} \tau_{90} \right) \sim \varepsilon \tau_{90}, \tag{14.122}$$

where use has been made of the approximation $\theta_c \approx \pi/2 - (2r/R_0)^{1/2}$. Observe that the effective collision time is reduced by an amount of order $\varepsilon$ from the full $90°$ collision time.

### The trapped particle neoclassical transport coefficients

All the separate components are now in place to evaluate the trapped particle neoclassical transport coefficients by means of the random walk model. The diffusion coefficients are given by the ratio of $(\Delta l)^2$ to $\tau_{\text{eff}}$, multiplied by the fraction $f$ of trapped particles (since only this portion of the particles is involved in the transport). Mathematically this is equivalent to

$$D = f \frac{\langle (\Delta r)^2 \rangle}{\tau_{\text{eff}}}. \tag{14.123}$$

Consider first particle diffusion. Ambipolarity again holds for trapped particles, implying that particle diffusion is caused by electron–ion collisions. This leads to the following expression for the particle diffusion coefficient:

$$D_n^{(NC)} = 5.2\, q^2 \left( \frac{R_0}{r} \right)^{3/2} \left( \frac{2 m_e T_e}{e^2 B_0^2 \bar{\tau}_{ei}} \right) = 5.2\, q^2 \left( \frac{R_0}{r} \right)^{3/2} D_n^{(CL)}. \tag{14.124}$$

An elegant, self-consistent kinetic theory of neoclassical transport has been formulated by Rosenbluth, Hazeltine, and Hinton that leads to the same scaling relation for $D_n^{(NC)}$ but with a corrected numerical coefficient. They find

$$D_n^{(NC)} = 2.2\, q^2 \left( \frac{R_0}{r} \right)^{3/2} D_n^{(CL)}. \tag{14.125}$$

Observe that for $q = 3$, $R_0 = 5$, and $r \approx a/2 = 1$, the neoclassical transport due to trapped particles is a factor of 220 larger than the classical value.

A similar analysis holds for the thermal diffusivities, which are again determined by like particle collisions. The same multiplying factor of $q^2 (R_0/r)^{3/2}$ appears in each diffusivity coefficient. The diffusivities with the correct numerical coefficients obtained by Rosenbluth, Hazeltine, and Hinton are given by

$$\chi_e^{(NC)} = 0.89\, q^2 \left( \frac{R_0}{r} \right)^{3/2} \chi_e^{(CL)} = 4.3 \times 10^{-3} q^2 \left( \frac{R_0}{r} \right)^{3/2} \left( \frac{n_{20}}{B_0^2 T_k^{1/2}} \right) \text{m}^2/\text{s},$$

$$\chi_i^{(NC)} = 0.68\, q^2 \left( \frac{R_0}{r} \right)^{3/2} \chi_i^{(CL)} = 0.068\, q^2 \left( \frac{R_0}{r} \right)^{3/2} \left( \frac{n_{20}}{B_0^2 T_k^{1/2}} \right) \text{m}^2/\text{s}. \tag{14.126}$$

Note that the neoclassical ion diffusivity is enhanced by a factor of about 68 for the test example. In absolute units, $\chi_i \approx 0.12$ m$^2$/s for the simple reactor design parameters.

A final point to be considered in the theory of trapped particle neoclassical transport is the regime of validity. The underlying assumption in the random walk argument is that trapped particles have sufficient time to complete one bounce period before undergoing a collision. The condition for "banana" regime transport to be valid is thus given by $\nu_{\text{eff}} \ll \omega_B$ (with

$v_{\mathrm{eff}} = \tau_{\mathrm{eff}}^{-1}$), which is independent of the particle mass and can be written as

$$v_* \equiv \frac{v_{\mathrm{eff}}}{\omega_B} \sim \left(\frac{R_0}{r}\right)^{3/2} \left(\frac{q R_0}{v_T \tau_{90}}\right) \sim 0.01 \left(\frac{R_0}{r}\right)^{3/2} \left(\frac{q R_0 n_{20}}{T_k^2}\right) \ll 1. \qquad (14.127)$$

For the simple reactor $v_* \approx 0.01$, which clearly satisfies the low-collisionality requirement.

Overall, in terms of the applicability of neoclassical transport theory to current tokamak experiments the situation is as follows. The ion thermal diffusivity is somewhat lower than typical experimentally observed values, which are on the order of $\chi_i \sim 1$ m²/s. However, low-turbulence modes of operation have been discovered and in these situations the observed ion thermal diffusivity over portions of the plasma approaches the neoclassical value, which is an irreducible minimum. Still, for most high-performance operation the value of $\chi_i$ is anomalous because of plasma micro-instabilities. Ion thermal conduction represents the fastest loss of energy, exceeding that of electron thermal conduction and particle diffusion. However, experimentally $\chi_e$ and $D_n$ are only slightly smaller, by a factor on the order of 3, compared to $\chi_i$, and not the much larger reduction of $(m_e/m_i)^{1/2}$ expected from the theory. Thus, both electron heat conduction and particle diffusion have large anomalies because of micro-turbulence.

The conclusion is that neoclassical theory serves as a useful reference point for the lower limit on energy transport, but is still optimistic with respect to actual tokamak operation.

### 14.4.4 The bootstrap current

The bootstrap current $J_B$ is one of the most interesting and important predictions of neoclassical transport theory. It is important because it is generated by the natural radial transport in the plasma, thereby creating a potentially steady state toroidal plasma current in a tokamak without the need for expensive, external current drive. A tokamak without a substantial fraction of bootstrap current would very likely not be viable as a reactor for economic reasons.

The bootstrap current is also a quite subtle phenomenon since the final form of $J_B$ is independent of collision frequency but yet is a consequence of collisional transport. An intuitive picture of the origin of the bootstrap current is presented in this subsection. It is shown that the bootstrap current flows parallel and not anti-parallel to the main toroidal current. Also its magnitude can be quite substantial, theoretically capable of approaching 100% of the toroidal current. This is critical since bootstrap fractions on the order of $f_B > 0.7$ are probably required for economic viability.

The intuitive picture, which assumes for simplicity that the ions are infinitely massive, shows that three electron currents need to be considered. These are the magnetization current due to the trapped electrons, the magnetization current due to the passing electrons, and the current that flows because of the frictional momentum exchange between trapped and passing electrons. The final result demonstrates that the bootstrap current is carried

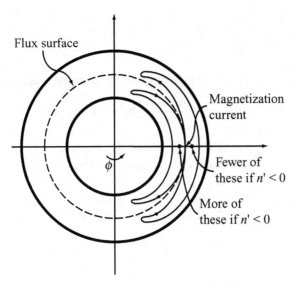

Figure 14.15 Top view of the tokamak showing the toroidal projection of two banana orbits. If $\partial n/\partial r < 0$, there are more inward that outward shifted banana orbits. This produces a net downward magnetization (for positive particles) at the point of tangency.

by a flow of passing electrons generated by collisional friction with the trapped electron magnetization current.

### The trapped electron magnetization current

A simple derivation of the magnetization current has been presented in Chapter 11. A similar analysis applies here, although it is necessary to evaluate the contributions from the trapped and passing particles separately. Consider first the trapped particles. The component of magnetization current that flows parallel to the magnetic field is most easily visualized by viewing the tokamak from the top as shown in Fig. 14.15. Since $B_\phi \gg B_\theta$, the projection of the guiding center motion from the top essentially traces out the parallel motion of the particles, the orbit of which also has the shape of a banana.

The magnetization current at the radius $r = r_0$ arises because of the gradient in the guiding center density (or temperature). Specifically, banana orbit electrons with a parallel velocity $v_\parallel > 0$ at $r = r_0$ have guiding centers shifted outward to $r = r_g^+ = r_0 + \Delta r/2$. These electrons produce a parallel current at $r = r_0$ approximately given by $J^+ = -ef_e(r_g^+, \mathbf{v})v_\parallel \, d\mathbf{v}$. Similarly, the banana orbits of electrons whose $v_\parallel$ is equal in magnitude but opposite in sign at $r = r_0$ have guiding centers shifted inward to $r = r_g^- = r_0 - \Delta r/2$. These electrons also produce a current at $r = r_0$. Its value is given by $J^- = -ef_e(r_g^-, \mathbf{v})v_\parallel \, d\mathbf{v} = +ef_e(r_g^-, \mathbf{v})|v_\parallel| \, d\mathbf{v}$.

If there are more inward than outward shifted guiding centers (i.e., a negative density gradient) there is a net magnetization current at $r = r_0$. The net current is obtained by summing the two contributions as follows:

$$J^+ + J^- = -e[f_e(r_g^+, \mathbf{v}) - f_e(r_g^-, \mathbf{v})]v_\parallel \, d\mathbf{v} \quad v_\parallel > 0. \tag{14.128}$$

This expression can be simplified by Taylor expanding assuming small $\Delta r$:

$$J^+ + J^- \approx -e\frac{\partial f_e(r_0, \mathbf{v})}{\partial r_0}\Delta r \, v_\parallel \, d\mathbf{v} \quad v_\parallel > 0. \tag{14.129}$$

Now, recall that for trapped particles $\Delta r \approx \langle(\Delta r)^2\rangle^{1/2} \approx q\,(R_0/r_0)^{1/2}\,r_{\mathrm{Li}}$. One substitutes this expression into Eq. (14.129) and then integrates over velocity space to determine the total magnetization current. In carrying out the $v_\parallel$ integration keep in mind that for trapped particles $0 < |v_\parallel| < (2r_0/R_0)^{1/2}\,v$. Thus, for a Maxwellian distribution function (with $T = $ const. for simplicity) the trapped particle magnetization current $J_t$ can be written as

$$
\begin{aligned}
J_t &= -\frac{m_e q}{B_0}\left(\frac{R_0}{r}\right)^{1/2}\int\frac{\partial F_M}{\partial r}v_\perp v_\parallel \, d\mathbf{v} \qquad v_\parallel > 0 \\[2mm]
&= -\frac{3}{2}q\left(\frac{R_0}{r}\right)^{1/2}\frac{T}{B_0}\frac{\partial n}{\partial r}\int_{\theta_c}^{\pi/2}\sin^2\theta\,\cos\theta\,d\theta \\[2mm]
&\approx -q\left(\frac{r}{R_0}\right)^{1/2}\frac{T}{B_0}\frac{\partial n}{\partial r}.
\end{aligned}
\tag{14.130}
$$

Here, $\cos\theta_c = (2r/R_0)^{1/2}$ is the critical angle defining the region of trapped particles. Also, for convenience the subscript "zero" has been dropped from $r_0$ and in the final expression the unimportant numerical multiplier has been ignored

Equation (14.130) is the desired expression. Observe that $J_t$ does not depend on collisions and is non-zero even though the distribution function for guiding centers is symmetric in $v_\parallel$: that is, $F_M(r_g, v_\parallel) = F_M(r_g, -v_\parallel)$. The current is generated solely because of the density (or temperature) gradient of the guiding centers. Formally, this contribution arises because of the $v_\parallel$ dependence of $r_g$; that is, $r_g = r + (v_\parallel/|v_\parallel|)\Delta r/2$, illustrating the fact that parallel and anti-parallel particles drift in opposite directions off the flux surface. A careful consideration of (1) the sign of the $\nabla B$ drift velocity, and (2) the orientation of the parallel direction as defined by the sign of $B_\theta/B_\phi$ shows that $J_t$ for a negative density gradient flows in the same direction as the $J_\phi$ current that generates $B_\theta$; in other words $J_t$ is parallel and not anti-parallel to $J_\phi$ for a negative density gradient.

The quantity $J_t$ plays a critical role in driving the bootstrap current but, as is shown shortly, by itself only represents a small fraction of $J_B$.

### The passing electron magnetization current

A completely analogous calculation applies to the passing particles as shown in Fig. 14.16. In this case the passing particle magnetization current can be expressed as

$$J_p \approx -e\int\frac{\partial f_e(r_0, \mathbf{v})}{\partial r_0}\Delta r \, v_\parallel \, d\mathbf{v} \quad v_\parallel > 0. \tag{14.131}$$

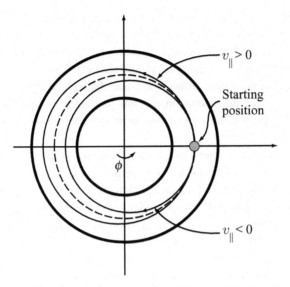

Figure 14.16 Top view of the tokamak showing the toroidal projection of two passing particle orbits, one with $v_\parallel > 0$, the other with $v_\parallel < 0$. Note that the radius of the $v_\parallel < 0$ orbit is the smaller, implying that if $\partial n/\partial r < 0$ there will be a downward magnetization current at the point of tangency.

The integral is easily evaluated by noting that for passing particles $\Delta r \approx q r_{Li}$ and that in the large aspect ratio limit the passing particles occupy almost the entire velocity space $(2r_0/R_0)^{1/2} v < |v_\parallel| < v$. Thus, one obtains

$$J_p = -\frac{m_e q}{B_0} \int \frac{\partial F_M}{\partial r} v_\perp v_\parallel \, d\mathbf{v} \qquad v_\parallel > 0$$

$$= -\frac{3}{2} q \frac{T}{B_0} \frac{\partial n}{\partial r} \int_0^{\theta_c} \sin^2 \theta \, \cos \theta \, d\theta$$

$$\approx -q \frac{T}{B_0} \frac{\partial n}{\partial r}. \qquad (14.132)$$

Equation (14.132) shows that the magnetization current arising from the passing particles also flows parallel and not anti-parallel to $J_\phi$ and is larger by $1/\varepsilon^{1/2}$ than the contribution from the trapped particles: $J_p \sim \varepsilon^{-1/2} J_t$. Even so, this larger contribution to the magnetization current does not represent the bootstrap current. As is shown next, the bootstrap current is larger still, and results from the requirement that electron–electron collisions conserve momentum.

### The collision-driven bootstrap current

The bootstrap current arises because of the collisional friction between the passing and trapped electrons. Recall that the derivation of the magnetization currents completely ignores the effect of collisions. Nevertheless, even if collisions are infrequent, when included in the steady state analysis, they impose a strong constraint on the electron currents. Specifically, the magnitudes of the trapped and passing particle currents must be such that the total momentum exchanged between all electrons is zero. In other words, when summing over both trapped and passing electrons momentum is exactly conserved in like particle collisions since the Coulomb interaction represents a purely elastic collision.

The first part of the derivation shows that when the trapped and passing particles are allowed to carry only the magnetization currents just calculated, collisional momentum balance is violated. The second part of the derivation shows how the currents must be altered to resolve the difficulty.

Consider first collisional momentum balance arising solely from the magnetization currents. The physics can be understood if one imagines that the electron fluid is divided into two species – trapped particles and passing particles. For the passing particles the average parallel momentum lost per particle per collision due to Coulomb interactions with the trapped particles is given by $\langle m_e v_\parallel \rangle_p = m_e u_p$, where $u_p = -J_p/en_p$ represents the macroscopic flow velocity. Now, the density of passing particles is $n_p$ and on average they lose their parallel momentum during a collision time $(\overline{\nu}_{pt})^{-1}$. Thus passing particles lose the following amount of momentum per cubic meter per second:

$$(\Delta P_\parallel)_p = (m_e u_p)(n_p)(\overline{\nu}_{pt}). \tag{14.133}$$

Since passing particles must scatter a full $90°$ to lose their momentum, this implies that $\overline{\nu}_{pt} \sim \overline{\nu}_{ee}$. Equation (14.133) reduces to

$$(\Delta P_\parallel)_p \approx -\frac{m_e}{e} J_p \overline{\nu}_{ee} \approx qT \frac{\partial n}{\partial r} \frac{\overline{\nu}_{ee}}{|\omega_{ce}|}, \tag{14.134}$$

where use has been made of the approximation $n_p \approx n$.

A similar estimate applies to the loss of trapped particle momentum $(\Delta P_\parallel)_t$. For the trapped particles $\langle m_e v_\parallel \rangle_t = m_e u_t = -m_e J_t/en_t$. The density of trapped particles is $n_t$ and they lose their momentum in a time $(\overline{\nu}_{tp})^{-1}$. Therefore, the momentum lost per cubic meter per second is given by

$$(\Delta P_\parallel)_t \approx -\frac{m_e}{e} J_t \overline{\nu}_{tp}. \tag{14.135}$$

This expression is simplified by recalling that $n_t \approx (r/R_0)^{1/2} n$ and that trapped particles lose their momentum in a much shorter time than passing particles since they only have to scatter through a small angle: $\overline{\nu}_{tp} \approx (R_0/r) \overline{\nu}_{ee}$. Equation (14.135) reduces to

$$(\Delta P_\parallel)_t \approx qT \left(\frac{R_0}{r}\right)^{1/2} \frac{\partial n}{\partial r} \frac{\overline{\nu}_{ee}}{|\omega_{ce}|}. \tag{14.136}$$

For collisional momentum balance one requires that the two losses be equal since the loss by one species represents a gain by the other. In other words, in steady state, conservation of total momentum in electron–electron collisions implies that the momentum transfer from passing to trapped particles is equal to the momentum transfer from trapped to passing particles: $(\Delta P_\parallel)_p = (\Delta P_\parallel)_t$. However, there is a basic scaling mismatch since

$$(\Delta P_\parallel)_p \approx \left(\frac{r}{R_0}\right)^{1/2} (\Delta P_\parallel)_t. \tag{14.137}$$

The conclusion is that momentum balance cannot be achieved solely by the magnetization currents.

What has happened and how can this difficulty be resolved? The problem is that while there are fewer trapped particles ($n_t/n_p \sim \varepsilon^{1/2}$) they lose the momentum associated with their magnetization flow at such a fast rate ($\bar{v}_{tp}/\bar{v}_{pt} \sim 1/\varepsilon$) that a momentum imbalance is created with respect to the passing particles. The difficulty is resolved by relaxing the constraint that in the guiding center reference frame the passing particles have a *pure stationary* Maxwellian distribution function. Instead, the passing electrons must have a net parallel flow due to their diamagnetism that can be approximately modeled by a *shifted* Maxwellian and that balances the excess momentum of the trapped electrons. The shift $u_B$ must be in the passing particles since the trapped particles are "trapped" and thus are not allowed to drift toroidally.

Mathematically, this requires the following replacement in the derivation of the passing particle current:

$$f_p(r_g, \mathbf{v}) = \frac{n(r_g)}{\pi^{3/2} v_T^3} \exp\left(-\frac{v_\perp^2 + v_\parallel^2}{v_T^2}\right) \rightarrow \frac{n(r_g)}{\pi^{3/2} v_T^3} \exp\left[-\frac{v_\perp^2 + (v_\parallel - u_B)^2}{v_T^2}\right]. \tag{14.138}$$

In the limit of small $\Delta r$ and small $u_B$ the distribution function can be Taylor expanded yielding

$$f_p(r_g, \mathbf{v}) \approx \frac{n(r)}{\pi^{3/2} v_T^3} \left[1 + \frac{v_\parallel}{|v_\parallel|}\left(\frac{1}{n}\frac{\partial n}{\partial r}\right)(\Delta r)_p + 2\frac{v_\parallel u_B}{v_T^2}\right] \exp\left(-\frac{v_\perp^2 + v_\parallel^2}{v_T^2}\right). \tag{14.139}$$

When calculating $J_\parallel$ one finds that the first term in the square bracket (i.e., the "1") averages to zero after multiplying by $v_\parallel$ and integrating over velocity space. The second term produces the contribution due to the passing particle magnetization current. The last term is a new contribution representing a passing particle flow driven by the collisional imbalance.

The net result of this modification is that the quantity $m_e u_p$ representing the average momentum lost per passing particle collision in Eq. (14.133) must be replaced by

$$m_e u_p = -m_e \left(\frac{J_p}{e n_p}\right) \rightarrow m_e \left(-\frac{J_p}{e n_p} + u_B\right). \tag{14.140}$$

This, in turn, implies that the passing particle collisional loss per cubic meter per second (Eq. (14.134)) becomes

$$(\Delta P_\parallel)_p \approx m_e \left(-\frac{J_p}{e} + n_p u_B\right) \bar{v}_{ee}. \tag{14.141}$$

Finally, the value of $u_B$ is obtained from the collisional momentum balance requirement $(\Delta P_\parallel)_p = (\Delta P_\parallel)_t$. In order for balance to occur one finds that $u_B \gg |J_p/en_p|$ leading to the following expression for $J_B = -en_p u_B$:

$$J_B \approx -q \left(\frac{R_0}{r}\right)^{1/2} \frac{T}{B_0} \frac{\partial n}{\partial r}. \tag{14.142}$$

The quantity $J_B$ is the bootstrap current. Note that it is $1/\varepsilon$ larger than the trapped particle magnetization current and $1/\varepsilon^{1/2}$ larger than the passing particle magnetization current.

Equation (14.142) actually represents only part of the total bootstrap current since in the derivation it has been assumed that the temperature is uniform and the ions are infinitely massive. Relaxing these constraints leads to additional contributions to $J_B$. Interestingly, the ion and electron density gradient contributions add together while the temperature gradient contributions tend to cancel. In any event all of these additional contributions are of the same order as that given by Eq.(14.142), so the basic scaling remains unchanged. General forms for the bootstrap current including all the above effects as well as an arbitrary cross section have been calculated self-consistently from kinetic theory. In the large aspect ratio, circular cross section limit the more exact form of the bootstrap current is given by

$$J_B = -4.71q \left(\frac{R_0}{r}\right)^{1/2} \frac{T}{B_0} \left[\frac{\partial n}{\partial r} + 0.04 \frac{n}{T} \frac{\partial T}{\partial r}\right]. \tag{14.143}$$

This is the desired low-collisionality expression to be used in future calculations. An important property to note is that the bootstrap current normally peaks off-axis since $n'/r^{1/2} \to 0$ as $r \to 0$.

The final point concerns the important question of the bootstrap fraction $f_B$. Since the total toroidal current flowing in the plasma is given by $\mu_0 J_\phi \approx (1/r) \partial r B_\theta/\partial r$ it follows that

$$f_B(r) \equiv \frac{J_B}{J_\phi} \approx -1.18G \left(\frac{r}{R_0}\right)^{1/2} \beta_p \sim \varepsilon^{1/2} \beta_p, \tag{14.144}$$

where $\beta_p(r) = 4\mu_0 n T/B_\theta^2$ is the local $\beta_p$ and $G(r)$ is a profile factor defined by

$$G(r) = (\ln n + 0.04 \ln T)' / (\ln r B_\theta)'. \tag{14.145}$$

Observe that the bootstrap fraction can be quite large. Recall that in the high-$\beta$ tokamak ordering $\beta_p \sim 1/\varepsilon$, implying that $f_B \sim 1/\varepsilon^{1/2} \gg 1$. The bootstrap current can theoretically overdrive the total current. In practice, however, the situation is more complicated. First, $\varepsilon^{1/2}$ is not that small. Second, the profile factor tends to be small for typical flat density profiles. Third, the collisionality may be low but it still leads to modifications of the numerical coefficients appearing in $J_B$ and a corresponding finite reduction in the bootstrap fraction. Fourth, the way to achieve high $\beta_p$ is through a combination of high pressure and low toroidal current. However, low toroidal current shortens the energy confinement time making it harder to achieve high pressure. The final bootstrap fraction therefore involves a number of tradeoffs and a careful analysis including profile effects. A simple example is described in Section 14.6.4.

For the moment, the main conclusion to be drawn from the analysis is that neoclassical trapped particle effects lead to a transport driven toroidal plasma current carried by the passing particles. This bootstrap current is capable of being maintained in steady state without the need of an ohmic transformer or external current drive. Furthermore, tokamak experiments indicate that the neoclassical prediction of $J_B$ is consistent with observations. There is no obvious "anomalous" degradation of $J_B$ due to micro-turbulence. This is indeed a favorable result as it opens up the possibility of steady state operation without the need for excessive amounts of external current drive power.

### 14.4.5 Summary

Neoclassical theory describes the effect of Coulomb collisions on various plasma transport phenomena in a toroidal geometry. The most striking difference between neoclassical and cylindrical transport theory is the effect of trapped particle banana orbits. It has been shown that particle and heat transport are enhanced by a factor of $q^2(R_0/r)^{3/2}$ for each species corresponding to nearly two orders of magnitude in practical situations.

However, experimental data show that the micro-turbulence-driven ion thermal conductivity represents the fastest loss of energy and is anomalously large by a factor on the order of 1–10 with respect to the neoclassical $\chi_i$. The electron heat conduction and particle diffusion coefficients are each anomalous by about two orders of magnitude. The end result is that in practice $\chi_i \sim \chi_e \sim D_n$. Despite the unreliability of ion neoclassical theory to predict experimental energy loss the model still serves as a firm reference point for understanding transport theory. Also, it has been discovered empirically that in certain modes of operation internal transport barriers can be formed which have in some cases led to ion transport approaching the neoclassical value.

Finally, one of the most important predictions of neoclassical theory is the existence of the bootstrap current. This is a natural current generated by the Coulomb friction between trapped and passing particles. The current is actually carried by the passing particles and is of a sufficiently large magnitude that it may be able to sustain the plasma in steady state operation with the addition of only a small amount of extra external current drive power. The experimental measurements and theoretical predictions of the neoclassical bootstrap current are in reasonably good agreement. This is a favorable result and is currently viewed as a critical element on the path to an economically viable tokamak reactor.

## 14.5 Empirical scaling relations

### 14.5.1 Introduction

The most important transport loss that one must understand and control on the path to a fusion rector is due to thermal conduction. Specifically, for a fusion reactor to sustain itself in a self-heated ignited state requires that the condition $p\tau_E \approx 8.3$ atm s be satisfied with $p \approx 7.2$ atm and $\tau_E \approx 1.2$ s. Thermal conduction is the dominant loss mechanism that sets the value of $\tau_E$.

The analysis thus far presented assumes that transport losses are the result of Coulomb collisions. This leads to the conclusion that $\tau_E \sim a^2/\chi_i^{(NC)}$, where $\chi_i^{(NC)}$ is the neoclassical ion thermal diffusivity. Unfortunately, the neoclassical value of $\tau_E$ is too optimistic as compared to experimental observations. Plasma micro-turbulence, driven largely by the ion temperature gradient, produces electric and magnetic field fluctuations that cause random perturbations in the guiding center orbits of the particles. The randomness of the fluctuations leads to a collision-like diffusion of particles and energy, usually referred to as "anomalous transport." Almost always, the anomalous heat transport is substantially larger than neoclassical heat transport.

Understanding anomalous transport is often considered to be a "grand challenge" of plasma physics. It involves both linear and non-linear analysis of sophisticated kinetic models in realistic geometries. Furthermore, there are usually several different classes of micro instabilities that can be simultaneously excited in a plasma and one must identify the most dangerous modes corresponding to the situation at hand. This, in turn, requires a knowledge of the non-linear saturated states driven by the micro-turbulence. The advent of high-speed, large-memory computers has led to a great improvement in the understanding of anomalous transport. Even so, the problem remains far from being completely solved. Obtaining the desired understanding will require a large number of numerical simulations, which when combined with analytic theory, will hopefully lead to a reasonably tractable, self-consistent mathematical form for the anomalous ion thermal conductivity. This highly desirable goal is still years away.

Based on these difficulties, one can then ask how plasma physicists have treated the problem of thermal transport in the past and how they are likely to treat it in the near-to-midterm future. As in many other fields of science and engineering, when a first principles theory is not available the necessary information is obtained by empirical scaling relations. In terms of energy transport the idea is to collect a large amount of data from many different experiments and then determine a best empirical fit to the data. These empirical fits usually do quite well when making predictions that interpolate within existing regimes of experimental operation. They are less trustworthy when extrapolating to new regimes or to large new experiments that lie beyond the existing database. Nevertheless, this is the best option currently available and the designs of large, next generation burning plasma experiments, such as ITER, are primarily based on empirical scaling relations when dealing with energy transport.

The goal of this section is to describe the method used to determine the empirical fit to $\tau_E$ and to present several specific forms corresponding to different regimes of operation. These forms are then compared with neoclassical thermal transport. It is again worth emphasizing that $\tau_E$ represents global thermal transport in the plasma core. However, there are also several important plasma edge transport phenomena that directly and indirectly affect core transport. These too are understood primarily on an empirical basis. As a prelude to the discussion of core transport a brief description is given of the main transport-related edge phenomena and how they affect $\tau_E$.

### 14.5.2 Edge transport phenomena in a tokamak

Described below are four important phenomena that directly impact the core transport of tokamak experiments. These are: (1) an upper limit on the density; (2) a low-to-high (i.e., L–H) transition boundary that produces a significant improvement in energy confinement; (3) the excitation of MHD modes at the edge of the plasma that can affect the energy confinement of the plasma; and (4) the appearance under certain conditions of internal transport barriers that slow down the flow of heat energy out of the plasma.

### The density limit

The discussion of MHD instabilities has shown that too high a value of current can cause major disruptions to occur in a tokamak. Depending upon the exact conditions in the plasma the unstable modes could be kink modes, ballooning-kink modes, or resistive versions of these modes. In any case, a major disruption leads to a catastrophic collapse of the plasma pressure and current, which clearly must be avoided in a power reactor.

In the practical operation of tokamaks there is an additional mechanism that causes major disruptions. Specifically, if the edge density of the plasma becomes too large, the plasma suffers a disruption. To avoid this situation, tokamak plasmas must operate below a critical density limit. This has a direct impact on core transport since $\tau_E$, as will be shown in Subsection 14.5.3, is an increasing function of plasma density. Therefore, there is a limit to how much $\tau_E$ can be improved by raising the density.

The physical mechanism driving the high-density disruption is usually associated with a radiation collapse near the low-temperature plasma edge caused by the presence of impurities from the first wall. Qualitatively, the explanation for an ohmically heated plasma is as follows. If the plasma edge density is increased at a fixed heating power, the edge temperature decreases by a comparable amount such that the pressure remains approximately constant. When the temperature becomes sufficiently small, on the order of 10 eV, there is a huge increase in the impurity radiation. The energy loss then becomes dominated by radiation rather than thermal conduction. Once this occurs, the plasma becomes essentially detached from the wall. The strong edge radiation region causes the core plasma radius to contract (i.e., $a$ becomes smaller, now limited by the radiation boundary rather than the wall). A fixed total current with a decreasing plasma radius causes a decrease in the value of $q(a)$ eventually leading to the onset of MHD instabilities and a disruption.

While similar phenomena occur for auxiliary heated tokamaks, the analogous theory is much more complicated, probably requiring the inclusion of edge turbulence. In fact, at present a first principles, self-consistent model for the auxiliary heated density limit does not exist. Instead, sufficient data have been collected from a large number of tokamaks, thereby allowing an empirical determination of the density limit. The analysis of this large volume of data was first carried out by Greenwald, who derived a remarkably simple empirical formula for the density limit which is usually referred to as the "Greenwald limit." This relation is given by

$$\bar{n}_{20} \leq n_G \equiv \frac{I_M}{\pi a^2}. \tag{14.146}$$

A set of experimental measurements demonstrating the onset of a disruption when the Greenwald limit is violated is shown in Fig. 14.17. Observe the rapid termination of the plasma internal energy and the plasma current.

Clearly, during the operation of existing experiments or when designing new ones, one must make sure that the desired number density lies below the Greenwald limit. For the simple fusion reactor operating at a maximum allowable current corresponding to $q_* = 2\pi a^2 \kappa B_0 / \mu_0 R_0 I = 2$, the resulting current (for $\kappa = 2$) has the value $I_M = 18.8$ MA. The value of the Greenwald density is then given by $n_G = 1.5 \, (10^{20}/\text{m}^3)$. This is just the value required for the reactor. Given the simplicity of the reactor model one should not take the fact that there is no safety margin as being a precise, unavoidable conclusion. On the other hand, the absence of a large safety margin suggests that the density limit must be considered seriously in future experimental designs.

### The L–H transition

Qualitatively there are two distinct modes of operation for tokamak experiments. These are the "L mode" referring to lower confinement and the "H mode" referring to higher confinement. Practically, $\tau_E$ for the H mode is about a factor of 2 higher than for the L mode.

Any given tokamak is capable of operation in either regime depending upon the detailed experimental conditions. The key features that determine which regime of operation prevails are the amount of external heating power supplied and the way in which the plasma makes contact with the first material surface. The situation is as follows. As the external power is increased in a tokamak experiment there is an abrupt transition from L mode confinement to H mode confinement. This transition was first observed on the ASDEX tokamak in Germany and has been subsequently observed on all other large tokamaks. A typical set of experimental measurements is illustrated in Fig. 14.18. Observe the abrupt increase in the plasma energy as the power exceeds a critical value.

In terms of contact of the plasma with the first material surface there are two widely used generic plasma–wall interfaces known as the "limiter" and the "divertor." These are illustrated schematically in Fig. 14.19. The idea behind the limiter is that, as the plasma slowly diffuses across the last closed flux surface (LCFS), both particles and energy are rapidly deposited on the limiter surface due to the enormously higher parallel transport. This isolates the first wall from the plasma. The limiter has the advantage of simpler, more compact construction, but its close proximity to the plasma almost always increases the number of impurities diffusing into the plasma.

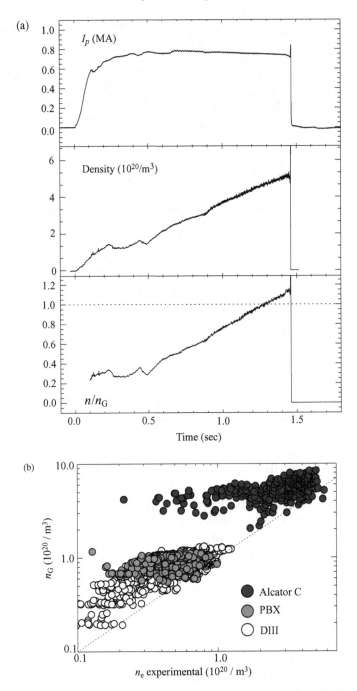

Figure 14.17 (a) Experimental data showing the onset of a disruption when the Greenwald limit is violated (courtesy of M. Greenwald). (b) Accumulated data from several tokamaks showing the experimental range of density operation vs. the Greenwald limit (Greenwald, M. (2002). *Plasma Physics and Controlled Fusion*, **44**, R27).

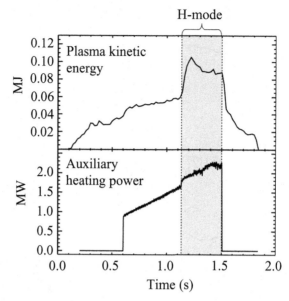

Figure 14.18 Experimental data showing the sudden transition from L to H mode confinement when the power threshold is reached (courtesy of M. Greenwald).

The divertor configuration has extra coils that produce a null point in the poloidal magnetic field near the edge of the plasma. As plasma diffuses across the separatrix there is a rapid loss of particles and energy along the field lines which is deposited on the target plates as shown in the diagram. The divertor, because of its remote location, does a better job of isolating the plasma from impurities and the first wall from the plasma, but takes up a larger volume and tends to focus the heat load onto a narrow area of the target plates. Most plasma experimentalists believe that impurity isolation is the dominant issue. Consequently, most tokamaks operate with some form of divertor. Returning to the question of the L–H transition it has been found that H modes are more easily accessible in divertor geometries.

Thus, a combination of high external power and divertor geometry is desirable for access to H mode operation. Again, edge transport physics associated with the interaction of the plasma with the first material surface has a direct impact on core transport, specifically whether $\tau_E$ corresponds to L mode or H mode operation.

Having established the conditions for the L–H transition, one can next ask how the improved H mode confinement affects the plasma profiles and what actually causes the abrupt transition. H mode profiles typically develop increases in the edge density and edge temperature. The density in particular becomes nearly flat across the entire profile. The end result is an increase in the edge pressure. The narrow transition layer between the plasma edge and the actual first material surface thus has the appearance of an edge pedestal in pressure. The ability of the plasma to support a substantial edge pressure suggests the formation of an edge "transport barrier" that prevents the rapid loss of energy. Overall, this

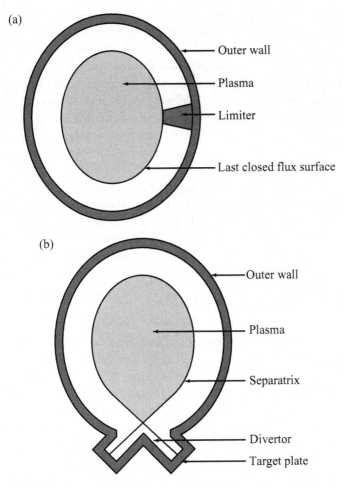

Figure 14.19 Schematic diagram of (a) a limiter and (b) a divertor.

transport barrier leads to an increase in the average density and a higher central temperature, both of which correspond to an improvement in $\tau_E$.

The reason for the L–H transition is not well understood physically. The prevailing belief is that at high auxiliary power levels strongly sheared flow velocities develop near the plasma edge that act to stabilize micro-turbulence. However, the theories are far from complete. The L–H transition thus remains an area of active fusion research.

Until such a first principles theory is developed, experimentalists and machine designers have to rely on an empirical scaling relation for the minimum threshold heating power for H mode operation. The analysis of a large experimental database has led to the following empirical threshold for the L–H transition:

$$P_{\text{LH}} = 1.38\, \bar{n}_{20}^{0.77}\, B_0^{0.92}\, R_0^{1.23}\, a^{0.76} \text{ MW}. \tag{14.147}$$

For the simple reactor design $P_{LH} \approx 100$ MW which, as is shown in Subsection 14.6.3, has the undesirable feature of being considerably higher than the actual auxiliary power required to heat the plasma to ignition once it is in H mode operation; that is, the threshold is very high. This suggests a more subtle path to reach ignition as follows: (1) start the plasma at a low density (e.g. $\bar{n}_{20} \approx 0.3$) for easier low-power access to H mode confinement; (2) heat the plasma to about 5–7 keV; and then (3) gradually raise the density to the desired operating value during which the alphas become the dominating heating source. In this way, if one assumes that $P_{LH} = P_h + P_\alpha$ (i.e., total heating equals auxiliary heating plus alpha heating) the entire evolution takes place with the H mode threshold condition satisfied.

In summary, the L–H transition is an important phenomenon in tokamak physics. There are two separate, although somewhat similar, scaling relations for $\tau_E$, corresponding to the different modes of confinement. These are presented shortly. The factor of 2 difference in magnitudes may not seem enormous, but perhaps surprisingly is critical in predicting the performance of experiments such as ITER. In fact, most researchers believe that ITER will not ignite in L mode but might just do so if operated in the H mode.

### Edge localized modes (ELMs)

The discovery of H mode confinement represents a major improvement in tokamak operation. A higher $\tau_E$ leads to a smaller, less costly ignition experiment and more closely approaches the value required in a reactor. However, H mode operation also has some potential disadvantages. If the buildup of edge density goes unchecked, eventually the Greenwald density limit may be violated leading to a disruption. Often, before this limit is reached, lower level, but nonetheless important, localized edge instabilities are excited in the plasma. These are known as ELMs. Plasma physicists believe these modes are MHD in nature, driven by the large edge pressure and current gradients associated with H mode operation. The situation is still not fully resolved theoretically and also remains an area of active research.

How do ELMs affect plasma performance? These modes qualitatively act as a pressure relief valve. When the edge pressure gradient becomes too high, a burst of ELMs is excited, thereby relieving the excess pressure. Importantly, impurities are also carried out of the plasma with these bursts of energy. The ELMs continue (i.e., the pressure relief valve remains open) until the pressure is reduced to a sufficiently low value (corresponding to the lower shut-off value of the pressure relief valve). In this way ELMs stabilize the time-averaged edge value of $\bar{p}(a, t)$. The presence of ELMs nominally might sound like an advantage, which it sometimes is, but there are different types of ELM behavior, most of which have some overall disadvantages. A summary of ELM behavior follows and is illustrated in Fig 14.20.

At one end of the spectrum there is ELM-free operation. This is normally a transient behavior leading to a large increase in edge density and a large increase in impurities. Eventually the impurities lead to a minor radiation collapse of the edge density and contaminate

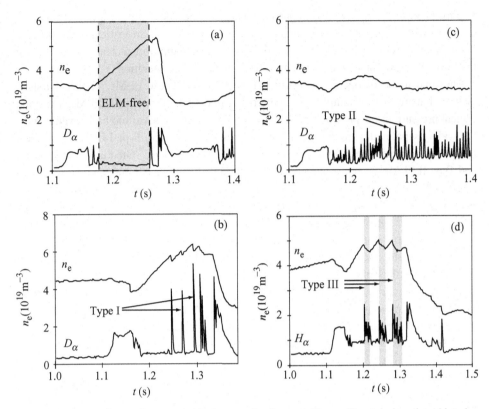

Figure 14.20 Experimentally measured electron density and $D_\alpha$ or $H_\alpha$ emission time histories showing: (a) ELM-free operation and (b) type I, (c) type II, and (d) type III ELMs (ASDEX Team. (1989). *Nuclear Fusion*, **29**, 1959).

the core plasma. The accumulation of impurities cancels one of the main proposed benefits for the divertor. ELM-free steady state operation is not a highly desirable goal for a fusion plasma if impurities are allowed to accumulate.

At the other end of the spectrum there is another mode of operation that has an almost continuous excitation of ELMs called Type III ELMs. In this case the time-averaged edge pressure is stabilized, but at a low level because of the continuous presence of ELMs. In terms of a mechanical analogy the upper and lower critical pressures of the pressure relief valve are quite close to each other and are set at a very low value, allowing a nearly continuous release of pressure. The problem with Type III ELMs is that the low average edge pressure eliminates the confinement benefits of H mode operation. This, too, is not a highly desirable mode of operation.

Type II ELMs represent an intermediate mode of operation. They produce moderate amplitude bursts of activity clearly separated in time. The average edge pressure equilibrates at a reasonably high level so that the benefits of H mode confinement can be realized. The value of $\tau_E$ is increased substantially over L mode confinement. Also, the net release of impurities from the first wall is kept to an acceptably low level because of their outward transport with the ELMs. The mechanical analog suggests that the critical upper pressure in the pressure relief valve has been set to an acceptable value, with a clear separation from the lower shut-off value. This is the most desirable mode of operation.

The final mode of operation involves Type I ELMs. These ELMs produce bursts of activity that are larger in amplitude and narrower in time than for Type II ELMs. The upper critical pressure in the pressure relief valve analogy has been set at too high a value and is too widely separated from the lower shut-off value. On average, Type I ELMs do not lead to a dramatic reduction in the overall energy confinement time. Stated differently, Type I and Type II ELMs are both characterized by improved H mode confinement. However, the large-amplitude short bursts of activity result in a high pulsed heat load on the divertor target plates, which is not acceptable from a thermal hydraulic cooling point of view. Consequently, Type I ELMs are also not a desirable mode of operation.

At present it is not possible to accurately predict the type and level of ELM activity in future experiments. Consequently, understanding ELMs is an area of active research in fusion plasma physics.

The overall conclusion is that ELMs play an important role in limiting the edge behavior of the plasma pressure and a moderate level of ELM activity is actually desirable. The empirical global $\tau_E$ associated with H mode operation is given shortly and is directly affected by ELMs. The expression presented corresponds to the desired situation with moderate Type II but no Type III ELM activity present.

### *Internal transport barriers*

The final phenomena concerns "internal transport barriers." This is an empirically discovered mode of tokamak operation that has several very desirable features. Primarily, there is a further improvement in $\tau_E$ over H mode operation. A second desirable feature is that when internal transport barriers are combined with AT operation the current profile often naturally overlaps with that which would be produced by a high fraction of bootstrap current. Internal transport barriers have been observed in many tokamaks, as long ago as 1984 on Alcator C at MIT.

As its name implies, an internal transport barrier is a region within the plasma core, although usually not far from the edge, where the local ion thermal conductivity is substantially reduced, approaching the ion neoclassical value. This produces a strong temperature gradient resulting in a high central temperature and a corresponding high value of $\tau_E$. Plots of experimentally measured values of $\chi_i$ and $\chi_e$ are illustrated in Fig. 14.21 and compared with $\chi_i^{(NC)}$. Observe the abrupt increase in $\chi_i$ just beyond the barrier and that $\chi_i$ becomes comparable to $\chi_i^{(NC)}$ over a large portion of the plasma core. Also, $\chi_i$ can become sufficiently low that its value is reduced below the value of $\chi_e$.

The two most noteworthy features of AT internal transport barrier discharges are that often the current profile is hollow and that there is substantial shear in the flow velocity. The hollow current profile often leads to a $q(r)$ with an off-axis minimum, and such plasmas are said to possess "reverse shear" (i.e., $dq/dr$ reverses sign at the minimum). Practically,

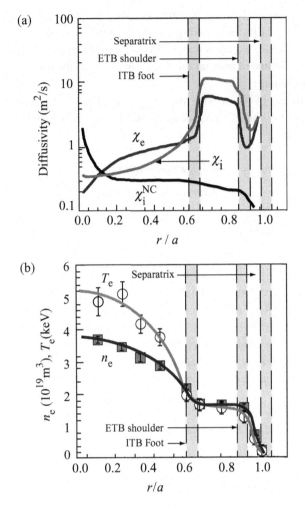

Figure 14.21  (a) Experimentally measured profiles of $\chi_e$ and $\chi_i$ illustrating the presence of an internal transport barrier. (b) Experimentally measured profiles of $n_e$ and $T_e$ for the same plasma discharges. (O'Connor *et al.* (2004). *Nuclear Fusion*, **44**, R1.)

internal transport barriers can be created experimentally by: (1) applying off-axis current drive or heating, (2) rapidly increasing the plasma current thereby producing a hollow profile by the skin effect, or (3) fueling the plasma by injecting high-density solid D pellets into the core of the discharge. There is no unequivocally accepted explanation for the initiation of internal transport barriers, although there is some indication that the high-flow shear may be effective in reducing or eliminating certain micro-turbulence. Also, in existing experiments internal transport barriers are extremely variable phenomena so it has not yet been possible to obtain empirical scaling relations for this regime analogous to those for L and H mode operation.

The existence of internal transport barriers holds great promise for AT operation. The required hollow current profiles for high $\tau_E$ are synergetic with the naturally occurring bootstrap profiles. Plasma experimentalists are actively carrying out research to try to learn how to extend the lifetime of internal transport barriers from the present transient behavior to steady state operation.

With the discussion of the density limit, L–H transition, ELMs, and internal transport barriers now complete, one can next focus attention on the empirical determination of $\tau_E$.

### 14.5.3 Empirical fit for $\tau_E$

#### Experimental procedure

The experimentally determined empirical fits for $\tau_E$ are based on a slightly simplified form of the single fluid, temperature equilibrated form of the energy balance equation given by

$$\frac{3}{2}\frac{\partial p}{\partial t} = \nabla \cdot (n\chi \nabla T) + S_\Omega + S_h. \tag{14.148}$$

Note that convection and compression are neglected as they usually play a small role in the discharges of interest. Also, radiation is neglected. It is usually a finite but rarely dominant contribution to the power balance. Furthermore, radiation is usually localized at the plasma edge, whereas $\tau_E$ is intended to model the overall plasma confinement. Statistical studies have shown that eliminating the radiation term yields a more reliable empirical fit to the data and from a practical point of view this is the main motivation for not including it in the analysis.

The general relation for the energy confinement time is obtained by integrating Eq. (14.148), which is valid for arbitrary geometry, over the entire plasma volume. One finds

$$dW/dt = \int n\chi \, (\mathbf{n} \cdot \nabla T) \, dS + P. \tag{14.149}$$

Here,

$$W = \int \tfrac{3}{2} p \, d\mathbf{r} \tag{14.150}$$

is the total stored energy in the plasma and

$$P = \int (S_\Omega + S_h) \, d\mathbf{r} \tag{14.151}$$

is the total ohmic plus auxiliary heating power supplied to the plasma. The energy confinement time is now defined in terms of the thermal conduction losses as follows:

$$W/\tau_E \equiv - \int n\chi (\mathbf{n} \cdot \nabla T) \, dS. \tag{14.152}$$

Combining terms leads to a useful expression for the energy confinement time given by

$$\tau_E = \frac{W}{P - \dot{W}}. \tag{14.153}$$

This expression is useful because each of the quantities on the right hand side is measured experimentally. Therefore, for a given plasma discharge, usually operating during the flat-top portion of the pulse, one has an experimental determination of $\tau_E$. Also measured are other plasma parameters such as $B_0$, $I$, $\bar{n}$, $a$, $R_0$, $\kappa$, $A$ (with $A$ the atomic mass of plasma ions). This information represents the combined data set for the discharge under consideration.

### Determining $\tau_E$

The empirical fit for $\tau_E$ is determined by first collecting a large number of data sets of the type just described from different discharges on the same device. Second, the complete data set from a given device is then combined with similar data sets from many other devices forming the overall database. These overall data are used to determine the empirical fit to $\tau_E$.

The pioneering work in this area was carried out by Goldston. He postulated that the overall data could be modeled by an empirical fit to $\tau_E$ of the form

$$\tau_E = C \, B_0^{\alpha_1} \, I^{\alpha_2} \, \bar{n}^{\alpha_3} \, a^{\alpha_4} \, R_0^{\alpha_5} \, \kappa^{\alpha_6} \, A^{\alpha_7} \, P^{\alpha_8}. \tag{14.154}$$

By means of a numerical regression analysis Goldston was able to determine values for the constant $C$ and the exponents $\alpha_j$. Since his original work the database has increased substantially. In fact, there is now a rather large database, containing thousands of data sets, for both L mode and H mode discharges. The unknown parameters are slowly but constantly improving in time as more data are included in the analysis.

Two forms for $\tau_E$ that have been widely accepted by the fusion community are, for complex historic reasons, designated as $\tau_E^{ITER89-P}$ for L mode discharges and $\tau_E^{IBP98(y,2)}$ for H mode discharges. For simplicity they are designated here $\tau_L$ and $\tau_H$ and are given by

$$\tau_L = 0.048 \frac{I_M^{0.85} R_0^{1.2} a^{0.3} \kappa^{0.5} \bar{n}_{20}^{0.1} B_0^{0.2} A^{0.5}}{P_M^{0.5}} \quad \text{s},$$

$$\tau_H = 0.145 \frac{I_M^{0.93} R_0^{1.39} a^{0.58} \kappa^{0.78} \bar{n}_{20}^{0.41} B_0^{0.15} A^{0.19}}{P_M^{0.69}} \quad \text{s},$$

(14.155)

using the standard practical units $[I_M(\text{MA}), P_M(\text{MW})]$. Observe that these two forms are qualitatively similar, at least to the extent that the same quantities appear in the numerator and denominator of each relation. Also, the exponents $\alpha_j$ are reasonably similar. One can test the approximate accuracy of the scaling relations by plotting the experimental values of $\tau_E$ from the database vs. the empirical predictions. As an example a plot of $\tau_E^{exp}$ vs. $\tau_E^{emp}$ is illustrated in Fig. 14.22 for the H-mode database. The agreement is quite reasonable.

The above forms of $\tau_E$ are particularly useful when applied to existing tokamaks where the dominant contribution to $P_M$ is due to auxiliary heating. They can also be used when extrapolating to ignition experiments or fusion reactors where the alpha heating becomes dominant. In this case $P_\alpha$ must be included in $P_M$. Since $P_\alpha$ is a strong function of temperature, the above forms, while correct, do not clearly show the dependence of $\tau_E$ on $T$.

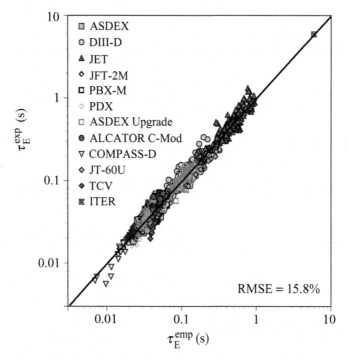

Figure 14.22 Comparison of experimental and empirical $\tau_E$ for the H-mode scaling relation (ITER Physics Experts Groups on Confinement Modelling and Database (1998). *Nuclear Fusion*, **39**, 2175).

This problem is addressed by means of an alternative representation of $\tau_E$, *valid only in steady state*, which is obtained by eliminating $P_M$ in terms of $T_k$ as follows. Equation (14.155) shows that the empirical $\tau_E$ for either L or H mode can be written as

$$\tau_E = K/P_M^\nu. \tag{14.156}$$

This relation is combined with the basic definition of $\tau_E$ assuming that the transient term is negligible:

$$\tau_E = \frac{W}{P - \dot{W}} \approx \frac{W}{P}. \tag{14.157}$$

Next, if the density profile is assumed to be approximately uniform, then this expression can be rewritten as

$$\tau_E = \frac{3nTV}{P} = 0.95\frac{\bar{n}_{20}\,\overline{T}_k\,R_0\,a^2\,\kappa}{P_M} = \frac{D}{P_M} \tag{14.158}$$

with $\overline{T}_k$ assumed to be the profile averaged temperature. The quantity $P_M$ can be eliminated from Eqs. (14.156) and (14.158) leading to a relation of the form $\tau_E = \tau_E(\overline{T}_k)$. This relation is given by

$$\tau_E = \left(\frac{K}{D^v}\right)^{\frac{1}{1-v}}. \tag{14.159}$$

Straightforward substitution yields the following expressions for $\tau_L$ and $\tau_H$:

$$\tau_L = 0.037 \frac{\varepsilon^{0.3}}{q_*^{1.7}} \frac{a^{1.7} \kappa^{1.7} B_0^{2.1} A}{\overline{n}_{20}^{0.8} \overline{T}_k} \quad s,$$

$$\tau_H = 0.28 \frac{\varepsilon^{0.74}}{q_*^{3}} \frac{a^{2.67} \kappa^{3.29} B_0^{3.48} A^{0.61}}{\overline{n}_{20}^{0.91} \overline{T}_k^{2.23}} \quad s. \tag{14.160}$$

In these expressions the $\varepsilon = a/R_0$ and $q_* = 2\pi a^2 \kappa B_0 / \mu_0 R_0 I$ dependence has been explicitly extracted as these are parameters that do not vary very much from tokamak to tokamak. Also it makes the comparison with neoclassical transport simpler to understand. Observe that again there is qualitative agreement between L and H modes with the same terms appearing in the numerator and denominator. However, there is a stronger variation in certain exponents.

Next, consider the predictions for the simple test reactor with $a = 2$, $B_0 = 4.7$, $\overline{n}_{20} = 1.5$, $\overline{T}_k = 15$, $\varepsilon = 0.4$, $\kappa = 2$, $q_* = 2$, and $A = 2.5$. One finds

$$\tau_L = 0.29 \text{ s}, \tau_H = 0.68 \text{ s}. \tag{14.161}$$

Both are below the required value of $\tau_E \approx 1.2$ s. The H mode value is about a factor of 1.8 too small. Even so, the strong dependence on $B_0$, $q_*$, and $\overline{T}_k$ suggests that relatively small changes in any of these quantities can remedy the situation. For instance lowering $q_* = 2 \rightarrow$ 1.7, lowering $\overline{T}_k = 15 \rightarrow 10$, raising $B_0 = 4.7 \rightarrow 5.7$, or any appropriate combination thereof, raises $\tau_E$ to the required value. Similarly, $\tau_E$ can be increased for the same parameters given above when profile effects are taken into account. As an example note that, for a peaked pressure profile $p = p_0(1 - r^2/a^2)^2$, it follows that $\langle p^2 \rangle = (9/4)(\overline{p})^2$. The factor 9/4 directly multiplies the value of $S_\alpha$ used for power balance.

The situation is considerably more difficult for L-mode scaling. Here, $\tau_L$ is too small by a factor of about 4.1. Also the exponents appearing on the various quantities are, in general, weaker than for the H mode. Consequently, the changes required in the basic reactor parameters may be too large from an engineering and economic point of view to result in a viable design.

To summarize, the strong dependencies in the H-mode scaling are advantageous in that small changes can produce the required energy confinement time. On the other hand, one must acknowledge that these strong variations are somewhat unsettling in view of the fact that the results are being applied to an extrapolated regime somewhat distant from where most of the data have been collected.

The final topic involves a comparison of the empirical scaling relations with the predictions of classical and neoclassical transport theory. Of particular interest is the temperature dependence. If one uses the relations $\tau_E \sim a^2 \kappa / \chi_i$ and $\chi_i \sim A^{1/2}$, then the comparisons (ignoring numerical coefficients) can be written as

$$\tau_E^{(CL)} \sim \frac{a^2 \kappa B^2 T^{1/2}}{n A^{1/2}},$$

$$\tau_E^{(NC)} \sim \frac{\varepsilon^{3/2}}{q^2} \frac{a^2 \kappa B^2 T^{1/2}}{n A^{1/2}},$$

$$\tau_L \sim \frac{\varepsilon^{0.3}}{q^{1.7}} \frac{a^{1.7} \kappa^{1.7} B^{2.1} A}{n^{0.8} T}, \qquad (14.162)$$

$$\tau_H \sim \frac{\varepsilon^{0.74}}{q^3} \frac{a^{2.67} \kappa^{3.29} B^{3.48} A^{0.61}}{n^{0.91} T^{2.23}}.$$

The main qualitative differences between Coulomb-driven transport and empirical transport are the opposite dependencies on $T$ and $A$. It is rather unfortunate that the optimistic scaling relations of classical and neoclassical theories that predict improvements in $\tau_E$ as $T$ increases do not hold for the empirical scaling relations. The empirically observed degradation in $\tau_E$ with increasing $T$ is a major reason why it is so difficult to ignite a plasma in a small, relatively inexpensive test experiment. One needs the large size of a reactor scale experiment to compensate the unfavorable scaling dependence on $T$.

### 14.5.4 Summary

The theoretical and experimental complexity associated with the turbulent behavior of thermal transport in a plasma has driven the fusion community to develop empirical scaling relations for $\tau_E$. These relations are based on a large database and provide a reasonably good guideline for predicting the performance of existing experiments that essentially lie in interpolated regimes of operation. The empirical scaling relations are also used to predict the performance of new, next generation burning plasma experiments. Here, one is not as confident about the reliability of the scaling relations since such experiments will operate in an extrapolated regime dominated by alpha heating. Even so, at present, the empirical relations remain the best option. Theoretical progress has increased substantially but a first-principles theory is still years away.

Analysis of tokamak data has shown that there are two basic modes of operation – the L mode and the H mode. The actual regime of operation of any given discharge depends upon the level of external heating power and whether or not the first plasma contact surface is a divertor or limiter. High-power divertor discharges usually operate in the H mode with a confinement time about a factor of 2 higher than for the L mode. The H mode confinement scaling relation predicts a value of $\tau_E$ which is close to that required in a reactor for a self-sustained alpha heated plasma.

Lastly, the earlier discovery of internal transport barriers coupled with the idea of AT operation leads to further improvements in confinement, approaching the ion neoclassical value, and may ultimately lead to a steady state ignited tokamak. The AT operation involves the use of hollow current profiles, which have the added advantage of closely overlapping the natural bootstrap current profile. AT operation is an area of current fusion research and experimentalists hope to discover ways to make such discharges operate for long periods of time and to develop a corresponding empirical AT scaling relation for $\tau_E$.

## 14.6 Applications of transport theory to a fusion ignition experiment

### 14.6.1 Introduction

Armed with the knowledge of the empirical scaling relations for $\tau_E$ and the neoclassical prediction of the bootstrap current, one can now more realistically investigate certain important aspects in the design of a tokamak fusion reactor or, alternatively, a tokamak ignition experiment. The applications described here focus on the nearer term objective of an ignition experiment. Three important topics are discussed.

First, the design of a self-sustained, superconducting ignition experiment is carried out using the empirical scaling relation in which $\tau_E = \tau_E(T)$. The design is constrained by several critical MHD stability limits. The analysis shows that the parameters of the final design are quite similar to those in an actual power reactor. In other words, the costs along the development path to a tokamak fusion reactor will be high, since it is difficult to construct a small-scale, low-cost ignition experiment to learn about alpha physics. Obtaining a large amount of alpha heating requires a reactor-scale experiment. This is an undesirable but not insurmountable consequence of tokamak physics.

Second, the evolution of the plasma from a cold initial state to the hot self-sustained final state in the ignition experiment is investigated. Of particular interest are the questions of the minimum auxiliary power required to reach ignition and the problem of thermal stability at the final operating point. It is shown that the temperature dependence of $\tau_E$ actually improves the situation as compared to the simple analysis presented in Chapter 4 where $\tau_E$ was assumed to be a constant.

Third, the question of the highest possible bootstrap current fraction is addressed. It is shown that achieving high bootstrap fractions on the order of $f_B > 0.75$ usually leads to a violation of the Troyon no-wall MHD $\beta$ limit. The implication is that the economic constraint of large $f_B$ leads to configurations in which the resistive wall mode is excited. Consequently, some form of resistive wall stabilization is required, probably feedback stabilization.

### 14.6.2 A superconducting ignition experiment

The discussion here closely follows that presented in Chapter 5. In the present case, however, attention is focused on a superconducting ignition experiment. The goal is to design the minimum cost experiment subject to the appropriate constraints. The cost is again assumed to be proportional to the combined volume of the blanket-and-shield and toroidal field coils. As for the simple reactor, the nuclear physics constraints require a blanket-and-shield thickness of $b = 1.2$ m, while the engineering constraints limit the magnetic field on the inside of the coil to be $B_{max} = 13$ T and assume the maximum allowable stress on the magnet support structure is $\sigma_{max} = 300$ MPa.

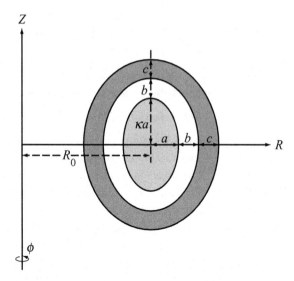

Figure 14.23 Schematic diagram of a fusion ignition experiment.

The main differences between the reactor and ignition experiment designs are as follows. In an ignition experiment the constraints of specifying the output power and maximum wall loading are removed. Instead, these are replaced by a new set of plasma physics constraints. Specifically, in the fully ignited state the plasma must satisfy the MHD Troyon $\beta$ limit, the MHD $q_*$ current limit, the MHD $n = 0$ vertical stability limit, and the Greenwald $n_G$ density limit. Also, and this point is crucial, in the ignition experiment the plasma is assumed to satisfy H-mode scaling ($\tau_E = \tau_H$) in contrast to the earlier reactor design in which the value of $\tau_E$ was determined as a required output rather than a specified input.

The analysis is relatively straightforward. The idea is to utilize the constraints in order to express all the unknown design parameters in terms of the temperature $T$. These relations are substituted into the expression for the device volume, which is then minimized over $T$ leading to the final design.

The model used in the analysis is slightly more sophisticated than the simple one used for the reactor in Chapter 4. Two modifications are introduced. First, the plasma cross section is assumed to be elongated from the outset. Second, rather than just setting all quantities equal to their average value, they are instead modeled with simple profiles representative of present day experimental observations. These profiles do not change the basic scaling relations but give slightly more accurate numerical coefficients when performing the averages to obtain the 0-D power balance relation. In general, more realistic peaked profiles improve the alpha particle heating in comparison to the thermal conduction losses.

### The device volume

A diagram of the simple elongated ignition experiment is shown in Fig. 14.23. For simplicity, the plasma cross section is assumed to be elliptical with an elongation $\kappa$. The cost of the experiment is assumed to be proportional to the combined volume of the blanket-and-shield and toroidal field coils given by

$$V = 2\pi^2 R_0[(a+b+c)(\kappa a + b + c) - \kappa a^2].\tag{14.163}$$

The quantity $V$ is the basic "cost function" that is to be minimized. At present, the only known quantity is $b = 1.2$ m based on nuclear physics constraints.

### The constraints

There are a number of constraints that the design must satisfy. First, the plasma elongation cannot be too large or else the $n = 0$ vertical instability would be excited. This requires that

$$\kappa \leq \kappa_{max} = 2.\tag{14.164}$$

To build in a reasonable safety margin, the elongation is chosen as $\kappa = 1.7$.

Second, the plasma current cannot be too large or else the external MHD kink mode would be excited. This condition can be written as

$$q_* = \frac{5a^2\kappa B_0}{R_0 I_M} \geq q_{*\min} = 1.7.\tag{14.165}$$

For a reasonable safety margin the value of safety factor is chosen as $q_* = 2$.

Third, the density cannot be too high or else the Greenwald density limit would be violated. This requires that one choose the average density to satisfy

$$\bar{n}_{20} = N_G \frac{I_M}{\pi a^2} = \left(\frac{5\kappa N_G}{\pi q_*}\right)\frac{B_0}{R_0} \qquad N_G \leq 1.\tag{14.166}$$

Here $N_G$ is a safety margin whose value is chosen to be 0.8.

Fourth, the plasma pressure cannot be too large or else the Troyon $\beta$ limit would be violated. This requires that

$$\beta = \beta_N \frac{I_M}{a B_0} \qquad \beta_N \leq \beta_{N\,max} = 0.03.\tag{14.167}$$

Equation (14.167) can be rewritten directly in terms of the average pressure by using the definition $\beta \equiv 2\mu_0 \bar{p}/B_0^2$ and then eliminating $I_M$ by means of Eq. (14.165). One finds (in practical units)

$$\bar{p}_a = \left(19.9\frac{\kappa\beta_N}{q_*}\right)\frac{a B_0^2}{R_0} \quad \text{atm.}\tag{14.168}$$

To insure a reasonable safety margin $\beta_N$ is set at 0.025.

The last constraint involves the coil thickness $c$, which is determined by the maximum allowable stress limit $\sigma_{max}$. The analysis is very similar to that presented in Chapter 5, the only difference being the need to account for the non-circularity of the coil. For the non-circular case the maximum magnetic force occurs along the elongated sides of the magnet.

A straightforward calculation then shows that

$$c = \frac{2\xi}{1 - \xi}(\kappa a + b) \qquad \xi = \frac{B_{max}^2}{4\mu_0\sigma_{max}} = 0.11. \tag{14.169}$$

The safety margins are already built into the values of $B_{max} = 13$ T and $\sigma_{max} = 300$ MPa. Also, $B_0$ is related to $B_{max}$ by the usual relation

$$B_0 = B_{max}\left(1 - \frac{a + b}{R_0}\right). \tag{14.170}$$

This completes the specification of the constraints. The best way to view the problem is to recognize that the constraint parameters $\kappa = 1.7$, $q_* = 2$, $N_G = 0.8$, $\beta_N = 0.025$, and $\xi = 0.11$ are now known quantities. The remaining unknown design parameters are $a$, $R_0$, $B_0$, $\overline{p}_a$, $\overline{n}_{20}$, $\overline{T}_k$. They can be expressed in terms of the constraint parameters and are determined by requiring power balance and minimizing the device volume.

### Power balance

The analysis of the power balance relation is somewhat involved, requiring a sequence of substitutions and simplifications. In following the discussion readers should keep in mind that the goal of the analysis is ultimately to determine two relationships expressing $R_0$ and $a$ as functions of $T$. Once these relations are derived they can be substituted into the expression for the device volume, which can then be minimized over $T$.

The steady state power balance in an ignition experiment requires that alpha heating balance the sum of the thermal conduction and Bremsstrahlung radiation losses. In steady state there is no ohmic heating power. Also, if the plasma is fully self-sustained, the auxiliary power, by definition, must be zero. Mathematically, the 0-D power balance requires that

$$\langle S_\alpha \rangle = \langle S_\kappa \rangle + \langle S_B \rangle, \tag{14.171}$$

where $\langle S \rangle$ denotes average over the volume.

The various contributions are evaluated as follows. To begin, note that the density and temperature are modeled by simple, experimentally motivated profiles rather than just being set to their average value. Specifically, the density and temperature are modeled as

$$n = \frac{4}{3}\overline{n}(1 - \rho^2)^{1/3},$$

$$T = \frac{5}{3}\overline{T}(1 - \rho^2)^{2/3}, \tag{14.172}$$

where

$$\rho^2 = \frac{x^2}{a^2} + \frac{y^2}{\kappa^2 a^2} \tag{14.173}$$

and the plasma surface is defined by $\rho = 1$. This implies that

$$\langle S \rangle = 2 \int_0^1 S \rho \, d\rho \tag{14.174}$$

and that $\langle n \rangle = \overline{n}$, $\langle T \rangle = \overline{T}$. Observe that the density profile is relatively flat. The temperature profile is slightly peaked. The fractional exponents provide a crude modeling of the edge pedestals characteristic of H-mode operation.

Using these profiles allows one to evaluate the various terms in the power balance relation. The first step is to determine the relationship between the average pressure, density and temperature: $\langle p \rangle = 2 \langle nT \rangle$. One finds that $\overline{p} = (10/9) \left( 2\overline{n}\,\overline{T} \right)$ or in practical units

$$\overline{p}_a = 0.356 \, \overline{n}_{20} \, \overline{T}_k \text{ atm.} \tag{14.175}$$

This expression can be further simplified by eliminating $\overline{n}_{20}$ by means of the Greenwald density limit given by Eq. (14.166)

$$\overline{p}_a = \left( 0.567 \frac{\kappa N_G}{q_*} \right) \frac{B_0 \overline{T}_k}{R_0} = 0.386 \frac{B_0 \overline{T}_k}{R_0} \quad \text{atm.} \tag{14.176}$$

The second step provides related information that will be required shortly to simplify the analysis. This information is obtained by equating the expressions for $\overline{p}_a$ in Eqs. (14.168) and (14.176). The result is an expression for the quantity $B_0 a$ in terms of $\overline{T}_k$:

$$B_0 a = \left( 0.0285 \frac{N_G}{\beta_N} \right) \overline{T}_k = 0.912 \overline{T}_k \quad \text{T m.} \tag{14.177}$$

Consider next the evaluation of the alpha power

$$\langle S_\alpha \rangle = \frac{1}{16} E_\alpha \left( 2 \int_0^1 p^2 \frac{\langle \sigma v \rangle}{T^2} \rho \, d\rho \right). \tag{14.178}$$

The integral can be evaluated by making the reasonably good approximation that $\langle \sigma v \rangle / T^2 \approx \langle \sigma v \rangle / T^2 |_{\overline{T}} = \text{const.}$ over the temperature regime of interest and noting that $\langle p^2 \rangle = (4/3) \overline{p}^2$. The factor $4/3$ is the gain due to using peaked profiles rather than simple average values. The value of $\langle S_\alpha \rangle$ in practical units is now given by

$$\langle S_\alpha \rangle = 1.82 \times 10^6 \overline{p}_a^2 \frac{\langle \sigma v \rangle_n}{\overline{T}_k^2} \quad \text{W/m}^3. \tag{14.179}$$

Here, the normalized $\langle \sigma v \rangle_n$ is equal to $\langle \sigma v \rangle$ measured in units of $10^{-22} \text{ m}^3/\text{s}$.

The Bremsstrahlung radiation loss is evaluated in a completely analogous manner. A short calculation assuming that $Z_{\text{eff}} = 1$ yields

$$\langle S_B \rangle = 4.84 \times 10^4 \frac{\overline{p}_a^2}{\overline{T}_k^{3/2}} \quad \text{W/m}^3. \tag{14.180}$$

The last quantity of interest is the thermal conduction loss. For H-mode confinement this is given by

$$\langle S_\kappa \rangle = 1.5 \times 10^5 \frac{\overline{p}_a}{\tau_H} \quad \text{W/m}^3. \tag{14.181}$$

These expressions are now substituted into the power balance relation yielding a requirement on $\overline{p}_a \tau_H$ that can be written as

$$\overline{p}_a \tau_H = 0.0824 \frac{\overline{T}_k^2}{\langle \sigma v \rangle_n - K_B \overline{T}_k^{1/2}} \quad \text{atm s}, \tag{14.182}$$

where $K_B = 0.0266$.

The next step in the analysis requires the simplification of $\overline{p}_a \tau_H$ by substitution of the actual empirical relation for $\tau_H$ given by Eq. (14.155) with $P = \langle S_\alpha \rangle V_p$. A straightforward, but somewhat tedious calculation leads to the following result.

$$\overline{p}_a \tau_H = \left( 0.0978 \frac{B_0^{0.02}}{a^{0.03}} \frac{\kappa^{1.05} N_G^{0.03}}{q_*^{0.96}} \right) \frac{(B_0 a)^{1.09}}{R_0^{0.26}} \frac{\overline{T}_k}{\langle \sigma v \rangle_n^{0.69}} \quad \text{atm s}. \tag{14.183}$$

The interesting feature in this expression is the coincidental fact that except for the very weak dependence on $B_0^{0.02}$ and $a^{0.03}$, the quantity $\overline{p}_a \tau_H$ depends only on the combination $B_0 a$. Consequently, the expression can be significantly simplified by substituting $B_0 \approx 6$ and $a \approx 2$ into the weakly dependent terms and substituting the expression in Eq. (14.177) for the $B_0 a$ combination term. A short calculation yields

$$\overline{p}_a \tau_H = \left( 2.05 \times 10^{-3} \frac{\kappa^{1.05} N_G^{1.12}}{q_*^{0.96} \beta_N^{1.09}} \right) \frac{\overline{T}_k^{2.09}}{R_0^{0.26} \langle \sigma v \rangle_n^{0.69}} \quad \text{atm s}$$

$$= 0.0799 \frac{\overline{T}_k^{2.09}}{R_0^{0.26} \langle \sigma v \rangle_n^{0.69}} \quad \text{atm s}. \tag{14.184}$$

The last step in the power balance relation requires setting the two expressions for $\overline{p}_a \tau_H$ in Eqs. (14.182) and (14.184) equal to each other. The result is an explicit expression for $R_0 = R_0 \left( \overline{T}_k \right)$:

$$R_0 = \left( 6.80 \times 10^{-7} \frac{\kappa^{4.04} N_G^{4.31}}{q_*^{3.69} \beta_N^{4.19}} \right) \overline{T}_k^{0.35} \langle \sigma v \rangle_n^{1.19} \left( 1 - K_B \overline{T}_k^{1/2} / \langle \sigma v \rangle_n \right)^{3.85} \quad \text{m},$$

$$= 0.886 \overline{T}_k^{0.35} \langle \sigma v \rangle_n^{1.19} \left( 1 - K_B \overline{T}_k^{1/2} / \langle \sigma v \rangle_n \right)^{3.85} \quad \text{m}. \tag{14.185}$$

The one remaining task is to derive a relationship between $a$ and $T$. This relationship is easily obtained by combining the expressions for $B_0 a$ in Eq. (14.177) with the relationship between $B_0$ and $B_{max}$ in Eq. (14.170). One finds

$$a \left( 1 - \frac{a+b}{R_0} \right) = 0.0285 \frac{N_G}{B_{max} \beta_N} \overline{T}_k. \tag{14.186}$$

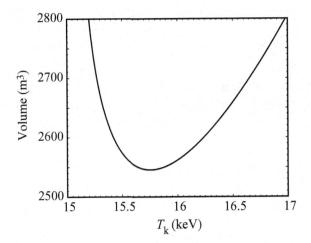

Figure 14.24 Volume vs. temperature for an ignition experiment. The optimum temperature corresponds to 15.7 keV.

Solving for $a$ yields

$$a = \frac{R_0 - b}{2} - \frac{1}{2}\left[(R_0 - b)^2 - K_M R_0 \overline{T}_k\right]^{1/2} \quad \text{m,} \tag{14.187}$$

where $K_M = 0.114 N_G / B_{max} \beta_N = 0.281$.

After a lengthy calculation one has finally obtained the desired relations for $R_0(\overline{T}_k)$ and $a(\overline{T}_k)$.

### *The minimum volume experiment*

The expressions for $R_0$ (Eq. (14.185)), $a$ (Eq. (14.187)) and $c$ (Eq. (14.169)) are now substituted in the expression for the device volume $V$ (Eq. (14.163)). The resulting expression is solely a function of $\overline{T}_k$ and is plotted in Fig. 14.24. Observe that the device volume has a minimum at $\overline{T}_k \approx 15.7$ keV. Using this value one can back substitute and evaluate all the design parameters. These are summarized in Tables 14.2 and 14.3, the first specifying the input parameters and the second the output design parameters.

A comparison of the design parameters between the ignition experiment and the simple reactor discussed in Chapter 5 shows that both devices are of comparable size. In fact, the ignition experiment is somewhat larger. The reason is that the value of $\tau_H$ obtained with the reactor parameters is about 0.68 s, nearly a factor of 2 smaller than that required in the design ($\tau_E = 1.2$ s). In the ignition experiment, self-sustained power balance is a required design goal, and this leads to a larger device to raise the smaller value of $\tau_H$ to the required value. Also, it is interesting to note that the simple model discussed here leads to parameters that are reasonably close to the proposed ITER design.

The simple analysis just presented shows how closely the size, and hence cost, of a fusion ignition experiment or a fusion reactor is tied to the achievable value of $\tau_E$. The factor of 2

Table 14.2. *Input parameters for the superconducting ignition experiment*

| Quantity | Symbol | Value |
|---|---|---|
| Blanket-and-shield thickness | $b$ | 1.2 m |
| Maximum field at the coil | $B_{max}$ | 13 T |
| Maximum magnet stress | $\sigma_{max}$ | 300 MPa |
| Elongation | $\kappa$ | 1.7 |
| Kink safety factor | $q_*$ | 2 |
| Greenwald density factor | $N_G$ | 0.8 |
| Troyon $\beta$ factor | $\beta_N$ | 0.025 |

Table 14.3. *Design parameters for the superconducting ignition experiment*

| Quantity | Symbol | Value |
|---|---|---|
| Average temperature | $\overline{T}_k$ | 15.7 keV |
| Major radius | $R_0$ | 7.1 m |
| Minor radius | $a$ | 2.0 m |
| Aspect ratio | $R_0/a$ | 3.5 |
| Coil thickness | $c$ | 1.1 m |
| Device volume | $V$ | 2600 m$^3$ |
| Plasma volume | $V_p$ | 960 m$^3$ |
| Plasma surface area | $S_p$ | 790 m$^2$ |
| Magnetic field on axis | $B_0$ | 7.1 T |
| Plasma current | $I_M$ | 17 MA |
| Average plasma pressure | $\overline{p}_a$ | 6.1 atm |
| Average plasma density | $\overline{n}_{20}$ | $1.1 \times 10^{20}$ m$^{-3}$ |
| H mode confinement time | $\tau_H$ | 1.2 s |
| Total alpha power | $P_\alpha$ | 760 MW |
| Total fusion power | $P_f$ | 3800 MW |
| Total Bremsstrahlung loss | $P_B$ | 28 MW |
| Wall loading | $P_W$ | 3.9 MW/m$^2$ |

increase required in $\tau_H$ has led to a device that is nearly twice as large as the simple reactor design. One can therefore easily appreciate why understanding, controlling, and improving transport is such an important area of fusion research.

The last point to reiterate is that the size of a superconducting tokamak ignition experiment is similar in scale to a full power reactor. The physics and engineering do not allow one to build a much smaller, less expensive, ignition experiment to investigate the science of burning plasmas. The implication is that the developmental costs for a tokamak fusion reactor will be high, since the ignition experiment leading up to the reactor is of comparable scale. This is an economic disadvantage, not insurmountable, but nonetheless undesirable.

### 14.6.3 Heating to ignition

The next application involves the time evolution of the plasma in the superconducting ignition experiment from a cold initial condition to its final steady state operating point. There are two main questions to consider. First, is the final operating point thermally stable or will some form of burn control be required? Second, how much external heating power is required to heat the plasma to a sufficiently high temperature so that the alphas dominate and complete the evolution to full ignition?

These questions were addressed in Chapter 4, where it was assumed that the energy confinement $\tau_E$ was a constant equal to the required value at ignition $\tau_E = 1.2$ s. The corresponding analysis showed that the steady state ignition point was thermally unstable and that the required auxiliary power was approximately equal to 25% of the ignited alpha power, quite a large fraction. In this subsection these questions are revisited using the empirically determined H mode confinement relation, which is temperature dependent: $\tau_E = \tau_H(T)$. The results, as shown below, are somewhat surprising.

#### Thermal stability

Both questions of interest can be answered by examining the time dependent, 0-D energy balance equation. The critical assumption in the analysis is that the entire evolution takes place at a constant density equal to the final desired value: $\bar{n}_{20} = 1.1$. This is not an unreasonable approximation since the particle confinement time is usually somewhat lower than the energy confinement time. Therefore, any external programming of the density would lag behind the faster evolving temperature, making control difficult.

The basic 0-D power balance includes the alpha power, external heating power, Bremsstrahlung loss, and thermal conduction loss. Ohmic heating is neglected as it makes a small contribution except at very low temperatures (as shown in the next chapter). The power balance equation for the assumed profiles can thus be written as

$$\frac{dW}{dt} = \langle S_\alpha \rangle - \langle S_B \rangle - \langle S_\kappa \rangle + \frac{P_h}{V_p}, \tag{14.188}$$

where

$$
\begin{aligned}
W &= \langle 3nT \rangle = 5.34 \times 10^4 \bar{n}_{20} \overline{T}_k \quad \text{J/m}^3, \\
\langle S_\alpha \rangle &= \langle E_\alpha n^2 \langle \sigma v \rangle / 4 \rangle = 2.31 \times 10^5 \bar{n}_{20}^2 \langle \sigma v \rangle_n \quad \text{W/m}^3, \\
\langle S_B \rangle &= \langle C_B n^2 T^{1/2} \rangle = 6.14 \times 10^3 \bar{n}_{20}^2 \overline{T}_k^{1/2} \quad \text{W/m}^3, \\
\langle S_\kappa \rangle &= \langle 3nT \rangle / \tau_H = 5.34 \times 10^4 \bar{n}_{20} \overline{T}_k / \tau_H \quad \text{W/m}^3.
\end{aligned}
\tag{14.189}
$$

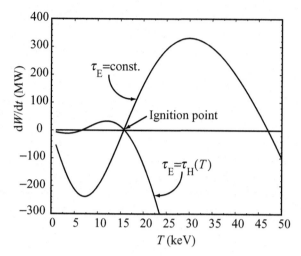

Figure 14.25  $dW/dt$ vs. $T$ for $\tau_E = \tau_E(T)$ and $\tau_E = $ const. $= 1.2$ assuming $P_h = 0$.

Also, $V_p = 960$ m$^3$ is the plasma volume, $P_h(\overline{T}_k)$ is the total externally supplied heating power in watts and $\tau_H$, repeated here for convenience, is given by

$$\tau_H = 0.145 \frac{I_M^{0.93} R_0^{1.39} a^{0.58} \kappa^{0.78} \overline{n}_{20}^{0.41} B_0^{0.15} A^{0.19}}{P_M^{0.69}} \quad \text{s}$$

$$= \frac{117}{P_M^{0.69}} \quad \text{s}. \tag{14.190}$$

Here, the second form of $\tau_H$ is obtained using the design parameters derived in the last subsection and $P_M$ is the total power deposited in the plasma measured in megawatts: $P_M(\overline{T}_k) = [\langle S_\alpha \rangle V_p + P_h]10^{-6}$.

The thermal stability of the system can now be ascertained by plotting the curve of $dW/dt$ vs. $\overline{T}_k$ for $P_h = 0$ and examining the slope at the equilibrium ignition point. Recall from the discussion in Chapter 4 that a positive slope is thermally unstable, while a negative slope is stable. This curve is plotted in Fig. 14.25 for the design value of density $\overline{n}_{20} = 1.1$. Also plotted for comparison is the equivalent curve of Chapter 4 which assumes that $\tau_E = 1.2$ s $= $ const.

With respect to the time dependence, $\overline{n}_{20} = $ const and $\overline{T}_k = \overline{T}_k(t)$ implying that $dW/dt \propto d\overline{T}_k/dt$.

Observe that, by construction, both curves intersect the axis at the same equilibrium ignition point $\overline{T}_k = 15.7$ keV. However, the $\tau_E = $ const. curve intersects with a positive slope indicating thermal instability while the $\tau_E = \tau_H$ curve intersects with a negative slope implying thermal stability. What has happened is that the degradation in $\tau_H$ with $\overline{T}_k$ has shifted the curve of $dW/dt$ vs. $\overline{T}_k$ to the left. The stable high-temperature intersection point of the $\tau_E = $ const. curve, which is not very interesting because the corresponding value of $\beta$ far exceeds the MHD stability limit, shifts down to a much more acceptable value when

$\tau_E = \tau_H$. The lower intersection point of the $\tau_E = \tau_H$ curve corresponds to the situation where the alpha power is balancing the Bremsstrahlung losses plus the relatively small thermal conduction losses (since $\tau_H$ is large at lower temperatures). There is not very much net power at this lower temperature so it is uninteresting from an energy point of view. Once the plasma crosses this temperature the thermal instability is excited, driving the plasma naturally to the stable higher equilibrium point $\overline{T}_k = 15.7$ keV. The conclusion is that if H mode scaling continues to apply to plasmas dominated by alpha heating, the point of self-sustained ignition is thermally stable, a highly desirable result which is exactly the opposite from that deduced from the $\tau_E = $ const. analysis in Chapter 4.

### The minimum power for ignition

Consider now the question of the minimum external power $P_h$ required to raise $\overline{T}_k$ to a sufficiently high value so that alpha heating becomes dominant. In the context of the $\tau_E = \tau_H$ curve one must add sufficient $P_h$ so that $dW/dt > 0$ for $0 < \overline{T}_k < 15.7$ keV. A positive $dW/dt$ implies that the temperature will continue to increase until the ignition point is reached. Clearly, $P_h$ must vanish at $\overline{T}_k = 15.7$ keV for the plasma to remain in equilibrium.

An examination of Fig. 14.25 suggests that external power must be added at low temperatures until the plasma passes the Bremsstrahlung equilibrium point at approximately $\overline{T}_k = 5$ keV. Beyond this point the alphas dominate and the external power can be gradually decreased to zero. A simple model for $P_h(\overline{T}_k)$ that has the desired properties is given by

$$
P_h = \begin{cases} P_0 & \overline{T}_k < T_B \\ P_0 \left[ 1 - \left( \dfrac{\overline{T}_k - T_B}{T_I - T_B} \right)^2 \right] & T_B < \overline{T}_k < T_I, \end{cases} \tag{14.191}
$$

where $T_B = 5$ keV is the Bremsstrahlung temperature, $T_I = 15.7$ keV is the ignition temperature, and $P_0$ is the maximum applied external heating power.

Curves of $dW/dt$ vs. $\overline{T}_k$ are illustrated in Fig. 14.26 for various values of $P_0$. Also illustrated is the curve of $P_h(\overline{T}_k)$. Figure 14.26 shows that the minimum value of $P_0$ for ignition is approximately $P_0 = 22.2$ MW. At this value the $dW/dt$ vs. $\overline{T}_k$ curve is everywhere positive with its minimum point just being tangent to the $dW/dt = 0$ axis. Realistically one wants a significantly higher value of $P_0$ in order to pass through the minimum point reasonably rapidly (i.e., hypothetically it would take an infinite time to reach ignition if one had to cross the point where $dW/dt = 0$). The curve corresponding to $P_0 = 40$ MW is thus a more realistic choice for $P_0$. If the power absorption efficiency is assumed to be 0.7, this implies that approximately 60 MW of external heating power must be injected into the plasma.

It is interesting to compare the value of $P_0 = 22.2$ MW with the value predicted in Chapter 4 using the $\tau_E = $ const. model. For this model the power absorbed by the plasma was shown to be about 25% of the alpha power at ignition, which for the superconducting design under consideration translates to $P_0 = 190$ MW. Using the H-mode confinement

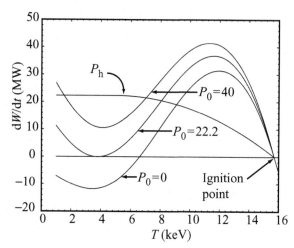

Figure 14.26 $dW/dt$ vs. $T$ for various $P_0$ assuming $\tau_E = \tau_E(T)$. Also shown is the curve of $P_h$ vs. $T$ for $P_0 = 22.2$ MW.

time has reduced the required auxiliary power by nearly a factor of 10, a highly desirable result. The reason is associated with the strong degradation of $\tau_H$ with temperature. This degradation is clearly undesirable when trying to reach high temperatures, but on the other hand implies a substantial increase in $\tau_H$ at low temperatures. To be specific, the confinement time increases by about a factor of 11 as the temperature decreases from 15 to 5 keV. The longer confinement time at lower $\overline{T}_k$ implies that considerably less external power is required to heat the plasma to the Bremsstrahlung temperature $\overline{T}_k \approx 5$ keV.

In summary, the undesirable effects associated with the degradation of $\tau_H$ with temperature are somewhat compensated by improvements in the dynamic evolution of the plasma to ignition. First, the final equilibrium ignition point is thermally stable. Second, a much lower value of external auxiliary power than one might have expected is sufficient to heat the plasma to ignition.

### 14.6.4 The bootstrap fraction

The final application involves the calculation of the bootstrap fraction $f_B$. The goal here is to evaluate the bootstrap fraction in order to determine the conditions under which $f_B \approx 0.75$ for economic viability. The calculation is relatively straightforward, involving the substitution of plausible experimental profiles into the expression for the bootstrap current density and then integrating over the plasma area to obtain the total bootstrap current. It is shown that the standard monotonic current density profile leads to low values of $f_B$, typically less than 0.4. Raising $f_B$ to the desired value of 0.75 requires AT operation characterized by hollow current profiles and a pressure profile quite likely exceeding the no-wall Troyon $\beta$ limit. In other words, the required higher values of $\beta$ lead to excitation of the resistive wall mode, which must then be feedback stabilized.

One point worth noting from the analysis is that the bootstrap fraction is quite sensitive to the density and temperature profiles. Slightly more optimistic profiles are used in this subsection than those used in the design of the ignition experiment in order not to obtain overly pessimistic results.

### Derivation of the bootstrap fraction

The bootstrap current density for a large aspect ratio, circular cross section tokamak in which the small contribution arising from the temperature gradient is neglected is repeated here for convenience:

$$J_B(r) = -4.71 \, q \left(\frac{R_0}{r}\right)^{1/2} \frac{T}{B_0} \frac{\partial n}{\partial r}. \tag{14.192}$$

To evaluate $J_B$ one needs to specify the profiles for $n, T, q$. In practice, it is simpler to specify the total toroidal current density profile $J$ from which it is then straightforward to calculate $q$.

Consider first the density, temperature, and pressure profiles. The pressure profile is assumed to be reasonably peaked with a peaking factor $p(0)/\overline{p} = 3$. This profile is held fixed for all cases studied. The density profile is flatter with a parameter $\nu$ that can be adjusted to vary the density peaking factor: $n(0)/\overline{n} = 1 + \nu$. Also, the larger cross sectional area associated with non-circularity is modeled by assuming a circular plasma with a minor radius given by $r_0 = \kappa^{1/2} a$. Under these assumptions the pressure, density, and temperature profiles can be written as

$$p(\rho) = 3 \, \overline{p}(1 - \rho^2)^2,$$
$$n(\rho) = (1 + \nu)\overline{n}(1 - \rho^2)^\nu, \tag{14.193}$$
$$T(\rho) = (3 - \nu)\overline{T}(1 - \rho^2)^{2-\nu},$$

where $\rho = r/r_0$ and $\overline{p} = (2/3)(1 + \nu)(3 - \nu)\overline{n}\overline{T}$.

The total current density is specified in terms of an arbitrary profile function $g(\rho)$:

$$J(\rho) = \frac{I}{\pi r_0^2} \frac{g(\rho)}{\overline{g}} \tag{14.194}$$

with

$$\overline{g} = 2 \int_0^1 g(\rho) \, \rho \, d\rho. \tag{14.195}$$

Different choices for $g(\rho)$ are made shortly to model standard operation and AT operation. Once $g(\rho)$ is specified the safety factor can be easily evaluated using the standard definition $q = r B_0 / R_0 B_\theta(r)$. A short calculation yields

$$q(\rho) = \frac{\pi r_0^2 B_0}{\mu_0 R_0 I} \frac{\rho^2 \overline{g}}{\int_0^\rho g(\rho) \, \rho \, d\rho} = q_* \frac{\rho^2 \overline{g}}{\int_0^\rho g(\rho) \, \rho \, d\rho}. \tag{14.196}$$

These profiles are next substituted into the expression for the bootstrap current density $J_B(\rho)$, which is then integrated over the plasma area to obtain the total bootstrap current. Dividing by the total plasma current leads to the desired expression for the bootstrap fraction $f_B$ which can be written as

$$f_B = \frac{I_B}{I} = 17.7\, G\, \frac{\nu\kappa^{1/4}\beta_N q_*}{\varepsilon^{1/2}}, \qquad (14.197)$$

where $G$ is a geometric factor defined by

$$G = \bar{g} \int_0^1 \left[ \rho^{5/2}(1 - \rho^2) \Big/ \int_0^\rho g\left(\rho'\right) \rho'\mathrm{d}\rho' \right] \rho\, \mathrm{d}\rho. \qquad (14.198)$$

The appearance of $\beta_N$ in Eq. (14.197) results from eliminating $\bar{p}$ in terms of the Troyon stability limit.

Equation (14.197) can now be used to estimate the bootstrap fraction for standard and AT current density profiles.

### Standard monotonic profiles

The standard current density profile for a tokamak is a monotonically decreasing function of radius that can be modeled by choosing $g(\rho)$ as follows:

$$g(\rho) = 1 - \rho^2. \qquad (14.199)$$

A simple numerical calculation shows that for this choice the geometric factor is given by $G = 0.225$. The resulting value for the bootstrap fraction is thus given by

$$f_B = 4.0 \frac{\kappa^{1/4}\beta_N q_* \nu}{\varepsilon^{1/2}}. \qquad (14.200)$$

For the design values $\kappa = 1.7$, $\beta_N = 0.025$, $q_* = 2$, $\varepsilon = 1/3.5$, $\nu = 1/3$, one finds that $f_B = 0.14$. This is far below the desired value of $f_B = 0.75$.

How can one increase the bootstrap fraction? There are several possible approaches, but all are fraught with difficulties. First, one can imagine having a more peaked density profile, for instance corresponding to $\nu = 1$. However, experimental density profiles tend to be rather flat when refueling takes place by gas puffing from the outside. This is particularly true for H-mode discharges. Fueling by internal pellet injection should lead to more peaked profiles, but the degree to which such peaking can be maintained over long periods of time is uncertain at present. Also, deep penetration into reactor grade plasmas is unlikely because of the high density and large size.

Another approach is to lower the current thereby raising $q_*$. This has the added advantage of providing a higher safety margin against current-driven disruptions. On the negative side, lower current implies poorer confinement since $\tau_H \sim I^{1.06}$. In practice the confinement issue dominates. Without good confinement the plasma could never be heated to a high enough temperature to ignite.

The third possibility is to raise $\beta_N$. The critical $\beta_N$ can be raised substantially if the plasma is surrounded by a perfectly conducting wall. In practice the wall must have a finite conductivity leading to the excitation of the resistive wall mode. There is a reasonable likelihood that the resistive wall mode could be feedback stabilized if the desired value of $\beta_N$ is not too far above the no-wall value. For the present case $\beta_N$ must be raised by a factor of over 5 to reach the value $f_B = 0.75$. This is too large a jump to be plausible.

These constraints have led to the discovery of the AT mode of operation. If successful AT operation can be achieved, then the likelihood of producing substantially higher bootstrap fractions is greatly improved. This is the next topic for discussion.

### AT profiles

AT operation corresponds to a mode of operation in which there is a substantial amount of profile control, primarily of the current density, by means of external power supplies. The goal is to achieve a hollow current profile. If achievable, a hollow current profile helps in two ways. First, a hollow current profile is quite similar in shape to the naturally induced bootstrap profile, implying a strong overlap. In other words, very little, if any, of the bootstrap current is wasted because of being produced in regions where it is not desired.

Second, the reduction of current in the core of the plasma implies a reduction in the total plasma current. Thus, even for a fixed bootstrap fraction, the total current to be driven is reduced. On a related point, the corresponding optimized profiles for the safety factor show that $q(\psi)$ should have an off-axis minimum (i.e., the shear should have an off-axis reversal point). Typically, for the optimized reversed shear profiles it has been found numerically that at the minimum point $q_{min}(\psi_{min}) > 2$. For these profiles the value of the kink safety factor must satisfy $q_* > 3$. Clearly the increased $q_*$ results in an increase in the magnitude of the bootstrap current. This is a desirable result. However, one might be concerned that the higher $q_*$, or equivalently lower current, may lead to a degradation in the confinement time since for H-mode scaling $\tau_H \sim I^{1.06}$. Interestingly, in the region of reversed shear experimental observations indicate that transport is improved to near neoclassical levels. The conclusion is that there may not be a reduction in $\tau_H$, although more experimental data are needed to confirm this important point.

For present purposes, one simply specifies a hollow current profile and then asks what must be done to generate a bootstrap fraction of $f_B = 0.75$. A simple model for the current profile is given in terms of the profile factor $g(\rho)$ as follows:

$$g(\rho) = (\rho^2 + \alpha)(1 - \rho^2), \qquad (14.201)$$

where $\alpha$ is a free parameter set to $\alpha = 0.2$ in the example below. A plot of $g(\rho)$ is illustrated in Fig. 14.27 for this case. A straightforward numerical calculation shows that the corresponding value of the geometric factor has been increased to $G = 0.34$. The bootstrap

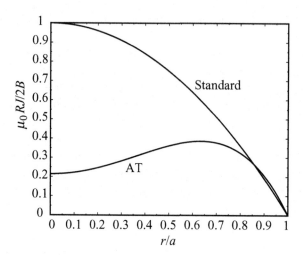

Figure 14.27 The profile functions $g(r/a)$ for the AT case with $\alpha = 0.2$ and for the standard case.

fraction for the hollow current profile is also increased and is given by

$$f_B = 5.9 \frac{\kappa^{1/4} \beta_N q_* \nu}{\varepsilon^{1/2}}. \tag{14.202}$$

Now, assume that for the hollow current profile $q_* = 3.5$. The other parameters are fixed at their standard values: $\kappa = 1.7$, $\beta_N = 0.025$, $\varepsilon = 1/3.5$, $\nu = 1/3$. This leads to $f_B = 0.37$, representing a substantial improvement but still below the desired value of $f_B = 0.75$.

The resolution of the shortfall in $f_B$ is to operate the plasma at a higher value of $\beta$, corresponding to $\beta_N \approx 0.05$. This violates the no-wall Troyon limit by a factor of about 2 leading to the excitation of the resistive wall mode. However, there is a reasonable expectation that a combination of feedback and/or plasma rotation should be capable of stabilizing the resistive wall mode to this more modest increase in $\beta_N$.

### 14.6.5 Summary

The need for steady state operation has led to the discovery of AT operation in which a large fraction of the toroidal current is sustained by the bootstrap current: $f_B \approx 0.75$. Unfortunately, it is not easy to generate such a large bootstrap fraction with standard profiles. Instead, one requires a highly tailored hollow current density profile with a corresponding reversed shear safety factor profile. Furthermore, it is necessary that the transport not be degraded because of the lower total current associated with the hollow current profile and that the resistive wall mode be stabilized at the higher values of $\beta_N$ required to achieve $f_B \approx 0.75$. This is a difficult set of constraints that has caused AT operation to be one of the important topics of tokamak research.

## 14.7 Overall summary

An energy confinement time on the order of $\tau_E \approx 1$ s is required to achieve a self-sustained ignited plasma in a fusion reactor. In practice, $\tau_E$ is dominated by thermal conduction losses. Experimentally, its value is orders of magnitude larger than that predicted by classical Coulomb transport in a 1-D cylinder, a consequence of micro-instability driven turbulence. Thus, while elegant and sophisticated first-principles theories have been developed for Coulomb transport in a cylinder and a torus, the current state of the art still relies heavily on experimentally determined empirical scaling relations for the design of new experiments. The empirical relations show that a tokamak operating in the H mode should be able to achieve self-sustained ignition in a device slightly larger than the simple reactor designed in Chapter 5. Looking to the future, it should be noted that great strides are being made in the development of a first-principles anomalous transport theory. When fully developed, this theory will substantially improve our confidence in predicting the performance of large new experiments. There is still a way to go before reaching this desirable goal.

The second main transport topic involving tokamak reactors is the generation of the bootstrap current. A simple derivation of $J_B$ has been presented which, when applied to experimental situations, shows that the corresponding bootstrap fraction $f_B$ is too low for economic viability under standard tokamak operation. High bootstrap fractions should be achievable with AT operation, but will likely lead to the excitation of the resistive wall mode, which would require feedback and/or rotational stabilization.

Lastly, thermal stability and the minimum power required to reach ignition were re-examined in the context of the empirical scaling laws. Here, the temperature dependence of $\tau_E(T)$ leads to more favorable results than those obtained in Chapter 4, where $\tau_E$ is assumed to be a constant. The new results indicate that the plasma will be thermally stable at the ignition point and that the auxiliary power required to heat the plasma to ignition is quite reasonable, nearly an order of magnitude less than the value predicted in Chapter 4.

Overall, transport in a tokamak still remains as the most difficult plasma physics problem on the path to ignition, but the progress made so far suggests that the performance required in a reactor should probably be achievable.

## Bibliography

There is a very large literature describing transport in a magnetized plasma because it is vitally important in the achievement of fusion energy. Listed below are several general references as well as references for specific transport phenomena.

### *General references*

Chen, F. F. (1984). *Introduction to Plasma Physics and Controlled Fusion*, second edn. New York: Plenum Press.
Hazeltine, R. D., and Meiss, J. D. (1992). *Plasma Confinement*. Redwood City, California: Addison-Wesley.

Helander, P., and Sigmar, D. J. (2002). *Collisional Transport in Magnetized Plasmas.* Cambridge, England: Cambridge University Press.

Hinton, F. L., and Hazeltine, R. D. (1976). Theory of plasma transport. *Reviews of Modern Physics,* **48**, 239.

ITER Physics Basis (1999), Chapter 2, Plasma confinement and transport, *Nuclear Fusion,* **39**, 2175.

Itoh, K., Itoh, I. S., and Fukuyama, A. (1999). *Transport and Structural Formation in Plasmas.* Bristol: Institute of Physics Publishing.

Spitzer, L. (1962). *The Physics of Fully Ionized Gases,* second edn. New York: Interscience.

Wesson, J. (2004). *Tokamaks,* third edn. Oxford: Oxford University Press.

## Neoclassical transport

Galeev, A. A., and Sagdeev, R. Z. (1968). Transport phenomena in a collisionless plasma in a toroidal magnetic system. *Soviet Physics JETP,* **26**, 233.

## Banana regime transport

Kadomstev, B. B., and Pogutse, O. P. (1971). Trapped particles in toroidal magnetic systems. *Nuclear Fusion,* **11**, 67.

Rosenbluth, M. N., Hazeltine, R. D., and Hinton, F. L. (1972). Plasma transport in toroidal confinement systems. *Physics of Fluids,* **15**, 116.

## The density limit

Greenwald, M., Terry, J., *et al.* (1988). A new look at density limits. *Nuclear Fusion,* **28**, 2199.

Greenwald, M. (2002). Density limits in toroidal plasmas. *Plasma Physics and Controlled Fusion,* **44**, R27.

## H mode

Wagner, F., Becker, G., *et al.* (1982). Regime of improved confinement and high beta in neutral beam heated divertor discharges in the Asdex tokamak. *Physical Review Letters,* **49**, 1408.

## Internal transport barriers

Greenwald, M., Gwinn, D., *et al.* (1984). Energy confinement of high density pellet fueled plasmas in the Alcator C tokamak. *Physical Review Letters,* **53**, 352.

Synakowski, E. J. (1998). Formation and structure of internal and edge transport barriers. *Plasma Physics and Controlled Fusion,* **40**, 581.

## Reversed shear

Levinton, F. M., Zarnstorff, M. C., *et al.* (1995). Improved confinement with reversed shear in TFTR. *Physical Review letters,* **75**, 4417.

### *Scaling relations*

Goldston, R. J. (1984). Energy confinement scaling in tokamaks: some implications of recent experiments with ohmic and strong auxiliary heating. *Plasma Physics and Controlled Fusion*, **26**, No. 1A, 87.

## Problems

14.1 This problem investigates the effect of the auxiliary heating deposition profile on confinement time. Consider a 1-D slab model of a plasma in which thermal conduction and auxiliary heating are the dominant contributions to plasma power balance. Assume the auxiliary heating power density is given by $S_h = (1 + \nu)\overline{S}(1 - x/a)^\nu$ with $0 < x < a$. The quantity $\nu$ is a profile parameter describing the deposition profile. Assume now that the thermal diffusivity $\chi$ and number density $n$ are given constants. The boundary conditions are $dT(0)/dx = 0$ and $T(a) = 0$. Derive an expression for the energy confinement time $\tau_E$ as a function of $\nu, n, \chi, a$ and compare the values for $\nu \to 0$ and $\nu \to \infty$.

14.2 Consider the two-fluid, steady state, 0-D power balance relations for the electrons and ions in an ohmically heated tokamak. For simplicity assume all profiles are uniform in space. The energy confinement times for each species are given by $\tau_{Ee}, \tau_{Ei}$ and are assumed to be known constants. Also, the tokamak is operated in the sawtooth regime so that the current density $J_0 = $ const. is independent of $T_e, T_i$. Recall now that the resistivity and energy equilibration time scale as $\eta = K_\eta/T_e^{3/2}$ and $\tau_{eq} = K_\tau T_e^{3/2}/n$ respectively. In the limit where $\tau_{eq}$ corresponds to a short but finite time, derive approximate expressions for the steady state values of $T_e$ and $1 - T_i/T_e$ in terms of $\tau_{Ee}, \tau_{Ei}, J_0, n, K_\eta, K_\tau$.

14.3 During the flat-top portion of a tokamak discharge the toroidal loop voltage $V_\phi = 2\pi R_0 E_\phi$ is measured to be 0.8 V. If the tokamak is operating in the sawtooth regime and the toroidal field is $B_0 = 4$ T, find the electron temperature on axis. Assume Spitzer resistivity: $\eta = 3.3 \times 10^{-8}/T_k^{3/2}$.

14.4 A cylindrical plasma of radius $a$ has an initial density profile $n(r, 0) = n_0$ for $0 < r < a$. The density profile satisfies the diffusion equation

$$\frac{\partial n}{\partial t} = \frac{D}{r}\frac{\partial}{\partial r}\left(r\frac{\partial n}{\partial r}\right),$$

where $D = $ const. Assume that all particles are absorbed at the edge of the plasma and that there are no sources present. Calculate the density profile after a large but finite time. Hint: review Fourier–Bessel series.

14.5 This problem describes a simple method for fueling a tokamak by external gas puffing. A series of gas jets injects a known flux of particles $\Gamma$ homogeneously around the surface of a cylindrical plasma. The particles entering the plasma are neutral D atoms. After penetrating a distance of about one mean free path the neutrals become ionized. This represents a distributed source $S_n(r)$ of new plasma particles which must be included in the conservation of mass equation. Note that near the center of the plasma $S_n(r) = 0$ since all the incoming particles have been ionized. A simple model for the source term is thus given by $S_n(r) = 0, 0 < r < a - \lambda$, and $S_n(r) = S_0, a - \lambda < r < a$, where for simplicity $\lambda = $ const. The 1-D steady state equation describing the

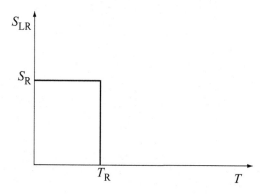

Figure 14.28

density profile is now given by

$$\frac{1}{r}\frac{\partial}{\partial r}\left(rD\frac{\partial n}{\partial r}\right) + S_n(r) = 0$$

with $D = \text{const.}$

(a) Relate the source amplitude $S_0$ to the input particle flux $\Gamma$.

(b) Assume a perfect sink condition at $r = a$. Find the steady state density profile and the density on axis in terms of $\Gamma, a, \lambda, D$.

14.6 Consider the energy transport in a 1-D, single-fluid, slab model of a plasma. The main contributions to energy balance are ohmic heating, thermal conduction losses, and externally applied auxiliary heating. For simplicity assume that the auxiliary heating power density is modeled by $S_h(x) = S_0$ and that the density $n(x) = n_0$. The goal of the problem is to determine the auxiliary power density $S_0$ required to achieve a specified temperature on axis $T_0$. Assume the plasma is in the steady state, flat-top portion of the discharge. The transport coefficients are modeled by resistivity $\eta = C_\eta / T^{3/2}$ and an anomalous thermal diffusivity $\chi_\perp = C_\chi T^{1/2}$. Furthermore, assume that the temperature and current density are related by $J = J_0 (T/T_0)^{3/2}$ where $J_0$ and $T_0$ are the values on the axis. The boundary conditions require that $T(x) = 0$ at $x = -a$ and $x = a$.

(a) Set $S_0 = 0$ and calculate $T_0 \equiv T_\Omega$, the temperature on axis due to pure ohmic heating.

(b) Derive an expression for the value of $S_0$ required to achieve a specified central temperature $T_0 > T_\Omega$.

14.7 This problem investigates the effect of line radiation at the edge of a plasma on thermal power balance. The geometry of interest is a 1-D plasma slab located between $0 < x < a$. The dominant contributions to power balance arise from external heating ($S_h = S_0$), thermal conduction ($\kappa = \kappa_0$), and line radiation loss $S_{LR}(T)$ as illustrated in Fig. 14.28. In practice $T_R \sim 10 - 30$ eV and $S_R \gg S_0$.

(a) For reference calculate the equilibrium temperature profile, neglecting the radiation loss. Assume $T'(0) = 0$ and $T(a) = 0$.

(b) Repeat part (a) for a finite radiation loss. Show that there is a critical value of $S_0$ such that the entire energy loss from the plasma is due to radiation. There is no thermal conduction loss. What happens when $S_0$ falls below this critical value?

14.8 Consider a large aspect ratio, circular cross section tokamak. During flat-top operation power balance is dominated by thermal conduction losses and auxiliary power heating.

(a) Assume the thermal diffusivity $\chi$, auxiliary power density $S_h$, and number density $n$ are uniform in space. Derive an expression for $T_e(r) = T_i(r) \equiv T(r)$ in kiloelectronvolts as a function of $\chi$, $n$, $R_0$, $a$ and the total auxiliary power $P_h(MW)$.

(b) Using the definition of global energy confinement time derive a relation between $\chi$ and $\tau_E$.

(c) Assume that $\tau_E$ is given by the following simplified form of L-mode scaling: $\tau_E \approx 0.06 I_M R_0^{3/2}/P_h^{1/2}$ s. Derive an expression for $\chi$ as a function of $n$, $a$, $R_0$, and $\overline{T}$. The answer should not be a function of $P_h$.

14.9 Consider a circular cross section tokamak with a major radius $R_0 = 0.6$ m, a minor radius $a = 0.2$ m, and a toroidal field on axis $B_0 = 10$ T. At what value of $q_a$ must the device be operated to provide 500 kW of ohmic heating power at a loop voltage of $V_\phi = 0.5$ V ? Assume the current density profile is given by $J_\phi = J_0(1 - r^2/a^2)$. Would you expect sawtooth oscillations to be excited? Explain.

# 15

# Heating and current drive

## 15.1 Introduction

For a plasma to be ignited in a steady state fusion reactor it must satisfy the ignition condition

$$p\tau_E = \frac{24}{E_\alpha} \frac{T^2}{\langle \sigma v \rangle}.$$

(15.1)

Minimizing the demands on pressure and confinement time requires that the reactor operate at $T = 15$ keV, the temperature that minimizes $T^2/\langle \sigma v \rangle$. Reaching this temperature is a two-stage process. First, a combination of ohmic and auxiliary heating must raise the plasma temperature to about 5–7 keV. During this evolution alpha power is negligible and the heating power must overcome the losses due to thermal conduction and Bremsstrahlung radiation. Above 5–7 keV alpha power becomes dominant, heating the plasma to its final ignition temperature of $T = 15$ keV.

The main purpose of this chapter is to investigate various methods of heating a plasma to reach the transition temperature of 5–7 keV.

Another purpose is to investigate the related problem of current drive. Recall that for a tokamak reactor to operate in steady state some form of non-inductive current drive is required. Most of the methods used to heat a plasma can also be used to drive current. It therefore makes sense to extend the discussion of heating to include current drive. The analysis of heating and current drive is focused on the tokamak where most of the research has been carried out. Even so, many of the results are applicable to other configurations as well.

It should be noted that heating and current drive cover a very broad range of topics, far too lengthy to include in the present book. The choice of topics discussed here is based on the strategy of focusing on methods that are likely to be relevant to ITER and fusion reactors. For heating these include ohmic heating, neutral beam heating, and electron and ion cyclotron heating. For current drive, the topic of interest is lower hybrid current drive.

To understand the heating issues, consider first ohmic heating. The toroidal current induced by the transformer in a tokamak produces ohmic heating. This is the simplest method in terms of the technology. However, the resistivity of a plasma decreases with temperature: $\eta \propto 1/T^{3/2}$. Thus, as the ohmic current increases the heating efficiency decreases. The

534

analysis shows that for typical parameters in a tokamak reactor the maximum temperature achievable by ohmic heating is about $T \lesssim 3$ keV. This is not high enough for the alpha power to dominate. Some other form of auxiliary heating is required.

The first option discussed is neutral beam heating. Here, a high-energy beam of either neutral deuterium or neutral tritium atoms is injected into the plasma. Heating takes place as follows. Neutral beam atoms are unaffected by the magnetic field. Thus beam atoms propagate in a straight line until they are ionized by collisions with the background plasma. Once ionized, the beam particles are confined by the magnetic field and gradually give up their energy to the plasma by Coulomb collisions. Since the heating mechanism depends on classical collisions it should reliably extrapolate to reactor grade plasmas. In terms of operation, the energy of the beam clearly must be much higher than the plasma temperature for good heating. The actual value of the energy is determined by the requirement that the beam be able to penetrate to the center of the plasma to produce central heating. This requirement poses the main problem for neutral beam heating and is technological in nature. Existing neutral beam systems are driven by positive ion sources which have good efficiency up to about 100 keV. This is sufficient for present day experiments. ITER and fusion reactors, because of their higher density and larger size, require 1 MeV beams for good penetration. This goal can be accomplished using a negative ion source to drive the neutral beam system. However, the technology is substantially more difficult and is not yet readily available. A major research and development program is underway to develop such negative ion sources and is expected to have been successful by the time they are needed for ITER.

A second option for auxiliary heating is the use of radio frequency (RF) waves. Here, high-frequency electromagnetic waves are launched into the plasma from an external source. The heating mechanism is similar to that in a microwave oven. When the applied frequency is carefully chosen to match the natural resonant frequency of the food, or in this case the plasma, there is a strong absorption of energy, which is converted into heat. There are several natural resonant frequencies of interest in a plasma: the cyclotron frequencies of the electrons and ions, and their cyclotron harmonics. Heating at the resonant frequencies of the electrons is known as electron cyclotron heating (ECH). For the ions it is ion cyclotron heating (ICH). An interesting feature of ECH and ICH is that the resonant absorption takes place by a mechanism known as "collisionless damping." Absorption does not depend on collisions as in a microwave oven. Collisionless damping is described in detail during the discussion of RF heating. While both ECH and ICH can produce a strong absorption of energy at the center of the plasma, both methods also face technological problems. For ECH the difficulty is that high-power, steady state gyrotron sources at the required frequency of 140 GH$_z$ are not yet readily available. For ICH the difficulty is that an antenna must be placed very close to the surface of the plasma to insure good coupling of the wave energy to the plasma. This leads to problems of arcing and plasma breakdown. Substantial research and development programs are underway and it is expected that successful solutions will have been found by the time they are needed for ITER.

Table 15.1. *Approximate relative cost per watt of auxiliary heating power options*

| Option | Requirement | Cost ($/W) |
|---|---|---|
| Negative ion beam | 1 MeV | 4 |
| ICH | 40 MHz | 2 |
| ECH | 140 GHz | 6 |
| LHCD | 3 GHz | 3 |

Overall, there is no clear winner in the choice of a heating system for experiments on the scale of ITER. Several options are available. All have good heating potential from the point of view of plasma physics. All have substantial technological issues to address, although successful solutions are expected. Based on this assessment, ITER has been designed with multiple heating options to determine experimentally which method ultimately will be most desirable in terms of plasma physics, reliability, and cost.

Consider now current drive. All of the methods just discussed can be used to drive a steady state non-inductive current. However, the discussion presented here focuses on lower hybrid current drive (LHCD) since this method has a higher efficiency (i.e., highest driven current per watt of applied power) than the other options. Here, too, a collisionless damping mechanism, known as Landau damping, leads to the possibility of a steady state non-inductive current. The idea in simple form is to launch waves into the plasma that propagate in one direction around the torus. These waves drag electrons with them in a somewhat analogous way to an ocean wave catching and pushing along a surfer. The electrons dragged with the wave produce the current drive. Sources are readily available in the microwave range to drive lower hybrid waves. Like ICH, LHCD does have the problem of requiring the launching structure to be near the plasma edge for good coupling. More importantly, the current-drive efficiency, while higher than for the other options, is still too low to drive all the current in a tokamak reactor in an economical way. The conclusion is that a tokamak reactor will have to depend on a substantial bootstrap current to lower the requirements on the current-drive system.

A final point to consider is the relative cost of the various auxiliary power options. Admittedly, this is a somewhat difficult and speculative task. Still, to obtain some idea of the economics examine the approximate costs given in Table 15.1, largely obtained from the ITER costing analysis. Observe that all the options involve a considerable amount of high-technology equipment leading to a substantial cost per watt. ICH has the lowest cost, while ECH has the highest cost, although all are comparable. It should be emphasized that none of these values include the high research and development costs involved. The values estimate only the direct hardware costs. Clearly, for the sake of economy, it is important to keep the auxiliary power requirements as low as possible.

The analysis that follows closely follows the order of the topics just described. Most of the results are derived from first principles, some requiring a relatively simple analysis (ohmic heating and neutral beams) and others requiring a more intensive analysis (ECH, ICH, LHCD).

## 15.2 Ohmic heating

The problem here is to calculate the maximum achievable plasma temperature resulting solely from ohmic heating, and to determine whether or not this temperature is sufficient to reach ignition without the need for auxiliary power. The analysis is carried out using the empirical $\tau_E$ and the neoclassical resistivity. It is shown that ohmic heating to ignition is extremely difficult, if not impossible, for typical engineering and plasma physics constraints. The biggest difficulty arises from the fact that the resistivity decreases with temperature: $\eta \sim 1/T^{3/2}$. Thus, as the plasma is heated, its resistivity decreases, implying a corresponding decrease in heating efficiency. In other words, as $I$ increases, the ohmic power ($P_\Omega = I^2 R$) increases slower than $I^2$ because the resistance $R$ decreases with $I$. The overall conclusion is that a substantial amount of external heating is required to reach ignition.

The specific goal of the analysis is to derive a relationship between the plasma temperature and the ohmic current, assuming no auxiliary power and no alpha power. An examination of this expression then leads to the above conclusion.

### 15.2.1 The ohmic heating model

A useful way to think about the problem is to assume ohmic ignition is possible and to then calculate how much current is required to heat the plasma half-way to the final ignition temperature of $T \approx 15\,\text{keV}$, ignoring radiation, auxiliary power, and alpha heating. Once $T \approx 7\,\text{keV}$ is achieved, one can assume that alpha heating becomes dominant raising the temperature to the final desired operating value $T \approx 15\,\text{keV}$. The current required to reach 7 keV must then be tested to guarantee that it does not violate the MHD stability limit. Stated mathematically, the goal of the analysis is to calculate $\overline{T}_k = \overline{T}_k(q_*)$ and to test whether $\overline{T}_k(q_* = 2) \geq 7\,\text{keV}$ for ohmic ignition.

### 15.2.2 Ohmic power balance

Under the above assumptions, steady state ohmic power balance is defined by

$$P_\kappa = P_\Omega, \tag{15.2}$$

where the thermal conduction losses and ohmic power are given by

$$P_\kappa = \frac{3}{\tau_L} \int nT \, d\mathbf{r},$$

$$P_\Omega = \int \eta J^2 \, d\mathbf{r}. \tag{15.3}$$

### 15.2.3 Thermal conduction losses

The thermal conduction losses are evaluated by assuming for simplicity that the density profile is uniform and that the temperature profile is modeled by

$$T = T_0 \left( 1 - \frac{x^2}{a^2} - \frac{y^2}{\kappa^2 a^2} \right)^{4/3}. \tag{15.4}$$

The temperature profile is slightly peaked and corresponds to an elliptical cross section. The exponent $4/3$ is chosen for convenience to allow certain integrals that arise shortly to be easily evaluated analytically. If one now introduces the transformation $x = a\rho \cos\theta$ and $y = \kappa a\rho \sin\theta$, then the differential area element becomes $dx\,dy = \kappa a^2 \rho\,d\rho\,d\theta$. A short calculation then shows that the relation between peak and average temperatures is $\overline{T} = (3/7)T_0$.

The thermal conduction losses are now evaluated as follows:

$$P_\kappa = 6V \frac{\overline{n}T_0}{\tau_L} \int_0^1 (1 - \rho^2)^{4/3} \rho\,d\rho = 3V \frac{\overline{n}\overline{T}}{\tau_L}, \tag{15.5}$$

where $V = 2\pi^2 R_0 a^2 \kappa$ is the plasma volume.

Next, note that ohmically heated tokamaks usually operate in the L-mode regime so that $\tau_L$ is the relevant empirical scaling relation. Substituting the value of $\tau_L$ from Eq. (14.154) leads to the desired expression for the thermal conduction losses, which in practical units can be written as

$$P_\kappa = \frac{25.6}{\varepsilon^{1.3}} \left( \frac{\overline{n}_{20}^{1.8} a^{1.3}}{\kappa^{0.7} B_0^{2.1} A} \right) q_*^{1.7} \overline{T}_k^2 \quad \text{MW}. \tag{15.6}$$

### 15.2.4 The ohmic power

The evaluation of the ohmic power requires several steps. The first step involves the relation between $J$ and $\eta$. In a large aspect ratio tokamak $J \approx J_\phi \approx J_\parallel$ and in steady state $E \approx E_\phi \approx E_\parallel$ with $E_\phi = E_0(R_0/R) \approx E_0 = \text{const}$. The parallel Ohm's law thus yields $\eta J \approx \eta_\parallel J_\parallel \approx E_0$ or equivalently $J_\parallel = E_0/\eta_\parallel$.

The second step involves the expression for $\eta_\parallel$. This value has already been given for a cylinder. However, there is a neoclassical modification that does not change the basic large aspect ratio scaling $\eta_\parallel \sim 1/T^{3/2}$, but does lead to a quantitative change in the numerical multiplier for realistic aspect ratios. A good approximation for the neoclassical resistivity is

$$\eta_\parallel^{(NC)} = \frac{1}{[1 - (r/R_0)^{1/2}]^2} \eta_\parallel^{(CL)}. \tag{15.7}$$

Note that the neoclassical value is higher. This is a consequence of the fact that trapped particles cannot carry parallel current. Thus, there are fewer current carriers and the resistivity increases. Interestingly, in contrast to predictions of ion thermal transport, the neoclassical resistivity appears to be in relatively good agreement with experimental observations.

Combining the above results leads to the following expression for $J_\parallel$:

$$J_\parallel = J_0(1 - \rho^2)^2(1 - \varepsilon^{1/2}\rho^{1/2})^2, \tag{15.8}$$

where $J_0$ is a new constant replacing $E_0$ that represents the current density on-axis. Actually, it is more convenient to express $J_0$ in terms of the total plasma current $I$ by the usual definition:

$$I = \int J_\parallel dx \, dy = 2\pi a^2 \kappa \int_0^1 J_\parallel \rho \, d\rho$$

$$= \frac{\pi a^2 \kappa J_0}{3}(1 - 1.31\varepsilon^{1/2} + 0.46\varepsilon). \tag{15.9}$$

The final step is to substitute these relations into the expression for the ohmic power given by Eqs. (15.3), recalling that $\eta_\parallel^{(CL)} = 3.3 \times 10^{-8}/\overline{T}_k^{3/2}$ $\Omega$ m. A short calculation yields

$$P_\Omega = \left(\frac{5.6 \times 10^{-2}}{1 - 1.31\varepsilon^{1/2} + 0.46\varepsilon}\right)\left(\frac{R_0 I_M^2}{a^2\kappa \overline{T}_k^{3/2}}\right) \text{MW}$$

$$= \left(\frac{1.4\varepsilon}{1 - 1.31\varepsilon^{1/2} + 0.46\varepsilon}\right)\left(\frac{a^2\kappa B_0^2}{q_*^2\overline{T}_k^{3/2}}\right) \text{MW}. \tag{15.10}$$

In the second expression the current has been eliminated in terms of the safety factor by the relation $I_M = 5a^2\kappa B_0/R_0 q_*$.

### 15.2.5 Ohmic power balance

The last step in the analysis is to equate the expressions for the thermal conduction losses (Eq. (15.6)) and ohmic heating power (Eq. (15.10)). A straightforward calculation then yields the desired relation $\overline{T}_k = \overline{T}_k(q_*)$:

$$\overline{T}_k = 1.1\frac{\varepsilon^{0.66}}{(1 - 1.31\varepsilon^{1/2} + 0.46\varepsilon)^{0.29}}\left(\frac{\kappa^{0.49}B_0^{1.17}A^{0.29}}{\overline{n}_{20}^{0.51}a^{0.09}}\right)\frac{1}{q_*^{1.06}} = \frac{6.9}{q_*^{1.06}} \text{ keV}. \tag{15.11}$$

The numerical value in the last expression has been obtained using the parameters in the simple fusion reactor and assuming $\kappa = 2$. Observe that the temperature scales approximately linearly with plasma current: $\overline{T}_k \sim I_M^{1.06}$. One might have hoped for a quadratic scaling $\overline{T}_k \sim I_M^2$ since $P_\Omega \sim I_M^2$. However, the decrease of resistivity with temperature weakens the dependence.

If one now assumes that $q_* \geq 2$ (corresponding to $I_M \leq 18.8$ MA) for MHD stability, then Eq. (15.11) predicts that ohmic heating can raise the plasma temperature to a maximum value of $\overline{T}_k = 3.3$ keV. Note that this value is smaller than the critical temperature required for alpha power to exceed Bremsstrahlung radiation ($\overline{T}_k > 4.4$ keV). The conclusion is that for typical parameters of a fusion reactor a substantial amount of auxiliary power is required to raise the plasma temperature to a high enough value so that alpha heating becomes the dominant heating source.

There is one final point to be made. If, along the path to a fusion reactor, one wants to study ignition physics on a relatively inexpensive experiment, it may be possible to do so using ultrahigh-field, pulsed copper magnets capable of producing $B_0 \sim 10$ T. According to Eq. (15.11) such a device should be capable of ohmically heating a plasma to a temperature greater than 6 keV. This is very close to the condition for ohmic ignition. This concept has been investigated in the Ignitor tokamak. Prototype components have been designed, constructed, and successfully tested in Italy, but a final commitment has not as yet been made to fully fund the construction of Ignitor.

## 15.3 Neutral beam heating

### 15.3.1 Overview

One highly successful method of raising the plasma temperature well above the maximum achievable ohmic value is by means of neutral beams. The idea is as follows. Assume for a moment the existence of a beam of high-energy neutral particles, for instance deuterium atoms. "High-energy" implies a beam energy much higher than the desired plasma temperature of 15 keV. The beam is now injected into the plasma. Since the particles are electrically neutral they are unaffected by the magnetic field and travel along straight-line trajectories until they are ionized by collisions with the background plasma. Once ionized the neutral beam particles become magnetically confined. In other words they become part of the plasma, corresponding to a high-energy tail on the deuterium distribution function. The high-energy tail slows down by Coulomb collisions, thus transferring its energy to the background plasma in the form of heat. Neutral beam heating as just described has been successful in achieving temperatures of approximately $T_i \approx 20$ keV in the JET experiment.

What are the problems of and prospects for using neutral beams to heat a plasma to ignition in a fusion reactor? The issues involve both physics and technology and can be summarized as follows.

Consider first the physics issues. The main requirement on the beam is a high flux, capable of penetrating to the center of the plasma before being ionized in order for the energy to be deposited where it is most needed. Intuitively, one expects the penetration depth to be proportional to the beam energy. Low-energy beams deposit most of their energy on the outside of the plasma, which is undesirable. If the beam energy is too high, the beam passes through the plasma and deposits its energy on the opposite wall, also an undesirable situation. One important physics issue, therefore, is to determine the optimum value of beam energy $E_b$ required to penetrate a distance $a$ to the center of the plasma; that is, to determine $E_b = E_b(a)$.

A second physics issue involves the transfer of the beam energy (once it has been ionized) to the background plasma, although this is relatively straightforward. The ionized beam slows down and deposits its energy by standard Coulomb collisions. As described in Chapter 9, a portion of the beam energy heats the plasma electrons, while the remainder heats the plasma ions. The actual fraction depends upon the ratio of the beam energy to the plasma temperature, with high beam energies favoring the electrons. In any event, in a fusion reactor the density is high enough to cause the electron and ion temperatures to rapidly equilibrate. On this basis it is assumed that virtually all of the neutral beam energy is deposited in the plasma, and is divided equally between electrons and ions.

Consider now the technological issues. For current experiments in which the density and minor radius are both smaller than that required in a reactor, the existing neutral beam technology is quite satisfactory. In particular, high-efficiency beams with an optimized energy on the order of 100 keV can be produced such that most of the beam energy is deposited in the center of the plasma. In a reactor, however, the physics relationship $E_b = E_b(a)$ implies that substantially higher beam energies (on the order of 1 MeV) will be required to achieve the increased penetration depths and this poses a difficult technological problem.

Specifically, most neutral beams currently in use are generated from an initial source of positive ions. The corresponding overall efficiencies are quite reasonable (i.e., $\eta =$ beam power/input power is reasonably high). However, the efficiency rapidly decreases with increasing beam energy. This unfortunate dependency leads to the widely accepted conclusion that the present positive ion technology will not successfully extrapolate into the reactor regime. To address the problem a different strategy has been developed that involves the production of neutral beams starting with an initial source of negative ions. Theory predicts that the overall efficiency using negative ion sources will remain high as the beam energy increases, although the technology for producing such sources is more complicated.

A further problem for both positive and negative ions is that the final neutral beam sources are physically quite large and involve substantial amounts of high-technology components. In other words, the cost per watt of neutral beam power is relatively high.

These issues have been analyzed in detail in *Tokamaks* by J. Wesson (Oxford: Oxford University Press, third edition, 2004) and in fact the discussion below largely consists of a simplified summary of his analysis of neutral beams. Three issues are examined here. The first involves a qualitative description of the operation of a neutral beam source. This is followed by a quantitative calculation of the function $E_b = E_b(a)$. Lastly, a short calculation is presented showing why positive ion beams are inefficient at high energies and how negative ion beams improve the situation. The main conclusion is that negative ion neutral beams offer a good method to heat a plasma. However, negative ion neutral beams need to be developed further before their usefulness as a heating method for fusion reactors can be fully assessed.

### 15.3.2 *How is a neutral beam produced?*

A neutral beam source is a four-stage device, illustrated schematically in Fig. 15.1. It works as follows. The purpose of the first stage is to produce a source of low-temperature ions. Positive ions can be created by standard techniques. The ions are produced as the positively charged species of a low-temperature plasma by one of several well-established methods

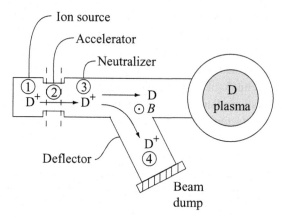

Figure 15.1 Schematic diagram of a neutral beam system: (1) ion source, (2) accelerating region, (3) neutralizer, (4) charged particle deflector.

(i.e., a Penning discharge, a pin cathode discharge, a "bucket source", etc.). One challenge that has been successfully addressed is minimization of the number of molecular $D_2^+$ and $D_3^+$ ions with respect to the number of atomic $D^+$ ions. The molecular forms, because of their heavier mass, ultimately produce neutrals with $1/2$ or $1/3$ the energy of the those produced by $D^+$, leading to low penetration, and a corresponding undesirable edge heating.

Negative ions, on the other hand, are more difficult to produce. They typically arise in low-temperature discharges in which electrons attach themselves to neutral molecules. This is one of the main technical challenges facing negative-ion-driven neutral beam sources. Even so, negative ions have one special property that is both desirable and crucial: the extra electron is only weakly bound to its neutral atom. This electron can therefore be relatively easily stripped away even at energies well in excess of 100 keV. The result is high efficiency at high energies, and represents the main advantage over positive-ion-driven sources.

For present purposes, assume that a source of low-temperature ions, of either positive or negative charge, has been produced.

The second stage of the device accelerates these ions to a high energy. The ions pass from one stage to the next through a thin surface containing a large number of small holes or some equivalent grid. They are then accelerated by means of a high voltage, negative for positive ions and positive for negative ions. This acceleration to high voltage is where most of the input electrical power is consumed. The output of the accelerator stage is a highly directed, nearly mono-energetic beam of high-energy ions.

The high-energy ions enter the third stage of the device, known as the neutralizer. This is essentially a long tube filled with a carefully chosen density of neutral particles, typically the same species as the source ions, for instance deuterium. For the case of positive ions, as they pass through the neutralizer they may undergo an inelastic collision known as a "charge exchange" collision. Here, a high-energy positive ion acquires an electron from a cold neutral; that is, the neutral and ion basically "exchange" roles with the outcome being a low-energy ion and a high-energy neutral. For the case of negative ions, they are

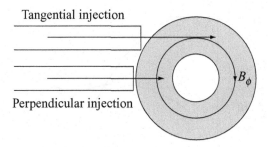

Figure 15.2 Schematic diagram of a top view of a tokamak showing tangential and perpendicular injection.

neutralized by an inelastic collision which strips the excess electron from the atom. It is worth again emphasizing that the neutralization efficiencies are quite different for the two types of ions. The efficiency of positive ion neutralization decreases with beam energy, while it remains approximately constant for negative ion neutralization. For either positive or negative ion sources it is the high-energy neutral that is the particle desired to inject into the plasma.

The last stage of the device is the magnetic deflector. This stage is necessary because the output of the neutralizer in general includes both high-energy neutral particles and high-energy ions which have escaped neutralization. Each typically carries a comparable amount of power. Because of their charge, the ions, if injected into the plasma, would have their straight-line orbits strongly altered by the magnetic field, and would most likely deposit their energy on the neutral beam entrance port, clearly an undesirable situation. To avoid this difficulty the combined beam passes through an applied magnetic field produced in the deflector region. Only the charged ions are affected. They are deflected, and their energy is collected on the beam dump as illustrated.

The remaining neutral particles are then injected into the plasma. Here they are re-ionized and deposit their energy in the background plasma, interestingly by reverse charge exchange collisions. Note that if possible it is advantageous to inject neutral beams parallel to the plasma axis as shown Fig. 15.2, even though this is more complicated geometrically and requires more beam energy because of the longer path. Perpendicular injection is simpler, but results in high-energy ionized particles with a large perpendicular velocity component. Such particles can be rapidly lost by means of neoclassical transport if the toroidal field ripple is too large. In actual experiments, injection is often at a compromise angle to allow for the multiple geometric and beam energy constraints that must be satisfied.

This completes the qualitative description of the operation of a neutral beam source. In summary, high-efficiency positive ion systems have been constructed and operated up to the 100 keV level, and have had great success in heating present day tokamaks to reactor level temperatures. The 1 MeV sources needed for the larger, higher-density plasmas in a reactor, however, require more technologically complicated negative ion systems. The development of negative-ion-driven neutral beam sources remains an important area of fusion technology research.

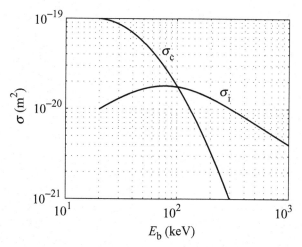

Figure 15.3 Cross sections for charge exchange and ionization (Wesson, J. (2004). *Tokamaks*, third edition. Oxford: Clarendon Press).

### 15.3.3 The physics problem – energy required for beam penetration

The main physics problem is to calculate how much beam energy is required for high-energy neutral particles to penetrate to the center of the plasma before being ionized. Specifically one wants to derive an expression for $E_b = E_b(a)$. The derivation is straightforward and depends primarily on the cross sections for various ionization mechanisms. The calculation proceeds as follows.

In general, there are three basic ways in which a high-energy neutral particle can be ionized in a plasma: charge exchange, ionization by ions, and ionization by electrons. Of these, electron ionization is the smallest effect and is neglected for simplicity. Consider now charge exchange collisions in which a high-energy neutral loses an electron to a colder ion. The charge exchange reaction for D–D collisions is

$$D_b + D_p^+ \rightarrow D_b^+ + D_p. \tag{15.12}$$

The notation here and below is as follows. The subscripts "b" and "p" refer to beam and plasma respectively. A superscript "+" refers to a plus charge. No superscript implies a neutral atom. Lastly, whether charged or not, a beam particle is always a high-energy particle, while a plasma particle is always a lower-energy particle whose energy is comparable to the plasma temperature. The cross section for the charge exchange process $\sigma_c$ is well known from atomic physics research and is illustrated in Fig. 15.3. Observe the rapid decrease starting at 50–100 keV.

The second main contribution to ionization occurs when a high-energy neutral has a strong collision with a plasma ion. In this case the neutral particle disassembles into an ion and an electron, both with a velocity comparable to the initial neutral velocity. Thus most of the energy is carried by the ion because of its heavier mass. The corresponding reaction is

$$D_b + D_p^+ \rightarrow D_b^+ + D_p^+ + e^-. \tag{15.13}$$

The cross section for this process $\sigma_i$ is also well known from atomic physics and is included in Fig. 15.3. Below 90 keV charge exchange is the dominant ionization mechanism. Furthermore, note that while the above reactions were written for deuterium neutrals colliding with deuterium ions, either the beam or plasma particles could be replaced with tritium with no significant change in the results.

Next, consider the relation between the cross sections and the penetration depth. Recall that as the neutral beam penetrates and is absorbed into the plasma its flux $\Gamma_b \equiv n_b v_b$ decays with distance in accordance with the relation

$$d\Gamma_b/dx = -n_p(\sigma_c + \sigma_i)\Gamma_b. \tag{15.14}$$

Here, $n_p$ is the background plasma density, which is in general a function of $x$. However, since the density profile is usually quite flat one can assume that $n_p \approx$ const. Since there is a one-to-one correspondence between the loss of high-energy neutral particles ($\Gamma_b$) and the gain in high-energy charged particles ($\Gamma_b^+$), the solution to Eq. (15.14) implies that

$$\Gamma_b^+(x) = \Gamma_b(0)(1 - e^{-x/\lambda}), \tag{15.15}$$

where the decay length $\lambda$ is given by

$$\lambda = \frac{1}{n_p(\sigma_c + \sigma_i)}. \tag{15.16}$$

The penetration depth is clearly of the order of the decay length. Focusing first on perpendicular heating, it follows that one should set $\lambda \sim a$ for central heating. A more quantitative definition is suggested by a simple line of reasoning. A short decay length (i.e., small $\lambda$) is undesirable since most of the energy is deposited on the outside of the plasma. A long decay length (i.e., large $\lambda$) leads to a high relative deposition in the center of the plasma but also results in much of the beam energy escaping from the far side of the plasma, also an undesirable situation. Intuitively, about the best that one can do is to choose $\lambda$ as large as possible subject to the constraint that negligible beam energy escapes from the far side of the plasma. Since (1) the beam traverses a distance $2a$ before leaving the plasma and (2) three decay lengths are required to reduce beam losses to an acceptably low level, it follows that the desired relation $E_b = E_b(a)$ can be written in implicit form as

$$a = \frac{3}{2}\lambda(E_b) \equiv a_p = \frac{1.5}{n_p(\sigma_c + \sigma_i)}. \tag{15.17}$$

A similar argument holds for parallel injection. The one main difference is that because of geometric effects the beam must propagate a longer distance. See Fig. 15.2. A simple calculation shows that in this case Eq. (15.17) is replaced by

$$a_p = 1.5\left(\frac{\varepsilon}{2+\varepsilon}\right)^{1/2} \lambda(E_b). \tag{15.18}$$

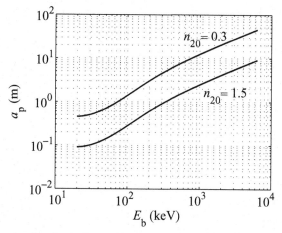

Figure 15.4 Beam penetration depths for a typical JET discharge with $n_{20} = 0.3$, $a = 1$, and a simple reactor with $n_{20} = 1.5$, $a = 2$ (Wesson, J. (2004). *Tokamaks*, third edition. Oxford: Clarendon Press).

For an aspect ratio of 2.5 the beam must penetrate about a factor of 2.4 further for parallel injection than for perpendicular injection.

Equation (15.17) is plotted in Fig. 15.4 for two different values of density. Observe that for a typical JET plasma with a density of $n_{20} = 0.3$ and minor radius $a = 1$ m, the penetration depth for a 100 keV neutral beam is $a_p = 1.3$ m, which is approximately the required depth for perpendicular injection. Purely parallel injection requires a penetration of about 3 m, corresponding to a beam energy of 200 keV, which is just beyond the present upper limits of high-efficiency positive ion technology. The conclusion, born out experimentally, is that positive ion technology is just adequate to efficiently heat a JET size plasma to reactor temperatures.

On the other hand, a fusion reactor plasma with a density of $n_{20} = 1.5$ requires a penetration depth of $a_p = 2$ m for perpendicular injection, which translates into a beam energy of 0.8 MeV. Parallel injection for $R_0/a = 2.5$ corresponds to a penetration depth of 6 m requiring a 3 MeV beam energy. These values are well beyond the range where positive ions can be used effectively. The conclusion is that high-energy negative ion beam technology must be developed if this form of auxiliary heating is to be used in ITER or a fusion reactor.

### 15.3.4 The technology problem – conversion efficiency in the neutralizer

It has been stated several times that positive-ion-driven neutral beams lose their efficiency for energies above 100 keV. In contrast negative-ion-driven sources maintain their efficiencies even at very high energies, on the order of 1 MeV. The goal of this subsection is to present a simple analysis that quantitatively demonstrates these points. The calculations are straightforward and are based on well-established cross sections for the various processes taking place. The specific mathematical task is to calculate the fraction of the incoming charged particle beam that is neutralized.

### Positive ions

The discussion begins with neutralization of positive ion beams. There are two important processes to consider. First, a high-energy positive ion can charge exchange with a background neutral particle in the neutralizer, producing the desired high-energy neutral particle. Second, this high-energy neutral can then be re-ionized also by colliding with a background neutral. This reaction is undesirable in that it reduces the number of neutralized particles. Consider now each of these reactions separately.

The charge exchange reaction is

$$D_b^+ + D_n \rightarrow D_b + D_n^+. \tag{15.19}$$

Here, $D_b$ is the desired high-energy neutral and $D_n$ is a low-temperature background diatomic molecule (i.e., $D_n$ is an abbreviation for $(D_2)_n$). The cross section for this reaction is the previously discussed quantity $\sigma_c$. Readers should not be suspicious of the fact that charge exchange converts high-energy charged particles to high-energy neutrals in the neutralizer, while doing the opposite, converting high-energy neutrals into high-energy charged particles in the plasma. An effective charge exchange process just requires that one species be neutral and the other charged, and that the high-energy species have a much lower density than the lower-temperature background species.

The second reaction involves re-ionization of the high-energy neutrals and can be written as

$$D_b + D_n \rightarrow D_b^+ + D_n + e^-. \tag{15.20}$$

The relevant cross section is the previously discussed quantity $\sigma_i$. Here too there may at first glance appear to be a contradiction. The cross section $\sigma_i$ was first used to describe the ionization of a high-energy neutral particle entering the plasma and colliding with a positively charged background ion. In the neutralizer, a high-energy neutral particle is ionized by colliding with an electrically neutral background molecule. There is no contradiction. Since both processes involve high-energy electrically neutral beam particles, it does not matter whether or not the background particles are charged; that is, the electric field of a charged background particle produces a negligible modification on the orbit of an electrically neutral beam particle.

Consider next the evolution of the high-energy charged particle beam as it propagates along the neutralizer. The equation determining the beam flux is

$$\frac{d\Gamma_b^+}{dx} = -n_n \sigma_c \Gamma_b^+ + n_n \sigma_i \Gamma_b, \tag{15.21}$$

where $n_n$ is the background density of the diatomic neutrals in the neutralizer. The first term on the right hand side represents the loss of the high-energy charged particle flux $\Gamma_b^+$. This loss is converted into the desired high-energy neutral beam flux $\Gamma_b$. The second term on the

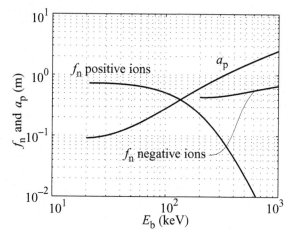

Figure 15.5 Neutralization fraction vs. beam energy for positive and negative ion beams. Also plotted is the penetration depth for $n_{20} = 1.5$. (Wesson, J. (2004). *Tokamaks*, third edition. Oxford: Clarendon Press).

right hand side represents a gain in $\Gamma_b^+$ due the re-ionization of the neutral beam flux $\Gamma_b$. The model is closed by noting that the conservation of beam particles requires that

$$\Gamma_b^+ + \Gamma_b = \Gamma_0 \equiv \Gamma_b^+(0). \tag{15.22}$$

Now, assume that neither neutral nor charged beam particles lose much energy as they progress along the neutralizer. Under this reasonably good assumption both $\sigma_c$ and $\sigma_i$ can be approximated as constants. One can then easily solve Eqs. (15.21) and (15.22) obtaining

$$\Gamma_b^+(x) = \Gamma_0 \left[ \frac{\lambda}{\lambda_i} + \left( 1 - \frac{\lambda}{\lambda_i} \right) e^{-x/\lambda} \right],$$

$$\Gamma_b(x) = \Gamma_0 \left( 1 - \frac{\lambda}{\lambda_i} \right) (1 - e^{-x/\lambda}), \tag{15.23}$$

where $\lambda = 1/n_n(\sigma_c + \sigma_i)$ and $\lambda_i = 1/n_n\sigma_i$.

The fraction of neutralization is now easily calculated by noting that the length of the neutralizer is typically much greater than $\lambda$. Consequently, at the output of the neutralizer, the exponential terms can be neglected. The neutral and charged particle fluxes reach equilibration with the fraction of neutralization being given by

$$f_n = \frac{\Gamma_b(x/\lambda \to \infty)}{\Gamma_0} = \frac{\sigma_c}{\sigma_c + \sigma_i}. \tag{15.24}$$

This is the desired result. The neutralization fraction is plotted as a function of energy in Fig. 15.5. Observe that $f_n$ decreases rapidly for energies above 100 keV because of the rapid decay of the charge exchange cross section. Also replotted is the perpendicular penetration depth vs. energy for $n_{20} = 1.5$. At this high density it is difficult for a positive-ion-driven

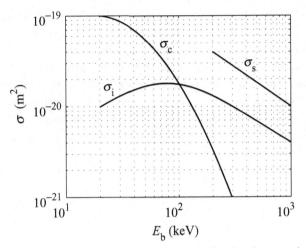

Figure 15.6 Stripping cross section for negative ions compared to the charge exchange and ionizing cross sections for positive ions. (Wesson, J. (2004). *Tokamaks*, third edition. Oxford: Clarendon Press).

neutral beam to penetrate more than 0.3 m when the conversion fraction, directly related to the efficiency, is required to be higher than 50%.

### Negative ions

A similar analysis applies to negative-ion-driven neutral beam sources. Again there are two main processes to consider. First, negative ions entering the neutralizer can have their excess electron "stripped" away by colliding with background neutrals. The process produces the desired high-energy neutral particles. The cross section $\sigma_s$ is different from that of positive ion charge exchange. The weak bonding of the extra electron leads to a much larger cross section for negative ions at high energy.

The second reaction involves ionization of high-energy neutrals by collisions with background neutrals and is identical to the positive ion case. The two reactions of interest can be written as

$$D_b^- + D_n \rightarrow D_b + D_n + e^-,$$
$$D_b + D_n \rightarrow D_b^+ + D_n + e^-. \tag{15.25}$$

The corresponding stripping cross section is illustrated in Fig. 15.6 in the interesting regime of high-beam energy. Observe that $\sigma_s$ is substantially larger than $\sigma_i$ at large energies.

The next step is to calculate the evolution of the high-energy negative ion and neutral beam fluxes as they progress along the neutralizer. The appropriate model is given by

$$d\Gamma_b^-/dx = -n_n \sigma_s \Gamma_b^-,$$
$$d\Gamma_b/dx = -n_n \sigma_i \Gamma_b + n_n \sigma_s \Gamma_b^-. \tag{15.26}$$

The right hand side of the first equation represents the loss of negative ions due to stripping collisions with the background neutrals. This loss appears as a gain in the flux of neutral particles corresponding to the second term on the right hand side of the second equation. The first term on the right hand side of the second equation represents the loss of high-energy neutrals due to ionization collisions with the background neutrals.

Note that unlike the case of positive ions, there is no replenishment of negative ions by ionization collisions, just the creation of a new species of energetic positive ions. Therefore, as one progresses along the neutralizer, both the high-energy negative ion and neutral fluxes simply decay to zero. There is no long-length limit of equilibration. The implication is that there is an optimum length for the neutralizer corresponding to the point where the fractional neutralization is a maximum.

This insight is confirmed by solving Eqs. (15.26). A short calculation yields

$$\Gamma_b^- = \Gamma_0 e^{-x/\lambda_s},$$

$$\Gamma_b = \Gamma_0 \frac{\lambda_i}{\lambda_s - \lambda_i} (e^{-x/\lambda_i} - e^{-x/\lambda_s}).$$

(15.27)

Here, $\Gamma_0 = \Gamma_b^-(0)$, $\lambda_s = 1/n_n \sigma_s$, and $\lambda_i = 1/n_n \sigma_i$. The neutralization fraction is a function of $x$ given by

$$f_n(x) = \frac{\Gamma_b(x)}{\Gamma_0} = \frac{\lambda_i}{\lambda_i - \lambda_s} (e^{-x/\lambda_i} - e^{-x/\lambda_s}).$$

(15.28)

This function has a maximum at

$$x_m = \frac{\lambda_i \lambda_s}{\lambda_i - \lambda_s} \ln \left( \frac{\lambda_i}{\lambda_s} \right) = \frac{1}{n_n(\sigma_s - \sigma_i)} \ln \left( \frac{\sigma_s}{\sigma_i} \right)$$

(15.29)

and corresponds to the optimum length of the neutralizer. The peak neutralization fraction thus occurs at the output of the neutralizer and has the value

$$f_n(x_m) = (\delta)^{\delta/(1-\delta)},$$

(15.30)

where $\delta = \sigma_i/\sigma_s$.

The function $f_n(x_m)$ is shown in Fig. 15.5. Observe that for a 1 MeV beam energy the neutralization efficiency remains high for negative ions, on the order of 60%. For positive ions the corresponding efficiency has plummeted well below 1%. This basic result is the reason why most researchers believe positive ion technology will not scale favorably into the reactor regime and why future research is focused on developing high-energy negative ion sources.

### 15.3.5 Summary

Neutral beam auxiliary heating systems have been very successful in heating many present day tokomaks to reactor level temperatures. These 100 keV beam systems have been driven by positive ion sources. However, typical reactor dimensions and plasma densities are significantly higher than in present day experiments. Reactor parameters require higher beam energies, on the order of 1 MeV. Unfortunately, positive ion efficiency decreases very rapidly above 100 keV. This has motivated a research and development program whose goal is to create high-energy negative ion sources that maintain high efficiency even at the 1 MeV level. The technological development of such sources is the most difficult problem facing neutral beam heating. Even so, there is a reasonable expectation that high-power, negative ion neutral beam systems will be available when needed for ITER.

## 15.4 Basic principles of RF heating and current drive

### *15.4.1 Overview*

Launching RF waves into a plasma is a successful technique for: (1) raising the temperature to the level required for ignition, or (2) driving a steady state non-inductive current. Furthermore, RF source technology appears to extrapolate favorably into the reactor regime. The idea is conceptually similar to using a microwave oven to cook food. In the case of fusion if the launching frequency is carefully chosen, the RF waves resonate with one of the natural frequencies of the plasma, leading to a large absorption of power that appears in the form of heat and additionally, under certain conditions, as non-inductive current.

The theory of RF heating involves some extremely interesting plasma physics and represents one of the great practical success stories in the fusion program. To understand RF heating and current drive one must learn about electromagnetic wave propagation in a plasma, a topic so broad and encompassing so many phenomena that several books have been devoted solely to this subject. The present discussion is narrower in scope and focuses primarily on using electromagnetic waves to heat a plasma and to drive a non-inductive current, the specific topics most relevant to a fusion reactor. As such, the discussion concentrates on describing: (1) the two most promising RF heating methods, ECH and ICH, and (2) the most efficient current drive method, LHCD. The goals are to answer the following questions.

- What types of RF sources and launching structures are needed to propagate waves into the plasma?
- At what frequency should the RF waves be launched in order to produce maximum absorption for heating or maximum current for current drive?
- How does one insure that the RF power for heating is absorbed in the center of the plasma where it is most needed?
- What fraction of the incident heating power is absorbed in the plasma and what is the absorption mechanism?
- How does one insure that the RF power for current drive is absorbed near the outer edge of the plasma where it is most needed to match the natural bootstrap profile?
- What is mechanism for driving such a current?
- How many watts of LHCD power are required to drive 1A of current?

The answers to these questions are essential in order to assess the desirability of RF heating and current drive with respect to economic viability and the recirculating power fraction in a fusion reactor.

The analysis of ECH, ICH, and LHCD for fusion applications is potentially quite complicated for three reasons. First, the plasma properties vary in space. Second, the plasma is situated in a complicated toroidal geometry. Third, the magnetic field causes highly anisotropic behavior in the plasma.

The plan of attack to overcome these difficulties involves several steps. The discussion begins with a brief description of the sources and launching structures used for ECH, ICH and LHCD in order to obtain an overview of the relevant technology. Next, some of the general principles of electromagnetic wave propagation in arbitrary media are reviewed. These principles are then applied to the "cold" plasma model which gives a surprisingly reliable description of the way waves propagate to the center of the plasma. However, the cold plasma model does not accurately describe the way in which waves are absorbed or drive current in a plasma. This requires an analysis of "collisionless" damping, which, although sounding like an oxymoron, is actually what happens in a plasma. Collisionless damping is, therefore, described in some detail by means of a simple model. Finally, basic wave theory is applied to ECH, ICH, and LHCD in fusion grade plasmas. One primary goal of practical importance for heating is to determine the fraction of the input power that is absorbed. For current drive the corresponding goal is to determine the amount of current driven per watt of LHCD power.

The main conclusions are that both ECH and ICH are capable of efficient plasma heating (i.e., a large fraction of absorption). For reactor parameters, the main difficulty with ECH is the lack of high-power, steady state sources at the requisite high frequencies. For ICH, there is no difficulty with sources. The main problem is the need for an antenna inside the vacuum chamber close to the plasma edge. The ICH issues involve electromagnetic shielding and the prevention of arcing near the antenna. Large research and development programs are underway to address both the ECH source problem and the ICH antenna problem and solutions should be available when needed for ITER.

For LHCD the sources and launching structures are readily available and the main technological problem is the need to place the launching structure very close to the plasma surface. The largest problem is that the conversion of power to current is not very efficient, implying the need for a substantial bootstrap current to achieve an economically viable steady state reactor.

### 15.4.2 *RF sources and launching structures*

Most of the analysis of RF heating involves learning how electromagnetic waves launched from the outer edge of the plasma propagate to a desired location in the plasma where they can be absorbed by collisionless damping: in the center for heating and near the edge for current-drive. Before proceeding along this path, however, it is instructive to begin with a brief discussion of the various RF sources currently available, and the methods by which energy is transmitted from these sources to the plasma edge.

The basic RF heating or current-drive configuration is illustrated in Fig. 15.7, which shows a source sending waves along a transmission path to a launching structure at the plasma edge.

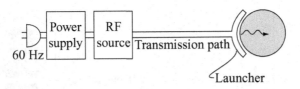

Figure 15.7 Schematic diagram of an RF heating or current-drive system.

The source is driven by a high-voltage power supply, which in turn is driven by standard 60 Hz power. The type of source utilized depends upon the operating frequency. Here it is useful to keep in mind that, as their names imply, ECH and ICH require RF waves at the electron and ion cyclotron frequencies respectively. For a magnetic field $B_0 = 5$ T, these frequencies are given by $f_{ce} = \omega_{ce}/2\pi = 140$ GHz and (for deuterium) $f_{ci} = \omega_{ci}/2\pi = 38$ MHz. Lower hybrid waves correspond to an intermediate frequency, typically on the order $f_{LH} = 3$ GHz. These frequencies must be matched with the typical frequencies generated by various types of RF sources as follows.

$$
\begin{aligned}
\text{high-power vacuum tubes}: &\quad f < 100 \text{ MHz} \\
\text{klystrons (microwaves)}: &\quad f \sim 1\text{–}10 \text{ GHz} \\
\text{gyrotrons (submillimeter waves)}: &\quad f \sim 10\text{–}300 \text{ GHz}
\end{aligned}
$$

Each type of source is illustrated in Fig. 15.8. Observe that ICH is driven by high-power vacuum tubes, while ECH requires gyrotrons. Klystrons drive intermediate frequencies which are required for LHCD. In terms of the technology, high-power, steady state vacuum tubes and klystrons are well developed and readily available. Gyrotrons are well on the way, but further development is still required. The goal is to develop a robust, reliable 140 GHz, 1 MW, steady state gyrotron.

Consider next the transmission path. There are three basic ways in which electromagnetic energy can be transmitted from the source to the plasma: standard electrical wire, a two-wire transmission line, and waveguides. The appropriate choice of a transmission method involves a comparison of the wavelength of the RF power with the characteristic dimensions of the transmitting circuit as shown in Fig. 15.9. The different regimes of applicability are as follows.

Normal household and industrial AC circuits require standard twisted-wire electrical cable to carry power. This is the appropriate choice when

$$\lambda \gg L_1 \gg L_t. \tag{15.31}$$

For example if $f = 60$ Hz, then $\lambda = c/f = 5000$ km, which clearly satisfies the above criterion. When analyzing AC circuits it is a good approximation to neglect the displacement current in Maxwell's equations.

Parallel wire or coaxial transmission lines are routinely used to guide TV and FM signals. This method is the appropriate choice when

$$L_1 \gtrsim \lambda \gg L_t. \tag{15.32}$$

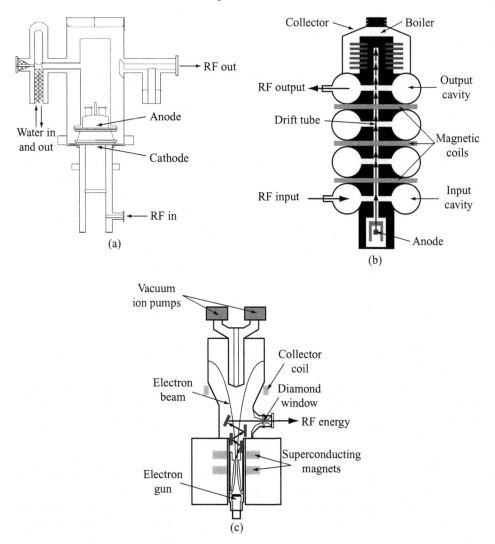

Figure 15.8 (a) High-power ICH vacuum tube (courtesy of V. L. Auslender); (b) LHCD klystron; (c) ECH gyrotron (courtesy of K. Fetch).

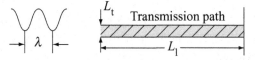

Figure 15.9 Comparison of three characteristic lengths: (1) the wavelength $\lambda$, (2) the transverse dimension of the transmission circuit, $L_t$, (3) the length of the transmission path $L_1$.

A typical ICH frequency $f = 40 \, MH_z$ corresponds to $\lambda = c/f = 7.5$ m, which satisfies the above inequality. The analysis of transmission line circuits shows that the displacement current must be maintained in Maxwell's equations. The RF power propagates as a pure transverse electromagnetic wave (TEM) whose field structure, as the name implies, has no component of electric or magnetic field parallel to the direction of propagation: $E_\parallel = B_\parallel = 0$. It is critical to use two wires to guide the electromagnetic waves.

Lastly, waveguides are used to transmit RF power when the wavelength lies in the range

$$L_l \gg L_t \sim \lambda. \tag{15.33}$$

For ECH a typical frequency in a reactor is $f = 140$ GHz and corresponds to $\lambda = c/f = 2.1$ mm. For lower hybrid waves $f = 3 \, GH_z$ and $\lambda = c/f = 10$ cm. Thus, ECH power is transmitted by means of a waveguide with a relatively small cross sectional dimension. Usually an "oversized waveguide" is used to minimize ohmic dissipation in the waveguide walls. Lower hybrid waves are transmitted using standard size waveguides.

For both of these cases the displacement current must be maintained in Maxwell's equations and the power propagates as either a transverse electric (TE) or transverse magnetic (TM) wave; that is, either $B_\parallel$ or $E_\parallel$ must be non-zero. At these high frequencies transmission of power requires only a hollow metal tube (i.e., a waveguide) which is usually rectangular or circular in cross section. No central conductor is needed.

In practice, transmission lines and waveguides have been studied, developed, and widely used for many, many years. The conclusion is that once an appropriate RF source is available, its power can be readily transmitted to the launcher at the plasma edge.

The last topic of interest involves the launcher itself. The launcher is a structure that acts as the interface between the transmission circuit and the edge of the plasma. Its form also depends critically on the wavelength being used. For ICH, waves are launched into the plasma by means of an antenna placed inside the vacuum chamber. For LHCD, the corresponding structure is a waveguide array. For ECH, an RF mirroring system is used. Examples of each are illustrated in Fig. 15.10. In general, for ICH, ECH, and LHCD it is highly desirable for reasons of geometric accessibility to launch waves from the outside of the plasma as illustrated in Fig. 15.11. There is just too much hardware in the center of most configurations to allow unimpeded access to the plasma edge. In addition to this general requirement, there are several specific issues facing each type of launcher, which are described below.

For ICH, the biggest problem is the proximity of the metal structure of the antenna to the plasma. High voltages are required to launch large amounts of power and these voltages can cause arcing and plasma breakdown near the antenna, both undesirable effects. One cannot just move the antenna further away from the plasma, because, as is shown shortly, close proximity is required or else the ICH waves do not couple strongly to the plasma. The design of a well-shielded antenna capable of preventing high-voltage arcing is an important technological problem for ICH. It is also worth noting that since the DC magnetic field in the plasma is tangential to the vacuum chamber, the geometric structure of the antenna determines the value of $k_\parallel$ of the electromagnetic waves as they enter the plasma. In other

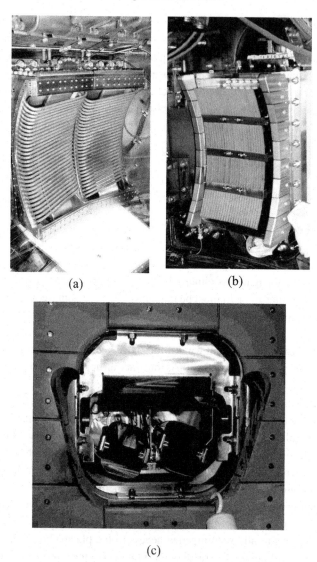

Figure 15.10 Launching structures for: (a) an ICH antenna (courtesy of E. Marmar), (b) an LHCD waveguide array (courtesy of E. Marmar), (c) an ECH mirroring system (Lohr, J. *et al.*, (2005). *Fusion Science and Technology*, **48**, 1226).

words, as the waves start to propagate into the plasma the frequency $\omega$ and parallel wave number $k_{\parallel}$ are known from the RF source and antenna structure respectively. This fact will be very important shortly when trying to understand how waves propagate to the center of the plasma.

For LHCD, the issues are somewhat different. A generic problem is to spread the RF power over a large enough area so that high-voltage breakdown problems do not occur, a

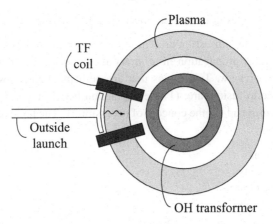

Figure 15.11 Schematic diagram of the top view of a tokamak showing an outside, low-field launch. Note the hardware congestion on the inside of the torus because of the converging geometry and the presence of the OH transformer.

situation somewhat driven by the small transverse dimensions of a single waveguide and the need to place the launching structure near the plasma edge for good coupling. This requires a large number of waveguides in the launching array. An equally difficult problem is maintaining an insulated vacuum interface between the plasma and the waveguides. The solution involves inserting sealed windows of a carefully chosen material at the end of each waveguide in the array to isolate the launcher from the plasma. The difficulty is that large amounts of RF power must pass through this window. Very little of it should be reflected back or else the efficiency would rapidly decrease. Very little of it should be absorbed in the window or else damage will occur because of thermal stresses. The development of high-power windows is an important technological problem for LHCD. Lastly, again as for ICH, the geometric arrangement and relative phases of the electromagnetic waves at the edge of the launching array set the value of $k_\parallel$ for the waves as they enter the plasma. Thus, here too both $\omega$ and $k_\parallel$ are known at the plasma edge.

ECH has a similar set of issues as LHCD, although there is good coupling even when the launching structure is moved away from the plasma. While the coupling problem is allevi-ated, other problems are exacerbated because of the very small size of the ECH waveguides.

ICH, ECH, and LHCD have been used to heat and drive current in present day plasma experiments. The technological problems described above have been satisfactorily resolved, although not without considerable effort, in these pulsed experiments. ICH is more prevalent than ECH because of the ready availability of high-power, reliable sources at reasonable cost. Even so ECH is being used in a number of existing experiments around the world. LHCD is gaining attention, although it has not been investigated as thoroughly as ICH. The reason is that current drive has not until recently been viewed as being as critical a problem as heating.

In the discussion that follows it is assumed that RF power at the desired frequency is available at the necessary power level. The task then is to understand how this power propagates to the center of the plasma and is absorbed.

### 15.4.3 Principles of electromagnetic wave propagation in a plasma

To start the discussion this subsection presents a brief description of the general principles of electromagnetic wave propagation in arbitrary media. For many readers much of the information will be a review. Two topics that are perhaps slightly less familiar and of particular importance to fusion, are: (1) the distinction between a wave resonance and a wave-particle resonance and (2) the concept of accessibility. The topics covered include:

(a) the dielectric tensor,
(b) the dispersion relation,
(c) phase and group velocity,
(d) cutoffs and wave resonances,
(e) wave-particle resonances,
(f) polarization,
(g) reflection, transmission, absorption, and mode conversion,
(h) accessibility.

### *The dielectric tensor*

Consider the propagation of electromagnetic waves in an arbitrary medium, for example a plasma. In general, the behavior of a plasma is far too complicated to be modeled by a simple $\varepsilon$ or $\mu$ as for dielectric or magnetic materials. Instead the plasma particles are assumed to be immersed in vacuum, characterized by $\varepsilon_0$ and $\mu_0$. The influence of the charged particles on the wave propagation characteristics is determined by explicitly calculating the current density $\mathbf{J}$ and charge density $\sigma$ by means of a suitable plasma model. The consequences of plasma behavior are then conveniently summarized by means of the resulting "dielectric tensor".

The starting point for the definition of the dielectric tensor is Maxwell's equations:

$$
\begin{aligned}
\nabla \times \mathbf{E} &= -\partial \mathbf{B}/\partial t, \\
\nabla \times \mathbf{B} &= \mu_0 \mathbf{J} + \frac{1}{c^2} \frac{\partial \mathbf{E}}{\partial t}, \\
\nabla \cdot \mathbf{E} &= \sigma/\varepsilon_0, \\
\nabla \cdot \mathbf{B} &= 0.
\end{aligned}
\tag{15.34}
$$

Next, note that most treatments of plasma waves assume small-amplitude perturbations. The implication is that wave propagation can be treated as a linearized perturbation about an equilibrium state. Furthermore, for the sake of simplicity, the equilibrium state is assumed to be infinite and homogenous, or at most slowly varying with respect to the wavelengths of interest. These rather gross approximations are surprisingly accurate and are discussed in more detail in the next section.

Using the small-amplitude assumption allows one to expand all quantities as

$$
Q(\mathbf{r}, t) = Q_0 + \tilde{Q}_1(\mathbf{r}, t)
\tag{15.35}
$$

with $Q_1 \ll Q_0$. The homogeneous assumption implies that Fourier analysis can be used in space and time:

$$
\tilde{Q}_1(\mathbf{r}, t) = Q_1 \exp\left(-i\omega t + i\mathbf{k} \cdot \mathbf{r}\right),
\tag{15.36}
$$

where $Q_1$ is now simply an amplitude. Fourier analysis, combined with linearization is useful because the time and space derivatives appearing in Maxwell's equations become

$$\partial/\partial t \rightarrow -i\omega,$$
$$\nabla \rightarrow i\mathbf{k}, \tag{15.37}$$

and all the exponential factors exactly cancel. This has the desirable effect of transforming the partial differential equations into a set of coupled algebraic equations with the amplitudes of the various field quantities being the unknowns.

The equations are obtained by linearizing Maxwell's equations about the homogeneous equilibrium, leading to

$$i\mathbf{k} \times \mathbf{E}_1 = i\omega \mathbf{B}_1,$$
$$i\mathbf{k} \times \mathbf{B}_1 = \mu_0 \mathbf{J}_1 - \frac{i\omega}{c^2}\mathbf{E}_1,$$
$$i\mathbf{k} \cdot \mathbf{E}_1 = \frac{\sigma_1}{\varepsilon_0}, \tag{15.38}$$
$$i\mathbf{k} \cdot \mathbf{B}_1 = 0.$$

The equations are simplified by eliminating $\mathbf{B}_1 = \mathbf{k} \times \mathbf{E}_1/\omega$ resulting in a single vector equation for $\mathbf{E}_1$:

$$\mathbf{k} \times \mathbf{k} \times \mathbf{E}_1 = -\frac{\omega^2}{c^2}\mathbf{E}_1 - i\omega\mathbf{J}_1. \tag{15.39}$$

The remaining two Maxwell equations can be shown to be redundant.

At this point one must assume that there is a fluid or kinetic model of the plasma that relates $\mathbf{J}_1$ to $\mathbf{E}_1$. This relation is not in general isotropic, implying a tensor relationship:

$$\mathbf{J}_1 = \overset{\leftrightarrow}{\sigma} \cdot \mathbf{E}_1, \tag{15.40}$$

where $\overset{\leftrightarrow}{\sigma}$ is defined as the conductivity tensor. Assume for now that $\overset{\leftrightarrow}{\sigma}$ is known. Equation (15.39) can thus be rewritten as

$$\mathbf{k} \times \mathbf{k} \times \mathbf{E}_1 = -\frac{\omega^2}{c^2}\mathbf{E}_1 - i\omega\mu_0\overset{\leftrightarrow}{\sigma} \cdot \mathbf{E}_1. \tag{15.41}$$

A more standard way to write this equation is to introduce the index of refraction

$$\mathbf{n} = \frac{c}{\omega}\mathbf{k} \tag{15.42}$$

and the dielectric tensor

$$\overset{\leftrightarrow}{\mathbf{K}} = \overset{\leftrightarrow}{\mathbf{I}} + \frac{i}{\varepsilon_0\omega}\overset{\leftrightarrow}{\sigma}. \tag{15.43}$$

Equation (15.41) reduces to

$$\mathbf{n} \times \mathbf{n} \times \mathbf{E}_1 + \overset{\leftrightarrow}{\mathbf{K}} \cdot \mathbf{E}_1 = 0. \tag{15.44}$$

This is the desired relation with the dielectric tensor $\overset{\leftrightarrow}{\mathbf{K}}$ defined by Eq. (15.43). It is worth keeping in mind that the role of $\overset{\leftrightarrow}{\mathbf{K}}$ is to describe the effects of the plasma on wave propagation and as such it is not explicitly known until a plasma model is introduced to determine $\overset{\leftrightarrow}{\sigma}$.

## The dispersion relation

One expects the solution to Maxwell's equations to yield a relationship between $\omega$ and $\mathbf{k}$, the details of which depend upon the properties of the plasma. For example, in the special case of no plasma (i.e., a vacuum) this relationship is given by $\omega = kc$. For the general case the relationship is known as the "dispersion relation" and is often written as $\omega = \omega(\mathbf{k})$ or more generically as $D(\omega, \mathbf{k}) = 0$ or equivalently $\hat{D}(\omega, \mathbf{n}) = 0$.

How does the dispersion relation arise? To answer this question note that Eq. (15.44) has the form of three coupled, linear, homogeneous algebraic equations for the three unknown amplitudes $E_{1x}$, $E_{1y}$, $E_{1z}$. In order for non-trivial solutions to exist the determinant of these equations must vanish. Setting the determinant to zero is equivalent to finding the eigenvalues of the system. The resulting eigenvalue relation is the dispersion relation $D(\omega, \mathbf{k}) = 0$.

The dispersion relation corresponding to Eq. (15.44) is easily found by writing down the three separate components of the vector equations. A short calculation then leads to the following determinant:

$$\begin{vmatrix} n_y^2 + n_z^2 - K_{xx} & -n_x n_y - K_{xy} & -n_x n_z - K_{xz} \\ -n_x n_y - K_{yx} & n_x^2 + n_z^2 - K_{yy} & -n_y n_z - K_{yz} \\ -n_x n_z - K_{zx} & -n_y n_z - K_{zy} & n_x^2 + n_y^2 - K_{zz} \end{vmatrix} = 0. \qquad (15.45)$$

This rather ambitious looking determinant will soon become noticeably simpler when the cold fluid model of a plasma is introduced. For present purposes, however, assume the $K_{ij}$ are known and the determinant has been set to zero to evaluate $D(\omega, \mathbf{k}) = 0$. A key point to recognize is that when solving the dispersion relation there will, in general, be multiple roots. Specifically, a given $\omega$, $n_y$, $n_z$ may lead to multiple solutions for $n_x$. Each root corresponds to an independent wave with different propagation characteristics. These multiple roots are important when trying to understand RF heating and current drive in a plasma.

## Phase and group velocity

Two important propagation properties of an electromagnetic wave are its phase and group velocities. These are defined as follows. The phase velocity $\mathbf{V}_p$ is the velocity at which an observer must travel in order for the phase of the wave to appear constant. In other words, since the phase of the wave is defined as $\phi = \omega t - \mathbf{k} \cdot \mathbf{r}$, then $\mathbf{V}_p$ is that velocity for which $d\phi/dt = 0$. The phase velocity is easily obtained from the dispersion relation $\omega = \omega(\mathbf{k})$ by noting that

$$d\phi/dt = \omega - \mathbf{k} \cdot d\mathbf{r}/dt = \omega - \mathbf{k} \cdot \mathbf{V}_p = 0. \qquad (15.46)$$

The phase velocity is therefore defined as

$$\mathbf{V}_p = \frac{\omega}{k}\mathbf{e}_k = \frac{\omega}{k}\left(\frac{k_x}{k}\mathbf{e}_x + \frac{k_y}{k}\mathbf{e}_y + \frac{k_z}{k}\mathbf{e}_z\right). \qquad (15.47)$$

Observe that the magnitude of $\mathbf{V}_p$ is $\omega/k$ and its direction points along $\mathbf{k}$. Also, the phase velocity can be greater or less than the speed of light. There is no contradiction with respect to relativity since the phase velocity of a single, monochromatic wave does not represent the propagation of a physical quantity such as information or energy.

The phase velocity is useful in understanding certain general properties of plasma waves. It is particularly important in the study of wave-particle resonances and the associated phenomenon of collisionless damping.

The next topic of interest is the group velocity. This is the velocity at which information or energy propagates and therefore must always be less than or equal to the speed of light. What exactly is meant by the "velocity at which information or energy propagates"? The question can be answered by considering the 1-D propagation of an amplitude modulated (AM) radio wave. Here there is a high-frequency carrier wave characterized by a frequency $\omega_0$ and wave number $k_0$. The electric field of the wave is given by $E = A \cos(\omega_0 t - k_0 x)$. Information is propagated by modulating the amplitude $A$ of the carrier wave. For simplicity $A$ is assumed to contain only a single harmonic. To maintain coherence of the information, the frequency $\omega_1$ and wave number $k_1$ of the modulating harmonic must be much lower than that of the carrier wave: $\omega_1 \ll \omega_0$ and $k_1 \ll k_0$. The electric field of the modulated wave can thus be written as

$$
\begin{aligned}
E &= A_0 \cos(\omega_1 t - k_1 x) \cos(\omega_0 t - k_0 x) \\
&= \frac{A_0}{2} \{\cos[(\omega_0 + \omega_1)t - (k_0 + k_1)x] + \cos[(\omega_0 - \omega_1)t - (k_0 - k_1)x]\}. \quad (15.48)
\end{aligned}
$$

The information is carried in the amplitude of the harmonic modulation and by definition propagates with a velocity given by $\omega_1/k_1$, known as the group velocity $V_g$. The value of $V_g$ can be expressed in terms of the dispersion relation by: (1) noting that the modulated wave can be written as the sum of two pure waves, one up-shifted and the other down-shifted, each of whose properties must satisfy the dispersion relation; and (2) exploiting the smallness of $\omega_1$ and $k_1$. If the dispersion relation is written as $\omega = \omega(k)$ it follows that

$$
\begin{aligned}
\omega_0 &= \omega(k_0), \\
\omega_0 \pm \omega_1 &= \omega(k_0 \pm k_1) \approx \omega(k_0) \pm \frac{d\omega}{dk_0} k_1.
\end{aligned} \quad (15.49)
$$

The group velocity is then given by

$$
V_g = \frac{\omega_1}{k_1} = \frac{d\omega(k_0)}{dk_0}. \quad (15.50)
$$

This relation is easily generalized to the 3-D case. One obtains

$$
\mathbf{V}_g = \nabla_k \omega = \frac{\partial \omega}{\partial k_x} \mathbf{e}_x + \frac{\partial \omega}{\partial k_y} \mathbf{e}_y + \frac{\partial \omega}{\partial k_z} \mathbf{e}_z, \quad (15.51)
$$

where for convenience the "0" subscript has been dropped from $k$.

Equation (15.51) is the desired definition of group velocity, a quantity that is important in understanding the flow of information and energy in electromagnetic waves propagating in a plasma. Physically, the group velocity must always satisfy $V_g < c$. Also, while $\mathbf{V}_p$ and $\mathbf{V}_g$ often point in the same direction, this is not a requirement. When two vector components

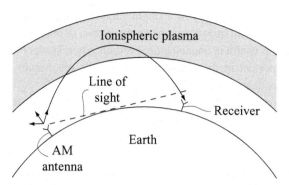

Figure 15.12 Schematic diagram of AM reflection from an ionispheric cutoff. Note that the reflection allows reception beyond the line of sight of the transmitting source.

point in opposite directions the wave is called a "backward wave," a situation that occurs for LHCD.

## Cutoffs and wave resonances

Cutoffs and wave resonances refer to two limiting values of $\mathbf{k}$ that dramatically alter the propagation characteristics of a wave. To understand the phenomena imagine an electromagnetic wave launched into a plasma whose properties (e.g. $n$, $B_0$, $T$) vary in space very slowly compared to the wavelength $\lambda = 2\pi/k$ of the wave. The propagation characteristics are governed by the dispersion relation $\omega = \omega(\mathbf{k}, \mathbf{r})$, the slow $\mathbf{r}$ dependence arising from the variation in plasma properties. The wave enters the outer edge of the plasma with specified values of $\omega$ and $k_z \equiv k_\parallel$ determined by the source and launcher. Both $\omega$ and $k_\parallel$ are real quantities for a propagating wave. For a 2-D geometry, the dispersion relation is then used to determine $k_x \equiv k_\perp$. The value of $k_\perp$ varies slowly in space in accordance with the slow variation of the plasma parameters. The critical point to keep in mind is that as long as $k_\perp$ remains real the wave will continue to propagate further into the plasma.

A cutoff occurs when the plasma parameters gradually reach values that make $k_\perp^2 = 0$ or equivalently $n_\perp^2 = 0$. At this point the phase velocity becomes infinite (although the group velocity still remains finite). Past this point, further changes in the plasma parameters cause $k_\perp^2$ to become negative. The roots are purely imaginary, and with no other sources present the wave exponentially decays (i.e., it is evanescent) away from the cutoff point. A classic example of a cutoff is the ionispheric reflection of AM radio waves as shown in Fig. 15.12. In summary, a cutoff occurs when $k_\perp^2$ changes sign passing through zero, preventing further penetration of the wave.

A wave resonance corresponds to the opposite limit. In this case $k_\perp^2$ changes sign passing through infinity. Specifically, a wave resonance occurs when $k_\perp^2 = \infty$ or equivalently $n_\perp^2 = \infty$. In this limit the phase and group velocity approach zero. The energy flow slows down and, hypothetically, the introduction of a small amount of dissipation would lead to a large absorption of energy. This is the conclusion that one would reach based on the analysis of the cold plasma dispersion relation which is described shortly. However, a more realistic plasma model shows that the cold plasma wave resonances vanish when kinetic effects are included. Instead of absorption, there is "mode conversion," a process by which the incoming plasma wave is transformed into a different type of plasma wave. This too is discussed shortly. The key point is that, in spite of the word "resonance" in the description, there is no direct absorption of RF power at a wave resonance.

Cutoffs and resonances play an important role in determining the best strategies to propagate RF waves to the center of the plasma for heating purposes.

### Wave-particle resonances

A wave-particle resonance is the basic mechanism by which the plasma absorbs energy from RF waves. It is this absorption that produces RF heating and current drive. The absorption involves a special group of particles whose particular velocities cause them to be in resonance with the wave. If the parameters of the plasma and the wave are set properly, a substantial number of particles can be in resonance leading to a large absorption of power.

The condition for a wave-particle resonance, stated here without proof, can be written as

$$\omega = k_\parallel v_\parallel + l\omega_c \qquad l = 0, \ 1, \ 2, \ \ldots \tag{15.52}$$

Resonance occurs when the parallel Doppler shifted frequency is equal to an exact harmonic of the cyclotron frequency. The basic physics underlying Eq. (15.52) is explained in the subsection on "collisionless damping."

The $l = 0$ resonance is known as "Landau damping" and is shown to be important in connection with LHCD. The $l = 1$ resonance corresponds to "heating at the fundamental frequency." It is important for ECH and sometimes for ICH as well. As its name implies the $l = 2$ resonance produces "second harmonic heating." It too is important for both ECH and ICH.

Wave-particle resonance is the key mechanism to keep in mind when trying to decide the best way to heat or drive current in a plasma by means of RF waves.

### Polarization

After one solves the dispersion relation it is straightforward to substitute back into the matrix equation for $\mathbf{E}_1$ to determine the relationship between any two components in terms of the third: for example $E_{1x}$ and $E_{1y}$ in terms of $E_{1z}$. The polarization describes the various possible relationships that can arise and helps in understanding which waves may be particularly effective in producing a wave-particle resonance.

As an example note that in optics the polarization of the wave is defined as the ratio of the electric field components perpendicular to $\mathbf{k}$. In plasma physics there are several different definitions that are widely used. Two definitions that are particularly useful for propagation perpendicular to the background magnetic field $\mathbf{B}_0 = B_0 \mathbf{e}_z$ are as follows. In the first, polarization is defined in terms of the perpendicular components of electric field

$E_{1x}$ and $E_{1y}$. The polarization $P$ of the wave is defined as

$$P = iE_{1x}/E_{1y}. \tag{15.53}$$

Its value depends upon the plasma model and specific wave under consideration. Special cases include:

$$
\begin{array}{ll}
P = 0 & \text{linearly polarized wave,} \\
P = \infty & \text{linearly polarized wave,} \\
P = \pm 1 & \text{circularly polarized wave.}
\end{array}
$$

A linearly polarized wave has a single Cartesian field component (i.e., $E_{1x}$) pointing in a constant direction. A circularly polarized wave has two field components 90° out of phase with each other. This creates a circularly rotating electric field as a function of time.

The second definition of polarization that is widely used in fusion physics distinguishes whether or not the wave of interest has a component of electric field parallel to the background magnetic field. Depending upon the answer, the wave is then either an "O" mode, denoting *ordinary* wave, or an "X" mode, denoting an *extraordinary* wave. The specific definitions of O and X are:

$$
\begin{array}{ll}
\text{O mode} & \text{has } E_\parallel \neq 0, \\
\text{X mode} & \text{has } E_\parallel = 0.
\end{array}
$$

A knowledge of the polarization is very important in understanding ECH, ICH, LHCD and is discussed in more detail as the analysis progresses.

### *Reflection, transmission, absorption, and mode conversion*

As a wave propagates into a plasma several different phenomena can occur with respect to the deposition of its wave energy: reflection, transmission, absorption, and mode conversion. These are illustrated schematically in Fig. 15.13 for the simple case of a plasma slab.

Consider, for example, the application of heating. Note that an incident wave is in general partially reflected and partially transmitted. Clearly for efficient heating the reflected wave should have a small amplitude. The transmitted wave may continue to propagate unattenuated deeper into the plasma. If it reaches a cutoff before reaching the center it will then be totally reflected, clearly an undesirable outcome. On the other hand, the wave may continue to propagate into the plasma until it reaches the center. If the plasma and wave parameters are properly chosen, there can be a strong wave-particle resonance at the center leading to a strong absorption. This is the desired result. A third possibility is that the wave first encounters a wave resonance which mode converts it into a different type of plasma wave. The new wave then continues to propagate into the plasma where it may reach either a cutoff (undesirable) or the center where there can be a wave-particle resonance (desirable). As one can see there are many possibilities.

At this point it is worth briefly discussing in slightly more detail the mechanism of mode conversion, since this may be a somewhat unfamiliar topic but it is important in certain heating schemes. Mode conversion involves two different plasma waves. For instance, both have

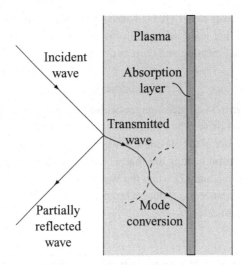

Figure 15.13 Schematic diagram of reflection, transmission, mode conversion, and absorption.

the same $\omega$ and $k_\parallel$ but different values of $k_\perp$. They are different roots of the same dispersion relation $k_\perp = k_\perp(\omega, k_\parallel)$. Mode conversion occurs when the two values $k_\perp$ just happen to coalesce at a certain point in the profile as the values $k_\perp$ evolve during their propagation into the plasma. Usually this occurs when one wave is at a cold plasma wave resonance. The other wave often arises as a new mode, introduced by a more sophisticated kinetic plasma model. At the coalescence point both waves lose their individual identity and there is strong mode coupling. The end result is that some of the energy from the externally excited first wave is transferred to the second wave which then propagates further into the plasma according to its own dispersion characteristics. This is the process of mode conversion.

The combination of reflection, transmission, absorption, and mode conversion plays an important role in determining the effectiveness of various RF heating methods.

### Accessibility

The issue of accessibility is crucial for successful RF heating and current drive. Accessibility takes into account all of the phenomena just described and asks whether or not a wave launched from the outside of the plasma can reach a desired location in the plasma where it can then be absorbed by a wave-particle resonance. For an outside launch the wave sees an increasing density as it propagates towards the center. It also sees a slightly increasing magnetic field because of the $1/R$ dependence of $B_\phi$. It is non-trivial for the wave to reach the center without encountering a cutoff region. Also, if a wave resonance is reached, one must know the coupling coefficient and the propagation characteristics of the secondary wave. Ideally, for good heating the wave should have $n_\perp^2(x) > 0$ from the edge to the desired absorption location in the plasma.

For both ECH and ICH clever methods have been devised for the wave energy to gain access to the center of the plasma. The situation is more complex, but also successfully resolved, with respect to off-axis LHCD. These methods are discussed in Sections 15.7–15.9. The conclusion is that good accessibility is a major requirement with respect to the ultimate desirability of RF heating and current drive.

## Summary

Some of the basic properties of electromagnetic theory have been reviewed in the context of RF heating and current drive. The conclusions are that waves launched from the outside propagate towards the center of the plasma according to the dispersion relation $D(\omega, \mathbf{k}) = 0$. The critical issues for central heating are that: (1) the wave should have good accessibility to the center of the plasma, and (2) when the wave arrives at the center there should be a strong wave-particle resonance to produce a large absorption of wave energy. A similar set of issues applies to current drive.

### 15.4.4 *Analysis of electromagnetic wave propagation in a plasma*

The next major task is to apply the basic electromagnetic wave principles just described to learn how to both heat and drive current in a plasma efficiently. One needs to know the frequencies and launcher wave numbers that allow waves to propagate undamped to a desired location in the plasma (i.e., the center for heating and off-axis for current drive), where conditions should be such that a strong wave-particle absorption takes place. There are several steps in the analysis which are outlined below.

The starting point is to recognize that a truly realistic model must include kinetic effects, toroidal geometry, and diffuse profiles, a relatively ambitious challenge. Over the years models have been developed to meet these requirements. They have been highly successful but as might be expected involve large numerical codes. For an introduction to RF heating and current drive it is more appropriate to focus instead on a simple model that can be solved analytically, thereby providing greater physical insight. The simple model described below semi-quantitatively reproduces the essential features of the more complete model.

To understand the simple model note that the crucial step in the analysis is the calculation of the dispersion relation, which for the general case has the form

$$D(\omega, \mathbf{k}) = D_r(\omega, \mathbf{k}) + iD_i(\omega, \mathbf{k}). \tag{15.54}$$

The functions $D_r$ and $D_i$ are real when $\omega$ and $\mathbf{k}$ are real. The first term $D_r$ basically describes the wave propagation characteristics (i.e., propagation, cutoffs, wave resonances). It thus determines the accessibility of a given type of wave to a desired location in the plasma. The second term, which usually satisfies $D_i \ll D_r$, represents the dissipation of the wave due to wave-particle resonances.

In the simplified analysis $D_r(\omega, \mathbf{k})$ is evaluated from the *two-fluid cold plasma model* in which all temperature effects are neglected. Furthermore, to simplify the geometry the plasma is treated as an infinite, homogeneous slab with $r \to x$. The resulting $D_r(\omega, \mathbf{k})$ is a function of the equilibrium density $n_0$ and magnetic field $B_0$, which, after the fact, are allowed to vary slowly in space: $n_0 = n_0(x)$ and $B_0 = B_0(x)$. The dispersion relation is

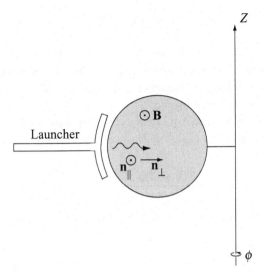

Figure 15.14 Schematic diagram of a low-field, outside launch in a toroidal geometry. Note the direction of $\mathbf{n}_\parallel$ and $\mathbf{n}_\perp$.

therefore a local function of spatial coordinate: $D_r(\omega, \mathbf{k}, x)$. These rather gross simplifications are difficult to justify a priori. The main justification comes from the hindsight obtained by solving the more general model numerically and then comparing the results with the predictions of the cold plasma model. The comparison shows that the cold plasma model is surprisingly reliable with respect to the question of accessibility.

Consider next the evaluation of $D_i(\omega, \mathbf{k}, x)$. This, in general, requires a full kinetic treatment in order to accurately evaluate the wave-particle resonant absorption. Fortunately, over the years, researchers have developed relatively simple models, based on single-particle motion, that reproduce the wave-particle damping results of kinetic theory. These models do not determine $D_i(\omega, \mathbf{k}, x)$ but instead directly determine the damping rate in time $\omega_i$, assuming that $\mathbf{k}$ is purely real. The model used to evaluate $\omega_i = \omega_i[\omega, k_\parallel, k_\perp(x)]$ is known as the *collisionless damping model*.

Thus, armed with a knowledge of $D_r(\omega, \mathbf{k}, x)$ and $\omega_i$ one is now in a position to investigate ECH, ICH, and LHCD. In each case the first step is to examine accessibility as determined by setting $D_r(\omega, \mathbf{k}, x) = 0$ in the cold plasma model. Towards this goal, consider the simple geometry illustrated in Fig. 15.14. The source sets the frequency $\omega$. The launcher sets the value[1] of $k_\parallel$. The wave propagates towards the center of the plasma with an axial (i.e., radial) wave number $k_\perp$. In the simple model the total wave number is thus a vector: $\mathbf{k} = k_\perp \mathbf{e}_x + k_\parallel \mathbf{e}_z$. Accessibility is investigated by inverting $D_r(\omega, \mathbf{k}, x) = 0$ to obtain

$$k_\perp = k_\perp(\omega, k_\parallel, x), \qquad (15.55)$$

where the $x$ dependence arises from the equilibrium variation in $n_0(x)$ and $B_0(x)$. Good accessibility for heating requires that $k_\perp$ be real from the outer edge of the plasma to the

---

[1] The launcher also sets the value of $k_y$, which is usually chosen as $k_y = 0$, so that the wave propagates directly towards the center of the plasma and not at an angle that would miss the center.

center, or at worse pass through only a narrow cutoff region where the wave can tunnel through.

Assuming good accessibility one must next calculate the small damping decrement in space $k_{\perp i}(x)$. This quantity is directly related to $\omega_i$, which is determined by the collisionless damping model. Once $k_{\perp i}(x)$ is known a straightforward power balance calculation determines the fraction of power absorbed.

The actual relationship between $k_{\perp i}(x)$ and $\omega_i$ is determined as follows. For a wave which decays in time characterized by $\omega_i \ll \omega_r$, $D_i \ll D_r$ and real $k_\perp$, $k_\parallel$, the Taylor expansion of the dispersion relation yields

$$D_r(\omega_r, k_{\perp r}, k_\parallel) + i\left[\frac{\partial D_r(\omega_r, k_{\perp r}, k_\parallel)}{\partial \omega_r}\omega_i + D_i(\omega_r, k_{\perp r}, k_\parallel)\right] \approx 0. \tag{15.56}$$

On the other hand, for a wave which damps in space with $k_{\perp i} \ll k_{\perp r}$ and real $\omega$, $k_\parallel$, one finds

$$D_r(\omega_r, k_{\perp r}, k_\parallel) + i\left[\frac{\partial D_r(\omega_r, k_{\perp r}, k_\parallel)}{\partial k_{\perp r}}n_{\perp i} + D_i(\omega_r, k_{\perp r}, k_\parallel)\right] \approx 0. \tag{15.57}$$

The relation between $k_{\perp i}$ and $\omega_i$ is obtained by subtracting Eqs. (15.56) and (15.57):

$$k_{\perp i} = \left[\frac{\partial D_r/\partial \omega_r}{\partial D_r/\partial k_{\perp r}}\right]\omega_i = -\frac{\omega_i}{(\partial \omega/\partial k_\perp)} = -\frac{\omega_i}{V_{g\perp}}. \tag{15.58}$$

The spatial and time damping decrements are related by the perpendicular group velocity. Note that to determine $k_{\perp i}$ requires a knowledge of the cold plasma form of $D_r$ and the value of $\omega_i$. One does not need an explicit expression for $D_i$.

A final point in the analysis involves the relationship between the Poynting vector and the stored energy in the wave. This relationship is not, in principle, required for the applications of interest. However, the relationship can greatly simplify the algebraic details involved in the calculation of $k_{\perp i}$, particularly for LHCD. The relationship is derived in Appendix D and is

$$P_\perp = \frac{\omega_i}{k_{\perp i}}U = V_{g\perp}U, \tag{15.59}$$

where

$$P_\perp = \mathbf{e}_x \cdot \mathbf{P} = \frac{1}{2\mu_0}\text{Re}(\mathbf{e}_x \cdot \mathbf{E}_1 \times \mathbf{B}_1^*) \tag{15.60}$$

is the perpendicular component of the Poynting vector and

$$U = \frac{1}{2}\left\{\frac{1}{2\mu_0}|\mathbf{B}_1^2| + \frac{\varepsilon_0}{2}\mathbf{E}_1^* \cdot \frac{\partial}{\partial \omega_r}[\omega_r \overset{\leftrightarrow}{\mathbf{K}}(\omega_r)] \cdot \mathbf{E}_1\right\} \tag{15.61}$$

is the total stored energy (i.e., wave energy plus plasma kinetic energy) in the system. Also $\overset{\leftrightarrow}{\mathbf{K}}(\omega)$ is the cold plasma dielectric tensor and the extra factors of "1/2" are introduced to express all quantities in terms of peak rather than rms values. The Poynting vector, which represents the flux of electromagnetic power, is equal to the product of the stored energy times the group velocity.

In summary, to evaluate the absorption fraction for ECH, ICH, or LHCD a series of steps must be carried out in the following sequence:

(a) calculate the cold plasma dispersion relation $D_r(\omega, k_\perp, k_\parallel) = 0$;
(b) analyze accessibility;
(c) calculate the collisionless damping rate $\omega_i$ and then $k_{\perp i}$;
(d) analyze power absorption.

The analysis below shows that ECH, ICH, and LHCD can each have a high absorption fraction.

## 15.5 The cold plasma dispersion relation

Recall from the basic properties of electromagnetic wave propagation that the dispersion relation is obtained by: (1) calculating the elements of the dielectric tensor, and (2) setting the resulting electric field determinant to zero. In this section the dispersion relation is calculated for the two-fluid cold plasma model. The end result is an explicit form for $\hat{D}_r(\omega, n_\perp, n_\parallel)$ which is equivalent to $D_r(\omega, k_\perp, k_\parallel)$. No accessibility analysis of $D_r = 0$ is presented here but it is instead described in detail in Subsections 15.7–15.9.

The starting point for the derivation of the dielectric tensor is the two-fluid cold plasma model. For a cold plasma only the momentum equations are needed which are given by

$$m_e \left( \frac{\partial \mathbf{u}_e}{\partial t} + \mathbf{u}_e \cdot \nabla \mathbf{u}_e \right) = -e(\mathbf{E} + \mathbf{u}_e \times \mathbf{B}),$$

$$m_i \left( \frac{\partial \mathbf{u}_i}{\partial t} + \mathbf{u}_i \cdot \nabla \mathbf{u}_i \right) = e(\mathbf{E} + \mathbf{u}_i \times \mathbf{B}).$$

(15.62)

The equilibrium corresponding to an infinite, homogeneous plasma can be written as

$$n_e = n_i \equiv n_0,$$
$$\mathbf{B} = B_0 \mathbf{e}_z,$$
$$\mathbf{u}_e = \mathbf{u}_i = \mathbf{E} = 0.$$

(15.63)

The fluid equations are now linearized about this equilibrium. The goal is to calculate $\mathbf{u}_{e1}$ and $\mathbf{u}_{i1}$ in terms of $\mathbf{E}_1$. From the velocities one can easily calculate $\mathbf{J}_1$ and then $\overset{\leftrightarrow}{\sigma}$ and finally $\overset{\leftrightarrow}{\mathbf{K}}$. The equation for the perturbed velocity is

$$m \frac{\partial \mathbf{u}_1}{\partial t} = q(\mathbf{E}_1 + \mathbf{u}_1 \times \mathbf{B}_0),$$

(15.64)

where $m$ represents $m_e$, $m_i$ and $q$ represents $\pm e$.

In accordance with the discussion in the previous section, the coordinate system is chosen so that $\mathbf{B}_0$ lies along the $z$ axis. Propagation into the plasma occurs along the $x$ axis. Thus, the wave vector has the form $\mathbf{k} = k_\perp \mathbf{e}_x + k_\parallel \mathbf{e}_z$. All perturbed quantities can then be Fourier analyzed in space and time as $Q_1(\mathbf{r}, t) = Q_1 \exp(-i\omega t + ik_\perp x + ik_\parallel z)$, where without loss in generality the frequency can hereafter always be considered to be positive: $\omega > 0$.

The components of the perturbed momentum equation become

$$-i\omega m u_{1x} = q(E_{1x} + u_{1y}B_0),$$
$$-i\omega m u_{1y} = q(E_{1y} - u_{1y}B_0),$$
$$-i\omega m u_{1z} = q E_{1z}.$$

$$(15.65)$$

One can easily solve these equations for $\mathbf{u}_1$. The result is

$$u_{1x} = \frac{q}{m}\frac{(i\omega E_{1x} - \omega_c E_{1y})}{\omega^2 - \omega_c^2},$$

$$u_{1y} = \frac{q}{m}\frac{(\omega_c E_{1x} + i\omega E_{1y})}{\omega^2 - \omega_c^2},$$

$$(15.66)$$

$$u_{1z} = \frac{iq}{m\omega}E_{1z},$$

where $\omega_c = qB_0/m$ is the gyro frequency and can have either sign. The current density is determined from the relation $\mathbf{J}_1 = e(n_i\mathbf{u}_i - n_e\mathbf{u}_e)_1 = en_0(\mathbf{u}_{i1} - \mathbf{u}_{e1})$. A short calculation then shows that the conductivity tensor $\overset{\leftrightarrow}{\sigma}$, defined by $\mathbf{J}_1 = \overset{\leftrightarrow}{\sigma}\cdot\mathbf{E}_1$, can be written as

$$\overset{\leftrightarrow}{\sigma} = \begin{vmatrix} \sigma_{xx} & \sigma_{xy} & 0 \\ \sigma_{yx} & \sigma_{yy} & 0 \\ 0 & 0 & \sigma_{zz} \end{vmatrix}$$

$$(15.67)$$

with

$$\sigma_{xx} = \sigma_{yy} = i\sum_j \frac{q_j^2 n_0}{m_j}\frac{\omega}{\omega^2 - \omega_{cj}^2},$$

$$\sigma_{xy} = -\sigma_{yx} = -\sum_j \frac{q_j^2 n_0}{m_j}\frac{\omega_{cj}}{\omega^2 - \omega_{cj}^2},$$

$$(15.68)$$

$$\sigma_{zz} = i\sum_j \frac{q_j^2 n_0}{m_j\omega}.$$

Finally, the dielectric tensor is found from the relation $\overset{\leftrightarrow}{\mathbf{K}} = \overset{\leftrightarrow}{\mathbf{I}} + (i/\omega\varepsilon_0)\overset{\leftrightarrow}{\sigma}$:

$$\overset{\leftrightarrow}{\mathbf{K}} = \begin{vmatrix} K_{xx} & K_{xy} & 0 \\ K_{yx} & K_{yy} & 0 \\ 0 & 0 & K_{zz} \end{vmatrix} = \begin{vmatrix} K_\perp & -iK_A & 0 \\ iK_A & K_\perp & 0 \\ 0 & 0 & K_\parallel \end{vmatrix}.$$

$$(15.69)$$

Here,

$$K_\perp = 1 - \sum_j \frac{\omega_{pj}^2}{\omega^2 - \omega_{cj}^2},$$

$$K_A = \sum_j \frac{\omega_{cj}}{\omega}\frac{\omega_{pj}^2}{\omega^2 - \omega_{cj}^2},$$

$$(15.70)$$

$$K_\parallel = 1 - \sum_j \frac{\omega_{pj}^2}{\omega^2},$$

and $\omega_{pj}^2 = n_0 q_j^2/m_j\varepsilon_0$ is the square of the plasma frequency.

Knowing the dielectric tensor, one can now calculate the dispersion relation by setting $\text{Det}|\mathbf{n} \times \mathbf{n} \times \mathbf{E}_1 + \overset{\leftrightarrow}{\mathbf{K}} \cdot \mathbf{E}_1| = 0$; that is, one sets

$$\begin{vmatrix} n_\parallel^2 - K_\perp & iK_A & -n_\perp n_\parallel \\ -iK_A & n_\perp^2 + n_\parallel^2 - K_\perp & 0. \\ -n_\perp n_\parallel & 0 & n_\perp^2 - K_\parallel \end{vmatrix} = 0. \tag{15.71}$$

A short calculation yields a quadratic equation for $n_\perp^2$:

$$\hat{D}_r(\omega, n_\parallel, n_\perp) \equiv An_\perp^4 + Bn_\perp^2 + C = 0,$$

$$A = K_\perp,$$

$$B = (K_\perp + K_\parallel)(n_\parallel^2 - K_\perp) + K_A^2, \tag{15.72}$$

$$C = K_\parallel\left[(n_\parallel^2 - K_\perp)^2 - K_A^2\right].$$

This somewhat complicated expression is the desired cold plasma dispersion relation which will be simplified and analyzed in detail in the subsections describing ECH, ICH, and LHCD.

## 15.6 Collisionless damping

Assume that an electromagnetic wave has been identified from the cold plasma dispersion relation that has good accessibility to the center of the plasma. The next step is to calculate how much power is deposited in the plasma by means of collisionless damping. Collisionless damping involves a transfer of wave energy to the plasma by means of a wave-particle resonant interaction. Recall that the condition for resonance is $\omega = k_\parallel v_\parallel + l\omega_c$. For effective heating the parameters of the plasma and the applied electromagnetic wave should be chosen such that resonance occurs at the center of the plasma. For current drive the resonance should occur at the desired off-axis position.

In this section a simple, single-particle model is described that predicts the temporal damping rate $\omega_i$ resulting from a wave-particle resonance. The analysis follows the early work of Stix. The idea is to calculate the trajectory of a charged particle in a small amplitude electromagnetic field. It is essential to carry out the calculation as an initial value problem. It is found that a small class of particles, whose parallel velocity satisfies the condition $\omega = k_\parallel v_\parallel + l\omega_c$, is resonant with the wave and can thus potentially absorb large amounts of power. The net power absorbed by the plasma is determined by averaging over all initial particle positions within a wavelength period and then integrating over all particle velocities. Finally, a simple power balance calculation yields the desired value of $\omega_i$.

Depending on the structure of the wave electric field, there are several different types of resonant interactions. First, when the wave electric field has a component along $\mathbf{B}_0$ resonance occurs for the O mode. The special $l = 0$, $n_\perp = 0$ O mode corresponds to a pure electrostatic wave and produces the simplest case of collisionless damping, known as "Landau damping." This is the first topic discussed below. Second, when the wave electric field rotates transversely to $\mathbf{B}_0$ resonance occurs for the $l \geq 1$ X mode. This type of resonance produces X-mode cyclotron damping and is the next topic discussed. The final topic involves a generalization of the Landau damping calculation to the $l \geq 1$, $n_\perp \neq 0$ O-mode resonance. This produces O-mode cyclotron damping.

### 15.6.1 Landau damping

Consider the motion of a charged particle in a small-amplitude electrostatic electric field whose direction is aligned parallel to the background magnetic field $\mathbf{B} = B_0 \mathbf{e}_z$; that is, $\mathbf{E} = E_1(z, t)\mathbf{e}_z = E_\| \cos(k_\| z - \omega t)\mathbf{e}_z$. Since $n_\perp = 0$ the perturbed magnetic field is zero. The equations of motion for an electron can then be written as

$$\frac{dv}{dt} = -\frac{e}{m_e} E_\| \cos(\omega t - k_\| z) \qquad v(0) = v_\|,$$
$$\frac{dz}{dt} = v \qquad\qquad\qquad\qquad z(0) = z_i. \tag{15.73}$$

Assuming the electric field is small, one can solve these equations by a straightforward expansion $v(t) = v_0(t) + v_1(t) + \cdots$, $z(t) = z_0(t) + z_1(t) + \cdots$.

The leading order solution corresponds to the unperturbed orbits and is

$$v_0 = v_\|,$$
$$z_0 = z_i + v_\| t. \tag{15.74}$$

The first order equations can be written as

$$\frac{dv_1}{dt} = -\frac{e}{m_e} E_\| \cos(\omega t - k_\| v_\| t - k_\| z_i) \qquad v_1(0) = 0,$$
$$\frac{dz_1}{dt} = v_1 \qquad\qquad\qquad\qquad\qquad\qquad z_1(0) = 0. \tag{15.75}$$

The solution satisfying the initial conditions is

$$v_1 = \frac{e}{m_e} E_\| \frac{\sin(k_\| z_i - \bar\omega t) - \sin(k_\| z_i)}{\bar\omega},$$
$$z_1 = \frac{e}{m_e} E_\| \left[ \frac{\cos(k_\| z_i - \bar\omega t) - \cos(k_\| z_i)}{\bar\omega^2} - \frac{t \sin(k_\| z_i)}{\bar\omega} \right], \tag{15.76}$$

where $\bar\omega = \omega - k_\| v_\|$.

The next task is to calculate the change in particle energy per unit time (i.e., the power gained), defined by

$$\frac{dW}{dt} = \frac{d}{dt}\left(\frac{m_e v^2}{2}\right) = -e\,\mathbf{v} \cdot \mathbf{E}, \tag{15.77}$$

which to second order has the form

$$\frac{dW}{dt} = -e\left[ v_0 E_1(z_0, t) + v_0 \frac{\partial E_1(z_0, t)}{\partial z_0} z_1 + v_1 E_1(z_0, t) \right]. \tag{15.78}$$

Since the initial position of an electron $z_i$ is assumed to be randomly located between any two successive peaks of the electric field, it is necessary to average over all $z_i$ to obtain the net change in particle energy corresponding to any given initial velocity $v_\parallel$. Specifically, one must evaluate

$$\frac{d\overline{W}}{dt} = \frac{k_\parallel}{2\pi} \int_0^{2\pi/k_\parallel} \frac{dW}{dt} dz_i. \tag{15.79}$$

A straightforward calculation shows that: (1) the term which is linear in amplitude averages to zero, and (2) the term which is quadratic in amplitude reduces to

$$\frac{d\overline{W}}{dt} = \frac{e^2 E_\parallel^2}{2m_e} \left( \frac{\omega}{\overline{\omega}^2} \sin \overline{\omega}t - \frac{\omega t}{\overline{\omega}} \cos \overline{\omega}t + t \cos \overline{\omega}t \right). \tag{15.80}$$

Observe that $d\overline{W}/dt$ is finite when $\overline{\omega} \to 0$. Also there are terms that grow linearly with $t$ which is indicative of a resonant response. Both of these conclusions are a consequence of solving the trajectory as an initial value problem. If the problem is instead solved by standard "AC sinusoidal steady state analysis" the last two terms, linearly proportional to $t$, vanish and the first term diverges for $\overline{\omega} \to 0$.

The analysis continues by integrating over the entire distribution of velocities. One multiplies $d\overline{W}/dt$ by $f_0(v_\parallel, v_\perp)$ and integrates over all $v_\parallel$ and $v_\perp$. The resulting expression gives the power gained per unit volume ($S_L$) due to all the particles. The derivation for $S_L$ is simplified by using the identity

$$\frac{\omega}{\overline{\omega}^2} \sin \overline{\omega}t - \frac{\omega t}{\overline{\omega}} \cos \overline{\omega}t + t \cos \overline{\omega}t = -\frac{\partial}{\partial \overline{\omega}} \left( \frac{\omega \sin \overline{\omega}t}{\overline{\omega}} - \sin \overline{\omega}t \right)$$

$$= \frac{1}{k_\parallel} \frac{\partial}{\partial v_\parallel} \left( \frac{\omega \sin \overline{\omega}t}{\overline{\omega}} - \sin \overline{\omega}t \right) \tag{15.81}$$

and integrating by parts over the velocity. The value of $S_L$ becomes

$$S_L(t) = -\frac{e^2 E_\parallel^2 n_0}{2m_e k_\parallel} \int_{-\infty}^{\infty} \left( \frac{\omega \sin \overline{\omega}t}{\overline{\omega}} - \sin \overline{\omega}t \right) \frac{\partial f_\parallel}{\partial v_\parallel} dv_\parallel, \tag{15.82}$$

where $f_\parallel(v_\parallel) = (2\pi/n_0) \int f_0 v_\perp dv_\perp$ is the normalized distribution function integrated over all perpendicular velocities.

Consider now the behavior of $S_L(t)$ for large $t$. In this limit all the initial transients have decayed away, leaving just the long-time asymptotic solution. For $t \to \infty$ the contribution to the integral from particles with $\overline{\omega} \neq 0$ (i.e., $v_\parallel \neq \omega/k_\parallel$) vanishes because of the rapidly oscillating integrand. This is the contribution of the non-resonant particles. The wave rapidly passes these particles, first giving them a small increase in energy over half the wave period and then having the energy returned during the second half period when the electric field has changed sign.

There is, however, a finite contribution to $S_L(t)$ from those particles with velocities near $\overline{\omega} \approx 0$ (i.e., $v_\parallel \approx \omega/k_\parallel$). These are the resonant particles. The contribution arises from the first term in Eq. (15.82). The second term makes a negligibly small contribution since the resonance region is narrow. In fact, for a Maxwellian, it can be shown that the second term can be exactly integrated for any $t$ leading to a term proportional to $t \cos(\omega t) \exp(-k_\parallel^2 v_T^2 t^2/4)$ which rapidly vanishes for $t \to \infty$.

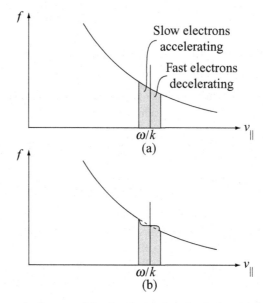

Figure 15.15 (a) Schematic diagram of the distribution function in the vicinity of the resonant particles. (b) The acceleration of slow particles plus the deceleration of fast particles tends to flatten the
distribution function near $v_\| = \omega/k_\|$.

The contribution from the first term is evaluated by noting that for large $t$ the range of res-
onant velocities is small: $\Delta v_\| \sim 1/k_\| t$. Therefore, within this small region of velocity space
one can approximate $\partial f_\|(v_\|)/\partial v_\| \approx \partial f_\|(v_\|)/\partial v_\||_{\omega/k_\|}$ and extract it from the integrand. This
yields

$$S_{\mathrm{L}}(t) = -\frac{e^2 E_\|^2 n_0 \omega}{2 m_e k_\|^2} \left(\frac{\partial f_\|}{\partial v_\|}\right)_{\omega/k_\|} \int \frac{\sin x}{x}\, dx, \qquad (15.83)$$

where $x = -\overline{\omega}t = (k_\| v_\| - \omega)t$. Since the main contribution to the integral occurs near
$x = 0$, a negligibly small error is made by considering the limits of integration to be $-\infty <
x < \infty$. One finally obtains

$$S_{\mathrm{L}}(t) = -\frac{\pi e^2 E_\|^2 n_0 \omega}{2 m_e k_\|^2} \left(\frac{\partial f_\|}{\partial v_\|}\right)_{\omega/k_\|}. \qquad (15.84)$$

Observe that $S_{\mathrm{L}}(t)$ is independent of $t$. More importantly, for the usual case where
$\partial f_\|/\partial v_\| < 0$, $S_L > 0$ indicating that the particles are gaining energy from the wave. This
can be understood physically as follows. (See Fig. 15.15.) Resonant particles have a velocity
$v_\| \approx \omega/k_\|$, which corresponds to the $l = 0$ harmonic of the general wave-particle resonance
condition $\omega = k_\| v_\| + l\omega_c$. These particles are resonant because in the wave reference frame
they are moving very slowly with respect to the wave and thus see a nearly DC electric
field. In particular, particles moving slightly slower than the phase velocity of the wave

are accelerated for a long period of time producing a gain in particle energy. Conversely, particles moving slightly faster than the wave are decelerated for a long period of time producing a loss of particle energy. If there are more slow particles than fast particles (i.e., $\partial f_\parallel / \partial v_\parallel < 0$), there is an overall gain in particle energy and the wave is damped.

This appealing picture, which is often compared to a surfer catching a wave, is oversimplified. In actuality a slow particle can either gain or lose energy from the wave depending upon its initial position $z_i$ with respect to the phase of the wave; that is, gain or loss depends upon whether the particle, at its initial position, sees a positive or negative electric field. Since the $z_i$ are randomly distributed, the net energy transfer to slow and fast particles tends to average to zero. Mathematically, this corresponds to the vanishing of the term linear in amplitude in the $z_i$ averaging in Eq. (15.79). The energy transfer is a smaller effect, second order in the amplitude of the wave. The simple model shows that, as might be expected, after averaging over $z_i$ there is a net gain in energy to the slower particles and a net loss in energy from the faster particles. Therefore, the discussion about $\partial f_\parallel / \partial v_\parallel < 0$ for energy absorption still holds but for slightly subtler reasons.

The last step in the analysis involves the calculation of the damping rate $\omega_i$ by means of a simple power balance argument. The argument is as follows. In general the total stored energy density in the wave $U$, averaged over one oscillation period in time, consists of the sum of the electric, magnetic, and plasma kinetic energy densities. Note that the plasma kinetic energy density arises from the bulk, non-resonant particles and is obtained from the cold plasma analysis. Without damping $U$ is thus given by

$$
U = \frac{\omega}{2\pi} \int_0^{2\pi/\omega} \left( \frac{\varepsilon_0 E^2}{2} + \frac{B^2}{2\mu_0} + \sum_j \frac{n_j m_j u_j^2}{2} \right) dt
$$

$$
= \frac{1}{2} \left( \frac{\varepsilon_0 |E_1|^2}{2} + \frac{|B_1|^2}{2\mu_0} + \sum_j \frac{n_0 m_j |u_{1j}|^2}{2} \right). \tag{15.85}
$$

The extra factor of $\frac{1}{2}$ in the second form arises because of the time averaging of sinusoidal functions: $\langle \sin^2(\omega t - \mathbf{k} \cdot \mathbf{r}) \rangle = \langle \cos^2(\omega t - \mathbf{k} \cdot \mathbf{r}) \rangle = \frac{1}{2}$. The quantity $U$ is thus expressed in terms of the peak (rather than rms) amplitudes.

If there is collisionless damping, then $U$ decays slowly in time (i.e., $\omega_i \neq 0$). Since $U$ is quadratic in the field amplitude, its explicit time dependence in the limit $\omega_i \ll \omega_r$ is proportional to terms of the form $E_{1z}^* E_{1z} \sim \exp(2\omega_i t)$ with $\omega_i < 0$ for damping. This implies that

$$
U(t) = \frac{1}{2} \left( \frac{\varepsilon_0 |E_1|^2}{2} + \frac{|B_1|^2}{2\mu_0} + \sum_j \frac{n_0 m_j |u_{1j}|^2}{2} \right) e^{2\omega_i t} = U_0 e^{2\omega_i t}. \tag{15.86}
$$

Thus, by definition one sees that

$$
dU/dt = 2\omega_i U. \tag{15.87}
$$

The quantity $2\omega_i U$ represents the energy loss rate of the wave. Power balance requires that this power loss be equal to the power gained by the resonant particles:

$$(2\omega_i U_0 + S_L)\, e^{2\omega_i t} = 0. \tag{15.88}$$

Consequently, the damping rate is given by

$$\omega_i = -S_L/2U_0. \tag{15.89}$$

The quantity $U_0$ can be explicitly evaluated from the cold plasma analysis. For the case of a generalized cold plasma wave (i.e., a wave with three non-zero components of electric field) one finds after a short calculation that $U_0 = (\varepsilon_0/4)\, \mathbf{E}_1^* \cdot \overset{\leftrightarrow}{\mathbf{M}} \cdot \mathbf{E}_1$, where

$$\overset{\leftrightarrow}{\mathbf{M}} = \begin{vmatrix} M_\perp + n_\parallel^2 & iM_A & -n_\parallel n_\perp \\ -iM_A & M_\perp + n^2 & 0 \\ -n_\parallel n_\perp & 0 & M_\parallel + n_\perp^2 \end{vmatrix} \tag{15.90}$$

and

$$M_\perp = 1 + \sum_j \frac{\omega_{pj}^2 \left(\omega^2 + \omega_{cj}^2\right)}{\left(\omega^2 - \omega_{cj}^2\right)^2},$$

$$M_A = 2\sum_j \frac{\omega_{pj}^2 \,\omega\,\omega_{cj}}{\left(\omega^2 - \omega_{cj}^2\right)^2}, \tag{15.91}$$

$$M_\parallel = 1 + \sum_j \frac{\omega_{pj}^2}{\omega^2}.$$

Note that by direct calculation one can also show that $U_0$ can be rewritten as

$$U_0 = \frac{\varepsilon_0}{4}\mathbf{E}_1^* \cdot \overset{\leftrightarrow}{\mathbf{M}} \cdot \mathbf{E}_1 = \frac{\varepsilon_0}{4}\left\{ c^2\left|\mathbf{B}_1^2\right| + \mathbf{E}_1^* \cdot \frac{\partial}{\partial\omega}[\omega\overset{\leftrightarrow}{\mathbf{K}}(\omega)] \cdot \mathbf{E}_1 \right\}, \tag{15.92}$$

explicitly demonstrating that the quantity $U$, introduced in connection with Poynting's theorem (i.e., Eq. (15.61)), is indeed the total wave energy.

For the simple electrostatic case under consideration $E_{1x} = E_{1y} = n_\perp = 0$, implying that $\mathbf{E}_1^* \cdot \overset{\leftrightarrow}{\mathbf{M}} \cdot \mathbf{E}_1 = M_\parallel E_\parallel^2$ and

$$U_0 = \frac{\varepsilon_0}{4}\left(1 + \frac{\omega_{pe}^2 + \omega_{pi}^2}{\omega^2}\right) E_\parallel^2. \tag{15.93}$$

Furthermore, the cold plasma dispersion relation reduces to $K_\parallel = 0$ or $\omega^2 = \omega_{pe}^2 + \omega_{pi}^2$. Therefore, $U_0 = \varepsilon_0 E_\parallel^2/2$. Using this value for $U_0$ and the expression for $S_L$ given in Eq. (15.84) leads to the desired expression for the damping rate:

$$\frac{\omega_i}{\omega_{pe}} = \frac{\pi\omega_{pe}^2}{2k_\parallel^2}\left(\frac{\partial f_\parallel}{\partial v_\parallel}\right)_{\omega_{pe}/k_\parallel} = -\pi^{1/2}\left(\frac{\omega_{pe}^3}{k_\parallel^3 v_{Te}^3}\right)\exp\left(-\frac{\omega_{pe}^2}{k_\parallel^2 v_{Te}^2}\right). \tag{15.94}$$

Here, the second form corresponds to a Maxwellian distribution function. This famous, classic result was first derived by Landau directly from kinetic theory and is widely known as "Landau damping". A key conclusion from Eq. (15.94) is that strong Landau damping only occurs when the phase velocity is comparable to the thermal velocity: $\omega/k_\parallel \sim v_T$. Waves with $\omega/k_\parallel \gg v_T$ produce a resonance on the tail of the distribution function, where there is an exponentially small number of particles and the corresponding damping is very weak. Landau damping is a very important plasma physics phenomena occurring in many fusion applications in addition to heating and current drive.

### 15.6.2 X-mode cyclotron damping

The calculation of X-mode cyclotron damping is similar to that of Landau damping, although it is more algebraically intensive because of the complications arising from the DC magnetic field. Still, the overall plan is the same: (1) calculate the gain in particle energy due to the wave, (2) average over initial positions, (3) multiply by the distribution function and integrate over velocities, and (4) determine $\omega_i$ by a simple power balance argument. In the analysis below it is useful to keep in mind that the plasma is assumed to be driven by a single frequency resonant with the $l$th cyclotron harmonic: $\omega \approx l\omega_c$.

The calculation starts by noting that X-mode cyclotron damping arises from the components of electric field perpendicular to $\mathbf{B}_0 = B_0 \mathbf{e}_z$. The parallel electric field makes a negligible contribution. Furthermore, rather than formulate the problem in terms of linearly polarized waves it is more convenient to use instead circularly polarized waves. Specifically, the wave electric field can, in general, be written as the superposition of two circularly polarized field components plus a parallel field component as follows:

$$\mathbf{E} = E_+(\cos\phi\,\mathbf{e}_x + \sin\phi\,\mathbf{e}_y) + E_-(\cos\phi\,\mathbf{e}_x - \sin\phi\,\mathbf{e}_y) + E_\parallel \cos\phi\,\mathbf{e}_z$$
$$\approx E_+(\cos\phi\,\mathbf{e}_x + \sin\phi\,\mathbf{e}_y), \tag{15.95}$$

where $E_{1x} = (E_+ + E_-)\,\mathrm{Re}(e^{i\phi})$, $E_{1y} = i(E_+ - E_-)\mathrm{Im}(e^{i\phi})$, and $\phi(t) = k_\perp x + k_\parallel z - \omega t$. The amplitudes $E_\pm$ correspond to the left and right circularly polarized waves which resonate with the ions and electrons respectively. The analysis below is carried out for ions, so that only the $E_+$ term is maintained for the collisionless damping analysis. The $E_-$ and $E_\parallel$ terms do not generate a strong wave-particle resonance for ions when $\omega \approx l\omega_{ci}$. However, $E_-$ and $E_\parallel$ as well as $E_+$ must be included when evaluating the wave energy $U_0$. In addition to the electric field, there is a wave magnetic field which can be neglected in the non-relativistic limit. Thus, $\mathbf{B} \approx B_0\mathbf{e}_z$.

Under these assumptions the single-particle equations of motion describing cyclotron damping reduce to

$$\frac{d\mathbf{v}}{dt} - \Omega_i \mathbf{v} \times \mathbf{e}_z = \frac{eE_+}{m_i}(\cos\phi\,\mathbf{e}_x + \sin\phi\,\mathbf{e}_y), \quad \mathbf{v}(0) = v_\perp\mathbf{e}_x + v_\parallel\mathbf{e}_z,$$

$$\frac{d\mathbf{r}}{dt} = \mathbf{v}, \qquad\qquad\qquad\qquad \mathbf{r}(0) = x_i\mathbf{e}_x + y_i\mathbf{e}_y + z_i\mathbf{e}_z, \tag{15.96}$$

with $\Omega_i = \omega_{ci} = eB_0/m_i$. Hereafter, in terms of notation, note that $\Omega_j = |\omega_{cj}| = |q_j B_0/m_j| > 0$.

The parallel motion is easily calculated: $v_z = v_\parallel$, $z = z_i + v_\parallel t$. The remaining two perpendicular vector components can be converted into a scalar representation by introducing complex quantities: $v_+ = v_x + iv_y$ and $r_+ = x + iy$. The perpendicular equations can now be written as

$$\frac{dv_+}{dt} + i\Omega_i v_+ = \frac{eE_+}{m_i} e^{i\phi}, \qquad v_+(0) = v_\perp,$$

$$\frac{dr_+}{dt} = v_+, \qquad r_+(0) = r_i, \tag{15.97}$$

where $r_i = x_i + iy_i$. Note that while $v_+$, $r_+$ are complex, $v_x$, $v_y$, $x$, $y$, $\phi$ are all real.

As in the case of Landau damping these equations are solved by assuming that $E_+$ is small and then introducing a small-amplitude expansion: $v_+(t) = v_0(t) + v_1(t) + \cdots$, $r_+(t) = r_0(t) + r_1(t) + \cdots$. This implies that $\phi(t) = \phi_0(t) + \phi_1(t) + \cdots$, where $\phi_0(t) = k_\perp x_0(t) + k_\parallel z(t) - \omega t$, $\phi_1(t) = k_\perp x_1(t)$ with $x_0(t) = \mathrm{Re}\,[r_0(t)]$, $x_1(t) = \mathrm{Re}\,[r_1(t)]$.

The zeroth order equations correspond to the unperturbed motion and are

$$\frac{dv_0}{dt} + i\Omega_i v_0 = 0, \qquad v_0(0) = v_\perp,$$

$$\frac{dr_0}{dt} = v_0, \qquad r_0(0) = r_i. \tag{15.98}$$

The solution is the familiar gyro motion which in the present notation has the form

$$v_0 = v_\perp e^{-i\Omega_i t},$$

$$r_0 = r_i + i\frac{v_\perp}{\Omega_i}(e^{-i\Omega_i t} - 1), \tag{15.99}$$

$$\phi_0 = \phi_i + (k_\parallel v_\parallel - \omega)t + \frac{k_\perp v_\perp}{\Omega_i} \sin\Omega_i t.$$

Here $\phi_i = k_\parallel z_i + k_\perp x_i$. The first order equations can be written as

$$\frac{dv_1}{dt} + i\Omega_i v_1 = \frac{eE_+}{m_i} e^{i\phi_0}, \qquad v_1(0) = 0,$$

$$\frac{dr_1}{dt} = v_1, \qquad r_1(0) = 0. \tag{15.100}$$

The quantity $\exp(i\phi_0)$ is somewhat complicated because of the appearance of $\sin\Omega_i t$ in the exponent. The term can be simplified by Fourier analysis using the identity

$$e^{i\alpha \sin\tau} = \sum_{-\infty}^{\infty} J_n(\alpha)\, e^{in\tau}, \tag{15.101}$$

where $J_n$ is the Bessel function of order $n$. It then follows that

$$e^{i\phi_0} = e^{i\phi_i} \sum_n J_n(w)\, e^{-i\bar{\omega}_n t}, \tag{15.102}$$

where $w = k_\perp v_\perp / \Omega_i$ and $\overline{\omega}_n = \omega - k_\| v_\| - n\Omega_i$. Using this relation one can solve the first order equations. The solution is

$$v_1 = i\frac{eE_+}{m_i}e^{i\phi_i}\sum_n \frac{J_n(w)}{\overline{\omega}_{n+1}}(e^{-i\overline{\omega}_n t} - e^{-i\Omega_i t}),$$

$$r_1 = -\frac{eE_+}{m_i}e^{i\phi_i}\sum_n \frac{J_n(w)}{\overline{\omega}_{n+1}}\left(\frac{e^{-i\overline{\omega}_n t} - 1}{\overline{\omega}_n} - \frac{e^{-i\Omega_i t} - 1}{\Omega_i}\right).$$

(15.103)

The next step in the analysis is to use the solutions for the trajectories to calculate the change in energy per unit time due to the resonant interaction with the wave. The change in particle energy is evaluated from the relation

$$\frac{dW}{dt} = \frac{d}{dt}\left(\frac{m_i v^2}{2}\right) = e\mathbf{v}\cdot\mathbf{E} = eE_+\mathrm{Re}(v_+^* e^{i\phi}).$$

(15.104)

As with Landau damping the calculation must be carried out to second order in amplitude. One obtains

$$\frac{dW}{dt} = eE_+\mathrm{Re}(v_0^* e^{i\phi_0} + v_1^* e^{i\phi_0} + i\phi_1 v_0^* e^{i\phi_0}).$$

(15.105)

The actual quantity of interest is the value of $dW/dt$ averaged over all initial positions:

$$\frac{d\overline{W}}{dt} = \frac{1}{2\pi}\int_0^{2\pi} \frac{dW}{dt} d\phi_i.$$

(15.106)

The term linear in wave amplitude again averages to zero. After a slightly tedious calculation the second and third terms can be averaged, yielding

$$\frac{d\overline{W}}{dt} = \frac{e^2 E_+^2}{m_i}\sum_{m,n} J_m(w) J_n(w)(T_1 + T_2),$$

$$T_1 = \frac{1}{\overline{\omega}_{n+1}}[\sin(m-n)\Omega_i t + \sin\overline{\omega}_{m+1}t],$$

(15.107)

$$T_2 = \frac{k_\perp v_\perp}{2}\left[\frac{\sin(m-n)\Omega_i t}{\overline{\omega}_n \overline{\omega}_{n+1}} + \frac{\sin\overline{\omega}_{m+2}t}{\Omega_i \overline{\omega}_{n+1}} - \frac{\sin\overline{\omega}_{m+1}t}{\Omega_i \overline{\omega}_n}\right].$$

This complicated looking expression can be simplified by recognizing that a strong wave-particle resonance only occurs for the particular value of $l$ corresponding to $\overline{\omega}_l = 0$, which is equivalent to driving the plasma with a frequency $\omega \approx l\Omega_i$. For $p \neq l$ the condition $\overline{\omega}_p = 0$ produces a resonance far out on the tail of the distribution function: $v_\|/v_{Ti} = (p-l)\Omega_i/k_\| v_{Ti} \sim c/n_\| v_{Ti} \gg 1$. A second simplification that arises is that only terms with a frequency $\sin\overline{\omega}_l t$ contribute to a strong resonant interaction. Terms of the form $(\sin\overline{\omega}_p t)/\overline{\omega}_l$ with $p \neq l$ oscillate very rapidly and therefore do not continuously add energy to the particle. To summarize, only those integers in the double sum that lead to terms of the form $(\sin\overline{\omega}_l t)/\overline{\omega}_l$ produce a strong resonant interaction. Extracting only these terms from the double sum leads to a simpler form of $d\overline{W}/dt$ given by

$$
\begin{aligned}
\frac{d\overline{W}}{dt} &= \frac{e^2 E_+^2}{m_i} \frac{\sin \overline{\omega}_l t}{\overline{\omega}_l} \left[ J_{l-1}^2 + \frac{w}{2} J_{l-1}(J_{l-2} - J_l) \right] \\
&= \frac{e^2 E_+^2}{2m_i} \frac{\sin \overline{\omega}_l t}{\overline{\omega}_l} \left[ \frac{1}{w} \frac{d}{dw} \left( w^2 J_{l-1}^2 \right) \right].
\end{aligned}
\tag{15.108}
$$

The analysis continues by multiplying $d\overline{W}/dt$ by the distribution function and integrating over all velocities. This yields the power absorbed per unit volume ($S_X$) by the particles. The calculation is carried out assuming for convenience a distribution function of the form $f_0(v_\parallel, v_\perp) = n_0 f_\parallel(v_\parallel) f_\perp(v_\perp)$. The expression for the power density can be written as

$$
S_X = \frac{\pi n_0 e^2 E_+^2}{m_i} \left[ \int_{-\infty}^{\infty} f_\parallel \frac{\sin \overline{\omega}_l t}{\overline{\omega}_l} dv_\parallel \right] \left[ \int_0^{\infty} \frac{f_\perp}{w} \frac{d}{dw} \left( w^2 J_{l-1}^2 \right) v_\perp dv_\perp \right].
\tag{15.109}
$$

For $t \to \infty$ the range of resonant parallel velocities is very narrow and, as for Landau damping, $f_\parallel$ can be extracted from the integrand and the remaining integral evaluated analytically. The $v_\perp$ integral is simplified by recalling that $w = k_\perp v_\perp / \Omega_i$ and integrating by parts. These simplifications yield

$$
S_X = -\frac{\pi^2 n_0 e^2 E_+^2}{m_i} \frac{f_\parallel(v_l)}{k_\parallel} \int_0^{\infty} v_\perp^2 J_{l-1}^2 \frac{\partial f_\perp}{\partial v_\perp} dv_\perp,
\tag{15.110}
$$

where $v_l = (\omega - l\Omega_i)/k_\parallel$.

The evaluation of $S_X$ is completed by assuming a Maxwellian distribution function for $f_\perp$ and making use of a perhaps surprising integral relation given by

$$
\begin{aligned}
\int_0^{\infty} x^3 e^{-x^2/2\alpha} J_p^2(x) dx &= 2\alpha^2 \frac{d}{d\alpha} \int_0^{\infty} x e^{-x^2/2\alpha} J_p^2(x) dx \\
&= 2\alpha^2 \frac{d}{d\alpha} [\alpha e^{-\alpha} I_p(\alpha)],
\end{aligned}
\tag{15.111}
$$

where $I_p$ is the modified Bessel function. The required form of $S_X$ is finally

$$
\begin{aligned}
S_X &= \frac{\pi n_0 e^2 E_+^2}{m_i} \frac{f_\parallel(v_l)}{k_\parallel} \frac{d}{db} [be^{-b} I_{l-1}(b)] \\
&= \frac{\pi^{1/2} n_0 e^2 E_+^2}{m_i k_\parallel v_{Ti}} \frac{d}{db} [be^{-b} I_{l-1}(b)] \exp\left[ -\left( \frac{\omega - l\Omega_i}{v_{Ti}} \right)^2 \right].
\end{aligned}
\tag{15.112}
$$

Here, $b = k_\perp^2 v_{Ti}^2 / 2\Omega_i^2 = k_\perp^2 r_{Li}^2 / 2$ and the second form is for a Maxwellian distribution function.

The final step in the analysis is to use power balance to determine the damping rate $\omega_i$. The same power balance argument applies to cyclotron damping as to Landau damping. Thus,

$$
\omega_i = -S_X/2U_0,
\tag{15.113}
$$

where the general expression for $U_0$ is given by Eq. (15.92): $U_0 = (\varepsilon_0/4)\, \mathbf{E}_1^* \cdot \overset{\leftrightarrow}{\mathbf{M}} \cdot \mathbf{E}_1$. The quantity $U_0$ for X-mode damping is evaluated by assuming that $n_\parallel = 0$ in the cold plasma dispersion relation, which, as is shown later, is a good approximation for the X mode. The condition $n_\parallel = 0$ also implies that $E_\parallel = 0$. Furthermore, note that $E_{1x} = E_+ + E_-$ and $iE_{1y} = E_+ - E_-$. A short calculation then leads to

$$U_0 = \frac{\varepsilon_0}{4} M_X |E_+^2|,$$

$$M_X = \frac{1}{|E_+^2|}[n_\perp^2 |E_+ - E_-|^2 + 2M_\perp(|E_+|^2 + |E_-|^2) + 2M_A(|E_+|^2 - |E_-|^2)].$$

(15.114)

The quantity $M_X$ can be explicitly evaluated when the dispersion relation is analyzed in detail to determine $n_\perp^2$ and the polarization $E_-/E_+$. For the moment, readers should just assume that $M_X$ is a known quantity.

The desired damping rate is thus given by

$$\frac{\omega_i}{\omega_{pi}} = -\frac{2\pi\,\omega_{pi}}{M_X} \frac{f_\parallel(v_l)}{k_\parallel} \frac{\mathrm{d}}{\mathrm{d}b}[be^{-b}I_{l-1}(b)].$$

(15.115)

For a Maxwellian $f_\parallel$ the damping rates for the lowest two harmonics in the small gyro radius limit reduce to

$$\frac{\omega_i}{\omega_{pi}} = -\frac{2\pi^{1/2}}{M_X} \frac{\omega_{pi}}{k_\parallel v_{Ti}} \exp\left[-\left(\frac{\omega - \Omega_i}{k_\parallel v_{Ti}}\right)^2\right], \qquad l = 1,$$

$$\frac{\omega_i}{\omega_{pi}} = -\frac{\pi^{1/2}}{M_X} \frac{\omega_{pi}}{k_\parallel v_{Ti}} \exp\left[-\left(\frac{\omega - 2\Omega_i}{k_\parallel v_{Ti}}\right)^2\right]\left(\frac{k_\perp v_{Ti}}{\Omega_i}\right)^2, \qquad l = 2. \quad (15.116)$$

Observe that the damping rate for second harmonic cyclotron heating is smaller than that of the fundamental by $k_\perp^2 r_{Li}^2$. Also, the maximum damping occurs when the applied frequency is equal to the local value of the gyro frequency harmonic: $\omega = l\Omega_i$. A similar set of relations exists for X-mode electron cyclotron resonance.

Looking ahead to the heating applications, one sees that it is actually $k_{\perp i}$ rather than $\omega_i$ that is the required quantity. Recall that $k_{\perp i}$ is related to $\omega_i$ by $k_{\perp i} = -\omega_i/V_{g\perp}$. Also $U_0$ and $V_{g\perp}$ are related to the Poynting vector $P_\perp$ by $P_\perp = V_{g\perp}U_0$. Consequently it follows that

$$k_{\perp i} = S_X/2P_\perp.$$

(15.117)

For the X mode, $P_\perp = E_{1y}B_{1z}^*/2\mu_0 = n_\perp|E_{1y}^2|/c$. Using these results leads to the desired expression for $k_{\perp i}$:

$$\frac{k_{\perp i}c}{\omega_{pi}} = \frac{\pi^{1/2}}{n_\perp}\left|\frac{E_+}{E_{1y}}\right|^2 \frac{\omega_{pi}}{k_\parallel v_{Ti}} \frac{\mathrm{d}}{\mathrm{d}b}[be^{-b}I_{l-1}(b)]\exp\left[-\left(\frac{\omega - l\Omega_i}{k_\parallel v_{Ti}}\right)^2\right], \quad (15.118)$$

where $E_{1y} = E_+ - E_-$.

Lastly it is of interest to develop a physical understanding of X-mode cyclotron damping. Consider first wave resonance at the fundamental, which has some similarities with the

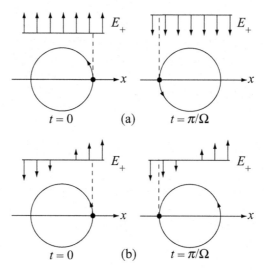

Figure 15.16 (a) Cyclotron resonance at the fundamental. The electric field is homogeneous in space and always in phase with the orbit. (b) Second harmonic resonance. For $\omega = 2\Omega$ the electric field has the same waveform at $t = 0$ and $t = \pi/\Omega$. However, the spatial dependence of $E_\parallel$ keeps the wave in phase with the particle.

Landau resonance. To begin assume that a circularly polarized wave has been launched that rotates in the same direction as the ion gyro motion. At a given point in space focus on those ions whose parallel velocity produces a Doppler shifted frequency that resonates with the local gyro frequency: $\omega - k_\parallel v_\parallel = \Omega_i$. For the fundamental the effects of $k_\perp$ are unimportant and can be neglected: $k_\perp \approx 0$. The equation describing the evolution of $v_+$ reduces to

$$\frac{dv_+}{dt} + i\Omega_i v_+ = \frac{eE_+}{m_i}e^{-i\Omega_i t}, \qquad v_+(0) = v_\perp. \tag{15.119}$$

Its solution is given by

$$v_+ = v_\perp e^{-i\Omega_i t} + \frac{eE_+}{m_i}te^{-i\Omega_i t}. \tag{15.120}$$

The second term clearly shows the secular behavior associated with a resonant interaction. In the Doppler shifted wave frame an ion sees an electric field that is always in phase with its gyro motion, thus leading to a constant absorption of energy. This is cyclotron damping at the fundamental harmonic.

The situation with higher harmonics is slightly more subtle. For instance, a Doppler shifted frequency corresponding to $\omega - k_\parallel v_\parallel = 2\Omega_i$ does not resonate with the gyro motion. However, when $k_\perp \neq 0$ the particles are not driven by a spatially uniform sine wave in time because of the finite perpendicular wavelength. In fact, when the perpendicular wavelength is comparable to the gyro radius, at a given instant of time the electric field reverses sign across the orbit as shown in Fig. 15.16. This spatially induced change in sign again brings the particle into resonance. The effect can be seen explicitly in the equation describing the evolution of $v_+$ in the limit of small $k_\perp r_L$.

$$\frac{dv_+}{dt} + i\Omega_i v_+ = \frac{eE_+}{m_i} e^{-i(2\Omega_i t - k_\perp x)}$$

$$\approx \frac{eE_+}{m_i} e^{-i2\Omega_i t} \left[ 1 + \frac{k_\perp v_\perp}{2\Omega_i} (e^{i\Omega_i t} - e^{-i\Omega_i t}) \right]. \qquad (15.121)$$

The spatial term proportional to $(k_\perp v_T / 2\Omega_i) \exp(i\Omega_i t)$ beats with the second harmonic driving frequency producing a finite gyro radius induced resonant interaction. It is by this mechanism that energy is absorbed at higher cyclotron harmonics.

### 15.6.3 O-mode cyclotron damping and generalized Landau damping

The final wave-particle resonant interaction of interest involves two closely related phenomena: O-mode cyclotron damping and generalized Landau damping, which are of interest to ECH and LHCD respectively. The corresponding analysis has similarities with both X-mode cyclotron damping and simple Landau damping. First, as its name implies, O-mode cyclotron damping occurs for an $l \geq 1$ resonance, characterized by a wave electric field with only a single component pointing along $\mathbf{B}_0$: $\mathbf{E} = E_\parallel \cos(k_\perp x + k_\parallel z - \omega t)\mathbf{e}_z$. For this case, the perturbed magnetic field is not zero and contains a transverse component: $\mathbf{B} = B_0\mathbf{e}_z + B_{1y} \cos(k_\perp x + k_\parallel z - \omega t)\mathbf{e}_y$. Since there is no applied $E_\pm$ it is the $v_\parallel B_{1y}$ term that produces an "effective transverse electric" field similar to that driving X-mode cyclotron damping. Second, generalized Landau damping corresponds to the $l = 0$ resonance and is also characterized by an electric field with a non-zero electric field parallel to the background magnetic field. However, unlike simple Landau damping, the general case includes non-zero perpendicular components of electric field $E_{1x}$, $E_{1y}$. Since $l = 0$, these components by assumption are not resonant with any cyclotron harmonics and thus do not produce any X-mode cyclotron damping.

With respect to O-mode cyclotron damping, note that similar $\mathbf{v}_\parallel \times \mathbf{B}_1$ terms also exist for X-mode cyclotron damping but have been neglected since they are small compared to the $E_\pm$ terms. The $\mathbf{v}_\parallel \times \mathbf{B}_1$ term is of comparable amplitude for O-mode cyclotron damping but in this case there is, by definition, no larger $E_\pm$ terms with which to make a comparison. The end result is that the O-mode cyclotron damping is weaker, at least at first glance, than that of the X mode because of the relative smallness of the O-mode $v_\parallel B_{1y}$ compared to the X-mode $E_\pm$.

Another point to keep in mind is that the interesting regime for O-mode cyclotron damping corresponds to propagation with $k_\parallel \to 0$. As will be shown, this maximizes the density at which heating can occur. Allowing $k_\parallel \to 0$ is a non-trivial mathematical limit. As such, the O-mode cyclotron damping analysis is carried out for finite $k_\parallel$ with the limit $k_\parallel \to 0$ explicitly evaluated in Section 15.7. However, in the cold plasma O-mode accessibility analysis one can set $k_\parallel = 0$ at the outset. For comparison, generalized Landau damping, which is applicable to LHCD, requires a finite $k_\parallel$. Thus $k_\parallel \neq 0$ must be assumed for both the damping and cold plasma accessibility analysis.

Since the collisionless damping calculations for both O-mode cyclotron damping and generalized Landau damping require a non-zero $k_\parallel$ as well as a finite $E_\parallel$, a single generalized analysis is all that is required. The details are very similar to those already presented for simple Landau damping and X-mode cyclotron damping. Consequently it suffices to simply outline the basic steps and state the relevant results. These are described below for electrons.

The analysis starts with the use of Faraday's law which relates $B_{1y}$ to $E_\parallel$. A simple calculation shows that $B_{1y} = -k_\perp E_\parallel/\omega$. The basic single-particle equations of motion for an electron then reduce to

$$\frac{d\mathbf{v}}{dt} + \Omega_e \mathbf{v} \times \mathbf{e}_z = -\frac{eE_\parallel}{m_e}\left[\frac{k_\perp v_z}{\omega}\mathbf{e}_x + \left(1 - \frac{k_\perp v_x}{\omega}\right)\mathbf{e}_z\right]\cos\phi,$$

$$\frac{d\mathbf{r}}{dt} = \mathbf{v}.$$

(15.122)

Here $\phi = k_\perp x + k_\parallel z - \omega t$, and the initial conditions are $\mathbf{v}(0) = v_\perp \mathbf{e}_x + v_\parallel \mathbf{e}_z$, $\mathbf{r}(0) = \mathbf{r}_i = x_i \mathbf{e}_x + y_i \mathbf{e}_y + z_i \mathbf{e}_z$.

The solution is again determined by an expansion in amplitude: $\mathbf{v} = \mathbf{v}_0 + \mathbf{v}_1 + \cdots$, $\mathbf{r} = \mathbf{r}_0 + \mathbf{r}_1 + \cdots$. The zeroth order solution corresponds to gyro motion and is given by

$$\mathbf{v}_0 = v_\perp(\cos\Omega_e t\, \mathbf{e}_x + \sin\Omega_e t\, \mathbf{e}_y) + v_\parallel \mathbf{e}_z,$$

$$\mathbf{r}_0 = \mathbf{r}_i + \frac{v_\perp}{\Omega_e}[\sin\Omega_e t\, \mathbf{e}_x - (\cos\Omega_e t - 1)\mathbf{e}_y] + v_\parallel t\, \mathbf{e}_z.$$

(15.123)

The first order solution requires a straightforward but slightly lengthy calculation. The quantities required for the energy calculation can be written as

$$v_{1x} = \frac{eE_\parallel}{m_e}\frac{k_\perp v_\parallel}{\omega}\sum_n J_n(w)\, V_x(t),$$

$$x_1 = \frac{eE_\parallel}{m_e}\frac{k_\perp v_\parallel}{\omega}\sum_n J_n(w)\, X(t),$$

$$v_{1z} = \frac{eE_\parallel}{m_e}\sum_n J_n(w)\, V_z(t),$$

$$z_1 = \frac{eE_\parallel}{m_e}\sum_n J_n(w)\, Z(t),$$

(15.124)

where $w = k_\perp v_\perp/\Omega_e$ and

$$V_x = \frac{\bar{\omega}_n \sin(\phi_i - \bar{\omega}_n t)}{\bar{\omega}_{n+1}\bar{\omega}_{n-1}} - \frac{\sin(\phi_i - \Omega_e t)}{2\bar{\omega}_{n+1}} - \frac{\sin(\phi_i + \Omega_e t)}{2\bar{\omega}_{n-1}},$$

$$X = \frac{\cos(\phi_i - \bar{\omega}_n t)}{\bar{\omega}_{n+1}\bar{\omega}_{n-1}} - \frac{\cos(\phi_i - \Omega_e t)}{2\bar{\omega}_{n+1}\Omega_e} + \frac{\cos(\phi_i + \Omega_e t)}{2\bar{\omega}_{n-1}\Omega_e},$$

$$V_z = \left(1 - \frac{n\Omega_e}{\omega}\right)\frac{\sin(\phi_i - \bar{\omega}_n t) - \sin\phi_i}{\bar{\omega}_n},$$

$$Z = \left(1 - \frac{n\Omega_e}{\omega}\right)\left[\frac{\cos(\phi_i - \bar{\omega}_n t) - \cos\phi_i}{\bar{\omega}_n^2} - \frac{t\sin\phi_i}{\bar{\omega}_n}\right].$$

(15.125)

The next step is to calculate the energy absorption rate for the electrons, averaged over initial positions. After a slightly tedious calculation one obtains

$$
\frac{d\overline{W}}{dt} = -\frac{1}{2\pi} \int_0^{2\pi} d\phi_i \frac{e}{m_e} \left( v_{0z} E_z + v_{0z} \frac{\partial E_z}{\partial x_0} x_1 + v_{0z} \frac{\partial E_z}{\partial z_0} z_1 + v_{1z} E_z \right)
$$

$$
= \frac{e^2 E_z^2}{2m_e} \left[ \frac{l k_\perp^2 v_\parallel^2}{\omega \Omega_e} \left( \frac{1}{w} \frac{dJ_l^2}{dw} \right) + \frac{\omega - l\Omega_e}{\omega} J_l^2 \frac{d}{d\overline{w}_l} \left( \frac{\overline{w}_l - \omega + l\Omega_e}{\overline{w}_l} \sin \overline{w}_l t \right) \right]. \quad (15.126)
$$

In the second form only the resonant contributions from the double sum have been maintained.

The analysis continues by multiplying $d\overline{W}/dt$ by the distribution function $f_0(\mathbf{v}) = n_0 f_\perp(v_\perp) f_\parallel(v_\parallel)$ and integrating over all velocities. This yields an expression for the power absorbed per unit volume $S_O$. A short calculation using a Maxwellian distribution function leads to

$$
S_O = \frac{\pi^{1/2} n_0 e^2 E_\parallel^2}{m_e} e^{-b} I_l(b) \frac{(\omega - l\Omega_e)^2}{k_\parallel^3 v_{Te}^3} \exp\left[ -\left( \frac{\omega - l\Omega_e}{k_\parallel v_{Te}} \right)^2 \right], \quad (15.127)
$$

where $b = k_\perp^2 v_{Te}^2 / 2\Omega_e^2$.

The last step is to calculate the damping rate $\omega_i$ from the power balance relation $\omega_i = -S_O/2U_0$. One finally obtains the desired relation

$$
\frac{\omega_i}{\omega_{pe}} = -\frac{2\pi^{1/2}}{M_O} \frac{\omega_{pe}}{k_\parallel v_{Te}} \left( \frac{\omega - l\Omega_e}{k_\parallel v_{Te}} \right)^2 e^{-b} I_l(b) \exp\left[ -\left( \frac{\omega - l\Omega_e}{k_\parallel v_{Te}} \right)^2 \right]. \quad (15.128)
$$

Here $M_O = M_\parallel + n_\perp^2 = 1 + n_\perp^2 + (\omega_{pe}^2 + \omega_{pi}^2)/\omega^2$ for O-mode cyclotron damping, where $E_{1x} = E_{1y} = 0$. For generalized Landau damping with $E_{1x} \neq 0$, $E_{1y} \neq 0$ the complete form $M_O = \mathbf{E}_1^* \cdot \overset{\leftrightarrow}{\mathbf{M}} \cdot \mathbf{E}_1 / E_\parallel^2$ must be used.

There are several important points to note, related to a comparison of the O-mode and X-mode electron cyclotron damping rates. First, for the fundamental frequency (i.e., $l = 1$) the X-mode damping rate is proportional to $I_0(b)$, while that of the O mode is proportional to $I_1(b)$. Thus, $\omega_i$ for the X mode is larger by a factor of $1/k_\perp^2 r_{Le}^2$ although, as is shown later, this benefit turns out to be elusive. Second, the O-mode damping rate has an additional multiplicative factor of $(\omega - l\Omega_e)^2 / k_\parallel^2 v_{Te}^2$ implying that exactly on cyclotron resonance ($\omega = l\Omega_e$) the damping is zero. The maximum damping occurs somewhat off resonance at two locations corresponding to $\omega = l\Omega_e \pm k_\parallel v_{Te}$. Lastly, observe that for $l = 0$ and $b \ll 1$, the generalized Landau damping rate reduces to the simple Landau damping rate, as expected.

The spatial damping rate $k_{\perp i}$, required for the heating and current drive applications, is again calculated from the relation $k_{\perp i} = S_O/2P_\perp$. A short calculation yields the desired results:

O-mode cyclotron damping:

$$\frac{k_{\perp i}c}{\omega_{pe}} = \frac{\pi^{1/2}}{n_\perp} \frac{\omega_{pe}}{k_\parallel v_{Te}} e^{-b} I_l(b) \left(\frac{\omega - l\Omega_e}{k_\parallel v_{Te}}\right)^2 \exp\left[-\left(\frac{\omega - l\Omega_e}{k_\parallel v_{Te}}\right)^2\right]. \quad (15.129)$$

Generalized Landau damping:

$$\frac{k_{\perp i}c}{\omega_{pe}} = \pi^{1/2} \frac{|E_\parallel^2|}{E_0^2} \frac{\omega_{pe}}{k_\parallel v_{Te}} e^{-b} I_0(b) \left(\frac{\omega}{k_\parallel v_{Te}}\right)^2 \exp\left[-\left(\frac{\omega}{k_\parallel v_{Te}}\right)^2\right],$$

$$E_0^2 = (n_\perp E_\parallel^* - n_\parallel E_{1x}^*)E_\parallel + n_\perp |E_{1y}|^2. \quad (15.130)$$

This completes the derivation of the damping rates for the various types of collisionless damping in a fusion plasma.

### 15.6.4 Summary

Landau and cyclotron damping have been calculated using a simple single-particle analysis. The main results required to analyze heating and current drive are the evaluations of the spatial damping rates. These are summarized below for Maxwellian distribution functions.

X-mode cyclotron damping:

$$\frac{k_{\perp i}c}{\omega_{pj}} = \frac{\pi^{1/2}}{n_\perp} \left|\frac{E_\pm}{E_{1y}}\right|^2 \frac{\omega_{pj}}{k_\parallel v_{Tj}} \frac{d}{db}[be^{-b}I_{l-1}(b)] \exp\left[-\left(\frac{\omega - l\Omega_j}{k_\parallel v_{Tj}}\right)^2\right]. \quad (15.131)$$

O-mode cyclotron damping:

$$\frac{k_{\perp i}c}{\omega_{pj}} = \frac{\pi^{1/2}}{n_\perp} \frac{\omega_{pj}}{k_\parallel v_{Tj}} e^{-b} I_l(b) \left(\frac{\omega - l\Omega_j}{k_\parallel v_{Tj}}\right)^2 \exp\left[-\left(\frac{\omega - l\Omega_j}{k_\parallel v_{Tj}}\right)^2\right]. \quad (15.132)$$

Generalized Landau damping:

$$\frac{k_{\perp i}c}{\omega_{pj}} = \pi^{1/2} \frac{|E_\parallel^2|}{E_0^2} \frac{\omega_{pj}}{k_\parallel v_{Tj}} e^{-b} I_0(b) \left(\frac{\omega}{k_\parallel v_{Tj}}\right)^2 \exp\left[-\left(\frac{\omega}{k_\parallel v_{Tj}}\right)^2\right],$$

$$E_0^2 = (n_\perp E_\parallel^* - n_\parallel E_{1x}^*)E_\parallel + n_\perp |E_{1y}|^2. \quad (15.133)$$

Here $b = (1/2)(k_\perp v_{Tj}/\Omega_j)^2 = (1/2)(v_{Tj}/c)^2(\omega/\Omega_j)^2 n_\perp^2$ and $j$ denotes ions or electrons. The upper and lower signs for X-mode cyclotron damping correspond to ions and electrons respectively. The explicit evaluation of the various ratios of electric fields as well as the value of $n_\perp^2$ requires a detailed analysis of the cold plasma dispersion relation.

All the separate pieces have now been calculated and one is therefore in a position to combine these results to study ECH, ICH, and LHCD.

## 15.7 Electron cyclotron heating (ECH)

ECH has proven to be a useful heating mechanism in present day tokamaks and may play an important role in a next generation ignition experiment such as ITER or in a fusion reactor. Because of the small wavelengths involved ECH provides the possibility of tailoring the profiles by controlled localized heating, a feature that is desirable for maximizing MHD $\beta$ limits, maximizing the bootstrap fraction, and suppressing localized resistive MHD instabilities. Its biggest drawback is a lack of high-power, steady state sources, although substantial development programs have been initiated to address this issue. The expectation is that such sources should be available when they are needed for ITER. The primary goal of this section is to develop an understanding of ECH and to calculate the absorption fraction, which is a critical parameter for reactor power balance and economics.

ECH can occur for either the O mode or the X mode. Two guiding principles that help decide the best option are as follows. First, there is a density limit above which ECH waves are reflected, implying that the accessibility becomes very poor beyond this point. This density limit can be expressed in terms of the plasma frequency and cyclotron frequency and is shown to be of the form $\omega_{pe}^2/\Omega_e^2 \leq K$, where $K$ is a constant of order unity, whose precise value depends upon the details of the ECH wave under consideration. Keep in mind that in the center of the simplified reactor the parameters corresponding to a density peaking factor of $n(0)/\bar{n} = 4/3$ are $\omega_{pe} = 8.0 \times 10^{11}$ s$^{-1}$, $\Omega_e = 8.3 \times 10^{11}$ s$^{-1}$ and $\omega_{pe}^2/\Omega_e^2 = 0.96$. These values are quite close to the ECH density limit. The first guiding principle, therefore, is to design the ECH system so that the density limit is as high as possible.

The second principle is that in general it is desirable to operate at as low a cyclotron harmonic as possible, with $l = 1$ being the optimum choice. The reason is that the damping rate is proportional to $(k_\perp^2 r_{Le}^2)^l$ for the O mode and $(k_\perp^2 r_{Le}^2)^{l-1}$ for the X mode. Since $k_\perp r_{Le} \ll 1$ in the regime of interest, the higher harmonic damping rates decrease rapidly with increasing values of $l$.

In terms of typical operation, ECH waves are usually launched from the outside of the torus for reasons of maximum geometric access. The waves are launched essentially perpendicular to the plasma implying that $n_\parallel \approx 0$. In practice there is a spectrum of $n_\parallel$ waves due to the finite structure of the launcher. Still, for present purposes it is a good approximation to set $n_\parallel = 0$ when discussing accessibility from the cold plasma dispersion relation, but to assume that $n_\parallel$ is small but finite when calculating the damping. The final power absorption fraction becomes independent of $n_\parallel$ in the limit $n_\parallel \to 0$.

The analysis below demonstrates the following points. There is reasonably good accessibility to the center of the plasma using O-mode ECH at the fundamental $l = 1$ harmonic. However, its damping rate is smaller by a factor of $k_\perp^2 r_{Le}^2$ than that of the $l = 1$ X mode. On the other hand, the $l = 1$ X mode only has good accessibility from an awkward high-field inside launch. An outside launch for X-mode ECH is only possible at the second harmonic (i.e., $l = 2$). The second harmonic X-mode damping rate is smaller by $k_\perp^2 r_{Le}^2$ from the fundamental, thus making it comparable to the damping rate of the O mode. One advantage of $l = 2$ X-mode heating is that the density limit is a factor of 2 higher than that of the O mode. One disadvantage is that it requires a driving frequency twice as high. For a reactor this corresponds to a frequency $f = 280$ GHz. High-power, steady sources at this frequency are at a very early stage of development, and one will have to wait to see how practical such sources become. In contrast, the development of $f = 140$ GHz gyrotrons

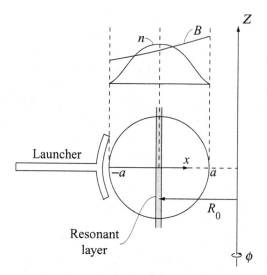

Figure 15.17 Geometry to study accessibility. The wave is launched at $x = -a$ and $x = 0$ is the center of the plasma. The resonant layer is a vertical strip corresponding to $\omega = \Omega_e(R_0)$.

for $l = 1$ resonance is much closer to fruition. In any event, even with the subtleties and issues described above ECH leads to a very high fraction of absorbed power, assuming that sources are available and that $n$ is below the density limit.

The strategy for the analysis is first to examine accessibility using the cold plasma dispersion relation and then to calculate the power absorption fraction using the collisionless damping rates. This procedure is carried out for both the O mode and X mode.

### 15.7.1 O-mode accessibility

The accessibility of the ECH O mode is easily determined from the cold plasma dispersion relation by assuming that $n_\parallel = 0$ and considering the wave polarization for which only $E_\parallel \neq 0$ (i.e., $E_+ = E_- = 0$). From Eq. (15.71) one sees that the O-mode dispersion relation is

$$n_\perp^2 = K_\parallel = 1 - \frac{\omega_{pe}^2}{\omega^2} - \frac{\omega_{pi}^2}{\omega^2} \approx 1 - \frac{\omega_{pe}^2}{\omega^2}. \tag{15.134}$$

A convenient way to understand the accessibility condition is as follows. Consider the geometry illustrated in Fig. 15.17. Waves are launched from the low-field side on the outside of the torus. As they propagate towards the center of the plasma (i.e., move to the right in the diagram) both the density and toroidal magnetic field increase. Next, observe that since $B_\phi \approx B_0 (R_0/R)$ is the dominant field in a tokamak, the cyclotron resonant absorption region corresponds to a narrow vertical strip in the plasma. Heating occurs where the ECH wave pattern overlaps with the cyclotron layer. For central heating the frequency must be chosen such that $\omega = \Omega_e(R_0)$ for the $l = 1$ fundamental.

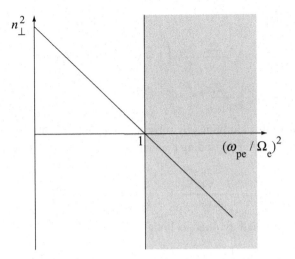

Figure 15.18 Dispersion relation for an outside launch using the ECH O mode. There is good accessibility for $\omega_{pe}^2/\Omega_e^2 < 1$.

This value of frequency is now substituted into Eq. (15.134) from which one can make a plot of $n_\perp^2$ vs. $\omega_{pe}^2/\Omega_e^2$ as shown in Fig. 15.18. (Note that for the O mode one can neglect toroidal effects and simply assume $\Omega_e = $ const. across the plasma.) Starting on the axis and moving to the right shows the value of $n_\perp^2$ as a function of increasing density as the waves propagate towards the plasma center. Good accessibility implies that $n_\perp^2 > 0$. Whenever, $n_\perp^2$ crosses a cutoff its value becomes negative and the accessibility is then very poor. Such regions are always shaded in the diagram. Thus, the region of good accessibility corresponds to the unshaded portion of the diagram starting from the axis on the left. The first transition to a shaded cutoff region represents the maximum density accessible to the ECH waves. This is the ECH density limit.

An examination of Fig. 15.18 shows that ECH with the O mode has good accessibility for densities satisfying the condition

$$\omega_{pe}^2/\Omega_e^2 \le 1. \tag{15.135}$$

This value is satisfactory for many present day experiments and is close to what is needed in a reactor.

### 15.7.2 O-mode absorption

The goal now is to calculate the fraction of input power absorbed as the wave passes through the resonant layer. The analysis involves substitution of the spatial damping rate $k_{\perp i}$ into a simple power balance relation, which then leads to the desired absorption fraction.

To obtain an explicit expression for $k_{\perp i}$, one sets $l = 1$ (heating at the fundamental) and substitutes the value of $n_\perp = (1 - \omega_{pe}^2/\omega^2)^{1/2}$ (from the accessibility analysis) into

Eq. (15.132), yielding

$$\frac{k_{\perp i} c}{\Omega_e} = \pi^{1/2} \frac{\Omega_e}{k_{\parallel} v_{Te}} \frac{\omega_{pe}^2}{\Omega_e^2} \left(1 - \frac{\omega_{pe}^2}{\omega^2}\right)^{-1/2} e^{-b} I_1(b) \zeta^2 e^{-\zeta^2},$$

$$\zeta = \frac{\omega - \Omega_e}{k_{\parallel} v_{Te}}. \tag{15.136}$$

This expression is further simplified by substituting $\omega = \Omega_e$ (for central heating) in all the multiplicative terms, Taylor expanding $\zeta$ about $x = 0$ assuming the resonance layer is very thin:

$$\zeta = \frac{\omega - \Omega_e(x)}{k_{\parallel} v_{Te}} \approx -\frac{\Omega_e' x}{k_{\parallel} v_{Te}} \approx -\frac{\Omega_e}{k_{\parallel} v_{Te}} \frac{x}{R_0}, \tag{15.137}$$

and noting that in the small gyro radius limit

$$e^{-b} I_1(b) \approx \frac{1}{4} \frac{k_{\perp}^2 v_{Te}^2}{\Omega_e^2} = \frac{n_{\perp}^2}{4} \frac{v_{Te}^2}{c^2} \frac{\omega^2}{\Omega_e^2} = \frac{1}{4} \left(1 - \frac{\omega_{pe}^2}{\Omega_e^2}\right) \frac{v_{Te}^2}{c^2}. \tag{15.138}$$

The desired expression for $k_{\perp i}$ reduces to

$$k_{\perp i} = \frac{\pi^{1/2}}{4} \frac{v_{Te} \Omega_e^2}{k_{\parallel} c^3} \frac{\omega_{pe}^2}{\Omega_e^2} \left(1 - \frac{\omega_{pe}^2}{\Omega_e^2}\right)^{1/2} \zeta^2 e^{-\zeta^2}. \tag{15.139}$$

The next step involves power balance. The reasoning is as follows. The flux of power flowing into the plasma from the launcher is given by the perpendicular component of the Poynting vector evaluated at $x = -a$:

$$P_{in} = \frac{1}{2\mu_0} \text{Re}(\mathbf{e}_x \cdot \mathbf{E} \times \mathbf{B}^*)_{x=-a} \quad \text{W/m}^2. \tag{15.140}$$

The power flux that propagates through the plasma without being absorbed and which ultimately escapes from the far side of the plasma is equal to the Poynting flux at $x = +a$:

$$P_{out} = \frac{1}{2\mu_0} \text{Re}(\mathbf{e}_x \cdot \mathbf{E} \times \mathbf{B}^*)_{x=+a} \quad \text{W/m}^2. \tag{15.141}$$

The difference $P_{in} - P_{out}$ represents the power flux absorbed by the plasma. This implies that the heating efficiency $\eta_h$, which is defined as the fraction of power absorbed, can be written as

$$\eta_h = \frac{P_{in} - P_{out}}{P_{in}}. \tag{15.142}$$

Now, note that a simple relationship exists between $P_{out}$ and $P_{in}$ as a consequence of the fact that the power flux is exponentially damped, with a damping rate equal to $2k_{\perp i}$, as it passes through the plasma. The factor "2" arises because the Poynting flux depends quadratically on the field amplitudes. The relation between $P_{out}$ and $P_{in}$ is thus

$$P_{out} = P_{in} \exp\left(-2 \int_{-a}^{a} k_{\perp i} dx\right). \tag{15.143}$$

The expression for the heating efficiency reduces to

$$\eta_h = 1 - e^{-\lambda},$$
$$\lambda = 2 \int_{-\infty}^{\infty} k_{\perp i} dx, \tag{15.144}$$

where for a thin layer the integration limits can, for simplicity, be extended to $\pm\infty$ with negligible error.

The parameter $\lambda$ is easily evaluated using the expression for $k_{\perp i}$ given by Eq. (15.139). A short calculation yields

$$\lambda = \frac{\pi}{4} \frac{v_{Te}^2}{c^2} \frac{\Omega_e R_0}{c} \frac{\omega_{pe}^2}{\Omega_e^2} \left(1 - \frac{\omega_{pe}^2}{\Omega_e^2}\right)^{1/2}. \tag{15.145}$$

An examination of Eq. (15.145) indicates that there is an optimum density to maximize absorption. The optimum density has the value

$$\omega_{pe}^2/\Omega_e^2 = \tfrac{2}{3}, \tag{15.146}$$

which is slightly below the maximum density limit. Substituting this value into Eq. (15.145) leads to the final, desired form of the coefficient $\lambda$:

$$\lambda = \frac{\pi}{6\sqrt{3}} \frac{v_{Te}^2}{c^2} \frac{\Omega_e R_0}{c}. \tag{15.147}$$

Note that $\lambda$, and hence the absorption, increases linearly with temperature: $\lambda \propto T$. For typical reactor parameters ($B_0 = 4.7$ T, $R_0 = 5$ m) one finds that for a starting temperature of $T = 1$ keV, the absorption coefficient $\lambda \approx 16$. All the power is absorbed and $\eta_h \approx 1$, indeed a favorable result.

Lastly, observe that while $k_\parallel$ has been assumed to be small but finite, the final value of $\lambda$ is independent of $k_\parallel$. However, setting $k_\parallel = 0$ at the outset of the calculation would lead to the result that $k_{\perp i} = 0$, because of the strong $k_\parallel$ dependence in the exponential factor. This, in turn, would imply that $\eta_h = 0$, corresponding to zero power absorbed. These two opposing conclusions with respect to $\eta_h$ are reconciled by a more sophisticated calculation in which relativistic effects are included. The relativistic calculation shows that there is a smooth limit as $k_\parallel \to 0$ with the damping rate identical to that derived for small but finite $k_\parallel$. Although the relativistic derivation is not presented here in order to keep the analysis simple, readers should keep in mind the necessity of including relativity in order to obtain a correct result for very small $k_\parallel$.

The overall conclusion is that O-mode ECH is a very effective way to heat a plasma. The electrons absorb all the wave energy in a narrow region of space centered about the electron cyclotron resonance layer. The energy is then shared with the ions by means of Coulomb collisions and rapidly equilibrates on each poloidal flux surface because of high parallel thermal conductivity. Since heating takes place at the $l = 1$ fundamental harmonic corresponding to $f \approx 140$ GHz, the high-power, steady state sources under development should be available by the time they are needed. The main drawback is that the maximum heating takes place at a peak density slightly lower than that required in a reactor.

### 15.7.3 X-mode accessibility

The X mode offers another option for using ECH to heat a plasma. In this case, accessibility is easily determined from Eq. (15.71), again assuming $n_\parallel = 0$ but choosing the polarization to satisfy $E_x \neq 0$, $E_y \neq 0$ and $E_\parallel = 0$. As for the O mode, $n_\parallel = 0$ is appropriate for accessibility but a small but finite $n_\parallel$ must be maintained for the absorption calculation. A simple calculation shows that the value of $n_\perp^2$ from the cold plasma dispersion relation is given by

$$n_\perp^2 = \frac{K_\perp^2 - K_A^2}{K_\perp}. \tag{15.148}$$

After substituting $K_\perp$ and $K_A$ from Eq. (15.70), and assuming $\omega \gg \Omega_i$, $\omega \gg \omega_{pi}$, one obtains

$$n_\perp^2 = \frac{\left(\omega^2 + \omega\Omega_e - \omega_{pe}^2\right)\left(\omega^2 - \omega\Omega_e - \omega_{pe}^2\right)}{\omega^2\left(\omega^2 - \Omega_e^2 - \omega_{pe}^2\right)}. \tag{15.149}$$

Consider now the possibility of X-mode heating at the $l = 1$ fundamental. Observe first that if toroidal effects are neglected (i.e., $\Omega_e = \text{const.}$) there is an indeterminate limit at the launcher position where $\omega = \Omega_e$ and $\omega_{pe} = 0$; both the numerator and denominator simultaneously vanish. The degeneracy is removed by including the toroidal variation of $\Omega_e$: $\Omega_e = \Omega_{e0}(R_0/R)$. When this modification is introduced one must distinguish between an inside launch and an outside launch.

To illustrate the point, assume that the density and magnetic field are given by $\omega_{pe}^2(\rho) = \omega_{pe0}^2(1 - \rho^2)$ and $\Omega_e = \Omega_{e0}(R_0/R) \approx \Omega_{e0}(1 \pm \varepsilon\rho)$. Here $\rho = |x|/a$, $0 \leq \rho \leq 1$, and $\rho = 0$ is the center of the plasma. The upper sign corresponds to a high-field inside launch, while the lower sign corresponds to a low-field outside launch. As a specific example assume that the plasma properties satisfy $\omega_{pe0} = \Omega_{e0}$, which is the same as the density limit for the O mode. Also, for X-mode ECH at the fundamental one must set $\omega = \Omega_{e0}$ for central heating. Under these assumptions it follows that $\omega_{pe}^2/\omega^2 = 1 - \rho^2$ and $\Omega_e/\Omega_{e0} = 1 \pm \varepsilon\rho$. As $\rho$ varies between 0 and 1 it is now straightforward to parametrically plot $n_\perp^2$ vs. $\omega_{pe}^2(\rho)/\Omega_{e0}^2$.

The curve corresponding to an inside launch is illustrated in Fig. 15.19(a) for the case $\varepsilon = 1/3$. Observe that there is good accessibility since $n_\perp^2 > 0$ from the launcher to the center of the plasma. However, the need for an inside launch is highly undesirable from a practical point of view and on this basis the option is rejected.

The more practical option of an outside launch is illustrated in Fig. 15.19(b), also for $\varepsilon = 1/3$. For this case note that there is a cutoff at the value of density for which

$$\frac{\omega_{pe}^2}{\Omega_{e0}^2} = \frac{\omega^2}{\Omega_{e0}^2} - \frac{\omega\Omega_e}{\Omega_{e0}^2} = 0.282, \tag{15.150}$$

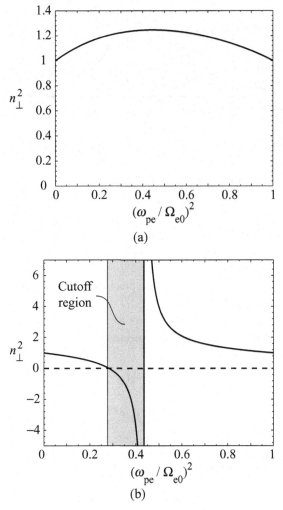

Figure 15.19 (a) Dispersion relation for the fundamental ECH X mode assuming a high-field, inside launch for the case $\varepsilon = 1/3$. There is good accessibility from the launch site to the center of the plasma. (b) The same dispersion relation for the low-field launch. Observe the cutoff and wave resonance at low $\omega_{pe}^2 / \Omega_{e0}^2$.

where the numerical value corresponds to the parameters in the example. There is also a wave resonance at the "upper hybrid resonance" at the value of density given by

$$\frac{\omega_{pe}^2}{\Omega_{e0}^2} = \frac{\omega^2}{\Omega_{e0}^2} - \frac{\Omega_e^2}{\Omega_{e0}^2} = 0.438. \qquad (15.151)$$

The implication is that the incoming wave sees a cutoff region for $0.282 < \omega_{pe}^2/\Omega_{e0}^2 < 0.438$. Since the wave cannot penetrate to the center of the plasma the low-field launch option must also be rejected because of poor accessibility.

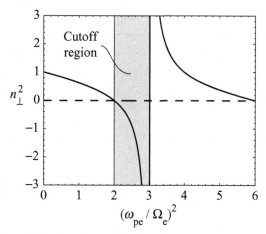

Figure 15.20 Dispersion relation for the second harmonic ECH X mode assuming a low-field, outside launch and $\varepsilon = 1/3$. Observe that there is good accessibility for $\omega_{pe}^2/\Omega_e^2 < 2$.

The conclusion is that X mode ECH is not a practical heating method at the fundamental $l = 1$ harmonic.

The situation is greatly improved when one considers second harmonic heating. In this case the toroidal variation in $\Omega_e$ can be neglected (i.e., $\Omega_e = \text{const.}$) and one must set $\omega = 2\Omega_e$. The dispersion relation reduces to

$$n_\perp^2 = \frac{\left(6 - \omega_{pe}^2/\Omega_e^2\right)\left(2 - \omega_{pe}^2/\Omega_e^2\right)}{4\left(3 - \omega_{pe}^2/\Omega_e^2\right)} \tag{15.152}$$

and is illustrated in Fig. 15.20. Observe that there is a cutoff at $\omega_{pe}^2/\Omega_e^2 = 2$ and a wave resonance (i.e., the upper hybrid resonance) at $\omega_{pe}^2/\Omega_e^2 = 3$.

The implication of Fig. 15.20 is that the second harmonic ECH X mode has good accessibility to the center of the plasma for densities below

$$\omega_{pe}^2/\Omega_e^2 \leq 2. \tag{15.153}$$

This limit is twice as high as the corresponding limit for the O mode. The higher value of critical density has the advantage of affording a larger margin of safety for typical reactor parameters, although the current lack of a 280 GHz gyrotron source is a serious technological challenge.

### 15.7.4 X-mode absorption

The X-mode absorption analysis is similar to that of the O mode. One first evaluates $k_{\perp i}$ and then calculates the absorption coefficient $\lambda$ from a simple power balance argument. The value of $\lambda$ determines the fraction of absorbed power $\eta_h$, which is the ultimate goal of the analysis.

The quantity $k_{\perp i}$ is evaluated from Eq. (15.131) as follows. First set $l = 2$ and $\omega = 2\,\Omega_e$ in all the multiplicative terms. Second, from Eq. (15.71) note that the polarization is given by

$$\frac{E_-}{E_{1y}} = \frac{i}{2}(E_{1x} - iE_{1y}) = \frac{i}{2}\left(\frac{K_A}{K_\perp} - 1\right) = -\frac{i}{4}\left(\frac{6-\alpha}{3-\alpha}\right), \tag{15.154}$$

where $\alpha = \omega_{pe}^2/\Omega_e^2$. Third, take the small gyro radius limit implying that

$$\frac{d}{db}[be^{-b}I_1(b)] \approx b = 2n_\perp^2 \frac{v_{Te}^2}{c^2}. \tag{15.155}$$

Fourth, eliminate $n_\perp^2$ using the dispersion relation

$$n_\perp^2 = \frac{(6-\alpha)(2-\alpha)}{4(3-\alpha)}. \tag{15.156}$$

A short calculation yields

$$k_{\perp i} = 2\pi^{1/2}\frac{v_{Te}\,\Omega_e^2}{c^3 k_\parallel}G(\alpha)e^{-\zeta^2}, \tag{15.157}$$

where in the limit of a thin resonant layer

$$\zeta^2 = \left(\frac{\omega - 2\Omega_e}{k_\parallel v_{Te}}\right)^2 \approx \left(\frac{2\Omega_e}{k_\parallel v_{Te}}\right)^2 \frac{x^2}{R_0^2} \tag{15.158}$$

and $G(\alpha)$ is a function of density given by

$$G(\alpha) = \frac{\alpha(2-\alpha)^{1/2}(6-\alpha)^{5/2}}{32(3-\alpha)^{5/2}}. \tag{15.159}$$

Now, a power balance argument identical to that for the O mode yields the following formula for the fraction of power absorbed ($\eta_h$):

$$\eta_h = 1 - e^{-\lambda}, \tag{15.160}$$

where

$$\lambda = 2\int_{-\infty}^{\infty} k_{\perp i}dx. \tag{15.161}$$

A simple calculation leads to

$$\lambda = 2\pi G(\alpha)\frac{v_{Te}^2}{c^2}\frac{\Omega_e R_0}{c}. \tag{15.162}$$

The function $G(\alpha)$ is illustrated in Fig. 15.21. One sees that $G(\alpha)$ has a maximum at a critical density given by

$$\alpha = \omega_{pe}^2/\Omega_e^2 = 1.75,$$
$$G(1.75) = 0.58. \tag{15.163}$$

At this optimum value

$$\lambda = 3.66\frac{v_{Te}^2}{c^2}\frac{\Omega_e R_0}{c}. \tag{15.164}$$

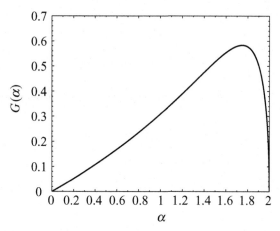

Figure 15.21 Curve of $G(\alpha)$ vs. $\alpha$ showing that the optimum density corresponds to $\alpha = \omega_{pe}^2/\Omega_e^2 = 1.75$.

Observe that the X-mode $\lambda$ has a scaling that is identical to that of the O mode. However, the multiplicative constant is more than a factor of 10 larger. The implication is that X-mode absorption is even stronger than that for the O mode, which is already very effective.

The overall conclusion is that second harmonic X mode ECH is potentially a very effective way to heat a plasma. There is good accessibility to the center of the plasma from an outside launch. All the energy is absorbed at the $\omega = 2\Omega_e$ resonant layer. The optimum density for absorption is about 2.5 times higher than for the O mode. A further practical advantage, applicable to both X mode and O mode, is that accessibility is good in the low-density region near the launcher. The implication is that the launcher can be shifted back from the plasma surface, thereby providing better protection against arcing and material damage due to interactions with the plasma edge. The one major drawback of X-mode ECH is the need for very-high-frequency sources, which have yet to be developed. Still, for many present day experiments that operate at lower fields, existing gyrotron sources do allow second harmonic X-mode ECH.

### 15.7.5 Summary

ECH can be a good way to heat a plasma. Both the $l = 1$ fundamental O mode and the $l = 2$ second harmonic X mode lead to essentially complete absorption of the power. The fraction of power absorbed can be written as

$$\eta_h = 1 - e^{-\lambda}, \tag{15.165}$$

where the absorption coefficient $\lambda$ is given by
O mode:

$$\lambda = 0.30\frac{v_{Te}^2}{c^2}\frac{\Omega_e R_0}{c} \quad \text{at} \quad \frac{\omega_{pe}^2}{\Omega_e^2} = \frac{2}{3}; \tag{15.166}$$

X mode:

$$\lambda = 3.66 \frac{v_{Te}^2}{c^2} \frac{\Omega_e R_0}{c} \quad \text{at} \quad \frac{\omega_{pe}^2}{\Omega_e^2} = 1.75. \tag{15.167}$$

In terms of a reactor the O mode has the disadvantage of a slightly low density limit while the X mode has the disadvantage of requiring high-frequency sources which have yet to be developed.

## 15.8 Ion cyclotron heating (ICH)

ICH is used in many present day tokamaks and will likely play an important role in next generation experiments and fusion reactors. One reason is the ready availability of high-power, steady state sources at a reasonable cost. A second reason is that ICH heats the ions directly. This may not be that critical in a reactor since even with electron heating, electron–ion equilibration is quite good at the high anticipated densities. However, for lower-density experiments, or even in reactor scenarios where heating to ignition takes place at low initial densities, too much electron heating could be an issue. The biggest drawback of ICH is the need for an internal antenna which, as is shown shortly, must be placed very close to the plasma surface. This leads to technological problems such as arcing and plasma breakdown. Substantial efforts are being made to improve antenna shielding. There is a reasonable expectation that satisfactory designs will be available when needed, for instance, on ITER.

As with ECH, the primary objective of this section is to develop an understanding of the physics of ICH, leading to an explicit evaluation of the absorption fraction, the critical parameter for reactor power balance and economics.

Intuitively one might expect it to be very difficult to heat a plasma with RF waves in the ion cyclotron frequency regime (e.g. typically $\Omega_i = 2.3 \times 10^8$ s$^{-1}$ corresponding to $f_i = 36$ MHz for deuterium in a reactor). The reason is that $\omega_{pe} \sim 8 \times 10^{11}$ s$^{-1}$, implying that $\Omega_i \ll \omega_{pe}$. A plasma tends to reflect waves whose frequency is much lower than the electron plasma frequency. In fact, for the ion cyclotron O mode this prediction is correct and such waves cannot penetrate into the plasma – accessibility is very poor. However, for the ion cyclotron X mode the large anisotropy in plasma properties produced by the background magnetic field opens up a window of accessibility which, interestingly, does *not* have a density limit as found with ECH. This accessibility is indeed a favorable property that makes heating with ion cyclotron waves a feasible option.

The physics of ICH accessibility is more complicated than that of ECH and in order to keep the analysis tractable attention is focused solely on the practical situation where waves are launched from the low-field, outside edge of the plasma. Even so, in order to understand the relevant issues three separate situations need to be discussed. First, an analysis is presented of the ICH X mode at the $l = 1$ fundamental $\omega = \Omega_i$. It is shown that, in principle, the accessibility is quite satisfactory. Unfortunately however, the polarization of the wave at the point of wave–particle resonance is exactly wrong; that is, the wave rotates in exactly the opposite direction to the ion motion and thus the plasma–wave coupling vanishes at the point of resonance. The resulting damping rate is too small for effective absorption.

Recognition of this problem leads to the second area of interest, second harmonic heating with the ICH X mode at $\omega = 2\,\Omega_i$. The analysis shows that good accessibility persists and that a reasonable portion of the wave has the proper in-phase polarization for strong wave–particle resonance. The calculated damping rate is large enough for fully satisfactory heating. One unfavorable feature is that the damping rate depends sensitively on temperature and density, and thus is not as robust and reliable as one might like.

A different option that alleviates this lack of robustness is the third area of interest. Fusion scientists have discovered that a clever way to improve ICH absorption is to add a small population of non-D–T particles to the plasma. The best accessibility and absorption occur when the cyclotron frequency of the new "minority" species is higher than that of deuterium. Thus hydrogen with $\Omega_H = 2\Omega_D$ and helium-3 with $\Omega_{He} = (4/3)\,\Omega_D$ are suitable minority candidates. The idea behind minority heating is to choose the ICH frequency to correspond to the $l = 1$ fundamental of the minority species. The difficulty with the "exactly wrong" polarization is overcome since the actual polarization is largely determined by the majority species. This appealing idea turns out to be not quite correct but even so the analysis shows that strong heating of the minority species does indeed occur. The minority species then transfers its energy to the bulk D–T plasma by Coulomb collisions. The calculations show that a 5% minority of helium-3 produces a substantial absorption of ICH power, sufficient to heat the plasma to ignition with high efficiency. Absorption using a hydrogen minority is even more efficient but for practical reasons, discussed shortly, is less desirable than helium-3.

With respect to the analysis itself, the essential features of accessibility are determined by using the cold plasma dispersion relation assuming that $n_\parallel^2 = 0$. As with ECH the absorption is calculated using the collisionless damping results assuming that $n_\parallel^2$ is small but finite. For X-mode second harmonic heating the final absorption results are independent of $n_\parallel^2$ in the limit $n_\parallel^2 \to 0$. For minority heating the situation is more complex. It is shown that one cannot take the limit $n_\parallel^2 \to 0$ in the absorption calculation due to the appearance of mathematical singularities. More importantly one does not want to let $n_\parallel^2 \to 0$ since the analysis shows that maximum absorption actually occurs for a finite value of $n_\parallel^2$. Assuming finite $n_\parallel^2$ thus maximizes absorption and eliminates the $n_\parallel^2 \to 0$ problem but then forces one to re-examine the $n_\parallel^2 = 0$ accessibility condition. This re-examination shows that accessibility is only weakly modified over most of the plasma core but that a small cutoff region always develops very near the edge of the plasma. It is this cutoff region that forces one to place the antenna as close to the plasma edge as possible to minimize reflection of the incoming waves.

### 15.8.1 X-mode fundamental accessibility and polarization

ICH accessibility is easily determined by examining the cold plasma dispersion relation under the assumption $n_\parallel^2 = 0$. The two roots, deduced from Eq. (15.71), reduce to

$$\text{O mode} \qquad n_\perp^2 = K_\parallel,$$
$$\text{X mode} \qquad n_\perp^2 = \frac{K_\perp^2 - K_A^2}{K_\perp}. \tag{15.168}$$

Simplified forms for the elements in the dielectric tensor are obtained for ICH by making use of the small electron mass expansion. For typical parameters in a fusion reactor, and in most present day experiments as well, the following expansion is applicable:

$$\omega_{pe}^2/\Omega_e^2 \sim 1,$$
$$\omega_{pi}^2/\Omega_i^2 \sim m_i/m_e,$$
$$\omega^2/\Omega_i^2 \sim 1, \qquad (15.169)$$
$$\omega_{pe}^2/\omega^2 \sim (m_i/m_e)^2.$$

From this ordering one immediately sees that the O-mode dispersion relation reduces to

$$n_\perp^2 = -\omega_{pe}^2/\omega^2. \qquad (15.170)$$

Since $n_\perp^2$ is large and negative, there is no accessibility to the center of the plasma and, consequently, no further discussion is warranted for the O mode.

Consider now the more interesting X mode, focusing for simplicity on a plasma with only a single ion species. The dielectric elements $K_\perp$ and $K_A$ are evaluated from Eq. (15.70). The dominant contribution to $K_\perp$, in the context of the mass ratio expansion, arises solely from the ion term and can be written as

$$K_\perp = -\frac{\omega_{pi}^2}{\omega^2 - \Omega_i^2}. \qquad (15.171)$$

The contributions to the anisotropic element $K_A$ arise from both the electron and ion terms. After invoking the charge neutrality condition $n_e = n_i$ one finds

$$K_A = \frac{\omega}{\Omega_i}\frac{\omega_{pi}^2}{\omega^2 - \Omega_i^2}. \qquad (15.172)$$

Substitution of these results into Eq. (15.168) leads to the desired dispersion relation

$$n_\perp^2 = \omega_{pi}^2/\Omega_i^2. \qquad (15.173)$$

Since $n_\perp^2 > 0$ there is good accessibility to the center of the plasma. Furthermore, there is no density limit as in ECH heating. These two features are highly desirable from the point of view of ICH.

As an aside, note that the dispersion relation can be rewritten as

$$\omega^2 = k_\perp^2 v_A^2, \qquad (15.174)$$

and thus corresponds to the compressional Alfvén wave. Similarly, the O-mode dispersion relation for $n_\parallel^2 \neq 0$ can be shown to reduce to the shear Alfvén wave: $\omega^2 = k_\parallel^2 v_A^2$. Since $k_\perp^2 \gg k_\parallel^2$ for most applications of interest the X mode for ICH, as described by Eq. (15.173), is usually referred to as the "fast wave" in the literature.

The next and crucial issue to investigate is the polarization of the fast wave at the resonant surface. The reason is that collisionless cyclotron damping of ions is proportional to the amplitude of the left circularly polarized wave ($E_+$). However, the fast wave that reaches the resonant surface is, in general, elliptically polarized, containing a combination of both left ($E_+$) and right ($E_-$) circularly polarized waves. The actual fraction of power absorbed therefore depends upon the ratio of $E_+/E_-$. This critical parameter is easily evaluated by back-substituting into Eq. (15.71). Its value is given by

$$\frac{E_+}{E_-} = \frac{K_A + K_\perp}{K_A - K_\perp} = \frac{\omega - \Omega_i}{\omega + \Omega_i}. \tag{15.175}$$

One sees that at the point of fundamental ion cyclotron resonance ($\omega = \Omega_i$) the left polarized damping amplitude $E_+ = 0$. In other words just at the point where a left polarized wave is needed for collisionless damping, its amplitude vanishes. The wave is entirely right polarized which corresponds to the "exactly wrong" polarization.

There is a small amount of damping slightly off resonance but this is far too weak to be of interest for heating. The conclusion is that because of the unfortunate vanishing of $E_+$ at the resonant surface, ICH at the fundamental harmonic is not a viable option for heating a plasma to ignition.

### 15.8.2 *Fast mode second harmonic accessibility*

The first idea to overcome the polarization problem for ICH at the fundamental is to switch to second harmonic heating. Since $E_+ = 0$ only at the fundamental resonance, there should be a finite $E_+$ at the second harmonic. As shown below, this is indeed the situation. In addition the accessibility remains good and there is no density limit. There is, however, a disadvantage to $\omega = 2 \Omega_i$ heating in that the damping is reduced by $k_\perp^2 r_{Li}^2$.

The analysis presented below considers the problem of accessibility followed by a calculation of the damping coefficient $\lambda$. For simplicity the analysis is carried out for a single-ion-species plasma. Results for a two-ion D–T plasma are discussed at the end of the analysis.

The critical relations determining accessibility for second harmonic ICH are the same as for the fundamental, the only change being the need to replace $\omega = \Omega_i$ with $\omega = 2 \Omega_i$. The expressions for the dispersion relation and the polarization can thus be written as

$$n_\perp^2 = \frac{\omega_{pi}^2}{\Omega_i^2},$$
$$\frac{E_+}{E_-} = \frac{K_A + K_\perp}{K_A - K_\perp} = \frac{\omega - \Omega_i}{\omega + \Omega_i} = \frac{1}{3}. \tag{15.176}$$

The first relation again clearly shows good accessibility. The second relation shows that the amplitude of the desired left hand polarized wave is $\frac{1}{3}$ that of the right polarized wave. The question now is whether the factor of $\frac{1}{3}$ coupled with the $k_\perp^2 r_{Li}^2$ reduction in damping still leads to sufficient absorption to make second harmonic ICH a viable option.

### 15.8.3 Fast mode second harmonic absorption

The task now is to calculate the spatial damping decrement $k_{\perp i}$. Once $k_{\perp i}$ is known it is then straightforward to calculate the absorption coefficient $\lambda$ and the corresponding heating efficiency $\eta_h$.

The quantity $k_{\perp i}$ is evaluated from Eq. (15.131) as follows: (1) set $l = 2$ and $\omega = 2\Omega_i$ in the multiplicative terms; (2) use the second harmonic polarization result

$$\frac{E_+}{E_{1y}} = i\frac{E_+}{E_+ - E_-} = -\frac{i}{2};\tag{15.177}$$

(3) take the small gyro radius limit

$$\frac{d}{db}[be^{-b}I_1(b)] \approx b = 2n_\perp^2\frac{v_{Ti}^2}{c^2};\tag{15.178}$$

and (4) substitute the dispersion relation $n_\perp = \omega_{pi}/\Omega_i$. A short calculation yields

$$k_{\perp i} = \frac{\pi^{1/2}}{2}\frac{v_{Ti}\omega_{pi}^3}{c^3\Omega_i k_\parallel}\exp(-\zeta^2),\tag{15.179}$$

where in the limit of a narrow resonance layer

$$\zeta = \frac{\omega - 2\,\Omega_i\,(x)}{k_\parallel v_{Ti}} \approx -2\frac{\Omega_i}{k_\parallel v_{Ti}}\frac{x}{R_0}.\tag{15.180}$$

As before, $x = 0$ corresponds to the center of the plasma and the launcher is located at $x = -a \approx -\infty$.

Finally, the heating efficiency is again defined as

$$\eta_h = 1 - e^{-\lambda}\tag{15.181}$$

with

$$\lambda = 2\int_{-\infty}^\infty k_{\perp i}dx.\tag{15.182}$$

A short calculation leads to the desired expression for $\lambda$ which can be written as

$$\lambda = \frac{\pi}{2}\left(\frac{\omega_{pi}R_0}{c}\right)\beta_i,\tag{15.183}$$

where $\beta_i = 2\mu_0 n T_i/B_0^2$ is the ion beta and all quantities are evaluated at the center of the plasma $(x = 0)$. Observe that $\lambda \sim n_i^{3/2}T_i/B_0^2$, showing a strong sensitivity on the plasma parameters.

In terms of numerical values consider a reactorlike deuterium plasma with an average density of $\bar{n} = 1.5 \times 10^{20}$ m$^{-3}$, a density peaking factor of $n_0/\bar{n} = \frac{4}{3}$, and a major radius of $R_0 = 5$ m. Assume that ohmic heating has raised the average temperature to $\bar{T}_i = 3$ keV with a temperature peaking factor of $T_{i0}/\bar{T}_i = 2$. The central magnetic field is $B_0 = 4.7$ T. With these parameters one finds that $\lambda = 7.6$, implying a good absorption efficiency $\eta_h \approx 1$. This favorable result is somewhat mitigated by the strong sensitivity on plasma parameters, implying a lack of robustness in the method. In fact, if one assumes that heating to ignition follows the more subtle low-density path described in Chapter 14, then for $\bar{n} = 0.3 \times 10^{20}$ m$^{-3}$, the value of $\lambda$ is reduced to 0.68, corresponding to $\eta_h = 0.49$. This is at best a marginally satisfactory result.

The results for a 50%–50% D–T reactor are quite similar to those just presented. The analysis in this case is straightforward, the main difference being the need to include two ion species in the analysis. Note that for a D–T plasma there are two possibilities for second harmonic resonance, one at $\omega = 2\Omega_D$ and the other at $\omega = 2\Omega_T = (4/3)\Omega_D$. The corresponding values of $\lambda$ are related to the single-species value given by Eq. (15.183) as follows

$$\begin{aligned}\lambda_{DT} &= 1.00\lambda_D & \omega &= 2\Omega_D, \\ \lambda_{DT} &= 0.094\lambda_D & \omega &= (4/3)\Omega_D.\end{aligned} \tag{15.184}$$

Here $\lambda_D$ is given by Eq. (15.183) but written in terms of $n_e$, still using deuterium as the reference fluid:

$$\lambda_D = \frac{\pi e \mu_0^{3/2} n_{e0}^{3/2} T_{D0} R_0}{m_D^{1/2} B_0^2} = 2^{3/2}\left[\frac{\pi}{2}\left(\frac{\omega_{pD} R_0}{c}\right)\beta_D\right]. \tag{15.185}$$

The factor $2^{3/2}$ appears because for a D–T plasma $n_D = n_e/2$.

Observe that the damping is smaller for second harmonic T than for second harmonic D by an order of magnitude. This is largely a consequence of poorer polarization. Even so, this should not really matter since in principle one can simply use ICH at second harmonic D to achieve a high heating efficiency. However, it is the strong density and temperature dependence that provides the motivation to seek a more robust alternative, particularly at low-density operation.

### 15.8.4 Fast wave minority accessibility

An interesting idea suggested by the RF heating community to improve ICH is to add a small minority of a non D–T species to the plasma. The wave frequency is chosen to produce wave–particle resonance at the fundamental of the minority. The hope is that the gain achieved by eliminating the $k_\perp^2 r_{Li}^2$ factor at the fundamental will compensate the reduction in collisionless absorption associated with the small density of minority particles.

At first glance, the problem with the "wrong polarization" at the fundamental would seem to be mitigated based on the idea that the actual polarization should be dominated by the majority species. Therefore, as long as the cyclotron frequency of the minority species is different from that of the majority species, a substantial component of left polarized wave should be present at the minority cyclotron frequency. This idea is not quite correct. The analysis shows that at both the majority and minority cyclotron frequencies, the polarization is exactly wrong: $E_+ = 0$ at $\omega = \Omega_{major}$ and $\omega = \Omega_{minor}$. However, the polarization changes very rapidly near the minority cyclotron frequency and as a result there is substantial damping near, but not exactly at, the resonant frequency.

A further point demonstrated in the analysis is that it is desirable for the minority species to have a higher cyclotron frequency than the majority species. This property allows good accessibility to the resonant layer from a low-field launch. The implication is that either hydrogen ($\Omega_H = 2\Omega_D$) or helium-3 ($\Omega_{He} = 4\Omega_D/3$) would be a suitable minority species.

For simplicity, the analysis below is carried out for a single-species deuterium plasma with a hydrogen minority. Results are also presented for a D–T plasma with a hydrogen or helium-3 minority. The analysis, even in the simple model under consideration, is somewhat complicated, involving a wave-particle resonance layer, a wave cutoff, and a wave resonance. Furthermore, it is shown that substantial damping occurs only for small but *finite* values of $k_\parallel$. This recognition forces a re-examination of accessibility, relaxing the assumption that $n_\parallel^2 = 0$ in the cold plasma dispersion relation. The main consequence is that there always exists an additional narrow cutoff region near the launcher. This imposes the important practical constraint that the launching antenna be situated as close as possible to the edge of the plasma to avoid a large edge reflection of wave energy.

Finally, it is worth noting that the analysis presented here represents only one of many options possible for minority heating. In general, kinetic effects and plasma inhomogeneities must be included to obtain accurate answers for these other minority options. Still, the particular analysis presented here describes one of the main options for using ICH in reactors and next generation ignition experiments.

The analysis begins with a discussion of accessibility. The dispersion relation is again calculated from the cold plasma model assuming initially that $n_\parallel^2 = 0$. For a two-species plasma consisting of deuterium (the majority species) and hydrogen (the minority species) the elements of the dielectric tensor in the frequency regime $\omega \sim \Omega_D$ are given by

$$K_\perp = -\frac{\omega_{pD}^2}{\omega^2 - \Omega_D^2} - \frac{\omega_{pH}^2}{\omega^2 - \Omega_H^2},$$

$$K_A = \frac{\omega}{\Omega_D}\frac{\omega_{pD}^2}{\omega^2 - \Omega_D^2} + \frac{\omega}{\Omega_H}\frac{\omega_{pH}^2}{\omega^2 - \Omega_H^2}. \tag{15.186}$$

These expressions can be simplified by noting that for most of the terms one can set $\omega = \Omega_H$ and ignore the spatial dependence of the magnetic field. Only in those terms that explicitly contain the factor $\omega - \Omega_H$ is the spatial variation of $B$ important. Such terms are singled out and left unchanged. Then $K_\perp$ and $K_A$ further simplify to

$$K_\perp \approx -\frac{\omega_{pD}^2}{\Omega_D^2}\left(\frac{2\xi - 3}{6\xi}\right),$$

$$K_A \approx \frac{\omega_{pD}^2}{\Omega_D^2}\left(\frac{4\xi - 3}{6\xi}\right). \tag{15.187}$$

Here,

$$\xi(x) = -\frac{\omega - \Omega_H(x)}{\Omega_D(0)}\frac{n_D}{n_H} \sim 1 \tag{15.188}$$

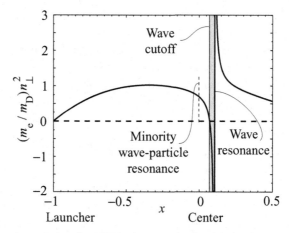

Figure 15.22 Dispersion relation for ICH hydrogen minority heating in a deuterium plasma for the case $\omega_{pD}(0)/\Omega_D(0) = (m_D/m_e)^{1/2}$, $\varepsilon = \frac{1}{3}$, $n_\parallel = 0$, and $n_H/n_D = 0.05$. Note that there is good accessibility from the launcher to the center of the plasma. A little further in there is a narrow cutoff region.

is a function that measures the small deviation away from hydrogen cyclotron resonance multiplied by the large density ratio factor, $n_D/n_H \gg 1$. Note that exact hydrogen cyclotron resonance occurs at $\xi = 0$. In the analysis that follows it will be convenient to express all quantities in terms of $\xi$.

The simplified expressions for $K_\perp$ and $K_A$ lead to the following cold plasma dispersion relation:

$$n_\perp^2 = \frac{K_\perp^2 - K_A^2}{K_\perp} = \frac{\omega_{pD}^2}{\Omega_D^2}\left(\frac{\xi - 1}{\xi - 3/2}\right). \tag{15.189}$$

Accessibility can now be investigated by noting that in the vicinity of the hydrogen cyclotron resonance

$$\xi(x) \approx 2\frac{n_D}{n_H}\frac{x}{R_0}, \tag{15.190}$$

where $B = B_0(R_0/R) \approx B_0(1 + x/R_0)$ with $x = 0$ corresponding to the center of the plasma and $x = -a$ corresponding to the location of the launcher. Good accessibility requires that $n_\perp^2 > 0$ for $\xi < 0$. An examination of Eq. (15.189) indicates that this requirement is indeed satisfied.

A further examination shows somewhat complicated behavior just beyond the resonance on the high-field side of the plasma. As shown in Fig. 15.22 a wave cutoff occurs at $\xi = 1$, followed by a wave resonance at $\xi = \frac{3}{2}$. These points are quite close to the resonant surface as can be seen by converting to real coordinates: $x/a = (2/\varepsilon)(n_H/n_D)\xi \ll 1$ for low minority concentrations. The impact of the cutoff and resonance is discussed shortly. For the moment readers should assume that there is good accessibility to the wave-particle resonant surface $\xi = 0$.

Consider now the question of the polarization. The goal is to determine under what conditions there is a substantial left polarized wave at the resonant surface. A straightforward calculation shows that

$$
\begin{aligned}
\frac{E_+}{E_-} &= \frac{K_A + K_\perp}{K_A - K_\perp} \\
&= \left[\frac{(\omega + \Omega_H) + (n_H/n_D)(\omega + \Omega_D)}{(\omega - \Omega_H) + (n_H/n_D)(\omega - \Omega_D)}\right]\left[\frac{(\omega - \Omega_D)(\omega - \Omega_H)}{(\omega + \Omega_D)(\omega + \Omega_H)}\right] \\
&\approx \frac{1}{3}\frac{\xi}{\xi - 1}.
\end{aligned}
\tag{15.191}
$$

The exact form shows that $E_+ = 0$ at both fundamental resonant frequencies, $\omega = \Omega_D$ and $\omega = \Omega_H$. The polarization is exactly wrong at both frequencies. The simpler approximate form focuses on the minority species and thus shows the wrong polarization only at the minority resonance $\xi = 0$. However, the simpler form also shows that the polarization becomes favorable just a short distance away; that is, $E_+/E_- \sim 1$ for $\xi \sim 1$. It is this rapid increase in the desired left hand polarization away from exact resonance that leads to a substantial collisionless absorption.

### 15.8.5 Fast wave minority absorption

The calculation of the ICH minority absorption coefficient is similar to previous calculations. One evaluates $k_{\perp i}$ from the collisionless damping results given by Eq. (15.131) and then integrates over the resonant layer to determine the damping coefficient $\lambda$ and the corresponding heating efficiency $\eta_h$.

The value of $k_{\perp i}$ for the $l = 1$ fundamental hydrogen resonance is calculated as follows. First, set $\omega = \Omega_H = 2\Omega_D$ in all the multiplicative terms. Second, note that in the vicinity of the minority cyclotron resonance, the wave polarization implies that

$$
\frac{E_+}{E_{1y}} = i\frac{E_+}{E_+ - E_-} = i\frac{\xi}{3 - 2\xi}.
\tag{15.192}
$$

Third, take the small gyro radius limit

$$
\frac{d}{db}[be^{-b}I_0(b)] \approx 1.
\tag{15.193}
$$

Fourth, substitute $n_\perp$ from the dispersion relation by Eq. (15.189). After a short calculation one obtains

$$
k_{\perp i} = \frac{\pi^{1/2}}{2}\frac{n_H}{n_D}\frac{\omega_{pD}\Omega_D}{ck_\parallel v_{TH}}\frac{\xi^2 e^{-\zeta^2}}{(\xi - 3/2)^{3/2}(\xi - 1)^{1/2}},
\tag{15.194}
$$

where

$$\zeta(x) = \frac{\omega - \Omega_H(x)}{k_\parallel v_{TH}} \approx -\frac{\Omega_H(0)}{k_\parallel v_{TH}} \frac{x}{R_0} = -\frac{x}{\Delta_\zeta},$$

$$\xi(x) = -\frac{\omega - \Omega_H(x)}{\Omega_D(0)} \frac{n_D}{n_H} \approx 2\frac{n_D}{n_H} \frac{x}{R_0} = \frac{x}{\Delta_\xi}.$$

(15.195)

Observe that two different small scale lengths enter the analysis. The first is $\Delta_\xi = (1/2)$ $(n_H/n_D)R_0$, representing the distance between the wave-particle resonant surface and the cutoff surface. The second is $\Delta_\zeta = R_0 k_\parallel v_{TH}/\Omega_H$, representing the finite width of the absorption layer due to the Doppler broadening associated with a small but finite $k_\parallel$.

To proceed further it is useful to introduce a new dimensionless wave number equal to the ratio of these two small scale lengths:

$$\hat{k}_\parallel = \frac{\Delta_\zeta}{\Delta_\xi} = 2\frac{n_D}{n_H} \frac{k_\parallel v_{TH}}{\Omega_H}.$$

(15.196)

This leads to the following expression for the damping coefficient:

$$\lambda = 2\int k_{\perp i}dx = \frac{\pi^{1/2}}{2}\frac{n_H}{n_D}\frac{\omega_{pD}R_0}{c}\int_{-\infty}^{\infty}\frac{\hat{k}_\parallel^2\zeta^2 e^{-\zeta^2}}{(\hat{k}_\parallel\zeta + 3/2)^{3/2}(\hat{k}_\parallel\zeta + 1)^{1/2}}d\zeta.$$

(15.197)

After this substantial amount of algebra one is now faced with two difficulties as alluded to earlier. Specifically, (1) the integral appears to diverge because of the singularity in the denominator of the integrand at $\hat{k}_\parallel\zeta = -\frac{3}{2}$ and (2) there are regions where the integrand becomes imaginary. It should be emphasized that these difficulties cannot be resolved in the context of the cold plasma model. A kinetic treatment is needed.

The results of such a kinetic treatment show that the singular, unphysical behavior is eliminated and that strong absorption occurs for $\hat{k}_\parallel\zeta \gg 1$. The regime $\hat{k}_\parallel\zeta \gg 1$ corresponds to the situation where the Doppler broadened absorption layer is much larger than the gap between resonance and cutoff. In this regime the cold plasma model can reproduce the kinetic results by (1) assuming that $\hat{k}_\parallel\zeta \gg 1$ in Eq. (15.197) and (2) simply ignoring the cutoff and resonance in the denominator. The physical reason why this works is associated with the narrowness of the cutoff–resonance gap. The narrow gap implies that very little energy is reflected at the cutoff and very little energy is mode converted at the resonance. Thus, the integrated contribution to the power absorption across the gap is small and can be ignored.

The result of this approximation is a simplified form of the integral which can be written as

$$\lim_{\hat{k}_\parallel\to\infty}\int_{-\infty}^{\infty}\frac{\hat{k}_\parallel^2\zeta^2 e^{-\zeta^2}}{(\hat{k}_\parallel\zeta + 3/2)^{3/2}(\hat{k}_\parallel\zeta + 1)^{1/2}}d\zeta \to \int_{-\infty}^{\infty}e^{-\zeta^2}d\zeta = \pi^{1/2}.$$

(15.198)

The desired final form of the damping coefficient is now given by

$$\lambda = \frac{\pi}{2}\frac{\omega_{pD}R_0}{c}\frac{n_H}{n_D}.$$

(15.199)

For $n_D = 1.5 \times 10^{20}$ m$^{-3}$, $R_0 = 5$ m, and $n_H/n_D = 0.05$ the damping coefficient has the value 15.0, implying that the efficiency $\eta_h = 1 - \exp(-\lambda) \approx 1$. This is clearly a favorable result. Furthermore, unlike second harmonic heating the damping rate is independent of temperature and only weakly dependent on density: $\lambda \propto n^{1/2}$. Thus, even along the low-density evolution to ignition described in Chapter 14, the value of $\lambda$ for a fixed $n_H/n_D = 0.05$ and $n_D = 0.3 \times 10^{20}$ m$^{-3}$ is only slightly reduced to $\lambda = 6.7$, again corresponding to $\eta_h \approx 1$.

Another point of interest is that the hydrogen minority resonance condition at the fundamental ($\omega = \Omega_H$) coincides with second harmonic heating of deuterium ($\omega = 2\Omega_D$). Consequently, the total damping is actually the sum of the two separate conditions. In other words, Eq. (15.199) should be replaced by

$$\lambda = \frac{\pi}{2} \frac{\omega_{pD} R_0}{c} \left( \frac{n_H}{n_D} + \beta_D \right). \tag{15.200}$$

There is one further issue to consider. Since efficient minority damping requires $\hat{k}_\| \zeta \gg 1$, it is of importance to re-examine the effect of a small but finite $k_\|$ on accessibility. As an example assume that $\hat{k}_\| = 1$, a value just on the border of satisfying the $\hat{k}_\| \zeta \gg 1$ condition. For the numerical example above, this translates into a value $k_\| = 10.5$ m$^{-1}$ and a corresponding value $n_\| = 7.0$. In the center of the plasma $n_\perp = 50.5$, so that the approximation $n_\|^2 \ll n_\perp^2$ is well satisfied. However, a difficulty arises at the edge of the plasma where the plasma density vanishes. Near the edge the dispersion relation reduces to the vacuum limit $n_\perp^2 = 1 - n_\|^2 \approx -n_\|^2 < 0$. The wave is damped just as it enters the plasma.

The effect can be quantified by returning to the $n_\|^2 \neq 0$ cold plasma dispersion relation given by Eq. (15.72). For ICH frequencies the dielectric element $K_\| \approx -\omega_{pe}^2/\omega^2$ is enormous compared to all other elements. This fact can be exploited to obtain a simple factoring of the dispersion relation. The branch corresponding to the fast wave is given by

$$n_\perp^2 \approx -\frac{C}{B} = \frac{(K_\perp + K_A - n_\|^2)(K_\perp - K_A - n_\|^2)}{K_\perp - n_\|^2}. \tag{15.201}$$

This function, when plotted against distance into the plasma for the case $n_\| = 7$, looks virtually identical to Fig. 15.22, which is the same curve except that $n_\| = 0$. However, if one focuses on the region near the launcher, as shown in Fig. 15.23, a narrow cutoff region appears just as the wave enters the plasma. Equation (15.201) implies that the cutoff region ends above a critical density after which the wave begins to propagate. The cutoff boundary is defined by the condition $K_\perp + K_A = n_\|^2$ and can be written as

$$\omega_{pD}^2 > 3n_\|^2 \Omega_D^2. \tag{15.202}$$

For the numerical example above, this condition reduces to the relatively low requirement $n_{20} > 0.087$. In addition, there is a vacuum gap between the actual launcher surface and the plasma edge, which also has $n_\perp^2 \approx -n_\|^2 < 0$, thereby extending the cutoff region. The practical implication is that the antenna surface must be placed as close to the plasma surface as possible to avoid a large reflection of the entering wave.

*Heating and current drive*

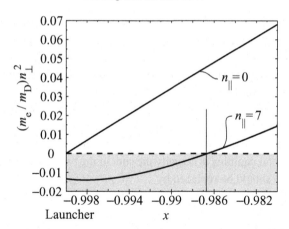

Figure 15.23 Expanded view of the ICH hydrogen minority dispersion relation near the launcher for $n_\parallel = 0$ and $n_\parallel = 7$. Observe that a finite $n_\parallel$ creates a narrow cutoff region near the launcher.

On a related practical point, note that even for second harmonic ICH $n_\parallel$ will likely be finite because of the finite structure of the port and the launcher. Here, though, finite $n_\parallel$ is a consequence of practicality but not a requirement arising from the plasma physics. For minority heating, the plasma physics requires a finite $n_\parallel$ for energy absorption.

The final topic of interest is the extension of minority heating results to a D–T plasma. A similar calculation to the one just presented shows that the damping coefficients for a hydrogen minority and a helium-3 minority are given by

$$\lambda_H = \lambda_M \left( 1.7 \frac{n_H}{n_e} + 2.0 \beta_D \right), \qquad \omega = \Omega_H = 2\Omega_D,$$

$$\lambda_{He} = \lambda_M \left( 0.5 \frac{n_{He}}{n_e} + 0.19 \beta_T \right), \qquad \omega = \Omega_{He} = 2\Omega_T, \tag{15.203}$$

where

$$\lambda_M = \frac{\pi}{2} \frac{e\mu_0^{1/2} n_e^{1/2} R_0}{m_D^{1/2}} = 2^{1/2} \left[ \frac{\pi}{2} \frac{\omega_{pD} R_0}{c} \right]. \tag{15.204}$$

Here, the second harmonic heating contributions due to deuterium for the hydrogen minority case and tritium for the helium-3 minority case have been added.

For typical parameters, both minorities produce a stronger absorption than the corresponding second harmonic damping. A hydrogen minority produces the larger damping coefficient. However, the helium-3 minority coefficient is also substantial. Both minority damping coefficients are comparable to the value $\lambda \approx 15$ for the simple case of a deuterium plasma with a hydrogen minority. Helium-3 is usually the favored choice for a practical reason not connected with heating. In general, one does not want either helium-3 or hydrogen in the plasma once it has been ignited. At ignition the minorities act like unwanted impurities diluting the fusion reactions. While helium-3 can be readily pumped out of the plasma it is very difficult practically to purge the discharge of all hydrogen. Thus, helium-3 is the favored minority species.

### 15.8.6 Summary

ICH is an effective way to heat a plasma to ignition. There are two basic approaches. First, one can utilize second harmonic heating of either deuterium or tritium. Second, one can use minority heating with either hydrogen or helium-3 serving as the minority species, with helium-3 being the favored choice for practical reasons. For reactor-like parameters minority heating is more effective and 5% helium-3 leads to full absorption of power.

The fraction of power absorbed can be written as

$$\eta_h = 1 - e^{-\lambda}, \tag{15.205}$$

where the damping coefficients $\lambda$ for the various options are

$$\lambda = \frac{\pi}{2} \frac{\omega_{pD} R_0}{c} \left( \frac{n_H}{n_D} + \beta_D \right) \qquad\qquad \text{D}+\text{H},$$

$$\lambda_H = 2^{1/2} \left[ \frac{\pi}{2} \frac{\omega_{pD} R_0}{c} \right] \left( 1.7 \frac{n_H}{n_e} + 2.0 \beta_D \right) \qquad \text{D–T}+\text{H}, \tag{15.206}$$

$$\lambda_{He} = 2^{1/2} \left[ \frac{\pi}{2} \frac{\omega_{pD} R_0}{c} \right] \left( 0.5 \frac{n_{He}}{n_e} + 0.19 \beta_T \right) \quad \text{D–T}+\text{He}^3.$$

In terms of reactor desirability, ICH has the advantage of readily available RF power at a reasonable cost. The main disadvantage is the need to place the antenna as close to the plasma edge as possible to minimize the effects of the edge cutoff region.

## 15.9 Lower hybrid current drive (LHCD)

### 15.9.1 Overview

The last major topic of the chapter involves non-inductive current drive, a necessity for the achievement of steady state operation in most fusion configurations. All of the previously discussed auxiliary heating methods (neutral beams, ECH, and ICH) can be modified to drive a steady state current. This section, however, focuses on a different method known as LHCD.

There are several reasons for this choice. First, in almost all situations of interest LHCD has the highest current-drive efficiency, defined as the number of amperes driven per watt of auxiliary power. Second, LHCD is very effective in driving off-axis currents, an important requirement for matching the natural bootstrap profiles and for establishing transport barriers at large minor radii, where they are most needed. Finally, the physics of LHCD is of scientific interest in its own right, representing a very clever way to solve a difficult problem.

The basic idea of LHCD is to launch RF waves into the plasma from the low-field side with a strongly asymmetric spectrum in $k_\parallel$. See Fig. 15.24. With such a spectrum, the

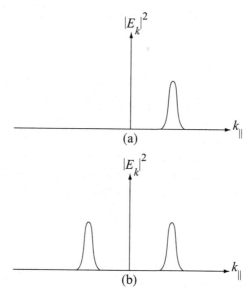

Figure 15.24 Fourier spectrum of the electric field energy for: the case of (a) current drive (asymmetric) and (b) heating (symmetric).

waves travel predominantly in one direction around the torus (against the direction of the ohmic current flow). For LHCD the wave frequency is typically between the electron and ion cyclotron frequencies: $\Omega_i \ll \omega \ll \Omega_e$. In this frequency range there is no cyclotron damping. However, for appropriately chosen parameters there can be a strong $l = 0$ Landau wave-particle resonance (at $\omega = k_\parallel v_\parallel$) acting on the electrons. At this resonance the electrons absorb energy from the wave. If the spectrum is asymmetric, there is also a net transfer of momentum to the electrons. This is in contrast to a heating spectrum, which is typically symmetric in $k_\parallel$ and consequently imparts no net momentum to the electrons; that is, electrons with $\pm v_\parallel$ absorb equal but opposite momentum at $\omega = \pm k_\parallel v_\parallel$. For an asymmetric spectrum the gain in electron momentum corresponds to a net toroidal electron fluid flow, or equivalently a net toroidal current. This is the desired current in "current drive."

Note that one must continually supply RF power to sustain the driven current. The reason is that the heating up of slow electrons and the slowing down of fast electrons produces a distortion of the distribution function away from a Maxwellian in the vicinity of the resonant velocity. Specifically, the distribution function becomes flattened near $v_\parallel = \omega/k_\parallel$ as has been shown in Fig. 15.15. Coulomb collisions act to restore the distribution function to a Maxwellian, reducing the current drive to zero. Only by continually driving the system with auxiliary power can the asymmetry in the distribution function and the corresponding current drive be maintained. The ratio of the driven current to the auxiliary power required to sustain the current is the critical parameter defining the efficiency of current drive. The analysis presented in this section shows that typically for LHCD the current-drive efficiency $\eta_{CD} \equiv$ driven current/auxiliary power $\sim 0.05$ A/W.

A subtle plasma physics point worth noting concerns the mechanism for driving current. The electrons involved in LHCD are shown to be those with velocities satisfying $v_\parallel \gtrsim 3v_{Te}$. They are somewhat out on the tail of the distribution function. As they heat and absorb momentum these electrons tend to enhance the tail of the distribution function as compared to a pure Maxwellian. Now, recall that the Coulomb collision frequency scales as $v_{ei} \sim 1/v_\parallel^3$. Therefore, since most of the current is carried by electrons with a higher $v_\parallel$, the Coulomb frictional drag is reduced, implying that there is less friction trying to restore the distribution function to a Maxwellian. The conclusion is that less power is required to sustain the current drive (i.e., the efficiency is higher) than one might have expected based solely on the classical collisional frequency $\bar{v}_{ei}$.

Also worth mentioning is the fact that the actual numerical value of $\eta_{CD}$ is a strong function of the density and temperature profiles and as a result there is no single, universal constant, determined from first principles, multiplying the scaling functions. Consequently, while the numerical value of $\eta_{CD}$ calculated in this section is plausible in terms of experimental observations, it should, nevertheless, be viewed only as a representative value.

The analysis itself is not too difficult, although it requires a considerable number of steps. A useful way to think about the problem is to assume that the plasma in a fusion reactor has been heated to the desired ignited state. Thus the temperature, density, and magnetic field profiles are known quantities. The task now is to design an LHCD system capable of driving a specified amount of non-inductive current to create steady state operation. Specifically, one wants to determine the following properties of the LHCD system: (1) the current drive efficiency $\eta_{CD}$, (2) the amplitude and radial profile of the driven current $J_{CD}(r)$, (3) the amplitude and radial profile of the lower hybrid power deposition $S_{LH}(r)$, (4) the frequency of the lower hybrid waves $\omega$, and (5) the parallel wave number of the lower hybrid waves $k_\parallel$.

As with RF heating, accessibility of lower hybrid waves to the region of interest is determined by the cold plasma dispersion relation. For LHCD it is crucial to keep $n_\parallel \neq 0$. The absorption of lower hybrid power by the Landau wave-particle resonance is readily deduced from the collisionless damping analysis already carried out.

Interestingly, for reactor-like parameters the damping is so strong that most of the lower hybrid power is absorbed in a relatively narrow penetration layer near the plasma edge, implying that current can only be driven at values of $r$ not too different from $a$. It is fortunate that the goal is to drive off-axis current, since strong Landau damping would not allow LHCD to be effective near the axis.

Lastly, the calculation of the driven current is carried out by a straightforward extension of the collisionless damping analysis to include the transfer of momentum as well as energy. A simple plasma momentum balance then determines the amplitude and profile of the driven current in terms of the wave properties.

The main conclusion from the analysis is that the current-drive efficiency is typically of order $\eta_{CD} \sim 0.05$, which, while higher than for other current-drive schemes, is still too low to economically drive all of the current in a tokamak reactor. This is the reason why the bootstrap current is so important.

### *15.9.2 Lower hybrid accessibility*

The analysis begins with a discussion of lower hybrid wave propagation and accessibility. The relevant results are obtained from the cold plasma dispersion relation assuming a frequency between the two cyclotron frequencies and a finite parallel wave number. With a finite $n_\parallel$ the wave has a combination of O-mode and X-mode electric field components, although it is the $E_\parallel$ component associated with the O mode that is responsible for the Landau damping and current drive.

In carrying out the analysis it is useful to treat the mass ratio $m_e/m_i$ as a small parameter in order to understand the basic scaling relations and to determine which terms must be kept and which can be neglected. The basic ordering that defines lower hybrid wave propagation in a plasma with reactor-like parameters is as follows:

$$\omega_{pe}/\Omega_e \sim \omega_{pi}/\omega \sim n_\parallel \sim 1,$$
$$\omega_{pi}/\Omega_i \sim \omega/\Omega_i \sim \Omega_e/\omega \sim n_\perp \sim (m_i/m_e)^{1/2} \gg 1. \tag{15.207}$$

Under this ordering scheme the elements of the dielectric tensor reduce to

$$K_\perp = 1 + \frac{\omega_{pe}^2}{\Omega_e^2} - \frac{\omega_{pi}^2}{\omega^2} \sim 1,$$

$$K_A = \frac{\omega_{pe}^2}{\omega\Omega_e} \sim (m_i/m_e)^{1/2}, \tag{15.208}$$

$$K_\parallel = -\frac{\omega_{pe}^2}{\omega^2} \sim m_i/m_e.$$

These results are substituted into the general solution for the cold plasma dispersion relation given by Eq. (15.72), leading to the following expression for $n_\perp^2$:

$$n_\perp^2 = -\frac{K_\parallel}{2K_\perp} \left\{ n_\parallel^2 - K_\perp + \frac{K_A^2}{K_\parallel} \pm \left[ \left( n_\parallel^2 - K_\perp + \frac{K_A^2}{K_\parallel} \right)^2 + \frac{4K_\perp K_A^2}{K_\parallel} \right]^{1/2} \right\}. \tag{15.209}$$

The interesting branch for LHCD corresponds to the $+$ sign for the square root. This choice results in the larger value for $n_\perp^2$ corresponding to the smaller value of $\omega/k_\perp$. Consequently, the lower hybrid wave is often referred to as the "slow wave" in the literature.

The task now is to analyze the somewhat complicated looking expression given in Eq. (15.209) to determine the conditions for accessibility. If one keeps in mind that $K_\parallel < 0$, then there are two requirements for good accessibility (i.e., $n_\perp^2 > 0$). First, $K_\perp$ must be positive everywhere in the region of interest in order to prevent the occurrence of a wave resonance, which if present opens up to an evanescent region where $n_\perp^2 < 0$. Avoiding this problem requires that $\omega$ be above a critical value. Second, the expression under the square root can become negative unless $n_\parallel^2$ is sufficiently large. For too small a value of $n_\parallel^2$, a mode conversion layer forms, where the slow wave is converted to a fast wave which then propagates back to the launcher. In other words there is no accessibility beyond the mode conversion layer. Lastly, it should be noted that once $n_\parallel^2$ is chosen to avoid mode conversion, this implies that the first term in the curly brackets in Eq. (15.209) is automatically positive. Thus, a combination of high $\omega$ and large $n_\parallel^2$ is required for good accessibility.

These conditions are now quantified as follows. The condition $K_\perp > 0$ can be written as

$$\omega^2 > \omega_{\text{LH}}^2(x_\text{r}) \equiv \frac{\omega_{\text{pi}}^2(x_\text{r})}{1 + \omega_{\text{pe}}^2(x_\text{r})/\Omega_\text{e}^2} \qquad (15.210)$$

or equivalently

$$\frac{\omega^2}{\Omega_\text{e}\Omega_\text{i}} > \frac{\omega_{\text{pe}}^2(x_\text{r})/\Omega_\text{e}^2}{1 + \omega_{\text{pe}}^2(x_\text{r})/\Omega_\text{e}^2}. \qquad (15.211)$$

Here, $\omega_{\text{LH}}(x)$ is the local lower hybrid frequency along the midplane. The resonant radius $x_\text{r}$ can lie anywhere in the range $-a \leq x_\text{r} \leq a$. The distance $a - |x_\text{r}|$ corresponds to the penetration depth of the wave from the launching structure at $x = -a$ into the plasma before reaching the lower hybrid resonance (neglecting the effects of Landau damping). For the moment, readers should assume that $x_\text{r}$ is an unknown quantity to be determined from the analysis. Also, it is not essential to maintain the $1/R$ toroidal dependence in the magnetic field. This allows $\Omega_\text{e}$ and $\Omega_\text{i}$ to be treated as constants.

Consider next the sign of the term under the square root. The analysis is somewhat involved. To begin note that a straightforward calculation shows that the condition for this term to be positive can be written as

$$n_\parallel^2 > \left[ K_\perp^{1/2} + \left( -\frac{K_A^2}{K_\parallel} \right)^{1/2} \right]^2. \qquad (15.212)$$

After substituting for the elements of the dielectric tensor one obtains

$$\begin{aligned} n_\parallel^2 &> [(1 - \gamma X)^{1/2} + X^{1/2}]^2, \\ X(x) &= \omega_{\text{pe}}^2(x)/\Omega_\text{e}^2, \\ Y &= \omega^2/\Omega_\text{e}\Omega_\text{i}, \\ \gamma &= (1 - Y)/Y. \end{aligned} \qquad (15.213)$$

This function has a maximum in space (i.e., $\partial n_\parallel^2/\partial X = 0$) at a radius $x_\text{m}$ corresponding to a density

$$X(x_\text{m}) = \frac{Y^2}{1 - Y}, \qquad (15.214)$$

which can be rewritten as

$$\frac{\omega_{\text{pe}}^2(x_\text{m})}{\Omega_\text{e}^2} = \frac{(\omega^2/\Omega_\text{e}\Omega_\text{i})^2}{1 - \omega^2/\Omega_\text{e}\Omega_\text{i}}. \qquad (15.215)$$

Furthermore, it can be shown that this maximum always occurs somewhere between the launcher and the lower hybrid resonance radius where $\omega^2 = \omega_{\text{LH}}^2(x_\text{r})$. That is, the maximum lies in the region where the wave is propagating: $|x_\text{r}| < |x_\text{m}| < a$. This is shown graphically

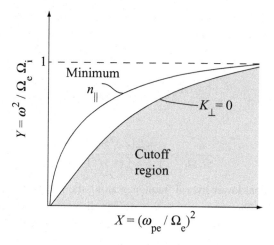

Figure 15.25 Curves of frequency vs. density. The lower curve is the boundary to avoid the wave resonance that occurs at $K_\perp = 0$. The upper curve corresponds to the minimum $n_\parallel$ required to avoid an evanescent layer.

in Fig. 15.25, in which two curves of $Y$ vs. $X$ are plotted. The first corresponds to $\omega^2 = \omega_{LH}^2$, which transforms to $Y = X/(1+X)$. One must be above this curve to avoid the wave resonance. The second curve corresponds to Eq. (15.214). Observe that as stated the maximum always lies in the region between the launcher and the resonant surface.

Based on this analysis, it follows that a necessary and sufficient condition on $n_\parallel^2$ can be obtained by substituting the maximizing value of $X$ into Eq. (15.213), yielding

$$n_\parallel^2 > n_{\parallel crit}^2 \equiv \frac{1}{1 - \omega^2/\Omega_e \Omega_i}. \qquad (15.216)$$

This is the desired result.

How does one now choose proper values for $\omega$ and $n_\parallel$? The choice is not obvious . Clearly $\omega$ and $n_\parallel$ must be chosen so that there is good accessibility over the entire region of interest, which for the moment is assumed to be the region between the launcher and the center of the plasma: $n_\perp^2 > 0$ for $a > |x| > 0$. There is also a second constraint that requires that $n_\parallel^2$ be as small as possible. This constraint is based on the fact (to be derived shortly) that the current-drive efficiency is maximized when $n_\parallel^2$ is minimized. Since $n_\parallel^2 > n_{\parallel crit}^2$ to avoid the mode conversion layer the best that one can do is to set $n_\parallel^2 = n_{\parallel crit}^2$.

The simplest approach that satisfies both requirements is to choose the frequency such that the lower hybrid resonance occurs at the center of the plasma, (i.e., $x_r = 0$) and to set $n_\parallel^2$ slightly above the critical value. With these choices it follows that $n_\perp^2 > 0$ for $a > |x| > 0$. The difficulty with this approach is that a strong non-linear, parametric-decay wave damping, not included in the analysis, occurs well before the launched wave reaches the center of the plasma.

A better approach that avoids this problem and does not lead to a large penalty in increased $n_\parallel^2$ is to choose a slightly higher frequency such that the mode conversion layer occurs at the center of the plasma (i.e., $x_m = 0$). As before, $n_\parallel^2$ is chosen to be slightly above the critical value. For these choices, the launched waves propagate to the center of the plasma without encountering a mode conversion layer, the lower hybrid resonance, or significant non-linear damping. There is good accessibility.

This would seem to resolve the choice for $\omega$ and $n_\parallel$ but there is one further subtlety. When Landau damping is included, it is shown that for reactorlike parameters the waves do not penetrate to the center of the plasma but only to some intermediate and as yet undetermined radius $x_c$. The above conclusions are still valid except that for an optimized result one should choose the frequency such that $x_m = x_c$ rather than $x_m = 0$. However, since the density profile is usually reasonably flat, one does not lose too much current-drive efficiency by using the simpler choice $x_m = 0$. This simpler assumption is made in the remainder of the analysis.

The points just raised with respect to the choice of $n_\parallel^2$ are elucidated by means of a graphical illustration of the dispersion relation, Eq. (15.209). Figure 15.26 shows three curves of $n_\perp^2$ vs. $\omega_{pe}^2/\Omega_e^2$, characterized by three different values of $n_\parallel^2 : n_{\parallel 1}^2 < n_{\parallel 2}^2 < n_{\parallel 3}^2$. The value of $n_{\parallel 2}^2$ corresponds to $n_{\parallel 2}^2 = n_{\parallel crit}^2$, the critical situation. The frequency is chosen so that for case (2) the $n_\parallel = n_{\parallel crit}$ mode conversion surface occurs at the center of the plasma: $x_m = 0$. The value of $\omega$ is then held fixed for the other two cases. To be specific, the parameters for $n_{\parallel 2}^2 = n_{\parallel crit}^2$ and $\omega_2^2/\Omega_e \Omega_i$ are determined by Eqs. (15.215) and (15.216), and are given by

$$\frac{\omega_2^2}{\Omega_e \Omega_i} = \frac{2}{1 + [1 + 4/\alpha]^{1/2}} = \frac{2}{1 + \sqrt{5}} \approx 0.618,$$

$$n_{\parallel 2}^2 = \frac{[1 + 4/\alpha]^{1/2} + 1}{[1 + 4/\alpha]^{1/2} - 1} = \frac{\sqrt{5} + 1}{\sqrt{5} - 1} \approx 2.618,$$

(15.217)

where $\alpha \equiv \omega_{pe}^2(0)/\Omega_e^2$ with $\alpha = 1$ the value used for each of the curves.

Observe the sensitivity of the results to the value of $n_\parallel^2$. A small decrease to $n_{\parallel 1}^2 = 0.995 n_{\parallel crit}^2$ results in a substantial mode conversion region where the mode does not propagate. Similarly, for $n_{\parallel 3}^2 = 1.005 n_{\parallel crit}^2$ the fast and slow wave are well separated and have good accessibility to the center of the plasma.

The final point worth noting is that Eq. (15.216) requires that $n_\parallel^2 > 1$, implying that $n_\perp^2 = 1 - n_\parallel^2 < 0$ at the edge of the plasma where the density vanishes. Consequently, as for ICH, there is a small cutoff region near the edge of the plasma that the waves must tunnel through. One is again forced to place the launcher near the plasma edge to avoid excessive wave reflection.

In summary, the lower hybrid slow wave has good accessibility up to the center of the plasma if $\omega = \omega_3$ and $n_\parallel = n_{\parallel 3}$. These are the values chosen in the analysis of lower hybrid power absorption and current drive.

### 15.9.3 Slow wave power absorption

As the lower hybrid slow wave propagates into the plasma two important wave-particle resonant phenomena occur. First, under the action of an asymmetric $n_\parallel$ spectrum, electrons

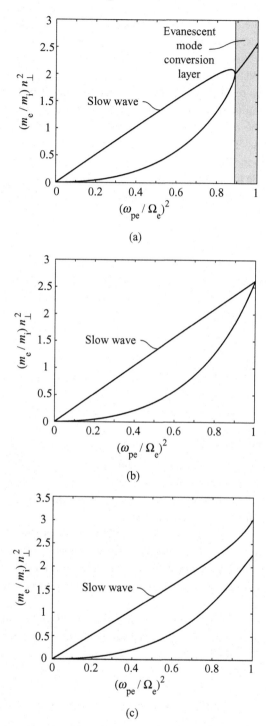

Figure 15.26 Dispersion relation for the fast and slow lower hybrid waves for the cases: (a) $n_\parallel = 0.995 n_{\mathrm{crit}}$, (b) $n_\parallel = n_{\mathrm{crit}}$, and (c) $n_\parallel = 1.005 n_{\mathrm{crit}}$.

acquire net momentum from the collisionless Landau resonance, producing the desired current drive. Second, the wave-particle resonance also produces Landau damping of the wave amplitude. Clearly current drive can only occur in the outer portion of the plasma where the wave amplitude has not as yet been damped. For reactorlike parameters, it is shown that this outer region is relatively narrow, a consequence of the fact that Landau damping is very strong near the plasma edge.

The purpose of this subsection is to present a derivation of the damping distance. Specifically, as the wave propagates from the outside edge into the plasma its amplitude has the form

$$|E_z^2| = E_\parallel^2 \exp[-\lambda(x)] = E_\parallel^2 \exp\left(-2\int_{-a}^x k_{\perp i}dx\right). \tag{15.218}$$

The goal is to evaluate $\lambda(x)$ and to then determine the critical penetration distance $x_c$ corresponding to $\lambda(x_c) \approx 3$. Beyond this point the wave is for all practical purposes completely damped and no further current can be driven.

In this connection it is worth pointing out one important difference between Landau damping and cyclotron damping. Cyclotron damping is confined to a narrow layer of the plasma centered about the local cyclotron resonance $x = x_\Omega$ (or its harmonics): $\omega = l\,\Omega(x_\Omega)$. Landau damping, on the other hand, occurs continuously over the entire profile as the wave propagates into the plasma and approaches a maximum when $\omega \sim k_\parallel v_{Te}$. This fact complicates the mathematical details of the analysis but does not alter the conceptual understanding of the physics.

The analysis proceeds along the by now familiar path of first calculating $k_{\perp i} = S_L/2P_\perp$ and then integrating over $x$ to determine the damping coefficient $\lambda(x)$. For lower hybrid waves the collisionless absorption arises from the generalized Landau damping relation given by Eq. (15.133). In the expression for $S_L$ one sets $l = 0$ and takes the small gyro radius limit $e^{-b}I_0(b) \approx 1$. The Poynting vector $P_\perp$ can be simplified by eliminating the magnetic field in terms of the electric field using Faraday's law, yielding

$$P_\perp = \frac{1}{2\mu_0 c}[(n_\perp E_\parallel^* - n_\parallel E_{1x}^*)E_\parallel + n_\perp |E_{1y}|^2]. \tag{15.219}$$

The quantities $E_{1x}$ and $E_{1y}$ are expressed in terms of $E_\parallel$ by means of the dispersion relation matrix given by Eq. (15.71):

$$n_\perp E_\parallel - n_\parallel E_{1x} = \frac{K_\parallel}{n_\perp}E_\parallel,$$

$$E_{1y} = i\frac{K_A(n_\perp^2 - K_\parallel)}{n_\perp^3 n_\parallel}E_\parallel. \tag{15.220}$$

Substitution of these relations leads to an intermediate form of $k_{\perp i}$ which can be written as

$$k_{\perp i} = \pi^{1/2}\frac{n_\perp \omega_{pe}^2}{\omega c}\frac{\zeta^3 e^{-\zeta^2}}{\left[K_\parallel + K_A^2\left(n_\perp^2 - K_\parallel\right)^2/n_\perp^4 n_\parallel^2\right]}, \tag{15.221}$$

$$\zeta = \frac{\omega}{k_\parallel v_{Te}}.$$

The required result is then obtained by substituting for the various quantities in Eq. (15.221), a task made simpler by the introduction of several normalized parameters:

$$\hat{\omega} = \omega/(\Omega_e \Omega_i)^{1/2},$$
$$\alpha = \omega_{pe}^2(0)/\Omega_e^2. \tag{15.222}$$

In the analysis that follows it is algebraically convenient to express all relationships in terms of the normalized frequency $\hat{\omega}$ rather than the normalized density $\alpha$. Thus, the LHCD choices for $\omega$ and $n_\parallel$ lead to

$$\alpha = \frac{\hat{\omega}^4}{1 - \hat{\omega}^2},$$
$$n_\parallel^2 = \frac{1}{1 - \hat{\omega}^2}. \tag{15.223}$$

The next step in the normalization procedure is to note that in order to evaluate $\lambda(x)$ one must specify profiles for the density and temperature. The normalized density $N(x)$ and temperature $\tau(x)$ are defined by

$$\omega_{pe}^2(x)/\Omega_e^2 = \alpha N(x),$$
$$\omega^2/k_\parallel^2 v_{Te}^2(x) = \zeta^2(x) = \zeta_0^2/\tau(x), \tag{15.224}$$
$$\zeta_0^2 = \omega^2/k_\parallel^2 v_{Te}^2(0) = c^2/n_\parallel^2 v_{Te}^2(0).$$

The functions $N(x)$ and $\tau(x)$ lie in the ranges $0 \leq N \leq 1$ and $0 \leq \tau \leq 1$ respectively.

The last quantity needed is $n_\perp^2$. A short calculation shows that

$$n_\perp = -\left(\frac{m_i}{m_e}\right)^{1/2} \frac{\hat{\omega}^2}{1 - \hat{\omega}^2}. \tag{15.225}$$

The minus sign in front of the right hand side is the consequence of the direction of energy flow. A calculation of $V_{g\perp}$ shows that the group velocity and the phase velocity have opposite signs, implying that the slow wave is a "backward wave." Thus, in order for energy to propagate *into* the plasma one must choose the negative sign for $n_\perp$: $n_\perp = -(n_\perp^2)^{1/2}$.

With these normalizations it is straightforward to evaluate $K_\perp$, $K_A$, $K_\parallel$. The results are then substituted in Eq. (15.221) leading to the desired form for $k_{\perp i}$:

$$k_{\perp i} = \pi^{1/2} \frac{\Omega_e}{c} \frac{\hat{\omega}^3}{1 - \hat{\omega}^2} \frac{\zeta^3 e^{-\zeta^2}}{(1 - N)D},$$
$$D = 1 + (1 - \hat{\omega}^4)N + (1 - \hat{\omega}^2)^2 N^2. \tag{15.226}$$

The information necessary to calculate $\lambda(x)$ is now available. All that is required is a specification of the profiles and evaluation of the integral determining $\lambda(x)$. A reasonable choice for the profiles is

$$\tau(\rho) = 1 - \rho^2,$$
$$N(\rho) = 1 - \rho^6, \tag{15.227}$$

where $\rho = x/a$ and $-1 \le \rho \le 1$. The temperature peaking factor has the usual value $T(0)/\overline{T} = 2$. The density profile is flatter, also in accordance with experimental observations, and its peaking factor has the value $n(0)/\overline{n} = 4/3$.

If these profiles and a change of integration variables $\rho^2 = 1 - 1/y$ are substituted into Eq. (15.226), then the expression for the damping coefficient reduces to

$$\lambda(y) = \pi^{1/2} \frac{\Omega_e a}{c} \frac{\hat{\omega}^2}{1 - \hat{\omega}^2} I(y),$$

$$I(y) = \zeta_0^3 \int_y^\infty \frac{y^3 e^{-\zeta_0^2 y} dy}{(y-1)^{7/2} D}. \tag{15.228}$$

The complicated integral $I(y)$ can be evaluated reasonably accurately by making two approximations based on the fact that for typical reactor parameters $\Omega_e a/c \sim 5 \times 10^3 \gg 1$ and $\zeta_0 \sim 2$. The first approximation assumes that $\zeta_0^2 \gg 1$, which is a reasonable, although not highly accurate approximation. The second approximation follows from the shape of the density profile, which is quite flat out to large radii, allowing one to set $N \approx 1$ in the expression for $D$ with only a small error. Under these assumptions the integral $I(y_c)$ and the damping coefficient $\lambda(y_c)$ simplify to

$$I(y_c) \approx \frac{\zeta_0^3}{3 - 2\hat{\omega}^2} \lim_{\zeta_0 \to \infty} \int_{y_c}^\infty \frac{y^3 e^{-\zeta_0^2 y} dy}{(y-1)^{7/2}} \approx \frac{\zeta_0}{(3 - 2\hat{\omega}^2)} \frac{y_c^3 e^{-\zeta_0^2 y_c}}{(y_c - 1)^{7/2}} \tag{15.229}$$

and

$$\lambda(y_c) = \pi^{1/2} \frac{\Omega_e a}{c} \frac{\hat{\omega}^2}{(1 - \hat{\omega}^2)(3 - 2\hat{\omega}^2)} \frac{\zeta_0 y_c^3 e^{-\zeta_0^2 y_c}}{(y_c - 1)^{7/2}} = 3. \tag{15.230}$$

Equation (15.230) should be viewed as a transcendental equation for $y_c$. Its solution and consequences are perhaps best illustrated by means of a practical numerical example corresponding to typical reactor parameters: $B_0 = 4.7$ T, $a = 2$ m, $T_k(0) = 2\overline{T}_k = 30$, and $n_{20}(0) = (4/3)\overline{n}_{20} = 2$. For these parameters it follows that $\alpha = 0.93$, $\hat{\omega}^2 = 0.61$ (equivalent to $f = 1.5$ GHz), and $n_\parallel^2 = 2.5$. Also, $\Omega_e a/c = 5.5 \times 10^3$ and $\zeta_0 = 1.8$. The numerical solution to Eq. (15.230) is easily calculated and is $y_c = 2.85$.

Finally, when this value is substituted back into the definition of $x_c$ one finds that

$$\frac{|x_c|}{a} = 0.81. \tag{15.231}$$

After an involved calculation, this is the desired information. The conclusion is that for typical reactor parameters the slow wave propagates only a short distance into the plasma before being fully attenuated. It is only in this narrow outer region that current can be driven.

### 15.9.4 LHCD

This subsection describes the most interesting part of the analysis, the calculation of the non-inductive current driven by the lower hybrid slow wave. The analysis consists of a straightforward extension of the collisionless damping calculations used to determine $\omega_i$ and $k_{\perp i}$ for Landau and cyclotron damping and which focused on resonant particle energy absorption. The present discussion extends this analysis to include the momentum absorption arising from an asymmetric $n_{\parallel}$ spectrum. Balancing the resonant particle momentum gain with the electron–ion collisional drag tending to restore the plasma to a Maxwellian leads to expressions for the magnitude of the driven current, the current-drive profile, and the current-drive efficiency. This information is the main goal of the analysis.

To begin, note that both the energy and momentum gain due to resonant wave–particle interactions are determined by solving the single-particle equations of motion as an initial value problem. For the generalized Landau resonance only the $E_{\parallel}$ component of electric field is important and the electron equations of motion reduce to

$$\frac{dv_z}{dt} = -\frac{e}{m_e} E_{\parallel} \cos(\omega t - k_{\perp} x - k_{\parallel} z)$$

$$\approx -\frac{e}{m_e} E_{\parallel} \cos(\omega t - k_{\parallel} z), \quad v_z(0) = v_{\parallel}, \tag{15.232}$$

$$\frac{dz}{dt} = v_z, \qquad\qquad z(0) = z_i.$$

Considerable algebraic simplification arises by neglecting the $k_{\perp} x$ term in the momentum equation. Keeping this term leads to the appearance of Bessel functions as in Eq. (15.133). For the $l = 0$ generalized Landau resonance in the small-gyro-radius limit the Bessel function corrections reduce to a value of unity. Therefore, a substantial simplification arises by neglecting these effects at the outset.

The solution of the equations of motion is obtained by an expansion in amplitude: $v_z = v_0 + v_1 + v_2 + \cdots$, $z = z_0 + z_1 + z_2 + \cdots$. The leading and first order solutions have already been found and are given by:

zeroth order

$$\begin{aligned} v_0 &= v_{\parallel}, \\ z_0 &= z_i + v_{\parallel} t; \end{aligned} \tag{15.233}$$

first Order

$$v_1 = \frac{e}{m_e} E_{\parallel} \frac{\sin(k_{\parallel} z_i - \overline{\omega} t) - \sin(k_{\parallel} z_i)}{\overline{\omega}},$$

$$z_1 = \frac{e}{m_e} E_{\parallel} \left[ \frac{\cos(k_{\parallel} z_i - \overline{\omega} t) - \cos(k_{\parallel} z_i)}{\overline{\omega}^2} - \frac{t \sin(k_{\parallel} z_i)}{\overline{\omega}} \right], \tag{15.234}$$

where $\overline{\omega} = \omega - k_{\parallel} v_{\parallel}$.

The change in energy and the new quantity, the change in momentum, can now be written as

$$\frac{d}{dt}\left(\frac{m_e v_z^2}{2}\right) = -e\left[ v_0 E_z(z_0, t) + v_0 \frac{\partial E_z(z_0, t)}{\partial z_0} z_1 + v_1 E_z(z_0, t) \right],$$

$$\frac{d}{dt}(m_e v_z) = -e\left[ E_z(z_0, t) + \frac{\partial E_z(z_0, t)}{\partial z_0} z_1 \right]. \tag{15.235}$$

Each of these quantities must now be averaged over initial position and then integrated over all $v_{\parallel}$. A short calculation shows that the initial position averaging leads to

$$\frac{d\overline{W}}{dt} = \frac{e^2 E_{\parallel}^2}{2m_e k_{\parallel}} \frac{\partial}{\partial v_{\parallel}} \left( \frac{\omega \sin \overline{\omega} t}{\overline{\omega}} - \sin \overline{\omega} t \right),$$

$$\frac{d\overline{P}}{dt} = \frac{e^2 E_{\parallel}^2}{2m_e} \frac{\partial}{\partial v_{\parallel}} \left( \frac{\sin \overline{\omega} t}{\overline{\omega}} \right),$$

(15.236)

where $W = m_e v_z^2/2$ and $P = m_e v_z$.

The power absorbed per unit volume by the resonant particles has already been calculated and for a Maxwellian is given by

$$S_L = \int \frac{d\overline{W}}{dt} f_M d\mathbf{v} = \pi^{1/2} \varepsilon_0 E_{\parallel}^2 \left( \frac{\omega_{pe}^2}{\omega} \right) \zeta^3 e^{-\lambda - \zeta^2}.$$

(15.237)

Here $\zeta = \omega/k_{\parallel} v_{Te}$. Also $E_{\parallel}^2$ has been replaced by $E_{\parallel}^2 \exp(-\lambda)$ with $\lambda(x)$ given by Eq. (15.230). This factor represents the decay of the amplitude due to Landau damping. The quantity $S_L$ will be required shortly to evaluate the current-drive efficiency.

The last quantity of interest is the driven current, which can be determined from a simple application of momentum balance. The argument is as follows. The momentum gained by resonant electrons must be balanced by the momentum lost due to collisional drag with the ions, which are tending to restore the distribution function to a Maxwellian. (The effect of electron–electron collisions is zero since such collisions exactly conserve momentum.) Mathematically, momentum balance has the form

$$d\overline{P}/dt = -m_e v_{ei} \Delta v_z,$$

(15.238)

where $\Delta v_z = \overline{v}_z - v_{\parallel}$ is the deviation of the initial-position-averaged parallel velocity from its Maxwellian value and $v_{ei}$ is the electron–ion collision frequency given by Eq. (9.49):

$$v_{ei}(v) = \left( \frac{1}{4\pi} \frac{e^4 n_i}{\varepsilon_0^2 m_e^2} \ln \Lambda \right) \frac{1}{v^3}$$

(15.239)

with $v^2 = v_{\perp}^2 + v_{\parallel}^2$.

Next, note that the steady state driven current arising from the integration over all particle velocities can be written as

$$J_{CD} = -e \int \Delta v_z f_M d\mathbf{v}.$$

(15.240)

Momentum balance then implies that

$$J_{CD} = \frac{e}{m_e} \int \left( \frac{1}{v_{ei}} \frac{d\overline{P}}{dt} \right) f_M d\mathbf{v}.$$

(15.241)

The quantity $J_{CD}$ is evaluated by substituting for the various quantities in the integrand and carrying out the velocity space integration. A short calculation involving an integration by parts plus a normalization of the velocities to the thermal velocities yields

$$J_{CD} = \frac{e^3 E_\parallel^2 n e^{-\lambda}}{\pi^{1/2} m_e^2 v_0 v_{Te}} \int \frac{\sin \overline{\omega} t}{\overline{\omega}} u u_\parallel (2u^2 - 3) e^{-u^2} u_\perp du_\perp du_\parallel, \qquad (15.242)$$

where $v_0 = v_{ei}(v = v_{Te})$. As with the energy absorption calculations, the $u_\parallel$ integration can be easily carried out in the limit $t \to \infty$ to obtain the steady state current drive. The quantity $\sin(\overline{\omega} t)/\overline{\omega}$ approaches a delta function and in all other terms one sets $u_\parallel = \omega/k_\parallel v_{Te} = \zeta$. The expression for $J_{CD}$ reduces to

$$J_{CD} = \pi^{1/2} \varepsilon_0 E_\parallel^2 \left(\frac{\omega_{pe}^2}{\omega}\right) \left(\frac{e}{m_e v_0 v_{Te}}\right) e^{-\lambda - \zeta^2} G(\zeta),$$
$$G(\zeta) = \zeta^2 \int_0^\infty (u_\perp^2 + \zeta^2)^{1/2} (2u_\perp^2 + 2\zeta^2 - 3) e^{-u_\perp^2} u_\perp du_\perp. \qquad (15.243)$$

A reasonable approximation for the function $G(\zeta)$ can be obtained by noting that from the previous analysis the regime of interest corresponds to $\zeta^2 \gg 1$. In this regime $G(\zeta) \approx \zeta^5$. Using this approximation leads to the desired form of $J_{CD}$:

$$J_{CD} = \pi^{1/2} \varepsilon_0 E_\parallel^2 \left(\frac{\omega_{pe}^2}{\omega}\right) \left(\frac{e}{m_e v_0 v_{Te}}\right) \zeta^5 e^{-\lambda - \zeta^2}. \qquad (15.244)$$

The first property of interest is the current-drive profile. Figure 15.27 is a plot of $J_{CD}/J_{CD_{max}}$ vs. $x/a$ for the numerical example under consideration. Observe that the current peaks sharply near the critical surface $x/a \approx x_c/a \approx 0.81$. The lower hybrid slow wave drives current near the outer edge of the plasma. This conclusion is unavoidable for reactor temperatures. Specifically, the definition of $x_c$ combined with Eq. (15.230) implies the following approximate scaling relation: $1 - x_c^2/a^2 \propto 1/T$. Thus, current can be driven near the center of the plasma only if the temperature is substantially lower than in a reactor. A more detailed calculation based on kinetic theory shows that only for $\overline{T}_k \lesssim 5$ keV is it possible to drive current near the axis.

Consider now the important question of the current-drive efficiency $\eta_{CD}$, which determines how much current is driven per watt of applied power. In terms of the present analysis, the definition is

$$\eta_{CD} = \frac{I_{CD}}{P_{CD}} = \frac{\int J_{CD} dA}{2\pi R_0 \int S_L dA}. \qquad (15.245)$$

The integration over the cross sectional area is accurately approximated by noting that if one plots $S_L$ vs. $x/a$ the curve almost exactly overlays the curve of $J_{CD}$ vs. $x/a$ illustrated in Fig. 15.27. The reason is that the strong exponential dependence $\exp(-\lambda - \zeta^2)$ is the same for both functions and dominates the behavior. Consequently,

$$\frac{\int J_{CD} dA}{\int S_L dA} \approx \frac{J_{CD}(x_J)}{S_L(x_J)} \approx \left.\frac{\zeta^2 e}{m_e v_0 v_{Te}}\right|_{x=x_J}, \qquad (15.246)$$

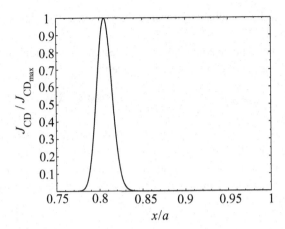

Figure 15.27 Normalized current profile as a function of normalized distance from the launcher. The profile peaks near the critical surface $x/a = 0.81$.

where $x_J \approx x_c$ corresponds to the location of the peak current density. Interestingly, in the ratio $\zeta^2/v_0 v_{T_e}$ the temperature profile exactly cancels and the only remaining spatial dependence involves the density appearing in $v_0$. Since the density is flat, one can set $n(x_J) \approx n(0) = (4/3)\bar{n}$ with only a small error. These simplifications lead to the desired expression for the current-drive efficiency

$$\eta_{CD} = \frac{I_{CD}}{P_{CD}} = \frac{1.17}{n_{\parallel}^2 R_0 \bar{n}_{20}} = 0.047 \quad \text{A/W.} \tag{15.247}$$

It takes about 20 W to drive 1 A of current in a reactor.

One sees that to drive the entire current in a tokamak reactor, which is on the order of 20 MA, would require 400 MW of delivered power. Clearly this is economically unacceptable in a 1000 MW reactor. It is for this reason that it is important to have a substantial bootstrap current flowing in the plasma. High bootstrap current, such as generated by AT operation, is essential for reducing the requirements on the current-drive system.

### 15.9.5 *Summary*

The lower hybrid slow wave is currently the most efficient method for driving a steady state non-inductive off-axis current in a tokamak reactor. High power at 1.5 GHz is obtained from klystron sources, which are readily available at a cost of about \$3/W. The biggest technological problem is the need to place the wave guide launching array near the plasma edge in order to prevent a large reflection because of the unavoidable edge cutoff region.

For typical reactor temperatures the accessibility properties force the current to be driven near the plasma edge. Lower plasma temperatures allow the current to penetrate further into the plasma. The off-axis current drive is, however, desirable for matching the natural bootstrap profile and for establishing transport barriers near the edge of the plasma where they are most needed.

The critical parameter in the context of reactor design is the current-drive efficiency. The analysis presented has shown this efficiency is given by

$$\eta_{CD} = \frac{I_{CD}}{P_{CD}} = \frac{1.17}{n_\parallel^2 R_0 \bar{n}_{20}} = 0.047 \quad A/W \tag{15.248}$$

for a typical tokamak reactor. Thus it takes about 20 W to drive 1 A. Although this efficiency is higher that those for all other methods, it is still too low to drive the entire current in a fusion reactor. A substantial bootstrap fraction is needed to reduce the requirements on the current drive system to a level more compatible with reactor economics. This will likely require AT operation.

## 15.10  Overall summary

A plasma must first be heated to a temperature of about 5–7 keV in order to reach the point where alpha power becomes dominant. During the rise to 5–7 keV alpha heating is negligible and a combination of ohmic and auxiliary power is required to overcome thermal conduction and Bremsstrahlung radiation losses. In addition, once ignition is achieved, non-inductive current drive is needed to maintain the plasma in steady state operation.

This chapter has described several heating methods. Ohmic heating is technologically the simplest method but, because the plasma resistivity decreases with temperature, can only achieve values of $T \lesssim 3$ keV for typical reactor parameters. Other heating methods are needed.

Neutral beam heating has been a very effective method for heating present day plasmas. The basic physics of the heating involves classical collisional processes and thus should extrapolate in a reliable way to future large devices such as ITER and fusion reactors. The main problem is technological. Present day positive ion driven neutral beam systems do not extrapolate well into the regime of large devices because the beam efficiency rapidly decreases with beam energy above 100 keV. Overcoming this difficulty requires the development of high-energy, negative-ion-driven, neutral beam systems. This is difficult technologically but is expected to be solved in time for ITER.

Various RF waves have also been proposed as sources of auxiliary heating. Both ECH and ICH should be capable of producing a high fraction of absorbed power in the center of a large, next generation experiment as well as in a reactor. Heating occurs by a process known as collisionless cyclotron damping, which depends on a strong wave–particle resonant interaction at the point where the applied frequency is equal to the local cyclotron frequency or its harmonics. ECH heating is the simpler of the two methods. It takes place very locally because of the small wavelengths involved. The main issue with ECH is the lack of high-power, steady state gyrotron sources. The successful development of a 140 GHz source, corresponding to resonance at the fundamental cyclotron frequency, is expected by the time it is needed for ITER. ICH involves much lower frequencies and high-power, steady state vacuum tube sources are readily available. The power absorption mechanism is complex, requiring a combination of minority heating and second harmonic heating. The main issue facing ICH is the need for an antenna inside the vacuum chamber that must be placed as close to the edge of the plasma as possible to insure good coupling.

Overall, no single heating method appears superior to all other methods. Each method described should be capable of good heating. Each method also faces technological problems associated with either sources or antennas. This assessment has led to an ITER design with multiple heating options to determine which method in fact is the most desirable from the point of view of reliability and cost.

The last topic discussed in the chapter was LHCD. This method has the highest efficiency and is capable of driving off-axis current which is required to match the natural bootstrap profiles and to create transport barriers near the plasma edge where they are most needed. LHCD is driven by the collisionless Landau resonance at frequencies between the electron and ion cyclotron frequencies. Klystron power sources are readily available. The main technological problem is the need to place the launching array near the plasma edge for good coupling. The main physics problem is that the current-drive efficiency, given approximately by $\eta_{CD} \approx 0.047$ A/W, is too low to economically drive all the current in a tokamak reactor. This conclusion motivates the need for a high bootstrap fraction and AT operation.

## Bibliography

Ohmic and neutral beam heating techniques have been well known for many years and are described in a number of books, several of which are listed below. RF heating and current-drive techniques have also been well known for many years. However, because of the relatively complex plasma physics involved, whole books as well as several major review articles have been written on the subject. Some of the major references are listed below.

### *Ohmic heating*

Chen, F. F. (1984). *Introduction to Plasma Physics and Controlled Fusion*, second edn. New York: Plenum Press.
Coppi, B. (1977). Compact experiment for $\alpha$-particle heating. *Comments on Plasma Physics and Controlled Fusion*, **III**, 47.
Wesson, J. (2004). *Tokamaks*, third edn. Oxford: Oxford University Press.

### *Neutral beam heating*

Sweetman, D. R., Cordey, J. G., and Green, T. S. (1981). Heating and plasma interactions with beams of energetic neutral atoms. *Philosophical Transactions of the Royal Society*, **A 300**, 589.

Stork, D. (1991). Neutral beam heating and current drive systems. *Fusion Engineering and Design*, **14**, 111.


## *RF general*

Brambilla, M. (1998). *Kinetic Theory of Plasma Waves, Homogeneous Plasmas*, International series on Monographs on Physics 96. Oxford: Oxford University Press.
Cairns, R. A. (1991). *Radio Frequency Heating of Plasmas*. Bristol: Adam Hilger
Hwang, D. Q. and Wilson, J. R. (1981). Radio frequency wave applications in magnetic fusion devices. *Proceedings of the IEEE*, **69**, 1030.
Porkolab, M. (1977). Review of RF heating. In *Theory of Magnetically Confined Plasmas. Proceedings of the International School of Plasma Physics, Varenna*. Oxford: Pergamon Press.
Stix, T. H. (1992). *Waves in Plasmas*, second edn. New York: American Institute of Physics.
Swanson, D. G. (1989). *Plasma Waves*. San Diego: Academic Press.
Wesson, J. (2004). *Tokamaks*, third edn. Oxford: Oxford University Press.


## *Landau damping*

Landau, L. D. (1946). On the vibrations of the electronic plasma. *Journal of Physics (USSR)*, **10**, 25.
Stix, T. H. (1992). *Waves in Plasmas*, 2nd edn. New York: American Institute of Physics.


## *ECH*

Manheimer, W. M. (1979). *Electron Cyclotron Heating in Tokamaks*, Vol. II (Button, K. J., editor). New York: Academic Press.
Bornatici, M., Cano, R., *et al.* (1983). Electron cyclotron emission and absorption in fusion plasmas. *Nuclear Fusion*, **23**, 1153.


## *ICH*

Hwang, D. Q. (1985). The theory of minority species fast magnetosonic wave heating in a tokamak. In *Wave Heating and Current Drive in Plasmas* (Granastein, V. L., and Colestock, P. L., editors). New York: Gordon and Breach.
Stix, T. H. (1975). Fast wave heating of a two component plasma. *Nuclear Fusion*, **15**, 737.
Swanson, D. G. (1985). Radio frequency heating in the ion-cyclotron range of frequencies. *Physics of Fluids*, **28**, 2645.


## *Lower hybrid heating and current drive*

Bonoli, P. (1985). Linear theory of lower hybrid waves in tokamak plasmas. In *Wave Heating and Current Drive in Plasmas* (Granastein, V. L. and Colestock, P. L., editors). New York: Gordon and Breach.

Fisch, N. J. (1978). Principles of current drive by Landau damping of traveling waves. *Physical Review Letters*, **41**, 873.

Fisch, N. J. (1987). Theory of current drive in plasmas. *Reviews of Modern Physics*, **59**, 175.

Porkolab, M. (1985). Lower hybrid wave propagation, heating and current drive experiments. In *Wave Heating and Current Drive in Plasmas* (Granastein, V. L., and Colestock, P. L., editors). New York: Gordon and Breach.

## Problems

15.1 A perfectly conducting cylindrical waveguide has a circular cross section of radius *a*. It is desired to propagate a TE wave with frequency $\omega$ and axial wave number $k$ along the wave guide. For simplicity consider azimuthally symmetric modes (i.e., $\partial/\partial\theta = 0$) with the following non-zero field components: $E_z$, $B_r$, $B_z$. All quantities are functions of $r, z, t$.

    (a) Derive a single ordinary differential equation describing the radial structure of the modes.

    (b) Determine the dispersion relation for wave propagation.

    (c) Calculate the lowest frequency that can propagate if $a = 0.03$ m.

15.2 The purpose of this problem is to calculate the power dissipated in the walls of the cylindrical wave guide described in Problem 15.1. Specifically, it is desired to calculate the length of waveguide required for the input power to decay by 10%. This gives an estimate of how far the RF power source can be placed with respect to the plasma chamber before appreciable input power is lost to dissipation. The exact problem is remarkably complicated to solve exactly. Instead an iterative approach is used whereby in lowest order the fields within the waveguide are assumed to be those determined in Problem 15.1, corresponding to an ideal perfectly conducting waveguide. The effects of finite conductivity are then calculated perturbatively.

    (a) Consider the fields in the waveguide. Assume the waveguide has a conductivity $\sigma$. It is thin enough that it can be represented by a slab geometry with a "radial" coordinate $x$ defined by $r = a + x$, but thick enough such that the fields completely decay to zero across the wall. Practically, these conditions require that $\delta \ll d \ll a$, where $\delta$ is the skin depth and $d$ is the waveguide thickness. The non-zero field components are again $E_z$, $B_x$, $B_z$ and vary as $\exp(-i\omega t + ikz - x/\delta)$. Using Maxwell's equations determine the fields in the waveguide wall and the skin depth $\delta$. Simplify your calculation by neglecting displace current and assuming $\delta^2 \ll k^2$, both of which are excellent approximations.

    (b) Connect the fields in the wall to the vacuum fields in the wave guide by making use of the critical fact that $B_z$ in the vacuum does not change very much whether the wall has $\sigma$ finite or infinite. Thus, the matching condition across the wave guide wall is $[\![B_z]\!] = 0$. Express each of the wave guide field components in terms of the vacuum amplitude of $B_z$.

    (c) Assume now that $|B_z|^2 = |B_z(0)|^2 \exp(-2z/L)$ is a slowly decreasing function of $z$ because of the dissipation losses. Calculate the differential loss in power transmitted in the waveguide $(dP_S/dz)$ by integrating the vacuum field Poynting vector over two cross sectional areas separated by an infinitesimal distance $dz$.

    (d) Calculate the differential power dissipated in the walls $(dP_D/dz)$ along the same segment $dz$.

(e) Use power balance to determine the damping length $L$. As a practical example evaluate $L$ for a lower hybrid system characterized by $f = 5 \times 10^9$ Hz, $a = 0.05$ m, and $\sigma = 5 \times 10^7$ mho/m.

15.3 ITER, FIRE, and IGNITOR are three proposed next generation experiments aimed at investigating ignition and alpha particle physics. The goal of this problem is to determine how much current is required to achieve $Q = 10$ operation solely by ohmic heating and to then check whether the resulting parameters are consistent with MHD stability limits.

The input information consists of the design parameters for each device as given in Table 15.2.

Table 15.2

|                              | ITER | FIRE | IGNITOR |
|------------------------------|------|------|---------|
| $I_{MA}$ (MA)                | 15   | 7.7  | 11      |
| $B_0$ (T)                    | 5.3  | 10   | 13      |
| $a$ (m)                      | 2.0  | 0.60 | 0.47    |
| $R_0$ (m)                    | 6.2  | 2.14 | 1.32    |
| $\kappa$ (elongation)        | 1.7  | 1.81 | 1.83    |
| $\overline{T}_e$ (keV)       | 19   | 11   | 10.5    |
| $\overline{T}_i$ (keV)       | 19   | 11   | 10.5    |
| $\overline{n}_{20}$ ($10^{20}$ m$^{-3}$) | 1.0 | 5.5 | 10 |

Also needed is the D–T fusion cross section. A very good analytic approximation is

$$\langle \sigma v \rangle = 10^{-6} \exp\left( \frac{a_{-1}}{T_i^\alpha} + a_0 + a_1 T_i + a_2 T_i^2 + a_3 T_i^3 + a_4 T_i^4 \right) \quad \text{m}^3/\text{s},$$

where $T_i = T_i$ (keV) and

| $\alpha$ | $a_{-1}$ | $a_0$ | $a_1$ | $a_2$ | $a_3$ | $a_4$ |
|----------|----------|-------|-------|-------|-------|-------|
| 0.2935 | −21.38 | −25.20 | $-7.101 \times 10^{-2}$ | $1.938 \times 10^{-4}$ | $4.925 \times 10^{-6}$ | $-3.984 \times 10^{-8}$ |

Lastly, for ohmically heated tokamaks the conventional wisdom is that energy confinement is determined by the L-mode energy confinement time:

$$\tau_L = 0.048 \frac{I^{0.85} R_0^{1.2} a^{0.3} \kappa^{0.5} n_{20}^{0.1} B_0^{0.2} A^{0.5}}{P^{0.5}} \quad \text{s}.$$

Here, $A = 2.5$ is the average mass number for a 50%–50% D–T fuel mixture, and $P = P(\text{MW})$ is the total heating power. For an ohmically heated high-$Q$ system set $P = P_\alpha + P_\Omega$. When calculating the ohmic heating power use the neoclassical value for $\eta \approx \eta_{\text{classical}}/(1 - \varepsilon)^2$.

Now, assume that the density and temperature profiles are given by

$$T_e = 2\overline{T}_e(1 - \rho^2),$$

$$T_i = 2\overline{T}_i(1 - \rho^2),$$

$$n = (1 + \nu)\overline{n}(1 - \rho^2)^\nu,$$

where $\rho^2 = x^2/a^2 + y^2/\kappa^2 a^2$, $\overline{Q}$ represents the area average of $Q$ assuming an elliptical cross section, and $\nu$ is a free parameter measuring the peaking factor of the density. Also, recall that for ohmic heating the current and temperature profiles are related by $J = J_0 (T_e/T_{e0})^{3/2}$.

Neglect Bremsstrahlung radiation and *ignore* the values of $I_{MA}$ given in Table 15.2. The tabulated current values are just for reference. Instead, calculate the required value of $I_{MA}$ for $Q = 10$ in each device as a function of the profile parameter $\nu$. You will need to do a numerical integration to complete this step. Knowing $I_{MA}$, calculate $q_*$ and $\beta_N$ for each device. Display the results as a set of curves of $I_{MA}$ vs. $\nu$, one curve for each device. Superimpose the corresponding design values from Table 15.2 for comparison. Next, plot $q_*$ vs. $\nu$ for all three devices on the same graph and superimpose the kink stability limit. Finally, plot $\beta_N$ vs. $\nu$ for all three devices on a single graph and superimpose the Troyon stability limit. What conclusions can be drawn about the likelihood of $Q = 10$ ohmic operation?

15.4 This problem involves the calculation of the self-consistent pressure profile of an ohmically heated cylindrical screw pinch operating in the "steady state" flat portion of the discharge. The plasma is described by the resistive MHD model. The non-trivial quantities of interest are $p = p(r)$, $\mathbf{E} = E_0\mathbf{e}_z$, and $\mathbf{B} = B_0\mathbf{e}_z + B_\theta\mathbf{e}_\theta$ where $E_0$, $B_0$ are known constants.
   (a) Derive a single differential equation for the pressure $p(r)$ by making use of the following information: Maxwell's equations, pressure balance, the parallel Ohm's law, the ideal gas law, classical resistivity $\eta = C_\eta/T^{3/2}$, and an assumed density–temperature relation $n(r) = n_0(T/T_0)^{1/2}$. The final result should be a second order ordinary differential equation. Its appearance can be simplified by introducing normalized variables $x$, $B(x)$, $P(x)$ defined by $r = ax$, $B_\theta(r) = B_I B(x)$, $p(r) = (B_I^2/\mu_0)P(x)$. Here $a$ is the characteristic plasma radius and $B_I = C_\eta n_0/a E_0 T_0^{1/2}$, both known constants.
   (b) The differential equation can be cast in a simpler form by introducing the transformation $x = \exp(-y)$, $P(x) = \exp[-V(y) + 2y]$. Solve this equation subject to the conditions $P(0) = P_0$, $P'(0) = 0$. Make a plot of $P(x)$ vs. $x$.

15.5 One novel suggestion to create a fusion plasma is to shine a powerful laser along a dense column of plasma. The laser energy heats the plasma by collisional absorption. To investigate this approach consider a plasma with infinitely massive ions and the following background equilibrium state: $n_e = n_i = n_0$, $p_e = p_i = p_0$, $\mathbf{E} = \mathbf{B} = \mathbf{v}_e = \mathbf{v}_i = 0$. For simplicity assume the laser propagation is 1-D along the column (i.e., neglect the transverse radial dependence of the wave).
   (a) Derive the dispersion relation for a TEM electromagnetic wave propagating in the plasma. Include the effect of electron–ion collisions in the electron momentum equation.
   (b) Assume that $\omega$ is given and calculate $k = k_r + ik_i$ in the limit $\overline{\nu}_{ei} \ll \omega$.

(c) Consider the case where $n_0 = 10^{24}$ m$^{-3}$ and $\bar{v}_{ei} = 4 \times 10^7$ s$^{-1}$. If the length of the plasma is set equal to the damping length of the waves (i.e., $L = 1/k_i$) what is the length of the column? Does this seem practical to you?

15.6 An electromagnetic wave with frequency $\omega$ and wave number $k$ propagates in the $z$ direction in a vacuum. The non-zero field components are $E_x$, $B_y$. The wave impinges perpendicularly on a semi-infinite slab of plasma whose interface is located at $z = 0$. The plasma is cold and homogeneous. It has a constant $n$, a negligible external DC magnetic field, and a small but finite electron–ion momentum exchange collision frequency. The wave frequency is sufficiently high that the ions can be considered to be infinitely massive.

(a) Calculate the dispersion relation for wave propagation in the plasma.

(b) Match across the interface to determine the amplitudes of the plasma fields and the wave reflected off the interface. Express your answers in terms of $E_0$, the amplitude of the incident electric field.

(c) Define the absorption efficiency as $\eta = \text{Re}(S_p)/\text{Re}(S_{in})$, where $S_{in}$ is the incident Poynting flux and $S_p$ is the Poynting flux just inside the plasma interface. Evaluate $\eta = \eta(\omega)$ and determine the frequency corresponding to maximum absorption. Is this equal to the plasma frequency?

15.7 A 1-D slab plasma chamber has a thickness of $a = 8$ cm. Initially the plasma chamber is empty – it is a vacuum. An 8 mm wavelength TEM wave is propagated through the chamber and the phase shift $\Delta\phi_1$ of the wave between $x = 0$ and $x = a$ is recorded. The experiment is repeated, this time with a plasma of density $n$ filling the chamber. The resulting phase shift is $\Delta\phi_2$. The difference in phase shifts between the two measurements is found to be $\Delta\phi_1 - \Delta\phi_2 = \pi/5$. What is the plasma density?

15.8 Consider the following simple model that examines the stability of a neutral beam heated tokamak. The electrons are an isothermal fluid with a temperature $T_e$ and a density $n_0$; that is, neglect electron inertia and collisions in the momentum equation. The bulk ions are a cold fluid with density $n_i$. The neutral beam is injected parallel to the magnetic field with a velocity $v_0$ and zero thermal spread. It quickly becomes ionized and for the short times of interest there is a negligible loss in energy or momentum. The beam density satisfies $n_b \ll n_i$.

(a) Calculate the dispersion relation for low-frequency electrostatic modes propagating parallel to the magnetic field.

(b) Calculate the condition on $v_0$ for instability to occur. Is this instability likely to occur for practical situations? Explain.

15.9 This problem investigates thermal effects on the cold plasma dielectric tensor in the context of electron cyclotron heating. Consider an infinite homogeneous plasma with an equilibrium magnetic field given by $\mathbf{B} = B_0\mathbf{e}_z$. The electron equilibrium satisfies $n = n_0$, $T = T_0$. The ions are assumed to be cold and infinitely massive. To calculate the dielectric tensor use the electron fluid equations. Neglect collisions but maintain electron pressure effects. Calculate the components of the dielectric tensor for the special case of interest to ECH corresponding to $k_\parallel = 0$.

15.10 Use the results of Problem 15.9 to show that the O-mode branch of the dispersion relation is unaffected by the inclusion of thermal effects.

15.11 Use the results of Problem 15.9 to investigate thermal effects on the X-mode branch of the dispersion relation. Assume an outside low-field launch, where the equilibrium quantities have the following midplane profiles:

$$B_0(r) = B_a \left( \frac{R_0}{R_0 + r} \right),$$

$$n_0(r) = n_a \left( 1 - \frac{r^2}{a^2} \right),$$

$$T_0(r) = T_a \left( 1 - \frac{r^2}{a^2} \right).$$

Assume $B_a = 5$ T, $n_a = 3.6 \times 10^{20}$ m$^{-3}$, $T_a = 5$ keV, $R_0 = 1$ m, and $a = 0.35$ m. Choose the wave frequency to correspond to second harmonic cyclotron resonance at the center of the plasma $r = 0$. Plot $n_\perp^2 \equiv (k_\perp c/\omega)^2$ vs. $r$ for the X-mode branch of the dispersion relation. Superimpose on this plot the curve corresponding to $T_a = 0$ in order to make comparisons. Is it still possible to heat at the second harmonic when thermal effects are included? Explain.

15.12 The purpose of this problem is to calculate the power dissipated at a wave resonance assuming that the cold plasma model, including a small collision frequency, accurately describes the physics. Consider for simplicity an isotropic plasma whose local index of refraction is given by $n^2 = n^2(\omega, x)$. At a certain point $x = x_0$ there is wave resonance implying that near this point

$$n^2(x) \approx \frac{C(x_0)}{\omega - \omega_R(x)},$$

where $\omega = \omega_R(x_0)$ is the resonant frequency.
(a) To calculate the power absorbed evaluate the conductivity $\sigma$ from the definition of the dielectric tensor for an electromagnetic wave with $\mathbf{E} \cdot \mathbf{k} = 0$.
(b) Using the relation $\mathbf{J} = \sigma \mathbf{E}$, calculate the power absorbed per unit area from the relation $P_D/A = \int \mathbf{E} \cdot \mathbf{J}^* dx$. Note that in the absence of collisions this expression has a singularity at $x = x_0$. Resolve this singularity by assuming a small collision frequency. This is most easily done by using the intuition that solutions which originally varied as $\exp(-i\omega t)$ now vary as $\exp(-i\omega t - \nu t)$. Thus, the singularity in the denominator of $n^2$ can now be resolved by the trick of replacing $\omega \to \omega - i\nu$. The calculation is further simplified by recognizing that for small collisionality the absorption layer is narrow. Thus the power integral can be easily evaluated by Taylor expanding all quantities about the point $x = x_0$. Show that in the limit of small collisionality the power dissipated is independent of $\nu$ and is given by

$$\frac{P_D}{A} = \frac{\pi \omega_R \varepsilon_0 |E|^2 C}{(d\omega_R/dx)} \bigg|_{x = x_0}.$$

15.13 This problem investigates another possible method for using waves in the ion cyclotron range of frequencies to heat a plasma. The idea is to generate a wave resonance in the plasma by means of two ion species. One assumes that the wave is mode converted at the resonance layer to a kinetic plasma wave which then continues to propagate until it is absorbed by collisionless damping. Since wave cutoffs often occur near wave resonances there is a crucial question regarding accessibility. Will the wave reach the resonance before it encounters the cutoff? The question can be addressed by the cold plasma dispersion relation. Consider a cold plasma with two different ion species heated by ion cyclotron waves.

(a) Show, using charge neutrality, that in the frequency regime of interest the dispersion relation for $n_\parallel^2 = 0$ can be written as

$$
n_\perp^2 = \frac{\displaystyle\sum_i \frac{\omega_{pi}^2}{\Omega_i\,(\omega + \Omega_i)} \sum_i \frac{\omega_{pi}^2}{\Omega_i\,(\omega - \Omega_i)}}{\displaystyle\sum_i \frac{\omega_{pi}^2}{(\omega^2 - \Omega_i^2)}},
$$

where the sum is only over the ions.

(b) Calculate the frequency $\omega_R$ at which the two-ion hybrid resonance occurs. Show that $\omega_R$ lies between the two cyclotron frequencies.

(c) Calculate the frequency $\omega_C$ at which the cutoff occurs. Show that $\omega_C$ lies between the two cyclotron frequencies.

(d) Show that $\omega_C > \omega_R$. If one wants to gain accessibility to the wave resonance before encountering the cutoff does this imply a low-field or a high-field launch?

# 16

# The future of fusion research

## 16.1 Introduction

The primary plasma physics issues facing the development of fusion energy have now been discussed: (1) macroscopic equilibrium and stability, (2) transport, and (3) heating and current drive. Armed with this knowledge one is in a position to assess the current status of fusion research and ask where the world fusion program should be directed in the future. These are the goals of Chapter 16. To begin, a brief overview is presented describing the present status of tokamak research in the context of the three main plasma physics building blocks listed above. This is followed by a discussion of the next major fusion experiment in the world fusion program: ITER. Assuming ITER is built and is successful, one can then project ahead to the design and construction of a demonstration fusion power plant (DEMO).

## 16.2 Current status of plasma physics research

Much of the material described in this book applies to many proposed fusion concepts. Even so, the discussion here is focused on the tokamak configuration as this is the clear leader on the path to a fusion reactor in terms of actual experimental plasma physics performance and the most detailed designs of potential fusion reactors.

The summary below describes the progress on plasma physics in tokamaks with respect to the basic ignition condition for a fusion reactor given by

$$p\tau_E = \frac{24}{E_\alpha} \frac{T^2}{\langle \sigma v \rangle}. \tag{16.1}$$

The progress has been substantial, although as shown below there still remain challenging, unanswered questions.

### 16.2.1 Macroscopic equilibrium and stability

Consider first the pressure $p$ appearing in the ignition condition. The maximum achievable pressure against major disruptions in a given fusion concept is largely determined by

633

macroscopic equilibrium and stability limits, which are well described by the MHD model. For the tokamak the theoretical predictions and experimental observations are, in general, in good agreement. The theory can therefore be used to reliably predict the pressure limits in next generation experiments.

In terms of actual performance, tokamak experiments operating in the standard mode have already achieved the values of $\beta$ required in a reactor. These values are close to the maximum no-wall $\beta$ limit. The corresponding plasma pressures are, however, less than those required in a reactor because the magnetic fields are smaller. A next generation, higher-field experiment should produce both high $\beta$ and high pressure. The main issue is that the high $\beta$ pressure and current profiles produce a bootstrap fraction that is too low to reduce the current-drive requirements to an acceptable level for economic viability in a reactor.

Avoiding this problem requires the achievement of high bootstrap fractions through AT operation. Unfortunately, the achievement of a high bootstrap fraction requires $\beta$ values that exceed the no-wall $\beta$ limit. A perfectly conducting wall can produce stability at these higher values of $\beta$. However, since a real wall has a finite resistivity, this leads to excitation of the resistive wall mode. Stabilization of the resistive wall mode is an important topic of research for both existing experiments and future large devices.

### 16.2.2 Transport

The dominant transport mechanism in tokamak plasmas is thermal conduction and is characterized by the energy confinement time $\tau_E$ in the ignition condition. Substantial progress in the basic understanding of core thermal transport has been made by a combination of analytic theory and large-scale computation. However, a first-principles prediction of $\tau_E$ is still not available and remains a grand challenge of present and future fusion research.

There are also several related transport problems involving the edge plasma that directly impact the core transport: the Greenwald density limit, the critical power threshold for the L–H transition, and ELMs. In addition, the physics of internal transport barriers, which may be important for the lower-current AT operation, is not well understood.

At present, the determination of $\tau_E$ as well as the relevant criteria for the edge phenomena is based on empirical scaling relations. These relations work reasonably well in existing experiments and hopefully will reliably extrapolate to future generation tokamaks. In fact, the size and cost of a next generation experiment are directly dependent on the prediction of the empirical scaling relation for $\tau_E$. Developing empirical scaling relations for thermal transport in the presence of a large population of energetic alphas is a major challenge for the future.

Lastly, it is worth noting that the theoretical prediction of the bootstrap current based on neoclassical transport theory seems to be in reasonably good agreement with experimental observations. An accurate prediction of $J_B$ is critical for determining the requirements on the current-drive system for steady state operation. Understanding the effects of alpha particles on the self-consistent bootstrap current is another important challenge for future research.

### 16.2.3 Heating and current drive

Minimizing the demands on the $p\tau_E$ product in the ignition condition requires achieving a plasma temperature that minimizes $T^2/\langle \sigma v \rangle$: $T \approx 15$ keV. Reaching $T \approx 15$ keV will be accomplished by a two-stage process, where initially auxiliary power heats the plasma to about $T \sim 5$–$7$ keV, after which the alphas dominate, completing the heating to $T \approx 15$ keV. Several methods of auxiliary power that provide central heating have been tested in existing experiments: neutral beam heating, ICH, and ECH. Temperatures well in excess of $T \sim 5$–$7$ keV have been achieved in existing tokamaks in pure deuterium plasmas, usually at lower densities than required in a reactor. Overall, the heating methods work reasonably well and are in good agreement with theoretical predictions. The implication is that a reasonable (in terms of power balance and cost) amount of auxiliary power should achieve the required heating mission in a next generation experiment or reactor. Also, based on present experimental experience, neutral beam heating is usually regarded as the simplest and most reliable heating method from a purely plasma physics point of view, and therefore will play a primary role in ITER.

The challenges of extrapolating heating methods to ITER and a reactor are largely techno-logical. Neutral beam heating requires the development of high-energy, negative ion sources to act as drivers. ECH requires the development of high-power, steady state gyrotron sources. ICH requires an antenna structure very close to the plasma edge. These are all topics of current and future research.

A related issue concerns current drive, which enters the ignition condition implicitly through the assumption of steady state operation. LHCD is the most efficient method presently available. Also it drives current off-axis, which is an advantage in matching to the natural bootstrap profiles. Even so, the absolute magnitude of the current-drive efficiency is too low to drive all the current in a tokamak reactor or ignition experiment. The conclusion is that a substantial bootstrap fraction will be required in order to reduce the current-drive requirements to a level compatible with reactor economics. Long-pulse current drive is thus an important research topic in present as well as next generation experiments.

### 16.2.4 Alpha particle plasma physics

The alpha particle physics discussed in the main text is primarily focused on issues of power balance and heating. There has been very little discussion of the plasma physics effects of the core on the alpha particles and vice versa. The main reason is that there are very few data involving alpha particles since only two experiments, TFTR and JET, have actually operated with tritium, and then only for a limited period of time.

Because of the lack of data, alpha particle plasma physics is often referred to as the next (and hopefully last) frontier in plasma physics. Learning about alpha particle plasma physics is one the most important physics goals of a next generation ignition experiment.

Of particular interest is whether the alpha pressure gradient will excite instabilities that would cause the alphas to be lost at an anomalously fast rate. This would be highly undesirable in that the alphas could be lost before transferring all their energy to the background plasma, thereby substantially increasing the difficulty of satisfying the steady state ignition condition.

Another issue involves the ability to externally control the pressure and current profiles by means of auxiliary heating and current drive. The difficulty here is that the alpha power in an ignited plasma completely dominates the auxiliary and current-drive powers. Studies are needed to determine how effectively these relatively "small" external power sources control the profiles.

A further topic of importance, not encountered in present experiments, is the removal of the alpha "ash". As the alphas build up due to fusion reactions, they replace D–T fuel because of the charge neutrality requirement. Too many alphas dilute the D–T fuel, leading to a reduction in fusion reactions, which adversely affects power balance. Rapidly removing the alphas is therefore an important challenge for future experiments.

Lastly, the issue of burn control needs to be addressed. The analysis in Chapter 14 shows that burn control should be automatic because of the shape of the $\dot{T}$ vs. $T$ with H-mode scaling. This needs to be demonstrated experimentally to show the viability of stable, steady state operation.

### 16.2.5 Fusion technology issues

As stated, great progress has been made in the basic understanding of *plasma physics* although important problems still remain that require investigation in a next generation ignition experiment. Equally importantly, an ignition experiment will have to start realistically addressing many of the *fusion technology* issues facing a reactor. Several of the these issues are summarized below.

A critical issue is the interaction of the first wall with the flux of 14.1 MeV neutrons. In the text it has been assumed that the neutron flux limits the wall loading on the first wall to $P_W \leq 4$ MW/m$^2$. This is probably an optimistic bound with respect to existing materials but not an unrealistic goal by the time of the first fusion reactor. Recall that the maximum neutron wall loading is a crucial design parameter, directly impacting the cost of a reactor. Unfortunately, there are only limited materials radiation data available because of the lack of 14 MeV sources. Most fusion researchers agree that progress towards a fusion reactor will require not only an ignition experiment such as ITER, but a dedicated materials testing facility to develop advanced materials capable of withstanding high wall loadings. This is the role of another future fusion facility known as the International Fusion Materials Irradiation Facility (IFMIF).

A second major technological problem involves the design of the divertor. Although a substantial number of data have been collected from existing tokamaks with divertors, the situation regarding ITER and fusion reactors is still not fully resolved. The reason is due to difficult tradeoff issues involving the choice of target materials, the durability of the target, the action of the target back on the plasma, and the need to robotically replace divertor modules as they wear out.

A third issue of importance is related to superconducting magnet technology. Although there has been a great deal of experience building high-field, superconducting magnets, no one has yet built magnets on the scale needed for ITER or a fusion reactor. The size, coupled with the need to use the more difficult to fabricate superconducting material niobium–tin to achieve high $B_{max}$, makes this a challenging technological problem.

A fourth technological issue involves the blanket. An experiment such as ITER will be the first to produce large amounts of fusion neutrons, thus requiring the presence of a blanket. Hopefully, the knowledge obtained from fission reactors will suffice with respect to the removal of neutron energy by means of a heat exchanger. However, there is almost no practical experience with respect to the breeding of tritium. This is an important technological issue since the world's supply of tritium is rather limited and it is very expensive to make in large quantities.

Fifth, there are the technological issues associated with plasma heating: the development of negative ion drivers for neutral beam heating, the development of high-power gyrotrons for ECH, and the development of robust antenna designs for ICH.

Finally, in closing this section it is worth noting that many researchers view the technological and plasma physics problems facing fusion to be of comparable difficulty. Although technological solutions seem conceptually possible, it is clear that a facility such as ITER must be built to test these ideas in an actual practical device.

## 16.3 ITER

The goals of the ITER experiment are to address the plasma physics and fusion technology issues just described. ITER thus has the crucial role of being the flagship facility for the world's fusion program for the next two decades. The project is enormously important in that progress towards a fusion reactor will be directly tied to the physics and technological performance of ITER.

To help understand ITER this section contains a brief history of the project and a description of the actual proposed experiment. The story unfolds below.

### 16.3.1 History

As early as the late 1970s fusion researchers around the world already recognized the importance of building a large-scale ignition experiment to investigate alpha physics and to start addressing many of the technological issues facing a reactor. An international collaboration was established to design such an experiment, which was named the International Tokamak Reactor (INTOR). The idea, which has a great deal of validity even today, is that a good way to learn about the issues facing an ignition experiment or a fusion reactor is to try to actually design one. The collaboration was highly successful in identifying many of the critical issues and suggesting important areas for future research. One difficulty faced by the INTOR group was that at the time of the design the world's large tokamaks had not yet been completed and as a consequence they did not have reliable scaling relations to predict the energy confinement time. The INTOR design, based on the best data available at the time, reached the conclusion that a plasma current of $I \approx 8$ MA would be sufficient to achieve its goals. Present understanding of energy confinement based on the H-mode scaling implies that $I \approx 20$ MA is required for ignition.

Researchers knew about the uncertainties in INTOR and so the design was never put forward for actual construction. However, INTOR had established a precedent for international collaboration. Thus, at the Geneva Summit Meeting in 1985, Soviet Leader Mikhail Gorbachov suggested to US President Ronald Reagan that the USA and the USSR should initiate an international collaboration to design and build a next generation fusion ignition experiment. The European Community and Japan quickly joined the collaboration. The project was called ITER.

The first step in the project was the development of a conceptual design, which was given the official name of the Conceptual Design Activity (CDA). The CDA started in 1989 and was completed in 1991. It was viewed as a success in that the researchers did indeed agree that such a device could be built and its aims would be achieved. Critical design parameters were also specified.

Based on this success, a second agreement was signed to develop an actual engineering design for ITER. This was called the Engineering Design Activity (EDA) and spanned the period 1992–8. At the end of this period a detailed engineering design was delivered. The huge effort devoted to the EDA resulted in a technologically successful final design: that is, the final design was deemed credible from both an engineering and plasma physics point of view by a large number of expert reviewers. The final design called for a 20 MA tokamak with a major radius of 8.1 m. It would cost about \$9B (in 2005 dollars) and take about 10 years to construct.

Although ITER was technologically credible, its cost was ultimately deemed too high by the various partners in the collaboration. This, coupled with the fact that energy was relatively inexpensive in the late 1990s, led to a situation in which none of the collaborators was willing at that time to put forward a site on which to construct ITER and to serve as the host, which involved a considerably higher cost.

A decision was, therefore, made to design a smaller version of ITER with a corresponding reduced mission and cost. The hope was that a lower cost, perhaps combined with a more favorable future economic climate for energy research, would lead to approval of the project. One of the main differences in missions is that the original ITER was designed to achieve full ignition (i.e., $Q = \infty$), while in the new version this requirement was relaxed, with the reduced goal being high but not infinite $Q$ (i.e., $Q = 10$). A further complication facing the reduced mission ITER was that the US Government decided to completely pull out of the ITER project in 1998.

The remaining partners continued their collaboration and by 2001 developed a successful engineering design for the reduced mission ITER. The new ITER design has a lower current of $I = 15$ MA and a smaller major radius of $R_0 = 6.2$ m. Its cost is about \$4B (in 2005 dollars) and should take 8–10 years to build. By the time the design was completed energy prices had begun to increase and the climate for energy research had improved. The US rejoined the collaboration. Also joining were two new partners, China and Korea. Canada also considered becoming a partner.

Another very positive result was that after the new ITER design was completed, four countries offered sites for construction of the facility: Canada, France, Japan, and Spain. The Canadian site was very attractive from a technological point of view. However, the population and corresponding tax base of Canada was too small to support the high level of funding required by the host country. Canada thus withdrew its offer of a site and is not at present an official member of the collaboration. The French and Spanish sites were both attractive technologically. Nevertheless, the European Union decided that a single entry would increase the likelihood of a European selection, and chose the French site at Cadarache as its official candidate. The Japanese proposed a technologically attractive site at Rokkasho at the north of Honshu Island.

The final competition was thus between the French and Japanese sites. Both the EU and Japan made very serious and attractive bids to become the host for the new ITER. An initial vote, leading to a longstanding stalemate, had three partners (the EU, Russia, and China) supporting the French site and the other three partners (Japan, the US, and Korea) supporting the Japanese site. After an arduous and torturous set of negotiations, an international agreement was finally reached (in July 2005) to construct the new ITER at the French site in Cadarache. This was indeed a major milestone.

The hope now, at the time this book is being written (October 2006), is that a final agreement will be signed by the end of 2006 allowing construction of the new ITER to begin in 2007.

### 16.3.2 The new ITER

It is now of interest to present a brief technical description of the new ITER, referred to hereafter simply as ITER. The discussion begins with a definition of the primary physics mission of ITER: to produce a stable, well-confined, $Q = 10$ plasma lasting for a sufficiently long duration to reach quasi-steady-state operation. A second physics mission is to achieve steady state operation using non-inductive current drive at $Q \gtrsim 5$. With respect to technology, the construction of ITER would demonstrate the viability of large superconducting magnets, various plasma facing materials, and large-scale remote handling. It would also test the effectiveness of the divertor design and begin to explore tritium breeding.

The actual ITER design is illustrated in a cutaway view in Fig. 16.1. An artist's sketch of the entire device is shown in Fig. 16.2. Note that ITER has a single null divertor and superconducting magnets constructed of niobium–tin. The magnetic field at the center of the plasma is $B_0 = 5.3$ T. To minimize the cost, the size of the machine has been minimized subject to the constraints of achieving $Q = 10$ operation with H-mode scaling in a plasma which is MHD stable without a conducting wall. This leads to a major radius $R_0 = 6.2$ m,

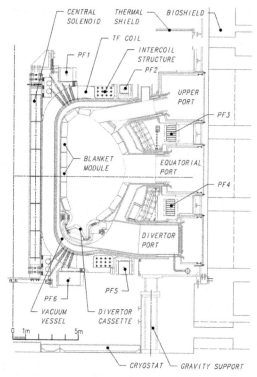

Figure 16.1  Cutaway drawing of the ITER design (*ITER Final Design Report* (2001). Vienna: IAEA).

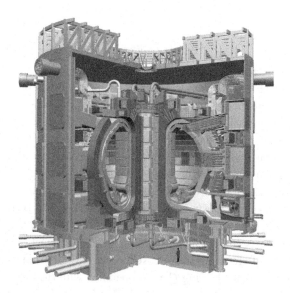

Figure 16.2  Artist's drawing of the entire ITER device (*ITER Final Design Report*. (2001). Vienna: IAEA).

Table 16.1. *Parameters for base operation of the ITER experiment*

| Parameters | Symbol | Units | ITER |
|---|---|---|---|
| Major radius | $R_0$ | m | 6.2 |
| Minor radius | $a$ | m | 2.0 |
| Aspect ratio | $R_0/a$ | | 3.2 |
| Elongation (95% flux surface) | $\kappa$ | | 1.7 |
| Plasma volume | $V_P$ | $m^3$ | 837 |
| Plasma surface area | $A_p$ | $m^2$ | 678 |
| Toroidal magnetic field | $B_0 = B(R_0)$ | T | 5.3 |
| Plasma current | $I$ | MA | 15 |
| Kink safety factor | $q_*$ | | 1.94 |
| Safety factor (95% flux surface) | $q_{95}$ | | 3.0 |
| Average temperature | $\overline{T}_e \approx \overline{T}_i \equiv \overline{T}_k$ | keV | 11.2 |
| Temperature peaking factor | $T(0)/\overline{T}$ | | 1.7 |
| Average electron density | $\overline{n}_{20}$ | $10^{20}$ m$^{-3}$ | 0.91 |
| Density peaking factor | $n(0)/\overline{n}$ | | 1.1 |
| Density/Greenwald density | $\overline{n}/n_G$ | | 0.85 |
| Energy confinement time | $\tau_E$ | s | 3.7 |
| Performance parameter | $\overline{p}\tau_E$ | atm s | 6.4 |
| Power gain | $Q = P_f/P_h$ | | 10 |
| Neutral beam power | $P_{NBI}$ | MW | 33 |
| ICH power | $P_{ICH}$ | MW | 20 |
| ECH power | $P_{ECH}$ | MW | 20 |
| Ohmic pulse length | $\tau_{pulse}$ | s | 400 |
| Toroidal beta | $\beta_t$ | | 0.026 |
| Normalized beta | $\beta_N = \beta_t/(I_M/aB_0)$ | | 0.018 |
| Cost | $C$ | $B (2005) | 4.3 |

a minor radius $a = 2$ m, and an aspect ratio $R_0/a = 3.1$. The current required to achieve the necessary confinement time is $I = 15$ MA. At $Q = 10$ operation the average density and temperature are $\overline{n}_{20} = 0.9$ and $\overline{T}_k = 11$.

For base operation, ITER will have three sources of auxiliary power: 33 MW of negative-ion-driven neutral beams, 20 MW of ICH, and 20 MW of ECH. The neutral beams and ICH will be used primarily for heating. The ECH will be used, at least initially, to stabilize a localized resistive MHD instability known as the neoclassical tearing mode, should this mode limit the achievable value of $\beta$.

ITER will operate for pulse durations of about $\tau_{pulse} \approx 400$ s, driven entirely by the ohmic transformer. The bootstrap current is expected to be small and no current drive is planned for base operation. If successful, ITER should produce a $Q = 10$ plasma corresponding to a fusion performance factor of $\overline{p}\tau_E = 6.4$ atm s.

An overall summary of the basic parameters characterizing ITER is given in Table 16.1. Observe that the parameters are comparable to the simple fusion reactor designed in

Chapter 5 and the ignition experiment designed in Chapter 14. ITER is close to being a full-scale prototype fusion reactor in terms of size and performance. The main difference is that ITER is still largely an experimental facility and therefore has not been designed to have the very high duty factor associated with a steady state power producing reactor.

In addition to the first stage of operation described in Table 16.1, there is a second stage of ITER operation that focuses on AT operation. The goal here is to obtain near steady state operation by means of substantial current drive and profile shaping. Several different scenarios are envisaged and the parameters below describe a representative example of AT operation.

For the AT experiments a combination of bootstrap current and external current drive should produce very, very long pulses (i.e. 3000 s) or even true steady state operation. The current drive will be provided by a combination of lower hybrid and electron cyclotron power. However, since ITER will not have sufficient current-drive power plus bootstrap fraction to achieve the entire 15 MA of base performance, the AT phase of the experiment will operate with somewhat reduced parameters. Specifically the total current will be reduced from 15 MA to about 9 MA, while the fusion gain will be reduced from $Q = 10$ to $Q = 5$.

Profile control should produce a hollow current density, a reversed shear safety factor, and bootstrap fractions on the order of $f_B \sim 0.4$. Also, the required confinement time to achieve $Q = 5$ will exceed the H-mode confinement time, implying the need for an improvement in transport due to the formation of internal transport barriers, again by means of profile control. Lastly, the anticipated value of $\beta$ will be very close to the Troyon no-wall stability limit. Quite possibly the resistive wall mode may be excited, and would need to be feedback stabilized.

A summary of the parameters for the representative AT operation of ITER just described is given in Table 16.2. Observe that while performance is somewhat reduced from the base values, AT operation still represents a major accomplishment in showing that a tokamak can be sustained in steady state in the presence of a large population of alpha particles.

The parameters in Tables 16.1 and 16.2 show that ITER is certainly a major international project with ambitious goals. Assuming that ITER is built and is successful in carrying out its goals, where to next? The answer is a full-scale demonstration power plant as described below.

## 16.4 A demonstration power plant (DEMO)

The current belief is that the information learned from ITER with respect to both plasma physics and fusion technology should be sufficiently complete and comprehensive to justify moving to the final step before commercialization. This final step is usually referred to as DEMO.

The transition from ITER to DEMO involves several important issues, mainly of a technological nature. The plasma physics issues should be similar to those for ITER since the devices will be of comparable size with comparable parameters. However, technologically

Table 16.2. *Parameters for AT operation of the ITER experiment*

| Parameters | Symbol | Units | ITER |
|---|---|---|---|
| Toroidal magnetic field | $B_0 = B(R_0)$ | T | 5.3 |
| Plasma current | $I$ | MA | 9.1 |
| Bootstrap current | $I_B$ | MA | 3.64 |
| Current drive current | $I_{CD}$ | MA | 5.46 |
| Bootstrap fraction | $f_B$ | | 0.4 |
| Kink safety factor | $q_*$ | | 3.2 |
| Safety factor (95% flux surface) | $q_{95}$ | | 5.0 |
| Minimum safety factor | $q_{min}$ | | 3.0 |
| Minor radius where $q = q_{min}$ | $r_{min}/a$ | | 0.7 |
| Ratio of $\tau_{required}$ to $\tau_H$ for $Q = 5$ | $H = \tau_E/\tau_H$ | | 1.4 |
| Toroidal beta | $\beta_t$ | | 0.024 |
| Normalized beta | $\beta_N = \beta_t/(I_M/aB_0)$ | | 0.028 |
| Power gain | $Q = P_f/P_h$ | | 5 |
| Lower hybrid power | $P_{LHCD}$ | MW | 40 |
| Electron cyclotron power | $P_{ECCD}$ | MW | 20 |
| Pulse length | $\tau_{pulse}$ | s | >3000 |

DEMO must be able to demonstrate full steady state operation in a safe, reliable, and maintainable way. While individual components may extrapolate in a straightforward manner, integration of all these components into a working power plant will be a major goal of DEMO.

Another major goal of DEMO will be to demonstrate tritium breeding with a recovery ratio greater than unity. This is crucial since the world's supply of tritium is very limited. There are huge reserves of lithium that can be used to breed tritium in the blanket but it is essential actually to demonstrate that more tritium can be produced than is consumed.

Lastly, DEMO, through the utilization of advanced materials developed during the interim period, should be able to demonstrate the attractiveness of fusion with respect to the environment in general and radioactive waste in particular. This is, after all, one of the primary advantages of a fusion reactor.

DEMO is clearly decades away. If built and successful, the step after DEMO is a commercial fusion power plant. Because of its complexity, the capital cost of a fusion power plant will likely be relatively high. However, its fuel and operating costs should be low. The net result is that the overall cost of electricity from a fusion power plant may indeed be competitive with other sources when such fusion plants become available. Time will tell.

In the immediate future the goal is to build and operate ITER. ITER is expensive but the ultimate attractiveness of fusion in terms of fuel supply and environmental impact suggests that this is a wise investment of research funds. Can the nations of the world afford not to try to harness this remarkable source of power?

## Bibliography

There are a large number of documents describing various aspects of the ITER experiment. Useful summaries are given in the references listed below.

Berk, H. L., Betti, R., *et al.* (2001). *Review of Burning Plasma Physics*, US Department of Energy Report DOE/SC-0041. Germantown, Maryland: US Department of Energy.

ITER Final Design Report (2001). ITER Documentation Series No. 22. Vienna: IAEA.

McCracken, G. and Stott, P. (2005). *Fusion, the Energy of the Universe*. London: Elsevier Academic Press.

Wesson, J. (2004). *Tokamaks*, third edn. Oxford: Oxford University Press.

# Appendix A

## Analytic derivation of $\langle \sigma v \rangle$

The goal here is to derive an analytic expression for $\langle \sigma v \rangle$, the definition of which is as follows:

$$\langle \sigma v \rangle = \frac{1}{n_1 n_2} \int f_1(\mathbf{v}_1) f_2(\mathbf{v}_2)\, \sigma(v)\, v\, d\mathbf{v}_1 d\mathbf{v}_2. \tag{A.1}$$

Here, $v = |\mathbf{v}_2 - \mathbf{v}_1|$. Each species is assumed to have a Maxwellian distribution function with the same temperature $T$:

$$
\begin{aligned}
f_1 &= n_1 \left( \frac{m_1}{2\pi T} \right)^{3/2} e^{-m_1 v_1^2/2T}, \\
f_2 &= n_2 \left( \frac{m_2}{2\pi T} \right)^{3/2} e^{-m_2 v_2^2/2T}.
\end{aligned}
\tag{A.2}
$$

Some general simplifications can be made before introducing an analytic approximation for $\sigma$. The first step is to replace the particle velocities $\mathbf{v}_1$ and $\mathbf{v}_2$ with a new set of independent variables representing the center of mass velocity $\mathbf{V}$ and the relative velocity $\mathbf{v}$:

$$
\begin{aligned}
\mathbf{V} &= \frac{m_2 \mathbf{v}_2 + m_1 \mathbf{v}_1}{m_2 + m_1}, \\
\mathbf{v} &= \mathbf{v}_2 - \mathbf{v}_1.
\end{aligned}
\tag{A.3}
$$

The inverse relations are given by

$$
\begin{aligned}
\mathbf{v}_2 &= \mathbf{V} + \frac{m_1}{m_2 + m_1} \mathbf{v}, \\
\mathbf{v}_1 &= \mathbf{V} - \frac{m_2}{m_2 + m_1} \mathbf{v}.
\end{aligned}
\tag{A.4}
$$

Some simple algebra leads to the following expressions for the Jacobian of the transformation and the sum of the kinetic energies appearing in the product of the Maxwellians:

$$
\begin{aligned}
d\mathbf{v}_1 d\mathbf{v}_2 &= d\mathbf{V} d\mathbf{v}, \\
m_1 v_1^2 + m_2 v_2^2 &= (m_1 + m_2) V^2 + m_r v^2,
\end{aligned}
\tag{A.5}
$$

where $m_r = m_1 m_2 / (m_1 + m_2)$ is the reduced mass. The expression for $\langle \sigma v \rangle$ reduces to

$$\langle \sigma v \rangle = \left( \frac{m_1}{2\pi T} \right)^{3/2} \left( \frac{m_2}{2\pi T} \right)^{3/2} \int d\mathbf{V} d\mathbf{v}\, \sigma(v)\, v\, e^{-[(m_1+m_2)V^2 + m_r v^2]/2T}. \tag{A.6}$$

Table A.1. *Values of $\sigma_m$ and $T_m$ for the analytic model of $\sigma(v)$*

| Reaction | $\sigma_m$ (barns) | $T_m$ (keV) |
|----------|--------------------|-------------|
| D–T | 5.03 | 296 |
| D–He$^3$ | 0.48 | 1970 |
| D–D | 0.029 | 246 |

The next step is to introduce spherical velocity variables

$$\mathrm{d}\mathbf{V} = V^2 \sin\theta \, \mathrm{d}V \, \mathrm{d}\theta \, \mathrm{d}\phi = 4\pi V^2 \mathrm{d}V,$$
$$\mathrm{d}\mathbf{v} = v^2 \sin\hat\theta \, \mathrm{d}v \, \mathrm{d}\hat\theta \, \mathrm{d}\hat\phi = 4\pi v^2 \mathrm{d}v. \tag{A.7}$$

Here, use has been made of the fact that none of the quantities in the integrand depends upon any of the angular velocity variables. This leads to the factor $4\pi$.

The integral over $V$ can now be easily evaluated analytically by making the substitution $x = [(m_1 + m_2)/2T]^{1/2}V$ and using the relation

$$\int_0^\infty x^2 \mathrm{e}^{-x^2} \mathrm{d}x = \pi^{1/2}/4. \tag{A.8}$$

The quantity $\langle \sigma v \rangle$ simplifies to

$$\langle \sigma v \rangle = 4\pi \, (m_r/2\pi T)^{3/2} \int_0^\infty \sigma(v) \, v^3 \, \mathrm{e}^{-m_r v^2/2T} \mathrm{d}v. \tag{A.9}$$

The final step in the general reduction of $\langle \sigma v \rangle$ is to transform the velocity variable from $v$ to the center of mass kinetic energy $W = m_r v^2/2$. This yields

$$\langle \sigma v \rangle = \frac{8\pi}{m_r^2} \left( \frac{m_r}{2\pi T} \right)^{3/2} \int_0^\infty W \, \sigma(W) \, \mathrm{e}^{-W/T} \mathrm{d}W. \tag{A.10}$$

This is the desired general expression for $\langle \sigma v \rangle$. To proceed further, one must introduce an explicit form for $\sigma(v)$. A useful approximation from the theory of nuclear physics is as follows:

$$\sigma(W) = \sigma_m \left( \frac{T_m}{W} \right) \mathrm{e}^{-2(T_m^{1/2}/W^{1/2}-1)}, \tag{A.11}$$

where $\sigma_m$ and $T_m$ are experimentally determined parameters modeling the specific reaction of interest. A sketch of $\sigma$ is presented in Fig. A.1. Note that $\sigma_m$ is the peak value of $\sigma$, while $T_m$ is the energy at which the peak occurs. Although qualitatively similar to the experimentally obtained curve, the analytic model is actually only accurate for energies well below the maximum. Best-fit experimental values of $\sigma_m$ and $T_m$ are given in Table A.1 for the fusion reactions of interest.

The evaluation of $\langle \sigma v \rangle$ proceeds by making the substitution $z = W/T_m$ in the integral and introducing the normalized temperature $T_* = T/T_m$. The integral reduces to

$$\langle \sigma v \rangle = \frac{2\sigma_m}{T_*^{3/2}} \left( \frac{2T_m}{\pi m_r} \right)^{1/2} \int_0^\infty \mathrm{d}z \, \exp\left( -\frac{z}{T_*} - \frac{2}{z^{1/2}} + 2 \right). \tag{A.12}$$

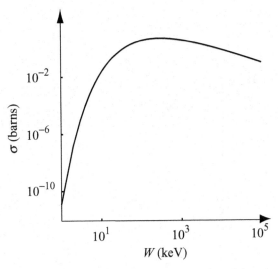

Figure A.1  Schematic illustration of the cross section $\sigma$ as a function of the energy $W$. The maximum value of $\sigma$ is $\sigma_m$ and occurs at $W = T_m$

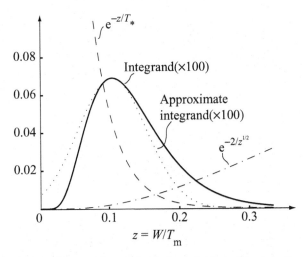

Figure A.2  Illustration of the integrand appearing in the evaluation of $\langle \sigma v \rangle$ as a function of normalized energy.

The integrand is sketched in Fig. A.2. Observe that the main contribution to the integral arises from the region $z \approx z_m$, corresponding to the maximum of the exponent. A simple calculation shows that this maximum occurs at $z_m = T_*^{2/3}$, which corresponds to

$$\frac{W_m}{T} = \left(\frac{T_m}{T}\right)^{1/3}.$$  (A.13)

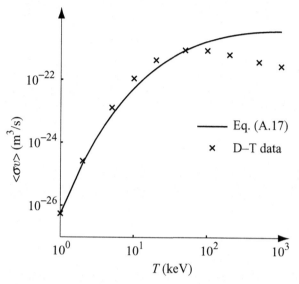

Figure A.3 Plot of the analytic $\langle \sigma v \rangle$ as a function of $T$. Also shown for comparsion are experimental data.

One consequence of this relation can be ascertained as follows. Since $T_m = 296$ keV for D–T, then for $T = 1$ keV, $W_m = 6.7$ keV. Similarly for $T = 10$ keV, $W_m = 31$ keV. Since $W_m > T$, the implication is that most of the fusion reactions occur for particles on the tail of the distribution function.

Returning to $\langle \sigma v \rangle$, one can obtain an accurate approximation for the integral by Taylor expanding the exponent about the point $z = z_m$. This yields

$$g(z) = -\frac{z}{T_*} - \frac{2}{z^{1/2}} + 2$$

$$\approx g(z_m) + g'(z_m)(z - z_m) + \frac{1}{2}g''(z_m)(z - z_m)^2 + \cdots$$

$$\approx -\frac{3}{T_*^{1/3}} + 2 - \frac{3}{4T_*^{5/3}}(z - z_m)^2 + \cdots \tag{A.14}$$

The approximate integrand is shown as the dotted curve in Fig. A.2. Observe that the exact and approximate integrands differ only in regions where the integrand is small and does not contribute much to the overall integral. Thus, the integral can be easily evaluated, with negligible error, by extending the range of integration to $-\infty < z < \infty$ as follows:

$$\int_0^\infty e^{g(z)}dz \approx e^{-3/T_*^{1/3}+2} \int_{-\infty}^\infty e^{-3(z-z_m)^2/4T_*^{5/3}} \, dz$$

$$\approx \frac{2\pi^{1/2}T_*^{5/6}}{\sqrt{3}} e^{-3/T_*^{1/3}+2}. \tag{A.15}$$

Here, use has been made of the relation

$$\int_{-\infty}^\infty e^{-x^2}dx = \pi^{1/2}. \tag{A.16}$$

Substituting Eq. (A.15) into Eq. (A.12) leads to the desired expression for $\langle \sigma v \rangle$:

$$\langle \sigma v \rangle = \frac{4\sigma_m}{\sqrt{3}} \left( \frac{2T_m}{m_r} \right)^{1/2} \left( \frac{T_m}{T} \right)^{2/3} e^{-3(T_m/T)^{1/3}+2}. \tag{A.17}$$

Equation (A.17) is plotted as a function of temperature in Fig. A.3. It is qualitatively similar to the experimentally determined curve. The maximum occurs at $T/T_m = 27/8$ and at this value

$$\langle \sigma v \rangle_{\max} = \frac{16}{9\sqrt{3}} \sigma_m \left( \frac{2T_m}{m_r} \right)^{1/2}. \tag{A.18}$$

One should keep in mind, however, that the analytic form of $\langle \sigma v \rangle$ is accurate only for $T \ll T_m$.

This completes the analytic derivation of $\langle \sigma v \rangle$.

# Appendix B

## Radiation from an accelerating charge

The purpose of Appendix B is to present a derivation of the formula for the power radiated by an accelerating charge. The derivation requires a complex analysis. An outline of the analysis is as follows. First, a careful definition is given of the "radiation" component of a specified electromagnetic field in the context of Maxwell's equations. Second, a derivation is presented of the formulas that determine the vector and scalar potentials $\mathbf{A}$ and $\phi$ arising from a general distribution of current and charge densities in the time dependent case. The resulting integral relations are generalizations of the well-known Biot–Savart law and Gauss' theorem from magnetostatics and electrostatics. Third, $\mathbf{A}$ and $\phi$, as well as the corresponding electromagnetic fields $\mathbf{E}$ and $\mathbf{B}$, are calculated assuming the charge and current densities correspond to a single accelerating charged particle. Finally, these results are combined to evaluate the outward Poynting flux on a surface far from the particle. The integral of the Poynting flux over this surface yields the desired expression for the power radiated by the accelerating charge.

### B.1 Definition of the radiation field

Assume the existence of a set of localized, time dependent charges and currents. Far from the sources the electric and magnetic fields decay away with distance. Part of these fields corresponds to the "radiation" field and part to the non-radiating "near-field." What is the precise definition that distinguishes the radiation component of the field from the components of the near-field? The definition is closely associated with the rate at which the fields decay with distance. In particular, consider the integral of the outward Poynting flux over the surface of a large sphere surrounding the sources. This integral represents the power leaving the system and scales as follows:

$$P = \int \mathbf{E} \times \mathbf{H} \cdot \mathbf{n}\, dA \sim |E|\,|B|\, r^2/\mu_0. \tag{B.1}$$

The radiation field is defined as that component of $\mathbf{E}$ and $\mathbf{B}$ that scales as

$$\begin{aligned} |\mathbf{E}| &\sim 1/r & r \to \infty, \\ |\mathbf{B}| &\sim 1/r & r \to \infty. \end{aligned} \tag{B.2}$$

For this scaling the integral of the Poynting flux approaches a constant value for large $r$, implying that the corresponding power is indeed lost from the system.

It can be shown that the remaining components of $\mathbf{E}$ and $\mathbf{B}$ decay faster than $1/r$. For these components, the integral of the Poynting flux approaches zero for large $r$. The implication is that the more rapidly decaying fields do not radiate away power but instead build up stored energy, which does not leave the system.

## B.2 Calculation of A and $\phi$ from a time dependent source

The task here is as follows. Consider a given localized, time dependent distribution of current density **J** and charge density $\sigma$. Knowing these quantities, derive a formula for the resulting vector potential **A** and scalar potential $\phi$. The derivation begins by recalling the relation between the fields and the potentials

$$\mathbf{E} = -\frac{\partial \mathbf{A}}{\partial t} - \nabla\phi, \tag{B.3}$$

$$\mathbf{B} = \nabla \times \mathbf{A}.$$

If one chooses the Lorentz gauge condition

$$\nabla \cdot \mathbf{A} + \frac{1}{c^2}\frac{\partial \phi}{\partial t} = 0, \tag{B.4}$$

then, as is well known, **A** and $\phi$ satisfy the 3-D inhomogeneous wave equations

$$\nabla^2 \mathbf{A} - \frac{1}{c^2}\frac{\partial^2 \mathbf{A}}{\partial t^2} = -\mu_0 \mathbf{J},$$

$$\nabla^2 \phi - \frac{1}{c^2}\frac{\partial^2 \phi}{\partial t^2} = -\frac{\sigma}{\varepsilon_0}. \tag{B.5}$$

Observe that in this formulation, $\phi$ and each of the rectangular components of **A** satisfy an equation of the form

$$\frac{\partial^2 \psi}{\partial x'^2} + \frac{\partial^2 \psi}{\partial y'^2} + \frac{\partial^2 \psi}{\partial z'^2} - \frac{1}{c^2}\frac{\partial^2 \psi}{\partial t'^2} = S(x', y', z', t'). \tag{B.6}$$

Here, primed coordinates have been introduced for convenience. They are tied to the geometric distribution of the sources. Ultimately, the total field is determined by integrating over the entire source (i.e., over the primed coordinates). Unprimed coordinates will also be introduced and these represent the location of an arbitrary observation point at which one wants to know the values of the potentials and fields.

The equation for $\psi$ is solved using a form of Green's theorem. The solution is based on Green's identity

$$\nabla \cdot (G \nabla \psi - \psi \nabla G) = G \nabla^2 \psi - \psi \nabla^2 G, \tag{B.7}$$

where $G$ is an arbitrary function satisfying the homogeneous wave equation

$$\nabla^2 G - \frac{1}{c^2}\frac{\partial^2 G}{\partial t^2} = \delta(x - x')\delta(y - y')\delta(z - z')g(t' + r'/c - t) \tag{B.8}$$

in which $r' = [(x' - x)^2 + (y' - y)^2 + (z' - z)^2]^{1/2}$ and $g$ is an arbitrary function. Integrating this equation over a closed volume $V$ using the divergence theorem yields

$$\int_V (G \nabla'^2 \psi - \psi \nabla'^2 G) \, dV' = \int_A (G \mathbf{n}' \cdot \nabla' \psi - \psi \mathbf{n}' \cdot \nabla' G) \, dA', \qquad (B.9)$$

where $\mathbf{n}'$ is the outward normal to the bounding surface $A$. A more convenient form of Eq. (B.8) is obtained by using the defining equations for $\psi$ and $G$:

$$\frac{1}{c^2} \frac{\partial}{\partial t'} \int_V \left( G \frac{\partial \psi}{\partial t'} - \psi \frac{\partial G}{\partial t'} \right) dV' + \int_V G S \, dV' - \psi(\mathbf{x}, t') g(t' - t)$$
$$= \int_A (G \mathbf{n}' \cdot \nabla' \psi - \psi \mathbf{n}' \cdot \nabla' G) \, dA'. \qquad (B.10)$$

The solution for $G$, satisfying Eq. (B.8), is given by

$$G = -\frac{1}{4\pi r'} g \left( t' + \frac{r'}{c} - t \right). \qquad (B.11)$$

Straightforward substitution accompanied by some slightly tedious algebra shows that $G$ indeed satisfies Eq. (B.8). The evaluation of $\psi$ is simplified by a particular choice for $g$:

$$g = \delta(t' + r'/c - t). \qquad (B.12)$$

Here, $\delta(x)$ is the standard impulse function of zero width, infinite height, and unit area.

The next step in the derivation is to substitute the special choice for $G$ into Eq. (B.10), the basic equation for $\psi$, and integrate over all time $-\infty < t' < \infty$. To define the time as well as spatial integrals uniquely one must also specify the volume. For the present case the volume is taken as a sphere of radius $r_2$ centered about the observation point. Ultimately the limit $r_2 \to \infty$ will be taken. With this choice for the volume the separate terms in Eq. (B.10) can now be evaluated.

The second term is evaluated as follows:

$$-\frac{1}{4\pi} \int_V d\mathbf{x}' \int_{-\infty}^{\infty} dt' \frac{1}{r'} S(\mathbf{x}', t') \delta(t' + r'/c - t) = -\frac{1}{4\pi} \int_V d\mathbf{x}' \frac{1}{r'} S(\mathbf{x}', t - r'/c). \quad (B.13)$$

The first term is a perfect differential with respect to $t'$, which can easily be integrated:

$$\frac{1}{c^2} \int_{-\infty}^{\infty} dt' \frac{\partial}{\partial t'} \int_V d\mathbf{x}' \left( G \frac{\partial \psi}{\partial t'} - \psi \frac{\partial G}{\partial t'} \right)$$
$$= \frac{1}{c^2} \int_V d\mathbf{x}' \left( G \frac{\partial \psi}{\partial t'} - \psi \frac{\partial G}{\partial t'} \right) \Big|_{-\infty}^{\infty} = 0. \qquad (B.14)$$

The integral vanishes because the delta functions in $G$ and $\partial G / \partial t'$ vanish at both end points. The contribution from the surface integral term vanishes if $r_2$ is chosen to be sufficiently large that the signal from the source, which travels at the speed of light, has not had time to reach the integration surface. Thus as $r_2 \to \infty$, $\psi$ and $\mathbf{n}' \cdot \nabla \psi$ both vanish on the integration surface. The remaining contribution is easily evaluated by noting that

$$\int_{-\infty}^{\infty} \psi(\mathbf{x}, t') \, \delta(t' - t) \, dt' = \psi(\mathbf{x}, t). \qquad (B.15)$$

Combining these results leads to the desired expression for $\psi$

$$\psi(\mathbf{x}, t) = -\frac{1}{4\pi} \int \frac{S(\mathbf{x}', t - r'/c)}{r'} \, d\mathbf{x}'. \tag{B.16}$$

Finally, applying this result to the vector and scalar potentials leads to

$$\mathbf{A}(\mathbf{x}, t) = \frac{\mu_0}{4\pi} \int \frac{\mathbf{J}(\mathbf{x}', t - r'/c)}{r'} \, d\mathbf{x}',$$

$$\phi(\mathbf{x}, t) = \frac{1}{4\pi\varepsilon_0} \int \frac{\sigma(\mathbf{x}', t - r'/c)}{r'} \, d\mathbf{x}'. \tag{B.17}$$

Observe that these expressions are nearly identical to the corresponding relations for pure magnetostatics and electrostatics. The only difference is that the source terms are evaluated at the retarded time $t - r'/c$. The interpretation is as follows. If one wants to know the values of $\mathbf{A}$ and $\phi$ at the time $t$, then the source terms must be evaluated at a slightly earlier $t$ since it takes a time $r'/c$ for the information to propagate from the source to the observation point.

## B.3 Application to a single accelerating charge

The next step in the derivation is to choose a current and charge density that correspond to a single accelerating charged particle and then evaluate the $\mathbf{A}$ and $\phi$ that result from Eq. (B.17). If a point particle has a charge $q$ and is moving along a trajectory $\mathbf{x}_0(t)$, then the equivalent $\sigma$ and $\mathbf{J}$ are given by

$$\sigma = q\,\delta(x' - x_0)\delta(y' - y_0)\delta(z' - z_0),$$

$$\mathbf{J} = q\,\mathbf{v}_0\,\delta(x' - x_0)\delta(y' - y_0)\delta(z' - z_0), \tag{B.18}$$

where $\mathbf{v}_0(t) = d\mathbf{x}_0(t)/dt$. The delta function integrals are not as trivial to evaluate as one might think. The reason is that the sources are evaluated at the retarded time implying for instance that $x_0(t - r'/c) = x_0(t, x', y', z', x, y, z)$; that is, $x_0$ itself is a function of the integration coordinates. This difficulty is overcome by introducing new integration coordinates as follows:

$$x' - x_0 = X,$$
$$y' - y_0 = Y, \tag{B.19}$$
$$z' - z_0 = Z.$$

After another slightly tedious calculation one can show that the Jacobian of the transformation can be written as

$$dX\,dY\,dZ = \left(1 - \frac{\mathbf{v}_0 \cdot \mathbf{e}_r}{c}\right) dx'\,dy'\,dz', \tag{B.20}$$

where

$$\mathbf{e}_r = \frac{(x - x')}{r'}\mathbf{e}_x + \frac{(y - y')}{r'}\mathbf{e}_y + \frac{(z - z')}{r'}\mathbf{e}_z. \tag{B.21}$$

With this transformation, the delta function integrals can now easily be evaluated, yielding

$$\phi(\mathbf{x}, t) = \frac{q}{4\pi\varepsilon_0} \frac{1}{r_0(1 - v_r/c)},$$

$$\mathbf{A}(\mathbf{x}, t) = \frac{\mu_0 q}{4\pi} \frac{\mathbf{v}_0}{r_0(1 - v_r/c)}, \tag{B.22}$$

Here, $v_r(t - r_0/c) = \mathbf{v}_0 \cdot \mathbf{e}_r$ and

$$r_0 = [(x - x_0)^2 + (y - y_0)^2 + (z - z_0)^2]^{1/2}. \tag{B.23}$$

The quantity $r_0$ represents the distance between the observation point and the charge at the retarded time.

## B.4 Calculation of E and B

The electric and magnetic fields are calculated by taking appropriate derivatives of $\mathbf{A}$ and $\phi$. The calculations are simplified by focusing on the radiation component of each field; that is, the component that scales like $1/r_0$ for large $r_0$. Since both $\mathbf{A}$ and $\phi$ are proportional to $1/r_0$ any derivatives with respect to $r_0$ produce contributions scaling like $1/r_0^2$, corresponding to non-radiating near-fields. Thus, the radiation field is determined only from various derivatives of the velocity $\mathbf{v}_0(\tau)$, where $\tau = t - r_0/c$. The derivatives leading to contributions to the radiation field are denoted with the superscript (R) and, in the non-relativistic limit are given as follows:

$$\frac{\partial \mathbf{A}^{(R)}}{\partial t} = \frac{\mu_0 q}{4\pi} \frac{1}{r_0(1 - v_r/c)} \frac{\partial \mathbf{v}_0}{\partial t} \approx \frac{\mu_0 q}{4\pi r_0} \dot{\mathbf{v}}_0,$$

$$\nabla \times \mathbf{A}^{(R)} = \frac{\mu_0 q}{4\pi} \frac{1}{r_0(1 - v_r/c)} \nabla \times \mathbf{v}_0 \approx \frac{\mu_0 q}{4\pi r_0} \nabla \tau \times \dot{\mathbf{v}}_0$$

$$\approx \frac{\mu_0 q}{4\pi r_0 c} \dot{\mathbf{v}}_0 \times \mathbf{e}_r, \tag{B.24}$$

$$\nabla \phi^{(R)} = \frac{q}{4\pi\varepsilon_0 r_0} \nabla \frac{1}{1 - v_r/c} = \frac{q}{4\pi\varepsilon_0 r_0} \frac{1}{(1 - v_r/c)^2} \nabla \frac{v_r}{c}$$

$$\approx \frac{\mu_0 q}{4\pi r_0} (\mathbf{e}_r \cdot \dot{\mathbf{v}}_0) \mathbf{e}_r.$$

Here $\dot{\mathbf{v}}_0 = d\mathbf{v}_0/d\tau$. From these relations it is straightforward to calculate the radiation components of the electric and magnetic fields:

$$\mathbf{E}^{(R)} = \frac{\mu_0 q}{4\pi r_0} (\dot{\mathbf{v}}_0 \times \mathbf{e}_r) \times \mathbf{e}_r,$$

$$\mathbf{B}^{(R)} = \frac{\mu_0 q}{4\pi r_0 c} (\dot{\mathbf{v}}_0 \times \mathbf{e}_r). \tag{B.25}$$

## B.5 Calculation of the power radiated

The power radiated is calculated by evaluating the normal component of the Poynting vector and then integrating over the surface of a large sphere surrounding the charged particle. The Poynting vector is given by

$$\mathbf{S} = \frac{1}{\mu_0} \mathbf{E}^{(R)} \times \mathbf{B}^{(R)} = \frac{1}{\mu_0 c} \left( \frac{\mu_0 q}{4\pi r_0} \right)^2 [(\dot{\mathbf{v}}_0 \times \mathbf{e}_r) \times \mathbf{e}_r] \times (\dot{\mathbf{v}}_0 \times \mathbf{e}_r)$$

$$= \left( \frac{\mu_0 q}{4\pi r_0} \right)^2 \frac{|\dot{\mathbf{v}}_0 \times \mathbf{e}_r|^2}{\mu_0 c} \mathbf{e}_r. \tag{B.26}$$

The final step is to integrate the Poynting vector over the surface area of a large sphere of radius $r_0$. To carry out the calculation, assume that the angle between $\dot{\mathbf{v}}_0$ and the outward normal $\mathbf{e}_r$ is defined as $\theta$: $|\dot{\mathbf{v}}_0 \times \mathbf{e}_r| = \dot{v}_0 \sin \theta$. Setting up a spherical coordinate system centered about $r_0 = 0$ then implies that the differential surface area on the sphere is given by

$$d\mathbf{A} = r_0^2 \sin \theta \, d\theta \, d\phi \, \mathbf{e}_r = 2\pi r_0^2 \sin \theta \, d\theta \, \mathbf{e}_r. \tag{B.27}$$

The power radiated can now be written as follows.

$$P(\tau) = \int \mathbf{S} \cdot d\mathbf{A} = \frac{\mu_0 q^2 \dot{v}_0^2}{8\pi c} \int_0^\pi \sin^3 \theta \, d\theta$$

$$= \frac{\mu_0 q^2 \dot{v}_0^2}{6\pi c} \quad W. \tag{B.28}$$

This is the desired relation. Note that for practical fusion applications in which the time scales are much longer than the retardation time one can accurately approximate $\tau \approx t$.

# Appendix C
## Derivation of Boozer coordinates

The derivation of Boozer coordinates requires several steps of analysis. First, a general transformation is introduced that converts the familiar laboratory coordinate system into a set of arbitrary flux coordinates. Second, by using the relationships $\mathbf{B} \cdot \nabla \psi = \mathbf{J} \cdot \nabla \psi = 0$ and $\nabla \cdot \mathbf{B} = \nabla \cdot \mathbf{J} = 0$, both $\mathbf{B}$ and $\mathbf{J}$ can be cast into a cross-product form in flux coordinates, close to the desired form of Boozer coordinates. Third, by means of the relation $\nabla \times \mathbf{B} = \mu_0 \mathbf{J}$, it is shown that $\mathbf{B}$ can also be written in a gradient form in flux coordinates, close to the desired form of Boozer coordinates. Fourth, it is shown how certain free functions appearing in the representation of $\mathbf{B}$ can be eliminated by means of an additional transformation of the angular flux coordinates $\chi, \zeta$. The new coordinates correspond to the actual Boozer coordinates. Fifth, the various free functions remaining in the expressions for $\mathbf{B}$ are rewritten in terms of physically recognizable quantities. Finally, the magnetic field expressed in Boozer coordinates is used to calculate the guiding center drifts of the particles.

### C.1 General coordinate transformation

The derivation starts with a general transformation from the familiar $r, \theta, \phi$ coordinates to the abstract $\psi, \chi, \zeta$ coordinates defined in Section 13.8 and repeated here for convenience:

$$\psi = \psi_0(r) + \sum_{m,n} \psi_{mn}(r) e^{i(m\theta + n\phi)},$$

$$\chi = \theta + \sum_{m,n} \theta_{mn}(r) e^{i(m\theta + n\phi)},$$

$$\zeta = -\phi + \sum_{m,n} \phi_{mn}(r) e^{i(m\theta + n\phi)}. \tag{C.1}$$

(The minus sign in front of $\phi$ is used to make $\psi, \chi, \zeta$ a right handed system.) Actually, since the derivation is rather formal in detail, it is convenient to switch from $r, \theta, \phi$ coordinates to the even simpler rectangular $x, y, z$ coordinates:
$x = (R_0 + r \cos \theta) \cos \phi$, $y = (R_0 + r \cos \theta) \sin \phi$, $z = r \sin \theta$. Thus, it is initially assumed that a general transformation exists (with appropriate periodicity properties), that is defined by

$$\psi = \psi(x, y, z),$$

$$\chi = \chi(x, y, z),$$

$$\zeta = \zeta(x, y, z). \tag{C.2}$$

Similarly, it is assumed that there is a well-defined inverse transformation given by

$$x = x(\psi, \chi, \phi),$$
$$y = y(\psi, \chi, \phi), \qquad \text{(C.3)}$$
$$z = z(\psi, \chi, \phi).$$

The Jacobian $J$ of the transformation is easily found by writing

$$d\psi \, d\chi \, d\zeta = J dx \, dy \, dz, \qquad \text{(C.4)}$$

where

$$J = \begin{vmatrix} \psi_x & \psi_y & \psi_z \\ \chi_x & \chi_y & \chi_z \\ \zeta_x & \zeta_y & \zeta_z \end{vmatrix} = \nabla\psi \cdot (\nabla\chi \times \nabla\zeta). \qquad \text{(C.5)}$$

It is now possible to define various sets of three independent basis vectors in terms of $\psi, \chi, \zeta$. One convenient choice is

$$\nabla\psi \times \nabla\chi \quad \nabla\zeta \times \nabla\psi \quad \nabla\chi \times \nabla\zeta. \qquad \text{(C.6)}$$

Note that both the magnetic field and current density can always be expressed in terms of these three basis vectors as follows:

$$\mathbf{B} = f_1(\psi, \chi, \zeta)\nabla\psi \times \nabla\chi + f_2(\psi, \chi, \zeta)\nabla\zeta \times \nabla\psi + f_3(\psi, \chi, \zeta)\nabla\chi \times \nabla\zeta,$$
$$\mathbf{J} = h_1(\psi, \chi, \zeta)\nabla\psi \times \nabla\chi + h_2(\psi, \chi, \zeta)\nabla\zeta \times \nabla\psi + h_3(\psi, \chi, \zeta)\nabla\chi \times \nabla\zeta. \qquad \text{(C.7)}$$

So far the discussion has been quite general. In the next section it is shown how these coordinates can be simplified into a cross-product form close to the desired form of Boozer coordinates.

## C.2  The partial simplification to the cross-product form of Boozer coordinates

The first step in the simplification is to make use of the fact that $\mathbf{B} \cdot \nabla\psi = 0$. Substitution into Eq. (C.7) yields $J f_3(\psi, \chi, \zeta) = 0$ or $f_3 = 0$. Similarly the relation $\mathbf{J} \cdot \nabla\psi = 0$ implies that $h_3 = 0$.

The second step requires substitution into the equation $\nabla \cdot \mathbf{B} = 0$. Using the relation

$$\nabla f(\psi, \chi, \zeta) = \frac{\partial f}{\partial \psi}\nabla\psi + \frac{\partial f}{\partial \chi}\nabla\chi + \frac{\partial f}{\partial \zeta}\nabla\zeta, \qquad \text{(C.8)}$$

one obtains

$$\nabla \cdot \mathbf{B} = J\left(\frac{\partial f_1}{\partial \zeta} + \frac{\partial f_2}{\partial \chi}\right) = 0. \qquad \text{(C.9)}$$

Equation (C.9) implies that $f_1$ and $f_2$ can always be written in terms of a stream function as follows:

$$f_1(\psi, \chi, \zeta) = \overline{f}_1(\psi) + \hat{f}(\psi)\zeta + \frac{\partial \tilde{f}(\psi, \chi, \zeta)}{\partial \chi},$$
$$f_2(\psi, \chi, \zeta) = \overline{f}_2(\psi) - \hat{f}(\psi)\chi - \frac{\partial \tilde{f}(\psi, \chi, \zeta)}{\partial \zeta}. \qquad \text{(C.10)}$$

In this expression, the non-oscillatory terms have been explicitly displayed. Note that the function $\hat{f}$ must satisfy $\hat{f} = 0$ since the corresponding contributions to $\mathbf{B}$ do not satisfy the periodicity requirement in $\chi$ or $\zeta$. The oscillatory part of the stream function $\tilde{f}(\psi, \chi, \zeta)$ must, by definition, satisfy $\langle \tilde{f} \rangle = 0$, where the average is over a period in either $\chi$ or $\zeta$. Furthermore, since $\nabla \cdot \mathbf{J} = 0$, a completely analogous argument applies to the current density leading to an identical set of conclusions with respect to the $h_j$ coefficients: $h_3 = 0$ and $h_1$, $h_2$ must have the same form as Eq. (C.10).

One now substitutes the forms for $f_1$, $f_2$, $h_1$, $h_2$ into the expressions for $\mathbf{B}$ and $\mathbf{J}$. A short calculation shows that the end results are cross-product representations of the magnetic field and current density given by

$$\mathbf{B} = \overline{f}_1 \nabla \psi \times \nabla \chi + \overline{f}_2 \nabla \zeta \times \nabla \psi + \nabla \psi \times \nabla \tilde{f},$$
$$\mathbf{J} = \overline{h}_1 \nabla \psi \times \nabla \chi + \overline{h}_2 \nabla \zeta \times \nabla \psi + \nabla \psi \times \nabla \tilde{h}. \tag{C.11}$$

These are the desired expressions. The key point that has been demonstrated is that $\overline{f}_1, \overline{f}_2, \overline{h}_1, \overline{h}_2$ are functions only of the flux $\psi$. It is shown shortly how the function $\tilde{f}$ can be eliminated by an additional transformation of the $\chi, \zeta$ coordinates.

### C.3 The partial simplification to the gradient form of Boozer coordinates

The next step in the derivation makes use of Ampère's law: $\nabla \times \mathbf{B} = \mu_0 \mathbf{J}$. First, the current density can easily be rewritten as follows

$$\mathbf{J} = \nabla \times [\overline{k}_1(\psi)\nabla \chi + \overline{k}_2(\psi)\nabla \zeta - \tilde{h}(\psi, \chi, \zeta)\nabla \psi], \tag{C.12}$$

where

$$d\overline{k}_1/d\psi = \overline{h}_1(\psi), \qquad d\overline{k}_2/d\psi = -\overline{h}_2(\psi). \tag{C.13}$$

Since $\nabla \times \mathbf{B} = \mu_0 \mathbf{J}$, one can then write

$$\mathbf{B} = \mu_0[\overline{k}_1\nabla \chi + \overline{k}_2\nabla \zeta - \tilde{h}\nabla \psi + \nabla \tilde{g}(\psi, \chi, \zeta)]. \tag{C.14}$$

Here, $\tilde{g}(\psi, \chi, \zeta)$ is an arbitrary free integration function. Without loss of generality it can be assumed that $\tilde{g}$ is periodic in $\chi$ or $\zeta$. Any non-periodic contribution must be of the form $\tilde{g} = C_1\chi + C_2\zeta$ with $C_1, C_2$ constants, and these can be absorbed in the free integration constants in $\overline{k}_1, \overline{k}_2$ arising from the integration of Eq. (C.13).

Equation (C.14) has many of the desired properties of the gradient form of Boozer coordinates. In particular, the functions $\overline{k}_1, \overline{k}_2$ are functions only of the flux $\psi$. It is shown shortly how the free function $\tilde{g}$ can be eliminated by an additional transformation of the $\chi, \zeta$ coordinates.

### C.4 Elimination of the free functions $\tilde{f}$ and $\tilde{g}$

At this point in the analysis it has been shown that the magnetic field can always be written in two alternative forms for arbitrary choices of the $\chi, \zeta$ coordinates. These forms are summarized below for convenience:

$$\mathbf{B} = \overline{f}_1 \nabla \psi \times \nabla \chi + \overline{f}_2\nabla \zeta \times \nabla \psi + \nabla \psi \times \nabla \tilde{f},$$
$$\mathbf{B} = \overline{k}_1 \nabla \chi + \overline{k}_2 \nabla \zeta - \tilde{h} \nabla \psi + \nabla \tilde{g}. \tag{C.15}$$

Assume now that an arbitrary but known choice of $\chi, \zeta$ coordinates has been specified. It is shown here how a transformation to a new set of coordinates $\chi', \zeta'$ leads to the elimination of the free functions $\tilde{f}, \tilde{g}$. The new coordinates are the Boozer coordinates.

The appropriate transformation of coordinates is defined by

$$\chi' = \chi + \overline{A}_1(\psi)\tilde{G}(\psi, \chi, \zeta) + \overline{C}_1(\psi)\tilde{F}(\psi, \chi, \zeta),$$
$$\zeta' = \zeta + \overline{A}_2(\psi)\tilde{G}(\psi, \chi, \zeta) + \overline{C}_2(\psi)\tilde{F}(\psi, \chi, \zeta), \qquad \text{(C.16)}$$
$$\psi' = \psi,$$

where $\overline{A}_1, \overline{A}_2, \overline{C}_1, \overline{C}_2$ and $\tilde{G}, \tilde{F}$ are arbitrary functions with $\langle\tilde{G}\rangle = \langle\tilde{F}\rangle = 0$. Clearly there is a great deal of freedom in the transformation because of the six free functions. A convenient choice for these free functions is

$$\chi' = \chi + \frac{1}{\overline{f}_1\overline{k}_2 + \overline{f}_2\overline{k}_1}(\overline{f}_2\tilde{g} + \overline{k}_2\tilde{f}),$$

$$\zeta' = \zeta + \frac{1}{\overline{f}_1\overline{k}_2 + \overline{f}_2\overline{k}_1}(\overline{f}_1\tilde{g} - \overline{k}_1\tilde{f}), \qquad \text{(C.17)}$$

$$\psi' = \psi.$$

With this choice of the transformation, one can show by a short calculation that the two forms of the magnetic field reduce to

$$\mathbf{B} = \overline{f}_1\nabla\psi \times \nabla\chi + \overline{f}_2\nabla\zeta \times \nabla\psi,$$
$$\mathbf{B} = \overline{k}_1\nabla\chi + \overline{k}_2\nabla\zeta + \tilde{k}\nabla\psi, \qquad \text{(C.18)}$$

where

$$\tilde{k}(\psi, \chi, \zeta) = -\tilde{h} + \frac{1}{\overline{f}_1\overline{k}_2 + \overline{f}_2\overline{k}_1}\left[(\overline{f}_2\tilde{g} + \overline{k}_2\tilde{f})\frac{\mathrm{d}\overline{k}_1}{\mathrm{d}\psi} + (\overline{f}_1\tilde{g} - \overline{k}_1\tilde{f})\frac{\mathrm{d}\overline{k}_2}{\mathrm{d}\psi}\right] \quad \text{(C.19)}$$

is a new free function satisfying $\langle\tilde{k}\rangle = 0$ that replaces $\tilde{h}$. For simplicity all the primes have been suppressed. Equation (C.18) is the desired representation of the magnetic field in Boozer coordinates. At this point there still remains considerable flexibility in defining the actual $\psi, \chi, \zeta$ coordinates since the condition $\mathbf{J} \times \mathbf{B} = \nabla p$ has not as yet been applied. However, for the present purpose of understanding single-particle confinement there is no need to carry out the pressure balance analysis. Instead, attention is next focused on the problem of relating the flux functions $\overline{f}_1, \overline{f}_2, \overline{k}_1, \overline{k}_2$ to more physically recognizable quantities.

## C.5 Introduction of physical quantities into the Boozer coordinates

In this section it is shown that the quantities $\overline{f}_1, \overline{f}_2, \overline{k}_1, \overline{k}_2$ are closely related to the magnetic fluxes and currents contained within a given pressure contour. The calculations are carried out by introducing the poloidal and toroidal differential surface areas required to calculate the fluxes and currents. These are illustrated in Fig. C.1. The mathematical expressions for the poloidal and toroidal differential surface areas are given by their usual definitions:

$$\mathrm{d}\mathbf{A}_\mathrm{p} = \frac{\partial\mathbf{r}}{\partial\zeta} \times \frac{\partial\mathbf{r}}{\partial\psi}\mathrm{d}\zeta\,\mathrm{d}\psi,$$

$$\mathrm{d}\mathbf{A}_\mathrm{t} = \frac{\partial\mathbf{r}}{\partial\psi} \times \frac{\partial\mathbf{r}}{\partial\chi}\mathrm{d}\chi\,\mathrm{d}\psi. \qquad \text{(C.20)}$$

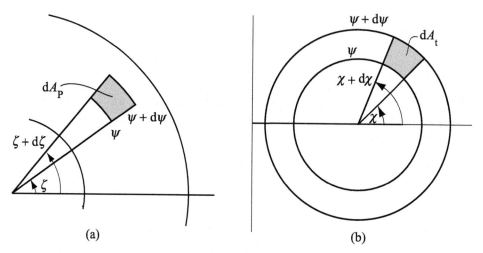

Figure C.1 Differential surface areas used to calculate: (a) the poloidal flux and current density, and (b) the toroidal flux and current density.

The various derivatives with respect to $\psi, \chi, \zeta$ can be conveniently taken in a rectangular coordinate system $x(\psi, \chi, \zeta), y(\psi, \chi, \zeta), z(\psi, \chi, \zeta)$. Thus, writing $\mathbf{r} = x(\psi, \chi, \zeta)\mathbf{e}_x + y(\psi, \chi, \zeta)\mathbf{e}_y + z(\psi, \chi, \zeta)\mathbf{e}_z$ one obtains

$$\partial\mathbf{r}/\partial\psi = x_\psi\mathbf{e}_x + y_\psi\mathbf{e}_y + z_\psi\mathbf{e}_z,$$
$$\partial\mathbf{r}/\partial\chi = x_\chi\mathbf{e}_x + y_\chi\mathbf{e}_y + z_\chi\mathbf{e}_z, \qquad (C.21)$$
$$\partial\mathbf{r}/\partial\zeta = x_\zeta\mathbf{e}_x + y_\zeta\mathbf{e}_y + z_\zeta\mathbf{e}_z.$$

Similarly the rectangular unit vectors can be written as

$$\mathbf{e}_x = \nabla x = x_\psi\nabla\psi + x_\chi\nabla\chi + x_\zeta\nabla\zeta,$$
$$\mathbf{e}_y = \nabla y = y_\psi\nabla\psi + y_\chi\nabla\chi + y_\zeta\nabla\zeta, \qquad (C.22)$$
$$\mathbf{e}_z = \nabla z = z_\psi\nabla\psi + z_\chi\nabla\chi + z_\zeta\nabla\zeta.$$

A straightforward calculation then shows that

$$\frac{\partial\mathbf{r}}{\partial\zeta} \times \frac{\partial\mathbf{r}}{\partial\psi} = J'\nabla\chi,$$
$$\frac{\partial\mathbf{r}}{\partial\psi} \times \frac{\partial\mathbf{r}}{\partial\chi} = J'\nabla\zeta, \qquad (C.23)$$

where $J' = [x_\psi(y_\chi z_\zeta - z_\chi y_\zeta) + x_\chi(y_\zeta z_\psi - z_\zeta y_\psi) + x_\zeta(y_\psi z_\chi - z_\psi y_\chi)]$. The quantity $J'$ is the inverse Jacobian of the coordinate transformation; that is, $J' = 1/J$. This follows from the definition

$$dx\, dy\, dz = J'd\psi\, d\chi\, d\zeta \qquad (C.24)$$

with

$$J' = \begin{vmatrix} x_\psi & x_\chi & x_\zeta \\ y_\psi & y_\chi & y_\zeta \\ z_\psi & z_\chi & z_\zeta \end{vmatrix}$$
$$= [x_\psi(y_\chi z_\zeta - z_\chi y_\zeta) + x_\chi(y_\zeta z_\psi - z_\zeta y_\psi) + x_\zeta(y_\psi z_\chi - z_\psi y_\chi)]. \qquad (C.25)$$

Combining these results leads to the following simple expressions for the differential surface areas:

$$d\mathbf{A}_p = \frac{\nabla \chi}{J} d\zeta \, d\psi,$$

$$d\mathbf{A}_t = \frac{\nabla \zeta}{J} d\chi \, d\psi.$$

(C.26)

Consider now the poloidal flux defined by

$$\Psi_p = \int \mathbf{B} \cdot d\mathbf{A}_p = \int_0^{2\pi} \int_0^\psi \frac{\mathbf{B} \cdot \nabla \chi}{J} d\zeta \, d\psi.$$

(C.27)

Using the cross-product representation of $\mathbf{B}$ it follows that

$$\mathbf{B} \cdot \nabla \chi = \overline{f}_2 (\nabla \zeta \times \nabla \psi) \cdot \nabla \chi = J \overline{f}_2.$$

(C.28)

The Jacobian factors cancel in the integrand and the $\zeta$ integral can be immediately evaluated yielding a factor of $2\pi$. The poloidal flux reduces to

$$\Psi_p = 2\pi \int_0^\psi \overline{f}_2 \, d\psi.$$

(C.29)

Recall that $\psi$ has been defined as $\psi = \Psi_p/2\pi$. Thus, differentiating Eq. (C.29) with respect to $\psi$ leads to the conclusion that

$$\overline{f}_2 = 1.$$

(C.30)

A completely analogous calculation for the toroidal flux shows that

$$\Psi_t = \int \mathbf{B} \cdot d\mathbf{A}_t = \int_0^{2\pi} \int_0^\psi \frac{\mathbf{B} \cdot \nabla \zeta}{J} d\chi \, d\psi = 2\pi \int_0^\psi \overline{f}_1 \, d\psi.$$

(C.31)

Again, recalling that $\psi_t = \Psi_t/2\pi$ and then differentiating with respect to $\psi$ leads to the following expression for $\overline{f}_1$:

$$\overline{f}_1(\psi) = \frac{d\psi_t(\psi)}{d\psi} \equiv q(\psi).$$

(C.32)

Here, $q(\psi)$ is the generalized definition of the safety factor.

The next quantities of interest are the currents flowing within a $\psi = $ const. contour. The net toroidal current is given by

$$I_t = \int \mathbf{J} \cdot d\mathbf{A}_t = \int_0^{2\pi} \int_0^\psi \frac{\mathbf{J} \cdot \nabla \zeta}{J} d\chi \, d\psi.$$

(C.33)

The quantity $\mathbf{J} \cdot \nabla \zeta$ is evaluated from $\mu_0 \mathbf{J} = \nabla \times \mathbf{B}$ using the gradient form of $\mathbf{B}$. A short calculation yields

$$\mu_0 \mathbf{J} \cdot \nabla \zeta = J \left( \frac{d\overline{k}_1}{d\psi} - \frac{\partial \widetilde{k}}{\partial \chi} \right).$$

(C.34)

Note that the $\chi$ integral involving $\widetilde{k}$ averages to zero. The remaining terms reduce to

$$\mu_0 I_t(\psi) = 2\pi \int_0^\psi \frac{d\overline{k}_1}{d\psi} d\psi.$$

(C.35)

One now defines $i_t(\psi) = \mu_0 I_t(\psi)/2\pi$ and directly integrates Eq. (C.35) leading to

$$\overline{k}_1(\psi) - \overline{k}_1(0) = i_t(\psi).$$

(C.36)

The integration constant $\bar{k}_1(0) = 0$. This follows from Eq. (C.18); that is, when the net toroidal current flowing in the plasma is zero, there is no average poloidal vacuum field.

The last quantity of interest is the poloidal current flowing within a $\psi = $ const. contour. A similar calculation to the one just presented shows that

$$I_p^{(\text{plasma})} = \int \mathbf{J} \cdot d\mathbf{A}_p = \int_0^{2\pi} \int_0^{\psi} \frac{\mathbf{J} \cdot \nabla \chi}{J} \, d\zeta \, d\psi. \tag{C.37}$$

The quantity $\mathbf{J} \cdot \nabla \chi$ is again found from the gradient form of the magnetic field and can be written as

$$\mu_0 \mathbf{J} \cdot \nabla \chi = -J \left( \frac{d\bar{k}_2}{d\psi} - \frac{\partial \tilde{k}}{\partial \zeta} \right). \tag{C.38}$$

The $\tilde{k}$ term averages to zero in the integral. The remaining contribution to $I_p$ reduces to

$$\mu_0 I_p^{(\text{plasma})}(\psi) = -2\pi \int_0^{\psi} \frac{d\bar{k}_2}{d\psi} \, d\psi. \tag{C.39}$$

Upon introducing the definition $i_p^{(\text{plasma})}(\psi) = \mu_0 I_p^{(\text{plasma})}(\psi)/2\pi$ and integrating Eq. (C.39), one obtains

$$\bar{k}_2(\psi) - \bar{k}_2(0) = -i_p^{(\text{plasma})}(\psi). \tag{C.40}$$

The integration constant $\bar{k}_2(0)$ is non-zero. Even when there is no poloidal plasma current there is still a vacuum toroidal magnetic field resulting from the net poloidal current in the coils (i.e., the $B_\phi = B_0 R_0/R$ contribution in a tokamak). The normalized coil current is defined as $i_p^{(\text{coil})} = \mu_0 I_p^{(\text{coil})}/2\pi = $ const. implying that $\bar{k}_2(0) = i_p^{(\text{coil})}$. Thus, Eq. (C.40) can be rewritten as

$$\bar{k}_2(\psi) = i_p^{(\text{coil})} - i_p^{(\text{plasma})}(\psi) \equiv i_p(\psi). \tag{C.41}$$

This discussion finally leads to a dual representation of the magnetic field in Boozer coordinates expressed in terms of the physical fluxes and currents. These expressions are

$$\begin{aligned} \mathbf{B} &= q(\psi) \nabla \psi \times \nabla \chi + \nabla \zeta \times \nabla \psi, \\ \mathbf{B} &= i_t(\psi) \nabla \chi + i_p(\psi) \nabla \zeta + \tilde{k}(\psi, \chi, \zeta) \nabla \psi. \end{aligned} \tag{C.42}$$

The Boozer dual representation has one crucial property that is essential in understanding particle orbits in a stellarator. This property is obtained by forming the dot product of the two representations:

$$J = J(\psi, B) = \frac{B^2}{i_t + q \, i_p}. \tag{C.43}$$

On any given flux surface the Jacobian of the coordinate transformation is only a function of $B$. It does not depend on the vector nature of the fields.

## C.6 The guiding center orbits in Boozer coordinates

The last step in the analysis involves the substitution of the Boozer coordinates into the equations describing the guiding center motion of the particles. To begin, note that two equivalent ways of locating a particle are by specifying the values of its $x$, $y$, $z$ coordinates or its $\psi$, $\chi$, $\zeta$ coordinates. If the particles are moving with their guiding center velocity, the equations describing the evolution of the orbits in Boozer coordinates, assuming steady state magnetic fields, can be written as

$$d\psi/dt = (\dot{\mathbf{r}}_g + v_\| \mathbf{b}) \cdot \nabla\psi,$$
$$d\chi/dt = (\dot{\mathbf{r}}_g + v_\| \mathbf{b}) \cdot \nabla\chi, \tag{C.44}$$
$$d\zeta/dt = (\dot{\mathbf{r}}_g + v_\| \mathbf{b}) \cdot \nabla\zeta.$$

Here, $\mathbf{v}_g = \dot{\mathbf{r}}_g + v_\| \mathbf{b}$ is the guiding center velocity, which in steady state includes the grad $B$ drift, the curvature drift, and parallel motion. In particular, the perpendicular contribution is given by

$$\frac{d\mathbf{r}_g}{dt} = \frac{v_\perp^2}{2\omega_c} \frac{\mathbf{B} \times \nabla B}{B^2} + \frac{v_\|^2}{\omega_c} \frac{\mathbf{R}_c \times \mathbf{B}}{R_c^2 B}. \tag{C.45}$$

The discussion now focuses on the evaluation of $d\psi/dt$ which represents the guiding center drift perpendicular to the flux surface. The reason is that the deviation of a particle's trajectory off a flux surface is the most critical quantity determining particle losses in multi-dimensional geometries. The justification for this statement is given in Section 14.4. For present purposes readers should just assume the statement is true and accept that good single-particle confinement follows when the flux surface deviation is small.

Since $\mathbf{b} \cdot \nabla\psi = 0$ the quantity to evaluate is $\dot{\mathbf{r}}_g \cdot \nabla\psi$. This step is easily carried out by using the gradient form of $\mathbf{B}$ and then noting that

$$\nabla\psi \cdot \mathbf{B} \times \nabla B = J \left( i_t \frac{\partial B}{\partial \zeta} - i_p \frac{\partial B}{\partial \chi} \right). \tag{C.46}$$

The curvature drift term is simplified by recalling that $\mathbf{R}_c/R_c^2 = -\kappa = -\mathbf{b} \cdot \nabla\mathbf{b}$ and using the relation

$$\begin{aligned}
\mathbf{b} \cdot \nabla\mathbf{b} &= \frac{\mathbf{B}}{B} \cdot \nabla\frac{\mathbf{B}}{B} = \frac{1}{B^2}\mathbf{B} \cdot \nabla\mathbf{B} - \frac{\mathbf{B} \cdot \nabla B}{B^3}\mathbf{B} \\
&= \frac{1}{B^2}\nabla\left(\mu_0 p + \frac{B^2}{2}\right) - \frac{\mathbf{B} \cdot \nabla B}{B^3}\mathbf{B} \\
&= \frac{1}{B^2}\frac{dp}{d\psi}\nabla\psi + \frac{\nabla B}{B} - \frac{\mathbf{B} \cdot \nabla B}{B^3}\mathbf{B}.
\end{aligned} \tag{C.47}$$

It then follows that

$$\nabla\psi \cdot \kappa \times \mathbf{B} = -\frac{J}{B}\left( i_t \frac{\partial B}{\partial \zeta} - i_p \frac{\partial B}{\partial \chi} \right). \tag{C.48}$$

The last simplification makes use of the fact that in guiding center theory a particle's energy and magnetic moment are conserved. Therefore the perpendicular and parallel velocities can be written as $mv_\perp^2/2 = \mu B$ and $mv_\|^2/2 = E - \mu B$. Combining these results and substituting for $J$ leads to the desired equation for the perpendicular guiding drift motion:

$$\frac{d\psi}{dt} = \frac{2E - \mu B}{eB}\left( \frac{i_t}{i_t + q i_p} \frac{\partial B}{\partial \zeta} - \frac{i_p}{i_t + q i_p} \frac{\partial B}{\partial \chi} \right). \tag{C.49}$$

The key feature of Eq. (C.49) is that the perpendicular guiding center drift has been shown to depend only upon $\psi$ and $B$ not upon the vector nature of $\mathbf{B}$. This result is used to understand the main motivation behind the design of the W7-X and NCSX stellarators.

# Appendix D
## Poynting's theorem

This appendix presents a derivation of the relationship between the Poynting vector and the total stored energy in an electromagnetic plasma wave.

The first step in the derivation is to form appropriate dot products of the linearized Maxwell equations (i.e., Eq. (15.38)) with $\mathbf{B}_1^*$ and $\mathbf{E}_1^*$ leading to Poynting's theorem:

$$\mathbf{k} \cdot \mathbf{P} = \frac{\omega}{2} \left( \frac{1}{2\mu_0} |\mathbf{B}_1^2| + \frac{\varepsilon_0}{2} \mathbf{E}_1^* \cdot \overset{\leftrightarrow}{\mathbf{K}} \cdot \mathbf{E}_1 \right). \tag{D.1}$$

$$\mathbf{P} \equiv \frac{1}{4\mu_0} (\mathbf{E}_1 \times \mathbf{B}_1^* + \mathbf{E}_1^* \times \mathbf{B}_1) = \frac{1}{2\mu_0} \mathrm{Re}(\mathbf{E}_1 \times \mathbf{B}_1^*).$$

Here $\mathbf{P}$ is the Poynting vector representing the flux of electromagnetic power. Assume now that the dielectric tensor is separated into real and imaginary parts: $\overset{\leftrightarrow}{\mathbf{K}} = \overset{\leftrightarrow}{\mathbf{K}}_r(\omega) + i\, \overset{\leftrightarrow}{\mathbf{K}}_i(\omega, \mathbf{k})$ with $\overset{\leftrightarrow}{\mathbf{K}}_r$ representing the cold plasma dielectric tensor and $\overset{\leftrightarrow}{\mathbf{K}}_i \ll \overset{\leftrightarrow}{\mathbf{K}}_r$ representing a small dissipation.

The desired relation is obtained by considering two separate situations. In the first, assume the plasma is driven by a source with a real frequency $\omega = \omega_r$. The resulting wave propagates into the plasma with a slight spatial damping due to the dissipation: $\mathbf{k} = k_\parallel \mathbf{e}_z + (k_{\perp r} + ik_{\perp i})\mathbf{e}_x = \mathbf{k}_r + ik_{\perp i}\mathbf{e}_x$. In the limit of small dissipation the real and imaginary parts of Poynting's theorem reduce to

$$\mathbf{k}_r \cdot \mathbf{P} = \frac{\omega_r}{2} \left( \frac{1}{2\mu_0} |\mathbf{B}_1^2| + \frac{\varepsilon_0}{2} \mathbf{E}_1^* \cdot \overset{\leftrightarrow}{\mathbf{K}}_r(\omega_r) \cdot \mathbf{E}_1 \right),$$

$$k_{\perp i} P_\perp \approx \frac{\omega_r \varepsilon_0}{4} \mathbf{E}_1^* \cdot \overset{\leftrightarrow}{\mathbf{K}}_i(\omega_r, \mathbf{k}_r) \cdot \mathbf{E}_1. \tag{D.2}$$

The first equation is automatically satisfied since it is equivalent to satisfying the cold plasma dispersion relation. The second gives a relationship between the Poynting flux and the dissipated power.

In the second situation, assume that a wave is propagating in the plasma with a real wave number $\mathbf{k} = k_\parallel \mathbf{e}_z + k_{\perp r}\mathbf{e}_x = \mathbf{k}_r$ but damps slowly in time due to the dissipation: $\omega = \omega_r + i\omega_i$. In this case the real and imaginary parts of Poynting's theorem reduce to

$$\mathbf{k}_r \cdot \mathbf{P} = \frac{\omega_r}{2} \left( \frac{1}{2\mu_0} |\mathbf{B}_1^2| + \frac{\varepsilon_0}{2} \mathbf{E}_1^* \cdot \overset{\leftrightarrow}{\mathbf{K}}_r(\omega_r) \cdot \mathbf{E}_1 \right),$$

$$0 \approx \omega_i U + \frac{\omega_r \varepsilon_0}{4} \mathbf{E}_1^* \cdot \overset{\leftrightarrow}{\mathbf{K}}_i(\omega_r, \mathbf{k}_r) \cdot \mathbf{E}_1, \tag{D.3}$$

where

$$U = \frac{1}{2} \left\{ \frac{1}{2\mu_0} |\mathbf{B}_1^2| + \frac{\varepsilon_0}{2} \mathbf{E}_1^* \cdot \frac{\partial}{\partial \omega_r} \left[ \omega_r \overset{\leftrightarrow}{\mathbf{K}}_r(\omega_r) \right] \cdot \mathbf{E}_1 \right\}. \tag{D.4}$$

The quantity $U$ represents the total stored energy (i.e., wave energy plus plasma kinetic energy) in the system. The first equation is again automatically satisfied, while the second gives a relation between the stored energy and the dissipation.

The desired relation is obtained by subtracting the two relations involving the dissipation terms. One obtains

$$P_\perp = \frac{\omega_i}{k_{\perp i}} U = V_{g\perp} U. \tag{D.5}$$

The Poynting flux is equal to the product of the stored energy and the group velocity.

# Index

Printed in the United States
By Bookmasters